D0938573

Plants and their Names
A Concise Dictionary

Plants and their Names
A Concise Dictionary

R. HYAM
R. PANKHURST

ROYAL
BOTANIC
GARDEN
EDINBURGH

OXFORD UNIVERSITY PRESS
1995

Oxford University Press, Walton Street, Oxford OX2 6DP
Oxford New York
Athens Auckland Bangkok Bombay
Calcutta Cape Town Dar es Salaam Delhi
Florence Hong Kong Istanbul Karachi
Kuala Lumpur Madras Madrid Melbourne
Mexico City Nairobi Paris Singapore
Taipei Tokyo Toronto
and associated companies in
Berlin Ibadan

Oxford is a trade mark of Oxford University Press

Published in the United States by
Oxford University Press Inc., New York

First published 1995

British Library Cataloguing in Publication Data
Data available

Library of Congress Cataloging in Publication Data
Pankhurst, R. J.
Plant and their Names: A Concise Dictionary / R. Pankhurst, R. Hyam.
p. cm.
1. Botany—Nomenclature. 2. Plant names, Popular—Dictionaries. 3. Botany—Dictionaries.
4. Botany, Economic—Dictionaries. 5. Botany—Great Britain—Dictionaries.
I. Hyam, R. (Roger) II. Title.
581'.014—dc20 QK96.P35 1995 94-5024
ISBN 0-19-866189-4

Typeset by Graphicraft Typesetters Ltd.
Printed in Great Britain by
Biddles Ltd.
Guildford and King's Lynn

FOREWORD

In the last few years a new word has entered the English language: biodiversity. It has become part of the vocabulary of the environmental movement. Politicians have found it important to refer to biodiversity in their manifestos and speeches. International conventions have been signed to protect it and, after some initial confusion, scientists have agreed, more or less, about what it is – the sum total of the variety of life on earth: diversity of genes, species, and ecosystems.

The book in front of you is a dictionary to help you talk about the biodiversity of plants. There is a special international language, governed by its own rules of grammar, use, and spelling, which is used in botany and horticulture to make communication about plant biodiversity as unambiguous and efficient as possible. It is derived from Latin in an obvious, though not always straightforward way, while the etymology of individual words can sometimes be a fascinating story, reflecting the social background or hobbies of botanists rather than having anything to do with the plants themselves. The language is sometimes inadequate and some people do not use it properly. Nevertheless, it has enabled mankind to catalogue and study the quarter of a million plants that have been discovered so far.

In 1989 Oxford University Press approached the Natural History Museum, London, about producing a dictionary for botanical names and Dr Richard Pankhurst took on the task. The first source of data was the *Oxford English Dictionary*, but progress was slow until 1992, when Dr Pankhurst, now at the Royal Botanic Garden, Edinburgh, was able to employ four young people with Lisztian fingers to enter new information into a specially designed database. Subsequently, Roger Hyam has laboured like a latter-day Hercules to enlarge and improve the database, under Dr Pankhurst's supervision. The result is before you and we hope it will be useful.

Dr David G. Mann
Deputy Director
Royal Botanic Garden, Edinburgh

INTRODUCTION

The plant kingdom is staggeringly diverse. Over 250,000 species of vascular plant have been named so far and many remain to be discovered. The aim of this dictionary is to provide the non-specialist with a reference source to some 16,000 of the more commonly occurring vernacular and Latin names. Technical terms and abbreviations have been kept to a minimum but those that have been used are included in a glossary at the back of the book. In selecting from the tens of thousands of plants that could have been included special attention was paid to those that are grown ornamentally in northern Europe and those that serve as sources of food or raw materials. All the plant families that are recognized by the Royal Botanic Gardens, Kew, and the Royal Botanic Garden, Edinburgh, are mentioned. Name derivations, descriptions, uses, and geographical distributions are given for most entries. Coverage extends to higher plants, that is to say, flowering plants, ferns, and gymnosperms, but not lower plants such as mosses, lichens, or algae.

The paragraphs below give an explanation of the different types of entry used in the dictionary and a little about how scientific names are structured. A list of principal sources and some suggestions for further reading are given after the main text.

HOW TO USE THE DICTIONARY

There are six types of entry in this dictionary. Five of them are to do with scientific names and one is for common names.

Scientific Names

Scientific names are designed to reflect how plants are arranged in a hierarchy of groups within groups. Each species has a two-word name, known as a binomial. The first of the two words is the name of the genus, in other words a group of closely related plants. The second name is termed the specific epithet and identifies a particular species within the genus. This is very similar to the way in which some common names are used. Buttercup is a general name for a group of plants but can be particularized by the addition of a qualifying word, Creeping Buttercup or Meadow Buttercup, for example. The scientific name for Creeping Buttercup is *Ranunculus repens* where '*Ranunculus*' is the genus name and '*repens*' is the specific epithet.

There are certain conventions that must be observed when writing down a scientific name. The genus name is always written first and starts with a capital letter, the specific name comes second and always starts with a lower-case letter. In the dictionary the whole name is in italics. When several species are mentioned from the same genus and there is no danger of confusion occurring, the genus name can be abbreviated to a single letter followed by a full stop. The scientific name for the Meadow Buttercup, which is also in the genus *Ranunculus*, can therefore be written *R. acris*. Specific epithets should never be used on their own to refer to a plant as they are adjectives and the same epithet may be used in combination with a number of different genus names. The epithet *repens*, for example, occurs sixteen times in this book. Genus names can be used alone to refer to a group of plants as they are technically nouns and also unique.

Sometimes further scientific research or the discovery of new species reveals that a plant has been placed in the wrong genus or that two or more names are being used for the same plant. In this case only one name is correct, usually the oldest, and the other names become synonyms and should no longer be used.

Just as species are collected into genera, so genera are grouped into families. In all but a few exceptions family names are based on the name of one of their genera with the ending -aceae added. Family names are not written in italics or abbreviated.

The five types of scientific entry in the dictionary are:

1. Family names. Brief descriptions of the 511 families recognized by the Royal Botanic Garden, Edinburgh, and the Royal Botanic Gardens, Kew, are given. Each entry is followed by a list of the genera in that family that are included elsewhere in the dictionary. A few families contain only one genus and where that genus is included in the dictionary these families are not described (the description would repeat the information contained in the genus entry).

2. Genus names. Brief descriptions of 2,169 genera are given. The name of the family to which that genus belongs is also mentioned. Each genus entry is followed by one or more species entries.

3. Species names. It is not possible to include all the species of every genus mentioned in a volume this size but brief mention has been made of at least one species for every genus, a total of around 5,500.

4. Common specific epithets. Specific epithets often have a meaning; *repens*, for example, actually means creeping. The meanings of 2,600 of the more commonly occurring epithets have therefore been given so that, even if a species is not mentioned under the genus entry, its epithet may be included. In combination the genus and epithet entries

will convey some understanding of the plant in question. Because epithets are adjectives they have to agree, under the rules of botanical Latin, with the genus name with which they are combined. Many of the entries are therefore given with a series of possible endings.

5. Synonyms. Some names that are included are no longer accepted as correct names and they are cross-referenced to the accepted name.

Carrying on up the hierarchy of groups, families (which are clusters of genera) are in turn gathered into higher groups. The general class of plants to which each family or genus belongs is indicated by a one-letter code at the end of the description. These codes are explained in the glossary at the back of the book.

Common Names

Scientific plant names are governed by a strict set of rules laid out in the International Code of Botanical Nomenclature. This code is designed to prevent the same name being used for different plants and different names being used for the same plant. Common names are not governed by any rules and so may mean different things to different people. As an example, Bluebell is commonly used for the woodland plant *Hyacinthoides non-scripta*. But in Scotland, Bluebell has traditionally been used for *Campanula rotundifolia*, a plant also known as the Harebell. Scientific names are also sometimes used as common names for other plants: see 'Nasturtium' as an example.

The 6,000 common names that have been included in this dictionary are those that were encountered in the botanical and horticultural literature and should reflect common usage. There may, however, be regional and cultural variations on the application of some names and there may well be others that are very familiar to the reader but have not been included.

ACKNOWLEDGEMENTS

The whole project was carried out at the Royal Botanic Garden, Edinburgh under the general supervision of Dr Richard Pankhurst. The dictionary was created from a computer database using Advanced Revelation database software on a Novell network of IBM-compatible personal computers at the Royal Botanic Garden, Edinburgh. The design and programming of the database was carried out by Richard Pankhurst. The text of the dictionary was generated from the database and sorted by programs which were especially written for the purpose. In the first phase of the project Zoë Gowler, Richard Parish, Peter Watkins, and Duncan Will entered family, genus, and species names. In the second phase of the project, entry of further data and the detailed editing of the database was carried out by Roger Hyam with help from Karen Sidwell. We are grateful for support and advice from Royal Botanic Garden staff in general, and especially to Dr Colin Will (Librarian), Mrs Gillian Lomas (Head of Administration), Dr Crinan Alexander (Editor of the *European Garden Flora*) and Dr David Mann (Deputy Director).

Roger Hyam
Richard Pankhurst

A

Aaron's beard *Hypericum calycinum*.

Aaron's rod *Verbascum thapsus*.

abaca *Musa textilis*.

abata *Cola acuminata*.

abele *Populus alba*.

Abelia (After Dr Clarke Abel (1780–1826), who collected *A. chinensis* while working as a physician in China.) A genus of 30 species of deciduous and evergreen shrubs with arching branches, simple, short stalked leaves and cymes of many small, bell- or funnel-shaped flowers. A number of species and hybrids are grown as ornamentals. *Distr.* Himalaya, E Asia, Mexico. *Fam.* Caprifoliaceae. *Class* D.

A. chinensis Spreading deciduous shrub. Flowers fragrant, white. *Distr.* China.

A. engleriana (After Heinrich Engler (1844–1930), German botanist) Deciduous shrub. Flowers borne in pairs, pink. *Distr.* China.

A. floribunda Evergreen shrub. Flowers pink to red, in pendulous clusters. *Distr.* Mexico.

A. × grandiflora A hybrid of *A. chinensis* × *A. uniflora*. Semi-evergreen shrub. Flowers white. *Distr.* Presumed garden origin.

A. rupestris See *A. x grandiflora*

A. schumannii (After K. M. Schumann (1851–1904), a German botanist.) Semi-evergreen shrub. Flowers solitary, pink. *Distr.* Central China.

Abeliophyllum (From the genus name *Abelia*, and the Greek, *phyllon*, a leaf, alluding to the similar leaves.) A genus of 1 species of sprawling, deciduous shrub with simple, opposite leaves and racemes of fragrant, white flowers that bear 4 petals and 2 stamens. *Distr.* Korea. *Fam.* Oleaceae. *Class* D.

A. distichum WHITE FORSYTHIA. Grown as a hardy ornamental.

Abelmoschus (From the Arabic *abu-l-mosk*, father of musk, alluding to the musk-scented seeds.) A genus of 15 species of annual and perennial herbs with large, palmately lobed or compound leaves and showy, regular flowers that bear 5, typically yellow petals and 10 stamens that are fused into a column below. The fruit is a 5-angled pod. *Distr.* Tropical regions of the Old World. *Fam.* Malvaceae. *Class* D.

A. esculentus OKRA, GUMBO, LADY'S FINGERS. Annual herb. Flowers white to bright yellow with a dark central blotch. The immature pods are eaten as a vegetable. *Distr.* Tropical regions of the Old World.

A. moschatus MUSK MALLOW. Perennial, roughly hairy herb to 2m high. Flowers white to yellow with a maroon centre. A number of ornamental hybrids have been raised from this species. *Distr.* Tropical Asia.

Abies (From the Latin *abire*, to rise, alluding to the great height some species attain.) FIR, SILVER FIR. A genus of about 40 species of evergreen coniferous trees with whorled branches and flattened, linear leaves. The woody, fruiting cones are borne on the upper branches and disintegrate at maturity. Several species are grown as ornamental specimen trees. The resin of some species is the source of Canada balsam. *Distr.* Eurasia, N Africa, North and Central America. *Fam.* Pinaceae. *Class* G.

A. alba CHRISTMAS TREE, SILVER FIR, WHITEWOOD. Tree to 60m high. Source of timber, notably for telegraph poles, and turpentine. *Distr.* Central and S Europe.

A. balsamea BALSAM FIR. Slender tree to 25m. The major source of the Canada balsam used in microscopic preparations. *Distr.* E North America.

A. concolor WHITE FIR, COLORADO FIR. Tree to 50m high. A number of ornamental cultivars have been raised from this species. *Distr.* W USA, Mexico.

A. grandis GIANT FIR. Tree to 75m high. *Distr.* W North America.

A. nobilis See *A. procera*.

A. pinsapo SPANISH FIR, HEDGEHOG FIR. Tree to 35m high. Now threatened in the wild. *Distr.* S Spain.

A. procera NOBLE FIR. Tree to 70m tall. *Distr.* W North America.

abnormis, -e: abnormal.

absinthe *Artemisia absinthium.*

Abutilon (From the Arabic name for a species of *Malva.*) FLOWERING MAPLE, INDIAN MAPLE, PARLOUR MAPLE. A genus of about 150 species of annual to perennial herbs, shrubs and small trees with simple to palmately lobed leaves and showy, bell-shaped flowers that bear their parts in 5s. Several species and numerous cultivars are grown ornamentally. Several cultivars have intricately variegated foliage caused by a viral infection. *Distr.* Tropical and subtropical regions. *Fam.* Malvaceae. *Class* D.

A. globosum See *A.* × *hybridum*

A.* × *hybridum CHINESE LANTERN. This name is assigned to a large group of cultivars that are typically shrubby with variegated foliage and solitary, red, orange, yellow or white flowers. *Distr.* Garden origin.

A. megapotamicum TRAILING ABUTILON. Slender, scrambling shrub to 2.5m high. Flowers pendent, yellow and red. *Distr.* Brazil.

A. pictum Shrub or small tree. Flowers yellow-orange with crimson veins. The source of a number of ornamental cultivars. *Distr.* Brazil, naturalized throughout Central and South America.

A. theophrasti CHINGMA, CHINESE HEMP, INDIAN MALLOW, MANCHURIAN JUTE, VELVET LEAF, BUTTER PRINT, CHINA JUTE. Annual herb. Flowers yellow, borne in small cymes. An important fibre plant in China. *Distr.* Tropical Asia, naturalized SE Europe, Mediterranean and USA.

abutilon, trailing *Abutilon megapotamicum.*

abyssinicus, -a, -um: of Africa.

Acacia (From the Greek *akis*, a sharp point, alluding to the thorns of many of the species.) WATTLE, MIMOSA. A genus of 700–1200 species of evergreen shrubs, trees, and a few lianas with alternate, bipinnate leaves and racemes or panicles of clusters of small, regular flowers that bear numerous stamens. The leaf stipules often develop as woody spines and in some species these act as homes for ants. The leaves are sometimes replaced by leaf-like, flat, green shoots known as phyllodes. This genus is ecologically very important in many of the drier areas of the world and is a source of both wood and fodder as well as numerous ornamental plants. *Distr.* Tropical and warm regions. *Fam.* Leguminosae. *Class* D.

A. baileyana (After F. M. Bailey (1827–1915), Australian botanist.) COOTAMUNDRA WATTLE, GOLDEN MIMOSA. Shrub or small, spreading tree. Flower-heads small, golden yellow. *Distr.* SW Australia.

A. cultriformis KNIFE ACACIA, KNIFE-LEAF WATTLE. Erect shrub. Leaves bright green. Flowers yellow-fragrant. *Distr.* SW Australia.

A. dealbata BLUE WATTLE, SILVER WATTLE, MIMOSA. Spreading tree. Flower-heads bright yellow, fragrant. *Distr.* Australia, Tasmania.

A. genistifolia SPREADING WATTLE. Shrub. Phyllodes needle-like. *Distr.* Australia, Tasmania.

A. longifolia SYDNEY GOLDEN WATTLE, SWALLOW WATTLE. Spreading tree. Phyllodes narrow, dark green. Flower clusters cylindrical. *Distr.* E Australia.

A. mearnsii BLACK WATTLE. Tree. Bark black. A source of tannin. *Distr.* SE Australia, Tasmania.

A. paradoxa KANGAROO THORN, HEDGE WATTLE. Spiny shrub. Phyllodes lance-shaped. *Distr.* SW Australia.

A. podalyrifolia MOUNT MORGAN WATTLE, QUEENSLAND SILVER WATTLE. Bushy shrub. Phyllodes silver-blue, sharply pointed. *Distr.* NE Australia.

A. pravissima OVENS WATTLE. Bushy shrub or small tree. Phyllodes silver-grey, spine tipped. *Distr.* SE Australia.

A. verticillata STAR ACACIA, PRICKLY MOSSES. Low shrub or small open tree. Phyllodes needle-like. Flower-heads bottle-brush-like. *Distr.* SW Australia, Tasmania.

acacia: bastard ~ *Robinia pseudoacacia* **false** ~ *R. pseudoacacia* **knife** ~ *Acacia cultriformis* **rose** ~ *Robinia hispida* **star** ~ *Acacia verticillata.*

acaciiformis from the genus name *Acacia*, and *formis*, in the form of.

Acaena (From the Greek *akaina*, a thorn, alluding to the spiny fruits.) A genus of 100 species of perennial herbs and subshrubs with alternate, pinnately divided leaves and spikes or heads of small, regular flowers that are followed by a barbed fruit. Several species are grown as ornamental ground cover. *Distr.* S hemisphere, N to California and Hawaii. *Fam.* Rosaceae. *Class* D.

A. buchananii (After John Buchanan (1819–98), botanist.) Rhizomatous perennial herb. Foliage grey-green. Flowers white. *Distr.* New Zealand (South Island), Australia.

A. microphylla Rhizomatous, perennial herb. Foliage bronze at first. Flowers surrounded by dull red bracts. *Distr.* New Zealand.

A. novae-zelandiae BIDGEE WID-GEE, BIDDY BIDDY. Creeping shrub. Branches rooting. Flowers in rounded heads. *Distr.* SE Australia, New Zealand, New Guinea, naturalized British Isles.

acajou *Anacardium occidentale.*

Acalypha (The Greek name for nettle, alluding to the similarity of the leaves to those of Stinging Nettle, *Urtica dioica.*) A genus of 430 species of herbs, shrubs and trees with simple, alternate leaves and small spikes or racemes of inconspicuous, red or green flowers. Several species are grown as ornamentals. *Distr.* Tropical and warm regions. *Fam.* Euphorbiaceae. *Class* D.

A. hispida Shrub. Leaves bright green. Flowers red, in pendent, tassel-like racemes. *Distr.* Malaya, New Guinea.

A. wilkesiana Shrub. Leaves copper coloured, marked red. Flowers in slender spikes. Several cultivars are available. *Distr.* Pacific Islands.

Acanthaceae The Acanthus family. About 300 genera and 3000 species of herbs, lianas and trees with opposite leaves and typically irregular, 2 lipped flowers that bear 4–5 sepals, 4–5 petals and 2–5 stamens. This family contains some ornamental and a few medicinal species. *Distr.* Widespread in tropical and subtropical regions, rarer in temperate zones. *Class* D.

See: Acanthus, Aphelandra, Asystasia, Barleria, Beloperone, Crossandra, Dicliptera, Fittonia, Hypoestes, Justicia, Mackaya, Pachystachys, Peristrophe, Pseuderanthemum, Ruellia, Ruttya, Sanchezia, Strobilanthes, Thunbergia.

acanthifolius, -a, -um: from the Greek *akanth-*, spiny, and the Latin *folius*, leaved.

acanthius, -a, -um: spiny.

Acanthocalyx See *Morina*

Acanthocereus (From the Greek *akanthos*, thorn, and the genus name *Cereus.*) A genus of 8 species of shrubs and tree-like cacti with jointed stems and white, funnel-shaped flowers that open at night. Several species are grown as ornamentals. *Distr.* SE North America to NE Brazil. *Fam.* Cactaceae. *Class* D.

A. pentagonus Fruit edible. *Distr.* NE Brazil.

A. tetragonus Shrub to 3m high. Stems arching and rooting at tips. Spines to 4cm long. *Distr.* S USA, West Indies, Mexico, Venezuela.

Acantholimon (From the Greek *akanthos*, thorn, and the genus name *Limonium.*) PRICKLY THRIFT. A genus of 120 species of tuft- or cushion-forming, perennial herbs and subshrubs with simple, often spiny leaves and spikes of small flowers that bear their parts in 5s. Several species are grown ornamentally. *Distr.* E Mediterranean to Central Asia. *Fam.* Plumbaginaceae. *Class* D.

A. androsaceum See *A. ulicinum*

A. glumaceum Cushion-forming subshrub. Flowers white. *Distr.* W Asia.

A. ulicinum Dense cushion-forming subshrub. Flowers pink. *Distr.* SE Europe, Turkey.

Acanthopanax See *Eleutherococcus*

acanthothamnos spiny shrub.

Acanthus (From the Greek *akanthos*, a thorn, referring to the spiny leaves and bracts.) BEAR'S BREECHES. A genus of 30 species of perennial herbs and subshrubs with pinnately lobed, spiny leaves and spikes of irregular flowers that have a 3-lobed lower lip but lack an upper lip. *Distr.* Tropical and warm regions of the Old World. *Fam.* Acanthaceae. *Class* D.

A. balcanicus To 1.5m high. Flowers pale-pink. Bracts red-purple. *Distr.* S Europe.

A. hungaricus See *A. balcanicus*

A. longifolius See *A. balcanicus*

A. mollis BEAR'S BREECHES. Clump-forming. Leaves to 1m long. Flowers purple and white. *Distr.* Europe, NW Africa.

A. spinosus Forms spreading clumps. Spikes erect, to 1m long. Flowers purple-white, bracts white. *Distr.* SE Europe.

acanthus Acanthaceae.

acaulis, -e: stemless.

Acer (From the Latin name for the MAPLE also meaning sharp, alluding to the hard wood.) MAPLE. A genus of 130–150 species of deciduous trees with simple-lobed or compound leaves noted for their spectacular autumn colours. The small flowers bear 4–5 sepals and 4–10 stamens and are followed by pairs of winged fruits. Some species are important timbers, used for flooring, furniture, musical instruments, and gunstocks. Several species are tapped for maple syrup and sugar. *Distr.* Temperate regions and mountainous areas of the tropics. *Fam.* Aceraceae. *Class* D.

A. campestre COMMON MAPLE, FIELD MAPLE, HEDGE MAPLE. Shrub or tree to 20m high. Leaves hairy below, turning red. Numerous ornamental cultivars available. Wood used for kitchen implements. *Distr.* Europe, N Africa, W Asia.

A. globosum See *A. platanoides*

A. negundo (From the local vernacular name for *Vitex negundo*, because of a supposed similarity of the leaf.) BOX ELDER. Leaves pinnate. Numerous cultivars available. *Distr.* Central and North America.

A. palmatum JAPANESE MAPLE. Small trees. Very numerous cultivars available. *Distr.* E Asia, Japan.

A. platanoides NORWAY MAPLE. An important timber tree. *Distr.* Europe, W Asia.

A. pseudoplatanus SYCAMORE, GREAT MAPLE, SCOTTISH MAPLE. An important timber tree as well as a source of early nectar for honey bees; resulting honey is green. Now often considered a weed in Britain. *Distr.* Europe, W Asia.

A. saccharum SUGAR MAPLE, STRIPED MAPLE. Grown as a timber tree as well as for the delicious maple syrup which is tapped from the tree in spring. *Distr.* Alaska and Canada to SE USA.

A. sieboldianum Small tree. *Distr.* Japan.

acer, acris, acre bitter, pungent.

Aceraceae The Maple family. 2 genera and 113 species of typically deciduous trees and shrubs with palmately lobed or divided leaves. The flowers are regular and typically bear their parts in 5s. The paired, winged fruits are characteristic of the family. Some species are planted as ornamentals, others harvested for timber or tapped for sweet sap. *Distr.* N temperate region and tropical mountains to Malaysia. *Class* D.
See: Acer, Dipteronia.

acerbus, -a, -um: bitter, sour.

acerifolius, -a, -um: maple-leaved, from the genus name *Acer*, and *folius*, leaved.

Aceriphyllum See *Mukdenia*

A. rossii See *Mukdenia rossii*

acerosus, -a, um: needle-shaped.

acetabulosus, -a, um: markedly concave, depressed.

acetosella: slightly acidic, an early name referring to plants with mildly acidic leaves.

acetosus, -a, um: acidic, an early name referring to plants with acidic leaves.

Achariaceae A family of 3 genera and 4 species of subshrubs and stemless or climbing herbs. This family is closely related to the Passifloraceae. *Distr.* South Africa. *Class* D.

Achatocarpaceae A family of 2 genera and 16 species of trees or shrubs. This family is closely related to the Phytolaccaceae. *Distr.* Tropical America. *Class* D.

Achillea (After Achilles, a hero of Greek mythology, who is said to have used it medicinally.) YARROW, MILFOIL. A genus of 85 species of perennial, sometimes mat-forming herbs with aromatic, typically pinnately divided leaves and dense umbels of daisy-like flower-heads that bear distinct rays. *Distr.* N temperate regions. *Fam.* Compositae. *Class* D.

A. abrotanoides To 40cm high. Rays white. *Distr.* W Balkans.

A. ageratifolia Leaves tuft-forming, grey-hairy. Rays white. *Distr.* Central Balkans.

A. argentea See *Tanacetum argenteum*

A. cartilaginea Leaves glandular. Rays round, white. *Distr.* Central Asia to E Germany and SW Romania.

A. chrysocoma Leaves hairy. Rays bright yellow. *Distr.* SE Europe.

A. clavennae (Named in honour of Niccola Chiavera (died 1617), Italian apothecary.) Leaves silky-hairy. Rays white. *Distr.* Alps, Balkans.

A. distans To 1.2m high. Rays white-pink. *Distr.* SW Alps to E Carpathians and Bulgaria.

A. erba-rotta Leaves simple or divided. Rays white. *Distr.* Alps, Apennines, Greece.

A. filipendulina Leaves densely hairy. Rays gold. *Distr.* W Asia, Caucasus.

A. grandifolia To 1m. Rays white. *Distr.* S and Central Balkans.

A. × jaborneggii A hybrid of *A. clavennae* × *A. erba-rotta*. *Distr.* Garden origin.

A. × kellereri A hybrid of *A. clypeolata* × *A. ageratifolia*. *Distr.* Garden origin.

A. × kolbiana A hybrid of *A. clavennae* × *A. umbellata*. *Distr.* Garden origin.

A. × lewisii A hybrid of *A. clavennae* × *A. clypeolata*. Evergreen. *Distr.* Garden origin.

A. millefolium MILFOIL, YARROW. Mat-forming. Rays white-pink. A weed of waste ground, pasture and hedgerow. It also has medicinal properties and has been used as a tobacco. *Distr.* Temperate Europe, W Asia, widely naturalized.

A. nana Leaves hairy, strongly aromatic. Rays white. *Distr.* Alps, Apennines.

A. nobilis Rays tinged yellow above. *Distr.* S and Central Europe to Central Asia.

A. ptarmica (From the Greek name for a plant used in snuff.) SNEEZEWEED, SNEEZEWORT. Leaves simple. Rays round, white. Used medicinally and as a salad vegetable. *Distr.* Europe.

A. sibirica Leaves long haired. Rays white. *Distr.* E Siberia.

A. tomentosa Silky-hairy. Rays yellow. *Distr.* SW Europe to Central Italy.

A. umbellata Leaves white-hairy. Rays white. *Distr.* S Greece.

A. × wilczekii A hybrid of *A. ageratifolia* x *A. lingulata*. Mat-forming. *Distr.* Garden origin.

Achimenes (Probably from the Greek *khemaino*, to suffer from cold, alluding to the tenderness of these plants.) HOT WATER PLANT. A genus of 25 species of rhizomatous perennial herbs with simple opposite or whorled leaves and 2-lipped, 5-lobed, funnel-shaped flowers that are borne singly or in cymes. Several species and numerous cultivars are grown as ornamentals. *Distr.* West Indies, Central America. *Fam.* Gesneriaceae. *Class* D.

A. antirrhina Stems erect, to 30cm high. Leaves in unequal pairs. Flowers yellow with red-purple markings. *Distr.* W Mexico, Guatemala.

A. bella Stems erect. Flowers solitary, violet. *Distr.* W Mexico.

A. dulcis Stems to 60cm high. Flowers solitary, white with a yellow throat. *Distr.* W Mexico.

A. erecta Bushy. Leaves narrow, in whorls of 3. Flowers red with a yellow throat. *Distr.* Central America, Mexico, Jamaica.

A. longiflora Stems hairy. Leaves in whorls of 3–4, marked red below. Flowers solitary, violet-blue. Numerous cultivars are available. *Distr.* Mexico to Panama.

achira *Canna indica*.

achyrostachys: from the Greek *achyron*, chaff or husks, and *stachys*, spike.

acicularifolius, -a, um: with needle-like leaves.

acicularis, -e: needle-like.

Acidanthera See *Gladiolus*

acinacifolius, -a, -um: with curved, sickle-shaped leaves.

Acinos (From the Greek *akinos*, a name used for an aromatic herb.) A genus of 10 species of small, annual and perennial herbs with simple, opposite leaves and spikes that bear whorls of 2-lipped flowers. Several species are grown as ornamentals. *Distr.* Mediterranean to Central Asia and Iran. *Fam.* Labiatae. *Class* D.

A. alpinus ALPINE CALAMINT. Densely branched perennial to 45cm high. Flowers violet with white markings. *Distr.* Mountains of Central and S Europe, N Africa.

A. arvensis BASIL THYME, HOME OF THYME. Short lived perennial or annual. Leaves somewhat aromatic. Flowers violet. *Distr.* Eurasia.

A. corsicus Creeping perennial. Flowers violet. *Distr.* Corsica.

acinosa: from the Greek *akinos*, an aromatic herb.

Aciphylla (From the Greek *akis*, point, and *phyllon*, leaf.) A genus of 39 species of evergreen, perennial herbs with basal rosettes of stiff, sharply pointed leaves and very large spike-like inflorescences of small, unisexual flowers. Several species are grown ornamentally. *Distr.* New Zealand, Australia. *Fam.* Umbelliferae. *Class* D.

A. aurea GOLDEN SPANIARD. Leaves long, lance-shaped. Flowers golden. *Distr.* New Zealand.

A. scott-thomsonii Forms large clumps. Leaves pinnate or bipinnate. Inflorescence prickly. Flowers yellow-cream. *Distr.* New Zealand.

A. squarrosa BAYONET PLANT, SPEARGRASS. Leaves pinnate with long pointed leaflets. Flowers yellow. *Distr.* New Zealand.

acmopetalum: with anvil-shaped petals.

acmosepalum: with anvil-shaped sepals.

aconite, winter *Eranthis, E. hyemalis.*

aconitifolius, -a, -um: aconite-leaved, from the genus name *Aconitus*, and *folius*, leaved.

Aconitum (The ancient Latin name for this genus, from the Greek name, *akoniton.*) MONK'S HOOD, WOLF'S BANE. A genus of about 100 species of annual, biennial and perennial herbs with tuberous roots, palmate leaves and hooded flowers that bear 5 petal-like sepals and 2–10 small, concealed petals. Many species and numerous cultivars are grown as ornamentals; all are poisonous and some have been used medicinally. *Distr.* N temperate regions. *Fam.* Ranunculaceae. *Class* D.

A. anthora Compact perennial. Flowers yellow. *Distr.* S Europe, W and Central Asia.

A. carmichaelii: (After J. R. Carmichael (1838–70).) Perennial to 2m high. Flowers deep purple. *Distr.* Central and W China.

A. fischeri See *A. carmichaelii*

A. hemsleyanum Scrambling and twining perennial. Flowers indigo to blue. *Distr.* Central and W China.

A. lycoctonum BADGER'S BANE, WOLF'S BANE. Erect perennial. Flowers purple, white or yellow. Formerly used as an animal poison. *Distr.* Europe, N Africa.

A. napellus MONK'S HOOD, GARDEN WOLF'S BANE, HELMET FLOWER, FRIAR'S CAP, SOLDIER'S CAP, TURK'S CAP, BEAR'S FOOT. Erect perennial. Flowers indigo-blue, borne in spike-like inflorescences. The source of the heart drug aconitine. *Distr.* Europe, Asia, America.

A. volubile Twining or scrambling perennial. Flowers straw yellow. *Distr.* Siberia, Mongolia, China, Japan.

A. vulparia See *A. lycoctonum*

Acoraceae A family of 1 genus and 2 species. *Class* M.
See: Acorus.

Acorus (A classical Greek name for a plant with an aromatic rhizome.) SWEET FLAG. A genus of 2 species of iris-like, herbs of wetlands. *Distr.* N temperate to tropical regions. *Fam.* Acoraceae. *Class* M.

A. calamus (From the Greek *calamus*, reed, referring to the appearance of the foliage.) CALAMUS, FLAGROOT, SWEET FLAG, MYRTLE FLAG. Used medicinally for toothache as a tonic and against dysentery since the 4th century BC, also used in holy ointment for anointing alters and sacred vessels in the Old Testament and as an effective insecticide. *Distr.* SE Asia and SE USA, very widely naturalized elsewhere.

Acradenia (From the Greek *akros*, at the tip, and *adenia*, a gland, alluding to the glands at the tips of the carpels.) A genus of 2 species of evergreen shrubs and trees with opposite trifoliate leaves and panicles of small, bisexual flowers that bear 5–6 petals and 10–12 stamens. *Distr.* E Australia. *Fam.* Rutaceae. *Class* D.

A. frankliniae (After Lady Franklin, wife of the Governor of Tasmania in 1842.) WHITEY WOOD. Foliage aromatic. Flowers star-shaped, white. Grown as a tender ornamental. *Distr.* W Tasmania.

acraeus: growing in high places.

acris, acre: see *acer.*

acrotrichum with terminal hairs.

Actaea (From the Greek name for ELDER, *aktea.*) BANEBERRY, COHOSH. A genus of about 8 species of rhizomatous herbs with pinnately divided, alternate leaves and racemes of small white flowers that are followed by fleshy poisonous fruits. Several species are grown ornamentally. *Distr.* N temperate regions. *Fam.* Ranunculaceae. *Class* D.

A. alba DOLL'S EYES, WHITE BANEBERRY. Clump-forming perennial. Flowers white. Berries white. *Distr.* E North America.

A. pachypoda See *A. alba*

A. rubra RED BANEBERRY, SNAKBERRY. Clump-forming perennial. Flowers white. Berries scarlet. *Distr.* North America.

A. spicata BANEBERRY, HERB CHRIS-TOPHER, BLACK COHOSH. Clump-forming perennial. Flowers white. Berries black. Formerly used medicinally. *Distr.* Europe to China.

Actinidia (From the Greek *aktinos*, a ray, alluding to the radiating styles.) A genus of 30 species of deciduous, climbing and twining shrubs with simple, alternate, long stalked leaves and solitary or clustered white flowers. The fruit is a many seeded berry. Some species have edible fruits, others have ornamental value. *Distr.* E Asia. *Fam.* Actinidiaceae. *Class* D.

A. arguta TARA VINE. Fruit similar to that of *A. deliciosa*. *Distr.* NE Asia.

A. chinensis See *A. deliciosa*

A. deliciosa KIWI FRUIT, CHINESE GOOSEBERRY, YANGTAO. Vigorous shrub with white-yellow flowers. Fruit densely hairy, brown with a green pith. Widely cultivated for fruit especially in New Zealand. *Distr.* China.

A. kolomikta (From the local vernacular name.) Erect shrub. Leaves variegated. *Distr.* China, Japan.

A. polygama SILVER VINE. Shrubby. Leaves silvery white to yellow. *Distr.* Central Japan.

Actinidiaceae The Kiwi Fruit family. 3 genera and about 350 species of trees, shrubs and lianas with simple leaves and flowers that usually bear 5 sepals, 5 petals, and numerous stamens. *Distr.* Warm temperate and tropical regions. *Class* D.
See: Actinidia, Saurauia.

Actiniopteridaceae A family of 1 genus and 5 species of small ferns that lack indusia. *Distr.* Africa, SW Asia to Australia. *Class* F.

actinophyllus, -a, -um: with radiating leaves, from the Greek *aktinos*, ray, and *phyllon*, leaf.

Actinotus (From the Greek *aktinotus*, furnished with rays, alluding to the showy spreading bracts.) A genus of 17 species of annual and perennial herbs with simple, toothed or compound leaves and umbels of small flowers. The umbels are subtended by a whorl of large bracts. *Distr.* Australia, New Zealand. *Fam.* Umbelliferae. *Class* D.

A. helianthi FLANNEL FLOWER. Erect perennial. Bracts white-woolly. Grown as an ornamental. *Distr.* Australia.

action plant *Mimosa pudica*.

aculeatus, -a, -um: prickly.

acuminatus, -a, -um: long-pointed, tapering to a narrow tip.

acutangulus, a, -um: from *acutus*, pointed, and *angulus*, angle or corner.

acutidens: from the Latin *acutus*, pointed, and *dens*, teeth.

acutifolius, -a, -um: from *acutus*, pointed, and *folius*, leaved.

acutilobus from *acutus*, pointed, and *lobus*, lobe.

acutissimus very sharply pointed.

acutus, -a, -um: pointed, narrowing to an angle of less than ninety degrees.

Adam's needle *Yucca, Y. smallia, Y. filamentosa.*

adder's meat *Stellaria holostea* **adder's mouth** *Malaxis*.

adder's tongue *Ophioglossum, Erythronium* **common** ~ *Ophioglossum vulgatum* **foetid** ~ *Scoliopus bigelovii* **Oregon foetid** ~ *S. hallii* **yellow** ~ ***Erythronium americanum.***

Adelocaryum See *Lindelofia*

adenocaula with a glandular stem.

adenogynus, -a, -um: with a glandular ovary.

Adenophora (From the Greek *aden*, gland, and *phoros*, bearing, alluding to the gland at the base of the style.) LADYBELLS, GLAND BELLFLOWER. A genus of 40 species of perennial herbs with fleshy roots, alternate or whorled, simple leaves and pendulous, 5-lobed, bell-shaped flowers. Several species are grown as ornamentals. *Distr.* Eurasia. *Fam.* Campanulaceae. *Class* D.

A. potaninii Rosette-forming perennial. Flowers lavender-blue. *Distr.* W China.

adenophyllus, -a, -um: with glandular leaves.

adenopodus, -a, -um: with glandular pedicels.

adenosus, -a, -um: with abundant glands.

adenothrix: with glandular hairs.

adenotrichum: with glandular hairs.

Adiantaceae The Maidenhair fern family. 33 genera and about 900 species of terrestrial, often very small ferns. The sori are borne along the veins or on the leaf margins and lack indusia. *Distr.* Widespread. *Class* F. *See: Adiantum, Cheilanthes, Cryptogamma, Dryopteris, Gymnopteris, Hemionitis, Onychium, Pellaea, Pityrogramma.*

adiantifolius, -a, -um: from the fern genus *Adiantum*, and *folius*, leaf.

adiantiformis from the fern genus *Adiantum* and *formis*, in the form of.

Adiantum (From the Greek *adiantos*, unwetted, alluding to the way the fronds repel water.) MAIDENHAIR FERN. A genus of 200 species of medium to large ferns with compound, occasionally simple leaves. Some species are grown ornamentally and some have medicinal uses. *Distr.* Cosmopolitan but especially tropical America. *Fam.* Adiantaceae. *Class* F.

A. capillus-veneris VENUS'S HAIR, SOUTHERN MAIDENHAIR FERN, COMMON MAIDENHAIR FERN. *Distr.* Widespread.

A. cuneatum See *A. raddianum*

A. pedatum NORTH AMERICAN MAIDENHAIR FERN, FIVE-FINGERED MAIDENHAIR FERN. Black leaf stalks used by American Indians in basketry. *Distr.* North America, E Asia.

A. raddianum (After Giuseppe Raddi (1770–1829).) DELTA MAIDENHAIR FERN. Several cultivars of this species are available. *Distr.* Brazil, W Indies.

Adlumia (After John Adlum (1759–1836), a grape breeder from Pennsylvania.) A genus of 1 species of biennial, climbing herb with pinnately divided, fern-like leaves and pendent panicles of tiny, tubular, white or purple flowers. *Distr.* E North America, Korea. *Fam.* Papaveraceae. *Class* D.

A. fungosa CLIMBING FUMITORY, MOUNTAIN FRINGE, ALLEGHENY VINE. Grown as a hardy ornamental.

Adonis (After the Greek god Adonis who was changed into a flower by Aphrodite, after he was killed by a wild boar.) PHEASANT'S EYE. A genus of about 20 species of annual and perennial herbs with pinnately divided leaves and large, solitary flowers that bear 3–30 yellow or red petals and numerous stamens. Several species are grown ornamentally. *Distr.* Temperate Eurasia. *Fam.* Ranunculaceae. *Class* D.

A. amurensis Clump-forming perennial. Flowers yellow. *Distr.* E Asia.

A. chrysocyathus Erect perennial. Flowers yellow. *Distr.* W Himalaya.

A. vernalis Clump-forming perennial. Flowers yellow. Formerly used medicinally, having a similar effect as *Digitalis*. *Distr.* SE and Central Europe, Siberia, Caucasus.

Adoxa (From the Greek *a*, without, and *doxa*, repute or glory, alluding to the flowers.) A genus of 1 species of delicate, rhizomatous herb with compound leaves that bear 3 leaflets each of which is 3-lobed. The flowers are inconspicuous, yellow-green and borne in a tiny head. *Distr.* N temperate regions. *Fam.* Adoxaceae. *Class* D.

A. moschatellina MUSKROOT, MOSCHATEL. Grown as a hardy ornamental.

Adoxaceae A family of 3 genera and 3 species of small perennial herbs with compound leaves that bear 3 leaflets. The flowers are inconspicuous and bear 2–3 sepals, 4–5 fused petals and an inferior ovary. *Distr.* N temperate regions. *Class* D.
See: Adoxa.

adoxoides: from the genus *Adoxa*, with the ending *-oides*, indicating resemblance.

adpressus, -a, -um: lying against, adpressed.

adscendens: ascending.

adsurgens: rising up, erect.

advenus, -a, -um: adventive.

Aechmea (From the Greek *aichme*, a point, alluding to the pointed sepals.) A genus of 172 species of stemless epiphytic herbs with rosettes of leathery, typically variegated leaves. The flowers are borne in branched or simple spikes on a tall stem and they are often somewhat concealed by brightly coloured bracts. Many species and cultivars are grown as ornamental house plants. *Distr.* Central and South America and the West Indies. *Fam.* Bromeliaceae. *Class* M.

A. distichantha Leaves rounded at tip with grey scales below and dark spines on margins. *Distr.* South America.

A. fasciata URN PLANT, SILVER VASE. Leaves rounded, grey-scaly, silver-banded. One of the most commonly cultivated species. *Distr.* Brazil.

A. fulgens CORAL BERRY. Leaves form a funnel-shaped tank. Flowers borne in pyramidal inflorescences and followed by showy red berries. *Distr.* Brazil.

A. sepiaria See *Poncirus trifoliata*

Aegopodium (From the Greek *aix*, a goat, and *pous*, a foot.) A genus of about 5 species of creeping rhizomatous herbs with compound leaves and compound umbels of yellow or white flowers. *Distr.* Temperate regions of Eurasia. *Fam.* Umbelliferae. *Class* D.

A. podagraria GROUND ELDER, GOUT WEED, BISHOP WEED, HERB GERARD, ASH WEED, GROUND ASH. Rhizome aromatic. Flowers white. Having been introduced to the British Isles in the middle ages as a medicinal and culinary herb it has now become a widespread weed, although there are ornamental cultivars available. *Distr.* Europe, naturalized in North America.

aegyptiacus, -a, um: of Egypt.

aemulus, -a, -um: very similar to, imitating.

Aeonium (The classical Latin name for these plants.) A genus of about 35 species of succulent herbs and shrubs with soft stems and simple leaves that are borne alternately or in rosettes at the ends of branches. The small regular flowers are very numerous and borne in cymes atop an erect stem. A number of species and hybrids are grown ornamentally. *Distr.* Macaronesia, Mediterranean, N Africa. *Fam.* Crassulaceae. *Class* D.

A. arboreum Perennial subshrub. Sap used to harden fishermen's lines. *Distr.* Morocco, naturalized S North and N South America.

A. balsamiferum Perennial sub-shrub smelling of balsam. *Distr.* Lanzarote, Cape Verde Islands.

A. canariense CANARY ISLAND AEONIUM, GIANT VELVET ROSE, VELVET ROSE. Perennial herb. *Distr.* Canary Islands (Tenerife).

A. haworthiae PINWHEEL. Low bushy perennial. Leaves in numerous rosettes. Flowers yellow, tinged pink. *Distr.* Canary Islands (Tenerife).

A. tabuliforme Biennial or perennial herb. Leaves in a very flat, regular rosette. Flowers yellow. *Distr.* Canary Islands (Tenerife).

A.× domesticum See *Aichryson x domesticum*

aeonium, Canary Island *Aeonium canariense*.

aequalis, -e: equal.

aeroplane propeller plant *Crassula falcata*.

Aeschynanthus (From the Greek *aischun*, shame, and *anthos*, flower, alluding to the red flowers.) BASKET PLANT. A genus of 100 species of climbing herbs and shrubs with simple leaves in somewhat unequal pairs. The 2-lipped, 5-lobed funnel-shaped flowers bear 4 stamens and are borne singly or in clusters. Numerous species and cultivars are grown as ornamentals. *Distr.* India to Malaysia. *Fam.* Gesneriaceae. *Class* D.

A. hildebrandii Dwarf, creeping subshrub. Flowers clustered, orange-red. *Distr.* Burma.

A. lobbianus Straggling, much branched shrub. Flowers red, hairy. *Distr.* Java.

A. marmoratus Trailing subshrub. Flowers green with dark brown markings. *Distr.* Burma, Thailand, Malaysia.

A. radicans LIPSTICK VINE. Creeping or epiphytic subshrub. Flowers bright red. *Distr.* Malaysia, Java.

A. speciosus Trailing subshrub. Flowers clustered, tubular, erect, orange-yellow. *Distr.* Malaysia.

Aesculus (The classical Latin name for an oak with edible acorns.) BUCKEYE, HORSE CHESTNUT. A genus of about 13 species of deciduous shrubs and trees with opposite, palmate leaves that bear 5–11 leaflets and leave a U- or V-shaped scar when they fall. The irregular flowers bear 5 fused sepals, 4–5 petals and 5–10 stamens and are borne in showy, erect, spike-like racemes. The fruit is a spiny, scaly or smooth capsule containing 1–6 brown seeds. *Distr.* SE Europe to E Asia, North America. *Fam.* Hippocastanaceae. *Class* D.

A. californica CALIFORNIAN BUCKEYE. Tree or spreading shrub. Flowers white, tinged pink. Fruit 1-seeded. Seeds much eaten by Californian Indians. *Distr.* USA (California).

A. flava YELLOW BUCKEYE, SWEET BUCKEYE. Tree to 30m high. Flowers yellow with 4 petals. *Distr.* USA.

A. glabra OHIO BUCKEYE. Shrub or tree to 30m high. Flowers yellow-green. Fruit 1–3 seeded, prickly. *Distr.* S USA.

A. hippocastanum HORSE CHESTNUT, CONKER TREE. Tree to 25m high. Flowers white with a red central patch. Fruit 1-seeded, spiny. The seeds are the conkers of children's games and may have some medicinal qualities. *Distr.* Balkans to Himalaya.

A. indica INDIAN HORSE CHESTNUT. Domed tree to 30m high. Flowers white with red markings. Fruit 1-seeded, smooth. Horse medicine. *Distr.* NE Himalaya.

A. octandra See *A. flava*

A. parviflora Bushy shrub. Flowers white. Fruit smooth. *Distr.* SE USA.

A. pavia (After Peter Paw (died 1616), Dutch botanist.) RED HORSE CHESTNUT, RED BUCKEYE. Shrub or small tree. Flowers red or occasionally yellow or red and yellow. Fruit smooth. *Distr.* North America.

A. splendens See *A. pavia*

A. turbinata JAPANESE HORSE CHESTNUT. Tree to 30m high. Flowers cream, marked red. Fruit pear-shaped, bumpy. *Distr.* Japan.

aestivalis, -e: referring to summer.

aestivus, -a, -um: of summer.

Aethionema (Possibly from the Greek *aethus*, unusual, and *nema*, a thread, alluding to the stamens.) STONE CRESS. A genus of 30–40 species of annual and perennial herbs with somewhat fleshy leaves and white to red-purple flowers. *Distr.* Europe, Mediterranean and SW Asia. *Fam.* Cruciferae. *Class* D.

A. armenum Densely tufted perennial with pink-white flowers. *Distr.* Turkey, Armenia and the Caucasus.

A. coridifolium Loosely branching perennial with pink flowers. *Distr.* Lebanon.

A. graecum See *A. saxatile*

A. grandiflorum Loosely branching perennial with pink flowers. *Distr.* Iran, Iraq, and the Caucasus.

A. iberideum Perennial with white flowers. *Distr.* Greece, Turkey, and the Caucasus.

A. saxatile Annual or perennial with purple-white flowers. *Distr.* S and S Central Europe.

A. schistosum Perennial with pink flowers. *Distr.* Turkey.

aethiopicus, -a, -um: of Africa.

aethusifolius, -a, -um: from the genus *Aethusa*, and *folius*, leaf.

aetnensis: of Mount Etna, Sicily.

aetolicus, -a, -um: of Aitolia, Greece.

Aextoxicaceae A family of 1 genus and 1 species of tree with simple leaves and racemes of flowers that typically bear their parts in 5s and have a superior, 2-chambered ovary. *Distr.* Chile. *Class* D.

afer: of Africa.

affinis, -e: with an affinity to, allied to.

afghanicus, -a, -um: of Afghanistan

afoliatus, -a, -um: without leaves.

africanus, -a, -um: of Africa.

Agapanthus (From the Greek *agape*, love, and *anthos*, a flower.) AFRICAN LILY. A genus of 10 species of perennial herbs with fleshy roots, linear leaves and umbels of tubular to bell-shaped flowers. Some species are cultivated as ornamentals. *Distr.* South Africa. *Fam.* Alliaceae. *Class* M.

A. africanus AFRICAN LILY, BLUE AFRICAN LILY, LILY OF THE NILE. Leaves evergreen. Flowers deep violet-blue. White-flowered garden forms are available.

A. campanulatus Leaves deciduous. Flowers bell-shaped, blue. A number of cultivars of this species are available. *Distr.* South Africa.

A. praecox Leaves evergreen, somewhat fleshy. Flowers blue or white. *Distr.* South Africa.

A. umbellatus See *A. campanulatus*

agapanthus, pink *Tulbaghia fragrans.*

Agapetes (From the Greek *agapetos*, desirable or beloved.) A genus of 95 species of evergreen shrubs with simple leathery leaves and attractive flowers. The leaves of some species are used in tea in India. *Distr.* Asia to N Australia. *Fam.* Ericaceae. *Class* D.

A. buxifolia An erect shrub with bright red, waxy flowers. *Distr.* Bhutan.

A. serpens A creeping shrub with numerous, bright red and crimson flowers. *Distr.* Nepal, Bhutan, Assam.

Agarista (Named after the beautiful daughter of Kliothenes in Greek mythology.) A genus of 20 species of small evergreen shrubs with leathery leaves. The flowers have 5 fused petals and 10 stamens. *Distr.* South America, Mauritius. *Fam.* Ericaceae. *Class* D.

A. neriifolia Flowers scarlet and pale pink. *Distr.* Brazil.

Agastache (From the Greek *agan*, very much, and *stachys*, a spike, alluding to the numerous flower spikes.) MEXICAN HYSSOP, GIANT HYSSOP. A genus of 20–30 species of aromatic, horizontal or erect, perennial herbs with simple opposite leaves and spikes that bear whorls of 2-lipped flowers. Several species are grown as ornamentals and some species are used as flavourings. *Distr.* Central and E Asia, North America, Mexico. *Fam.* Labiatae. *Class* D.

A. anethiodora See *A. foeniculum*

A. anisata See *A. foeniculum*

A. coelestis See *Felicia amelloides*

A. foeniculum GIANT HYSSOP, ANISE HYSSOP. Stems erect. Flowers blue or white. Used as the basis of a drink. *Distr.* North America.

A. mexicana Stems erect. Flowers pink or red. *Distr.* Mexico.

A. nepetoides YELLOW GIANT HYSSOP. Stems erect, to 2m high. Flowers yellow-green. *Distr.* E North America.

Agathis (From the Greek *agathis*, a ball of thread, alluding to the shape of the female cones.) DAMMAR PINE, KAURI PINE. A genus of 10–20 species of evergreen coniferous trees with flat leathery leaves and globose fruiting cones that take 2 years to ripen. *Distr.* Sumatra to New Zealand. *Fam.* Araucariaceae. *Class* G.

A. australis KAURI, KAURI PINE. Occasionally grown as a half-hardy ornamental. *Distr.* New Zealand (North Island).

Agavaceae The Sisal family. 12 genera and about 400 species of rhizomatous shrubs and climbers with fleshy, sharp-pointed leaves crowded at the base of the stem and regular flowers that bear 2 whorls of 3 petal-like tepals that are fused at their base. A number of species provide fibre, the sap of others is fermented as a beverage. *Distr.* Tropical and subtropical regions, particularly in arid areas. *Class* M.

See: Agave, Beschorneria, Bravoa, Cordyline, Furcraea, Hesperaloe, Manfreda, Polianthes, Yucca.

Agave (From the Greek *agave*, noble.) CENTURY PLANT, MAGUEY. A genus of about 300 species of herbs with short stems and rosettes of large, succulent leaves that bear sharp apical spines. The flowers are usually borne in very tall, paniculate inflorescences. These plants are an important source of a number of products including fibre from the leaves, food in the form of the flowers and young buds, and pulque from the fermented sap which is distilled to give mescal or tequila. *Distr.* North America to N South America. *Fam.* Agavaceae. *Class* M.

A. americana CENTURY PLANT, MAGUEY. Grown as an ornamental with numerous cultivars available and as a hedging plant. *Distr.* Mexico, widely naturalized in the Mediterranean area and elsewhere.

A. cantala CANTALA, MANILA MAGUEY. Widely cultivated, particularly in the Old World, for its fibre which is somewhat softer than sisal. *Distr.* Mexico.

A. filifera THREAD AGAVE. Terminal leaf-spine thread-like. *Distr.* Origin uncertain, probably E Mexico.

A. schottii Rosettes small. *Distr.* Arizona to Mexico (Sonora).

A. sisalana SISAL, BAHAMA HEMP. An important source of fibre. *Distr.* E Mexico.

A. striata Leaves striped dark green. *Distr.* USA, Baja California.

A. vera-cruz A commercial source of fructose. *Distr.* Mexico.

agave, thread *Agave filifera.*

agavifolius, -a, -um: from the genus *Agave*, and *folius*, leaved.

agavoides: from the genus *Agave*, with the ending *-oides*, indicating resemblance.

Agdestidaceae A family of 1 genus and 1 species. *Class* D.
See: Agdestis.

Agdestis A genus of 1 species of tuberous vine with a much branched red stem, simple leaves and panicles of star-shaped fragrant flowers. *Distr.* Central America. *Fam.* Agdestidaceae. *Class* D.

A. clematidea Grown as an ornamental.

Ageratina A genus of 230 species of annual to perennial herbs and shrubs with daisy-like flower-heads that lack distinct rays. *Distr.* Central and W South America, E USA, naturalized in the Old World. *Fam.* Compositae. *Class* D.

A. altissima WHITE SNAKE ROOT. Perennial, to 2m high, blue-grey, hairy. Flowers white. *Distr.* E USA.

A. aromatica Perennial to 1.5m. Flowers white. *Distr.* E USA.

A. glechonophylla Subshrub or perennial herb. Flowers white-pink. *Distr.* Chile.

A. ligustrina Large shrub. Flowers fragrant. *Distr.* Central America.

aggregatus, -a, -um: clustered.

Aglaomorpha (From the Greek *aglaos*, bright, and *morphe*, shape.) A genus of 10 species of epiphytic ferns with creeping fleshy rhizomes and pinnately divided leaves. Several species are grown as tender ornamentals. *Distr.* Tropical Asia. *Fam.* Polypodiaceae. *Class* F.

A. meyeniana BEAR'S PAW FERN. Fronds to 80cm long, pinnately divided at the tip but simple at the base. *Distr.* Taiwan, Philippines.

Aglaonema (From the Greek *aglaos*, bright, clear or manifest, and *nema*, a thread, alluding to the appearance of the stamens.) CHINESE EVERGREEN. A genus of 21 species of evergreen, tufted perennial herbs with simple, variegated leaves and small flowers that are borne on a cylindrical or club-shaped spadix subtended by an ovate spathe. Several species are cultivated as ornamentals, chiefly for their variegated foliage. *Distr.* Tropical Asia. *Fam.* Araceae. *Class* M.

A. commutatum Leaves with bar-shaped variegation. Frequently cultivated as an ornamental with numerous cultivars available. *Distr.* Central Malaysia.

A. crispum PAINTED DROP-TONGUE. *Distr.* Philippines (Mount Bulusan, Luzon).

A. pictum Leaves with irregular silver blotches. Frequently grown as an ornamental. *Distr.* Sumatra.

A. roebelinii See *A. crispum*

agrestis, -e: of fields or cultivated land.

Agrimonia (Possibly a misrendering of the genus name *Argemonia* used by the Greeks and derived from *argema*, a fleck in the eye.) AGRIMONY, HARVEST LICE, COCKLEBUR. A genus of 15 species of rhizomatous perennial herbs with alternate, pinnately divided leaves and spike-like racemes of small yellow flowers. *Distr.* N temperate regions. *Fam.* Rosaceae. *Class* D.

A. eupatoria AGRIMONY. Glandular hairy herb. Flowers golden yellow. Formerly used medicinally and as a source of yellow dye. *Distr.* Europe, N and S Africa, N Asia.

A. odorata See *A. repens*

A. procera FRAGRANT AGRIMONY. Erect perennial. Flowers fragrant. *Distr.* Europe.

A. repens Spreading perennial herb. Leaves leathery. Flowers golden yellow. *Distr.* Turkey, Iraq.

agrimony *Agrimonia*, *A. eupatoria* **fragrant** ~ *A. procera* **hemp** ~ *Eupatorium cannabinum.*

Agropyron (From the Greek *agros*, field, and *puros*, wheat.) WHEATGRASS, DOG GRASS. A genus of about 40 species of rhizomatous, perennial grasses with spike-like inflorescences. Several species are pernicious weeds of cultivated ground. *Distr.* Temperate and cool regions. *Fam.* Gramineae. *Class* M.

A. cristatum CRESTED WHEATGRASS, FAIRWAY CRESTED WHEATGRASS. Used as a drought resistant pasture grass in central North America. *Distr.* Eurasia, naturalized North America.

Agrostemma (From the Greek *agros*, field and *stemma*, a crown or garland.) CORN COCK-

LE. A genus of 2–3 species of annual herbs with simple, opposite, linear leaves and large flowers that bear 5 sepals, that are fused at the base but have spreading leaf-like tips, and 5 petals that have spreading notched limbs. *Distr.* S Europe, W Asia, introduced North America. *Fam.* Caryophyllaceae. *Class* D.

A. coronaria See *Lychnis coronaria*

A. githago CORN COCKLE. Previously a common weed of cereal crops, this beautiful flower has become increasingly rare since the introduction of modern herbicides but is now being grown as a garden ornamental. *Distr.* E Mediterranean region, widely naturalized.

Agrostis (From the Greek *agrostis*, a kind of grass.) BENT. A genus of 120 species of annual and perennial, stoloniferous or tufted grasses with flat to rounded leaves and tall branched inflore scences. Several species are important for lawns and pastures. *Distr.* Cosmopolitan, especially in N temperate regions. *Fam.* Gramineae. *Class* M.

A. canina VELVET BENT, RHODE ISLAND BENT, BROWN BENT. Leaves narrow, to 6cm long. Commonly used in fine lawns such as putting greens. *Distr.* Europe.

A. nebulosa CLOUD GRASS. Tufted annual. Leaves few. Panicle loose. *Distr.* Spain, Portugal and Morocco.

aguacate *Persea americana.*

ague root *Aletris farinosa.*

Aichryson (The Greek name for *Aeonium arboreum.*) A genus of 15 species of succulent, annual and perennial herbs and subshrubs with simple leaves borne along branched stems or in loose rosettes, and yellow star-shaped flowers that are borne in panicles. Several species are grown as ornamentals. *Distr.* Macaronesia to Morocco and Portugal. *Fam.* Crassulaceae. *Class* D.

A. × domesticum YOUTH AND OLD AGE. A hybrid of *A. tortuosum* × *A. punctatum.* Perennial subshrub. Leaves clustered near branch tips. *Distr.* Canary Islands.

ailanthoides from the genus *Ailanthus*, with the ending *-oides*, indicating resemblance.

Ailanthus (From *ailanto*, TREE OF HEAVEN, the Indonesian name for *A. moluccana.*) TREE OF HEAVEN. A genus of 5 species of deciduous shrubs and trees with alternate, pinnate leaves and panicles of small flowers that bear 5 petals and 10 stamens and are followed by winged seeds. Several species are grown as ornamentals. *Distr.* Asia to Australia. *Fam.* Simaroubaceae. *Class* D.

A. altissima TREE OF HEAVEN. Tree to 30m high. Flowers green-white, malodorous. Grown as a street tree and used in soil conservation programmes. *Distr.* China, naturalized in North America, Central and S Europe.

A. glandulosa See *A. altissima*

ailantifolius, -a, -um: from the genus *Ailanthus*, and *folius*, leaved.

Aizoaceae A family of 128 genera and 2500 species of succulent, annual or perennial herbs and small shrubs with simple leaves. The flowers are usually bisexual with around 5 sepals and numerous petals; they often appear superficially daisy-like. In some genera the plants are reduced to just two stone-like leaves. Many ornamental species are found in this family including *Mesembryanthemum* and the Living Stones (*Lithops*). *Distr.* Tropical and subtropical regions, especially South Africa. *Class* D.

See: Aptenia, Carpobrotus, Delosperma, Dorotheanthus, Drosanthemum, Faucaria, Lampranthus, Lithops, Machairophyllum, Malephora, Mesembrianthemum, Mesembryanthemum, Meyerophytum, Mitrophyllum, Monilaria, Neohenricia, Odontophorus, Oophytum, Ophthalmophyllum, Oscularia, Pleiospilos, Rhinephyllum, Ruschia, Tetragonia.

aizoides resembling the genus *Aizoon.*

aizoon resembling the genus *Aizoon.*

Ajania (Of Ajan, E Asia.) A genus of about 30 species of perennial herbs or shrubs with daisy-like flower-heads that bear distinct rays. *Distr.* Central and E Asia. *Fam.* Compositae. *Class* D.

A. adenantha Strongly scented, silky hairy herb. Flowers orange-yellow. Grown as a hardy ornamental. *Distr.* W China.

Ajuga (From the Greek *a*, not, and *zugon*, yoke, alluding to the regular sepals.) BUGLE. A genus of 40–50 species of rhizomatous herbs with 4-angled stems, simple, opposite leaves and whorls of 2-lipped flowers. Several species are grown ornamentally. *Distr.* Temperate regions of the Old World, naturalized S Africa and Australia. *Fam.* Labiatae. *Class* D.

A. genevensis BLUE BUGLE, UPRIGHT BUGLE. Erect perennial to 45cm. Flowers bright blue. *Distr.* Eurasia.

A. metallica See *A. pyramidalis*

A. pyramidalis PYRAMID BUGLE. Evergreen carpet-forming perennial. Flowers blue, somewhat concealed by purple-tinged bracts. *Distr.* Europe.

A. reptans COMMON BUGLE. Carpet-forming perennial with leafy stolons. Flowers azure, subtended by purple-tinted bracts. Numerous ornamental cultivars have been raised from this species. *Distr.* Europe, W Asia.

Akaniaceae A family of 1 genus and 1 species of small tree with spirally arranged, pinnate leaves and panicles of flowers that bear 5 sepals, 5 petals, 8 stamens and a superior ovary. *Distr.* NE Australia. *Class* D.

Akebia (The Latinized version of the Japanese name, *akebi.*) CHOCOLATE VINE. A genus of 4–5 species of twining woody climbers with alternate palmate leaves and pendulous racemes of unisexual, violet-brown flowers in which the petals have been replaced by nectaries and the sepals have become petal-like. Several species are grown ornamentally. *Distr.* Japan, China, Korea. *Fam.* Lardizabalaceae. *Class* D.

A.×pentaphylla A hybrid of *A. quinata* × *A. trifoliata*. Leaves bear 4–5 leaflets. Flowers slightly fragrant. *Distr.* Japan.

A. quinata Leaves deciduous or semi-evergreen, bearing 5 leaflets. Flowers vanilla-scented. *Distr.* China, Japan, Korea, naturalized E USA.

A. trifoliata Leaves deciduous, bearing 3 leaflets. Flowers not scented. *Distr.* China, Japan.

alabamensis, -e: of Alabama, USA.

alabaster plant *Dudleya virens*.

Alangiaceae A family of 1 genus and 17 species. *Class* D.
See: Alangium.

Alangium (From the Tamil name *alangi.*) A genus of 17 species of shrubs, climbers and trees with simple or lobed leaves and cymes of regular flowers that bear 4–10 fused sepals, 4–10 reflexed petals and a superior ovary. Some species are important for their timber or medicinal qualities and some are grown as ornamentals. *Distr.* Tropical Africa, China to E Australia. *Fam.* Alangiaceae. *Class* D.

A. platanifolium Shrub. Leaves deciduous. Flowers white, occasionally orange. Grown as an ornamental. *Distr.* Japan, Korea.

alaskana: of Alaska.

alatus, -a, -um: winged.

albanicus, -a, -um: of Albania.

albanus, -a, -um: of Albana (now Daghestan), Caucasus.

albescens: whitish, turning white.

albicans: somewhat off-white.

albicaulis: white-stemmed.

albidus, -a, -um: whitish.

albiflorus, -a, -um: white-flowered.

albiflos: with white flowers.

albifrons: with white fronds.

Albizia (After F. degli Albizzi, 18th century Italian naturalist.) A genus of 150 species of deciduous shrubs, trees and climbers with bipinnate leaves and clusters or heads of regular flowers that bear numerous stamens. Several species are grown as ornamentals. *Distr.* Tropical regions, chiefly in the Old World. *Fam.* Leguminosae. *Class* D.

A. juilibrissin SILK TREE. Small tree. Inflorescences showy, green-white or pink. *Distr.* Asia, naturalized North America.

A. lophantha PLUME ALBIZIA. Small tree or shrub. Flowers pale-green to golden yellow. *Distr.* Australia.

albizia, plume *Albizia lophantha*.

albo-picta: from *albus*, white, and *picta*, coloured, painted.

albomarginatus, -a, -um: with a white margin.

albopilosus, -a, -um: with white hairs.

albopurpurescens: from *albus*, white, and *purpurescens* purplish.

alboroseus, -a, -um: from *albus*, white, and *roseus*, pure red; the flowers change from white to red.

albostriatus, -a, -um: from *albus*, white and *striatus*, marked with fine lines or streaks.

Albuca (From the Latin *albus*, white.) A genus of 30 species of perennial bulbiferous herbs with flat or rounded leaves and loose racemes of open green-white or yellow flowers that bear 2 whorls of 3 free tepals. Several species are grown as ornamentals. *Distr.* S Africa to Arabia. *Fam.* Hyacinthaceae. *Class* M.

A. canadensis Flowers yellow, occurring in spring. *Distr.* South Africa.

A. humilis Flowers few, white with a green stripe on the outer tepals. *Distr.* South Africa.

A. nelsonii Flowers white with a red or green stripe on each tepal. *Distr.* South Africa.

albus, -a, -um: white.

Alcea (From Greek name *alkaia*, used for a kind of MALLOW.) HOLLYHOCK. A genus of 60 species of biennial and perennial herbs with erect stems, palmately lobed or divided leaves and racemes of showy regular flowers that bear their parts in 5s. Several species are grown ornamentally. *Distr.* Mediterranean to Central Asia. *Fam.* Malvaceae. *Class* D.

A. ficifolia FIG LEAVED HOLLYHOCK, ANTWERP HOLLYHOCK. Biennial or perennial herb. Leaves 6-lobed. Flowers yellow or orange. *Distr.* Siberia.

A. rosea HOLLYHOCK. Biennial or perennial herb to 3m high. Flowers purple to white. Numerous cultivars of this species have been raised. *Distr.* Origin unknown, probably W Asia.

Alchemilla (From *alkemelych*, the Arabic name for these plants.) LADY'S MANTLE. A genus of 250–300 species of perennial herbs with palmately lobed or divided leaves and cymes of small flowers that lack petals. Several species are grown ornamentally. *Distr.* Eurasia, rarer in North America and in mountainous regions of the tropics. *Fam.* Rosaceae. *Class* D.

A. alpina ALPINE LADY'S MANTLE. Mat-forming perennial. Leaves densely, silky-hairy beneath. *Distr.* Mountainous regions of W and Central Europe, and N Europe.

A. mollis Robust perennial to 80cm high. Foliage densely hairy throughout. *Distr.* E Carpathians, Turkey, Caucasus.

alchimilloides: from the genus *Alchemilla* with the ending *-oides*, indicating resemblance.

alder *Alnus* **black** ~ *Ilex verticillata*, *Alnus glutinosa* **common** ~ *A. glutinosa* **European** ~ *A. glutinosa* **green** ~ *A. viridis* **grey** ~ *A. incana* **hazel** ~ *A. serrulata* **Italian** ~ *A. cordata* **Japanese** ~ *A. japonica* **Manchurian** ~ *A. hirsuta* **mountain** ~ *A. tenuifolia* **Oregon** ~ *A. rubra* **red** ~ *A. rubra* **smooth** ~ *A. serrulata* **speckled** ~ *A. rugosa* **thinleaf** ~ *A. tenuifolia* **white** ~ *Clethra*, *Alnus incana*.

alecost *Tanacetum balsamita*.

Alectryon (From the Greek *alektryon*, cock, alluding to the clusters of fruit in some species, which resemble a cock's comb.) A genus of 15–20 species of evergreen trees with pinnately divided, alternate leaves and panicles of small flowers that lack petals. *Distr.* W Pacific to New Zealand. *Fam.* Sapindaceae. *Class* D.

A. excelsus TITOKI. Small tree. Bark black. Flowers dark red. The tough wood is used for tool handles. *Distr.* New Zealand.

alehoof *Glechoma hederacea*.

alerce *Fitzroya cupressoides*.

Aletris (From the Greek *aletris*, the female slave who ground meal, alluding to the appearance of the flowers.) STARGRASS, COLIC ROOT. A genus of about 12 species of rhizomatous perennials with grass-like leaves and spikes of small, tubular, wrinkled flowers. *Distr.* E North America and E Asia. *Fam.* Melanthiaceae. *Class* M.

A. farinosa COLIC ROOT, AGUE ROOT, CROW ROOT, UNICORN ROOT. Leaves ribbed, yellow-green. Flowers white. Reported to have several medicinal properties notably as a diuretic; also grown as an ornamental. *Distr.* SE USA.

aleuticus, -a, -um: of the Aleutian Islands, North Pacific.

alexanders *Smyrnium olusatrum* **golden** ~ *Zizia aurea*.

Alexander's surprise *Phlox subulata*.

alfa *Stipa tenacissima.*

alfalfa *Medicago sativa.*

algaroba *Prosopis chilensis.*

algarvensis, -e: of the Algarve.

algeriensis, -e: of Algeria, North Africa.

algidus, -a, -um: cold.

alienus, -a, -um: alien or foreign.

Alisma (The Greek name for a water plant.) WATER PLANTAIN. A genus of 9 species of aquatic or marginal, perennial herbs with ovate or heart-shaped leaves and small, white-green or purple flowers. Several species are grown as ornamental bog and water garden plants. *Distr.* N Temperate regions. *Fam.* Alismataceae. *Class* M.

A. lanceolatum Flower purple-pink. *Distr.* Europe, Asia and N Africa.

A. plantago-aquatica WATER PLANTAIN. Reported to be edible and to have medicinal qualities, but the base is believed to be poisonous. *Distr.* N temperate regions.

Alismataceae The Water Plantain family. 12 genera and about 90 species of aquatic or amphibious, annual and perennial herbs with strap-shaped submerged leaves and broad floating leaves. The flowers have 3 free, short-lived petals. Some species are cultivated as ornamentals and a few have edible corms. *Distr.* Widespread but centred in the New World. *Class* M.
See: Alisma, Echinodorus, Luronium, Sagittaria.

alismoides from the genus *Alisma*, with the ending *-oides*, indicating resemblance.

alkanet *Alkanna tinctoria* **bastard** ~ *Lithospermum arvense.*

Alkanna (From the Arabic name for this plant, *al-hinna*.) A genus of 25–30 species of glandular, annual and perennial herbs with numerous, alternate, simple leaves and cymes of funnel- or salver-shaped flowers. Several species are grown as ornamentals. *Distr.* S Europe, Mediterranean to SW Asia. *Fam.* Boraginaceae. *Class* D.

A. lehmannii See *A. tinctoria*

A. orientalis Perennial herb. Flowers fragrant, white or yellow. *Distr.* S Europe, SW Asia.

A. tinctoria ALKANET. Perennial herb. Flowers bright blue. The root is a source of red dye as a colouring for food and alcohol thermometers. *Distr.* S Europe.

Allamanda (After Fredrick Allamanda (1735–*c.* 1795), Swiss botanist who introduced *Allamanda cathartica* to Europe.) A genus of 12 species of erect or climbing shrubs with simple leaves and cymes of large, yellow to mauve, funnel-shaped flowers. Several species are grown as ornamentals. *Distr.* Tropical America. *Fam.* Apocynaceae. *Class* D.

A. blanchetii PURPLE ALLAMANDA. Flowers purple with a dark throat. *Distr.* South America.

A. cathartica GOLDEN TRUMPET, COMMON ALLAMANDA. Vigorous climbing shrub. Flowers yellow. *Distr.* South America.

A. schottii BUSH ALLAMANDA. Erect shrub. Flowers yellow, marked red-orange. *Distr.* South America.

A. violacea See *A. blanchetii*

allamanda: bush ~ *Allamanda schottii* **common** ~ *A. cathartica* **purple** ~ *A. blanchetii.*

Allardia A genus of about 5 species of perennial herbs with pinnately divided leaves and large daisy-like flower-heads that bear distinct rays. *Distr.* Himalaya, Central Asia. *Fam.* Compositae. *Class* D.

A. tomentosa Leaves hairy. Rays numerous, pink-white. *Distr.* W Himalaya.

all good *Chenopodium bonus-henricus.*

Alliaceae The Onion family. 31 genera and about 850 species of herbs with bulbs, corms or rhizomes and linear leaves that often smell of ONION when crushed. The flowers are borne in few- to many-flowered umbels. Many species are grown as ornamentals, others, notably ONION and GARLIC, are of great economic importance. *Distr.* Widespread. *Class* M.
See: Agapanthus, Allium, Bloomeria, Brodiaea, Caloscordum, Dichelostemma, Ipheion, Leucocoryne, Milla, Muilla, Nectaroscordum, Nothoscordum, Tristagma, Triteleia, Tulbaghia.

alliaceus, -a, -um: from the genus *Allium*, with the ending *-aceus*, indicating resemblance.

Alliaria A genus of 5 species of biennial or perennial herbs with kidney-shaped leaves and white flowers. *Distr.* Europe and temperate Asia. *Fam.* Cruciferae. *Class* D.

A. officinalis See *A. petiolata*

A. petiolata JACK BY THE HEDGE, HEDGE GARLIC, GARLIC MUSTARD. Biennial herb. Used as a flavouring. *Distr.* Europe.

Allium (The Greek name for GARLIC, possibly from the Celtic *all*, hot.) ONION. A genus of about 700 species of biennial and perennial herbs with foliage that has a characteristic onion smell especially when crushed. The linear leaves arise from solitary or clustered bulbs and the flowers are borne in umbels. A large number of species are cultivated as ornamentals, others are important food plants. *Distr.* Europe, Asia and America, most abundant in Central Asia. *Fam.* Alliaceae. *Class* M.

A. albopilosum See *A. christophii*

A. ampeloprasum WILD LEEK, LEVANT GARLIC, KURRAT. Flowers very numerous, cup-shaped, borne in a spherical head. *Distr.* S Europe, Caucasus to N Africa.

A. angulosum MOUSE GARLIC. Flowers cup-shaped, white-purple in a hemispherical umbel. *Distr.* Europe to Siberia.

A. bulgaricum See *Nectaroscordum siculum*

A. canadense CANADA GARLIC, MEADOW LEEK, ROSE LEEK. Flowers bell-shape, white-pink. Bulbs consumed by American Indians. *Distr.* North America.

A. carinatum KEELED GARLIC. Bulbs small, flowers cup-shaped, purple. *Distr.* Europe to Turkey and C Asia.

A. cepa SHALLOT, TREE ONION, ONION. A variable biennial, often with large bulbs, a hollow stem and white-green flowers. There are a number of varieties and many cultivars of this species, the edible bulbs of which are, perhaps, the most important flavouring in the majority of the world's kitchens. *Distr.* Not known in the wild.

A. cernuum LADY'S LEEK, NODDING ONION, WILD ONION. Flowers cup-shaped, in a pendulous umbel. Edible (strong). *Distr.* North America from Canada to Mexico.

A. chinense RAKKYO. Similar to CHIVES. but not so stiffly erect. Ingredient in pickles. *Distr.* Central and E China.

A. christophii STAR OF PERSIA. Flowers star-shaped, violet. *Distr.* Iran to Turkey.

A. cirrhosum See *A. carinatum*

A. cowanii See *A. neapolitanum*

A. fistulosum WELSH ONION, JAPANESE LEEK, JAPANESE BUNCHING ONION. Leaves hollow, round and sheathing the stem. The leaves are eaten in salads. *Distr.* Originated in the Far East, but not known in the wild; naturalized on turf roofs in Norway.

A. flavum SMALL YELLOW ONION. Stem to 30cm high. Flowers yellow, sweet scented. *Distr.* Central Europe to W Asia.

A. glaucum See *A. senescens*

A. moly MOLY, LILY ONION, YELLOW ONION. Flowers numerous, star-shaped, yellow. *Distr.* E Spain and SW France.

A. neapolitanum DAFFODIL GARLIC, FLOWERING GARLIC, NAPLES GARLIC. Flowers numerous, large, cup-shaped, white. *Distr.* S Europe, N Africa, Turkey.

A. odorum See *A. ramosum*

A. paradoxum FEW-FLOWERED LEEK. Bulb only bears a single leaf. *Distr.* Caucasus, Iran.

A. pulchellum See *A. carinatum*

A. ramosum FRAGRANT-FLOWERED GARLIC. Tepals white with a red midrib. *Distr.* Central Asia.

A. roseum ROSY GARLIC. Flowers bell-shaped, white-pink. *Distr.* S Europe, N Africa, Turkey.

A. sativum GARLIC. Bulbs clustered. Flowers small, white-pink, rarely produced. The second most important *Allium* crop after ONION, GARLIC plays a major roll as a flavouring in many cuisines, has numerous medicinal qualities and is of importance in folklore. *Distr.* Not known in the wild.

A. schoenoprasum CHIVES. Leaves cylindrical, hollow. Flowers white, lilac or purple. Leaves used as a flavouring. *Distr.* Eurasia, Europe, North America.

A. scorodoprasum SPANISH GARLIC, GIANT GARLIC, SAND LEEK, ROCAMBOLE. Bulbs are used as a flavouring in a similar way to GARLIC. but have a milder flavour. *Distr.* Central and E Europe, Caucasus.

A. senescens GERMAN GARLIC. Flowers cup-shaped, violet. *Distr.* Europe to Siberia.

A. siculum See *Nectaroscordum siculum*

A. sphaerocephalon ROUND-HEADED GARLIC, ROUND-HEADED LEEK. Bulblets red. Flowers lilac to dark purple. Bulbs edible. *Distr.* Europe, N Africa, W Asia.

A. tanguticum LAVENDER GLOBE LILY. Flowers numerous, purple. *Distr.* W China.

A. triquetrum TRIQUETROUS LEEK, THREE-CORNERED LEEK. Stem 3-angled. Flowers white, pendent. *Distr.* W Mediterranean, naturalized in Britain.

A. tuberosum CHINESE CHIVES, GARLIC CHIVES, ORIENTAL GARLIC. Bulbs clustered on a rhizome. Flowers numerous, fragrant. Eaten in Asia. *Distr.* SE Asia.

A. ursinum RAMSONS, WILD GARLIC, WOOD GARLIC. Leaves broad, stalked. Flowers star-shaped, white. *Distr.* Europe, Caucasus.

A. vineale CROW GARLIC, FALSE GARLIC, STAG'S GARLIC. Leaves hollow. Flowers white or pink. *Distr.* Europe, N Africa, W Asia.

A. violaceum See *A. carinatum*

allspice *Pimenta dioica* **Californian** ~ *Calycanthus occidentalis* **Carolina** ~ *C. floridus*, **wild** ~ *Lindera.*

almond *Prunus dulcis* **Barbados** ~ *Terminalia catappa* **dwarf flowering** ~ *Prunus glandulosa* **earth** ~ *Cyperus esculentus* **green** ~ *Pistacia vera* **Indian** ~ *Terminalia catappa* **wild** ~ *T. catappa.*

alnifolius, -a, -um: from the genus *Alnus*, and *folius*, leaved.

Alnus (The Latin name for ALDER.) ALDER. A genus of 35 species of deciduous shrubs and trees with simple, toothed leaves and small, unisexual, wind pollinated flowers. The male flowers are borne in clusters of pendent catkins and the females in small, spreading or pendent, cone-like catkins that become woody in fruit. Some species are grown as ornamentals and some are important for their timber. *Distr.* N hemisphere, Andes of South America. *Fam.* Betulaceae. *Class* D.

A. cordata ITALIAN ALDER. Tree to 30m high. Leaves heart-shaped at base. *Distr.* Corsica and S Italy.

A. glutinosa BLACK ALDER, COMMON ALDER, EUROPEAN ALDER. Tree to 20m high. Young shoots sticky. The wood is good for carving and has been used in the manufacture of musical instruments, clogs, boats and furniture; it is also used in the manufacture of charcoal and in construction, and is reported to have been used in some of the early piles in Venice. The bark has been used in the tanning industry, having the tendency to turn leather red. There are numerous ornamental cultivars available. *Distr.* Europe, W Asia, mountains of N Africa.

A. hirsuta MANCHURIAN ALDER. Tree, to 20m high. *Distr.* E Asia.

A. incana WHITE ALDER, GREY ALDER. Tree, to 20m high. Leaves white-grey, hairy on the underside. Numerous ornamental cultivars are available. *Distr.* Europe, W Siberia.

A. japonica JAPANESE ALDER. Conical tree, to 25m high. Grown as an ornamental as well as being as source of charcoal for gunpowder. *Distr.* NE Asia, Philippines.

A. oregona See *A. rubra*

A. rubra RED ALDER, OREGON ALDER. Tree, to 30m high. Young shoots red. *Distr.* W North America.

A. rugosa SPECKLED ALDER. Shrub or small tree, to 6m high. *Distr.* E North America, naturalized Central Europe.

A. serrulata HAZEL ALDER, SMOOTH ALDER. Shrub or small tree, to 4m high. *Distr.* E USA.

A. tenuifolia THINLEAF ALDER, MOUNTAIN ALDER. Tree, to 10m high. *Distr.* W North America.

A. viridis GREEN ALDER. Shrub to 3m high. *Distr.* Mountains of Central and E Europe, Corsica.

Aloaceae The Aloe family. 5 genera and about 300 genera species of rosette-forming herbs, shrubs and trees. This family is sometimes included within Liliaceae. *Distr.* Arabia, Africa, Madagascar and Mascarenes. *Class* M. *See: Aloe, Gasteria, Haworthia, Poellnitzia.*

Alocasia (From the Greek *a*, without or not, and *Calocasia*, a related genus.) ELEPHANT'S EAR PLANT. A genus of 70 species of perennial herbs with large, typically peltate, darkly marked leaves on long leaf-stalks. The flowers are borne on a spike-like spadix which has a sterile apical appendage and is surrounded by a green-yellow spathe. Several species and numerous hybrids are grown as ornamentals, chiefly for their foliage. *Distr.* Tropical S and SE Asia. *Fam.* Araceae. *Class* M.

A. macrorrhiza GIANT TARO. Cultivated both as a ornamental and for its edible rhizomes. *Distr.* India to Malaysia.

A. picta See *A. veitchii*

A. veitchii Leaves arrow-head shaped, dark green above, red-purple below. Often grown as a pot plant. *Distr.* Borneo.

Aloe (Possibly from the Arabic *alloch*, or the Hebrew *allal*, bitter.) A genus of about 300 species of perennial herbs, shrubs and trees typically with rosettes of succulent leaves and racemes or panicles of red or yellow, bird-pollinated flowers that are borne on tall stems. The yellow juice of leaves has medicinal qualities. *Distr.* Mostly S Africa; also Madagascar, the Canary Islands, E Africa, S Arabian peninsula. *Fam.* Aloaceae. *Class* M.

A. arborescens CANDELABRA PLANT, TORCH PLANT, OCTOPUS PLANT. Large, branched shrub producing showy racemes of red flowers. *Distr.* South Africa, Zimbabwe, Mozambique, Malawi.

A. aristata LACE ALOE, TORCH PLANT. Rosettes stemless, clustered. Inflorescence simple with 20–30 flowers. *Distr.* Cape Province, Orange Free State, Natal.

A. barbadensis See *A. vera*

A. brevifolia Rosettes stemless, clustered. Inflorescence somewhat conical. *Distr.* Cape Province.

A. camperi Rosettes on erect or decumbent stems. *Distr.* Ethiopia.

A. ciliaris CLIMBING ALOE. Stems slender, climbing to 5m. *Distr.* Cape Province.

A. ferox CAPE ALOE. Rosettes on tall branched stems that are covered in old leaves. Source of Cape aloes. *Distr.* Cape Province.

A. humilis SPIDER ALOE, CROCODILE JAWS, HEDGEHOG ALOE. Rosettes stemless and aggregated into clumps. *Distr.* Cape Province.

A. karasbergensis Rosettes short-stemmed. Inflorescences much branched, pyramidal. *Distr.* Cape Province.

A. mitriformis Rosettes on long procumbent stems. *Distr.* Cape Province.

A. saponaria SOAP ALOE. Rosettes stemless or on short stems. *Distr.* South Africa, Zimbabwe, Botswana.

A. striata CORAL ALOE. Stems decumbent and clothed in dead leaves below the terminal rosette. *Distr.* Cape Province.

A. vera The commerical source of aloe, the leaves being used in the manufacture of cosmetics. *Distr.* Mediterranean, Cape Verde Islands to W Indies and Central America.

aloe *Aloaceae* **Cape** ~ *Aloe ferox* **climbing** ~ *A. ciliaris* **cobweb** ~ *Haworthia arachnoidea* **coral** ~ *Aloe striata* **green** ~ *Furcraea foetida* **hedgehog** ~ *Aloe humilis* **lace** ~*A. aristata* **soap** ~ *A. saponaria* **spider** ~ *A. humilis*.

aloides: from the genus *Aloe*, with the ending *-oides*, indicating resemblance.

aloifolius, -a, -um: from the genus *Aloe*, and *folius*, leaved.

Alonsoa (After Alonzo Zanoni, Secretary of State of Colombia in the 18th century.) MASK FLOWER. A genus of 6 species of herbs and shrubs with simple leaves and terminal racemes of scarlet or orange, more or less regular, funnel-shaped flowers that bear 5 fused petals and 4 stamens. Several species are grown as ornamentals. *Distr.* Tropical America. *Fam.* Scrophulariaceae. *Class* D.

A. acutifolia Small shrub. Leaves sharply toothed. Flowers orange-red. *Distr.* Peru, Bolivia.

A. warscewiczii (After Joseph Warscewicz (1812–66) who collected in South America.) MASK FLOWER. Perennial herb or shrub. Stems tinged red. Flowers scarlet. Often grown as a pot plant or as an annual. *Distr.* Peru.

alopecuroides: from the genus *Alopecurus*, with the ending *-oides*, indicating resemblance.

Alopecurus (From *alopekouros*, the Greek name used for foxtail-like grasses.) FOXTAIL, MEADOW GRASS. A genus of 25 species of annual and perennial grasses with flat leaves and cylindrical, spike-like inflorescences. Several species are grown as ornamentals or as meadow grasses; others are aggressive weeds. *Distr.* Temperate regions of the N hemisphere. *Fam.* Gramineae. *Class* M.

A. alpinus Low-growing perennial. *Distr.* Scotland, Arctic Europe.

A. lanatus Perennial. Leaves and stems densely white-hairy. An attractive ornamental. *Distr.* E Mediterranean.

A. pratensis MEADOW FOXTAIL. Often grown as an ornamental, with several variegated cultivars available. *Distr.* Europe and N Asia, naturalized in North America.

Aloysia (After Maria Louis (died 1819), Princess of Parma.) A genus of 37 species of aromatic deciduous shrubs with simple, opposite or occasionally whorled leaves and spikes or racemes of small tubular flowers. *Distr*. The New World. *Fam*. Verbenaceae. *Class* D.

A. citriodora See *A. triphylla*

A. triphylla LEMON VERBENA, LIMO-NETTO, CIDRON. Shrub to 3m high. Leaves in whorls of 3 or 4. Flowers white. Grown as an ornamental for its fragrant foliage which is also used as a flavouring and herbal tea. *Distr*. Argentina, Chile.

alpestris, -e: of lower mountains.

alpicola: growing on mountains.

Alpinia (After Prospero Alpino (1553–1616), Italian botanist.) GINGER LILY. A genus of 200–250 species of rhizomatous herbs with reed-like stems, 2 ranks of linear leaves and racemes or panicles of 2-lipped flowers. The rhizomes are ginger-scented. *Distr*. Asia and Australasia. *Fam*. Zingiberaceae. *Class* M.

A. officinarum GALANGAL. Rhizome a source of essential oil. *Distr*. China.

A. purpurata RED GINGER. Floral bracts red. *Distr*. Islands of SE Pacific.

A. sanderae See *A. vittata*

A. vittata VARIEGATED GINGER. Leaves striped cream. *Distr*. Islands of the SE Pacific.

A. zerumbet PINK PORCELAIN LILY, SHELL GINGER. *Distr*. Tropical and subtropical Asia.

alpinus, -a, -um: alpine.

Alseuosmia (From the Greek *alsos*, grove, and *euosmia*, a pleasing fragrance.) A genus of 5 species of evergreen shrubs with simple alternate leaves and fragrant funnel-shaped flowers. *Distr*. New Zealand. *Fam*. Alseuosmiaceae. *Class* D.

A. macrophylla Shrub to 2m high. Flowers 2cm long, red or cream. Grown as a tender ornamental.

Alseuosmiaceae A family of 3 genera and 8 species of evergreen shrubs with simple leaves and regular fragrant flowers that typically bear their parts in 5s. *Distr*. Australia, New Zealand, New Caledonia. *Class* D.
See: Alseuosmia, Wittsteinia.

Alsobia See *Episcia*

A. dianthiflora See *Episcia dianthiflora*

Alstroemeria (After Baron Claus Alstroemer (1736–94), Swedish naturalist and pupil of Linnaeus.) PERUVIAN LILY, LILY OF THE INCAS. A genus of 50 species of tuberous or rhizomatous herbs with leafy stems and umbels of irregular, attractively marked flowers. Some species are grown as garden ornamentals or as long lasting cut flowers; the roots of others are used as a source of starch. *Distr*. South America. *Fam*. Alstroemeriaceae. *Class* M.

A. aurantiaca See *A. aurea*

A. aurea Flowers yellow or orange with red flecks. *Distr*. Chile.

A. brasiliensis Flowers yellow-red, flecked brown. *Distr*. Central Brazil.

A. haemantha Flowers deep red to orange, striped purple. *Distr*. Chile.

A. hookeri Flowers pink and yellow with maroon flecks. *Distr*. Chile, Peru.

A. pelegrina Flowers cream, tinged pink with maroon flecks. *Distr*. Peru.

A. psittacina Flowers green, tinged with red-purple and marked with red. *Distr*. N Brazil.

A. pulchella See *A. psittacina*

A. pygmaea Dwarf. Flowers yellow, spotted red. *Distr*. Argentina.

Alstroemeriaceae A family of 4 genera and 152 species of rhizomatous herbs and climbers with somewhat irregular, often showy flowers. The flowers bear 2 whorls of 3 tepals and 6 stamens. This family is sometimes included within the family Liliaceae. Numerous species are cultivated both as hardy and as tender ornamentals and for cut flowers. *Distr*. Mexico to South America. *Class* M.
See: Alstroemeria, Bomarea.

altaicus, -a, -um: of the Altai Mountains, Central Asia.

altaiensis, -e: of the Altai mountains, Central Asia.

alternans: alternating.

alternifolius, -a, -um: with alternate leaves.

Althaea (From the Greek *althaine*, to heal or cure, alluding to the medicinal qualities of some of the species.) A genus of 12 species of annual and perennial herbs with palmately lobed leaves and racemes or panicles of regular bisexual flowers that bear 5 entire or notched petals and 10 fused stamens. Several species are grown ornamentally. *Distr.* Europe to Siberia. *Fam.* Malvaceae. *Class* D.

A. officinalis MARSH MALLOW, WHITEMALLOW. Grey-hairy perennial herb to 2m high. Flowers lilac-pink. *Distr.* Europe, naturalized E USA.

A. rosea See *Alcea rosea*

Altingiaceae See Hamamelidaceae

altissimus, -a, -um: tallest, from the Latin *altus*, tall.

altus, -a, -um: tall.

aluminium plant *Pilea cardierei.*

alum root *Heuchera.*

Alyogyne A genus of 4 species of shrubs with simple to palmately divided leaves and showy, regular, bisexual flowers that bear 5 lilac petals and 10 fused stamens. Several species are grown ornamentally. *Distr.* S and W Australia. *Fam.* Malvaceae. *Class* D.

A. huegelii LILAC HIBISCUS. Sparsely hairy shrub to 2.5m high. Flowers with a dark central spot. *Distr.* SW Australia.

Alyssoides A genus of 3 species of perennial herbs with small yellow flowers. *Distr.* Eurasia. *Fam.* Cruciferae. *Class* D.

A. utriculata Leaves densely crowded and hairy. *Distr.* Alps, Balkans.

alyssoides from the genus *Alyssum*, with the ending *-oides*, indicating resemblance.

Alyssum (From the Greek *a*, not, and *lyssa*, madness; it was said to cure rabies.) MADWORT. A genus of 168 species of annual and perennial herbs and subshrubs with simple leaves and small flowers. *Distr.* Mediterranean to Siberia. *Fam.* Cruciferae. *Class* D.

A. argenteum Dense perennial with yellow flowers. *Distr.* S Europe.

A. cuneifolium Tufted perennial with notched yellow petals. *Distr.* S Europe.

A. idaeum Diffuse perennial. *Distr.* Greece and Crete.

A. moellendorfianum Tufted perennial with notched yellow petals. *Distr.* Balkans, Yugoslavia.

A. montanum Tufted perennial with bright yellow, notched petals. *Distr.* Europe.

A. murale YELLOW TUFT. Tufted perennial with yellow flowers. *Distr.* E Europe.

A. pulvinare Cushion-forming perennial with lemon yellow flowers. *Distr.* Bulgaria.

A. repens Diffuse to erect perennial with golden yellow, notched petals. *Distr.* S and Central Europe to SW Asia.

A. saxatile *Distr.* S and Central Europe, Turkey.

A. serpyllifolium Procumbent perennial with pale yellow flowers. *Distr.* SW Europe.

A. spinosum Small, somewhat spiny shrub with white flowers. *Distr.* S France, Spain.

A. stribrnyi Diffuse perennial with orange-yellow notched petals. *Distr.* E Balkans to W Turkey.

A. tortuosum Prostrate to erect perennial. *Distr.* SE Europe.

A. wulfenianum Erect to prostrate perennial with pale yellow flowers. *Distr.* S Europe and Turkey.

Alzateaceae A family of 1 genus and 1 species of tree with opposite simple leaves and very small regular flowers. *Distr.* Peru, Bolivia. *Class* D.

amabilis, -e: beautiful.

Amana See *Tulipa*

amanum: of the Amanus mountains, Turkey.

Amaranthaceae The Cockscomb family. 70 genera and 800 species of herbs and shrubs with opposite or alternate, simple leaves and very small flowers that lack petals. Some species are grown as ornamentals, others as potherbs and vegetables. *Distr.* Widespread. *Class* D.

See: Amaranthus, Celosia, Iresine.

amaranthoides: from the genus *Amaranthus*, with the ending *-oides*, indicating resemblance.

Amaranthus (From the Greek *a-*, without, and *marain*, to wither, alluding to inflorescences of some species that appear to be everlasting.) A genus of 60 species of erect or prostrate, annual herbs with alternate, long-stalked leaves and catkin-like cymes of small, unisexual, red or green, flowers. A number of species are grown as ornamentals; others are weeds or are eaten as vegetables or as grain. *Distr.* Cosmopolitan. *Fam.* Amaranthaceae. *Class* D.

A. caudatus LOVE LIES BLEEDING, VELVET FLOWER, TASSEL FLOWER, INCA WHEAT. Erect annual or short-lived perennial. Inflorescences tassel-like, pendent. Flowers red or green. Widely grown as a grain crop in Central and South America before being replaced by European cereals in the 16th century. A number of ornamental varieties are available for the garden. *Distr.* South America, Africa, India.

A. tricolor TAMPALA, CHINESE SPINACH. Leaves green or purple. Inflorescences numerous. Flowers pale green or red. Grown as an ornamental for its coloured foliage, with a number of different cultivars available; also eaten as a leaf vegetable. *Distr.* Africa, India, China.

amarissimus, -a, -um: very bitter, from *amarus*, bitter.

×*Amarygia* (From the names of the parent genera.) A genus of hybrids between members of the genera *Amaryllis* and *Brunsvigia*. *Fam.* Amaryllidaceae. *Class* M.

×*A. parkeri* A hybrid of *A. belladonna* × *B. josephinae* with red or white flowers. A number of cultivars of the hybrid are grown ornamentally. *Distr.* Garden origin.

Amaryllidaceae The Daffodil family. 65 genera and about 1000 species of herbs with bulbs or more rarely rhizomes and strap-shaped leaves. The flowers have 2 whorls of 3 tepals that may be fused or free. In many genera a corona (crown) is also present and is most marked in *Narcissus*. Many horticulturally important plants occur in this family including *Narcissus* (DAFFODIL), *Galanthus* (SNOWDROP), *Leucojum* (SNOW-FLAKE) and *Hippeastrum* (AMARYLLIS). *Distr.* Widespread but mostly warm temperate and subtropical regions. *Class* M.

See: × *Amarine*, × *Amarygia*, *Zephyranthes*, *Amaryllis*, *Brunsvigia*, *Chlidanthus*, *Clivia*, *Crinum*, *Cybistetes*, *Cyrtanthus*, *Elisena*, *Eucharis*, *Eustephia*, *Galanthus*, *Griffinia*, *Habranthus*, *Haemanthus*, *Hippeastrum*, *Hymenocallis*, *Lapiedra*, *Leucojum*, *Lycoris*, *Narcissus*, *Nerine*, *Pamianthe*, *Pancratium*, *Paramongaia*, *Phaedranassa*, *Rhodophiala*, *Scadoxus*, *Sprekelia*, *Stenomesson*, *Sternbergia*, *Urceolina*, *Vagaria*, *Worsleya*.

Amaryllis (After a shepherdess, *Amaryllis*, who occurs in Greek mythology.) A genus of 1 species of bulbiferous perennial herb with 2 ranks of strap-shaped leaves and an umbel of pink to purple, funnel-shaped flowers. *Distr.* South Africa. *Fam.* Amaryllidaceae. *Class* M.

A. belladonna JERSEY LILY, BELLADONA LILY. A frequently cultivated ornamental with numerous cultivars available. Specially prepared bulbs are often produced to flowers at or around Christmas.

amaryllis *Hippeastrum* **blue** ~ *H. procerum*.

amazonicus, -a, -um: of the Amazon river basin, Brazil.

Amazon sword plant *Echinodorus paniculatus* **dwarf** ~ *E. magdalenensis*.

ambiguus, -a, -um: ambigous

amberbell *Erythronium americanum*.

amboinensis: of Ambon, Moluccas, Indonesia.

Amborellaceae A family of 1 genus and 1 species of evergreen shrub with simple, spirally arranged leaves and flowers that bear 5–8 tepals and numerous stamens. The carpels are not fused into an ovary and are open at the tip. *Distr.* New Caledonia. *Class* D.

Amelanchier (From *amélanchier*, the French name for *Amelanchier ovalis*.) SERVICE BERRY, SUGAR PLUM, SHAD, SHAD BUSH, JUNE BERRY. A genus of 6 species of deciduous shrubs and trees with simple, alternate leaves and racemes of regular, white flowers that are followed by purple globose fruits. Several species are grown ornamentally and some have edible fruits which are eaten fresh, dried or made into preserves. *Distr.* N temperate regions, especially North America. *Fam.* Rosaceae. *Class* D.

A. alnifolia ALDER-LEAVED SERVICE BERRY. Shrub or small tree. Flowers small, fragrant, cream-white. Grown as a windbreak and to control soil erosion. *Distr.* NW North America.

A. asiatica Shrub or spreading tree. Flowers star-shaped, white, borne in profusion in late spring. Fruits juicy, edible. *Distr.* Japan, China, Korea.

A. canadensis Dense shrub. Flowers star-shaped. Leaves white-hairy at first, turning red in autumn. Fruits juicy, edible. *Distr.* E North America.

A. florida See *A. alnifolia*

amelloides: from the genus *Amellus*, with the ending *-oides*, indicating resemblance.

amellus: resembling the genus *Amellus*.

americanus, -a, -um: of America.

amethystinus, -a, -um: amethyst-coloured, deep violet.

Amicia (After Giovanni Battista Amici (1786–1863), professor of astronomy and microscopy in Florence.) A genus of 7 species of herbs and shrubs with bipinnate leaves and racemes of irregular pea-like flowers which bear 10 stamens. *Distr.* Tropical America. *Fam.* Leguminosae. *Class* D.

A. zygomeris Perennial herb. Flowers yellow, marked with purple. Grown as an ornamental. *Distr.* E Mexico.

amoenus, -a, -um: pleasant.

Amomum (From the Greek *a*, without, and *momos*, harm, alluding to their use as an antidote.) A genus of 90 species of rhizomatous herbs with reed-like stems, linear leaves and pyramidal inflorescences of irregular flowers. Several species are used as spice plants. *Distr.* Asia and Australia. *Fam.* Zingiberaceae. *Class* M.

A. cardamomum See *A. compactum*

A. compactum ROUND CARDAMOM. Flowers yellow and purple. Grown as an ornamental. *Distr.* Indonesia, Malaysia.

Amomyrtus A genus of 2 species of aromatic shrubs and trees with opposite, evergreen, leathery leaves and small regular flowers that are borne singly or in small clusters and bear their parts in 5s. *Distr.* S South America. *Fam.* Myrtaceae. *Class* D.

A. luma LUMA, PALO MADRONO, CAUCHAO. Shrub or tree to 20m high. Grown as a half-hardy ornamental. *Distr.* Chile, Argentina.

Amorpha (From the Greek *amorphos*, shapeless or deformed, alluding to the flowers.) A genus of 15 species of deciduous shrubs with pinnate leaves and panicles or spike-like racemes of irregular flowers that bear 5 fused sepals, a single, erect, 2-lobed petal and 10 fused stamens. Several species are grown as ornamentals. *Distr.* North America. *Fam.* Leguminosae. *Class* D.

A. canescens LEAD PLANT. Shrub to 1m high. Petals violet with a ragged margin. *Distr.* North America.

A. fruticosa BASTARD INDIGO, FALSE INDIGO. Upright shrub to 4m high. Petals dark purple. *Distr.* S USA.

Amorphophallus (From the Greek *amorphos*, shapeless or deformed, and *phallus*, in allusion to the form and shape of the inflorescence.) DEVIL'S TONGUE, SNAKE PALM. A genus of 90 species of perennial herbs producing an often gigantic inflorescence, the tiny unisexual flowers being borne on a spadix that is up to 4.5m long and surrounded by a funnel-shaped spathe. A single very large, divided leaf is produced from the substantial corm when flowering is complete. The corms of some species are edible. Several species are grown as ornamentals for their foul-smelling inflorescences and dramatic foliage. *Distr.* Tropical and subtropical regions of the Old World. *Fam.* Araceae. *Class* M.

A. brooksii Considered to have the largest inflorescence of any herbaceous plant, the spadix reaching a length of over 4m. *Distr.* Sumatra.

A. bulbifer Bears small bulbils on the leaf margins. *Distr.* NE India and Burma.

A. paeoniifolius ELEPHANT YAM, TELINGO POTATO. Corms eaten after boiling to remove toxins. *Distr.* India to Australasia.

A. rivierei DEVIL'S TONGUE, UMBRELLA PALM, SNAKE PALM. *Distr.* Indonesia to Japan.

A. titanum TITAN ARUM. One of the largest herbs. Corm to 50cm across and weighing over 50kg. Leaf to over 4.5m high and 4m across. Inflorescence to 2.5m long. *Distr.* Sumatra.

Ampelodesmus A genus of 1 species of robust, clump-forming, perennial grasses with solid stems to 3m high, wiry leaves and loose pendulous panicles. *Distr.* Mediterranean. *Fam.* Gramineae. *Class* M.

A. mauritanicus MAURITANIA VINE REED, DISS GRASS. Used locally as a source of fibre and of fodder when young. Grown as an attractive half-hardy ornamental.

Ampelopsis (From the Greek *ampelos*, grape, and *opsis*, resembling, alluding to the similarity to the genus *Vitis*.) A genus of 2 species of deciduous woody climbers with simple or compound leaves and cymes of small green flowers which are followed by berries. Several species are grown as ornamentals. *Distr.* Temperate and subtropical America, Asia. *Fam.* Vitaceae. *Class* D.

A. megalophylla Vine to 13m high. Leaves bipinnate, to 60cm across. Fruit dark purple. *Distr.* W China.

A. tricuspidata See *Parthenocissus tricuspidata*

amphibius, -a, -um: amphibious, growing in water or on land.

amplectens: clasping.

amplexicaulis, -e: with stemclasping leaves, from *amplectens*, clasping, and *caulis*, stem.

ampliatus, -a, -um: amplified, enlarged.

amplus, -a, -um: ample, large, abundant.

ampullaceus, -a, -um: flask-shaped, enlarged at the base, narrow above.

Amsonia (After Dr Charles Amson, an 18th-century Virginian physician.) BLUESTAR. A genus of 20 species of perennial herbs and subshrubs with milky sap, simple alternate leaves and panicles of tubular blue flowers. Several species are grown as ornamentals. *Distr.* S Europe, Turkey, Japan, North America. *Fam.* Apocynaceae. *Class* D.

A. tabernaemontana Clump-forming herb. Flowers small, pale blue. *Distr.* SE USA.

amurensis, -e: of the Amur river, NE Asia.

Amygdalus See *Prunus*

Anacardiaceae The Cashew family. 68 genera and about 800 species of trees, shrubs and lianas often with tissues secreting irritating resins. The leaves are usually pinnately compound although sometimes simple. The flowers are regular, bisexual and have 5 sepals, 5 free petals and 5 or 10 stamens. Some species are important for their edible fruits, notably CASHEW and PISTACHIO; others have an irritant effect, notably POISON IVY. *Distr.* Widespread but most species are tropical. *Class* D.

See: Anacardium, Cotinus, Pistacia, Rhus, Schinus.

Anacardium (From the Greek *ana*, up, and *kardia*, heart.) A genus of 8 species of shrubs and trees with simple leaves and panicles or corymbs of small flowers. The fruit consists of a single seed in a hard shell that often contains a poisonous juice. It is suspended below a large fleshy receptacle. *Distr.* Tropical America. *Fam.* Anacardiaceae. *Class* D.

A. occidentale CASHEW NUT, ACAJOU. The source of the cashew nut of commerce and cashew-nut-shell-liquid (CNSL), used in the manufacture of clutch and break linings, the tree is also a source of various gums and an indelible ink. *Distr.* Tropical America; now widely cultivated throughout the tropics.

Anacyclus (From the Greek, *an*, without, *anthos*, a flower, and *kuklos*, a ring, alluding to the arrangement of the ovaries.) A genus of 9 species of annual and perennial herbs with pinnately divided leaves and daisy-like flower-heads that bear distinct rays. A number of species, particularly the perennial ones, are cultivated ornamentally. *Distr.* Mediterranean and Europe. *Fam.* Compositae. *Class* D.

A. pyrethrum Rosette-forming perennial. Rays white with a red stripe below. This species has been used medicinally and as a flavour for liqueurs. *Distr.* Spain, Algeria, Morocco.

Anagallis (From the Greek, *anagelao*, to laugh; these herbs were believed to dispel sadness.) PIMPERNEL. A genus of 20 species of annual and perennial herbs with simple leaves and flat or funnel-shaped flowers that bear their parts in 5s. Several species are grown ornamentally. *Distr.* Europe, mountains of Africa, South America. *Fam.* Primulaceae. *Class* D.

A. arvensis COMMON PIMPERNEL, SCARLET PIMPERNEL, POOR MAN'S WEATHER GLASS, SHEPHERD'S WEATHER GLASS, SHEPHERD'S CLOCK. Annual, biennial, short-lived perennial herb with red or white flowers. This species was formerly used medicinally. *Distr.* Europe, widely naturalized elsewhere.

A. linifolia See *A. monelli*

A. monelli BLUE PIMPERNEL. Perennial herb. Flowers blue-red. *Distr.* Mediterranean.

A. tenella BOG PIMPERNEL. Perennial, semi-aquatic herb. Flowers pink. *Distr.* W Europe.

Ananas (From *nana*, the South American Tupi Indian name for a PINEAPPLE.) A genus of 8 species of stemless or short-stemmed, evergreen herbs with dense rosettes of thin, rigid leaves. The flowers are borne in a terminal spike which becomes a compound fruit or syncarp when the ovaries fuse together. *Distr.* South America. *Fam.* Bromeliaceae. *Class* M.

A. comosus PINEAPPLE. Leaves usually spiny, although spineless in some cultivars. Flowers red. Fruit very juicy. PINEAPPLE is an important fruit crop in Hawaii, Malaysia and E Africa; it is of less importance as a source of crowa, a fine leaf fibre. Numerous commercial cultivars are available as well as some variegated ornamentals. *Distr.* A seedless cultigen only known in cultivation but probably originating in either Paraguay or Brazil.

Anaphalis (From the Greek name for a similar plant.) PEARLY EVERLASTING. A genus of about 100 species of perennial herbs with simple leaves and small daisy-like heads of flowers which do not bear distinct rays. *Distr.* N temperate regions and tropical mountains. *Fam.* Compositae. *Class* D.

A. cinnamomea See *A. margaritacea*

A. margaritacea PEARLY EVERLASTING. Flower-heads numerous, in dense corymbs. *Distr.* North America, Europe, NE Asia.

A. nepalensis Small. Flower-heads solitary or few. *Distr.* Himalaya, W China.

A. nubigena See *A. nepalensis*

A. sinica Flower-heads in a globose cluster. *Distr.* Japan.

A. trinervis Leaves woolly below. Flower-heads few. *Distr.* New Zealand.

A. triplinervis Flower-heads numerous, in dense, domed clusters. *Distr.* Himalaya, China, Afghanistan.

Anarrhinum (From the Greek *a-*, without, and *rhis*, snout, alluding to the short-spurred petal tube.) A genus of 12 species of biennial and perennial herbs and subshrubs typically with basal rosettes of simple leaves and tall stems that bear a raceme or panicle of tubular, 2-lipped flowers. Several species are grown as ornamentals. *Distr.* Mediterranean region. *Fam.* Scrophulariaceae. *Class* D.

A. bellidifolium Biennial or perennial. Basal rosette compact. Flowering stem to 80cm high, slender. Flowers pale lilac-blue. *Distr.* SW Europe.

Anarthriaceae A family of 1 genus and 6 species of rhizomatous perennial herbs with small regular flowers. This family is sometimes included within the family Restionaceae. *Distr.* W Australia. *Class* M.

anatolicus, -a, -um: of Anatolia, Turkey.

anceps: doubtful; the country of origin may be uncertain.

Anchusa (From the Greek *ankhousa*, a skin paint; some species are a source of pigment.) BUGLOSS. A genus of about 25 species of annual, biennial and perennial, hairy herbs with simple alternate leaves and cymes of cup- or funnel-shaped, typically blue flowers. Several species and a number of cultivars are grown as ornamentals. *Distr.* Europe, N and S Africa, W Asia. *Fam.* Boraginaceae. *Class* D.

A. azurea Perennial, to 1.5m high. Flowers violet-blue. Several cultivars are available. *Distr.* Europe, N Africa, W Asia, naturalized North America.

A. caespitosa Tuft-forming perennial. Leaves narrow, borne in basal rosettes. Flowers deep blue with a white centre. *Distr.* Crete.

A. capensis Erect biennial. Flowers blue with a white centre. Several cultivars are available. *Distr.* S Africa.

A. italica See *A. azurea*

A. myosotidiflora See *Brunnera macrophylla*

A. officinalis Perennial or occasionally biennial. Flowers dark blue, red or sometimes white. *Distr.* Europe, Turkey.

A. sempervirens See *Pentaglottis sempervirens*

Ancistrocladaceae A family of 1 genus and 12 species of lianas and a few shrubs with simple alternate leaves and a coiled tip to the end of each branch. The flowers are small, regular, with 5 sepals, 5 petals and typically

10 stamens. *Distr.* Tropical Africa, tropical Asia.
Class D.

andinus, -e: of the Andes Mountains, South America.

androgynus, -a, -um: androgynous, with female and male flowers together on the plant, from the Greek *andro*, male, and *gyna*, female.

Andromeda (After Andromeda in Greek mythology, daughter of Cepheus and Cassiope, who was rescued from the sea monster by Perseus.) BOG ROSEMARY. A genus of 2 species of low-growing, evergreen shrubs with simple leathery leaves that are white-hairy below. The flowers have their parts in 5s and are borne in terminal umbels. *Distr.* N temperate and Arctic regions. *Fam.* Ericaceae. *Class* D.

A. glaucophylla BOG ROSEMARY. Wiry shrub with pink or white flowers. *Distr.* NE USA, Arctic, Canada.

A. polifolia COMMON BOG ROSEMARY, MARSH ANDROMEDA. Small, wiry shrub. Leaves and twigs are used for tanning in Russia. *Distr.* N temperate regions.

andromeda, marsh *Andromeda polifolia.*

Andropogon (From the Greek *aner*, a man, and *pogon*, a beard, alluding to the hairs on the spikelets of some species.) BEARD GRASS. A genus of about 120 species of annual and perennial rhizomatous grasses to 2.5m high with palmately branched, spreading inflorescences. Some species are used for thatching and erosion control whilst others are important ornamentals, grown for their large inflorescences. *Distr.* Warm temperate and tropical regions. *Fam.* Gramineae. *Class* M.

A. gerardii Clump-forming perennial. Leaves grey-green, turning bronze-purple in autumn. *Distr.* Canada to Mexico.

Androsace (From the Greek *andros*, male and *sakos*, a buckle, alluding to the shape of the anther.) ROCK JASMINE. A genus of about 100 species of alpine, annual and perennial herbs with simple, often tuft-forming leaves and clusters of showy flowers that bear 5 basally fused petals. This genus differs from the closely related *Primula* in that the petal tube is shorter than the sepals and somewhat constricted at the mouth so that the flowers resemble those of the FORGET ME NOT. Many species and cultivars are grown ornamentally. *Distr.* N temperate regions. *Fam.* Primulaceae. *Class* D.

A. carnea Perennial forming loose tufts. Flowers white or pink. *Distr.* Alps, Pyrenees.

A. chamaejasme DWARF JASMINE. Perennial forming loose mats. Flowers with long silky hairs on the margins. Flowers white or pink. *Distr.* S Europe to Siberia.

A. jacquemontii See *A. villosa*

A. lanuginosa Densely silky white-hairy perennial. Flowers white with a yellow throat. *Distr.* Himalaya.

A. limprichtii See *A. sarmentosa*

A. mollis See *A. sarmentosa*

A. primuloides See *A. sarmentosa*

A. sarmentosa Stoloniferous perennial. Flowers deep pink. *Distr.* Himalaya.

A. villosa Densely tufted, cushion-forming perennial. Flowers white to pink. *Distr.* Europe, W Asia.

A. vitaliana See *Vitaliana pri-muliflora*

A. watkinsii See *A. sarmentosa*

androsaceus, -a, -um: resembling the genus *Androsace*.

androsaemum: resembling the genus *Androsaemum*.

Andryala A genus of 25 species of annual and perennial, hairy herbs with simple or pinnate leaves and dandelion-like heads of flowers. Several species are grown as ornamentals. *Distr.* Mediterranean. *Fam.* Compositae. *Class* D.

A. agardhii Perennial with solitary flower-heads. *Distr.* S Spain.

A. lanata See *Hieracium lanatum*

Anemia (From the Greek *aneimon*, naked or unclad, alluding to the bare panicles of sporangia.) FLOWERING FERN. A genus of about 75 species of terrestrial ferns with pinnately divided sterile leaves and paniculate fertile leaves that are densely covered in sori. Several species are grown as ornamentals. *Distr.* Tropical and warm regions, especially America and South Africa. *Fam.* Schizaeaceae. *Class* F.

A. phyllitidis Sterile leaves to 30cm long, arching. Fertile leaves stiffly erect. *Distr.* Central and South America.

Anemone (From the Greek *anemos*, wind.) A genus of 120 species of rhizomatous herbs with palmately cut or divided leaves and regular, saucer-shaped flowers that bear 5–20 petal-like tepals. A number of species and numerous cultivars are grown as ornamentals. *Distr.* Widespread, especially in the N hemisphere. *Fam.* Ranunculaceae. *Class* D.

A. blanda Spreading perennial. Flowers white or blue. *Distr.* SE Europe, Cyprus, Turkey, Caucasus.

A. × flugens A hybrid of *A. pavonina* x *A. hortensis*. Flowers scarlet. Grown for the cut flower industry. *Distr.* Central Mediterranean.

A. hepatica See *Hepatica nobilis*

A. × hybrida A complex group of hybrids that has given rise to over 20 ornamental cultivars. *Distr.* Garden origin.

A. japonica See *A.* × *hybrida*

A. nemorosa WOOD ANEMONE. Creeping perennial. Flowers white, tinged purple-pink. Numerous cultivars have been raised from this species. *Distr.* Europe.

A. pavonina Tuberous perennial. Flowers scarlet or purple, rarely yellow. *Distr.* S Europe, Mediterranean region.

A. pulsatilla See *Pulsatilla vulgaris*

A. rivularis Flowers small, white, borne on branched stems. *Distr.* India, SW China.

A. superba See *A. x hybrida*

A. vernalis See *Pulsatilla vernalis*

anemone: false rue ~ *Isopyrum* **tree** ~ *Carpenteria californica* **wood** ~ *Anemone nemorosa*.

Anemonella (The diminutive of the genus name *Anemone*.) A genus of 1 species of tuberous perennial herb with compound basal leaves and umbels of regular white flowers that bear 5–10 petal-like tepals. *Distr.* E North America. *Fam.* Ranunculaceae. *Class* D.

A. thalictroides RUE ANEMONE. Grown as an ornamental. The tubers are reported to be edible.

anemoneus: resembling the genus *Anemone*.

anemoniflorus, -a, um: from the genus *Anemone*, and *florus*, flowers.

anemonoides: from the genus *Anemone*, with the ending *-oides*, indicating resemblance.

Anemonopsis (From the genus name *Anemone*, and the Greek *opsis*, likeness or similarity.) A genus of 1 species of rhizomatous herb with compound leaves and panicles of pale blue, waxy flowers that bear 7–10 petal-like sepals, 10 small, nectar-secreting petals and numerous stamens. *Distr.* Japan (Honshu). *Fam.* Ranunculaceae. *Class* D.

A. macrophylla Grown as an ornamental.

anethifolius, -a, -um: from the genus *Anethum*, and *folius*, leaved.

Anethum (From the Greek name for DILL, *anethon*.) A genus of 2 species of fragrant annual and biennial herbs with leaves that are finely divided into needle-like segments and yellow flowers that are borne in compound umbels. *Distr.* Warm regions of the Old World. *Fam.* Umbelliferae. *Class* D.

A. graveolens DILL. Erect, blue-green annual. A culinary herb used, in particular, with smoked salmon. *Distr.* W Asia, naturalized Europe, North America.

Angelica (From the Latin *angelus*, alluding to the 'angelic' medicinal properties.) A genus of about 50 species of biennial and perennial herbs with deep taproots, finely divided leaves and compound umbels of white, green or pink flowers. Several species are grown as ornamentals and some are used as medicinal and culinary herbs. *Distr.* N hemisphere, New Zealand. *Fam.* Umbelliferae. *Class* D.

A. archangelica ANGELICA, ARCH-ANGEL, WILD PARSNIP, GARDEN ANGELICA. Biennial herb that dies after it has flowered. Flowers cream, tinged green. The leaves are eaten as a vegetable or crystallized as a confectionery or decoration; they are also used medicinally and as a flavouring for the liqueur Benedictine. *Distr.* Eurasia, naturalized in Great Britain, Greenland.

angelica *Angelica archangelica* **garden** ~ *A. archangelica*.

angelica tree *Aralia elata* **American** ~ *A. spinosa* **Japanese** ~ *A. elata*.

angel's eyes *Veronica chamaedrys*.

angel's fishing rod *Dierama*, *D. pendulum*.

angel's tears *Billbergia nutans*, *Soleirolia soleirolii*, *Narcissus triandrus*, *Billbergia x windii*.

angel's trumpet, red *Brugmansia sanguinea*.

angel wings *Caladium*.

anglicus, -a, -um: of England.

angularis, -e: angled.

angulatus, -a, -um: angled.

angustatus, -a, -um: narrowed.

angustifolius, a, -um: narrow-leaved.

angustipetalus, -a, -um: with narrow petals.

angustisectum: narrowly divided, from the Latin *angustus*, narrow, and *sectus*, cut.

angustissimus, -a, -um: very narrow.

angustus, -a, -um: narrow.

Anigozanthos (From the Greek *anoigo*, to open, and *anthos*, flower; the flowers appear as if they have been ripped open.) KANGAROO PAW, AUSTRALIAN SWORD LILY, CAT'S PAW. A genus of about 12 species of rhizomatous, clump-forming herbs with sword-shaped leaves and panicles of irregular tubular flowers. Several species are grown as ornamentals. *Distr.* SW Australia. *Fam.* Haemodoraceae. *Class* M.

A. flavidus TALL KANGAROO PAW, EVERGREEN KANGAROO PAW, YELLOW KANGAROO PAW. Flowers yellow-green. This is the parent species of numerous garden hybrids.

A. manglesii RED AND GREEN KANGAROO PAW. Inflorescence simple, red. Flowers green. *Distr.* Australia.

A. preissii ALBANY CAT'S PAW. Inflorescence red, hairy. Flowers yellow to orange.

anisatus, -a, -um: smelling or tasting of aniseed (like licorice).

anise *Pimpinella anisum* **Chinese** ~ *Illicium verum* **Japanese** ~ *I. anisatum* **purple** ~ *I. floridanum* **star** ~ *I. verum* **tree** ~ *I. floridanum*

anise tree *Illicium*.

Anisodontea (From the Greek *anisos*, unequal and *odon*, a tooth, alluding to the teeth on the fruits.) A genus of 19 species of subshrubs and shrubs with decumbent or ascending stems, simple to palmately lobed leaves and regular bisexual flowers that bear 5 red to pink petals and 10 fused stamens. Several species and hybrids are grown ornamentally. *Distr.* S Africa. *Fam.* Malvaceae. *Class* D.

A. capensis Bushy subshrub. Flowers bowl-shaped, magenta. *Distr.* South Africa (Cape Province).

A. scabrosa HAIRY MALLOW. Rough-hairy shrub to 2m high. Flowers purple to magenta, darkening towards the centre. *Distr.* South Africa (Cape Province).

Anisophylleaceae A family of 4 genera and 30 species of shrubs and trees with spirally arranged, simple leaves and panicles or racemes of regular flowers that bear their parts in 4s. *Distr.* Tropical regions. *Class* D.

Anisotome (From the Greek *anisos*, unequal, and *tome*, cut, alluding to unequally divided leaves.) A genus of 13 species of perennial herbs with fern-like, pinnate leaves and large, compound umbels of unisexual, white, red, or purple flowers. The umbel is often subtended by a whorl of bracts. Several species are grown as ornamentals. *Distr.* New Zealand, subantarctic islands. *Fam.* Umbelliferae. *Class* D.

A. aromatica To 50cm high. Leaves leathery. Flowers white. *Distr.* New Zealand.

A. imbricata Cushion-forming. *Distr.* New Zealand (South Island).

annatto Bixaceae, *Bixa orellana*.

Annona (One Latin word for corn or food.) CUSTARD APPLE. A genus of about 100 species of shrubs and trees with alternate, simple, evergreen leaves, solitary or clustered regular flowers and large fleshy fruits. *Distr.* Tropical America and Africa. *Fam.* Annonaceae. *Class* D.

A. cherimola CUSTARD APPLE, CHERIMOYA. A large bush or small tree. The large fleshy fruits are widely eaten. *Distr.* Mountain valleys of Peru and Ecuador, cultivated throughout the tropics.

A. squamosa SUGAR APPLE, SWEETSOP. Similar to *A. cherimolia* but with narrow leaves. Fruit eaten.

Annonaceae A family of 125 genera and about 2000 species of shrubs and trees with simple leaves arranged in 2 ranks. The flowers are fragrant and have 3 whorls of 3 tepals. The fruits are large and made up of a number of berries fused together. Some species are cultivated for their edible fruits or aromatic oils. *Distr*. Widespread tropical and subtropical regions, rare warm-temperate areas. *Class* D.
See: Annona, Mitrella, Monodora.

annulatus, -a, -um: with a ring (*annulus*) or rings.

annuus, -a, -um: annual.

Anoda (From the Greek *a-*, without, and the Latin *nodus*, joint.) A genus of 10 species of annual and perennial herbs and subshrubs with simple to palmately lobed leaves and regular bisexual flowers that bear 5 petals and 10 fused stamens. *Distr*. Tropical America, mainly Mexico, naturalized in the Old World. *Fam*. Malvaceae. *Class* D.

A. cristata Annual or short-lived perennial herb to 1.5m high. Flowers purple-blue to white, borne singly or in pairs. Grown as an ornamental. *Distr*. SW USA, Mexico.

Anoiganthus See *Cyrtanthus*

anomalus, -a, -um: abnormal.

Anomatheca (From the Latin *anomalus*, abnormal, and *theca*, case, alluding to the unusual fruit.) A genus of 4–6 species of perennial herbs with corms, lance-shaped basal leaves and irregular, somewhat 2-lipped, tubular flowers. Several species are grown as ornamentals. *Distr*. S and Central Africa. *Fam*. Iridaceae. *Class* M.

A. cruenta See *A. laxa*

A. laxa Flowers red, occasionally blue-white. Grown as a garden ornamental. *Distr*. S Africa.

anopetalus, -a, -um: with upright petals.

Anopterus (From the Greek *ano*, up-wards, and *pteron*, wing, alluding to the winged seeds.) A genus of 2 species of evergreen shrubs and trees with simple alternate leaves and racemes of tubular flowers. *Distr*. SE Australia, Tasmania. *Fam*. Grossulariaceae. *Class* D.

A. glandulosus TASMANIAN LAUREL. Shrub or small tree. Flowers white. Grown as a half-hardy ornamental. *Distr*. Tasmania.

Anredera A genus of 5–10 species of twining herbs with fleshy heart-shaped or oval leaves and spikes or racemes of small flowers that bear 5 tepals and 5 stamens. *Distr*. Tropical America, naturalized S Europe. *Fam*. Basellaceae. *Class* D.

A. cordifolia MADEIRA VINE, MIGNO-NETTE VINE. Tuberous vine. Leaves heart-shaped. Flowers, white, fragrant, borne in pendent racemes. Grown as an ornamental and as a vegetable. *Distr*. Paraguay to S Brazil and N Argentina.

anserinifolius, -a, -um: from the genus *Anserinus* (now *Chenopodium*), and *folius*, leaved.

anserinus, -a, -um: to do with geese.

antarcticus, -a, -um: of Antarctic regions.

antelope bush *Purshia*.

Antennaria (From the Latin *antenna*; the pappus hairs of the male flowers have swollen tips resembling a butterfly's antennae.) EVERLASTING, PUSSY TOES, LADIES' TOBACCO. A genus of about 45 species of perennial hairy herbs with rosettes of simple leaves and small daisy-like heads of flowers that lack distinct rays. Several species are cultivated as rock garden ornamentals. *Distr*. Temperate and warm regions (excluding Africa). *Fam*. Compositae. *Class* D.

A. alpina Mat-forming with cream flowers. *Distr*. Mountains of North America, N Europe to NW Central Asia.

A. aprica See *A. parvifolia*

A. dioica CAT'S FOOT. Mat-forming. Leaves densely hairy. *Distr*. Cold regions of the N hemisphere.

A. neglecta Leaves to 6cm. *Distr*. W North America.

A. parvifolia Mat-forming. Leaves grey-hairy. *Distr*. W North America.

A. rosea Tufted or mat-forming. Flower-heads cream to rose. *Distr*. Mountains of W North America.

anthemifolius, -a, -um: from the genus *Anthemis*, and *folius*, leaved.

Anthemis (The Greek name for *Chamaemelum nobile*, probably derived from *anthemon*, flower.) DOG FENNEL. A genus of about

100 species of perennial herbs and small shrubs with hairy, aromatic foliage and heads of daisy-like flowers which may or may not bear distinct rays. Several species are grown ornamentally, others are considered weeds. *Distr.* Europe, Mediterranean to Iran. *Fam.* Compositae. *Class* D.

A. biebersteinii See *A. marschalliana*

A. cretica Cushion-forming herb with yellow flowers and white rays. *Distr.* Mountains of S Europe to Turkey.

A. frutescens See *Argyranthemum frutescens*

A. marschalliana Herb with densely downy foliage and solitary flower-heads. *Distr.* NE Anatolia, S Caucasia.

A. nobilis See *Chamaemelum nobile*

A. punctata Foliage glabrous or hairy. Rays absent. *Distr.* Europe, NW Africa.

A. rudolphiana See *A. marschalliana*

A. sancti-johannis (Referring to St John's Day, 24th June; the flowers appear in mid summer.) Tuft-forming. *Distr.* Bulgaria.

A. tinctoria DYER'S CHAMOMILE, YELLOW CHAMOMILE. Perennial. A source of yellow dye. *Distr.* Central and S Europe, naturalized in Great Britain and North America.

A. tuberculata Woolly, short-lived perennial. *Distr.* The mountains of Central and S Spain.

anthemoides from the genus *Anthemis*, with the ending *-oides*, indicating resemblance.

Anthericaceae A family of 29 genera and about 500 species of rhizomatous herbs with linear leaves and flowers that bear 2 whorls of 3 tepals and 6 stamens. *Distr.* Old World, especially Australasia, America. *Class* M.

See: Anthericum, Arthropodium, Chlorophytum, Diuranthera, Herpolirion, Leucocrinum, Pasithea, Thysanotus.

Anthericum (From the Greek name, *anthericos*, for the flower of the ASPHODEL (*Asphodelus*).) ST BERNARD'S LILY. A genus of 50 species of rhizomatous perennial herbs with rosettes of grass-like leaves and racemes or panicles of showy white flowers. Several species are cultivated as ornamentals. *Distr.* Africa, S Europe and tropical America. *Fam.* Anthericaceae. *Class* M.

A. algeriense See *A. liliago*

A. coccinea See *Crocosmia paniculata*

A. liliago ST BERNARD'S LILY. Stem erect, to 1m high. Flowers in racemes. *Distr.* Europe, Turkey.

A. paniculata See *Crocosmia paniculata*

A. ramosum Standing to 60cm tall. Flowers in panicles. *Distr.* Europe, Turkey.

Anthoxanthum (From the Greek *anthos*, a flower, and *xanthos*, yellow.) VERNAL GRASS. A genus of 15 species of annual and perennial grasses to around 75cm high with aromatic leaves and spike-like, compressed inflorescences. Some species are used as fodder grasses but have a poor food value. *Distr.* Europe, N Asia, North and South America and the mountains of E Africa. *Fam.* Gramineae. *Class* M.

A. gracile Annual. Occasionally grown as a fragrant ornamental. *Distr.* Mediterranean.

A. odoratum SWEET VERNAL GRASS, SPRING GRASS. Perennial. Sometimes grown as a fragrant ornamental and for use in dried flower arrangements. *Distr.* Eurasia.

Anthriscus (From the Greek name for CHERVIL, *anthriskos*.) A genus of about 12 species of annual, biennial and perennial herbs with finely divided leaves and compound umbels of small white flowers. The umbels may or may not be subtended by a whorl of bracts. *Distr.* Eurasia, N Africa. *Fam.* Umbelliferae. *Class* D.

A. cerefolium CHERVIL. Grown as an annual, aromatic, culinary herb. *Distr.* SE Europe, W Asia.

A. sylvestris COW PARSLEY, KECK. Biennial or perennial. Occasionally grown as an ornamental but more familiar as a hedgerow plant producing masses of white flowers in early summer. *Distr.* Europe, W Asia, N Africa.

Anthurium (From the Greek *anthos*, a flower, and *oura*, a tail, alluding to the tail-like spadix of the inflorescence.) FLAMINGO FLOWER, TAILFLOWER. A genus of about 700 species of terrestrial or epiphytic herbs with entire or palmate leaves and small flowers that are borne on a long, often tapering, spadix that is subtended by a flat or hooded spathe. A number of species and hybrids are grown as ornamentals both for their foliage and inflorescences. The dried leaves of some species are used to perfume tobacco in South America.

Distr. Tropical regions of the New World. *Fam.* Araceae. *Class* M.

A. andraeanum (After Edouard François André (1840–1911), French landscape architect.) TAILFLOWER, FLAMINGO FLOWER, PAINTER'S PALETTE. Spathe bright red, c15cm long, spadix white, c9cm long. Frequently grown as a house plant with numerous cultivars available. *Distr.* Colombia, Ecuador.

A. × ferrierense A complex collection of hybrids, involving *A. andraeanum* and *A. nymphalaeifolium* in their parentage, from which numerous cultivars have arisen.

A. scherzerianum (After Herr Scherzer of Vienna, who discovered it) FLAMINGO FLOWER, FLAME PLANT. Spathe c12cm long, bright red-orange, spadix c8cm long, orange yellow, spirally contorted. Frequently grown as a house plant, with numerous cultivars available. *Distr.* Central Costa Rica.

Anthyllis A genus of 20–25 species of annual and perennial herbs or shrubs with pinnate leaves and dense racemes of pea-like flowers that bear 10 stamens. Several species are grown as ornamentals. *Distr.* Mediterranean, W Europe, Macaronesia, NE Africa. *Fam.* Leguminosae. *Class* D.

A. barba-jovis JUPITER'S BEARD. Evergreen shrub. Flowers cream, enclosed by silvery bracts. *Distr.* Spain.

A. hermanniae Diffuse shrub. Leaves reduced to 1–3 leaflets. Flowers yellow. *Distr.* Mediterranean region.

A. montana Clump-forming perennial herb. Flowers purple, red or white. *Distr.* S Europe, Alps.

A. vulneraria KIDNEY VETCH, LADY'S FINGERS. Variable annual or short-lived perennial herb. Flowers borne in umbels. Sepals forming a somewhat inflated tube. Petals white to pink. *Distr.* Europe, N Africa, W Asia.

Antigonon (From the Greek *anti*, against or opposed to, and *gonia*, an angle, alluding to the flexible stems.) CORAL VINE. A genus of 2–3 species of perennial climbers with tendrils, alternate simple leaves and racemes of irregular flowers that bear 5 sepals and 8 stamens. *Distr.* Central America, Mexico. *Fam.* Polygonaceae. *Class* D.

A. leptopus CORAL VINE, ROSA DE MONTANA, CHAIN OF LOVE, CORALLITA, CONFEDERATE VINE, MEXICAN CREEPER. Fast-growing vine to 12m high. Flowers coral pink. Grown as a tender ornamental. The tubers are edible. *Distr.* Mexico.

antioquiensis: of Antioquia, Colombia.

antipodus, -a, -um: of the Antipodes, Australasia.

antirrhiniflorus, -a, -um: from the genus *Antirrhinum*, and *florum*, flowers.

antirrhinoides: from the genus *Antirrhinum*, with the ending *-oides*, indicating resemblance.

Antirrhinum (From the Greek *anti*, opposite, and *rhis*, a snout, alluding to the appearance of the flowers.) A genus of 42 species of annual and perennial herbs and subshrubs with simple leaves and showy 2-lipped flowers that are borne singly or in terminal racemes. The petal tube is usually closed at the mouth and somewhat inflated at the base. Several species and a number of cultivars are grown as ornamentals. *Distr.* Temperate regions of the N hemisphere. *Fam.* Scrophulariaceae. *Class* D.

A. glutinosum See *A. hispanicum*

A. hispanicum Dwarf shrub. Flowers white or pink with yellow markings. *Distr.* SE Spain.

A. majus SNAPDRAGON. Erect or scrambling perennial herb. Flowers typically purple or pink. Numerous ornamental cultivars of this species have been raised. *Distr.* SW Europe, naturalized throughout temperate regions.

antrorsus, -a, -um: directed upwards.

apennina of the Apennine mountains, the backbone of Italy.

apertus, -a, -um: open.

apetalon without petals.

Aphelandra (From the Greek *apheles*, simple, and *aner*, male, alluding to the one-celled anthers.) A genus of 170 species of evergreen subshrubs and shrubs with simple stalked leaves and 4-sided spikes that bear 2-lipped flowers and colourful bracts. Several species are grown as house plants for their showy bracts and foliage. *Distr.* Tropical America. *Fam.* Acanthaceae. *Class* D.

A. squarrosa SAFFRON SPIKE, ZEBRA PLANT. Leaves dark green with white veins. Flowers bright yellow. Bracts yellow, marked red. *Distr.* Brazil.

Aphyllanthaceae A family of 1 genus and 1 species. *Class* M.
See: Aphyllanthes.

Aphyllanthes (From the Greek *a*, without, *phyllos*, leaf, and *anthos*, flower, alluding to the rush-like leafless stems.) A genus of 1 species of perennial herb with slender rush-like leaves and simple deep blue flowers. *Distr.* Portugal to Italy and N Africa. *Fam.* Aphyllanthaceae. *Class* M.

A. monspeliensis Grown as a half-hardy ornamental.

aphyllanthes: flowering without leaves.

aphyllus, -a, -um: without leaves.

Apiaceae See Umbelliferae

apiculatus, -a, -um: ending in a short point.

apifera: bee-bearing; the flowers resemble bees.

apiifolia: from the genus *Apium*, and *folius*, leaved.

Apios (From the Greek *apios*, pear, alluding to the shape of the tubers.) A genus of 10 species of tuberous, twining, perennial herbs with alternate pinnate leaves and short racemes of pea-like flowers. *Distr.* E Asia, North America. *Fam.* Leguminosae. *Class* D.

A. americana POTATO BEAN, INDIAN POTATO, WILD BEAN. Sweet, edible tubers, formerly an important food for North American Indians. *Distr.* North America.

A. tuberosa See *A. americana*

Apium (The classical Latin name for CELERY and PARSNIP.) A genus of 20 species of biennial herbs with long-stalked, pinnate leaves and compound umbels of small white flowers. *Distr.* N temperate regions of Europe snd Asia. *Fam.* Umbelliferae. *Class* D.

A. graveolens WILD CELERY, CELERY, CELERIAC. Biennial. Leaf-stalks large, ridged. This species has given rise to a number of cultivars that either have large, edible leaf stalks (CELERY) or a swollen edible rootstock (CELERIAC). *Distr.* Coastal regions of Europe.

Apocynaceae The Periwinkle family. 168 genera and about 1500 species of tall trees, shrubs, and a few perennial herbs. The leaves are simple, typically opposite and may be evergreen or deciduous. The flowers are often large, showy and fragrant and bear 5 fused sepals, 5 fused petals and 5 stamens. This family is a source of cardiac drugs, latex, and a number of garden ornamentals. *Distr.* Widespread, centred in tropical regions. *Class* D.
See: Allamanda, Amsonia, Beaumontia, Carissa, Catharanthus, Elytropus, Mandevilla, Nerium, Pachypodium, Parsonsia, Plumeria, Rhazya, Thevetia, Trachelospermum, Vinca.

apodus, -a, -um: stalkless, from the Greek *a-*, not, and *podus*, footed.

Aponogeton (From the Latin name of the healing springs at Aquae Aponi, Italy, and the Greek *geiton*, a neighbour.) CAPE PONWEED. A genus of 44 species of aquatic, often floating, perennial herbs with branched spikes of small, frequently scented flowers. Some species are grown as ornamentals for the water garden or aquarium and some have edible tubers. *Distr.* Old World tropics, South Africa. *Fam.* Aponogetonaceae. *Class* M.

A. distachyos CAPE PONDWEED, WATER HAWTHORN. Leaves oval, floating. Flowers yellow or white, scented. *Distr.* South Africa.

A. madagascariensis LACE LEAF, LATTICE LEAF. Leaves submerged, lace-like. Frequently grown as an ornamental in aquaria. *Distr.* Madagascar.

Aponogetonaceae A family of 1 genus and 44 species. *Class* M.
See: Aponogeton.

appendiculatus, -a, -um: with small appendages.

apple *Malus*, *M. pumila*, **balsam** ~ *Clusia major* **belle** ~ *Passiflora laurifolia* **bitter** ~ *Citrullus colocynthis* **custard** ~ *Annona*, *A. cherimolia* **downy thorn** ~ *Datura inoxia* **European crab** ~ *Malus sylvestris* **Indian** ~ *Datura inoxia* **jew's** ~ *Solanum melongena* **kangaroo** ~ *S. aviculare* **love** ~ *Lycopersicon esculentum* **mad** ~ *Solanum melongena* **May** ~ *Passiflora incarnata*, *Podophyllum peltatum* **pitch** ~ *Clusia major* **Siberian crab** ~ *Malus baccata*

sugar ~ *Annona squamosa* **thorn** ~ *Datura stramonium, Crataegus.*

apple of Peru *Nicandra physalodes.*

apples, devil's *Mandragora officinarum.*

approximatus, -a, -um: close together.

apricot *Prunus armeniaca.*

apricus sun-loving.

Aptenia (From the Greek *apten*, unfledged or unwinged, alluding to the absence of wings on the fruit.) A genus of 2 species of small succulent subshrubs with lance shaped or heart-shaped leaves and solitary or clustered pink-purple terminal flowers. *Distr.* Transvaal and the E coastal districts of South Africa, naturalized Australia. *Fam.* Aizoaceae. *Class* D.

A. cordifolia Prostrate. Leaves heart-shaped. Flowers pink. *Distr.* South Africa (Cape Province), naturalized S Europe, Channel Islands.

apterus, -a, -um: without leaves.

aquaticus, -a, um: growing in water.

aquatilis: growing in water.

Aquifoliaceae The Holly family. 2 genera and about 400 species of shrubs and trees with leathery, simple, often evergreen leaves. The flowers are regular, green-white and inconspicuous with 4–8 fused, often absent petals. The fruit is a berry. As well as having a number of ornamental species this family is a source of light-coloured wood. *Distr.* Widespread. *Class* D.
See: Ilex.

aquifolium: Classical Latin name for holly.

Aquilegia (From the Latin *aquila*, an eagle, alluding to the form of the petals.) COLUMBINE, GRANNY'S BONNET. A genus of about 70 species of perennial herbs with erect, branched stems, compound leaves and showy, nodding flowers. The flowers bear 5 petal-like sepals and 5 petals each of which is extended backwards into a long, nectar-secreting spur. Many species and cultivars are grown as ornamentals. *Distr.* N temperate regions, especially in mountainous areas. *Fam.* Ranunculaceae. *Class* D.

A. alpina Short-lived perennial to 50cm high. Flowers clear blue. *Distr.* Alps.

A. baicalensis See *A. vulgaris*

A. canadensis MEETING HOUSES, HONEYSUCKLE, CANADIAN COLUMBINE. To 1m high. Flowers yellow with red spurs. *Distr.* E North America.

A. clematiflora See *A. vulgaris*

A. ecalcarata See *Semiaquilegia ecalcarata*

A. jonesii Tuft-forming perennial to 10cm high. Flowers erect, purple-blue. *Distr.* W North America (Rocky Mountains).

A. longissima To 1m high. Flowers pale yellow. *Distr.* S USA, N Mexico.

A. nevadensis See *A. vulgaris*

A. scopulorum Clump-forming perennial to 20cm high. Flowers pale blue to pink, with a yellow centre. *Distr.* W USA.

A. stellata See *A. vulgaris*

A. vulgaris COLUMBINE, GRANNY'S BONNET. Clump-forming perennial to 1m high. Flowers blue. Numerous cultivars have been raised from this species with differing flower colours. *Distr.* Europe, naturalized British Isles, North America.

aquilegiifolium: from the genus *Aquilegia*, and *folius*, leaved.

arabicus, -a, -um: of Arabia.

Arabis ROCKCRESS, WALL CRESS. A genus of 120 species of annual to perennial herbs with simple leaves and small, white to deep-purple flowers. *Distr.* N temperate regions to tropical African mountains, typically in rocky places. *Fam.* Cruciferae. *Class* D.

A. albida See *A. caucasica*

A. alpina Perennial with white flowers. *Distr.* Europe to African mountains and Himalaya.

A. androsacea Densely tufted perennial with white flowers. *Distr.* Turkey.

A. × arendsii Hybrid of *A.aubrietioides* x *A. caucasica*.

A. aubrietioides Dwarf tufted perennial with pink-purple flowers. *Distr.* Turkey (Cilician Taurus).

A. blepharophylla Perennial with round rose-purple petals. *Distr.* USA (California).

A. bryoides Densely tufted perennial with white flowers. *Distr.* Greece, Balkan peninsula.

A. caucasica Grey-green perennial with white flowers. *Distr.* SE Europe to Iran.

A. ferdinandi-coburgii Stoloniferous perennial with large white flowers. *Distr.* Bulgaria.

A. × kellereri Hybrid of *A. bryoides* x *A. ferdinandi-coburgii*. *Distr.* Garden origin.

A. muralis Perennial with large, rose-purple to white flowers. *Distr.* SE and Central Europe.

A. soyeri Perennial with white flowers. *Distr.* Pyrenees, Alps, W Carpathians.

A. stricta Perennial with leathery leaves and yellow flowers. *Distr.* W Europe.

A. × suendermannii Hybrid of *A. ferdinandi-coburgii* × *A. procurrens*. *Distr.* Garden origin.

Araceae The Arum family. 105 genera and about 3000 species of herbs, scrambling shrubs, climbers and free floating aquatics, many containing a milky or watery sap. The leaves are simple or compound. The flowers are small and borne on a characteristic spike-like spadix subtended by a leaf-like spathe. Some members of the family are cultivated as ornamentals, others are important as water weeds. *Distr.* Widespread, especially in tropical regions. *Class* M.
See: Aglaonema, Alocasia, Amorphophallus, Anthurium, Arisaema, Arisarum, Arum, Biarum, Caladium, Calla, Cercestis, Colocasia, Cyrtosperma, Dieffenbachia, Dracunculus, Eminium, Epipremnum, Lysichiton, Monstera, Nephthytis, Orontium, Peltandra, Philodendron, Pinellia, Pistia, Rhektophyllum, Rhodospatha, Sauromatum, Scindapsus, Spathicarpa, Spathiphyllum, Symplocarpus, Syngonium, Xanthosoma, Zantedeschia.

Arachis (From the Greek *a*, without, and *rachos*, branch.) A genus of 22 species of annual or perennial herbs with pinnate leaves and dense spikes of irregular pea-like flowers. After pollination the inflorescence arches over so that the fruits can develop below ground level. *Distr.* South America. *Fam.* Leguminosae. *Class* D.

A. hypogaea PEANUT, GROUNDNUT, MONKEY NUT. Erect or trailing annual herb. The seeds are a rich source of B complex vitamins and are eaten or crushed for oil that is used in margarines and cooking, the residue being fed to animals. *Distr.* Paraguay and S Brazil, grown in most tropical and subtropical countries.

Arachniodes (From the Greek *arachniodes*, like a spider's web, alluding to the soft indusia.) A genus of 20–40 species of medium-sized terrestrial ferns with divided leaves. Several species are grown as tender ornamentals. *Distr.* E Asia, Malaysia, New Zealand. *Fam.* Dryopteridaceae. *Class* F.

A. aristata Leaves to 40cm long, bipinnately divided. *Distr.* Asia, Polynesia.

arachnoideus, -a, um: with hairs like a spiders web.

aragonensis: of Aragon, northern Spain.

Aralia (From the Canadian-French name *aralie*.) A genus of about 36 species of herbs, shrubs and trees with pinnate, deciduous leaves and clusters of small, yellow, white or green flowers that are followed by fleshy black berries. Some species are edible or have medicinal qualities. *Distr.* North America, E Asia, Malaysia. *Fam.* Araliaceae. *Class* D.

A. californica ELK CLOVER, SPIKENARD. Perennial herb. *Distr.* W North America.

A. continentalis MANCHURIAN SPIKENARD. Perennial herb. Leaves pinnate. *Distr.* China, Manchuria, Korea.

A. cordata UDO, JAPANESE SPIKENARD. Herbaceous perennial. *Distr.* China, Japan.

A. elata JAPANESE ANGELICA TREE, ANGELICA TREE. Clump-forming, deciduous tree or shrub. *Distr.* China, Manchuria, Korea, Japan.

A. racemosa AMERICAN SPIKENARD, PETTY MOREL, LIFE OF MAN. Large perennial herb. *Distr.* SW North America.

A. sieboldii See *Fatsia japonica*

A. spinosa HERCULES CLUB, DEVIL'S WALKINGSTICK, AMERICAN ANGELICA TREE. Spiny shrub or small tree. *Distr.* E North America.

aralia, fern-leaf *Polyscias filicifolia.*

Araliaceae The Ivy family. 47 genera and about 700 species of herbs, shrubs, trees and climbers with simple or compound leaves and compound umbels of small regular flowers that typically bear 5 sepals, 5 petals and an inferior ovary. A number of species are grown as ornamentals, others have medicinal properties notably GINSENG, *Panax quinquefolia*. *Distr.* Widespread but mostly in tropical regions. *Class* D.

See: Aralia, Eleutherococcus, Fatsia, Hedera, Kalopanax, Meryta, Oreopanax, Panax, Polyscias, Pseudopanax, Schefflera, Tetrapanax, X Fatshedera.

Aralidiaceae A family of 1 genus and 1 species of shrub with palmately lobed leaves and cymes of small regular flowers. This family is intermediate between the families Araliaceae and Cornaceae and is sometimes included within one of them. *Distr.* Tropical Asia. *Class* D.

aralioides: from the genus *Aralia*, with the ending *-oides*, indicating resemblance.

araratica: of Mount Ararat, Eastern Turkey.

araucanus, -a, -um: of Arauco, Chile.

Araucaria: (After the Araucani Indians who live where *Araucaria araucana* grows.) MONKEY PUZZLE. A genus of 19 species of evergreen coniferous trees with whorled branches and whorled, flat or needle-like leaves. The fruiting cones are large and woody, take 2.5 years to mature and break up at maturity. Several species are cultivated ornamentally and some are a source of timber. *Distr.* New Caledonia, New Zealand, Australia, New Guinea and South America. *Fam.* Araucariaceae. *Class* G.

A. araucana MONKEY PUZZLE, CHILE PINE, CHILE NUT TREE. Tree to 30m high. Leaves leathery, broad-triangular. Frequently planted as an ornamental. Seeds edible (Chile nuts). *Distr.* Chile, Argentina.

A. augustifolia BRAZILIAN PINE, PARANA PINE. Tree to 35m high. Branches ascending. Leaves broad, leathery. *Distr.* S Brazil.

A. excelsa See *A. heterophylla*

A. heterophylla NORFOLK ISLAND PINE, HOUSE PINE. Tree to 60m high. Leaves scale-like. *Distr.* New Zealand (Norfolk Island).

A. imbricata See *A. araucana*

Araucariaceae The Monkey Puzzle family. 2 genera and 31 species of evergreen conifers with broad to needle-like leaves and large female cones that shatter at maturity. The family is an important source of timber. *Distr.* South America, the Pacific, Australasia, SE Asia. *Class* G.

See: Agathis, Araucaria.

Araujia (From the Brazilian name for the plant.) A genus of 2–3 species of evergreen climbing shrubs with simple leaves and cup- or bell-shaped, white-pink, fragrant flowers. Several species are grown as tender ornamentals. *Distr.* South America, naturalized in Australia. *Fam.* Asclepiadaceae. *Class* D.

A. grandiflora Yellow-hairy climber. Foliage bad-smelling when crushed. *Distr.* Brazil.

A. sericofera CRUEL PLANT. Robust climber. Flowers cup-shaped, white. Pollinated by night-flying moths which are held by their proboscises until morning, hence the vernacular name. *Distr.* South America, naturalized in Australia.

arborescens: becoming woody, from the Latin *arbor*, tree.

arboreus, -a, -um: tree-like.

arboricola growing on trees.

arbor vitae *Thuja* **American** ~ *T. occidentalis* **western** ~ *T. plicata*.

arbusculus, -a, -um: like a small tree.

arbutifolius, -a, -um: from the genus *Arbutus*, and *folius*, leaved.

Arbutus (The Latin for STRAWBERRY TREE, probably originally applied to *A. unedo*.) MANZANIA, MANZANITA, MADRONA, STRAWBERRY TREE. A genus of about 14 species of evergreen trees and shrubs with smooth flaking bark and simple leathery leaves. The flowers have their parts in 5s with the corolla fused into an urn-shaped tube. The fruit is a red or orange berry. *Distr.* W North America, W Europe to Mediterranean and Turkey. *Fam.* Ericaceae. *Class* D.

A. andrachne Tree to 12m high with orange fruit. *Distr.* SE Europe and Turkey.

A. × andrachnoides A naturally-occurring hybrid of *A. unedo* x *A. andrachne* which rarely sets fruit.

A. menziesii (After Archibald Menzies (1754–1842), naval surgeon and botanist.) MADRONA, MADRONE. A large tree with white flowers and orange-red fruit. The bark is used in tanning and the wood is used as timber. *Distr.* W North America.

A. unedo STRAWBERRY TREE. Tall shrub or small tree with scarlet fruits reminiscent of strawberries. The bark is used in tanning and the fruit is used in preserves and to flavour liqueurs, especially in Portugal. *Distr.* S Europe, Turkey and Ireland.

arbutus, trailing *Epigaea repens.*

archangel *Angelica archangelica*
yellow ~ *Lamium galeobdolon.*

archangelica: after the archangel Raphael.

Archontophoenix (From the Greek *arkhon*, a chieftain, and the genus name *Phoenix*.) KING PALM, BANGALOW PALM. A genus of 2 species of large palms with unbranched stems, feather-shaped leaves and lilac or yellow flowers in pendent panicles. Both species are grown as tender ornamentals. *Distr.* E Australia. *Fam.* Palmae. *Class* M.

A. alexandrae ALEXANDRA PALM, NORTHERN BANGALOW PALM. To 25m high. Flowers yellow-white. *Distr.* Australia (N Queensland).

A. cunninghamiana ILLAWARA PALM, PICCABBEN PALM. To 22m high. Flowers lilac. *Distr.* Australia (New South Wales, Queensland).

Arctanthemum (From the Greek *arktos*, North, and *anthemon*, flower.) A genus of about 5 species of perennial coastal herbs with simple or pinnate leaves and daisy-like heads of flowers that bear distinct rays. *Distr.* Arctic and subarctic. *Fam.* Compositae. *Class* D.

A. arcticum Low-growing herb with white rays. Grown as an ornamental in the rock garden. *Distr.* Arctic Siberia, Alaska.

Arcterica See *Pieris*

arcticus, -a, -um: of the Arctic.

Arctostaphylos (From the Greek *arktos*, a bear, and *staphyle*, a bunch of grapes; bears are said to eat the fruit.) BEARBERRY, MANZANITA. A genus of about 50 species of typically evergreen, prostrate shrubs or small trees with sinuous twigs and small pendulous flowers. The fruit is a berry-like drupe. *Distr.* N and Central America, 2 species are circumpolar. *Fam.* Ericaceae. *Class* D.

A. glandulosa EASTWOOD MANZANITA. Erect or spreading shrub with red bark, white flowers and red-brown fruit. *Distr.* USA (Oregon to S California).

A. × media A hybrid of *A. columbiana* × *A. uva-ursi* with a horizontal main stem and erect or ascending branches. *Distr.* NW USA.

A. myrtifolia IONE MANZANITA. Low, often mat-forming shrub. *Distr.* USA (California).

A. nummularia FORT BRAGG MANZANITA. Rounded or mat-forming shrub with green fruit. *Distr.* California.

A. patula GREEN MANZANITA. Erect shrub with pink flowers and dark brown fruit. *Distr.* California (Sierra Nevada).

A. pumila DUNE MANZANITA. Mat-forming shrub with brown fruits. *Distr.* USA (California).

A. stanfordiana Erect shrub with pink fruit. *Distr.* USA (California).

A. uva-ursi BEARBERRY, CREASHAK, BEAR'S GRAPE, MOUNTAIN BOX, SAND BERRY, KINNIKINICK, HOG CRANBERRY, MEAL BERRY, COMMON BEARBERRY. Mat-forming shrub with rooting branches and globose, scarlet, mealy fruit. Leaves used in tanning Russian leather and in tea in Russia. *Distr.* Cold regions of the N hemisphere.

Arctotis (From the Greek *arktos*, a bear, and *otus*, an ear; the pappus scales are said to resemble a bear's ears.) AFRICAN DAISY. A genus of 50 species of annual and perennial herbs with basal rosettes of white-hairy leaves and daisy-like heads of flowers which bear distinct rays. *Distr.* South Africa to Angola. *Fam.* Compositae. *Class* D.

A. × hybrida A hybrid of *A. venusta* × *A. fastuosa*. Grown as a frost-tender bedding plant. *Distr.* Garden origin.

A. venusta BLUE-EYED AFRICAN DAISY. Annual. Flowers blue-red. *Distr.* S Africa.

arcuatus, -a, -um: arched.

Ardisia (From the Greek *ardis*, a point, alluding to the pointed anthers.) A genus of 250 species of evergreen shrubs and small trees with simple, leathery, spirally arranged leaves and small fused flowers that bear 5-fused sepals and are followed by red berries. Several species are grown as ornamentals and some have medicinal uses. *Distr.* Tropical and warm regions excluding Africa. *Fam.* Myrsinaceae. *Class* D.

A. crenata CORALBERRY, SPICEBERRY. Upright open shrub. Flowers white, star-shaped. Fruits bright red. *Distr.* NE India to Japan.

Areca (From a vernacular name used in Malabar.) A genus of 50 species of large palms with feather-shaped, short-stalked leaves and spikes or panicles of small flowers that are followed by rounded fruits. *Distr.* India to tropical Australia, Solomon Islands. *Fam.* Palmae. *Class* M.

A. catechu (From the local vernacular name.) BETEL NUT, CATECHU, BETEL PALM, PINANG. Seed cut into slices and chewed in a wad of betel paper (*Piper betle*) with lime, causes saliva to turn red, dulls appetite and acts as a mild narcotic. Used in vast quantities throughout Asia. *Distr.* Origin unknown, widely cultivated from India through SE Asia to Pacific Islands.

A. triandra Grown as an ornamental. *Distr.* E India, Sumatra, Borneo and Philippines.

Arecaceae See Palmae

Arecastrum See *Syagrus*

Arenaria (From the Latin *arena*, sand; many species grow on sandy soils.) SANDWORT. A genus of 150–160 species of low-growing perennial herbs with opposite, linear to rounded leaves and small flowers that bear 5 free, typically white petals and 10 stamens. A few species are grown as rock garden ornamentals. *Distr.* N temperate and Arctic regions. *Fam.* Caryophyllaceae. *Class* D.

A. balearica Prostrate mat-forming perennial. Flowers minute, white. *Distr.* W Mediterranean islands, widely naturalized from the gardens of England.

A. montana Mat-forming. Flowers relatively large, rounded, white. *Distr.* SW Europe, temperate Asia.

A. purpurascens Mat-forming. Petal lobes narrow, pink. *Distr.* Pyrenees, mountains of N Spain.

A. tetraquetra Cushion-forming. Stems 4-angled. Flowers numerous, star-shaped, white. *Distr.* Pyrenees, mountains of E Spain.

arenarius, -a, -um: growing in sandy places.

Arenga (From *areng*, the Malayan name for this palm.) A genus of about 17 species of medium to large palms with simple or feather-shaped, spiny stalked leaves and erect or pendent inflorescences. Several species are grown as ornamentals. *Distr.* Tropical Asia to N Australia. *Fam.* Palmae. *Class* M.

A. pinnata SUGAR PALM, GOMUTI, EJOW. Grown as a source of palm sugar, sago and fibre. *Distr.* Malaysia.

areolatus, -a, -um: marked with small patches.

aretioides from the genus *Aretia*, with the ending *-oides*, indicating resemblance.

Argemone (From the Latin *argema*, cataract of the eye, alluding to the former medicinal use of these plants.) BRITTONIA, PRICKLY POPPY, ARGEMONY. A genus of 28 species of annual and perennial herbs and shrubs with orange or yellow latex, spiny leaves and regular flowers that bear 3–6 spine-tipped sepals and 2 whorls of 3 petals. Several species are grown ornamentally. *Distr.* North to South America, West Indies, Hawaii. *Fam.* Papaveraceae. *Class* D.

A. mexicana MEXICAN POPPY, DEVIL'S FIG, PRICKLY POPPY. Annual herb to 1m high. Latex bright yellow. Flowers yellow. *Distr.* Central America, West Indies, S USA.

argemony *Argemone*.

argenteus, -a, -um: silvery.

argophyllus, -a, -um: with silver leaves.

argutifolius, a, -um: from the Latin *argutus*, sharply-toothed, and *folius*, leaved.

argutus, -a, -um: sharp-toothed.

argyraea, argyrea. silvery.

Argyranthemum (From the Greek *argyro*, silver, and *anthemos*, flower.) A genus of 23 species of perennial herbs and subshrubs with simple or pinnate leaves and daisy-like heads of flowers that bear distinct rays. Several species are grown as pot or tender bedding plants. *Distr.* Macaronesia. *Fam.* Compositae. *Class* D.

A. broussonetii Flowers yellow, rays white. *Distr.* Canary Islands.

A. foeniculaceum Leaves tripinnate. Flowers yellow, rays white. *Distr.* Canary Islands (Tenerife).

A. frutescens Leaves pinnate, leathery. Flowers yellow, rays white. *Distr.* Canary Islands.

A. gracile Flower-heads numerous. Flowers yellow, rays white. *Distr.* Canary Islands (Tenerife).

A. maderense To 1m high. Flowers yellow. *Distr.* Canary Islands (Lanzarote).

A. ochroleucum See *A. maderense*

argyrophyllus, -a, -um: with silvery leaves.

argyrotricha: with silvery hairs.

arietinus, -a, -um: resembling the horns of a ram.

Arisaema (From the genus name *Arum*, and the Greek *aima*, blood-red, in allusion to the red blotches on the leaves of some species.) JACK IN THE PULPIT, INDIAN TURNIP, DRAGON ARUM. A genus of 150 species of tuberous or rhizomatous herbs with divided leaves and small flowers borne at the base of a spadix that is surrounded by a tubular spathe with a hooded tip. Several species are grown as ornamentals for their delicate foliage, large hooded spathes and bright red fruits. *Distr.* Predominantly E Asia and the Himalayan region, but also North America and E Africa. *Fam.* Araceae. *Class* M.

A. candidissimum Spathe with a striking white hood marked with pink stripes. *Distr.* S and W China.

A. griffithii Spathe large, purple, netted green and white outside and pale green with purple eye-shaped blotches on the inside. *Distr.* E Himalaya.

A. helleborifolium See *A. tortuosum*

A. nepenthoides COBRA PLANT. Spathe mottled brown, striped white, with the limb bent forward. *Distr.* Himalaya to SW China, Burma.

A. ringens Spathe helmet-shaped, green or purple. *Distr.* S Japan, S Korea and E China.

A. tortuosum Spathe purple-green, hooded with a protruding contorted spadix. *Distr.* Himalaya (Kashmir to SW China).

A. triphyllum JACK IN THE PULPIT, INDIAN TURNIP. Leaves usually 2, each bearing 3 leaflets. Spathe green-purple, striped green-white. *Distr.* E North America.

Arisarum (From the Greek name, *arisaron*, used for *Arisarum vulgare*.) A genus of 3 species of tuberous or rhizomatous herbs with arrow-shaped, long-stalked leaves and small flowers borne on a spadix that is surrounded by a tubular spathe. *Distr.* Mediterranean region and the Atlantic islands. *Fam.* Araceae. *Class* M.

A. proboscideum MOUSE-TAIL PLANT, MOUSE PLANT. Spathe hooded, the tip extended into a long, tail-like twisting tip. *Distr.* Central and S Italy, S Spain.

A. vulgare FRIAR'S COWL. Spathe striped purple-brown. Tubers have been eaten in times of famine. *Distr.* Mediterranean, Canaries, Azores.

aristatus, -a, -um: awned, with long pointed tip.

Aristea (From the Greek *arista*, a point, alluding to the leaves or possibly from *aristos*, best, pleasing.) A genus of 50 species of perennial herbs with lance-shaped basal leaves and spike-like in-florescences of blue or occasionally white saucer-shaped flowers. Several species are grown as ornamentals. *Distr.* S Africa to N tropical Africa, Madagascar. *Fam.* Iridaceae. *Class* M.

A. ecklonii Flowers in loose panicles. *Distr.* South Africa, Swaziland.

A. major Flowering stem to 1.5m high. Flowers pale blue. *Distr.* South Africa (Cape Peninsula).

Aristolochia (From the Greek *aristos*, best, and *lochia*, childbirth, alluding to the curved form of the flower with base and top together recalling the human foetus in the correct position for birth.) DUTCHMAN'S PIPE, SNAKE ROOT, BIRTHWORT. A genus of 200–300 species of herbs, lianas, scrambling and erect shrubs with tuberous roots, simple or palmately lobed leaves and bad-smelling flowers that consist of a straight or curved sepal-tube without petals. The flowers are pollinated by the capture and release of flies (cf. *Arum*). A number of species are grown ornamentally and many have medicinal properties. *Distr.* Tropical and temperate regions. *Fam.* Aristolochiaceae. *Class* D.

A. clematitis BIRTHWORT. Herb to 1m high. Rhizome creeping. Flowers with a swollen base and slightly curved tube, borne in clusters. Used as an abortifacient. *Distr.* Europe, naturalized in British Isles and E North America.

A. durior See *A. macrophylla*

A. elegans See *A. littoralis*

A. gigas See *A. grandiflora*

A. grandiflora PELICAN FLOWER, SWAN FLOWER. Liana. Flowers large, white, solitary with a curved tube opening into a heart-shaped lip. *Distr.* Central America, Caribbean.

A. littoralis CALICO FLOWER. Fast growing climber. Flowers solitary, tube spreading into a broad, purple limb with white marbling. *Distr.* South America, naturalized Central America.

A. macrophylla DUTCHMAN'S PIPE. Deciduous, fast-growing climber. Flowers solitary, with an upward curving tube. *Distr.* E North America.

A. sempervirens Evergreen climber. *Distr.* S Greece, Italy.

A. sipho See *A. macrophylla*

Aristolochiaceae The Birthwort family. 8 genera and 410 species of herbs and shrubs, many of them twining lianas. The flowers have 3–4 sepals fused into a S-shaped tube and usually lack petals. Many species are cultivated as ornamentals. *Distr.* Tropical to warm temperate regions. *Class* D.
See: Aristolochia, Asarum.

Aristotelia (After Aristotle (384–322 BC), Greek philosopher.) A genus of 5 species of evergreen or deciduous shrubs and small trees with simple or occasionally pinnate leaves and clusters of typically unisexual flowers with their parts in 4s or 5s. Several species are grown as ornamentals and for their fruit. *Distr.* E Australia, New Zealand, S South America. *Fam.* Elaeocarpaceae. *Class* D.

A. chilensis MACQUI. Evergreen shrub or small tree. Fruit edible, used in preserves and to colour wine. *Distr.* Chile.

A. macqui See *A. chilensis*

A. serrata NEW ZEALAND WINE BERRY. Deciduous shrub or tree. *Distr.* New Zealand.

arizelus, -a, -um: notable.

arizonicus, -a, -um: of Arizona, USA.

arkansanus, -a, -um: of Arkansas, USA.

armatus, -a, -um: spiny.

armeniacus, -a, -um: apricot-coloured.

Armeria (The classical Latin name for a *Dianthus*.) THRIFT, SEA PINK. A genus of 80 species of tuft- or cushion-forming, perennial herbs and subshrubs with rosettes of simple leaves and dense heads of small flowers that bear their parts in 5s. Several species and numerous cultivars have been grown as ornamentals. *Distr.* Europe, Turkey, N Africa, W North and South America. *Fam.* Plumbaginaceae. *Class* D.

A. alliacea JERSEY THRIFT. Loosely branched herb. Flowers white to pale pink. *Distr.* W Europe.

A. maritima THRIFT, SEA PINK. Tufted herb or dwarf shrub. Flowers red, pink or white. Numerous ornamental cultivars have been raised from this species. Formerly used medicinally to treat obesity. *Distr.* Europe, W Asia, N Africa, North America.

armerioides: from the genus *Armeria*, with the ending *-oides*, indicating resemblance.

armillaris: encircled.

Armoracia (Classical Greek name for HORSERADISH.) A genus of 3 species of perennial herbs with small white flowers. *Distr.* SE Europe to Siberia. *Fam.* Cruciferae. *Class* D.

A. rusticana HORSERADISH, RED COLE. Root the source of relish typically eaten with beef or oysters. *Distr.* SE Europe, naturalized throughout Europe and North America.

Arnebia (From the Arabian name for these plants.) A genus of 25 species of annual and perennial, stiffly hairy herbs with simple leaves and cymes of funnel- or salver-shaped flowers. Several species are grown as ornamentals. *Distr.* Mediterranean, tropical Africa, Himalaya. *Fam.* Boraginaceae. *Class* D.

A. echioides See *A. pulchra*

A. pulchra PROPHET FLOWER. Perennial. Flowers yellow. *Distr.* Armenia, Caucasus, N Iran.

Arnica (From the Greek *arnakis*, lambskin, alluding to the texture of the leaves.) A genus of 32 species of perennial rhizomatous herbs with simple leaves and daisy-like heads of flowers that bear distinct rays. *Distr.* N temperate regions and the Arctic. *Fam.* Compositae. *Class* D.

A. angustifolia To 50cm high. Leaves basal. *Distr.* North America, Arctic, N Eurasia.

A. chamissonis To over 1m. Rays pale yellow. *Distr.* Alaska to New Mexico.

A. fulgens The roots and flower-heads have medicinal qualities similar to those of *A. montana*. *Distr.* N America.

A. longifolia Stems much branched. Rays yellow. *Distr.* Mountains of W USA.

A. montana MOUNTAIN TOBACCO, ARNICA. To 75cm high. Rays toothed. The roots and flower-heads have medicinal qualities. *Distr.* Central and N Europe, W Asia.

A. unalaschensis Stems simple. Flower-heads solitary. *Distr.* Japan and neighbouring islands.

arnica *Arnica montana.*

aromaticus, -a, -um: aromatic, scented.

Aronia (From *aria*, the Greek name for *Sorbus aria*.) CHOKEBERRY. A genus of 2 species of deciduous shrubs with simple alternate leaves and clusters of small white flowers that are followed by red, purple or black fruits. *Distr.* NE America. *Fam.* Rosaceae. *Class* D.

A. arbutifolia RED CHOKEBERRY. Grown as an ornamental for its spring flowers, red fruits and brilliant autumn colours. *Distr.* E North America.

arrayan *Luma apiculata.*

Arrhenatherum (From the Greek *arren*, male, and *anther*, a bristle, alluding to the spiky male inflorescences.) OAT GRASS. A genus of 6 species of perennial, occasionally tuberous grasses to 1.5m high with flat leaves and narrow panicles. *Distr.* Europe, N Africa and N and W Asia. *Fam.* Gramineae. *Class* M.

A. elatius FALSE OAT, FRENCH RYE. Tussock-forming meadow grass. Sub-species *bulbosum* is grown as a curiosity because of the bulbs formed along the length of its stem. Variegated cultivars are also available. *Distr.* Europe.

arrowhead *Sagittaria, S. sagittifolia* **Old World** ~ *S. sagittifolia.*

arrowroot *Maranta arundinacea*, Marantaceae **East Indian** ~ *Tacca leontopetaloides* **Florida** ~ *Zamia* **Guyana** ~ *Dioscorea alata* **Portland** ~ *Arum* **Queensland** ~ *Canna indica* **Tahiti** ~ *Tacca leontopetaloides.*

arrow wood *Viburnum, V. acerifolium, V. dentatum*, **southern** ~ *V. dentatum.*

Artemisia (After the Greek goddess Artemis.) WORMWOOD, MUGWORT, SAGE BRUSH. A genus of about 300 species of annual, biennial or perennial herbs, shrubs and subshrubs typically with pinnately divided, silky-hairy leaves and pendent daisy-like heads of flowers that lack rays. Members of this genus are used as medicinal and culinary herbs. *Distr.* W South America, S Africa, Russian steppes. *Fam.* Compositae. *Class* D.

A. abrotanum SOUTHERNWOOD, LAD'S LOVE, OLD MAN. Shrub. Foliage strongly aromatic. Used to make tea and reported to have medicinal qualities. *Distr.* S Europe.

A. absinthium ABSINTHE, WORMWOOD. Aromatic perennial herb. Flowers yellow. Used medicinally. *Distr.* Europe.

A. alba Perennial aromatic subshrub to 1m tall. *Distr.* S Europe, N Africa.

A. annua SWEET WORMWOOD. Erect annual, to 1m high. Used as a treatment for Malaria in China. *Distr.* Eurasia, naturalized N America.

A. arborescens Glabrous perennial to 1m high with woody stems. *Distr.* Mediterranean.

A. armeniaca Rhizomatous perennial. *Distr.* Caucasus to Central Asia and Iran.

A. assoana See *A. pedemontana*

A. camphorata See *A. alba*

A. canariensis Shrub to 1m high. *Distr.* Canary Islands.

A. chamaemelifolia Glabrous aromatic perennial. *Distr.* Pyrenees, Alps, Bulgaria, Caucasus.

A. douglasiana Perennial with herbaceous stems to 3m high. *Distr.* W North America.

A. dracunculus TARRAGON, ESTRAGON. Much-branched, aromatic perennial. An important culinary herb, especially as flavouring with fish. *Distr.* Eurasia.

A. eriantha Tufted hairy perennial to 30cm high. *Distr.* Mountains of Central Europe.

A. frigida Tufted hairy perennial to 50cm high. *Distr.* SE Russia.

A. genipi Tufted, more or less hairy perennial to 40cm high. *Distr.* Alps.

A. glacialis Tufted, long-hairy perennial to 20cm high. *Distr.* SW Alps.

A. gmelinii Annual or biennial subshrub. *Distr.* E Europe to Siberia and NE Asia.

A. gnaphalodes See *A. ludoviciana*

A. gracilis See *A. scoparia*

A. granatensis Tufted, long-haired perennial. *Distr.* S Spain.

A. lactiflora WHITE MUGWORT. Perennial herb to 2m high. Flowers white. *Distr.* China.

A. lanata See *A. pedemontana*

A. ludoviciana WESTERN MUGWORT, WHITE SAGE, CUDWEED. Rhizomatous perennial. *Distr.* W North America to Mexico.

A. maritima See *Seriphidium maritimum*

A. nutans See *Seriphidium nutans*

A. palmeri See *Seriphidium palmeri*

A. pedemontana Tufted, woolly-hairy perennial. *Distr.* Central Spain to Ukraine.

A. pontica ROMAN WORMWOOD. Rhizomatous aromatic perennial. *Distr.* E and Central Europe.

A. purshiana See *A. ludoviciana*

A. rupestris Shrub with horizontal branches. *Distr.* Baltic to Central Asia.

A. schmidtiana Tufted rhizomatous perennial. *Distr.* Japan.

A. scoparia Biennial with red or yellow flowers. *Distr.* Europe to SW Asia and Siberia.

A. splendens Mound-forming perennial. *Distr.* Caucasus, SW Asia.

A. stelleriana BEECH WORMWOOD, OLD WOMAN, DUSTY MILLER. Densely white hairy, rhizomatous perennial. *Distr.* NE Asia, E North America.

A. tridentata See *Craspedia*

A. vallesiaca See *Seriphidium vallesiacum*

A. vulgaris MUGWORT. Tufted aromatic perennial. Leaves used as a condiment and also believed to have magical properties. *Distr.* Europe and N Africa to Siberia.

artemisioides: from the genus *Artemisia*, with the ending *-oides*, indicating resemblance.

Arthropodium (From the Greek *arthron*, a joint, and *podion*, a little foot, alluding to the jointed pedicels.) A genus of 12 species of rhizomatous, perennial, often tuft-forming herbs with linear sheathing leaves and racemes or panicles of small flowers. Several species are cultivated as ornamentals. *Distr.* Australia, New Zealand, New Guinea, New Caledonia and Madagascar. *Fam.* Anthericaceae. *Class* M.

A. cirrhatum An evergreen perennial with large panicles of star-shaped white flowers. *Distr.* New Zealand.

artichoke; Chinese ~ *Stachys affinis* **French** ~ *Cynara scolymus* **globe** ~ *Helianthus tuberosus*, *Cynara scolymus* **Japanese** ~ *Stachys affinis* **Jerusalem** ~ *Helianthus tuberosus*.

articulatus, -a, -um: articulated, jointed.

artillery plant *Pilea microphylla*.

Arum (From *aron*, the classical Greek name for these plants.) LORDS AND LADIES, CUCKOO PINT, JACK IN THE PULPIT, WAKE ROBIN, PORTLAND ARROWROOT. A genus of about 15 species of tuberous herbs with arrow-shaped or simple leaves and small flowers born on a spadix that is surrounded by a spathe. Pollination occurs via a capture and release mechanism. Flies are attracted to the appendage at the tip of the spadix and fall into the chamber created by the spathe. They are trapped there by a ring of hairs until covered in pollen at which point the hairs wither releasing the flies to visit another inflorescence. Several species are cultivated as ornamentals and the tubers have been used as a source of arrowroot or to starch linen. *Distr.* Europe to W Asia. *Fam.* Araceae. *Class* M.

A. cornutum See *Sauromatum venosum*

A. creticum Spathe yellow-green. *Distr.* Crete and Karpathos.

A. dioscoridis Spathe variable, usually with an erect sail-like limb. *Distr.* Cyprus to Turkey, Israel and Iraq.

A. dracunculus See *Dracunculus vulgaris*

A. idaeum See *Arum maculatum*

A. italicum Spathe pale, sometimes completely white. Several subspecies and a number of cultivars grown as ornamentals. *Distr.* S and W Europe.

A. maculatum CUCKOO PLANT, LORDS AND LADIES, JACK IN THE PULPIT. Leaves often black-spotted. Spathe pale green. *Distr.* Europe to the Ukraine.

A. pictum Leaves white-veined. *Distr.* W Mediterranean (Balearics, Corsica, Sardinia).

arum Araceae **arrow** ~ *Peltandra* **black-throated** ~ *Zantedeschia albomaculata* **bog** ~ *Calla palustris* **dragon** ~ *Arisaema*, *Dracunculus vulgaris* **golden** ~ *Zantedeschia elliottiana* **green arrow** ~ *Peltandra virginica* **hairy** ~ *Dracunculus muscivorus* **pink** ~ *Zantedeschia rehmannii* **spotted** ~ *Z. albomaculata* **titan** ~ *Amorphophallus titanum* **water** ~ *Calla palustris* **white arrow** ~ *Peltandra sagittifolia*.

Aruncus (The Greek name.) GOAT'S BEARD. A genus of 2 species of rhizomatous herbs with pinnately divided, alternate leaves and panicles of small white flowers. *Distr.* N temperate and subarctic regions. *Fam.* Rosaceae. *Class* D.

A. *dioicus* GOAT'S BEARD. Erect herb to 2m high. Flowers small, malodorous, borne in large branching plumes. *Distr.* W and Central Europe, Caucasia.

A. *sylvester* See *A. dioicus*

arundinacea reed-like, from the Latin *arundo*, reed, with the ending *-aceus*, indicating resemblance.

Arundinaria (From the Latin, *arundo*, a reed.) BAMBOO. A genus of about 1 species of creeping, rhizomatous bamboo with erect stems to 10m high. *Distr.* N America. *Fam.* Gramineae. *Class* M.

A. *auricoma* See *Pleioblastus auricoma*

A. *disticha* See *Pleioblastus pygmaeus*

A. *fastuosa* See *Semiarundinaria fastuosa*

A. *fortunei* See *Pleioblastus variegatus*

A. *gigantea* GIANT CANE, SWITCH CANE, GIANT REED, CANE REED. Edible young shoots, eaten by Indians and early settlers, mature stems used as canes. Now grown ornamentally. *Distr.* SE USA.

A. *japonica* See *Pseudosasa japonica*

A. *marmorea* See *Chimonobambusa marmorea*

A. *palmata* See *Sasa palmata*

A. *pygmaea* See *Pleioblastus pygmaeus*

A. *quadrangularis* See *Chimono-bambusa quadrangularis*

A. *spathiflora* See *Thamnocalamus spathiflorus*

A. *variegata* See *Pleioblastus variegatus*

A. *veitchii* See *Sasa veitchii*

A. *viridistriata* See *Pleioblastus auricoma*

Arundo (From the Latin *arundo*, a reed.) GIANT REED. A genus of 3 species of half hardy, rhizomatous grasses of damp areas with reed-like stems to 6m high, broad-linear leaves and large, terminal, feathery panicles. *Distr.* Mediterranean region to China and Japan. *Fam.* Gramineae. *Class* M.

A. *donax* (From the Greek name for a kind of reed.) GIANT REED. Sections of the stem have been used for over 5000 years in reed instruments which today include clarinets, saxophones, oboes and bassoons. The stems are also used for walking sticks and fishing rods and as a source of cellulose. *Distr.* Mediterranean region.

arvensis, -e: of cultivated land or fields.

asafoetida *Ferula assa-foetida*

asarabacca *Asarum*, *A. europaeum*.

asarifolius, -a, um: from the genus *Asarum*, and *folius*, leaved.

Asarina (The Spanish vernacular name for the genus *Antirrhinum*.) TWINING SNAPDRAGON. A genus of 16 species of climbing or spreading, perennial herbs with scattered, simple leaves and 2-lipped flowers that bear 5 sepals, 5 petals and 4 stamens. Several species are grown as ornamentals. *Distr.* Mexico and SW North America, S Europe. *Fam.* Scrophulariaceae. *Class* D.

A. *antirrhiniflora* VIOLET TWINING SNAPDRAGON. Stems diffuse, twining. Flowers pale purple with a yellow boss on the lower lip. *Distr.* SW USA to S Mexico.

A. *erubescens* CREEPING GLOXINIA. Stems twining, glandular-hairy. Flowers pink. *Distr.* Mexico.

A. *hispanica* See *Antirrhinum hispanicum*

asarina from a local Spanish name for *Antirrhinum*.

Asarum (From *asaron*, the classical Greek name.) WILD GINGER, ASARABACCA. A genus of about 75 species of rhizomatous herbs with long stalked, kidney- or heart-shaped leaves, and solitary flowers that typically bear 3 sepals fused into a cup and 2 whorls of 6 stamens. Several species are grown ornamentally and some are used medicinally. *Distr.* N temperate regions. *Fam.* Aristolochiaceae. *Class* D.

A. canadense WILD GINGER. The rhizomes are used medicinally and as a substitute for GINGER. *Distr.* North America.

A. caudatum Prostrate herb. Flowers red-brown with long thin lobes. *Distr.* W North America.

A. europaeum ASARABACCA. Creeping herb. Leaves heart-shaped, dark, glossy green. Used medicinally as a hangover cure. *Distr.* Europe.

A. hartwegii Flowers black with long tail-like lobes. *Distr.* W USA.

ascendens: ascending.

Asclepiadaceae The Milkweed family. 315 genera and about 3000 species of perennial herbs, shrubs, woody climbers and trees usually with a milky sap and simple leaves. The flowers are regular with their parts in 5s and often bear extra appendages on the stamens or corolla that form a corona (crown). *Distr.* Widespread tropical, subtropical and occasionally temperate areas. *Class* D.
See: Araujia, Asclepias, Ceropegia, Cionura, Dregea, Gomphocarpus, Hoya, Marsdenia, Matelea, Oxypetalum, Periploca, Stephanotis, Tweedia, Vincetoxicum, Wattakaka.

asclepiadeus, -a, -um: resembling the genus *Asclepias*.

Asclepias (From the Greek *Asklepios*, god of medicine, alluding to its medicinal properties.) MILKWEED, SILKWEED. A genus of about 120 species of tuberous annual and perennial herbs, subshrubs and shrubs with milky sap, simple leaves and cymes of numerous small flowers that are followed by pods of hairy seeds. A number of species are grown as ornamentals. *Distr.* North America, naturalized in the Old World. *Fam.* Asclepiadaceae. *Class* D.

A. curassavica BLOOD FLOWER, SWALLOW WORT, MATAC, INDIAN ROOT, BASTARD IPECACUANUA, MATAL. Short-lived evergreen subshrub. Flowers orange-red with a yellow centre. Grown as an ornamental and used medicinally. *Distr.* South America, naturalized throughout the tropics.

A. incarnata SWAMP MILKWEED. *Distr.* E USA.

A. physocarpa SWAN PLANT. Slender, deciduous shrub. Flowers cream-white. *Distr.* South Africa.

A. purpurascens PURPLE SILKWEED. Flowers pink-purple. *Distr.* E North America.

A. syriaca Perennial herb. Flowers purple-pink. Cultivated for fibre from the stems and seeds. *Distr.* E North America, naturalized Europe.

A. tuberosa BUTTERFLY WEED, TUBER ROOT, CHIEGER FLOWER. Erect, finely hairy herb with yellow-red flowers. Used medicinally as well as frequently being grown ornamentally. *Distr.* North America.

ash *Fraxinus* **Alpine** ~ *Eucalyptus delegatensis* **American** ~ *Fraxinus pennsylvanica, F. americana* **American mountain** ~ *Sorbus americana* **Canadian** ~ *Fraxinus pennsylvanica, F. americana* **common** ~ *F. excelsior* **common mountain** ~ *Sorbus aucuparia* **European** ~ *Fraxinus excelsior* **flowering** ~ *F. ornus* **green** ~ *F. pennsylvanica* **ground** ~ *Aegopodium podagraria* **manna** ~ *Fraxinus ornus* **mountain** ~ *Sorbus aucuparia* **prickly** ~ *Zanthoxylum* **red** ~ *Fraxinus pennsylvanica* **stinking** ~ *Ptelea trifoliata* **water** ~ *P. trifoliata* **white** ~ *Fraxinus pennsylvanica, F. americana* **yellow** ~ *Cladrastis lutea*.

asiaticus, -a, -um: of Asia.

Asparagaceae A family of 1 genus and about 100 species. *Class* M.
See: Asparagus.

asparagoides: from the genus *Asparagus*, with the ending *-oides*, indicating resemblance.

Asparagus (A classical Greek name for these plants.) A genus of about 100 species of rhizomatous perennial herbs, shrubs and climbers with green branches. The leaves are reduced to small, often spiny scales. The inconspicuous flowers are borne in the branch axils or in simple inflorescences and have their parts in 6s. Cultivated ornamentally and for their edible young shoots. *Distr.* Old World. *Fam.* Asparagaceae. *Class* M.

A. asparagoides Stem twining. Branches flattened, oval, leaf-like. *Distr.* South Africa (Cape Province), naturalized in the Mediterranean area.

A. densiflorus Stems somewhat woody, arising from a tuber. Branches flattened, linear. Flowers pink. Berries red. *Distr.* S Africa.

A. officinalis GARDEN ASPARAGUS, ASPARAGUS. Widely cultivated for the edible young shoots which are considered a delicacy. It has been in cultivation in Europe for over 2000 years and was originally thought to have medicinal properties. *Distr.* S Europe, N Africa, SW Asia.

A. setaceus ASPARAGUS FERN. An evergreen scrambler or climber with small sharp spines. The fine feathery foliage is much used by florists, typically as a backing for buttonholes. *Distr.* S Africa.

asparagus Asparagaceae, *Asparagus officinalis* **Bath** ~ *Ornithogalum pyrenaicum* **garden** ~ *Asparagus officinalis*.

aspen *Populus, P.tremula.*

asperata: rough texture.

Asperula (From the Latin *asper*, rough, alluding to the rough hairy stems.) WOODRUFF. A genus of 90–100 species of annual and perennial herbs and small shrubs with simple, opposite or whorled leaves and panicles of heads of bell- to cup-shaped, 4-lobed flowers. Several species are grown as ornamentals. *Distr.* Eurasia, Australia. *Fam.* Rubiaceae. *Class* D.

A. odorata See *Galium odoratum*

A. suberosa Clump-forming perennial. Flowers pink, very numerous. *Distr.* Greece, Bulgaria.

A. tinctoria DYER'S WOODRUFF. Erect or prostrate, stoloniferous perennial. Flowers white. *Distr.* Europe.

asperuloides: from the genus *Asperula*, with the ending *-oides*, indicating resemblance.

asperus, -a, -um: rough.

asphodel *Asphodelus* **bog** ~ *Narthecium, N. ossifragum* **false** ~ *Tofieldia* **giant** ~ *Eremurus* **mountain** ~ *Xerophyllum asphodeloides* **Scotch** ~ *Tofieldia pusilla* **white** ~ *Asphodelus albus* **yellow** ~ *Asphodeline lutea.*

Asphodelaceae A family of 11 genera and about 200 species of rhizomatous annual and perennial herbs with tufts or rosettes of typically linear leaves. The flowers are borne in spikes or panicles and have 2 whorls of three tepals that may or may not be fused. This family is sometimes included within the family Liliaceae. Several genera contain important ornamental species. *Distr.* Chiefly temperate regions of the Old World but with several species occurring in New Zealand and Mexico. *Class* M.

See: Asphodeline, Asphodelus, Bulbine, Bulbinella, Eremurus, Kniphofia, Paradisea, Simethis.

Asphodeline (Like the genus *Asphodelus.*) JACOB'S ROD. A genus of 10–20 species of rhizomatous perennial or biennial herbs with linear basal leaves and spikes of numerous, somewhat irregular, fragrant flowers. Several species are cultivated as ornamentals. *Distr.* Mediterranean area to the Caucasus. *Fam.* Asphodelaceae. *Class* M.

A. lutea YELLOW ASPHODEL, KING'S SPEAR. Perennial to 1.3m high. Flowers yellow and fragrant. Often grown as an ornamental with several cultivars available. Bulbs edible. *Distr.* Mediterranean area.

asphodeloides: from the genus *Asphodelus*, with the ending *-oides*, indicating resemblance.

Asphodelus (The Greek name for *Asphodelus ramosus.*) ASPHODEL. A genus of 12 species of annual and perennial herbs with swollen rhizomes and linear basal leaves. The regular white flowers are borne in tall racemes or panicles and bear tepals with a dark-striped midrib. Several species are cultivated as ornamentals. *Distr.* Mediterranean to Himalaya. *Fam.* Asphodelaceae. *Class* M.

A. acaulis Flowering stem absent or very short. *Distr.* Algeria, Morocco.

A. aestivus Stems to 2m high. Bulbs a source of yellow dye. *Distr.* S Europe, N Africa.

A. albus WHITE ASPHODEL. Flowering stem to 1m high, sometimes branched. Fermented as a source of alcohol. *Distr.* S and Central Europe.

A. fistulosus Annual or short-lived perennial. Bulbs edible. *Distr.* SW Europe, SW Asia, India.

A. luteus See *Asphodeline lutea*

A. microcarpus See *A. aestivus*

Aspidistra (From the Greek *aspidion*, a small round shield, alluding to the shape of the stigma.) A genus of 8 species of rhizomatous herbs with basal, elliptic to lance-shaped, leathery leaves that narrow to the leaf-stalk and are usually held vertically. The inconspicuous flowers are borne directly on the rhizome at soil level and are said to be pollinated by slugs and snails. Several species are grown for their ornamental foliage. *Distr.* E Asia, Himalaya to Japan. *Fam.* Convallariaceae. *Class* M.

A. elatior CAST-IRON PLANT, BAR ROOM PLANT. A robust house plant that was particularly popular during the 19th century. *Distr.* China, Japan

A. lurida See *A. elatior*

Aspleniaceae The Spleenwort family. 9 genera and 700 species of terrestrial or epiphytic, small to medium-sized ferns. The leaves are simple to several times pinnately divided. The sporangia are gathered into sori on the veins of the leaf. *Distr.* Widespread. *Class* F.
See: Asplenium, Camptosorus, Quercifilix.

aspleniifolius: from the genus *Asplenium*, and *folius*, leaved.

Asplenium (From the Greek *a*, not, and *splen*, spleen, alluding to supposed medicinal properties.) SPLEENWORT. A genus of 650–700 species of terrestrial and epiphytic ferns with divided leaves. Several species are grown as ornamentals and some have medicinal properties. *Distr.* Cosmopolitan. *Fam.* Aspleniaceae. *Class* F.

A. adiantum-nigrum BLACK SPLEENWORT. *Distr.* Europe, Africa, Asia.

A. bulbiferum MOTHER SPLEENWORT, HEN AND CHICK FERN. Small plantlets are produced on the margins of the leaves. *Distr.* SW Pacific regions.

A. nidus BIRD'S NEST FERN. Epiphytic. Leaves simple, borne in rosettes. *Distr.* Tropical regions of the Old World.

A. ruta-muraria WALL RUE. *Distr.* North America, Europe, tropical Asia.

A. scolopendrium HART'S TONGUE FERN. Leaves simple, erect. *Distr.* North America, Europe, Asia.

A. trichomanes MAIDENHAIR SPLEENWORT. Terrestrial. Leaves pinnate. *Distr.* N temperate, tropical mountains.

assurgentiflorus, -a, -um: from the Latin *assurgens*, rising upwards, and *florus*, flowers.

Astelia (From the Greek *a*, without, and *stele*, pillar, alluding to the inability of the pendent species to support themselves.) A genus of about 25 species of tufted epiphytic and terrestrial herbs with rosettes of hairy or scaly leaves and panicles of inconspicuous flowers. Several species are grown as ornamentals for their striking foliage. *Distr.* Australia, New Zealand, Polynesia, Reunion, Mauritius and the Falkland Islands. *Fam.* Asteliaceae. *Class* M.

A. nervosa BUSH FLAX. Terrestrial. Leaves to 2m long, rigid. The berries are edible and the leaves are a source of fibre. *Distr.* New Zealand.

Asteliaceae A family of 4 genera and about 40 species of herbs, shrubs and trees. This family is sometimes included within the family Liliaceae. *Distr.* Mascarenes, New Guinea, Australasia, Pacific, Chile. *Class* M.
See: Astelia.

Aster (From the Latin *aster*, a star, alluding to the flower-heads.) A genus of about 250 species of annual, biennial and perennial herbs with simple leaves and flowers in daisy-like heads that may or may not bear distinct rays. Many species and hybrids are grown as ornamentals. *Distr.* America, Eurasia, Africa. *Fam.* Compositae. *Class* D.

A. acris See *A. sedifolius*

A. albescens Much-branched shrub. Rays lilac. *Distr.* Himalaya, W China.

A. alpinus Perennial herb. Flowers yellow, rays violet-blue. *Distr.* Europe (Alps, Pyrenees), Asia.

A. amelloides See *Felicia amelloides*

A. amellus Perennial herb. Flowers yellow, rays usually purple. *Distr.* Eurasia, North America.

A. asper See *A. bakerianus*

A. bakerianus Tuberous perennial. Rays narrow. *Distr.* South Africa.

A. capensis See *Felicia amelloides*

A. carolinianus Straggling shrub. *Distr.* E USA (Florida to N Carolina).

A. coelestis See *Felicia amelloides*

A. cordifolius Perennial herb to 1.5m high. *Distr.* E North America.

A. corymbosus See *A. divaricatus*

A. diffusus See *A. lateriflorus*

A. divaricatus Rhizomatous perennial with purple stems. *Distr*. E North America.

A. dumosus Perennial herb. *Distr*. SE North America.

A. ericoides Perennial herb with thin, much-branched stems. *Distr*. North America, Mexico.

A. farreri Tufted perennial herb. *Distr*. W China, Tibet.

A. foliaceus Perennial rhizomatous herb. *Distr*. W North America.

A.×frikartii (After Carl Ludwig Frikart (1879–1964).) A hybrid of *A. amellus* × *A. thomsonii*. *Distr*. Garden origin.

A. himalaicus Small perennial herb to 20cm. *Distr*. Nepal, China.

A. hybridus See *X Solidaster luteus*

A. laevis Perennial herbs with a stout rootstock. *Distr*. California (Sierra Nevada).

A. lateriflorus Perennial herbs with a stout rootstock. Several cultivars are available. *Distr*. North America, naturalized in Europe.

A. linosyris GOLDILOCKS. Perennial herb. Flower-heads numerous, yellow, rays absent. *Distr*. S and SE Europe.

A. macrophyllus Tuft-forming perennial herb. *Distr*. North America, naturalized in Europe.

A. natalensis See *Felicia rosulata*

A. novae-angliae Perennial herb with overlapping leaves. Rays deep purple. *Distr*. E North America.

A. novi-belgii MICHAELMAS DAISY. Perennial herb. Flowers yellow, rays deep purple. *Distr*. E North America, especially coasts, widely naturalized in Europe.

A. paniculatus See *A. novi-belgii*

A. pappei See *Felicia amoena*

A. petiolatus See *Felicia petiolata*

A. pilosus Perennial herb. Rays white, turning purple. *Distr*. E North America.

A. ptarmicoides Perennial herb. Rays white. *Distr*. Central North America.

A. pyrenaeus Perennial herb. Flower-heads solitary, rays blue. *Distr*. Pyrenees.

A. rotundifolius See *Felicia amelloides*

A. scandens See *A. carolinianus*

A. sedifolius Perennial, rarely annual herb. *Distr*. S Europe.

A. sibiricus Small perennial. Rays violet. *Distr*. Arctic Asia.

A. thomsonii (After Thomas Thomson (1817–78), Scottish physician and superintendent of Calcutta Botanic Garden.) Perennial herb with thin stems. Rays purple. *Distr*. W Himalaya (Pakistan to Uttar Pradesh).

A. tongolensis Small stoloniferous perennial with solitary flower-heads. *Distr*. W China to India.

A. tradescantii See *A. pilosus*

A. tripolium SEA ASTER, MICHAELMAS DAISY. Annual or short lived perennial. Rays blue or absent. *Distr*. Eurasia, including Great Britain.

A. turbinellus Tall perennial. Rays violet. *Distr*. E Central North America.

A. umbellatus Tall, rhizomatous, perennial herb. *Distr*. E North America.

A. vahlii Short, rhizomatous, perennial herb. *Distr*. Falkland Islands, Tierra del Fuego.

A. yunnanensis Rhizomatous perennial with glandular hairs. *Distr*. W China to SE Tibet.

aster: beach ~ *Erigeron glaucus* **golden** ~ *Chrysopsis* **sea** ~ *Aster tripolium* **Stoke's** ~ *Stokesia laevis*.

Asteraceae See Compositae

Asteranthera (From the Latin *aster*, a star, and *anthera*, an anther, alluding to the star-like arrangement of the anthers.) A genus of 1 species of evergreen climbing shrub with simple leaves and pairs of bright red, 2-lipped, funnel-shaped flowers. *Distr*. Chile, Argentina. *Fam*. Gesneriaceae. *Class* D.

A. ovata Grown as a half-hardy ornamental.

Asteriscus (From the Greek *asteriskos*, little star.) A genus of 3–4 species of annual to perennial herbs with simple leaves and flowers in daisy-like heads that bear distinct rays. *Distr*. Mediterranean, Cape Verde Islands, Canary Islands. *Fam*. Compositae. *Class* D.

A. maritimus Tuft-forming herb. Rays yellow-orange. *Distr*. W Mediterranean, S Portugal, Canary Islands, Greece.

asteroides: resembling a star.

Asteropeiaceae A family of 1 genus and 5 species of vines and trees. This family is sometimes included within the family Theaceae. *Distr.* Madagascar. *Class* D.

Astilbe (From the Greek *a*, without and *stilbe*, brilliance, alluding to the small flowers.) SPIRAEA. A genus of 12 species of perennial clump-forming herbs with numerous, small, white to dark red flowers. *Distr.* E Asia, USA. *Fam.* Saxifragaceae. *Class* D.

A. × arendsii (After George Arends of Ronsdorf (1862–1952), who hybridized *Astilbes.*) A number of ornamental hybrids and cultivars with large panicles of flowers. *Distr.* Garden origin.

A. astilboides A name that is usually misapplied to *A. japonica*.

A. biternata FALSE GOATSBEARD. Flowers creamy-white to yellow, profuse. *Distr.* North America.

A. chinensis Flowers white, tinged red or magenta and borne in dense clusters. *Distr.* China, Japan.

A. × crispa A number of cultivars with finely divided leaves are available under this name. *Distr.* Garden origin.

A. glaberrima Dwarf perennial with mauve flowers.

A. grandis Perennial with pyramids of white flowers. *Distr.* China.

A. japonica SPIRAEA. Tall perennial with white flowers. Numerous ornamental cultivars of this species are available. *Distr.* Japan.

A. rivularis Perennial with leaves woolly below and flowers cream-white. *Distr.* Nepal.

A. simplicifolia Perennial with star-like white flowers. *Distr.* Japan.

Astilboides (From the genus name *Astilbe*, and the Greek *oides*, indicating resemblance.) A genus of 1 species of perennial herb with numerous small white flowers. *Distr.* N China. *Fam.* Saxifragaceae. *Class* D.

A. tabularis A cultivated ornamental. *Distr.* N China.

Astragalus GOAT'S THORN, MILK VETCH. A genus of about 2000 species of annual and perennial herbs and shrubs typically with pinnate leaves and spikes, racemes or clusters of irregular pea-like flowers that bear 10 stamens. This is the largest genus of the flowering plants and is of great ecological importance as well as containing many ornamental species and being the source of tragacanth gum that is use in icecream, cosmetic lotions, and pharmaceuticals. *Distr.* N temperate regions. *Fam.* Leguminosae. *Class* D.

A. glycyphyllos WILD LIQUORICE. Perennial herb. Flowers pale yellow. Used as fodder and a herbal tea. *Distr.* Europe.

A. purshii Mat-forming perennial herb. Flowers white to pink or purple. *Distr.* Rocky mountains.

Astrantia (Derivation obscure, possibly from the Latin *aster*, a star, alluding to the inflorescence.) MASTERWORT. A genus of 10 species of perennial herbs with palmate leaves and simple umbels of pink or white flowers. The umbels are subtended by a whorl of fused bracts that form a showy cup or star that is often larger than the inflorescence. Several species and numerous cultivars are grown ornamentally. *Distr.* Central and SW Europe, W Asia. *Fam.* Umbelliferae. *Class* D.

A. major GREATER MASTERWORT, ASTRANTIA. To 80cm high. Flowers papery, white, flushed with green or pink. This species has given rise to a number of garden cultivars. *Distr.* Europe.

A. rubra See *A. major*

astrantia *Astrantia major*.

asturiensis: of Asturia, Spain.

Asyneuma A genus of 50 species of biennial and perennial herbs with simple leaves and spikes or panicles of lilac or indigo flowers. This genus is closely related to the genus *Phyteuma*. Several species are grown as ornamentals. *Distr.* S Europe, Caucasus, E Asia. *Fam.* Campanulaceae. *Class* D.

A. canescens Stems erect, to 90cm high. Flowers borne in panicles, pale lilac. *Distr.* S Europe.

A. limoniifolium Stems erect or ascending, to 1.2m high. Flowers in spikes or panicles, violet. *Distr.* S Europe, Turkey.

Asystasia (From the Greek *asystasia*, confusion.) A genus of about 70 species of perennial herbs and shrubs with simple leaves and spikes of irregular tubular flowers. *Distr.* Tropical regions of the Old World. *Fam.* Acanthaceae. *Class* D.

A. bella See *Mackaya bella*

A. gangetica A weed throughout the tropics that is occasionally grown as ground cover. *Distr.* Tropical regions of the Old World, now widely naturalized.

Athamanta (After Mount Athamas in Sicily, where some of the species grow.) A genus of 15 species of perennial herbs with much divided leaves and umbels of small, white or yellow flowers. *Distr.* Mediterranean region. *Fam.* Umbelliferae. *Class* D.

A. cretensis CANDY CARROT. Leaves with fine needle-like segments. Flowers white, tinged red. This species is used in flavouring for liqueurs as well as being grown as an ornamental. *Distr.* S Europe.

Atherosperma (From the Greek *atheros*, barb, and *sperma*, seed.) A genus of 1 species of large, evergreen, aromatic tree with simple opposite leaves and solitary cream flowers that bear 8–10 tepals and 10–18 stamens. *Distr.* SW Australia. *Fam.* Monimiaceae. *Class* D.

A. moschatum Grown as a tender ornamental.

Atherospermataceae See Monimiaceae

Athrotaxis (From the Greek *athroos*, crowded together, and *taxis*, arrangement, alluding to the densely arranged leaves.) A genus of 2–3 species of evergreen coniferous trees with spirally arranged, scale-like leaves and woody fruiting cones. *Distr.* Tasmania. *Fam.* Taxodiaceae. *Class* G.

A. selaginiodes KING WILLIAM PINE. Tree to 30m high. Occasionally grown as an ornamental. *Distr.* W Tasmania.

Athyrium (Derivation obscure, possibly from the Greek *anthoros*, breeding well, alluding to the diverse forms of sori, or from *anthyros*, doorless, alluding to the late-opening indusia in some species.) LADY FERN. A genus of about 150 species of terrestrial ferns with erect or horizontal rhizomes and pinnately lobed to tripinnately divided, leathery leaves. Several species are grown as ornamentals. *Distr.* Widespread. *Fam.* Woodsiaceae. *Class* F.

A. filix-femina (The Latin for Lady-fern, alluding to the delicate leaves as compared with the MALE FERN, *Dryopteris filixmas*.) COMMON LADY FERN. Rhizome creeping. Leaves to 1m long. Numerous ornamental cultivars have been raised from this species. *Distr.* Temperate regions of the N hemisphere.

atlanticus, -a, -um: of the Atlantic.

Atragene See *Clematis*

Atraphaxis (From *atraphaxys*, the Greek name for a species of *Atriplex*.) A genus of about 20 species of deciduous, frequently spiny shrubs with simple alternate leaves and racemes of small, white or pink flowers. Several species are grown as ornamentals. *Distr.* SE Europe and N Africa to Central Asia. *Fam.* Polygonaceae. *Class* D.

A. frutescens Upright spineless shrub to 1m high. Flowers pink to white. *Distr.* S Europe to Central Asia.

Atriplex (The classical Latin name for these plants.) ORACH, SALT BUSH. A genus of about 200 species of annual and perennial herbs and shrubs with simple, lobed or toothed leaves and small unisexual flowers. Several species are grown as bedding plants, others are eaten as vegetables and some are used as hedging. *Distr.* Temperate and tropical regions. *Fam.* Chenopodiaceae. *Class* D.

A. canescens Shrubs. Leaves hairy. *Distr.* W North America.

A. halimus TREE PURSLANE. Erect spreading shrub. *Distr.* Mediterranean area, W Asia.

A. hortensis ORACH. Erect annual. Cultivated as a vegetable, the leaves being used like spinach. *Distr.* Asia, naturalized Central and S Europe.

atriplicifolius, -a, -um: from the genus *Atriplex*, and *folius*, leaved.

atrocyaneum: dark blue.

Atropa (After Atropos, one of the three Fates in Greek mythology, whose role it was to snip the thread of life, alluding to the toxic nature of these plants.) A genus of 4 species of erect perennial herbs with simple, typically alternate leaves and bell- or funnel-shaped, 5-lobed flowers. *Distr.* W Europe to Himalaya. *Fam.* Solanaceae. *Class* D.

A. belladonna DEADLY NIGHTSHADE, BELLADONNA. Stems erect, to 2m high. Flowers pendent, purple-brown or green. Fruit a purple-black berry. A highly poisonous plant that is occasionally used medicinally. The sap was formerly used as a cosmetic to make the pupils of the eyes dilate, hence the name, meaning beautiful lady. *Distr.* Eurasia, Mediterranean.

atropurpureus, -a, -um: dark purple.

atrorubens: dark red.

atrosanguineus, -a, -um: dark, somewhat brownish red.

atrovaginata: with a dark or black sheath, from the Latin *atro-*, dark, and *vaginatus*, sheathed.

atroviolaceus: dark violet.

atroviridipetalus: with dark green petals.

attenuatus, -a, -um: drawn out.

aubergine *Solanum melongena*.

aubretia *Aubrieta*.

Aubrieta (After Claude Aubriet (1668–1743), French botanical artist.) AUBRETIA, AUBRIETIA. A genus of about 12 species of low, evergreen, perennial herbs with white or violet-purple flowers. *Distr.* S Europe to Iran. *Fam.* Cruciferae. *Class* D.

A. canescens Tufted perennial with large lilac-violet flowers. *Distr.* Turkey.

A. gracilis Cushion- or mat-forming perennial with purple flowers. *Distr.* Balkan Peninsula.

A. pinardii Flowers large and purple. *Distr.* Turkey.

aubrietia *Aubrieta*.

aubrietioides from the genus *Aubrieta*, with the ending *-oides*, indicating resemblance.

Aucuba (From the Japanese name.) A genus of 3–4 species of evergreen shrubs and small trees with simple leathery leaves and cymes of small star-shaped flowers which bear 4 petals and 4 stamens. This genus is sometimes placed in the family Theaceae. *Distr.* Himalaya to Japan. *Fam.* Aucubaceae. *Class* D.

A. japonica JAPANESE LAUREL, SPOTTED LAUREL. Bushy shrub. Flowers drab green. Fruit bright red. Grown as a very hardy ornamental with several cultivars available. *Distr.* China, Taiwan, S Japan.

Aucubaceae A family of 1 genus and 4 species. *Class* D.
See: Aucuba.

aucuparius, -a, -um: from the Latin *avis*, a bird, and *capere*, to catch; the fruit attract birds.

augustifolius, -a, -um: with majestic leaves.

aulicum furrowed.

aurantiacus, -a, -um: orange-coloured.

aurantiifolius, -a, -um: from the Latin *aurantius*, orange-coloured, and *folius*, leaved.

aurantius, -a, -um: orange-coloured.

auratus, -a, -um: golden yellow.

aureiflorus, -a, -um: with golden yellow flowers.

aureispina: with golden yellow spines.

aureus, -a, -um: golden yellow.

auricomus, -a, -um: with golden hair.

auricula *Primula auricula*.

auricula: auricle, ear.

auriculatus, -a, um: with an ear-like appendage.

aurihamata: with golden hooks, referring to the yellow hooked spines.

auritus, -a, -um: large-eared.

australasicus, -a, -um: Australian.

australis, -a: southern.

austriacus, -a, -um: of Austria.

Austrobaileyaceae A family of 1 genus and 1 species of large climbing shrubs and lianas with opposite simple leaves and large solitary flowers that bear c12 tepals and numerous stamens. *Distr.* Queensland, Australia. *Class* D.

Austrocedrus (From the Latin *australis*, southern, and the genus name *Cedrus*, CEDAR.) A genus of 1 species of evergreen coniferous tree with scale-like leaves and small fruiting cones that open to release their seeds. *Distr.* W Argentina, Chile. *Fam.* Cupressaceae. *Class* G.

A. chilensis CHILEAN CEDAR. Occasionally grown as an ornamental.

austromontana: of southern mountains.

autograph tree *Clusia major.*

autumnalis, -e: of autumn.

auxillaris: auxiliary.

avellana: slightly red-brown, hazel-coloured.

avellanidens: with hazel-coloured teeth.

Avena (From the Latin *avena*, OAT.) OAT. A genus of 10–15 species of annual grasses with flat leaves and very loosely branched panicles of florets. *Distr.* Temperate regions. *Fam.* Gramineae. *Class* M.

A. barbata SLENDER WILD OAT. Grown as a graceful ornamental bedding plant. *Distr.* SW Europe to Asia.

A. candida See *Helictotrichon sempervirens*

A. sativa COMMON OAT. Eaten as a high energy cereal typically in porridge, muesli and oatcakes and also used as an animal fodder. *Distr.* Europe and the Middle East, cultivated since antiquity.

A. sterilis ANIMATED OAT. Awns of some varieties twist and untwist with changes in humidity. Grown as an ornamental and for use in dried flower arrangements. *Distr.* Mediterranean region, SW Asia.

avenaceus, -a, -um: from the genus *Avena*, oat, with the ending *-aceum*, indicating resemblance.

avens *Geum* **Alpine** ~ *G. montanum* **creeping** ~ *G. reptans* **mountain** ~ *Dryas octopetala* **purple** ~ *Geum rivale*, *G. triflorum* **water** ~ *G. rivale* **wood** ~ *G. urbanum* **yellow** ~ *G. macrophyllum.*

Avenula See *Helictotrichon*

Avicenniaceae A family of 1 genus and 14 species of tall mangrove-forming trees. This family is sometimes included within the family Verbenaceae. *Distr.* Tropical to subtropical coastal regions. *Class* D.

avicenniifolius, -a, -um: from the genus *Avicennia*, and *folius*, leaved.

aviculare: of small birds.

avium: of birds.

axillaris: axillary, borne in the leaf axils.

Azalea See *Rhododendron*

azalea: Alpine ~ *Loiseleuria procumbens* **mountain** ~ *L. procumbens.*

Azara (After J. N. Azara (1731–1804), Spanish scientist.) A genus of 10 species of evergreen shrubs and trees with simple leaves and clusters of small, yellow-green, fragrant flowers. Cultivated ornamentally. *Distr.* Chile, NW Argentina, Bolivia, Uruguay. *Fam.* Flacourtiaceae. *Class* D.

A. lanceolata Shrub or small tree. Leaves narrow. Flowers yellow, in rounded clusters. *Distr.* Chile, Argentina.

A. microphylla Shrub or small tree. Leaves small, dark green. Flowers yellow, vanilla scented. *Distr.* Chile, Argentina.

A. serrata Erect shrub or tree. Flowers prolific, fragrant, yellow. *Distr.* Chile.

Azolla (From the Greek *azo*, to dry, and *ollua*, to kill; they are killed by drying.) MOSQUITO FERN, WATER FERN, FAIRY MOSS. A genus of 6–8 species of free floating aquatic ferns. The leaves have a basal cavity which frequently contains a living blue-green alga which is able to convert atmospheric nitrogen into a form that can be used by the fern. *Distr.* Tropical and warm regions. *Fam.* Azollaceae. *Class* F.

Azollaceae A family of 1 genus and 6 species. *Class* F.
See: Azolla.

Azorella (The diminutive of Azores.) A genus of 70 species of tuft- or carpet-forming perennial herbs with simple or palmately divided leaves and simple subsessile umbels of yellow to brown flowers. *Distr.* Andes to temperate South America, Falkland Islands, Antarctic Islands. *Fam.* Umbelliferae. *Class* D.

A. glebaria See *A. trifurcata*

A. trifurcata Leaves crowded, glossy-green, 3-lobed, leathery. Flowers yellow. Grown as a rock garden ornamental. *Distr.* Chile, Argentina.

azoricus, -a, -um: of the Azores, in the Atlantic Ocean.

Azorina (Of the Azores.) A genus of 1 species of evergreen subshrub with robust scarred stems, simple toothed leaves and long racemes of nodding, bell-shaped, waisted, pink or white flowers. *Distr.* Azores. *Fam.* Campanulaceae. *Class* D.

A. vidalii Grown as a tender ornamental.

azureus, -a, -um: dark blue.

B

Babiana (Latinized version of an Afrikaans word, *babiaan*, baboon; baboons were reported to eat the corms.) BABOON ROOT. A genus of about 60 species of perennial herbs with deeply buried corms, folded, lance-shaped leaves and spikes of regular flowers. *Distr.* Sub-Saharan Africa, from Zambia S to Cape Province. *Fam.* Iridaceae. *Class* M.

B. plicata Flower violet-blue with a yellow patch at the base. Corms eaten by South African settlers. *Distr.* South Africa.

B. rubrocyanea Flowers red and blue. *Distr.* South Africa.

B. stricta Leaves forming a fan. Flowers funnel-shaped, purple to cream or yellow. *Distr.* South Africa (Cape Province).

baboon root *Babiana.*

baby blue eyes *Nemophila menziesii.*

babylonicus, -a, -um: of Babylon.

baby's breath *Gypsophila paniculata*
false ~ *Galium mollugo.*

baby's tears *Hypoestes phyllostachya, Soleirolia soleirolii.*

baccans: becoming quite pulpy, juicy or berry-like.

baccatus, -a, -um: pulpy or juicy, from the Latin *bacca*, berry.

baccharifolius: from the genus *Baccharis*, and *folius*, leaved.

Baccharis (After Bacchus, god of wine.) A genus of about 350 species of shrubs that often lack leaves but have flattened green stems and clusters of daisy-like flower-heads that lack distinct rays. Some species have medicinal properties. *Distr.* New World. *Fam.* Compositae. *Class* D.

B. halimifolia CONSUMPTION WEED, TREE GROUNDSEL, BUSH GROUNDSEL, COTTON SEED TREE. Shrub to 3m with blunt leaves. *Distr.* E North America, W Indies.

B. magellanica Prostrate or erect shrub with sessile leaves. *Distr.* S South America, Falkland Islands.

B. patagonica Leaves very small. *Distr.* S Chile, S Argentina.

badger's bane *Aconitum lycoctonum.*

Baeckea (After Dr Abraham Baeck (1713–1795), Swedish naturalist and physician.) A genus of about 90 species of upright and prostrate shrubs with small, simple, opposite leaves and small, solitary or clustered, white to pink flowers that bear their parts in 5s. Several species are grown as half-hardy ornamentals for their heath-like foliage. *Distr.* Tropical Asia, Australia, New Caledonia. *Fam.* Myrtaceae. *Class* D.

B. virgata TALL BAECKEA, TWIGGY BAECKEA. Prostrate shrub or small tree. Branches spreading and sometimes weeping. Flowers white. *Distr.* Australia.

baeckea: tall ~ *Baeckea virgata*
twiggy ~ *B. virgata.*

bag flower *Clerodendrum thomsoniae.*

baicalensis, -e: of Lake Baikal, East Siberia.

Balanitaceae A family of 1 genus and 25 species of spiny trees and shrubs with leaves made up of 2-leaflets. The flowers are borne in clusters and bear 5 sepals, 5 petals and a superior ovary. This family is sometimes included within the family Zygophyllaceae. *Distr.* Tropical Africa to Jordan, India, Burma. *Class* D.

Balanopaceae A family of 1 genus and 9 species of evergreen shrubs and trees with male and female flowers borne on separate plants and acorn-like drupes for fruits. *Distr.* Queensland, New Caledonia, Fiji. *Class* D.

Balanophoraceae A family of 17 genera and about 40 species of parasitic herbs with large tuberous underground parts. The above ground parts consist of a fleshy club-shaped inflorescence that resembles a fungus and bears some of the smallest known flowers. *Distr.* Tropical and subtropical regions. *Class* D.

balcanicus, -a, -um: of the Balkans.

bald money *Meum athamanticum.*

baldschaunicus, -a, -um: of Balzhuan, central Asia.

balearicus, -a, -um: of the Balearic Islands.

balloon flower *Platycodon grandiflorus.*

Ballota (The Greek name for *B. nigra.*) A genus of 35 species of perennial herbs and subshrubs with simple opposite leaves and whorls of 2-lipped, funnel-shaped flowers. The fruits have been used as floating wicks in olive oil lamps. Several species are occasionally grown ornamentally. *Distr.* Europe, Mediterranean, W Asia. *Fam.* Labiatae. *Class* D.

B. acetabulosa Erect perennial herb. Leaves densely hairy. Flowers small, white with purple-pink markings. *Distr.* E Mediterranean region.

B. nigra BLACK HOREHOUND. Erect perennial to 130cm. Flowers white, blue or pink. A source of essential oils and formerly used medicinally. *Distr.* Eurasia, naturalized North America.

B. pseudodictamnus Mound-forming perennial. Leaves rounded, woolly. Flowers small, pink. *Distr.* Greece, Crete.

balm: bastard ~ *Melittis melissophyllum* **bee** ~ *Monarda didyma, Melissa officinalis* **canary** ~ *Cedronella canariensis* **field** ~ *Glechoma hederacea* **horse** ~ *Collinsonia canadensis* **lemon** ~ *Melissa officinalis*

Balm of Gilead *Liquidambar orientalis.*

balmony *Chelone glabra.*

balsa *Ochroma lagopus.*

balsam *Impatiens*, Balsaminaceae **Alpine** ~ *Erinus alpinus* **Himalayan** ~ *Impatiens glandulifera* **liver** ~ *Erinus alpinus.*

balsameus, -a, -um: resembling balsam.

balsamiferus, -a, -um: balsambearing.

Balsaminaceae The Balsam family. 2 genera and 850 species of annual and perennial herbs with watery translucent stems and simple toothed leaves. The flowers are irregular and bear 3 or 5 sepals and 5 petals. *Distr.* Widespread except Australia and South America. *Class* D.
See: Impatiens.

Balsamita See *Tanacetum*

balsamita resembling the genus *Balsamita* (now *Chrysanthemum*).

bamboo *Chimonobambusa, Chusquea, Phyllostachys, Arundinaria* **black** ~ *Phyllostachys nigra* **Calcutta** ~ *Dendrocalamus strictus* **dwarf fern-leaf** ~ *Pleioblastus pygmaeus* **dwarf white-striped** ~ *P. variegatus* **fish-pole** ~ *Phyllostachys aurea* **fountain** ~ *Sinarundinaria nitida* **giant** ~ *Dendrocalamus, D. giganteus, Bambusa* **giant timber** ~ *Phyllostachys bambusoides;* **golden** ~ *P. aurea* **heavenly** ~ *Nandina domestica* **hedge** ~ *Bambusa multiplex* **male** ~ *Dendrocalamus strictus* **Mexican** ~ *Polygonum japonicum* **Muriel** ~ *Thamnocalamus spathaceus* **Narihira** ~ *Semiarundinaria fastuosa* **umbrella** ~ *Thamnocalamus spathaceus* **yellow groove** ~ *Phyllostachys aureosulcata.*

Bambusa (Latinized version of the Malayan vernacular name.) GIANT BAMBOO. A genus of about 100 species of large bamboos with hollow stems, numerous branches and small leaves. *Distr.* Tropical and subtropical Asia, Africa and America. *Fam.* Gramineae. *Class* M.

B. glaucescens See *B. multiplex*

B. multiplex HEDGE BAMBOO. Used as a hedging plant locally. Several cultivars available for the garden. *Distr.* S China.

B. vulgaris Grown throughout the tropics for stems and pulp.

bambusoides: from the genus *Bambusa*, with the ending *-oides*, indicating resemblance.

banana *Musa*, Musaceae **Abyssinian** ~ *Ensete ventricosum* **desert** ~ *Musa x paradisiaca* **edible** ~ *M. x paradisiaca*

Ethiopian ~ *Ensete ventricosum* **Japanese** ~ *Musa basjoo.*

banaticus, -a, -um: of Banat, North Romania.

baneberry *Actaea, A. spicata* **red** ~ *A. rubra* **white** ~ *A. alba.*

Banksia (After Sir Joseph Banks (1743–1820), botanist, explorer, and President of the Royal Society.) AUSTRALIAN HONEYSUCKLE. A genus of 50–70 species of shrubs and trees with woody tubers, simple leathery leaves and dense terminal spikes of 4-lobed tubular flowers. Numerous species are grown as ornamentals and a few are used as a source of timber. *Distr.* Australia, New Guinea. *Fam.* Proteaceae. *Class* D.

B. baueri WOOLLY BANKSIA. Low shrub. Inflorescence orange-brown. *Distr.* W Australia.

B. baxteri BAXTER'S BANKSIA. Spreading shrub. Inflorescences rounded. *Distr.* W Australia.

B. canei MOUNTAIN BANKSIA. Spreading or rounded shrub. Inflorescencep cylindrical, yellow. *Distr.* Australia (Victoria, New South Wales).

B. grandis BULL BANKSIA. Tree to 10m high. Inflorescences cylindrical, yellow. *Distr.* SW and W Australia.

B. integrifolia COAST BANKSIA. Low shrub to large tree. Inflorescences cylindrical, pale yellow. *Distr.* Australia (E seaboard).

B. marginata SILVER BANKSIA. Low shrub. Inflorescences lemon yellow. *Distr.* SW Australia.

B. occidentalis RED SWAMP BANKSIA. Spreading shrub. Inflorescences red. *Distr.* S coastal Australia.

banksia: Baxter's ~ *Banksia baxteri* **bull** ~ *B. grandis* **coast** ~ *B. integrifolia* **mountain** ~ *B. canei* **red swamp** ~ *B. occidentalis* **silver** ~ *B. marginata* **woolly** ~ *B. baueri.*

banyan, Indian *Ficus benghalensis*

banyan tree *Ficus benghalensis.*

Baptisia (From the Greek *bapto*, to dye.) FALSE INDIGO, WILD INDIGO. A genus of 17 species of perennial herbs typically with trifoliate leaves and racemes of irregular pea-like flowers. Several species are grown as ornamentals. *Distr.* E USA. *Fam.* Leguminosae. *Class* D.

B. australis FALSE INDIGO, BLUE FALSE INDIGO. Erect or spreading to 1.5m high. Flowers indigo or light purple with a yellow keel. Used as an indigo substitute. *Distr.* E USA.

B. tinctoria WILD INDIGO, HORSEFLY WEED, RATTLE WEED. Erect, much-branched herb. Flowers numerous, yellow. Formerly used as a dye and medicinally. *Distr.* SE USA.

Barbados pride *Caesalpinia pulcherrima.*

Barbarea (From its old name, Herba Sanctae Barbarae.) ST BARBARA'S HERB, WINTERCRESS. A genus of 12 species of biennial or perennial herbs with small, yellow to white flowers. *Distr.* N temperate regions. *Fam.* Cruciferae. *Class* D.

B. praecox See *B. verna*

B. verna LANDCRESS, EARLY WINTER CRESS, BELLE ISLE CRESS, NORMANDY CRESS, AMERICAN CRESS. Perennial or biennial, grown as an annual, with yellow flowers. Eaten as a salad vegetable. *Distr.* Europe.

B. vulgaris YELLOW ROCKET, ROCKET CRESS, BITTER CRESS. Yellow-flowered perennial. Grown as an ornamental in Europe but a noxious weed in the USA. *Distr.* Europe, naturalized North America.

barbarus, -a, -um: foreign.

barbatus, -a, -um: bearded.

barberry Berberidaceae, *Berberis* **common** ~ *B. vulgaris* **European** ~ *B. vulgaris.*

Barbeuiaceae A family of 1 genus and 1 species of woody climber. This family is sometimes included within the family Phytolaccaceae. *Distr.* Madagascar. *Class* D.

Barbeyaceae A family of 1 genus and 1 species of small tree with simple leaves and clusters of small, regular, wind-pollinated flowers that bear 4–5 sepals and lack petals. *Distr.* NE tropical Africa, Arabia. *Class* D.

barbigera: bearded.

barbinervis, -e: with bearded nerves.

Barleria (After Jacques Barrelier (1606–73), French monk and botanist.) A genus of

250 species of shrubs and perennial herbs with simple leaves and spikes of showy, tubular, blue or white flowers. Several species are grown as ornamentals and a few are said to have medicinal qualities. *Distr.* Tropical and subtropical regions of the Old World. *Fam.* Acanthaceae. *Class* D.

B. cristata PHILIPPINE VIOLET. Stiff-hairy throughout. Flowers violet-blue. Tender ornamental, often used for hedging. *Distr.* India and Burma.

B. obtusa Flowers mauve, borne on spikes in the axils of the upper leaves. *Distr.* S Africa.

barley *Hordeum* **four-row** ~ *H. polystichum* **fox-tail** ~*H. jubatum* **six-rowed** ~ *H. polystichum* **squirrel-tail** ~ *H. jubatum* **two-rowed** ~ *H. distichum.*

barrenwort *Epimedium, E. alpinum.*

bar-room plant *Aspidistra elatior.*

Bartsia (After J. Bartsch, a Prussian botanist.) A genus of 60 species of parasitic herbs with simple opposite leaves and racemes of 2-lipped, 5-lobed flowers that bear 4 stamens. These plants typically parasitise the roots of grasses. *Distr.* N temperate regions and tropical mountains. *Fam.* Scrophulariaceae. *Class* D.

B. alpina VELVET BELLS, ALPINE BARTSIA. Densely hairy annual. Flowers dark-purple, subtended by purple-tinged bracts. *Distr.* Europe.

bartsia, Alpine *Bartsia alpina.*

Basellaceae The Madeira Vine family. 5 genera and 15 species of vines with simple, fleshy, alternate leaves. The flowers are small and regular with 5 tepals and 5 stamens; they are usually subtended by 2 bracts. *Distr.* Tropical and subtropical regions. *Class* D. *See: Anredera, Boussingaultia.*

basil *Ocimum basilicum* **hoary** ~ *O. americanum* **holy** ~ *O. tenuiflorum* **sacred** ~ *O. tenuiflorum*; **sweet** ~ *O. basilicum* **wild** ~ *Clinopodium vulgare.*

basilaris, -e: basal, relating to the base.

basilicum: royal or princely.

basket flower *Hymenocallis narcissiflora*

basket plant *Aeschynanthus.*

basswood *Tilia* **American** ~ *T. americana.*

Bataceae A family of 1 genus and 2 species of shrubs with simple, opposite, succulent leaves and spikes of very small, unisexual flowers. *Distr.* New Guinea, Queensland, Pacific Islands, coasts of North and South America. *Class* D.

bat flower *Tacca integrifolia, T. chantrieri.*

bat plant *Tacca integrifolia.*

bay: bull ~ *Magnolia grandiflora* **california** ~ *Umbellularia californica* **loblolly** ~ *Franklinia lasianthus* **rose** ~ *Nerium oleander* **sweet** ~ *Laurus nobilis, Magnolia virginiana* ~ **tree** *Laurus nobilis.*

bayberry *Myrica pensylvanica.*

bayonet: Spanish ~ *Yucca baccata*

bayonet plant *Aciphylla squarrosa.*

bead plant *Nertera granadensis.*

bead tree *Melia azedarach.*

bean *Phaseolus* **algarroba** ~ *Ceratonia siliqua* **black** ~ *Kennedia nigricans* **bog** ~ *Menyanthes trifoliata* **broad** ~ *Vicia faba* **buck** ~ *Menyanthes trifoliata* **canellini** ~ *Phaseolus vulgaris* **carob** ~ *Ceratonia siliqua* **Dutch cane knife** ~ *Phaseolus coccineus* **Egyptian** ~ *Cicer arietinum* **English** ~ *Vicia faba* **field** ~ *V. faba* **French** ~ *Phaseolus vulgaris* **garbanzo** ~ *Cicer arietinum* **haricot** ~ *Phaseolus vulgaris* **kidney** ~ *P. vulgaris* **Lima** ~ *P. lunatus* **locust** ~ *Ceratonia siliqua* **lucky** ~ *Thevetia peruviana* **potato** ~ *Apios americana* **runner** ~ *Phaseolus coccineus* **smoking** ~ *Catalpa bignonioides* **soja** ~ *Glycine max* **soya** ~ *G. max* **wild** ~ *Apios americana.*

bean plant, string *Hoya longifolia.*

bean tree *Laburnum* **Indian** ~ *Catalpa bignonioides.*

bearberry *Arctostaphylos, A. uva-ursi* **common** ~ *A. uva-ursi.*

beargrass *Nolina microcarpa.*

bear's breeches *Acanthus*, *A. mollis*.

bear's ears *Primula auricula*.

bear's foot *Helleborus foetidus*, *Aconitum napellus*.

bear-tongue *Clintonia*.

Beaucarnea A genus of about 6 species of trees and subshrubs of dry areas, with robust linear leaves and panicles of small, unisexual, cream flowers. *Distr.* S USA and Mexico. *Fam.* Dracaenaceae. *Class* M.

B. recurvata BOTTLE PALM, ELEPHANT'S FOOT TREE, PONY TAIL. Trunk bottle-shaped, to 2m across at base. Leaves channelled and recurved. *Distr.* Mexico.

Beaufortia (After Mary Somerset (1630–1714), Duchess of Beaufort, English amateur botanist.) A genus of 16 species of prostrate to erect, densely-branched shrubs with small crowded leaves and brush-like heads or spikes of small regular flowers that bear prominent stamens. Several species are grown as half-hardy ornamentals. *Distr.* SW Australia. *Fam.* Myrtaceae. *Class* D.

B. sparsa SWAMP BOTTLEBRUSH, GRAVEL BOTTLEBRUSH. Medium-sized spreading shrub. Flower-heads bright red, occasionally orange to yellow. *Distr.* SW Australia.

Beaumontia (After Lady Diana Beaumont (died 1831), of Bretton Hall, Yorkshire.) A genus of 8 species of lianas with opposite simple leaves and cymes of fragrant, funnel- or bell-shaped flowers. *Distr.* China, India to Malaysia. *Fam.* Apocynaceae. *Class* D.

B. grandiflora HERALD'S TRUMPET, EASTER LILY VINE, NEPAL TRUMPET FLOWER. Flowers to 13cm long, white. Grown as a tender ornamental. *Distr.* India to Vietnam.

beauty, Californian *Fremontodendron*

beauty bush *Kolkwitzia amabilis*.

Beauverdia See *Leucocoryne*

beaver wood *Celtis occidentalis*.

bedstraw *Galium* **hedge** ~ *G. mollugo* **lady's** ~ *G. verum* **Our Lady's** ~ *G. verum* **white** ~ *G. mollugo* **yellow** ~ *G. verum*.

beech Fagaceae, *Fagus* **Antarctic** ~ *Nothofagus antarctica* **Australian** ~ *N. moorei* **black** ~ *N. solandri* **Chinese** ~ *Fagus engleriana* **common** ~ *F. sylvatica* **copper** ~ *F. sylvatica* **European** ~ *F. sylvatica* **Japanese** ~ *F. crenata* **rauli** ~ *Nothofagus procera* **red** ~ *N. fusca* **roble** ~ *N. obliqua* **silver** ~ *N. menziesii* **southern** ~ *Nothofagus* **weeping** ~ *Fagus sylvatica*.

beef plant *Iresine herbstii*

beefsteak plant *Iresine herbstii*.

beef suet tree *Shepherdia argentea*.

beefwood *Casuarina equisetifolia*.

beesianus, -a, -um: after Bees Nursery, Cheshire, from the nickname of its founders, A.K. Bulley and his sister, who were known as the 'busy Bs'.

beet *Beta* **fodder** ~ *B. vulgaris* **sea kale** ~ *B. vulgaris* **spinach** ~ *B. vulgaris* **sugar** ~ *B. vulgaris*.

beetle weed *Galax urceolata*.

beetroot *Beta vulgaris*.

beggartick *Bidens*.

Begonia (After Michael Begon (1683–1710), Governor of French Canada.) A genus of about 1200 species of perennial herbs, shrubs and climbers with asymmetric leaves and white-pink, unisexual flowers, both sexes occurring in the same inflorescence. The male flowers have 2 whorls of 2 tepals and the females have 2–6 tepals. The fruits are 2–3 winged and bear numerous, very small seeds. Many species and hybrids are cultivated as indoor ornamentals and summer bedding plants. *Distr.* Widespread in tropical and warm temperate regions. *Fam.* Begoniaceae. *Class* D.

B. albo-picta GUINEA-WING BEGONIA. Roots fibrous. Flowers numerous, green-white. *Distr.* Brazil.

B. angularis Shrub with numerous white flowers. *Distr.* Brazil.

B. bowerae EYELASH BEGONIA. Leaves pale with dark markings. Flowers few, small, pink. *Distr.* Mexico.

B. cucullata Succulent to 1m high. Flowers white. *Distr.* South America (Brazil to Argentina).

B. × erythrophylla RED-LEAVED BEEFSTEAK BEGONIA, KIDNEY BEGONIA, WHIRLPOOL BEGONIA, POND-LILY BEGONIA. A hybrid of *B. hydrocotylifolia* × *B. manicata*. Leaves round, leathery, red-purple below. Several cultivars are available with different leaf forms. *Distr.* Garden origin.

B. feastii See *Begonia* x *erythrophylla*

B. foliosa FERN BEGONIA, FERN-LEAVED BEGONIA. Plant shrubby. Leaves 2-ranked. Flowers white. *Distr.* Colombia, Venezuela.

B. fuchsioides FUCHSIA BEGONIA, CORAZON DE JESUS. Leaves flushed red. Flowers few-many, red-pink. *Distr.* Venezuela.

B. glaucophylla See *Begonia radicans*

B. grandis HARDY BEGONIA. Tuberous perennial. Stems annual. Flowers fragrant. Can be grown as a hardy plant in some areas. *Distr.* China.

B. haageana See *Begonia scharffii*

B. × hiemalis WINTER FLOWERING BEGONIA. A hybrid of *B. socotrana* × *B. tuberhybrida*. *Distr.* Garden origin, Great Britain.

B. limmingheana See *Begonia radicans*

B. listada Roots fibrous. Leaves small. Flowers white with red hairs. *Distr.* Brazil, Paraguay, Argentina.

B. masoniana (After Mr L. Maurice Mason who introduced it from Singapore in 1957.) IRON CROSS BEGONIA. Leaves covered in cone-shaped projections. Flowers small. *Distr.* China.

B. procumbens See *Begonia radicans*

B. radicans SHRIMP BEGONIA. Stems climbing or scrambling. Flowers in dense cymes, pink-red. *Distr.* Brazil.

B. rex KING BEGONIA, PAINTED-LEAF BEGONIA, BEEFSTEAK GERANIUM. Numerous cultivars of this species and of the hybrids it forms with other Asian species are available. They are grown chiefly for their 2 or 3 coloured leaves; the flowers are often inconspicuous. *Distr.* India, Himalaya.

B. scharffii (After Carl Scharff, who collected in Brazil about 1888.) ELEPHANT'S EAR BEGONIA. Leaves red below. Flowers numerous, in large clusters. *Distr.* Brazil.

B. semperflorens Strictly this name is a synonym of *B. cucullata* but is widely applied to a large range of hybrids and cultivars used as bedding plants. *Distr.* SE Brazil, NE Argentina.

B. serratipetala Roots fibrous. Leaves small. Flowers few, pink. *Distr.* Papua New Guinea.

B. socotrana Much used as a parent in the production of ornamental cultivars. *Distr.* Socotra.

B. solananthera Stems scrambling or climbing. Flowers fragrant. *Distr.* Brazil.

B. sonderiana Leaves very lopsided, 3-lobed. *Distr.* S Africa.

B. stipulacea Leaves dull red beneath. Flowers white. *Distr.* Brazil.

B. sutherlandii Tuberous perennial. Stems annual. Flowers orange-red. *Distr.* Natal to Tanzania to Congo.

begonia Begoniaceae **elephant's ear** ~ *Begonia scharffii* **eyelash** ~ *B. bowerae* **fern** ~ *B. foliosa* **fern-leaved** ~ *B. foliosa* **fuchsia** ~ *B. fuchsioides* **guinea-wing** ~ *B. albo-picta* **hardy** ~ *B. grandis* **iron cross** ~ *B. masoniana* **kidney** ~ *B. x erythrophylla* **king** ~ *B. rex* **painted-leaf** ~ *B. rex* **pond-lily** ~ *B. x erythrophylla* **red-leaved beefsteak** ~ *B. x erythrophylla* **shrimp** ~ *B. radicans* **whirlpool** ~ *B. x erythrophylla* **winter-flowering** ~ *B. x hiemalis*.

Begoniaceae The Begonia family. 3 genera and about 1400 species of perennial herbs and shrubs typically with succulent, jointed stems and asymmetric leaves. The flowers are unisexual, the sexes being borne in the same inflorescence. Most species belong to the genus *Begonia*. Many species are grown as ornamentals. *Distr.* Tropical and subtropical regions. *Class* D.
See: Begonia.

beharensis: of Behara, Madagascar.

Belamcanda (From a local vernacular name.) BLACKBERRY LILY, LEOPARD FLOWER. A genus of 2 species of bulbiferous herbs with fan- or sword-shaped leaves and branched stems that bear a succession of flat orange-red flowers. *Distr.* E Asia. *Fam.* Iridaceae. *Class* M.

B. chinensis LEOPARD LILY, BLACKBERRY LILY. Inflorescence to 1m high. Flowers yellow-red, spotted maroon. Grown as a short-lived garden ornamental. *Distr.* Eurasia.

bell: golden ~ *Forsythia suspensa* **rock** ~ *Wahlenbergia* **silver** ~ *Halesia, H. tetraptera*

belladonna *Atropa belladonna.*

belladonna: beautiful lady.

bellatulus, -a, -um: pretty.

Bellevalia (After Pierre Richer de Belleval (1564–1632).) A genus of 45–50 species of bulbiferous herbs with linear leaves and racemes of bell-shaped or tubular flowers. Several species are grown as ornamentals. *Distr.* Mediterranean and Black Sea regions to Afghanistan and Tadzhikistan. *Fam.* Hyacinthaceae. *Class* M.

B. paradoxa See *B. pycnantha*

B. pycnantha Inflorescences conical. Flowers blue-black with yellow-green edges. *Distr.* Caucasus, E Turkey, N and W Iran, N Iraq.

B. romana ROMAN HYACINTH. Flowers wide bell-shaped, white. *Distr.* Mediterranean area, SW France.

bellflower *Campanula*, Campanulaceae **bearded** ~ *Campanula barbata* **Canary Island** ~ *Canarina canariensis* **Chilean** ~ *Lapageria rosea*, *Nolana* **Chimney** ~ *Campanula pyramidalis* **Chinese** ~ *Platycodon grandiflorus* **climbing** ~ *Littonia modesta* **clustered** ~ *Campanula glomerata* **giant** ~ *Ostrowskia magnifica* **gland** ~ *Adenophora* **Italian** ~ *Campanula isophylla* **ivy-leaved** ~ *Wahlenbergia hederaceae* **nettle-leaved** ~ *Campanula trachelium* **ring** ~ *Symphyandra* **trailing** ~ *Cyananthus* **tussock** ~ *Campanula carpatica* **willow** ~ *C. persicifolia.*

bellidifolius, -a, -um: from the genus *Bellis*, and *folius*, leaved.

bellidioides: from the genus *Bellis*, with the ending *-oides*, indicating resemblance.

Bellis (From the Latin *bellus*, pretty.) DAISY. A genus of 7 species of annual or perennial herbs with simple leaves, often borne in a basal rosette, and numerous small flowers (florets) in dense heads (capitula), the outer flowers being differentiated into petal-like structures (ray florets or rays). Some species have medicinal properties. *Distr.* Europe and the Mediterranean. *Fam.* Compositae. *Class* D.

B. perennis DAISY, COMMON DAISY. A very common plant of lawns and short grassland, often considered a weed but sometimes cultivated as an ornamental, several cultivars having been raised. *Distr.* Europe, W Asia.

B. rotundifolia Cultivated as a half-hardy rock garden plant. *Distr.* S Europe.

Bellium (From the name of a related genus, *Bellis*.) A genus of 4 species of annual and perennial herbs with simple basal leaves and conical daisy-like heads of flowers that bear distinct white rays. Several species are cultivated as rock garden plants. *Distr.* Mediterranean and Europe. *Fam.* Compositae. *Class* D.

B. bellidioides Stoloniferous perennial. *Distr.* W Mediterranean Islands.

B. crassifolium Perennial with fleshy leaves. *Distr.* Sardinia.

B. minutum Annual. *Distr.* Greece, W Asia.

bells: Canterbury ~ *Gloxinia perennis*, *Campanula medium* **Chile** ~ *Lapageria rosea* **Christmas** ~ *Sandersonia aurantiaca*, *Blandfordia*, *Chironia baccifera* **coral** ~ *Heuchera* **fairy** ~ *Disporum* **grassy** ~ *Edraianthus* **honey** ~ *Freylinia lanceolata*, *Cephalanthus occidentalis* **monastery** ~ *Cobaea scandens* **Nippon** ~ *Shortia uniflora* **peach** ~ *Campanula persicifolia* **temple** ~ *Smithiantha*, *S. cinnabarina* **velvet** ~ *Bartsia alpina* **yellow** ~ *Tecoma*, *T. stans*

bells of Ireland *Moluccella lavis.*

bellus, -a, -um: pretty.

bellwort *Uvularia.*

Beloperone See *Justicia*

B. guttata See *Justicia brandegeana*

benghalensis: of Bengal, India.

benjamin, stinking *Trillium*, *T. erectum.*

benjamin bush *Lindera benzoin.*

benjamin tree *Ficus benjamina.*

Bensoniella (After G. T. Benson (1896–1928), botanist and author.) A genus of 1 species of perennial herb with heart-shaped basal leaves and racemes of small green-yellow flowers. *Distr.* USA (California, Oregon). *Fam.* Saxifragaceae. *Class* D.

B. oregona A cultivated ornamental.

bent *Agrostis* **brown** ~ *A. canina* **Rhode Island** ~ *A. canina* **sand** ~ *Mibora mimima* **silky** ~ *Cynosurus* **velvet** ~ *Agrostis canina.*

Berberidaceae The Barberry family. 18 genera and about 600 species of shrubs and perennial herbs. The leaves may be simple or compound, evergreen or deciduous, and in *Berberis* may be replaced by spines. The flowers are bisexual and regular with several whorls of tepals which may or may not be differentiated into sepals and petals. Some species are grown as ornamentals whilst others have medicinal properties. *Distr.* Widespread in N temperate areas and the mountains of South America. *Class* D.
See: Berberis, Bongardia, Caulophyllum, Diphylleia, Epimedium, Gymnospermium, Jeffersonia, Leontice, ×Mahoberberis, Mahonia, Nandina, Podophyllum, Ranzania, Vancouveria.

Berberidopsis (From the genus name *Berberis*, and the Greek *opsis*, indicating resemblance.) A genus of 1 species of evergreen woody climber with alternate, simple, dark green leaves and pendent clusters of globular red flowers. *Distr.* Chile, E Australia. *Fam.* Flacourtiaceae. *Class* D.

B. corallina CORAL PLANT. Grown as an ornamental. *Distr.* Chile.

Berberis (From *berberys*, the Arabic name for the fruit.) BARBERRY. A genus of about 500 species of deciduous and evergreen shrubs with yellow wood, spiny branches and simple leaves. The yellow to orange-red flowers are regular with numerous tepals and 6 stamens; they are followed by yellow, red or blue berries. Many species are cultivated ornamentally, especially as hedging plants. A number of plants have edible fruits. *Distr.* Temperate parts of Old and New Worlds. *Fam.* Berberidaceae. *Class* D.

B. aggregata Deciduous shrub to 2m. Leaves giving spectacular autumn colour. Flowers pale yellow. *Distr.* China.

B. bealei See *Mahonia japonica*

B. buxifolia Evergreen to 3m high. Leaves leathery. Flowers orange-yellow. Low growing cultivars are often grown in rock gardens. *Distr.* Chile and Argentina.

B. darwinii (After Charles Darwin (1809–82), who discovered it in 1835.) Evergreen to 2m high. Leaves leathery, pale below. Flowers yellow, tinged red. Several cultivars available. Used as a hedging plant. *Distr.* Chile and Argentina.

B. empetrifolia Evergreen. Stems prostrate, red. Flowers deep yellow. *Distr.* Chile, Argentina.

B. gagnepainii (After Francois Gagnepain (1866–1952).) Evergreen to 1.5m high. Leaves yellow below. Flowers bright yellow. A good hedging plant. *Distr.* China.

B. × hybrido-gagnepainii A hybrid of *B. verruculosa* × *B. gagnepainii.* A source of numerous evergreen cultivars. *Distr.* Garden origin.

B. linearifolia Evergreen, little branched, to 1.5m high. Flowers clustered, to 2cm across, orange. *Distr.* Chile and Argentina.

B. × stenophylla A hybrid of *B. darwinii* × *B. empetrifolia.* The source of a wide range of robust deciduous cultivars. *Distr.* Garden origin.

B. thunbergii (After Carl Peter Thunberg (1743–1828), professor of botany at Uppsala University, who travelled in E Asia.) Deciduous, to 1m high. Flowers clustered, yellow, tinged red. Grown as a rock garden plant. *Distr.* Japan.

B. vulgaris COMMON BARBERRY, EUROPEAN BARBERRY. Deciduous. Flowers yellow. Fruit dull red. A source of wood for turning, dyestuff for wool and leather and edible fruit used in preserves. *Distr.* Europe, North America, Turkey, Caucasus, naturalized in British Isles.

Berchemia (After M. Berchem, an early 17th-century French botanist.) A genus of about 12 species of deciduous climbing subshrubs and trees with small simple leaves and racemes or panicles of minute white flowers that bear their parts in 5s. Several species are grown as ornamentals. *Distr.* E Africa, E Asia, NW America. *Fam.* Rhamnaceae. *Class* D.

B. racemosa Scrambling subshrub to 4m high. *Distr.* Japan, Taiwan.

bergamot *Monarda* **sweet** ~ *M. didyma* **wild** ~ *M. fistulosa.*

Bergenia (After Karl August von Bergen (1704–59), German botanist.) ELEPHANT'S EAR, SIBERIAN SAXIFRAGE. A genus of 6–7 species of rhizomatous perennials with leathery leaves in open rosettes and cymes of white to red flowers. Several species are grown as ground cover. *Distr.* Temperate and subtropical regions, E Asia. *Fam.* Saxifragaceae. *Class* D.

B. ciliata Leaves ciliate on the margin, the flowers few, weakly fragrant, flushed rose. *Distr.* W Pakistan and N India to SW Nepal.

B. cordifolia Petals almost round, rose to dark pink. Several ornamental cultivars of this species are available. *Distr.* Siberia, Mongolia.

B. crassifolia Flowers nodding, magenta to crimson. *Distr.* Siberia, Mongolia.

B. delavayi See *B. purpurascens*

B. purpurascens Leaves tinged purple, flowers pink to maroon. *Distr.* Himalaya, China.

B. × schmidtii A hybrid of *B. ciliata* × *B. crassifolia* The most widely cultivated and vigorous of the garden BERGENIAS. *Distr.* Garden origin.

B. × spathulata A naturally occurring hybrid of *B. ciliata* × *B. stracheyi* that has also been produced in cultivation. *Distr.* India.

B. stracheyi (After Lieutenant-General Sir Richard Strachey (1817–1908), who collected in the Himalaya.) Leaves red in winter. Flowers yellow-pink, fragrant. *Distr.* Himalaya.

Berkheya (After Jan Le Francq van Berhey (1729–1812), Dutch botanist.) A genus of about 100 species of perennial herbs, subshrubs or shrubs, with pinnately divided leaves and daisy-like heads of flowers which may or may not bear distinct rays. The shrubs are often spiny. *Distr.* South Africa. *Fam.* Compositae. *Class* D.

B. armata Rhizomatous perennial herb with yellow flowers. *Distr.* Cape Province.

B. barbata Subshrub or shrub with spiny leaves. *Distr.* Cape Province.

B. grandiflora See *B. barbata*

bermudiana: of Bermuda.

berry: alder-leaved service ~ *Amelanchier alnifolia* **buffalo** ~ *Shepherdia C S. argentea.* **China** ~ *Melia azedarach* **Christmas** ~ *Photinia* **coral** ~ *Aechmea fulgens* **crake** ~ *Empetrum nigrum* **curlew** ~ *E. nigrum* **June** ~ *Amelanchier* **lemonade** ~ *Rhus integrifolia* **liver** ~ *Streptopus amplexifolius* **male** ~ *Lyonia ligustrina* **meal** ~ *Arctostaphylos uva-ursi* **New Zealand wine** ~ *Aristotelia serrata* **oso** ~ *Oemleria cerasiformis* **partridge** ~ *Gaultheria procumbens*, *Mitchella repens* **pigeon** ~ *Duranta erecta*, *Phytolacca americana* **salmon** ~ *Rubus parviflorus* **sand** ~ *Arctostaphylos uva-ursi* **sapphire** ~ *Symplocos paniculata* **scoot** ~ *Streptopus amplexifolius* **service** ~ *Amelanchier* **silver** ~ *Elaeagnus angustifolia*, *E. commutata* **snow** ~ *Gaultheria hispida* **soap** ~ *Sapindus* **thimble** ~ *Rubus parviflorus*, *R. odoratus* **two-eyed** ~ *Mitchella repens* **velvet** ~ *Rubiaceae scabra.*

Bertholletia (After L. C. Berthollet (1748–1822), French chemist.) A genus of 1 species of large deciduous tree with simple leathery leaves and erect panicles of large flowers that bear 2 sepals, 6 petals and numerous stamens. The fruit consists of a large, very hard, round capsule that may weigh as much as 2kg and contain as many as 24 angular seeds. *Distr.* Tropical S America. *Fam.* Lecythidaceae. *Class* D.

B. excelsa BRAZIL NUT, PARA NUT. The edible seeds are consumed in large quantities around Christmas.

Beschorneria (After Friedrich Beschorner (1806–73), German amateur botanist.) A genus of 10–12 species of perennial herbs closely related to the genus *Agave* but differing in having shorter flower tubes. Several species are cultivated as somewhat frost tender ornamentals. *Distr.* Mexico. *Fam.* Agavaceae. *Class* M.

B. tubiflora Stem very short. Leaves around 60cm long.

B. yuccoides Stem prominent. Leaves blue-green. The most hardy species available for cultivation.

Beta (From the Celtic *bett*, red.) BEET. A genus of 6 species of annual and biennial herbs with a robust tap root, long stalked, simple leaves and spikes of small bisexual flowers. About half the world's sugar supplies are derived from beet. *Distr.* Mediterranean region, especially coastal areas. *Fam.* Chenopodiaceae. *Class* D.

B. vulgaris BEETROOT, SUGAR BEET, FODDER BEET, MANGEL, SPINACH BEET, SEA KALE BEET. Perennial herbs with a whorl of leaves and a large woody or fleshy tap root; cultivated forms are biennial. Numerous agricultural cultivars are available with a number of different uses. Leaves are eaten as vegetables, tap roots are used as animal fodder, as vegetables or a source of refined sugar. *Distr.* Coastal Europe.

betel nut *Areca catechu*.

Betonica See *Stachys*

B. officinalis See *S. officinalis*

betonicifolius, -a, -um: from the genus *Betonica*, now *Stachys*, and *folius*, leaved.

betonicoides: from the genus *Betonica*, now *Stachys*, with the ending *-oides*, indicating resemblance.

betony *Stachys*, *S. officinalis* **marsh** ~ *S. palustris* **water** ~ *Scrophularia auriculata* **wood** ~ *Stachys officinalis* **woolly** ~ *S. byzantina*.

Betula BIRCH. A genus of about 60 species of deciduous shrubs and trees with simple alternate leaves and small, unisexual, wind-pollinated flowers. The male flowers are borne in pendent catkins that appear in the autumn and open to release their pollen in the spring. The female flowers are borne in smaller catkins that break up to release their narrowly winged seeds in autumn. The trees are grown as a source of timber and pulp as well as ornamentally. Bundles of thin branches were formerly used as an implement of corporal punishment known as the birch. *Distr.* Temperate and Arctic regions of the N hemisphere. *Fam.* Betulaceae. *Class* D.

B. alba See *Betula pubescens*

B. albosinensis CHINESE RED BIRCH. Tree, to 25m. Bark orange-red, peeling. *Distr.* China.

B. alleghaniensis YELLOW BETULA, YELLOW BIRCH. Tree, to 30m high. Leaves turning yellow in autumn. Good timber. *Distr.* E North America.

B. costata See *Betula ermanii*

B. ermanii (After Adolf Erman.) GOLD BIRCH, RUSSIAN ROCK BIRCH, ERMAN'S BIRCH. Tree or large shrub, to 25m high. *Distr.* E Asia.

B. fontinalis WATER BIRCH, AMERICAN RED BIRCH. Shrub, to 7m. Bark dark red. *Distr.* W North America.

B. grossa JAPANESE CHERRY BIRCH. Tree, to 25m high. *Distr.* Japan.

B. jacquemontii See *Betula utilis*

B. lenta AMERICAN BLACK BIRCH, CHERRY BIRCH, SWEET BIRCH. Tree, to 25m high. Bark dark brown. Distilled bark gives a medicinal oil, birch beer is made from fermenting sap. *Distr.* E USA.

B. lutea See *Betula alleghaniensis*

B. maximowicziana MONARCH BIRCH. Tree, to 18m high. Bark not peeling. Leaves large, turning bright yellow in autumn. *Distr.* Japan, Kurile Islands.

B. nana DWARF BETULA, DWARF BIRCH. Shrub, to 50cm high. Stems growing horizontally or ascending. *Distr.* Circumpolar, through subarctic Eurasia, North America and mountains of N temperate areas.

B. nigra RED BETULA, RIVER BIRCH. Tree, to 20m high. Bark dark red-brown, almost black. *Distr.* E USA.

B. occidentalis See *Betula fontinalis*

B. papyrifera PAPER BIRCH, WHITE BIRCH, CANOE BIRCH. Tree, to 30m high. Bark peeling, white to red-brown. The bark is impervious to water and was formerly used in the manufacture of baskets and canoes. *Distr.* N America.

B. pendula SILVER BETULA, COMMON BIRCH, SILVER BIRCH, EUROPEAN WHITE BIRCH, SILVER BIRCH, WARTY BIRCH. Tree, to 25m high. Bark silver-white and peeling when young, grey and fissured on older trees. A number of products are derived from this species. The timber is used for furniture, the bark may be eaten, and the shoots and wood are a source of dye. *Distr.* Europe.

B. pubescens EUROPEAN BIRCH, DOWNY BIRCH, WHITE BIRCH. A very similar tree to *B. pendula*. Leaves downy. *Distr.* Europe and W Asia.

B. schmidtii Tree, to 35m high with a thick trunk. The wood of this species is so dense that it will sink in water. *Distr.* Japan, Korea, Manchuria.

B. utilis HIMALAYAN BIRCH. Tree, to 18m. Thin peeling bark formerly used as a writing and packing material. *Distr.* SW China to the Himalaya and Afghanistan.

B. verrucosa See *Betula pendula*

betula: dwarf ~ *Betula nana* **red** ~ *B. nigra* **silver** ~ *B. pendula* **yellow** ~ *B. alleghaniensis.*

Betulaceae The Birch family. 2 genera and about 100 species of trees and shrubs with simple, alternate, deciduous leaves. The male flowers are borne in hanging catkins and the females in stiff, often erect, cone-like catkins. Both sexes occur on the same plant. This family is an important source of timber. *Distr.* N temperate regions and mountains in tropical South America. *Class* D.
See: Alnus, Betula.

betulifolius, -a, -um: from the genus *Betula*, birch, and *folius*, leaved.

betulinus, -a ,-um: resembling the genus *Betula*, birch.

betulus: of *Betula*, birch.

bhutanensis, -e: of Bhutan.

bhutanicus, -a, -um: of Bhutan.

Biarum (A classical Greek name for an unknown plant.) A genus of 12–15 species of tuberous herbs with simple leaves and small flowers borne on a short spadix that is surrounded by a spathe. The fruit consists of small clusters of white to red berries borne at ground level. Several species are grown as ornamentals for their interesting inflorescences and fruit. *Distr.* Mediterranean region and W Asia. *Fam.* Araceae. *Class* M.

B. tenuifolium Spathe acrid, dark purple above, borne before leaves. *Distr.* Mediterranean region from Greece W to Portugal.

bicallosus, -a, -um: with two hardened thickenings.

bicolor: two-coloured.

biconvexa: bi-convex.

bicostata: with two main veins, rather than a single midrib.

bictoniensis, -e: of Bicton, Devon.

biddy biddy *Acaena novae-zelandiae.*

Bidens (From the Latin *bis*, twice, and *dens*, a tooth, alluding to the two teeth on the fruit.) BUR MARIGOLD, PITCHFORK, SPANISH NEEDLES, TICKSEED, BEGGARTICK, STICK TIGHT, CUCKOLD. A genus of 233 species of annual or perennial herbs and shrubs with opposite, simple or pinnate leaves and daisy-like heads of yellow flowers that bear distinct rays. The fruits bear barbed bristles and so stick to clothing. Several species and numerous cultivars are grown as ornamentals. *Distr.* Cosmopolitan, especially Mexico. *Fam.* Compositae. *Class* D.

B. atrosanguinea See *Cosmos atrosanguineus*

B. aurea Perennial herb to 1m high. *Distr.* S USA to Guatemala.

B. ferulifolia Annual or perennial herb to 1m high. *Distr.* S USA, Mexico, Guatemala.

B. tripartita Annual herb to 70cm high. Rays absent. *Distr.* Temperate Eurasia.

bidgee widgee *Acaena novae-zelandiae.*

biennis, -e: biennial.

bifidus, -a, -um: divided into two.

bifloriformis: in the form of two flowers.

biflorus, -a, -um: two-flowered.

bifolius, -a, -um: two-leaved.

bifurcatus, -a, -um: forked into two.

bifurcus, -a, -um: forked into two.

bigarade *Citrus aurantium.*

big boy *Lyonia ligustrina.*

biglandulosus, -a ,-um: with two glands.

Bignonia (After Abbé Jean Paul Bignon (1662–1743), librarian to Louis XIV of France.) A genus of 1 species of evergreen woody climber with pinnately lobed leaves that bear a 3-branched tendril. The orange-red, 2-lipped, trumpet-shaped flowers are borne in axillary cymes. *Distr.* SE USA. *Fam.* Bignoniaceae. *Class* D.

B. capreolata CROSS VINE, TRUMPET FLOWER, QUARTERVINE. Grown as a hardy ornamental. *Distr.* SE USA.

B. unguis-cati See *Macfadyena unguis-cati*

Bignoniaceae A family of 111 genera and 725 species of trees, shrubs and lianas with compound or simple leaves and showy, bell- or funnel-shaped flowers that bear 5 fused sepals, 5 fused petals and typically 2

pairs of fertile stamens. *Distr.* Tropical and warm-temperate regions. *Class* D.
See: Bignonia, Campsis, Catalpa, Clytostoma, Distictis, Eccremocarpus, Incarvillea, Jacaranda, Macfadyena, Markhamia, Pandorea, Podranea, Pyrostegia, Tecoma, Tecomanthe, Tecomaria.

bignonioides: from the genus *Bignonia*, with the ending *-oides*, indicating resemblance.

bilberry *Vaccinium membranaceum, V. myrtillus* **bog** ~ *V. uliginosum* **dwarf** ~ *V. caespitosum* **Kamchatka** ~ *V. praestans* **red** ~ *V. parvifolium* **thin-leaved** ~ *V. membranaceum.*

Billardiera (After J. J. H de Labillardière (1755–1834), French botanist who worked on the Australian flora.) A genus of 8 species of evergreen climbers with simple alternate leaves and pendulous bell-shaped flowers that bear their parts in 5s. The fruit is a many-seeded berry. *Distr.* Australia. *Fam.* Pittosporaceae. *Class* D.

B. longiflora BLUEBERRY. Flowers solitary, green-yellow. Fruit blue. Grown as a tender ornamental. *Distr.* Tasmania.

Billbergia (After J. G. Billberg (1772–1884), Swedish botanist.) A genus of 54 species of evergreen, typically epiphytic herbs with funnel-shaped rosettes of leathery, often cross-banded leaves and bright coloured, tubular flowers that are borne in pendent spike-like inflorescences on long arching stems. Several species and many hybrids are grown as ornamentals. *Distr.* E Brazil and Central America. *Fam.* Bromeliaceae. *Class* M.

B. nutans ANGEL'S TEARS, QUEEN'S TEARS, FRIENDSHIP PLANT. Leaves strap-shaped. Flowers yellow-green, marked purple on the edge. Bracts pink. A popular house plant. *Distr.* Brazil, Paraguay, Uruguay and Argentina.

B. pyramidalis SUMMER TORCH. Inflorescence erect or nearly so, pyramidal. Flowers bright red to pink. *Distr.* E Brazil, naturalized Central and South America.

B. × windii (After Wind, a gardener.) ANGEL'S TEARS. A hybrid of *B. decora* x *B. nutans* that is similar to *B. nutans* but with grey-green leaves and a longer flowering period. *Distr.* Garden origin.

bills, mosquito *Dodecatheon hendersonii.*

bilobus, -a, -um: two-lobed.

bindweed *Calystegia*, Convolvulaceae, *Convolvulus* **field** ~ *C. arvensis* **rough** ~ *Smilax aspera.*

bine *Humulus lupulus.*

bipennifolius, -a, -um: with bipinnate leaves.

bipinnatifidus, -a, -um: twice pinnately divided.

bipinnatus, -a, -um: twice pinnate.

birch *Betula*, Betulaceae **American black** ~ *Betula lenta* **American red** ~ *B. fontinalis* **canoe** ~ *B. papyrifera* **cherry** ~ *B. lenta* **Chinese red** ~ *B. albosinensis* **common** ~ *B. pendula* **downy** ~ *B. pubescens* **dwarf** ~ *B. nana* **Erman's** ~ *B. ermanii* **European** ~ *B. pubescens* **European white** ~ *B. pendula* **gold** ~ *B. ermanii* **Himalayan** ~ *B. utilis* **Japanese cherry** ~ *B. grossa* **monarch** ~ *B. maximowicziana* **paper** ~ *B. papyrifera* **river** ~ *B. nigra* **Russian rock** ~ *B. ermanii* **silver** ~ *B. pendula* **sweet** ~ *B. lenta* ~ **warty** *B. pwndula.* **water** ~ *B. fontinalis* **white** ~ *B. papyrifera, B. pubescens* **yellow** ~ *B. alleghaniensis.*

bird-catcher tree *Pisonia umbellifera.*

bird of paradise *Caesalpinia gilliesii* Strelitziaceae **false** ~ *Heliconia.*

bird of paradise flower *Strelitzia reginae.*

bird's eye *Veronica, V. chamaedrys.*

bird's eyes *Gilia tricolor.*

birds of paradise *Strelitzia.*

birthroot *Trillium.*

birthwort *Aristolochia, A. clematitis* Aristolochiaceae.

Biscutella (From the Latin *bis*, twice, and *scutella*, a small flat dish, in allusion to the form of the fruits.) A genus of 40 species of herbs or small shrubs with yellow flowers. *Distr.* S and Central Europe, Turkey. *Fam.* Cruciferae. *Class* D.

B. frutescens White hairy perennial. *Distr.* SW Spain, Balearic Islands.

biserratus, -a, -um: twice serrated.

bisetaea: with two setae or bristles.

bishop's cap *Mitella* ~ **hat** *Epimedium* ~ **mitre** *Epimedium* ~ **wort** *Stachys macrantha, S. officinalis.*

bispinosus,-a, -um: with two spines.

bistort *Polygonum bistorta.*

Bistorta See *Polygonum*

bistorta: twice twisted.

bit: frog's ~ *Hydrocharis morsusranae*, Hydrocharitaceae **sheep's** ~ *Jasione, J. montana.*

biternatus, -a, -um: twice divided into threes.

bithynicus, -a, -um: of Bithynia.

bitter root *Lewisia rediviva.*

bittercress *Cardamine* **coral root** ~ *C. bulbifera.*

bitter nut *Carya cordiformis.*

bittersweet *Solanum dulcamara* **American** ~ *Celastrus scandens* **shrubby** ~ *Celastrus.*

bitterweed *Helenium amarum, Picris.*

biuncinatum: with two hooks.

bivalve: with two valves.

bivittatus: with two stripes.

Bixa (From *biche*, the local name.) A genus of 1 species of small tree with simple leaves that taper to a long point. The regular white flowers are around 5cm across, bear 5 free sepals, 5 free petals and are borne in terminal panicles. The fruit is a showy red capsule. *Distr.* Tropical America. *Fam.* Bixaceae. *Class* D.

B. orellana ANNATTO, LIPSTICK TREE. Grown as a tender, ornamental hedging plant. A red dye is obtained from the fruits that has been used as body paint and food colouring.

Bixaceae A family of 1 genus and 1 species. *Class* D.
See: Bixa.

black root *Veronicastrum virginicum* ~ **sarana** *Fritillaria camschatcensis.*

blackberry *Rubus, R. fruticosus.*

black-eyed Susan *Rudbeckia, Rudbeckia hirta, Thunbergia alata.*

Blackstonia (After John Blackstone (died 1753), English apothecary and botanist.) A genus of 5–6 species of annual herbs with simple leaves and trumpet-shaped, 6- to 8-lobed, yellow flowers. *Distr.* Europe. *Fam.* Gentianaceae. *Class* D.

B. perfoliata YELLOW WORT. Stems erect, to 45cm high. Flowers to 1.5cm across. *Distr.* Europe, SW Asia, N Africa.

blackthorn *Prunus spinosa.*

bladder pod *Lesquerella.*

bladdernut *Staphylea, S. pinnata*, Staphyleaceae.

bladderpod *Physaria.*

bladderseed *Levisticum officinale.*

bladderwort *Utricularia* **greater** ~ *U. vulgaris.*

blaeberry *Vaccinium myrtillus.*

Blandfordia (In honour of George Spencer-Churchill (1766–1840), Marquis of Blandford, later 5th Duke of Marlborough, who had a celebrated garden at Whiteknights, Reading.) CHRISTMAS BELLS. A genus of 4 species of rhizomatous herbs with grass-like linear leaves and stems to 1m high. The funnel- or bell-shaped flowers are borne on a showy raceme. All 4 species are grown as ornamentals. *Distr.* E Australia, Tasmania. *Fam.* Blandfordiaceae. *Class* M.

B. punicea TASMANIAN CHRISTMAS BELLS. Flowers red-pink outside, yellow within. *Distr.* Tasmania.

Blandfordiaceae A family of 1 genus and 4 species. *Class* M.
See: Blandfordia.

blandus, -a, -um: charming.

blanket flower *Gaillardia.*

blazing star *Chamaelirium luteum, Mentzelia lindleyi, Liatris.*

Blechnaceae A family of 9 genera and 250 species of terrestrial or lithophytic ferns and tree ferns with pinnately divided or occasionally entire leaves. The sori are long and thin or rounded and their indusia opens towards the midrib, a character not occurring in other ferns. *Distr.* Widespread. *Class* F. *See: Blechnum, Doodia, Sadleria, Stenochlaena, Woodwardia.*

blechnifolius, -a, -um: from the fern genus *Blechnum*, and *folius*, leaved.

Blechnum (From the classical Greek *blechnon*, fern.) HARD FERN. A genus of 200–220 species of terrestrial and epiphytic ferns. Several species are grown as ornamentals. *Distr.* Cosmopolitan but especially S hemisphere. *Fam.* Blechnaceae. *Class* F.

B. gibbum MINIATURE TREE FERN. Trunk erect, to 1m high. *Distr.* W Pacific.

B. spicant HARD FERN, DEER FERN. Sterile leaves spreading, broad. Fertile leaves, narrow, erect. *Distr.* Europe, W Asia, Japan, W North America.

bleeding heart, wild *Dicentra formosa.*

bleeding hearts *Dicentra spectabilis.*

blepharophyllus, -a, -um: with fringed leaves, from the Greek *blepharo-*, relating to eyelashes, and *phyllon*, leaf.

Bletilla (Named after Louis Blet (died 1794), Spanish apothecary.) A genus of 10 species of terrestrial deciduous orchids. The sepals and petals are similar, the lip is 3-lobed with central lobe reflexed and the laterals enveloping the column. *Distr.* E Asia. *Fam.* Orchidaceae. *Class* M.

B. hyacinthina See *B. striata*

B. striata Leaves sometimes white-variegated. Flowers rose-magenta. *Distr.* W China (Yunnan).

blood flower *Scadoxus multiflorus, Asclepias curassavica* **Mexican** ~ *Distictis buccinatoria.*

blood leaf *Iresine.*

bloodroot *Sanguinaria canadensis.*

blood-twig *Cornus sanguinea.*

bloodwort Haemodoraceae **burnet** ~ *Sanguisorba officinalis.*

Bloomeria (After Dr H. G. Bloomer (1821–74), Californian botanist.) A genus of 3 species of bulbous perennial herbs with linear leaves and umbels of many small flowers. *Distr.* SW USA and N Mexico. *Fam.* Alliaceae. *Class* M.

B. crocea Leaves solitary. Flowers golden yellow. *Distr.* S California, Baja California.

blowballs *Taraxacum.*

bluebeard *Clintonia borealis, Salvia viridis.*

bluebell *Hyacinthoides non-scripta* **Australian** ~ *Sollya heterophylla* **Californian** ~ *Phacelia campanularia* **English** ~ *Hyacinthoides non-scripta* **New Zealand** ~ *Wahlenbergia albomarginata* **Scottish** ~ *Campanula rotundifolia* **Spanish** ~ *Hyacinthoides hispanica.*

bluebells *Mertensia, Hyacinthoides.*

blueberry *Billardiera longiflora, Vaccinium corymbosum* **American** ~ *V. corymbosum* **box** ~ *V. ovatum* **evergreen** ~ *V. myrsinites* **highbush** ~ *V. corymbosum* **late sweet** ~ *V. angustifolium* **low-bush** ~ *V. angustifolium* **low sweet** ~ *V. angustifolium* **male** ~ *Lyonia ligustrina* **mountain** ~ *Vaccinium membranaceum* **New Zealand** ~ *Dianella nigra* **rabbit-eye** ~ *Vaccinium virgatum* **southern black** ~ *V. virgatum* **swamp** ~ *V. corymbosum.*

bluebottle *Centaurea cyanus.*

blue-eyed Mary *Omphalodes verna.* ~ **skyflower** *Thunbergia grandiflora.*

bluegrass *Festuca.*

bluestar *Amsonia.*

blue stem, Caucasian *Bothriochloa caucasica* **yellow** ~ *B. ischaemum.*

bluets *Houstonia caerulea* **creeping** ~ *Hedyotis michauxii.*

boaria: of cattle.

Bocconia (After Paolo Boccone (1633–1703), Italian monk and physician.) A genus of 9 species of herbs, shrubs and trees with yellow latex, pinnately lobed leaves and panicles of regular flowers that lack petals. *Distr.* Tropical and warm regions of the New World. *Fam.* Papaveraceae. *Class* D.

B. cordata See *Macleaya cordata*

B. frutescens TREE CELANDINE. Small tree to 7m high. Flowers purple-green. Grown as an ornamental. *Distr.* Central America.

Boehmeria (After George Rudolf Boehmer (1723–1803), professor of botany and anatomy at Wittenberg, Germany.) CHINA GRASS, FALSE NETTLE. A genus of 50–100 species of herbs, shrubs and small trees with simple, opposite or alternate leaves and clusters of small, unisexual, green flowers. *Distr.* E Asia. *Fam.* Urticaceae. *Class* D.

B. nivea CHINA GRASS, CHINESE SILK PLANT, RAMINE. Perennial herb to 2m high. Grown as the source of the longest of all known vegetable fibres that is used in cordage and weaving. *Distr.* Japan to Malaysia.

Boenninghausenia (After Clemens N. F. Boenninghausen (1785–1864), German physician and botanist.) A genus of 1 species of deciduous subshrub with alternate, pinnately divided leaves and large panicles of white flowers. *Distr.* India to Japan. *Fam.* Rutaceae. *Class* D.

B. albiflora Grown as an ornamental.

boerboon *Schotia* **weeping** ~ *S. brachypetala.*

bogbean Menyanthaceae.

bog plant *Saururus.*

Bolax (From the Greek *bolax*, clod.) A genus of 2–3 species of tuft- or carpet-forming perennial herbs with simple crowded leaves and simple umbels of small green-white flowers. *Distr.* South America. *Fam.* Umbelliferae. *Class* D.

B. glebaria See *Azorella trifurcata*

B. gummifera Leaves 3-lobed, leathery, to 1cm long. Umbels with 3–20 flowers. Grown as a rock garden ornamental. *Distr.* W Argentina, Chile, Falkland Islands.

boliviensis, -e: of Bolivia, South America.

Boltonia (After James Bolton (1758–99), British botanist.) FALSE CHAMOMILE. A genus of about 8 species of tall perennial herbs with simple leaves and panicles of daisy-like flowerheads that bear distinct rays. Grown as autumn flowering ornamentals and for cut flowers. *Distr.* E Asia, North America. *Fam.* Compositae. *Class* D.

B. asteroides To 3m tall. Leaves large. Rays white to purple. Used as a leaf vegetable in Japan. *Distr.* E USA.

B. incisa See *Kalimeris incisa*

Bolusanthus (After Dr Harry Bolus (1834–1911), South African botanist.) A genus of 1 species of small tree with pinnate leaves and crowded racemes of irregular pea-like flowers. *Distr.* S Central Africa, introduced North America. *Fam.* Leguminosae. *Class* D.

B. speciosus WILD WISTERIA, ELEPHANT WOOD, SOUTH AFRICAN WISTERIA. Grown as an ornamental and for timber.

Bomarea (After Jacques Christophe Valmont de Bomare (1731–1807), French patron of science.) A genus of 100–120 species of rhizomatous twining herbs with lance-shaped leaves and umbels of brightly marked flowers similar to those of *Alstroemeria*. Several species are grown as tender ornamentals. *Distr.* New World, from Mexico to the Andes. *Fam.* Alstro-emeriaceae. *Class* M.

B. caldasii (After Francisco Jose de Caldas (1771–1816), botanical explorer.) Outer tepals red-brown, inner tepals orange-yellow. *Distr.* Andes of Colombia and Ecuador.

B. multiflora Inflorescence many-flowered. Flowers red-yellow. *Distr.* Colombia, Venezuela.

Bombacaceae The Durian family. 30 genera and 250 species of deciduous, often bottle-shaped trees with palmate leaves and large showy flowers that typically open whilst the tree is leafless. The commercially important timber BALSA is derived from *Ochroma pyramidale* and the fibre KAPOK. is produced from the pods as *Ceiba pentandra* and *Bombax* species. *Distr.* Tropical and subtropical regions. *Class* D.

See: Bombax, Ceiba, Chorisia, Durio, Ochroma.

Bombax (From the Greek word for cotton, alluding to the fibre in the capsules.) SILK COTTON TREE. A genus of 8 species of large deciduous trees with palmate leaves and showy cup-shaped flowers. The fruit is a wooden capsule containing fibre-coated seeds. Several species are grown for their soft timber and the fibre from the capsules which is used in packing. *Distr.* Tropical regions of the Old World. *Fam.* Bombacaceae. *Class* D.

B. buonopozense GOLD COAST BOMBAX. Tree to 35m high. Trunk spiny. Flowers red. Grown commercially for its timber. *Distr.* Tropical W Africa.

B. ceiba RED SILK COTTON TREE, SIMUL. To 30m. Trunk spiny when young. Flowers red. Exploited for its timber as well as the fibrous seed which are considered inferior to kapok. *Distr.* SE Asia.

B. malabaricum See *B. ceiba*

bombax, Gold Coast *Bombax buonopozense.*

bombyciferum: bearing silk.

bombycinus, -a,-um: silky.

boneset *Eupatorium perfoliatum.*

Bongardia (After August Gustav Heinrich Bongard (1786–1839), German botanist.) A genus of 1 species of rhizomatous herb with pinnate, red-marked, basal leaves and a panicle of regular yellow flowers that are followed by red-tinged capsules. *Distr.* Greece, SW Asia. *Fam.* Berberidaceae. *Class* D.

B. chrysogonum Grown as an ornamental.

borage *Borago officinalis.*

Boraginaceae The Forget-me-not family. 131 genera and about 2200 species of annual and perennial herbs, shrubs, trees and lianas with simple entire leaves, and regular, salver- or bell-shaped flowers that typically bear 5 sepals, 5 petals, 5 stamens and a superior ovary. Several genera are important in the garden; others are a source of dye or have medicinal properties. *Distr.* Widespread. *Class* D.

See: Adelocaryum, Alkanna, Anchusa, Arnebia, Borago, Brunnera, Buglossoides, Cerinthe, Cynoglossum, Echium, Ehretia, Eritrichium, Heliotropium, Lindelofia, Lithodora, Lithospermum, Mertensia, Moltkia, Myosotidium, Myosotis, Omphalodes, Onosma, Pentaglottis, Pseudomertensia, Pulmonaria, Symphytum, Trachystemon.

Borago (Possibly from the Latin *burra*, a hairy garment, alluding to the leaves which are usually densely hairy.) A genus of 3 species of annual and perennial herbs with erect or horizontal stems, simple alternate leaves and cymes of bell-shaped or flattened flowers. *Distr.* Mediterranean, Europe, Asia. *Fam.* Boraginaceae. *Class* D.

B. laxiflora See *B. pygmaea*

B. officinalis BORAGE, TAILWORT. Annual. Stems erect, branched. Flowers wheel-shaped, blue, occasionally white. An important medicinal and culinary herb. The young leaves are rich in calcium and potassium and are eaten raw or cooked. This herb is also grown as an ornamental although it may become weedy. *Distr.* Europe and Mediterranean.

B. pygmaea Perennial. Stems slender, horizontal. Flowers bell-shaped, pale blue. Grown as a rock garden ornamental. *Distr.* Europe.

borealis, -e: northern.

Bornmuellera (After Joseph Bornmuller (1862–1948), traveller, collector and curator of the Haussknecht Herbarium at Weimar.) A genus of 6 species of perennial herbs with white flowers. *Distr.* Balkans, Turkey. *Fam.* Cruciferae. *Class* D.

B. cappadocica Perennial herb with a shrubby stock. *Distr.* Turkey.

Boronia (After Francesco Borone (1769–94), Italian botanist.) A genus of 96 species of evergreen aromatic shrubs with simple or compound, opposite leaves and cymes or umbels of typically fragrant flowers that bear 4 petals and 8 stamens. A number of species are grown as half-hardy ornamentals. *Distr.* Australia, New Caledonia. *Fam.* Rutaceae. *Class* D.

B. heterophylla RED BORONIA. Flowers pink to scarlet, slightly fragrant. *Distr.* W Australia.

B. megastigma SCENTED BORONIA, BROWN BORONIA. Flowers purple-brown to yellow, strongly fragrant. Grown commercially for scent. *Distr.* W Australia.

boronia: brown ~ *Boronia megastigma* **red** ~ *B. heterophylla* **scented** ~ *B. megastigma.*

bosniacus, -a, -um: of Bosnia.

Boswellia (After James Boswel (1740–95), biographer of Samuel Johnson.) A genus of

24 species of deciduous, resinous shrubs and trees with pinnate leaves and racemes of small regular flowers that bear 5 sepals, 5 petals and 10 stamens. *Distr.* Tropical regions of the Old World. *Fam.* Burseraceae. *Class* D.

B. sacra FRANKINCENSE, GUM OLIBANUM. Source of the resins used in the production of frankincense.

B. serrata A source of timber. *Distr.* India.

Bothriochloa (From the Greek *bothrion*, a shallow pit, and *chloe*, grass.) A genus of about 28 species of perennial rhizomatous grasses with flat leaves and erect raceme-like inflorescence. Several species are cultivated for fodder and as ornamentals. *Distr.* Tropical and warm temperate regions. *Fam.* Gramineae. *Class* M.

B. caucasica PURPLE BEARD GRASS, CAUCASIAN BLUE STEM. Inflorescence tinged purple. An attractive ornamental. *Distr.* India to the Caucasus.

B. ischaemum YELLOW BLUE STEM. Tussock-forming. Occasionally grown as an ornamental. *Distr.* S Europe.

botrys: a bunch of grapes.

bottlebrush, Alpine *Callistemon sieberi* **Cape** ~ *Greyia sutherlandii* **crimson** ~ *Callistemon citrinus* **gravel** ~ *Beaufortia sparsa* **lemon** ~ *Callistemon pallidus* **Natal** ~ *Greyia* **stiff** ~ *Callistemon rigidus* **swamp** ~ *Beaufortia sparsa* **weeping** ~ *Callistemon viminalis*.

~ **bush** *Callistemon*.

bottle tree, Chinese Firmiana simplex.

Bougainvillea (After Louis Antoine de Bougainville (1729–1811), French explorer.) A genus of 18 species of lianas, shrubs, and small trees with simple alternate leaves and 1 to several, small, tubular flowers that are subtended by large colourful bracts. Several species and numerous cultivars are grown as tender ornamentals particularly around the Mediterranean. *Distr.* Tropical and subtropical South America. *Fam.* Nyctaginaceae. *Class* D.

B. glabra PAPER FLOWER. Fast climber. Bracts white to magenta. *Distr.* Brazil.

B. spectabilis Strong climber or scrambler. Bracts purple-pink. *Distr.* Brazil.

bougainvillea Nyctaginaceae.

bouncing Bet *Saponaria officinalis*.

Boussingaultia See *Anredera*

Bouteloua (After two Spanish brothers, Claudio (1774–1842) and Esteban (1776–1813) Boutelou, both botanists.) GRAMA GRASS. A genus of 40–50 species of annual and perennial, stoloniferous or rhizomatous, clump-forming grasses with flat or folded leaves. One of the more important genera of grasses of the Great Plains of North America; several species are also grown as ornamentals. *Distr.* Warm temperate and tropical America. *Fam.* Gramineae. *Class* M.

B. curtipendula SIDE OATS GRASS. Rhizomatous perennial to 80cm high. *Distr.* Canada to Argentina.

B. gracilis BLUE GRAMAGRASS, MOSQUITO GRASS. Rhizome short, stout. Leaves rough. *Distr.* Central and S USA and Mexico.

Bouvardia (After Dr Charles Bouvard (1572–1658), keeper of the Jardin du Roi, Paris.) A genus of 20–30 species of perennial herbs and evergreen shrubs with simple, opposite or whorled leaves and tubular to funnel-shaped, 4-lobed flowers. Several species are grown as ornamentals. *Distr.* Tropical America. *Fam.* Rubiaceae. *Class* D.

B. longiflora SWEET BOUVARDIA. Shrub. Flowers tubular, white, highly scented. *Distr.* Mexico.

B. scabrida Herb or shrub. Flowers pink or white, unscented. *Distr.* Mexico.

B. ternifolia SCARLET TROMPETILLA. Herb or shrub. Flowers red. *Distr.* Texas, Mexico.

B. triphylla Shrub. Flowers orange-red, unscented. *Distr.* S USA.

bouvardia, sweet *Bouvardia longiflora*.

Bowenia (After Sir George Ferguson Bowen (1821–99), Governor of Queensland.) A genus of 2 species of palm-like conifers with an underground stem and pinnate leaves. The cones are borne on separate specialised branches. *Distr.* Australia (Queensland). *Fam.* Boweniaceae. *Class* G.

B. serrulata BYFIELD FERN. Occasionally grown as a tender ornamental. *Distr.* Australia (SE Queensland).

Boweniaceae A family of 1 genus and 2 species. *Class* G.
See: Bowenia.

bower plant *Pandorea jasminoides.*

Bowiea (After James Bowie (1789–1869), of the Royal Botanic Gardens, Kew, who collected in Brazil and South Africa.) A genus of 2–3 species of bulbiferous herbs with annual, succulent, twining stems, small leaves and racemes of small, green-white or yellow flowers. *Distr.* S Africa. *Fam.* Hyacinthaceae. *Class* M.

B. volubilis CLIMBING ONION. Occasionally grown as an ornamental. Bulbs toxic and used as a medicine locally. *Distr.* SW Africa.

Bowkeria (After James Henry Bowker (died 10900), and his sister, Mrs Mary Elizabeth Barber (died 1899), South African botanists.) A genus of 5 species of evergreen shrubs and trees with simple, sessile, typically whorled leaves and clusters of 2-lipped, 5-lobed flowers that bear 4 stamens. *Distr.* S Africa. *Fam.* Scrophulariaceae. *Class* D.

B. gerrardiana Shrub to 3m high. Flowers white, spotted red within. Grown as a half-hardy ornamental. *Distr.* South Africa.

bowman's root *Veronicastrum virginicum.*

box Buxaceae, *Buxus* **bastard** ~ *Polygala chamaebuxus* **Chinese** ~ *Murraya paniculata* **Christmas** ~ *Sarcococca* **common** ~ *Buxus sempervirens* **European** ~ *B. sempervirens* **Himalayan** ~ *B. wallichiana* **Japanese** ~ *B. microphylla* **mountain** ~ *Arctostaphylos uva-ursi* **running** ~ *Mitchella repens* **sweet** ~ *Sarcococca* **Venezuelan** ~ *Casearia praecox* **Victorian** ~ *Pittosporum undulatum.*

boxthorn *Lycium.*

boxwood *Buxus* **African** ~ *Myrsine africana* **Balearic** ~ *Buxus balearica* **Oregon** ~ *Paxistima myrtifolia.*

Boykinia (After Dr Samuel Boykin (1786–1846), American botanist, naturalist, and physician.) A genus of 8 species of perennial rhizomatous herbs with alternate leaves and crowded flowers. *Distr.* North America, Japan. *Fam.* Saxifragaceae. *Class* D.

B. aconitifolia Flowers white. *Distr.* E USA (Appalachians).

B. heucheriformis Flowers violet-purple. *Distr.* W North America.

B. jamesii Flowers crimson. *Distr.* NW USA.

B. rotundifolia Basal leaves round. *Distr.* USA (S California).

B. tellimoides See *Peltoboykinia tellimoides*

boysenberry *Rubus.*

brachiatus, -a, -um: branching.

brachyandrus: with short stamens.

brachyanthus, -a, -um: with short flowers.

brachybotrys: with a short clustered inflorescence, from the Greek *brachy*, short, and *botrys*, a bunch of grapes.

brachycalyx: with short sepals.

brachycarpus, -a, -um: with short fruit.

brachycaulos: short-stemmed.

brachycerus, -a, -um: with short horns.

Brachyglottis (From the Greek *brachys*, short, and *glotta*, a tongue, alluding to the short rays.) A genus of 29 species of evergreen shrubs, vines, or perennial herbs with simple leaves and fragrant flowers in daisy-like heads that may or may not bear rays. *Distr.* New Zealand, Chatham Islands, Tasmania. *Fam.* Compositae. *Class* D.

B. bidwillii Shrub with leathery leaves. *Distr.* New Zealand (North Island).

B. buchananii Robust shrub. Leaves silver-hairy beneath. *Distr.* New Zealand.

B. compacta Shrub with glabrous leaves. *Distr.* New Zealand (North Island).

B. elaeagnifolia Large shrub. Leaves silver-hairy beneath. *Distr.* New Zealand (North Island).

B. greyi Shrub. Leaves leathery. *Distr.* New Zealand (North Island).

B. hectorii Large shrub with spreading branches. *Distr.* New Zealand (South Island).

B. huntii Large shrub or tree. Rays recurved, yellow. *Distr.* New Zealand (Chatham Islands).

B. kirkii Shrub. Rays spreading, white. *Distr.* New Zealand (North Island).

B. laxifolia Shrub. Rays spreading, yellow. *Distr.* New Zealand (South Island).

B. monroi Much branched shrub. Leaves leathery. *Distr.* New Zealand (South Island).

B. repanda Large shrub or tree. Rays absent. *Distr.* New Zealand (North Island).

B. rotundifolia Large shrub or tree. Leaves leathery. Rays absent. *Distr.* New Zealand (South Island).

brachylobus, -a, -um: with short lobes.

brachypetalus, -a, -um: with short petals.

brachyphyllus, -a, -um: with short leaves.

Brachypodium (From the Greek *brachys*, short, and *pous*, foot, alluding to the very short pedicels.) A genus of about 16 species of annual or perennial, rhizomatous grasses to 1.2m high with flat or curled leaves and spike-like racemes. *Distr.* Temperate regions of the N hemisphere. *Fam.* Gramineae. *Class* M.

B. pinnatum TOR GRASS. Tussock-forming perennial with stiff leaves. The locally dominant species in some areas of calcareous grassland.

B. sylvaticum SLENDER FALSE BROME, WOOD FALSE BROME. Tuft-forming perennial. Leaves soft. A woodland grass.

brachypus: short-stalked.

brachysiphon: with a short tube.

brachystachys: with a short spike.

Brachystachyum See *Semiarun-dinaria*

brachytricha: with short hairs.

brachytyla: with short swellings, from the Greek *brachys*, short, and *tylos*, swelling or lump.

bracken *Pteridium aquilinum.*

bracteatus, -a, um: with conspicuous bracts.

bracteosus, -a, -um: with conspicuous bracts.

bractescens: with bracts.

Brahea (After Tycho Brahe (1546–1601), Danish astronomer.) SOYA PALM, HESPER PALM, ROCK PALM. A genus of 12 species of medium-sized palms with fan-shaped leaves and branched inflorescences of bisexual flowers. Some species have edible fruit and oils. *Distr.* Mexico, Central America. *Fam.* Palmae. *Class* M.

B. armata BLUE HESPER PALM, GREY GODDESS, BLUE FAN PALM, MEXICAN BLUE FAN PALM. To 15m. Leaves blue-green, waxy, with spiny stalks. Grown as an ornamental. *Distr.* Mexico, Baja California.

B. brandegeei SAN JOSE HESPER PALM. Leaves with spiny stalks. *Distr.* Baja California.

brain plant *Calathea makoyana.*

brake *Pteris* **Cretan** ~ *P. cretica* **spider** ~ *P. multifida.*

bramble *Rubus fruticosus, R. ulmifolius* **Arctic** ~ *R. arcticus* **crimson** ~ *R. arcticus.*

brandy bottle *Nuphar lutea.*

brasiliensis, -e: of Brazil.

Brassaia See *Schefflera*

Brassica (Latin name for CABBAGE.) A genus of about 30 species of annual, biennial and occasionally perennial herbs with yellow or white flowers. A genus of great importance containing a number of major vegetables and flavourings as well as animal fodder and oil seeds. *Distr.* Mediterranean. *Fam.* Cruciferae. *Class* D.

B. campestris FIELD MUSTARD, TURNIP RAPE. Grown as an oil seed in areas not suitable for cultivation of *Brassica napus. Distr.* Central and S Europe.

B. chinensis PAK CHOI. A vegetable of Chinese cooking. *Distr.* E Asia, widely cultivated in China and Japan.

B. japonica See *B. juncea*

B. juncea INDIAN MUSTARD, CHINESE MUSTARD, BROWN MUSTARD. Seeds are the principal source of the condiment mustard in Europe. *Distr.* Eurasia.

B. napus SWEDE, KALE, FORAGE RAPE, OILSEED RAPE, COLZA. Roots eaten as vegetable and as animal fodder, oil expressed from seeds has wide range of uses from lubrication to cooking, seedlings now used as the mustard in Mustard and Cress as a substitute for *Sinapis alba*. *Distr.* Not known in the wild, derived from *B. oleracea* and *B. rapa* in cultivation.

B. nigra BLACK MUSTARD. Original plant used in the production of mustard (the condiment). *Distr.* Eurasia, cultivated in Mediterranean since antiquity.

B. oleracea CABBAGE, CURLY KALE, BRUSSEL SPROUTS, CAULIFLOWER, BROCCOLI, CALABRESE, CATTLE CABBAGE, SAVOY CABBAGE. A very variable annual or perennial herb with a large number of cultivars. This species has given rise to very many familiar vegetables. *Distr.* Mediterranean and SW Europe, cultivated since antiquity.

B. pekinensis PE TSAI. Vegetable of Chinese cooking, a salad vegetable in Europe. *Distr.* E Asia, Widely cultivated in N China.

B. rapa TURNIP. Root widely eaten as vegetable. Also grown as both a fodder and forage crop. *Distr.* Derived in cultivation from *Brassica campestris*.

Brassicaceae See Cruciferae

Braya (After Count Francisco Gabrieli von Bray (1765–1831), German botanist.) A genus of 20 species of tufted perennial herbs with small white-purple flowers. *Distr.* N circumpolar, Alps, Central Asia, Himalaya. *Fam.* Cruciferae. *Class* D.

B. alpina Loosely tufted perennial. *Distr.* E Alps.

braziliensis: of Brasil.

Brazil nut Lecythidaceae.

bread, St John's *Ceratonia siliqua*.

~ **of India tree** *Elaeocarpus sphaericus*.

breath of heaven *Diosma ericoides*.

Bretschneideraceae A family of 1 genus and 1 species of deciduous tree with spirally arranged, pinnate leaves and racemes of slightly irregular, 5-petalled flowers. *Distr.* China, Vietnam, Thailand. *Class* D.

brevicalcaratus, -a, -um: with a short spur on the flower, from the Latin *brevis*, short, and *calcaratus*, spurred.

brevicaudatus, -a, um: with a short tail-like appendage, from the Latin *brevis*, short, and *caudatus*, caudate, ending in a tail.

brevicaulis, -e: with a short stem.

breviflorus, -a, -um: with short flowers.

brevifolius, -a, -um: short-leaved.

brevilobis with short lobes.

brevipedicellatus, -a, -um: with short flower stalks, from the Latin *brevis*, short, and *pedicellatus*, pedicelled.

brevipedunculatus, -a, -um: with short inflorescence stalks, from the Latin *brevis*, short, and *pedunculatus*, peduncled.

breviracemosus, -a, -um: with short inflorescences, from the Latin *brevis*, short, and *racemosus*, racemose, a form of inflorescence.

breviserratus, -a, -um: shortly serrated, referring to the margin of the leaves.

brevispinus, -a, -um: with short spines.

brevistylus, -a, -um: with short styles.

Breynia (After Jacob Breyne (1637– 1697), and his son, Johann Philipp Breyne (1680–1764), botanists from Danzig.) A genus of about 25 species of shrubs and small trees with simple alternate leaves, small unisexual flowers and red berries. *Distr.* China to New Caledonia and Australia. *Fam.* Euphorbiaceae. *Class* D.

B. nivosa SNOW BUSH. Slender shrub. Leaves variegated. Grown as an ornamental with several cultivars available. *Distr.* Pacific Islands.

bridal veil *Gibasis pellucida*.

bridal wreath *Francoa ramosa*, *F. sonchifolia*, *Spiraea*, *Stephanotis floribunda*.

brier: Austrian ~ *Rosa foetida* **bull** ~ *Smilax rotundifolia* **cat** ~ *Smilax* **common** ~ *Rosa canina* **dog** ~ *R. canina* **green** ~ *Smilax* **horse** ~ *S. rotundifolia* **sweet** ~ *Rosa rubiginosa*.

Briggsia (After Munro Briggs Scott (1889–1917), botanist.) A genus of 23 species of rhizomatous herb with basal rosettes of stalked simple leaves and solitary or clustered, 2-lipped, tubular flowers. *Distr.* E Himalaya, China, Burma. *Fam.* Gesneriaceae. *Class* D.

B. muscicola Flowers clustered, tube swollen at mouth, yellow-orange with purple marks. *Distr.* Bhutan, Tibet, W China.

Brimeura (After Maria de Brimeur, a 16th-century lover and grower of flowers.) A genus of 2 species of small bulbiferous herbs with narrow leaves and slender racemes of bell-shaped flowers. *Distr.* S Europe. *Fam.* Hyacinthaceae. *Class* M.

B. amethystina Flowers bright blue, narrow bell-shaped. Occasionally grown as an ornamental. *Distr.* Pyrenees and NE Spain, NW Yugoslavia.

brinjal *Solanum melongena.*

bristoliensis: of Bristol, England.

brittonia *Argemone.*

Briza (The Greek name for one of the food grains, possibly rye.) QUAKE, SHAKING GRASS, QUAKING GRASS. A genus of 12 species of annual and perennial grasses to 1m high with flat leaves and dense or loose panicles of pendent spikelets. Several species are grown as ornamentals, their inflorescences being particularly suited for dried flower arrangements. *Distr.* Temperate regions, mainly in the N hemisphere. *Fam.* Gramineae. *Class* M.

B. maxima PEARL GRASS, GREAT QUAKING GRASS. Annual. Panicles nodding. *Distr.* Mediterranean region.

B. media JIGGLE JOGGLE, DIDDER, TOTTER, COMMON QUAKING GRASS. Perennial. Panicle erect, spreading. *Distr.* Europe, Asia.

B. minor LESSER QUAKING GRASS. Annual. Panicle with loose ascending branches. *Distr.* W Europe and the Mediterranean region.

broccoli *Brassica oleracea.*

Brodiaea (After James Brodie (1744–1824), Scottish botanist.) A genus of 15 species of slender perennial herbs arising from a corm. The leaves are linear and the inflorescence is an umbel of small flowers. The corms of some species are eaten by North American Indians others are grown as ornamentals. *Distr.* W North and South America. *Fam.* Alliaceae. *Class* M.

B. capitata See *Dichelostemma pulchellum*

B. elegans Flowers few, deep-mauve. *Distr.* USA (California, Oregon).

B. laxa See *Triteleia laxa*

B. peduncularis See *Triteleia peduncularis*

B. terrestris Flowers mauve-pink. *Distr.* USA (California, S Oregon).

brome *Bromus* **slender false** ~ *Brachypodium sylvaticum* **wood false** ~ *B. sylvaticum.*

Bromeliaceae The Pineapple family. 51 genera and about 2000 species of typically short-stemmed herbs with rosettes of leaves, or occasionally large subshrubs or highly reduced epiphytes such as *Tillandsia usneoides*. The regular flowers have 3 sepals, 3 petals and 2 whorls of 3 stamens. Some species are a source of fibre, others have edible fruit (*Ananas comosus*, PINEAPPLE.). *Distr.* Tropical regions of the New World, 1 species in W tropical Africa. *Class* M.

See: Aechmea, Ananas, Billbergia, Cryptanthus, × *Cryptbergia, Fascicularia, Fosterella, Guzmania, Neoregelia, Nidularium, Ochagavia, Pitcairnia, Puya, Quesnelia, Tillandsia, Vriesea.*

bromeliad, bird's nest *Nidularium innocentii* **blushing** ~ *N. fulgens, Neoregelia carolinae.*

bromeliifolius, -a, -um: with bromeliad-like leaves.

bromelioides: bromeliad-like.

Bromus (From the Greek *bromos*, OATS.) BROME GRASS, BROME, CHESS. A genus of about 100 species of annual, biennial or perennial grasses to 2m high with flat or in-rolled leaves and erect or nodding panicles. Several species are important as forage crops, others are grown ornamentally. *Distr.* Temperate areas and in mountainous regions of the tropics. *Fam.* Gramineae. *Class* M.

B. lanceolatus See *B. macrostachys*

B. macrostachys Annual. Panicles dense, erect. Grown for ornamental panicles. *Distr.* Mediterranean.

B. ramosus Perennial. Panicle very loose, pendulous. *Distr.* Europe, N Africa, SW Asia.

bronchialis, -e: relating to use of the plant for the treatment of bronchitis.

brooklime *Veronica beccabunga* **European** ~ *V. beccabunga.*

broom *Genista, Cytisus* **butcher's** ~ *Ruscus, R. aculeatus,* **climbing butcher's** ~ *Semele androgyna*

common ~ *Cytisus scoparius* **coral** ~ *Corallospartium crassicaule* **hedgehog** ~ *Erinacea anthyllis* **Mount Etna** ~ *Genista aetnensis* **Scotch** ~ *Cytisus scoparius* **Spanish** ~ *Spartium junceum* **weaver's** ~ *S. junceum* **weeping tree** ~ *Chordospartium stevensonii* **white** ~ *Retama raetam.*

Broussonetia (After P. M. A. Broussonet (1761–1807), professor of botany at Montpellier.) A genus of 8 species of deciduous, latex-containing trees with simple, large, alternate leaves and small, unisexual, wind-pollinated flowers that are borne in pendulous catkins. *Distr.* E Asia, Polynesia. *Fam.* Moraceae. *Class* D.

B. papyrifera PAPER MULBERRY. Tree to 15m high. Grown as an ornamental. The bark is used to make paper in Japan. *Distr.* China, Japan, Korea, Polynesia.

Browallia (After John Browall (1707–55), Bishop of Abo, Sweden, and a botanist.) A genus of 2 species of annual and perennial herbs with simple, typically alternate leaves and solitary, irregular, blue or white flowers. *Distr.* Tropical South America. *Fam.* Solanaceae. *Class* D.

B. speciosa BUSH VIOLET. Somewhat shrubby perennial to 1.5m high. Flowers deep purple-blue. Grown as a tender ornamental. *Distr.* Colombia.

browallia, orange *Streptosolen jamesonii.*

brownies *Scoliopus bigelovii.*

Bruckenthalia (After Samuel and Michael von Bruckenthal, 18th-century Austrian noblemen.) A genus of 1 species of evergreen, heath-like, mat-forming shrubs with narrow leaves and racemes of small, pink, bell-shaped flowers. *Distr.* SE Europe, Turkey. *Fam.* Ericaceae. *Class* D.

B. spiculifolia SPIKE HEATH. Grown as a hardy ornamental.

Brugmansia (After Sebald Justin Brugmans (1763–1819), professor of natural history at Leiden.) A genus of 5 species of shrubs and trees with simple alternate leaves and very large, solitary, 5-lobed, funnel-shaped flowers. *Distr.* S America, especially the Andes. *Fam.* Solanaceae. *Class* D.

B. arborea ANGEL'S TRUMPET, MAIKOA. Shrub or small tree to 4m high. Flowers large, white. *Distr.* Ecuador, N Chile.

B. × insignis A hybrid of *B. suaveolens* x *B.versicolor*. *Distr.* E Andes.

B. sanguinea RED ANGEL'S TRUMPET. Shrub or small tree. Flowers red. Used as a hallucinogen by American Indians. *Distr.* Colombia to Chile.

B. suaveolens Shrub. Flowers white, yellow or pink to 30cm long, fragrant at night. *Distr.* SE Brazil.

B. versicolor Shrub or small tree. Flowers to 50cm long, white, becoming pink. *Distr.* Ecuador.

Brunelliaceae A family of 1 genus and 52 species of trees with simple or compound, often densely hairy leaves and panicles of regular bisexual flowers which lack petals. *Distr.* Mexico to Peru and the West Indies. *Class* D.

Brunfelsia (After Otto Brunfels (1489–1534), German botanist.) A genus of about 40 species of evergreen shrubs and small trees with simple alternate leaves and large, tubular or bell-shaped, 5-lobed flowers. Several species are grown ornamentally and some are used medicinally. *Distr.* Tropical America. *Fam.* Solanaceae. *Class* D.

B. americana LADY OF THE NIGHT. Shrub to 5m high. Flowers white, purple in the centre, to 2cm across. *Distr.* West Indies.

B. calycina See *B. pauciflora*

B. pauciflora YESTERDAY TODAY AND TOMORROW. Shrub to 3m high. Flowers blue and white, to 7cm across. *Distr.* Brazil.

B. undulata WHITE RAINTREE, RAINTREE. Shrub or small tree to 6m high. Flowers in many-flowered cymes, 20cm long, white. *Distr.* Jamaica.

Bruniaceae A family of 12 genera and 70 species of heath-like shrubs with small needle-like leaves and typically clusters or spikes of regular bisexual flowers which bear their parts in 4s or 5s. *Distr.* S Africa. *Class* D.

bruniifolius, -a, -um: with brown leaves.

Brunnera (After Samuel Brunner (1790–1844), Swiss botanist.) A genus of 3 species of rhizomatous herbs with erect stems, simple leaves and panicles of small, star-shaped, purple or blue flowers. *Distr.* Mediterranean to W Siberia. *Fam.* Boraginaceae. *Class* D.

B. macrophylla SIBERIAN BUGLOSS. Clump-forming perennial. Flowers bright blue, appearing before the leaves. Grown as a hardy ornamental. *Distr.* Caucasus, W Siberia.

brunneus, -a, -um: brown.

Brunsvigia (After Carl Wilhelm Ferdinand (1713–1780), Duke of Brunswick-Luneburg.) A genus of about 20 species of large bulbiferous perennial herbs with strap-shaped leaves and umbels of regular or irregular funnel-shaped flowers. Several species are grown as tender ornamentals. *Distr.* South Africa. *Fam.* Amaryllidaceae. *Class* M.

B. josephinae Bulbs borne on soil surface. Flowers scarlet. *Distr.* Cape Province.

B. orientalis Bulbs subterranean. Flowers pink to scarlet. *Distr.* Cape Province.

Bryanthus (From the Greek *bryon*, moss, and *anthos*, flower, referring to the low mossy habit.) A genus of 1 species of procumbent evergreen shrubs with linear, moss-like, leathery leaves and small pink flowers. *Distr.* Kamchatka, Japan. *Fam.* Ericaceae. *Class* D.

B. gmelinii Grown as a hardy ornamental.

bryoides: moss-like.

Bryonia (From the Greek *bruein*, to burgeon.) BRYONY. A genus of about 12 species of climbing perennial herbs with tuberous roots, palmately lobed leaves and unbranched tendrils. The white or yellow, bell-shaped flowers are unisexual, the males and the females typically being borne on separate plants. The fruit is a small berry. *Distr.* Eurasia, N Africa, Canary Islands. *Fam.* Cucurbitaceae. *Class* D.

B. alba WHITE BRYONY. Flowers white. Fruit black. A common hedgerow plant. *Distr.* Europe to W Asia.

B. dioica RED BRYONY, WILD HOP. Flowers tinged green. Fruit red. A common hedgerow plant. *Distr.* Eurasia, N Africa.

bryonioides: resembling bryony.

bryony *Bryonia* **black** ~ *Tamus communis* **red** ~ *Bryonia dioica* **white** ~ *B. alba*.

bucharicus, -a, -um: of Bokhara, Turkestan.

buckeye *Aesculus* **Californian** ~ *A. californica* **Ohio** ~ *A. glabra* **red** ~ *A. pavia* **sweet** ~ *A. flava* **yellow** ~ *A. flava*.

buckhorn *Osmunda cinnamomea*.

buckthorn Rhamnaceae, *Rhamnus* **alder** ~ *R. frangula* **common** ~ *R. cathartica* **European** ~ *R. cathartica* **purging** ~ *R. cathartica* **sea** ~ *Hippophae, H. rhamnoides*.

buckwheat *Fagopyrum esculentum* **wild** ~ *Eriogonum*.

buddleia, olive *Buddleja saligna*.

buddleioides: from the genus *Buddleja*, with the ending *-oides*, indicating resemblance.

Buddleja (After Revd Adam Buddle (1660–1715), amateur botanist.) BUTTERFLY BUSH. A genus of about 100 species of deciduous or evergreen herbs, shrubs and trees with simple, typically opposite leaves and clusters of small, often fragrant, tubular flowers. The flower clusters are frequently arranged along the branches so as to produce a spike-like inflorescence. Many species are grown as ornamentals and some are used medicinally. *Distr.* Tropical and subtropical Asia, America and Africa. *Fam.* Buddlejaceae. *Class* D.

B. alternifolia Deciduous weeping shrub or small tree. Leaves alternate. Flowers pink-purple. *Distr.* China.

B. asiatica Evergreen shrub or small tree. Leaves narrow. Flowers white, very fragrant. A source of fish poison. *Distr.* India to Malaysia.

B. colvilei (After Sir James Colvile (died 1890).) Deciduous shrub or small tree. Flowers relatively large, deep purple-red with a white centre. *Distr.* Himalaya.

B. crispa Deciduous bushy shrub. Flowers small, pale purple-pink with a white centre, fragrant. *Distr.* Himalaya.

B. davidii (After Armand David (1826–1900), French missionary who introduced it into cultivation.) BUTTERFLY BUSH, SUMMER LILAC, ORANGE EYE. Deciduous shrub. Flowers abundant, purple-blue, with an orange centre. A weed of waste ground in the British Isles, this species is often planted in gardens to attract butterflies. Numerous cultivars are available with differing habits and flower colours. *Distr.* China and Japan, naturalized in many parts of the world.

B. fallowiana (After George Fallow (1891–1915), gardener at the Royal Botanical Garden, Edinburgh.) Deciduous shrub. Flowers light blue-purple, orange in the centre, fragrant. *Distr.* Burma, W China.

B. globosa ORANGE BALL TREE. Semi-evergreen shrub. Flowers in many spherical heads, bright orange-yellow, fragrant. *Distr.* Argentina, Chile, Peru.

B. madagascariensis Evergreen half-hardy shrub. Flowers orange-yellow. *Distr.* Madagascar.

B. nicodemia See *B. madagascariensis.*

B. saligna OLIVE BUDDLEIA, BASTARD OLIVE. Shrub or small tree, to 12m high. Flowers in rounded heads, light blue to cream, with an orange centre. *Distr.* S Africa.

B. salviifolia SOUTH AFRICAN SAGE WOOD. Shrub or small tree. Exploited for timber. *Distr.* Tropical and S Africa.

B. × weyeriana (After Van de Weyer, who raised it in 1914.) A hybrid of *B. davidii* × *B. globosa*. Deciduous shrub. Flowers in spherical heads, blue-yellow.

Buddlejaceae The Butterfly Bush family. 8 genera and 122 species of herbs, shrubs and trees with simple, typically opposite leaves that are often covered in star-shaped hairs. The flowers are born in clusters and typically have 4 fused sepals, and a 4-lobed petal tube. *Distr.* Widespread but mostly tropical to warm-temperate regions. *Class* D.
See: Buddleja.

buddlejifolius, -a, -um: from the genus *Buddleja*, and *folius*, leaved.

bugbane *Cimicifuga* **foetid** ~ *C. foetida.*

bugle *Ajuga* **blue** ~ *A. genevensis* **common** ~ *A. reptans* **pyramid** ~ *A. pyramidalis* **upright** ~ *A. genevensis.*

bugloss *Anchusa* **Siberian** ~ *Brunnera macrophylla.*

Buglossoides (From the old genus name *Buglossum*, and the Greek *-oides*, resembling.) A genus of about 15 species of annual and perennial herbs with simple leaves and cymes of blue or white, funnel-shaped flowers. *Distr.* Europe, Asia. *Fam.* Boraginaceae. *Class* D.

B. purpuro-caerulea Perennial. Stems erect, little branched. Flowers red-blue, becoming bright blue. Grown as an ornamental. *Distr.* Europe, W Asia.

bulbiferum: producing bulbs.

bulbilifera: producing small bulbs.

Bulbine (From the Greek *bolbine*, star-flower.) A genus of 30–40 species of succulent annual and perennial herbs with soft fleshy leaves that die back during the dry season in some species. The small, yellow, white or orange flowers are borne in a terminal raceme. Many species are important medicinal plants in South Africa. *Distr.* S and E Africa, Australia. *Fam.* Asphodelaceae. *Class* M.

B. annua Annual. Flowers bright yellow. *Distr.* South Africa (Cape Province).

B. caulescens See *B. frutescens*

B. frutescens Branched subshrub with variable, white, orange or yellow flowers. Several ornamental cultivars are available. *Distr.* South Africa (Cape Province).

Bulbinella (Diminutive of the genus name *Bulbine*) CAT'S TAIL. A genus of 10–20 species of perennial herbs with rosettes of succulent leaves and numerous small flowers borne in terminal racemes. Several species are grown as ornamentals. *Distr.* Africa, New Zealand. *Fam.* Asphodelaceae. *Class* M.

B. hookeri Racemes cylindrical. Flowers yellow. *Distr.* New Zealand.

bulbocodioides from the genus name *Bulbocodium*, with the ending *-oides*, indicating resemblance.

Bulbocodium (From the Greek *bolbos*, a bulb, and *kodya*, capsule) A genus of 2 species of herbs with linear leaves arising from a corm. The stalkless flowers usually appear before the leaves and have 6 spreading to erect tepals and 6 stamens. This genus is closely related to the genus *Colchicum* but differs in the shape of the style. Both species are grown as spring-flowering ornamentals. *Distr.* S and E Europe. *Fam.* Colchicaceae. *Class* M.

B. vernum Flowers usually solitary, rose to magenta, occasionally white. *Distr.* Pyrenees, Alps.

B. versicolor Very similar to *B. vernum* but with smaller flowers and narrower leaves. *Distr.* E Europe to W Asia.

bulbosus, -a, -um: bulbous.

bulgaricus, -a, -um: of Bulgaria.

bullatus, -a, -um: with puckered, blistered leaves.

bulrush *Typha latifolia, Scirpur* **lesser** ~ *T. angustifolia.*

bunchberry *Cornus canadensis.*

bunchflower *Melanthium virginicum.*

bunny ears *Opuntia microdasys.*

Buphthalmum (From the Greek *bous*, an ox, and *ophthalmos*, an eye, alluding to the appearance of the flower.) OX-EYE DAISY. A genus of 2 species of perennial herbs with simple alternate leaves and yellow flowers in large, solitary, daisy-like heads that bear distinct yellow rays. *Distr.* Europe, W Asia. *Fam.* Compositae. *Class* D.

B. salicifolium To 1.5m high. Rays to 2cm long, fine. *Distr.* Central Europe.

B. speciosum See *Telekia speciosa.*

Bupleurum (From the Greek name for a similar plant.) THOROW WAX. A genus of 70–100 species of annual and perennial herbs and shrubs with simple leaves and umbels of typically yellow flowers. The umbels are often subtended by a whorl of leafy bracts. Several species are grown as ornamentals. *Distr.* Eurasia, N Africa, Canary Islands, arctic North America, South Africa. *Fam.* Umbelliferae. *Class* D.

B. fruticosum SHRUBBY HARE'S EAR. Evergreen bushy shrub. Leaves dark blue-green. Flowers yellow. *Distr.* S Europe, Mediterranean region.

B. rotundifolium THOROW WAX. Annual or short-lived perennial. Flowers green-yellow. *Distr.* Europe, naturalized North America.

Burchardia (After H. Burchard, an 18th-century botanist.) MILKMAIDS. A genus of 1–3 species of rhizomatous herbs with linear leaves and umbels of fragrant white flowers. *Distr.* Temperate Australia. *Fam.* Colchicaceae. *Class* M.

B. umbellata Occasionally grown as a tender ornamental.

burhead *Echinodorus.*

burmanicus, -a, -um: of Burma.

Burmanniaceae A family of 15 genera and about 130 species of small, annual and perennial herbs many of which are saprophytic with much reduced leaves. The flowers are usually regular and bear 2 whorls of 3 fused, colourful tepals. This family is closely related to the Orchids (Orchidaceae). *Distr.* Widespread throughout the tropics. *Class* M.

burnet *Sanguisorba, S. minor, S. officinalis* **Canadian** ~ *S. canadensis* **garden** ~ *S. minor* **great** ~ *S. officinalis.*

burnet saxifrage *Pimpinella saxifraga* **greater** ~ *P. major.*

burning bush *Dictamnus albus.*

burrawong *Macrozamia communis.*

Bursaria (From the Latin *bursa*, a purse, alluding to the shape of the fruits.) A genus of 3 species of spiny evergreen shrubs and small trees with simple alternate leaves and panicles of small flowers that bear their parts in 5s. The fruit is a flattened capsule. *Distr.* Australia. *Fam.* Pittosporaceae. *Class* D.

B. spinosa Shrub or small tree. Flowers white, fragrant. Occasionally grown as a half-hardy ornamental. *Distr.* Australia (New South Wales, Tasmania).

Burseraceae The Frankincense family. 18 genera and 540 species of trees and shrubs rich in resin. The leaves are spirally arranged, pinnate, and usually clustered at the branch tips. The flowers are small and have their parts in 3s, 4s or 5s. Some species are a source of timber. FRANKINCENSE and MYRRH are extracted from members of *Boswellia* and *Commiphora* respectively. *Distr.* Widespread in tropical regions. *Class* D. *See: Boswellia, Commiphora.*

burweed *Medicago.*

busy lizzie *Impatiens, I. walleriana.*

butcher's broom Ruscaceae.

Butia (The Brazilian name for *B. capitata*.) YATAY PALM, JELLY PALM. A genus of 8–12 species of medium-sized palms with feather-shaped leaves and spikes of male and female, red, yellow or purple flowers. Several species are grown as ornamentals. *Distr.* Tropical and subtropical South America. *Fam.* Palmae. *Class* M.

B. monosperma BASTARD TEAK, FLAME OF THE FOREST. Flowers bright red-orange; used as a red dye. *Distr.* India to Burma.

Butomaceae A family of 1 genus and 1 species. *Class* M.
See: Butomus.

Butomus (From the Greek, *bous*, an ox, and *temmo*, to cut, alluding to the sharp-edged leaves that cannot be eaten by cattle.) A genus of 1 species of aquatic, rhizomatous herb with twisted, 3-angle leaves that are up to 1.5m long. The pink, fragrant flowers are borne in a large umbel. The root-stocks are reported to be edible. *Distr.* Temperate regions of Europe and Asia. *Fam.* Butomaceae. *Class* M.

B. umbellatus FLOWERING RUSH, GRASSY RUSH, WATER GLADIOLUS. Rhizomes powdered for bread in N Eurasia. Often grown as a water garden ornamental.

butter and eggs *Linaria vulgaris.*

butterbur *Petasites, P. hybridus.*

buttercup Ranunculaceae, *Ranunculus* **Colombian** ~ *Oncidium cheirophorum* **creeping** ~ *Ranunculus repens* **meadow** ~ *R. acris.*

butterfly bush Buddlejaceae, *Buddleja, B. davidii*

butterfly weed *Asclepias tuberosa.*

butternut *Juglans cinerea.*

butter nut Caryocaraceae

butter print *Abutilon theophrasti.*

butterwood tree *Platanus.*

butterwort Lentibulariaceae, *Pinguicula, P. vulgaris.*

button bush *Cephalanthus occidentalis.*

button flower *Hibbertia.*

button: bachelor's ~ *Centaurea cyanus, Tanacetum parthenium, Craspedia, C. glauca* **blue** ~ *Knautia arvensis, Succisa pratensis* **brass** ~ *Cotula, C. coronpifolia* **common billy** ~ *Craspedia glauca* **frog's** ~ *Lemna* **golden** ~ *Tanacetum vulgare* **red** ~ *Hoya pubicalyx* **yellow bachelor's** ~ *Ranunculus acris.*

buttonwood *Platanus occidentalis.*

Buxaceae The Box family. 5 genera and 60 species of evergreen shrubs plus a few herbs and small trees with simple, leathery leaves and small regular, unisexual flowers which lack petals. Several species are grown ornamentally or as a source of wood. *Distr.* Widespread, except Australasia and the Pacific. *Class* D.
See: Buxus, Pachysandra, Sarcococca.

buxifolius: from the genus name *Buxus*, and Latin *folius*, leaved.

Buxus (The classical Latin name for *B. sempervirens*.) BOX, BOXWOOD. A genus of 30 species of evergreen shrubs and trees with simple, opposite, leathery leaves and short racemes of small, unisexual, yellow-green flowers. Several species are used for ornamental hedging and topiary. Some species have medicinal qualities. *Distr.* Mediterranean, temperate E Asia, West Indies, Central America. *Fam.* Buxaceae. *Class* D.

B. aurea See *B. sempervirens*

B. balearica BALEARIC BOXWOOD. Shrub or small tree. Leaves oval, bright green. *Distr.* Balearic Islands, SW Spain, N Africa.

B. microphylla JAPANESE BOX. Compact rounded shrub. Twigs 4-angled. Leaves thin. Numerous garden cultivars are available. *Distr.* Origin confused, possibly China or Korea; cultivated in Japan since 15th century. Not known in the wild.

B. sempervirens EUROPEAN BOX, COMMON BOX. Shrub or small tree to 8m high. Source of the hardest and heaviest timber in N Europe now largely replaced by the use of VENEZUELAN BOX, *Casearia praecox*. Numerous garden cultivars are available. *Distr.* Europe, Mediterranean, N Africa, W Asia.

B. wallichiana HIMALAYAN BOX. Shrub or small tree. Leaves dark green. *Distr.* NW Himalaya, India.

Byblidaceae A family of 1 genus and 2 species of insectivorous herbs with spirally arranged, linear leaves and flowers which bear 5 sepals, 5 free or basally fused petals and 5 stamens. *Distr.* New Guinea, Australia. *Class* D.

byzantinus, -a, -um: of Istanbul (Byzantium), Turkey.

C

cabbage *Brassica oleracea* **cattle** ~ *B. oleracea*; **John's** ~ *Hydrophyllum virginianum* **palm** ~ *Sabal palmetto* **Savoy** ~ *Brassica oleracea* **skunk** ~ *Lysichiton, L. americanus. Symplocarpus foetidus.*

cabbage tree *Sabal palmetto, Cordyline.*

Cabombaceae A family of 2 genera and 8 species of rhizomatous, aquatic herbs with finely divided underwater leaves and simple floating leaves. The solitary flowers typically bear 3 sepals, 3–18 stamens and 2–18 free carpels that mature into leathery capsules. This family is closely related to the family Nymphaceae but differs in not having the carpels fused into an ovary. *Distr.* Tropical to warm temperate regions. *Class* D.

cacao: the ancient Aztec name for *Theobroma cacao.*

cachemiricus, -a,-um: of Kashmir, on the India, Pakistan borders.

Cactaceae The Cactus family. 100 genera and about 1500 species of subshrubs, shrubs and trees of dry areas which are nearly always succulent and usually leafless. The leaves are often replaced by spines which appear to arise from a cushion, the areole, which is thought to represent a condensed lateral branch. The areoles are often borne on ridges or atop small mounds (tubicles). The flowers are solitary, sessile and bisexual and bear numerous sepals, petals and stamens. The fruit is typically a juicy berry. Many species are grown ornamentally. *Distr.* New World. *Class* D.
See: Acanthocereus, Discocactus, Lepismium, Maihuenia, Mammillaria, Matucana, Melocactus, Mila, Monvillea, Myrtillocactus, Neolloydia, Neoporteria, Neoraimondia, Neowerdermannia, Nopalxochia, Notocactus, Nyctocereus, Obregonia, Opuntia, Oreocereus, Oroya, Pachycereus, Parodia, Pediocactus, Pelecyphora, Peniocereus, Pereskia, Pfeiffera, Pterocactus, Thelocactus.

cactus Cactaceae **bird** ~ *Pedilanthus tithymaloides* **cane** ~ *Opuntia cylindrica* **chain-link** ~ *O. imbricata* **coral** ~ *Mammillaria heyderi* **corncob** ~ *Euphorbia mammillaris* **cotton pole** ~ *Opuntia vestita* **Eve's pin** ~ *O. subulata* **feather** ~ *Mammillaria plumosa* **gold lace** ~ *M. elongata* **old woman** ~ *M. hahniana* **ribbon** ~ *Pedilanthus tithymaloides* **rose** ~ *Pereskia grandifolia* **silver cluster** ~ *Mammillaria prolifera* **snowball** ~ *M. bocasana.*

caerulescens: becoming blue.

caeruleus, -a -um: dark blue.

Caesalpinia (After Andrea Cesalpini (1519–1603), Italian botanist physician to Pope Clement VIII.) A genus of about 100 species of herbs, shrubs, trees and climbers with alternate, bipinnate leaves and irregular flowers that bear 10 prominent stamens. Several species are grown ornamentally and a few are a source of timber. *Distr.* Tropical and subtropical regions. *Fam.* Leguminosae. *Class* D.

C. gilliesii BIRD OF PARADISE. Shrub or small tree. Flowers yellow, in erect racemes. Stamens red. *Distr.* Argentina, Uruguay.

C. pulcherrima BARBADOS PRIDE, PEACOCK FLOWER, FLAMBOYANT TREE, PARADISE FLOWER. Shrub or small tree. Flowers yellow-orange. Stamens very long, red. *Distr.* Tropical America.

caesiiglauca: lavender-blue coloured, with a slight bloom.

caesius, -a, -um: a shade of lavender blue.

caespititius tufted.

caespitosus, -a, -um: growing in tufts.

caffrum: of South Africa.

calabash, sweet: *Passiflora maliformis.*

calabrese *Brassica oleracea.*

calabricus, -a, -um: of Calabria, Italy.

calabriensis, -e: of Calabria, Italy.

Caladium (Latinized form of the Malay plant name *kaladi.*) ANGEL WINGS, ELEPHANT'S EARS. A genus of 7–8 species of tuberous herbs with large, variegated leaves and small flowers borne on a spadix that is surrounded and exceeded by a green-white spathe. A number of species are cultivated as house plants for their brightly coloured foliage. *Distr.* N South America, naturalized throughout the tropics. *Fam.* Araceae. *Class* M.

C. bicolor This species has given rise to numerous cultivars with a great variety of foliage, shape and colour. *Distr.* Australia, Brazil.

C. × hortulanum See *C. bicolor*

caladium, black *Colocasia esculenta.*

Calamagrostis (From the Greek *kalamos*, a reed, and *agrostis*, a kind of grass.) REED GRASS. A genus of about 250 species of rhizomatous, perennial grasses of damp areas with long, channelled leaves and dense, compressed panicles of spikes. *Distr.* Temperate regions. *Fam.* Gramineae. *Class* M.

C. × acutiflora FEATHER REED. A hybrid of *C. arundinacea* and *C. epigejos*. Grown as an ornamental for its bronze inflorescences.

calamint *Calamintha* **Alpine** ~ *Acinos alpinus* **common** ~ *Calamintha sylvatica* **cushion** ~ *Clinopodium vulgare* **large-flowered** ~ *Calamintha grandiflora* **lesser** ~ *C. nepeta* **wood** ~ *C. sylvatica*

Calamintha (From the Greek *kallis*, beautiful, and *mintha*, mint.) CALAMINTHA. A genus of about 7 species of aromatic, perennial herbs with simple, opposite leaves and stalked cymes of red-violet, 2-lipped flowers. Several species are grown ornamentally. *Distr.* Europe to Central Asia. *Fam.* Labiatae. *Class* D.

C. alpina See *Acinos alpinus*

C. clinopodium See *Clinopodium vulgare*

C. cretica Densely hairy, prostrate perennial. Flowers white. *Distr.* Crete.

C. grandiflora LARGE-FLOWERED CALAMINT. Rhizomatous perennial. Flowers large, pink. *Distr.* S and SE Europe.

C. nepeta LESSER CALAMINT. Erect perennial. Leaves strongly aromatic. Flowers small, mauve. *Distr.* S and W Europe.

C. sylvatica WOOD CALAMINT, COMMON CALAMINT. Erect perennial to 60cm high. Leaves mint scented. Flowers small, pink marked with purple. *Distr.* Europe.

calaminthifolius, -a, -um: from the genus *Calamintha*, and *folius*, leaved.

calamondin × *Citrofortunella microcarpa.*

calamus *Acorus calamus.*

calamus: from the Greek *calamus*, reed.

Calandrinia (After Jean Louis Calandrini (1703–1758), professor of mathematics and philosophy at Geneva.) ROCK PURSLANE. A genus of about 150 species of annual and perennial herbs with narrow, alternate leaves and racemes or panicles of red, purple or white flowers that bear 2 sepals and 5–7 petals. Several species are grown as ornamentals and some are edible. *Distr.* South America, Australia and W North America. *Fam.* Portulacaceae. *Class* D.

C. discolor Annual or perennial herb. Flowers light purple. *Distr.* Chile.

C. grandiflora Perennial herb to 1m high. Flowers purple to magenta, borne in racemes. *Distr.* Chile.

calandrinioides: from the genus *Calandrinia*, with the ending *-oides*, indicating resemblance.

Calanthe (From the Greek *kalos*, beautiful, and *anthos*, a flower.) A genus of about 120 species of terrestrial and a few epiphytic orchids with small pseudobulbs. The sepals and petals are elliptic and similar, and the lip is 3-lobed with a deeply notched central lobe. An important ornamental group. Some of the first artificial ORCHID., hybrids were made within this genus. *Distr.* Madagascar to Himalaya, SE Asia, Japan and Australia. *Fam.* Orchidaceae. *Class* M.

C. bicolor See *C. discolor*

C. discolor Flowers to 5cm across, maroon to white, lip pale pink. *Distr.* Japan, Korea and Taiwan.

C. sieboldii See *C. striata*

C. striata Flowers to 6.5cm across, yellow-bronze. *Distr.* Japan.

C. tricarinata Flowers yellow-green with a red-brown lip. *Distr.* Himalaya, China, Japan and Taiwan.

Calathea (From the Greek *kalathos*, a basket.) A genus of 300 species of large, rhizomatous or tuberous herbs with elliptic to ovate, frequently patterned leaves and spikes of small flowers. Some species have edible tubers and some are grown for their ornamental foliage. *Distr.* Tropical America. *Fam.* Marantaceae. *Class* M.

C. makoyana PEACOCK PLANT, CATHEDRAL WINDOWS, BRAIN PLANT. Leaves ovate, dark green with pale green or cream markings. *Distr.* E Brazil.

C. oppenheimiana See *Ctenanthe oppenheimiana*

C. zebrina ZEBRA PLANT. Leaves velvety dark green with pale green-yellow veins. *Distr.* E Brazil.

calcaratus, -a, -um: spurred.

calcareus, -a, -um: of chalky places.

Calceolaria (From the Latin *calceolus*, a slipper, alluding to the shape of the lower lip of the flower.) SLIPPER FLOWER. A genus of about 300 species of annual to perennial herbs and shrubs with simple or pinnate, stalked leaves and irregular cymes of 2-lipped flowers in which the lower lip is inflated into a large pouch. Numerous species and cultivars are grown ornamentally. *Distr.* Central and South America. *Fam.* Scrophulariaceae. *Class* D.

C. arachnoidea Clump-forming perennial. Leaves covered in long, white, downy hairs. Flowers dull purple. *Distr.* Chile.

C. bicolor Subshrub. Flowers pale yellow, white and sulphur yellow. *Distr.* Peru.

C. biflora Perennial. Flowers yellow. *Distr.* Chile, Argentina.

C. darwinii Clump-forming, short-lived perennial. Flowers yellow with dark brown spots. *Distr.* Tierra del Fuego.

C. integrifolia Upright subshrub, grown as an annual. Flowers yellow to red-brown. *Distr.* Chile.

calceolus, -a, -um: slipper-shaped.

calcicola: growing in chalky places.

calderiana: from a hot place.

Calectasiaceae A family of 1 genus and 1 species of rhizomatous subshrubs. *Distr.* Australia. *Class* M.

Calendula (From the Latin *calendae*, the day of the month on which interest is paid; the flowers are reputed to last a month.) A genus of 20 species of annual and perennial herbs with simple, alternate, often aromatic leaves, and flowers in daisy-like heads that bear distinct rays. Several species and numerous cultivars, particularly of *C. officinalis*, are grown as ornamental. *Distr.* Mediterranean, Macaronesia. *Fam.* Compositae. *Class* D.

C. officinalis COMMON MARIGOLD, RUDDLES, POT MARIGOLD, SCOTCH MARIGOLD. Annual to perennial herb. Flowers yellow, rays yellow-orange. An important summer bedding plant which also has medicinal qualities, being effective against chilblains and warts, and is used to colour butter and thicken soups. *Distr.* Origin unknown.

calendulaceus, -a -um: resembling the genus *Calendula*.

calico bush *Kalmia latifolia*

calico flower *Aristolochia littoralis*.

californicus, -a, -um: of California, USA.

Calla (From the Greek *kallos*, beauty.) A genus of 1 species of aquatic or marginal herb of bogs and wetlands with oval to kidney-shaped leaves on long leaf-stalks. The small flowers are borne on a short spadix that is subtended by a leaf-like spathe. *Distr.* N temperate regions. *Fam.* Araceae. *Class* M.

C. aethiopica See *Zantedeschia aethiopica*

C. palustris WATER ARUM, BOG ARUM, WILD CALLA, WATER DRAGON. Grown as an ornamental.

calla: red ~ *Sauromatum venosum*
wild ~ *Calla palustris*.

Calliandra (From the Greek *kalos*, beautiful, and *aner*, man, alluding to the showy stamens.) POWDER PUFF TREE. A genus of 200 species of perennial herbs, shrubs and small trees with bipinnate, alternate leaves and clusters or heads of regular flowers which bear 10 to many showy stamens. Several species are grown as ornamentals. *Distr.* Tropical

America, Madagascar, India. *Fam.* Leguminosae. *Class* D.

C. eriophylla FAIRY DUSTER. Shrub to 1m high. Flowers minute, forming pompon-like, pink heads. *Distr.* SW USA, N Mexico.

C. haematocephala Shrub or small tree. Flowers in pink heads 6cm across. *Distr.* South America.

Callianthemum (From the Greek *kallos*, beauty, and *anthemon*, a flower.) A genus of about 10 species of rhizomatous herbs with pinnately divided, fern-like basal leaves and regular flowers that bear 5 sepals, 5–20 showy petals and numerous stamens. Several species are grown ornamentally. *Distr.* Central Europe, Central and E Asia. *Fam.* Ranunculaceae. *Class* D.

C. coriandrifolium Prostrate perennial. Leaves blue-green, divided many times. Flowers white with a yellow centre. *Distr.* S Europe.

callianthus, -a, -um: with beautiful flowers.

Callicarpa (From the Greek *kallos*, beauty, and *karpos*, fruit.) A genus of about 140 species of shrubs and trees with simple, opposite leaves and cymes of small, tubular or bell-shaped, 4-lobed flowers. Several species are grown as ornamentals. *Distr.* Tropical and subtropical regions. *Fam.* Verbenaceae. *Class* D.

C. bodinieri Shrub to 3m high. Flowers lilac. Used medicinally and as a fish poison. *Distr.* China.

callimorphus, -a, -um: beautifully shaped.

Callirhoe (After Callirhoe, in Greek mythology, daughter of the river god *Achelous*.) POPPY MALLOW. A genus of 8 species of annual and perennial herbs with palmately lobed or divided leaves and regular, bisexual flowers that bear 5 petals and 10 fused stamens. Several species are grown as ornamentals. *Distr.* North America, Mexico. *Fam.* Malvaceae. *Class* D.

C. digitata FRINGED POPPY MALLOW. Erect perennial herb. Flowers pink to purple. *Distr.* S USA.

C. involucrata PURPLE POPPY MALLOW. Procumbent or ascending perennial. Flowers red or purple, to 5cm across. *Distr.* S USA.

Callisia (From the Greek *kallos*, beauty.) A genus of 20 species of perennial or occasionally annual herbs with succulent leaves and regular white or pink flowers. Several species are grown as ornamentals, chiefly for their foliage. *Distr.* Tropical America. *Fam.* Commelinaceae. *Class* M.

C. repens Creeping perennial with rounded, often white-banded leaves.

callistegioides from the genus *Calystegia*, with the ending *-oides*, indicating resemblance.

Callistemon (From the Greek *kallos*, beauty, and *stemon*, stamen.) BOTTLE-BRUSH BUSH. A genus of about 20 species of evergreen shrubs and trees with narrow, alternate, often leathery leaves and spike-like clusters of small, regular flowers that bear 5 sepals, 5 petals and numerous prominent stamens. A number of species are grown as ornamentals. *Distr.* Australia, New Caledonia. *Fam.* Myrtaceae. *Class* D.

C. citrinus CRIMSON BOTTLEBRUSH. Large shrub or tree. Flowers green or red. Stamens bright red.

C. pallidus LEMON BOTTLEBRUSH. Arching shrub. Flowers in spikes. Stamens yellow.

C. paludosus See *C. sieberi*

C. rigidus STIFF BOTTLEBRUSH. Bushy shrub. Flowers in stiffly erect spikes. Stamens deep red.

C. sieberi ALPINE BOTTLEBRUSH. Dwarf shrub. Leaves sharply pointed. Flowers in spikes. Stamens yellow.

C. viminalis WEEPING BOTTLEBRUSH. Arching shrub. Flowers in spikes. Stamens bright red.

Callitrichaceae A family of 1 genus and 17 species. *Class* D.
See: Callitriche.

Callitriche STARWORT. A genus of 17 species of delicate aquatic herbs, some of which are entirely submerged, with opposite or rosette-forming leaves. The flowers are minute, unisexual and solitary. Some species are sensitive to impurities in the water and have been used in pollution detection. *Distr.* Widespread. *Fam.* Callitrichaceae. *Class* D.

C. stagnalis WATER STARWORT. Grown as herbarium ornamental. *Distr.* Europe and the Mediterranean.

callitrichoides: from the genus *Callitriche*, with the ending *-oides*, indicating resemblance.

Callitris (From the Greek *kallos*, beauty, and *tris*, three, alluding to the whorls of 3 leaves.) CYPRESS PINE. A genus of 14 species of evergreen coniferous trees with needle-like juvenile leaves borne in whorls of 4, and scale-like adult leaves borne in whorls of 3. Some species are used as a source of timber and resin. *Distr.* Australia. *Fam.* Cupressaceae. *Class* G.

C. rhomboidea ILLAWARA PINE, PORT JACKSON PINE. Shrub or tree to 15m high. *Distr.* SE Australia.

callizonus: with beautiful markings.

callosus, -a, -um: calloused; with a hardened or thickened area.

Calluna (From Greek kallunein, to beautify, referring to stems used as brooms.) A genus of 1 species of low, evergreen shrub with small leaves that are tightly appressed to the stem, and small, pink, 4-lobed flowers. *Distr.* Europe, Turkey, naturalized in North America. *Fam.* Ericaceae. *Class* D.

C. vulgaris SCOTTISH HEATHER, LING. An important ornamental with over 1000 cultivars available. This species is of ecological importance as an understorey shrub and on managed moors. It has also been used as a source of yellow dye.

Calocedrus (From the Greek *kalos*, beautiful, and *kedros*, a cedar.) A genus of 3 species of evergreen, coniferous trees with scale-like leaves borne in whorls of 4. *Distr.* E Asia, W North America. *Fam.* Cupressaceae. *Class* G.

C. decurrens INCENSE CEDAR. Slender tree to 45m high. A source of timber. *Distr.* NW USA.

Calocephalus (From the Greek *kallos*, beautiful, and *kephale*, head.) A genus of 18 species of annual and perennial herbs and small shrubs with simple leaves and flowers in daisy-like heads that lack distinct rays. *Distr.* Temperate Australia. *Fam.* Compositae. *Class* D.

C. brownii Small, much-branched shrubs. Flowers white. Grown as a tender ornamental. *Distr.* Coastal temperate Australia.

Calochortus (From the Greek *kalos*, beautiful, and *chortos*, grass.) MARIPOSA LILY, FAIRY LANTERN, CAT'S EAR. A genus of 57 species of bulbiferous herbs with sword-shaped leaves and cymes of nodding or erect, saucer-shaped flowers that bear 3 sepals, 3 petals and 6 stamens. Some species are grown as ornamentals and some have edible flowers. *Distr.* W North and Central America. *Fam.* Liliaceae. *Class* M.

C. albus Flowers nodding, white, globose. *Distr.* USA (California).

C. amabilis GOLDEN FAIRY LANTERN, GOLDEN GLOBE TULIP. Flowers nodding, fringed, deep yellow. *Distr.* USA (California).

C. luteus Flowers open bell-shaped, petals yellow with a brown blotch, reflexed. *Distr.* W USA.

C. splendens Flowers erect, pale purple. *Distr.* USA (California), Mexico (Baja California).

C. superbus Flowers erect, cream to lilac marked brown. *Distr.* USA (California).

C. venustus Flowers erect, white with an orange-brown centre. *Distr.* USA (California).

Calomeria (From the Greek *kalos*, beautiful, and *meris*, portion.) A genus of 14 species of annual to perennial herbs with simple, overlapping leaves and flowers in daisy-like heads that lack distinct rays. *Distr.* Africa, Madagascar, S Australia. *Fam.* Compositae. *Class* D.

C. amaranthoides INCENSE PLANT, PLUME BUSH. Aromatic biennial or perennial to 2m high. *Distr.* Australia (New South Wales, Victoria).

calophytum: beautiful plant.

caloptera: beautifully winged.

Caloscordum (From the Greek *kalos*, beautiful, and *scordum*, GARLIC.) A genus of 1 species of bulbiferous herb with linear leaves and an umbel of star-shaped, pink flowers. *Distr.* N China to SE Siberia. *Fam.* Alliaceae. *Class* M.

C. neriniflorum A cultivated ornamental.

calostrotus, -a, -um: with beautiful covering.

Calothamnus (From the Greek *kalos*, beautiful, and *thamnos*, bush.) A genus of 24 species of evergreen shrubs with simple, rigid leaves, and spikes or clusters of small, regular flowers that bear 4–5 petals and 4–5 prominent stamens. Several species are grown as tender ornamentals. *Distr.* W Australia. *Fam.* Myrtaceae. *Class* D.

C. villosus WOOLLY NETBUSH. Bushy shrub. Leaves needle-like. Flowers in dense clusters. Stamens red. *Distr.* W Australia.

caloxanthum: a beautiful yellow colour.

Caltha (From the classical Latin name for a plant with a yellow flower.) MARSH MARIGOLD, KINGCUP. A genus of about 10 species of perennial, water-side herbs with simple, alternate, stalked leaves and regular, bisexual flowers that bear 5 petal-like sepals and numerous stamens but lack true petals. Several species are grown as ornamentals. *Distr.* Temperate regions of both hemispheres. *Fam.* Ranunculaceae. *Class* D.

C. leptosepala Flowers white with a yellow centre. Sepals narrow. *Distr.* W North America.

C. palustris KINGCUP, MARSH MARIGOLD, MEADOW BRIGHT, MAYBLOB. Flowers bright golden-yellow. *Distr.* N temperate regions.

calthifolium: from the genus *Caltha*, and *folius*, leaved.

caltrop Zygophyllaceae **water** ~ *Trapa natans*.

calvescens: becoming bald.

Calycanthaceae A family of 3 genera and 9 species of deciduous or evergreen shrubs and small trees with simple opposite leaves and solitary, fragrant flowers that bear 15–30 petal-like tepals, 5–30 stamens and 5–35 free carpels. Some species are grown as ornamentals, others are used locally as spices. *Distr.* S USA and China. *Class* D.
See: Calycanthus, Chimonanthus.

Calycanthus (From the Greek, *kalyx*, calyx (the sepals), and *anthos*, a flower, alluding to the lack of differentiation between sepals and petals.) CAROLINA ALLSPICE. A genus of 2–5 species of deciduous shrubs with aromatic bark, simple, opposite leaves and solitary, regular flowers that bear numerous petal-like tepals. The fruit is a hard, round capsule. Several species are grown as ornamentals. *Distr.* North America and Australia. *Fam.* Calycanthaceae. *Class* D.

C. floridus CAROLINA ALLSPICE, STRAWBERRY SHRUB. Bushy shrub. Leaves oval. Flowers red-brown. The bark has been used medicinally and as a substitute for CINNAMON. *Distr.* E North America.

C. occidentalis CALIFORNIAN ALLSPICE. Bushy shrub. Leaves large. Flowers purple-red. *Distr.* USA (California).

Calyceraceae A family of 6 genera and 55 species of annual and perennial herbs with alternate, entire to pinnate leaves, and heads of small flowers that are surrounded by a whorl of bracts. The flowers typically bear 5 free sepals, 5 fused petals, 5 stamens and an inferior ovary. *Distr.* Tropical and temperate South America. *Class* D.

calycinus, -a, -um: with a well-developed whorl of sepals.

Calyptridium umbellatum See *Spraguea umbellata*

Calystegia (From the Greek *kalyx*, calyx and *stege*, cover, alluding to the two bracts that subtend the flowers.) BIND-WEED. A genus of about 25 species of scrambling and climbing, rhizomatous herbs with simple, stalked leaves and large funnel-shaped flowers. Several species are grown as ornamentals. *Distr.* Cosmopolitan. *Fam.* Convolvulaceae. *Class* D.

C. hederacea Leaves narrowly arrowhead-shaped. Flowers red-pink. *Distr.* E Asia, naturalized North America.

C. macrostegia Leaves somewhat fleshy. Flowers white, aging to pink. *Distr.* SW North America.

Calytrix (From the Greek *kalyx*, bud, and *thrix*, hair.) A genus of about 40 species of prostrate and erect shrubs with small leaves and solitary flowers that bear 5 sepals, 5 petals, and numerous stamens. Several species are grown as half-hardy ornamentals. *Distr.* Australia. *Fam.* Myrtaceae. *Class* D.

C. alpestris SNOW MYRTLE. Shrub to 2.5m high. Flowers white to pink. *Distr.* S Australia.

camas *Camassia* **death** ~ *Zigadenus nuttallii* **poison** ~ *Z. nuttallii* **white** ~ *Z. elegans*.

Camassia (From the North American Indian name, *Quamash*.) CAMAS, BEAR GRASS.

A genus of 5 species of bulbiferous herbs with narrow, channelled leaves and racemes of star-shaped, white to purple flowers. Several species are grown as ornamentals. *Distr.* North and South America. *Fam.* Hyacinthaceae. *Class* M.

C. esculenta See *C. quamash*

C. fraseri See *C. scilloides*

C. leichtlinii Flowers usually white, becoming twisted as they wither. *Distr.* W North America.

C. quamash QUAMASH, CAMOSH. Bulbs edible. *Distr.* W North America.

C. scilloides WILD HYACINTH, INDIGO SQUILL. Bulb with a black tunic. Flowers blue. *Distr.* Central and E North America.

cambricus, -a, -um: of Wales.

Camellia (After Georg Kamel (1661–1706), Moravian Jesuit and botanist who travelled in Asia.) A genus of 82 species of evergreen shrubs and trees with simple, alternate leaves and regular flowers that bear 5–6 sepals, 5–12 petals and numerous stamens. Cultivated forms often bear large, double flowers. This is a very important cultivated genus with many species and numerous cultivars that are grown as ornamentals. *C. sinensis* is the source of the beverage Tea. *Distr.* India to Indonesia. *Fam.* Theaceae. *Class* D.

C. japonica Shrub or small tree to 15m high. Petals 5–6, red. Over 2000 cultivars have been raised from this species with a great diversity of flower colour and form. The seed is a source of hair oil (tsubaki oil) used by Japanese women. *Distr.* Korea, Japan, Taiwan.

C. reticulata Loosely branching shrub to 10m high. Numerous ornamental cultivars have been raised from this species including those with the largest flowers of the genus. *Distr.* W China.

C. sinensis TEA. Shrub or small tree to 7m high. Leaves dark green, leathery. Flowers white. The young leaves are hand picked and dried to provided Tea, the most important caffeine-rich beverage in the world. *Distr.* S and E Asia.

C. thea See *Camellia sinensis*

camellia Theaceae **mountain** ~ *Stewartia ovata* **silky** ~ *S. malacodendron.*

camomile: German ~ *Matricaria recutita* **scentless false** ~ *Tripleurospermum* **sweet false** ~ *Matricaria recutita* **wild** ~ *M. recutita.*

camosh *Camassia quamash.*

campaniflorus, -a, -um: with bell-shaped flowers.

Campanula (From the diminutive of the Latin *campana*, bell, alluding to the flowers.) BELLFLOWER. A genus of 300 species of annual, biennial and perennial herbs with simple, alternate, sessile leaves and panicles or racemes of delicate bell-shaped flowers which bear 5 fused sepals, 5 fused petals, 5 free stamens and an inferior ovary. Many species are grown as ornamentals. *Distr.* N temperate regions especially around the Mediterranean. *Fam.* Campanulaceae. *Class* D.

C. alaskana See *C. rotundifolia*

C. alliariifolia Perennial. Leaves heart-shaped. Flowers nodding, white, borne in spikes. *Distr.* Turkey, Caucasus.

C. americana Annual or biennial. Stems erect, to 2m high. Flowers pale blue. Has been considered as a possible source of oil and rubber. *Distr.* E North America.

C. barbata BEARDED BELLFLOWER. Perennial. Leaves in a basal rosette. Flowers small, nodding, in 1-sided racemes, lavender, ciliate. *Distr.* Alps, Norway.

C. carpatica TUSSOCK BELLFLOWER. Clump-forming perennial. Grown as a rock-garden plant. Cultivars are available in a number of flower colours. *Distr.* Carpathian Mountains.

C. cochleariifolia FAIRIES' THIMBLES. Mat-forming perennial. Leaves small, round borne in rosettes. Flowers clustered, white or blue. Several cultivars are available. *Distr.* Mountainous areas of Europe.

C. eriocarpa See *C. latifolia*

C. glomerata CLUSTERED BELLFLOWER. Stems rigid, erect. Flowers in crowded spikes, violet-blue or white. *Distr.* Europe, temperate Asia.

C. isophylla ITALIAN BELLFLOWER, STAR OF BETHLEHEM, FALLING STARS. Dwarf, trailing perennial. Flowers large, erect, borne in loose clusters, violet to white. *Distr.* N Italy.

C. lactiflora Stems robust, erect, to 1.5m high. Flowers erect, borne in broad panicles, blue. Several cultivars are available. *Distr.* Caucasus.

C. latifolia Perennial. Stems, simple, erect, robust, to 1m high. Leaves softly hairy. Flowers pale blue-purple. *Distr.* Eurasia.

C. medium CANTERBURY BELLS, CUP AND SAUCER. Biennial. Lower leaves in rosettes. Stems much branched. Flowers typically blue. Numerous ornamental cultivars have been raised from this species. *Distr.* S Europe.

C. muralis See *C. portenschlagiana*

C. nitida See *C. persicifolia*

C. persicifolia WILLOW BELLFLOWER, PEACH BELLS. Perennial. Stems erect. Flowers nodding, borne in slender racemes or solitary. Very many cultivars are available. *Distr.* Europe, N Africa, W Asia.

C. planiflora See *C. persicifolia*

C. portenschlagiana Finely hairy, tuft-forming perennial. Flowers numerous, erect, purple. *Distr.* S Europe.

C. poscharskyana Tuft-forming perennial. Flowers almost star-shaped. *Distr.* E Alps.

C. pusilla See *C. cochleariifolia*

C. pyramidalis CHIMNEY BELLFLOWER. Short-lived perennial. Flowers fragrant, borne in a very large pyramidal panicle that can reach 1.5m high. *Distr.* S Central Europe.

C. rapunculoides A noxious weed with brittle rhizomes. *Distr.* Eurasia.

C. rapunculus RAMPION. Biennial. Stems erect, to 1m high. Flowers small, white or pale blue. The parsnip-like roots were formerly eaten as a vegetable but the most suitable cultivars seem to have been lost in the early 19th century. *Distr.* Eurasia.

C. rhomboidalis See *C. rapunculoides*

C. rotundifolia HAIRBELL, SCOTTISH BLUEBELL. Small perennial. Rhizomes creeping. Flowers pendulous, white to deep blue. *Distr.* N temperate regions.

C. trachelium NETTLE-LEAVED BELLFLOWER, THROATWORT. Perennial. Leaves rough, pointed, toothed. Flowers in panicles, purple-blue. *Distr.* S Europe, N Africa.

C. versicolor Perennial. Stem erect. Flowers fragrant, blue and white. The somewhat leathery leaves are occasionally eaten. *Distr.* SE Europe.

C. vidalii See *Azorina vidalii*

Campanulaceae The Bellflower family. 84 genera and about 1800 species of annual to perennial herbs, shrubs, small trees and climbers, with simple or occasionally pinnate leaves. The flowers are typically regular, bell-shaped and bear 5 fused sepals, 5 fused petals and 5 stamens. Many of the genera contain ornamental species and a few that have medicinal uses. *Distr.* Widespread. *Class* D. *See: Adenophora, Asyneuma, Azorina, Campanula, Canarina, Codonopsis, Cyananthus, Diosphaera, Edraianthus, Hypsela, Jasione, Laurentia, Lobelia, Michauxia, Musschia, Ostrowskia, Physoplexis, Phyteuma, Platycodon, Pratia, Symphyandra, Trachelium, Wahlenbergia.*

campanularia: referring to the genus *Campanula*.

campanulatus, -a, -um: bell-shaped.

campanuloides: from the genus *Campanula*, with the ending *-oides*, indicating resemblance.

campestris, -e: of fields or meadows.

camphor *Cinnamomum camphora*.

camphoratus, -a, -um: resembling camphor.

campion *Silene* **Alpine** ~ *Lychnis alpina* **bladder** ~ *Silene vulgaris* **moss** ~ *S. acaulis* **red** ~ *S. dioica* **rose** ~ *Lychnis coronaria, L.* **sea** ~ *Silene uniflora* **white** ~ *S. latifolia*.

Campsis (From the Greek *kampsis*, bending, alluding to the curved stamens.) TRUMPET CREEPER. A genus of 2 species of deciduous woody climbers with pinnate leaves and curved, funnel- or bell-shaped, orange flowers. Both species and several cultivars are grown as hardy ornamentals. *Distr.* E Asia. *Fam.* Bignoniaceae. *Class* D.

C. grandiflora CHINESE TRUMPET FLOWER. Flowers orange outside, yellow inside. Used medicinally since ancient times. *Distr.* Japan, China.

C. radicans TRUMPET CLIMBER. Flowers pale orange with a yellow lining to the tube and red lobes. *Distr.* E USA.

C. × tagliabuana A hybrid of *C. grandiflora* × *C. radicans*. *Distr.* Garden origin.

camptotrichus, -a, -um: with curved hairs, from the Greek *camptos*, curved, and *trichos*, hair.

campylocarpus, -a, -um: with a curved fruit.

campylogynus, -a, -um: with a curved ovary.

Campylotropis (From the Greek *kamptos*, curved, and *tropis*, keel, alluding to the form of the flowers.) A genus of 65 species of subshrubs and shrubs with trifoliate leaves and racemes of irregular, pea-like flowers. *Distr.* E Asia. *Fam.* Leguminosae. *Class* D.

C. macrocarpa Shrub to 1m high. Flowers purple. *Distr.* N and Central China.

camtschatcensis: of Kamchatka, Siberia.

camtschaticum: of Kamchatka, Siberia.

canadensis, -e: of Canada or NE North America.

canaliculatus, -a, -um: channelled; with a longitudinal groove.

canariensis, -e: of the Canary Islands.

Canarina (Of the Canary Islands.) A genus of 3 species of tuberous, perennial, scrambling herbs with opposite, simple leaves and bell-shaped flowers that bear 6 fused sepals, 6 fused petals and 6 stamens. The fruit is an edible berry. *Distr.* Tropical E Africa, Canary Islands. *Fam.* Campanulaceae. *Class* D.

C. canariensis CANARY ISLAND BELLFLOWER. Flowers waxy, orange with red veins. Grown as a tender ornamental. *Distr.* Canary Islands.

C. campanula See *C. canariensis*

canary bird bush *Crotalaria agatiflora.*

canary bird flower *Tropaeolum.*

cancellatus, -a, -um: latticed.

candelabra plant *Aloe arborescens.*

candelabrus, -a, -um: like a candelabra.

candicans: becoming pure white.

candidissimus, -a, -um: very white.

candidus, -a, -um: pure glossy white.

candle plant *Plectranthus oertendahlii.*

candleberry *Myrica pennsylvanica* **swamp** ~ *M. pennsylvanica.*

candles, little *Mammillaria prolifera.*

Candollea See *Hibbertia*

candytuft *Iberis* **florist's** ~ *I. amara* **Gibraltar** ~ *I. gibraltarica* **rocket** ~ *I. amara.*

cane: giant ~ *Arundinaria gigantea* **switch** ~ *A. gigantea.*

Canellaceae A family of 6 genera and 16 species of aromatic trees with spirally arranged, leathery, glandular leaves and regular, bisexual flowers which bear fused stamens. A tonic is produced from the bark of some species. *Distr.* E Africa, Madagascar, S North America and South America. *Class* D.

canescens: rather greyish white.

canina: of dogs, from the Latin *canis*, dog.

Canna (From the Greek *kanna*, a reed.) A genus of about 25 species of rhizomatous herbs with large, spirally arranged leaves and racemes or panicles of irregular, showy flowers. Some species are grown for their starchy tubers, others as tender ornamentals. *Distr.* Tropical regions of the New World. *Fam.* Cannaceae. *Class* M.

C. edulis See *C. indica*

C.×generalis CANNA LILY. A hybrid that has given rise to many of the large number of ornamental cultivars grown today. *Distr.* Garden origin.

C. indica INDIAN SHOT, QUEENSLAND ARROWROOT, ACHIRA. Flowers red or orange. Grown for its starchy edible rhizomes, the starch grains being particularly suited to infants and invalids. Also cultivated as an ornamental for the heated greenhouse in temperate zones. *Distr.* Naturalized throughout tropics.

C. iridiflora Flowers orange-red. A robust ornamental. *Distr.* Peru.

C. lutea See *C. indica*

Cannabaceae The Hemp family. 2 genera and 3 species of perennial, glandular hairy herbs and lianas with palmate or palmately lobed leaves and small, unisexual, wind-pollinated flowers that bear 5 sepals, no petals, 5 stamens and a superior ovary. *Distr.* Temperate Europe to SE Asia, North America. *Class* D.
See: Cannabis, Humulus.

Cannabidaceae See *Cannabaceae*

Cannabis (From the Greek name for this plant, *kannabis*.) A genus of 1 species of roughly hairy, annual herb to 4m high with fibrous stems, palmate leaves and small, unisexual, wind-pollinated flowers. *Distr.* Central Asia. *Fam.* Cannabaceae. *Class* D.

C. sativa HEMP, MARIHUANA, POT, GANJA, HASHISH. Fibre from the stems is used in the production of canvas, twines and ropes. The flowers and seeds are a source of a narcotic drug and an oil used in paints and varnishes. Because of its narcotic uses cultivation of this plant is carefully licensed in many Western countries.

Cannaceae A family of 1 genus and 25 species. *Class* M.
See: Canna.

Canotiaceae A family of 2 genera and 3 species of shrubs and small trees with simple leaves and small flowers. *Distr.* S USA, Mexico. *Class* D.

cantabricus, -a, -um: of Cantabria, Spain.

cantala *Agave cantala.*

cantua (The Quechua vernacular name for *Cantua buxifolia*.) A genus of 6 species of evergreen shrubs and trees with simple, alternate leaves and corymbs of tubular flowers that bear their parts in 5s. Several species are grown as ornamentals. *Distr.* Andes. *Fam.* Polemoniaceae. *Class* D.

C. buxifolia SACRED FLOWER OF THE INCAS, MAGIC FLOWER, CANTUA. Small, hairy shrub. Inflorescence pendent. Flowers pink-purple with yellow stripes. Grown as a half-hardy ornamental. *Distr.* Peru, Bolivia, N Chile.

cantua *Cantua buxifolia.*

canus, -a, -um: very pale grey.

canyon *Quercus chrysolepis.*

caoutchouc tree *Hevea brasiliensis.*

cap, sailor's *Dodecatheon hendersonii.*

capensis: of the Cape of Good Hope, South Africa.

caper Capparaceae.

caperatus, -a, -um: wrinkled.

capers *Capparis spinosa.*

capillaris, -e: hair-like.

capitatus, -a, -um: forming a dense head.

capitellatus, -a, -um: having a small rounded head (usually of flowers).

capitulatus, -a, um: having a rounded head (usually of flowers).

cappadocica: of Cappadocia, Turkey.

Capparaceae The Caper family. 42 genera and about 650 species of herbs, shrubs, trees and a few lianas with simple or palmate leaves. The flowers are regular or irregular with 4–6 or no petals. *Distr.* Widespread in tropical regions, some in warm-temperate areas. *Class* D.
See: Capparis, Cleome.

Capparidaceae See Capparaceae

Capparis (From the Greek name *kapparis*, capers.) A genus of 250 species of evergreen, occasionally spiny shrubs and small trees with simple, spirally arranged leaves. The small white or yellow flowers are borne in racemes or panicles and bear 4 sepals, 4 petals and numerous stamens. *Distr.* Warm regions. *Fam.* Capparaceae. *Class* D.

C. spinosa CAPERS. Cultivated for the flower buds which are pickled as a relish. *Distr.* Mediterranean.

caprea: of goats.

capreolata: bearing tendrils.

Caprifoliaceae The Honeysuckle family. 13 genera and 450 species of herbs, shrubs, small trees and lianas with opposite, simple leaves. The bisexual, regular or 2-lipped flowers typically bear 5 fused sepals, 5 fused petals, 5 stamens and an inferior ovary. The family contains a number of hardy ornamental shrubs. *Distr.* Chiefly N temperate and occasionally tropical regions. *Class* D.
See: Abelia, Diervilla, Dipelta, Heptacodium, Kolkwitzia, Leycesteria, Linnaea, Lonicera, Sambucus, Symphoricarpos, Triosteum, Viburnum, Weigela.

caprifolius, -a, -um: used as goat fodder, from the Latin *capra*, goat, and *folius*, leaved.

Capsicum (From the Greek *kapto*, to bite,

alluding to the taste of the fruits.) PEPPER. A genus of 10 species of annual to perennial herbs and subshrubs with simple leaves and 5-lobed, typically white flowers. Fruit is a 2–3 chambered berry. Several species are grown ornamentally and the fruits of some are used as flavourings and as vegetables. *Distr.* Tropical America. *Fam.* Solanaceae. *Class* D.

C. annuum SWEET PEPPER, CHILLI PEPPER, CHILLI, CAYENNE PEPPER, CAPSICUM, BELL PEPPER, CHERRY PEPPER. Annual. Flowers solitary. The fruits of this species have been used as a flavouring for thousands of years. Numerous cultivars have been produced that have fruits with a range of flavours and sizes from the smaller hotter CHILLI and CAYENE PEPPERS to the larger and sweeter BELL and SWEET PEPPERS. *Distr.* Not known in the wild but thought to have originated in tropical America.

C. frutescens TABASCO PEPPER, HOT PEPPER, SPUR PEPPER. Perennial herb to 1.5m high. Flowers clustered. The fruits of this species are used as a flavouring in a similar way to those of *C. annuum* but are generally smaller and hotter tasting. *Distr.* Tropical America.

capsicum *Capsicum annuum.*

capsularis: bearing capsules.

Caragana (From *karaghan*, the Mongolian name for *Caragana arborescens.*) PEA TREE. A genus of 80 species of deciduous shrubs and small trees with alternate, pinnate leaves and solitary or clustered irregular, pea-like flowers. Several species are grown as ornamentals and some are important as a source of fuel wood. *Distr.* E Europe, Central and E Asia. *Fam.* Leguminosae. *Class* D.

C. arborescens SIBERIAN PEA TREE. Shrub to 6m high. Flowers clustered, pale yellow. Grown as a windbreak and for the bark which is used to make ropes and the young pods which are eaten. *Distr.* Siberia, Manchuria.

C. aurantiaca Shrub to 1m high. Flowers solitary, orange-yellow. *Distr.* Central Asia.

C. franchetiana Spiny shrub. Flowers clustered, yellow, marked with red. *Distr.* SW China.

C. frutex RUSSIAN PEA TREE. Erect shrub. Flowers large, yellow, clustered. Several ornamental cultivars of this species have been produced. *Distr.* Turkestan, Siberia.

caraway *Carum carvi.*

Cardamine BITTERCRESS. A genus of 130 species of annual or perennial herbs with white to purple or, rarely, pink flowers. Some species are eaten like WATERCRESS. *Distr.* Temperate regions. *Fam.* Cruciferae. *Class* D.

C. asarifolia See *Pachyphragma macrophyllum*

C. bulbifera CORAL ROOT BITTERCRESS. Rhizomatous perennial with large purple flowers. *Distr.* Europe.

C. californica Rhizomatous perennial with large pink flowers. *Distr.* California.

C. enneaphyllos Rhizomatous perennial with nodding, white flowers. *Distr.* W Carpathians, E Alps, S Italy.

C. heptaphylla Rhizomatous perennial with large purple-white flowers. *Distr.* W and Central Europe.

C. kitiabelii Rhizomatous perennial with pale yellow flowers. *Distr.* Central Europe.

C. latifolia See *C. raphanifolia*

C. pentaphyllos Rhizomatous perennial with lilac flowers. *Distr.* W and Central Europe.

C. pratensis CUCKOO FLOWER, LADY'S SMOCK, MEADOW CRESS, SPINKS. Rhizomatous perennial with white-lilac flowers. *Distr.* N temperate regions.

C. raphanifolia Rhizomatous perennial, typically with purple flowers. *Distr.* S Europe.

C. trifolia Rhizomatous perennial with pink or white flowers. *Distr.* Central and S Europe.

cardamom *Elettaria cardamomum* **Ceylon** ~ *E. cardamomum* **Malabar** ~ *E. cardamomum* **round** ~ *Amomum compactum.*

cardiacus, -a, -um: referring to the heart.

cardinal flower *Lobelia cardinalis, Sinningia cardinalis* **blue** ~ *Lobelia siphilitica.*

cardinalis, -e: scarlet.

cardinal's guard *Pachystachys coccinea.*

Cardiocrinum (From the Greek, *kardia*, heart, and *krinon*, lily, alluding to the heart-shaped leaves.) A genus of 3 species of large bulbiferous herbs with heart-shaped, long-stalked leaves and racemes of funnel-shaped, green-white flowers that bear 6 tepals and 6 stamens. All three species are cultivated as ornamentals. *Distr.* Himalaya, China and Japan. *Fam.* Liliaceae. *Class* M.

C. giganteum GIANT LILY. Stems to 4m high. Flowers large, white, striped maroon inside, fragrant. The most popular cultivated species. *Distr.* Himalaya.

cardiophyllus, -a, -um: with heart-shaped leaves, from the Greek *cardio-*, heart, *phyllos*, leaf.

Cardiopteridaceae A family of 1 genus and 2 species of twining herbs with milky sap, spirally arranged, simple leaves and panicles of small flowers that usually bear 5 fused petals and a superior ovary. Some species are eaten locally as a vegetable. *Distr.* India to Solomons. *Class* D.

Cardiospermum (From the Greek *kardia*, heart, and *sperma*, seed, the seed is heart-shaped.) A genus of 14 species of climbing herbs and shrubs with tendrils, compound leaves, and small white flowers. *Distr.* Tropical regions, especially America. *Fam.* Sapindaceae. *Class* D.

C. grandiflorum HEARTSEED. Climbing herb. Flowers cream-white, scented. The leaves are eaten as a vegetable and the black seeds are used as beads. *Distr.* Tropical America and Africa.

cardoon *Cynara cardunculus.*

Carduncellus (The diminutive of the genus name *Carduus.*) A genus of 28–30 species of perennial, often spiny herbs with pinnate leaves and flowers in daisy-like heads that lack distinct rays but bear numerous bracts. *Distr.* Mediterranean. *Fam.* Compositae. *Class* D.

C. rhaponticoides Stemless perennial. Leaves in a basal rosette. Flowers blue. *Distr.* Algeria.

Carduus (The name used by Virgil for a THISTLE.) THISTLE. A genus of 91 species of annual to perennial herbs with simple to pinnate, spiny-toothed leaves and flowers in dandelion-like heads that are surrounded by spiny bracts. *Distr.* Eurasia, Mediterranean, Mountains of E Africa. *Fam.* Compositae. *Class* D.

C. benedictus See *Cnicus benedictus.*

C. nutans SCOTCH THISTLE, MUSK THISTLE. Biennial to perennial, spiny herb. The dried flowers were formerly used to curdle milk and the pith is edible when cooked. *Distr.* Eurasia, naturalized in North America.

Carex (From Greek *kairo*, to cut, alluding to the sharp edges of the leaves of some species.) SEDGE. A genus of 1000–1500 species of grass-like, clump-forming, rhizomatous herbs with 3-angled aerial stems, 3-ranked leaves and spikes of very small, wind-pollinated flowers. A number of species are cultivated as ornamentals, with numerous cultivars and variegated forms available; others are used as a source of fibre or for packing and weaving. *Distr.* N temperate regions. *Fam.* Cyperaceae. *Class* M.

C. buchananii LEATHERLEAF SEDGE. Leaves semicircular in cross-section, red-green below, cream above. *Distr.* New Zealand.

C. elata TUFTED SEDGE. Forming tussocks to 1m high. Leaves folded in cross-section. Several ornamental cultivars are available. *Distr.* Europe to N Africa and Caucasus.

C. grayi MACE SEDGE. Leaves bright green. Female spikes mature to a knobbly brown fruit. *Distr.* E North America.

C. nigra COMMON SEDGE. Clump-forming. Leaves very narrow. *Distr.* Europe.

C. pendula PENDULOUS SEDGE. Flowering spikes pendulous. *Distr.* Europe, N Africa SW Asia.

C. riparia GREATER POND SEDGE. Leaves flat, 3-angled. Often grown in bog gardens but can be invasive. *Distr.* Europe, N Africa, W Asia.

C. stricta See *C. elata*

caribaea: of the Lesser Antilles, West Indies.

Carica A genus of 22 species of shrubs and thick-stemmed trees with palmately lobed, alternate leaves and unisexual, tubular to salver-shaped flowers. The fruit is a large, pendent, many-seeded berry. *Distr.* Subtropical and tropical America. *Fam.* Caricaceae. *Class* D.

C. papaya PAWPAW, PAPAYA, MELON TREE. Thick-stemmed tree. Cultivated throughout the tropics for edible fruits and extraction of papain, a meat tenderizer. It is also grown as a tender ornamental. *Distr.* Tropical South America.

C. pubescens MOUNTAIN PAWPAW. Small fruits can be candied or preserved. *Distr.* Andes.

carica: of Caria, West Asia.

Caricaceae The Pawpaw family. 4 genera and 31 species of small trees with palmately lobed or compound leaves and milky latex. The flowers are typically unisexual with their parts in 5s. The fruits are many-seeded berries. Some species have edible fruits, notably *Carica papaya* (PAWPAW.). *Distr.* Tropical and subtropical America and tropical W Africa. *Class* D.
See: Carica.

carinatus, -a, -um: keeled.

carinthiacus, -a, -um: of Carinthia, Austria.

Carissa (From the Indian vernacular name for these plants.) A genus of 37 species of shrubs and small trees with opposite, simple, leathery leaves and clusters of funnel-shaped, fragrant flowers. Grown for hedging and tart fruit. *Distr.* Warm regions of the Old World. *Fam.* Apocynaceae. *Class* D.

C. grandiflora NATAL PLUM. Grown as a spiny hedging plant in frost-free areas and as a tub plant in cooler areas. The fruits are eaten. *Distr.* South Africa.

Carlemanniaceae A family of 2 genera and 5 species of herbs and subshrubs with simple, opposite leaves. This family is sometimes included within the family Caprifoliaceae. *Distr.* Tropical Asia. *Class* D.

Carlina (After Charlemagne, whose army is said to have been cured of plague by a species of this genus.) CARLINE THISTLE. A genus of 28 species of annual to perennial herbs with simple to pinnately lobed leaves, and flowers in large, solitary, daisy-like heads that lack distinct rays but may have showy, spiny bracts. *Distr.* Macaronesia, Mediterranean, W Asia. *Fam.* Compositae. *Class* D.

C. acanthifolia Stemless perennial. Flowers lilac. Bracts yellow. *Distr.* SE and Central Europe.

C. acaulis CARLINE THISTLE. Stemless perennial. Flowers white-purple. Bracts white-pale pink, shiny. *Distr.* Europe, common in Alps.

C. vulgaris COMMON CARLINE THISTLE. Biennial to 70cm high. *Distr.* Eurasia.

Carludovica (From Charles IV of Spain.) and Queen Louisa. A genus of 3 species of short stemmed, rhizomatous, perennial herbs with fan-shaped leaves and spikes of small, unisexual flowers. The leaves are a source of fibre. *Distr.* Mainland tropical America. *Fam.* Cyclanthaceae. *Class* M.

C. palmata PANAMA HAT PALM, TOQUILLA. Young leaves used in the production of Panama Hats; 6 leaves are required to make each hat, a million of which are exported from Ecuador alone each year. Older leaves are used to make baskets and mats. *Distr.* Central America to Bolivia.

Carmichaelia (After Captain Dugald Carmichael (1722–1827), Scottish army officer and botanist.) A genus of 40 species of shrubs and small trees with pinnate leaves in juvenile plants, replaced by flattened stems (phyllodes) in adults. The numerous, small, pea-like flowers are borne on clustered racemes and are often fragrant. Several species are grown as ornamentals. *Distr.* New Zealand, Lord Howe Islands. *Fam.* Leguminosae. *Class* D.

C. enysii Mound-forming shrub. Flowers violet. *Distr.* New Zealand.

carmineus, -a, -um: carmine, a vivid crimson red.

× ***Carmispartium*** (From the names of the parent genera.) A genus of species of hybrids between members of the genera *Carmichaelia* and *Notospartium*. *Fam.* Leguminosae. *Class* D.

× *C. astens* See *X. hutchinsii*

× *C. hutchinsii* A hybrid of *C. astonii* × *N. glabrescens*. Shrub or small tree with pendulous, green branches, sparse, simple leaves and racemes of purple-white flowers. Grown as an ornamental. *Distr.* Garden origin.

carnation *Dianthus*, *D. caryophyllus*.

carneus, -a, -um: flesh-coloured.

carniolica: of Carniola, an old Austrian territory in the north of the former Yugoslavia.

carnosulus, -a, -um: slightly fleshy.

carnosus, -a, -um: fleshy, soft, succulent.

carolinianus, -a, -um: of Carolina, USA.

carota from the Greek name for carrot.

carpaticus, -a, -um: of the Carpathian Mountains, Central Europe.

Carpenteria (After Professor William Carpenter (1811–48) of Louisiana.) A genus of 1 species of evergreen shrub with opposite, simple leaves and fragrant, white flowers that bear 5–7 petals and are borne singly or in small cymes. *Distr.* SW USA. *Fam.* Hydrangeaceae. *Class* D.

C. californica TREE ANEMONE. Grown as a hardy ornamental.

carpinifolius, -a, -um: from the genus *Carpinus*, and *folius*, leaved.

Carpinus (The classical Latin name for these plants.) HORNBEAM. A genus of 35 species of typically deciduous shrubs and trees with slender winter buds and alternate, simple leaves that are borne in 2 ranks. The tiny, unisexual flowers are borne in separate catkins and appear in spring. The fruit is a small winged seed. Several species are grown as ornamentals and some are important for their timber which is particularly good for turning. *Distr.* N temperate regions, chiefly in E Asia and North America. *Fam.* Corylaceae. *Class* D.

C. betulus EUROPEAN HORNBEAM. Tree to 30m high. The timber was formerly used for mill cogwheels and is still important in the mechanisms of pianos. *Distr.* Europe, SW Asia.

C. caroliniana AMERICAN HORNBEAM. Large shrub or tree, to 13m high. *Distr.* USA, NE Mexico.

C. tschonoskii Rounded tree to 15m high. Branches with drooping tips. *Distr.* NE China, Korea, Japan.

C. turczaninowii Spreading shrub or small tree. Leaves orange in autumn. *Distr.* N China, Korea, Japan.

Carpobrotus (From the Greek *karpos*, fruit, and *brotos*, edible.) A genus of 20–30 species of small perennial subshrubs with creeping stems, succulent leaves and solitary, large, many-petalled flowers. The fruit is a juicy, cone-shaped capsule. A number of species are grown as ornamentals, some have edible fruit. *Distr.* South Africa, some naturalized on cliffs in Europe. *Fam.* Aizoaceae. *Class* D.

C. deliciosus Fruit edible, tastier than *C. edulis*. *Distr.* South Africa (Cape Province).

C. edulis HOTTENTOT FIG. Leaves tapering to the point, curved. Flowers yellow, turning purple. Fruit edible. *Distr.* South Africa (Cape Province), naturalized Europe.

carrot Umbelliferae, *Daucus carota* **candy** ~ *Athamanta cretensis* **wild** ~ *Daucus carota*.

Carthamus (From the Hebrew *qarthami*, to paint, alluding to the dye obtained from the flowers of *C. tinctorius*.) A genus of 14 species of annual and occasionally perennial herbs with pinnate, spiny leaves and flowers in solitary, daisy-like heads that lack distinct rays. *Distr.* Mediterranean, Asia. *Fam.* Compositae. *Class* D.

C. tinctorius SAFFLOWER, FALSE SAFFRON. Annual. Grown as a garden ornamental. The flowers were formerly used in dyeing and it has been considered a potential oilseed crop. *Distr.* Garden origin, possibly originating from the Near East.

cartwheel flower *Heracleum mantegazzianum*.

Carum A genus of about 30 species of aromatic, biennial, and perennial herbs with much-divided leaves and compound umbels of white or pink flowers. *Distr.* Temperate and warm regions. *Fam.* Umbelliferae. *Class* D.

C. carvi CARAWAY. Biennial herb to 60cm high. Grown for its fruits (caraway seeds) which are used as a flavouring for breads and liqueur; also for its leaves which may be used as a culinary herb and its roots which are eaten as a vegetable. *Distr.* Mediterranean.

Carya (From the Greek *karya*, a nut-bearing tree.) HICKORY, PECAN. A genus of 17 species of large deciduous trees with alternate, pinnate leaves and unisexual flowers. The male flowers occur in a pendent, branched catkin and the females in an erect spike. The fruit is a rounded drupe containing a nut. A source of timber and nuts (pecans). Strips of outer bark were formerly used in the construction of chairs. *Distr.* North America, Mexico, E Asia. *Fam.* Juglandaceae. *Class* D.

C. amygdaliferum SWARRI NUT. Tree to 55m high. Flowers yellow-green. Seeds a source of edible fat. *Distr.* Panama, Colombia.

C. aquatica BITTER PECAN, WATER HICKORY. Found in wet places. Nut astringent. *Distr.* SE USA.

C. cordiformis BITTER NUT, SWAMP HICKORY. Found in wet places. Used as a rootstock for other species. *Distr.* E North America.

C. glabra PIGNUT, HOGNUT, SMALL FRUITED HICKORY, BROOM HICKORY. Fruit small, astringent. *Distr.* E North America and Central Florida.

C. glabrum SOAPWOOD. Tree to 25m high. Inner bark used as a source of soap. *Distr.* West Indies, Tropical South America.

C. illinoinensis PECAN. Grown for its delicious edible nuts and oil which is used in cosmetics. The timber is of low quality. Numerous cultivars have been produced. *Distr.* USA, Mexico.

C. nuciferum BUTTER NUT, SOUARI NUT. Tree to 45m high. Flowers large, purple or deep red. Seeds a source of oil. *Distr.* Guyana, West Indies.

C. ovata SHAGBARK HICKORY. Bark shaggy. Grown for timber. *Distr.* E and Central North America.

C. tomentosa MOCKERNUT. Grown for its excellent timber. *Distr.* E North America.

Caryocaraceae The Butter Nut family. 2 genera and 24 species of trees and a few shrubs with trifoliate leaves and racemes of regular flowers that bear 5 sepals, 5 petals and a superior ovary. *Distr.* Tropical America. *Class* D.
See: Caryocar.

Caryocar (From the Greek *karyon*, nut, alluding to the fruit.) BUTTER NUT. A genus of 10 species of shrubs and trees with opposite, palmate, leathery leaves and racemes of regular flowers that bear 5–6 sepals, 5–6 petals and numerous stamens. The fruits are large, weighing as much as 11kg, and bear numerous edible nuts. *Distr.* Tropical America.

Caryophyllaceae The Pink family. 66 genera and about 2000 species of annual or perennial herbs and a few subshrubs with simple, opposite leaves. The flowers are regular and usually bisexual with 4–5 sepals and petals; cultivated forms are often doubles. Many garden plants are found in this family, notably the PINKS and CARNATIONS, derived from the genus *Dianthus*. *Distr.* Widespread but centred in temperate regions. *Class* D.
See: Agrostemma, Arenaria, Cerastium, Colobanthus, Dianthus, Drypis, Gypsophila, X Lycene, Lychnis, Minuartia, Moehringia, Petrocoptis, Petrorhagia, Sagina, Saponaria, Silene, Spergula, Spergularia, Stellaria, Vaccaria, Viscaria.

Caryopteris (From the Greek *karyon*, nut, and *pteron*, wing.) A genus of 6 species of herbs and deciduous shrubs with simple, opposite, aromatic leaves and panicles or cymes of 5-lobed flowers. The fruit is a winged nutlet. *Distr.* E Asia. *Fam.* Verbenaceae. *Class* D.

C. × clandonensis A hybrid of *C. incana* × *C. mongholica*. Shrub to 1.5m high. Foliage silvery. Flowers bright blue. *Distr.* Garden origin.

Caryota (From the Greek *karyon*, a nut.) FISHTAIL PALM. A genus of 12 species of large palms with bipinnate leaves and pendent inflorescences. *Distr.* Asia. *Fam.* Palmae. *Class* M.

C. mitis BURMESE FISHTAIL PALM, CLUSTERED FISHTAIL PALM, TUFTED FISHTAIL PALM. Stems narrow, clustered. Flowers cream. Fruit yellow to dark red. *Distr.* SE Asia.

Casearia A genus of 180 species of shrubs and trees with simple alternate leaves and clusters of small bisexual flowers. *Distr.* Tropical regions. *Fam.* Flacourtiaceae. *Class* D.

C. praecox VENEZUELAN BOX. The main source of box wood, largely replacing EUROPEAN BOX, *Buxus sempervirens*. *Distr.* Venezuela.

cashew Anacardiaceae.

cashmeriana: of Kashmir, on the India, Pakistan borders.

caspica: of the region of the Caspian Sea.

cassandra *Chamaedaphne calyculata.*

Cassia (From Greek.) A genus of about 30 species of annual and perennial herbs, trees and shrubs with pinnate leaves that are sometimes reduced to a single leaflet and large irregular flowers with 5 overlapping petals and typically 10 stamens. Several species are grown as ornamentals and some have medicinal properties. *Distr.* Tropical and warm temperate regions, excluding Europe. *Fam.* Leguminosae. *Class* D.

C. artemisioides See *Senna artemsioides*

C. didymobotrya See *Senna didymobotrya*

C. fistula GOLDEN SHOWER TREE, PURGING CASSIA, INDIAN LABURNUM, PUDDING PIPE TREE. Deciduous or semi-evergreen tree to 20m high. Flowers fragrant, in pendent racemes. *Distr.* SE Asia, Central and South America, N Australia.

C. javanica Deciduous tree to 25m high. Flowers red-pink, in rigid racemes. *Distr.* SE Asia.

C. marilandica See *Senna marilandica*

cassia *Cinnamomum aromaticum* **feathery** ~ *Senna artemsioides* **purging** ~ *Cassia fistula* **solver** ~ *Senna artemisioides.*

Cassinia (After Viscomte Alexandre Henri Gabriel Cassini (1781–1832), French botanist.) A genus of about 20 species of heather-like shrubs and perennial herbs with alternate leaves and flowers in numerous small, daisy-like heads that lack distinct rays. *Distr.* Australia, New Zealand. *Fam.* Compositae. *Class* D.

C. leptophylla Shrub to 2m. Flowers white. *Distr.* New Zealand.

C. retorta Shrub to 5m. *Distr.* New Zealand.

cassinoides: resembling the species *Ilex cassine.*

Cassiope (After Cassiope, in Greek mythology mother of Andromeda.) A genus of 12 species of evergreen, dwarf, sprawling shrubs with small leaves arranged in 4 distinct ranks and small, solitary, white to pink flowers. *Distr.* Himalaya, N temperate regions. *Fam.* Ericaceae. *Class* D.

C. fastigiata Flowers bell-shaped, pendulous. The parent of a number of ornamental hybrids and cultivars. *Distr.* Himalaya.

C. lycopodioides IWA-HIGE. Flowers tubular to bell-shaped, pink or white. The parent of a number of ornamental hybrids and cultivars. *Distr.* NE Asia and Japan.

C. mertensiana WHITE HEATHER. Mound-forming shrub with bell-shaped, usually white flowers. *Distr.* Mountains of W North America.

C. selaginoides Flowers with a 5-lobed corolla. *Distr.* W China and the Himalaya.

C. stelleriana Flowers with a 4-lobed, conspicuously cleft corolla. *Distr.* NW North America and NE Asia.

C. tetragona Much used as a fuel by Eskimos. *Distr.* The Arctic.

Castanea (The classical Latin name for CHESTNUT, after the town of Castania in Thessaly. The Greek for CHESTNUT is *kastana.*) A genus of 12 species of deciduous shrubs and trees with simple, toothed leaves and catkins of small, musky-scented, unisexual flowers. The fruit consists of a spiny case (cupule) enclosing 1–4 large brown nuts. Several species are grown as ornamentals and some are a source of edible nuts, timber or bark for tanning. *Distr.* N hemisphere temperate regions, introduced into Britain by the Romans. *Fam.* Fagaceae. *Class* D.

C. mollissima CHINESE CHESTNUT. Tree to 20m high. Grown for its seeds, several cultivars being available. *Distr.* China, Korea.

C. sativa SWEET CHESTNUT, SPANISH CHESTNUT. Tree to 40m high. Widely grown for its edible nuts which are often sold freshly roasted on the streets of large cities. *Distr.* S Europe, SW Asia, N Africa, thought to have been introduced into Britain by the Romans.

castaneifolius, -a, -um: from the genus *Castanea*, the sweet chestnut, and *folius*, leaved.

Castanopsis (From the genus name *Castanea* and the Greek suffix *-opsis*, resembling.) A genus of 110 species of evergreen shrubs and trees with leathery, 2-ranked or spirally arranged leaves and erect catkins of small flowers. The fruit consists of a spiny case (cupule) surrounding 3 nuts. The seeds of some species are eaten roasted like true chestnuts. *Distr.* W North America. *Fam.* Fagaceae. *Class* D.

C. cuspidata Much planted in Japanese parks and gardens; the leaves were formerly used as rice bowls and the nuts are edible. *Distr.* Japan, China.

Castanospermum (From the Latin *castanea*, chestnut, and *sperma*, seed.) A genus of 1 species of large tree with pinnately divided leaves and clusters of large, yellow, pea-like flowers that turn orange-red as they mature. The seeds are large and black. *Distr.* Australia, New Caledonia. *Fam.* Leguminosae. *Class* D.

C. australe MORETON BAY CHESTNUT, AUSTRALIAN CHESTNUT. Grown as an ornamental shade tree and for its timber. It yields compounds that are under investigation for possible use against the AIDS virus.

cast-iron plant *Aspidistra elatior.*

castor bean plant *Ricinus communis.*

castor oil plant *Ricinus communis* **false** ~ *Fatsia japonica.*

Casuarina (The drooping branches are said to resemble the feathers of a cassowary bird, *Casuarinus.*) IRONWOOD, SHE OAK, SWAMP OAK. A genus of about 80 species of evergreen trees and shrubs with weeping branches and minute scale-like leaves. The male flowers are borne in catkins and the females in cone-like structures. *Distr.* Tropical regions of the Old World. *Fam.* Casuarinaceae. *Class* D.

C. cunninghamiana Tree to 35m high. *Distr.* E Australia.

C. equisetifolia HORSETAIL TREE, SOUTH SEA IRONWOOD, MILE TREE, BEEFWOOD, BULL OAK, WHISTLING PINE. Tree to 35m high. Used for hedging, windbreaks, and as a source of timber. A pioneer tree of the sea shore. *Distr.* India, Malaysia, Australia; widely planted.

C. torulosa Tree to 20m high. Foliage bronze-tinged. *Distr.* E Australia.

Casuarinaceae The She Oak family. 4 genera and 70 species of tall trees with a weeping habit. The leaves are reduced to many-toothed sheaths surrounding the pendent stems. The male flowers are borne in spikes at the top of the tree whilst the females are borne in round clusters lower down. The fruits are surrounded by small hard bracts so that the whole cluster somewhat resembles a pine cone. A source of hardwood for furniture and other uses. *Distr.* Tropical Asia to Australia and the Pacific. *Class* D.
See: Allocasuarina, Casuarina.

Catalpa (From *catawba*, the North American Indian name for this plant.) A genus of 11 species of deciduous trees with large, simple or 3-lobed leaves and panicles or racemes of 2-lipped, bell- or funnel-shaped flowers that are followed by long cylindrical pods. Most of the species are grown as ornamentals, some are used for timber. *Distr.* E Asia, North America. *Fam.* Bignoniaceae. *Class* D.

C. bignonioides INDIAN BEAN TREE, SMOKING BEAN, INDIAN CIGAR, COMMON CATALPA. Tree to 15m high. Leaves malodorous when crushed. Flowers numerous, white with yellow and purple marks in throat. A source of durable timber. *Distr.* SE USA.

C. × erubescens A hybrid of *C. ovata* × *C. bignonioides*. The source of a number of ornamental cultivars including doubles. *Distr.* Garden origin.

C. ovata Spreading. Leaves 3-lobed, tinged purple. Flower white, yellow within. *Distr.* China.

C. speciosa CATAWBA, CIGAR TREE. Tree to 30m high. Leaves occasionally lobed. Flowers large, white, in dense heads. *Distr.* USA.

catalpa, common *Catalpa bignonioides.*

Catananche (From the Greek *katanangke*, a powerful incentive, alluding to its use in love potions.) A genus of 5 species of annual and perennial herbs with linear leaves and flowers in dandelion-like heads that are often surrounded by papery bracts. *Distr.* Mediterranean. *Fam.* Compositae. *Class* D.

C. caerulea CUPID'S DARTS. Perennial herb to 1m high. Grown as an ornamental. *Distr.* S Europe.

cataractae: of waterfalls.

cataria: of cats.

catawba *Catalpa speciosa.*

catawbiense: of the Catawba River, USA.

catberry *Ribes grossularioides.*

catchfly *Lychnis, Silene* **German** ~ *Lychnis viscaria* **nodding** ~ *Silene pendula.*

catechu *Areca catechu.*

catechu: from a local name for these plants in Vietnam.

Catharanthus (From the Greek *katharos*, pure, and *anthos*, flower.) PERIWINKLE. A genus of 8 species of annual and perennial herbs with simple opposite leaves and funnel-shaped flowers that bear 5 stamens fused to the inside of the tube. *Distr.* Madagascar. *Fam.* Apocynaceae. *Class* D.

C. roseus MADAGASCAR PERIWINKLE, OLD MAID, CAYENNE JASMINE. A source of numerous alkaloids some of which are reported to be useful in the treatment of leukaemia. There are a number ornamental cultivars available. *Distr.* Madagascar, widely naturalized throughout the tropics.

cathartica: purging.

cathedral windows *Calathea makoyana.*

catmint *Nepeta N. cataria*

catnip *N. cataria.*

cat's claw *Macfadyena unguiscati.*

cat's ear *Hypochoeris, Calochortus* **spotted** ~ *Hypochoeris radicata.*

cat's foot *Antennaria dioica.*

cat's paw *Anigozanthos* **Albany** ~ *A. preissii.*

cat's tail *Bulbinella, Typha latifolia* **meadow** ~ *Phleum pratense.*

cat's whiskers *Tacca chantrieri.*

cat tail *Typha.*

caucasicus: of the Caucasus Mountains.

cauchao *Amomyrtus luma.*

caudatifolius, -a, -um: with leaves ending in a slender tail.

caudatus, -a, -um: with a slender tail.

caulescens: with a slight stem.

cauliflower *Brassica oleracea.*

Caulophyllum (From the Greek *kaulos*, a stem, and *phyllon*, a leaf.) BLUE COHOSH. A genus of 2 species of rhizomatous herbs with a solitary 3-lobed leaf borne directly below a raceme or panicle of small, green or yellow flowers. *Distr.* E North America and Japan. *Fam.* Berberidaceae. *Class* D.

C. thalictroides BLUE COSH, PAPOOSE ROOT, SQUAW ROOT. Herb to 75cm high. Flowers yellow, appearing before the leaf opens. Grown as an ornamental. Dried rhizomes used as a diuretic. *Distr.* E North America.

cauticola: growing in rocky places.

cautleoides: from the genus *Cautleya*, with the ending *-oides*, indicating resemblance.

Cautleya (After Major-General Sir P. T. Cautley (1802–71), British naturalist.) A genus of about 5 species of rhizomatous herbs with reed-like stems, 2 ranks of oblong to lance-shaped leaves and terminal inflorescences of irregular yellow flowers. *Distr.* Himalaya. *Fam.* Zingiberaceae. *Class* M.

C. spicata Grown as an tender ornamental. *Distr.* Himalaya.

cavus, -a, -um: hollow.

Ceanothus (From a Greek name used by Theophrastus for a spiny plant.) CALIFORNIAN LILAC. A genus of 55 species of subshrubs, shrubs and small trees with simple, somewhat leathery leaves and clusters of small, white, blue or purple flowers that bear their parts in 5s. A number of species and numerous hybrids and cultivars are grown as ornamentals. *Distr.* North America, especially W California. *Fam.* Rhamnaceae. *Class* D.

C. americanus NEW JERSEY TEA, WILD SNOWBALL. Deciduous shrub to 1m high. Flowers white. This species plays an important part in the ecology of chaparral vegetation. The leaves can be used to make tea. *Distr.* E North America.

C. gloriosus Evergreen shrub. Flowers dark blue to purple. *Distr.* W USA.

Cecropiaceae A family of 6 genera and 200 species of shrubs, trees and lianas with aerial or stilt roots, spirally arranged, simple to palmately lobed leaves and cymes of small unisexual flowers. *Distr.* Widespread in tropical regions. *Class* D.

cedar *Cedrus* **Atlantic** ~ *C. atlantica* **Chilean** ~ *Austrocedrus chilensis* **Cyprus** ~ *Cedrus brevifolia* **deodar** ~ *C. deodara* **incense** ~ *Calocedrus decurrens* **Japanese** ~ *Cryptomeria japonica* **Oregon** ~ *Chamaecyparis lawsoniana* **red** ~ *Thuja* **salt** ~ *Tamarix* **western red** ~ *Thuja plicata* **western white** ~ *T. occidentalis* **white** ~ *T. occidentalis* **yellow** ~ *Chamaecyparis nootkatensis.*

Cedar of goa *Cupressus lusitanica.*

Cedar of lebanon *Cedrus libani*

Cedrela sinensis See *Toona sinensis*

Cedronella (From the Greek *kedros*, CEDAR, alluding to the resinous smell of the leaves.) A genus of 1 species of tall perennial herb with trifoliate leaves and whorls of pink to white, funnel-shaped, 2-lipped flowers. *Distr.* Canary Islands. *Fam.* Labiatae. *Class* D.

C. canariensis CANARY BALM. Grown as an ornamental for its scented leaves which have been used to make a kind of tea.

C. mexicana See *Agastache mexicana*

C. triphylla See *C. canariensis*

Cedrus (From the classical Greek name for these trees, *kedros.*) CEDAR. A genus of 4 species of evergreen coniferous trees with needle-like leaves that are borne singly along the longer shoots and in whorls on short, lateral shoots. The woody fruiting cones disintegrate at maturity. These trees are a source of timber and scented oil. *Distr.* W Himalaya, Mediterranean and Morocco. *Fam.* Pinaceae. *Class* G.

C. atlantica ATLANTIC CEDAR. Tree to 40m high. Foliage blue-green or grey-green. *Distr.* Morocco (Atlas Mountains).

C. brevifolia CYPRUS CEDAR. Tree to 20m high. *Distr.* Restricted to 2 forests on Mount Phabos, Cyprus.

C. deodara DEODAR CEDAR. Tree to 35m high. A fine source of timber. *Distr.* W Himalaya.

C. libani CEDAR OF LEBANON. Tree to 40m high. A source of hard fragrant timber, possibly that used to build Solomon's temple. *Distr.* Syria, Lebanon, SE Turkey.

Ceiba (From a South American vernacular name for these plants.) A genus of 4 species of large deciduous trees with palmate leaves. The flowers are solitary or in clusters and are pollinated by bats and other mammals. The seeds are embedded in fibre within pods. *Distr.* Tropical regions of the New World and Africa. *Fam.* Bombacaceae. *Class* D.

C. pentandra KAPOK. The fibre from the pods constitutes the kapok of commerce, the seeds are a source of oil, and the wood is used for making matches.

celandine: greater ~ *Chelidonium majus* **lesser** ~ *Ranunculus ficaria* **tree** ~ *Bocconia frutescens.*

Celastraceae The Spindle family. 85 genera and about 1300 species of trees, shrubs and climbers with simple, often leathery leaves. The small, green, regular flowers may be unisexual or bisexual and typically bear 5 sepals, 5 petals, and a superior ovary. The seeds are often covered by a bright red aril. Some species are important for timber, others are a source of dye. *Distr.* Widespread. *Class* D.
See: Celastrus, Euonymus, Maytenus, Paxistima, Tripterygium

Celastrus (From the Greek plant name *kelastros*, used for *Phillyrea latifolia*) SHRUBBY BITTERSWEET. A genus of 31 species of scrambling or twining, deciduous or evergreen shrubs with simple, alternate leaves. The small unisexual flowers are borne in large panicles or cymes. The seeds are covered by a showy orange or red aril. Several species are grown as ornamental climbing shrubs, particularly for their long-lasting, showy fruits. *Distr.* Tropical to warm temperate regions. *Fam.* Celastraceae. *Class* D.

C. orbiculatus Leaves small, rounded. Flowers green. Fruits small, splitting to reveal yellow and red interior. *Distr.* E Asia, naturalized in North America.

C. scandens AMERICAN BITTERSWEET, STAFF TREE, WAXWORK. Twining shrub. Leaves oval. Flowers yellow. Fruit splits to reveal orange and scarlet interior. *Distr.* North America.

celendine, tree *Macleaya cordata.*

celeriac *Apium graveolens.*

celery *A. graveolens* **wild** ~ *A. graveolens.*

Celmisia (After Celmisius, in Greek mythology, son of the nymph Alciope.) A genus of 61 species of perennial herbs and subshrubs with tufted or overlapping leaves and flowers in daisy-like heads that bear distinct rays. *Distr.* Australia, New Zealand. *Fam.* Compositae. *Class* D.

C. argentea Cushion forming subshrub. *Distr.* New Zealand.

C. bellidioides Creeping, mat-forming herb. *Distr.* New Zealand.

C. coriacea Tufted herb. *Distr.* New Zealand.

C. densiflora Subshrub. Rays white. *Distr.* New Zealand.

C. gracilenta Tufted herb. *Distr.* New Zealand.

C. hectorii Much branched subshrub to 1m high. *Distr.* New Zealand.

C. hookeri Tufted perennial herb. *Distr.* New Zealand.

C. lyallii Tufted perennial herb. *Distr.* New Zealand.

C. ramulosa Stout, small shrub. *Distr.* New Zealand.

C. walkeri Sprawling shrub. *Distr.* New Zealand.

C. webbiana See *C. walkeri*

Celosia (From the Greek *keleos*, burning, alluding to the brilliant colour of some of the flowers.) COCKSCOMB. A genus of 50 species of annual and perennial herbs with simple, alternate leaves and spike-like cymes of small flowers. *Distr.* Warmer regions of Africa and America though widely grown throughout tropical and subtropical regions. *Fam.* Amaranthaceae. *Class* D.

C. argentea COCKSCOMB. Much grown as a perennial pot plant and as an annual bedding plant in cooler regions. Numerous cultivars are available in different colours. *Distr.* Tropical regions.

C. cristata See *C. argentea*

Celsia See *Verbascum*

× Celsioverascum See *Verbascum*

Celtis HACKBERRY, SUGARBERRY, NETTLE TREE. A genus of about 65 species of evergreen and deciduous shrubs and trees with simple, alternate leaves and clusters of inconspicuous, 5-lobed, green flowers that lack petals. Several species have edible fruits and some are a source of dye or timber. *Distr.* N temperate and tropical regions. *Fam.* Ulmaceae. *Class* D.

C. australis MEDITERRANEAN HACKBERRY, EUROPEAN NETTLE TREE, LOTE TREE. Deciduous tree to 20m high. Fruit red-brown. Widely planted as a street tree in Mediterranean regions. *Distr.* S Europe, N Africa and W Asia.

C. occidentalis BEAVER WOOD, HACKBERRY. Deciduous tree to 25m high. A source of timber for fences; also used as fuel. *Distr.* E North America.

Centaurea BLUEBOTTLE, KNAPWEED, STAR THISTLE. A genus of about 450 species of annual and perennial herbs with simple or pinnate leaves and flowers in daisy-like heads that lack distinct rays but bear spiny or pointed bracts. *Distr.* Mediterranean, Eurasia, North America, Australia. *Fam.* Compositae. *Class* D.

C. bella Low cushion-forming perennial. Flowers purple. *Distr.* Caucasus.

C. cana See *C. triumfettii*

C. candidissima See *C. cineraria*

C. cineraria Perennial to 90cm high. *Distr.* Italy.

C. cyanus BACHELOR'S BUTTONS, CORNFLOWER, BLUEBOTTLE. Annual to biennial herb. Flowers purple and white. *Distr.* Eurasia.

C. cynaroides See *Leuzea centauroides*

C. dealbata Perennial to 1m high. Flowers bright pink. *Distr.* Caucasus.

C. debeauxii Perennial to 90cm high. Flowers orange. *Distr.* W Europe.

C. glastifolia Perennial to 90cm. Flowers yellow. *Distr.* Turkey, Caucasus.

C. gymnocarpa See *C. cineraria*

C. hypoleuca Perennial to 50cm. Flowers pink. *Distr.* Turkey to Iran and the Caucasus.

C. jacea Perennial. Flowers pink. *Distr.* Europe to W Russia.

C. macrocephala Perennial to 1m high. Bracts with a large appendage. *Distr.* Caucasus.

C. montana PERENNIAL CORNFLOWER. Creeping rhizomatous herb with violet and blue flowers. *Distr.* Central Europe.

C. nigra HARDHEADS, LESSER KNAPWEED. Perennial to 1m high. Bracts with black-tipped appendages. Used medicinally as a diuretic. *Distr.* Eurasia, naturalized in North America.

C. orientalis Perennial to 1m. Flowers cream. *Distr.* SE Europe to W Asia.

C. pulcherrima Woody-based perennial to 40cm high. Flowers rose-purple. *Distr.* Caucasus.

C. rutifolia Perennial to 90cm high. Flowers bright pink. *Distr.* SE Europe.

C. scabiosa GREATER KNAPWEED. Perennial to 1.5m high. Flowers purple. Bracts brown-tipped. *Distr.* Europe.

C. simplicicaulis Tufted perennial to 40cm. Flowers rose-coloured. *Distr.* Caucasus.

C. triumfettii Rhizomatous perennial. Flowers purple. *Distr.* S and Central Europe.

C. uniflora Small perennial. Flowers violet. *Distr.* Mountains of Central and SE Europe.

Centaurium (From the Greek *kentaur*, centaur; the centaur Chiron in Greek mythology was said to know the medicinal value of plants.) CENTAURY. A genus of 30 species of annual to perennial herbs with simple paired leaves and broad, flat-topped clusters of funnel-shaped flowers. *Distr.* N hemisphere, 1 species in Australia, 1 species in Chile. *Fam.* Gentianaceae. *Class* D.

C. scilloides Tuft-forming perennial. Flowers bright pink to white. Grown as an ornamental and reported to have medicinal properties. *Distr.* Azores, W Europe.

centaury *Centaurium.*

Centradenia (From the Greek *kentron*, spur, and *aden*, gland, alluding to the spur-like gland borne by the anthers.) A genus of 4 species of perennial herbs and shrubs with unequal pairs of simple, thin leaves and panicles or racemes of irregular, pink or white flowers that bear 4 petals. *Distr.* Central America. *Fam.* Melastomataceae. *Class* D.

C. floribunda Evergreen, rounded shrub. Flowers in large panicles, lilac-pink, white inside. Grown as an ornamental. *Distr.* Guatemala, Mexico.

Centranthus (From the Greek *kentron*, spur, and *anthos*, flower.) A genus of 9 species of annual and perennial herbs and subshrubs with opposite, simple or pinnately divided leaves and dense heads of small, funnel-shaped, spurred flowers. *Distr.* Mediterranean and Europe. *Fam.* Valerianaceae. *Class* D.

C. ruber RED VALERIAN, JUPITER'S BEARD, FOX'S BRUSH. Herb to 80cm high. Flowers red, spurred. Grown as an ornamental. *Distr.* Mediterranean, Europe, naturalized in Great Britain and California.

Centrolepidaceae A family of 4 genera and about 30 species of small annual or perennial herbs with grass-like leaves and a spike-like inflorescence. *Distr.* Tropical Asia to Pacific and temperate South America. *Class* M.

century plant *Agave*, *A. americana.*

Cephalanthus (From the Greek *kephale*, head, and *anthos*, flower.) A genus of 6 species of deciduous and evergreen shrubs and trees with simple, whorled or opposite leaves and dense, rounded heads of 4- or 5-lobed, tubular or funnel-shaped flowers. *Distr.* Tropical regions, North America. *Fam.* Rubiaceae. *Class* D.

C. occidentalis BUTTON BUSH, HONEY BELLS, BUTTON WILLOW. Deciduous shrub or tree. Flowers cream-white, fragrant. Grown as an ornamental and previously used medicinally as a laxative. *Distr.* North America.

cephalanthus, -a, -um: with flowers in heads.

Cephalaria (From the Greek *kephale*, head, alluding to the inflorescences.) A genus of about 65 species of annual and perennial herbs with opposite, pinnately lobed leaves and rounded heads of tubular flowers that bear their parts in 4s. Several species are grown as ornamentals. *Distr.* Mediterranean to Central Asia, Ethiopia, South Africa. *Fam.* Dipsacaceae. *Class* D.

C. gigantea Robust perennial to 2m high. Flowers cream-yellow. *Distr.* Caucasus to Siberia.

C. tatarica See *C. gigantea*

cephalonicus, -a -um: of Cephalonia (now Kefallinia), a Greek island.

Cephalotaceae A family of 1 genus and 1 species. *Class* D.
See: Cephalotus.

Cephalotaxaceae A family of 1 genus and 4 species. *Class* G.
See: Cephalotaxus.

Cephalotaxus (From the Greek *kephale*, a head, and the genus name *Taxus*.) PLUM YEW. A genus of 4 species of evergreen, coniferous shrubs and small trees with soft, linear leaves and head-like inflorescences. The fruit is a small, olive-like drupe. Several species are grown as ornamentals. *Distr.* E Himalaya to Japan. *Fam.* Cephalotaxaceae. *Class* G.

C. harringtonia JAPANESE PLUM YEW. Shrub or small tree to 5m high. *Distr.* Japan, China and Korea.

Cephalotus A genus of 1 species of perennial herb with some of the rosette leaves modified into pitcher-shaped insect traps, an example of parallel evolution with the genera *Nepenthes* and *Sarracenia*. The flowers are small, regular and bisexual and are borne in racemes on a tall leafless stem. *Distr.* SW Australia. *Fam.* Cephalotaceae. *Class* D.

C. follicularis Occasionally grown as an ornamental.

cephalotus, -a, -um: with a small head.

ceraceus, -a, -um: waxy.

cerasiferus: bearing cherry-like fruits.

cerasiformis: cherry-shaped, usually referring to the fruit.

cerastioides: from the genus *Cerastium*, with the ending *-oides*, indicating resemblance.

Cerastium (From the Greek *keras*, a horn, alluding to the shape of the seed capsules.) MOUSE-EAR CHICKWEED, CHICKWEED. A genus of 60 species of annual and perennial, often tuft- or cushion-forming herbs with simple, opposite leaves and white flowers that bear 5 petals which are typically deeply notched. Several species are grown as ornamentals. *Distr.* N temperate and Arctic regions. *Fam.* Caryophyllaceae. *Class* D.

C. alpinum ALPINE MOUSE-EAR. Mat-forming perennial. Flowers solitary or in small clusters. Petals deeply divided. *Distr.* Arctic regions, Mountains of Europe.

C. fontanum COMMON MOUSE-EAR CHICKWEED. Short-lived perennial or annual with creeping rootstock. A common garden weed. *Distr.* N temperate regions.

C. tomentosum SNOW IN SUMMER. Dense mat-forming perennial. Leaves grey-white, hairy. Petals white, notched. Used as ground cover in the garden. Numerous cultivars are available. *Distr.* Mountains of S and Central Europe, W Asia.

cerasus: the Latin name for cherry.

ceratocarpa: with horned fruit, from the Greek *cerato-*, horned, and *carpos*, fruit.

Ceratonia (From the Greek *keras*, a horn, alluding to the shape of the fruit pod.) A genus of 1 species of evergreen shrub or tree with alternate, pinnate, leathery leaves and racemes of minute flowers that bear 5 fused sepals and 5 stamens but lack petals. The fruit is a leathery pod containing numerous shiny seeds. *Distr.* Arabia and Somalia, naturalized North America. *Fam.* Leguminosae. *Class* D.

C. siliqua CAROB BEAN, LOCUST BEAN, ST JOHN'S BREAD, ALGARROBA BEAN. Trees planted for shade and timber. Seeds used as chocolate substitute and sweets; previously used as weights because of their uniformity.

Ceratophyllaceae A family of 1 genus and 2 species. *Class* D.
See: Ceratophyllum.

Ceratophyllum (From the Greek *keras*, a horn, and *phyllon*, a leaf.) HORNWORT. A genus of about 30 species of submerged, aquatic herbs entirely lacking roots. The leaves are thin, forked, and borne in whorls of 3–10. The flowers are tiny, borne in the leaf axils, and giving rise to a spiny fruit. *Distr.* Cosmopolitan. *Fam.* Ceratophyllaceae. *Class* D.

C. demersum Occasionally grown as an ornamental. *Distr.* Tropical Africa, S Europe.

Ceratostigma (From the Greek *keras*, a horn, and *stigma*, a stigma.) A genus of 8 species of perennial herbs and subshrubs with simple, alternate, somewhat bristly leaves and heads of small flowers that bear their parts in 5s. Several species are grown as ornamentals. *Distr.* Tropical Africa, Tibet, China, SE Asia. *Fam.* Plumbaginaceae. *Class* D.

C. willmottianum JAPANESE PLUMBAGO. Shrub to 1m high. Flowers purple-red. *Distr.* W China, Tibet.

Ceratotheca (From the Greek *keras*, horn, and *theke*, case, alluding to the horned fruit.) A genus of 5 species of annual herbs with simple, toothed leaves and solitary, 5-lobed, tubular flowers. *Distr.* Tropical and S Africa. *Fam.* Pedaliaceae. *Class* D.

C. triloba Erect herb to 2m high. Flowers mauve. Grown as an ornamental. *Distr.* S Africa, naturalized in North America.

cercidifolius from the genus *Cercidium*, and *folius*, leaved.

Cercidiphyllaceae A family of 1 genus and 1 species. *Class* D.
See: Cercidiphyllum.

Cercidiphyllum (From the Greek *kerkis*, Judas tree (*Cercis*), and *phyllon*, a leaf.) A genus of 1 species of tree with deciduous leaves of two types, palmately veined on the short shoots and pinnately veined on the longer shoots. The unisexual flowers are small, lack petals and the males and females are borne on separate plants. *Distr.* China, Japan. *Fam.* Cercidiphyllaceae. *Class* D.

C. *japonicum* KATSURA TREE. Grown as a hardy ornamental. *Distr.* China and Japan.

Cercis (From the Greek *kerkis*, a weaver's shuttle, alluding to the large woody fruits.) A genus of 6 species of deciduous shrubs and trees with simple, alternate, kidney-shaped leaves and clusters or racemes of irregular, pea-like flowers. Several species are grown as ornamentals. *Distr.* N temperate regions. *Fam.* Leguminosae. *Class* D.

C. *canadensis* REDBUD. Large shrub or small tree. Flowers red or lilac. Flowers used in salads and pickles. *Distr.* E USA.

C. *chinensis* CHINESE REDBUD. Large erect shrub. Flowers red to lavender. *Distr.* Central China.

C. *griffithii* Spreading shrub or small tree. Flowers mauve or lilac. *Distr.* W and C Asia.

C. *occidentalis* WESTERN REDBUD, CALIFORNIAN REDBUD. Large shrub. Flowers pink-red. *Distr.* W USA.

C. *siliquastrum* JUDAS TREE, LOVE TREE. Large shrub or small tree. Flowers magenta or pink. Grown for its colourful timber which is used in veneers. *Distr.* S Europe, E Mediterranean regions.

Cercocarpus (From the Greek *kerkos*, tail, and *karpos*, fruit, alluding to the plume-like style which remains attached to the fruit.) A genus of 6 species of evergreen shrubs and small trees with simple, alternate leaves and clusters of small, cup-shaped flowers which lack petals. The fruits bear a distinct hairy tail formed from the remains of the style. *Distr.* W and SW North America. *Fam.* Rosaceae. *Class* D.

C. *breviflorus* See *C. montana*

C. *montana* Shrub to 4m high. Flowers clustered. Fruit tails to 6cm long. Occasionally grown as an ornamental and minor source of timber. *Distr.* W North America.

cerefolium: waxen-leaved, from the Greek *keras*, wax, and Latin *folius*, leaved.

cerifera: wax-bearing.

ceriman *Monstera deliciosa*.

Cerinthe (From the Greek *keros*, wax, and *anthos*, flower; bees are alleged to make wax from the flowers.) A genus of 10 species of annual, biennial, and perennial herbs with simple leaves and pendent cymes of more or less tubular, yellow flowers. Several species are grown as ornamentals. *Distr.* Mediterranean. *Fam.* Boraginaceae. *Class* D.

C. *glabra* Biennial or perennial to 50cm high. Flowers yellow, marked with purple within. *Distr.* Central and S Europe.

cernuus, a, um: nodding.

Ceropegia (From the Greek *keros*, wax, and *pege*, fountain, alluding to the shape of the waxy flowers.) A genus of 160 species of perennial, sometimes succulent herbs and climbers with simple, often narrow leaves. The petals are fused to form a tube with a swollen base and 5 spreading lobes above. In some species the tips of the lobes are fused so as to form a lantern-like structure. Many species are grown as ornamentals. *Distr.* Tropical Old World, Australia, warm Africa including the Canary Islands. *Fam.* Asclepiadaceae. *Class* D.

C. *barkleyi* Tuberous perennial. Stems trailing or erect. Flowers to 2.5cm long, green, striped maroon. *Distr.* South Africa (Cape Province).

C. *linearis* STRING OF HEARTS, HEART VINE, HEARTS ENTANGLED. Stems thin, twining, pendent. Leaves heart-shaped. Flowers to 2cm long, dull pink. A popular house plant, particularly subspecies *woodii*. *Distr.* S Africa.

Ceroxylon (From the Greek *keros*, wax, and *xylon*, wood, alluding to the waxy coating of the trunk.) WAX PALM. A genus of about 15 species of very large palms with feather-shaped, stalked leaves and 2m-long, arching spikes of small, somewhat fleshy flowers. *Distr.* Andes. *Fam.* Palmae. *Class* M.

C. *alpinum* To 60m tall, tallest of all the palms. Wax from trunk used to manufacture candles. *Distr.* Colombia.

Cestrum (From the Greek *kestron*, a name applied to an unknown plant.) BASTARD JASMINE. A genus of 175 species of evergreen or deciduous shrubs with simple, alternate leaves and funnel-shaped to tubular, 5-lobed flowers that are followed by succulent berries. Several species are grown as ornamentals. *Distr.* Tropical America. *Fam.* Solanaceae. *Class* D.

C. aurantiacum Evergreen scrambling shrub. Flowers bright orange, borne in large clusters. *Distr.* Guatemala.

C. elegans Evergreen shrub or small tree. Flowers purple-red, borne in dense racemes. *Distr.* Mexico.

C. nocturnum LADY OF THE NIGHT, NIGHT JASMINE. Spreading evergreen shrub. Flowers yellow-green. Used medicinally. *Distr.* W Indies.

C. parqui WILLOW-LEAVED JASMINE. Deciduous shrub. Flowers yellow-green, fragrant at night. Toxic to sheep and cattle. *Distr.* S South America.

Ceterach See *Asplenium*

Chaenactis (From the Greek *chaino*, to gape, and *aktis*, ray, alluding to the shape of the outer flowers.) A genus of about 25 species of annual and low-growing perennial herbs with alternate leaves and flowers in daisy-like heads that lack distinct rays. *Distr.* W North America. *Fam.* Compositae. *Class* D.

C. alpina Small perennial. Leaves in terminal rosettes. Flowers white. *Distr.* USA (Oregon, Montana, Colorado).

C. glabriuscula Annual. Leaves pinnately divided. Flowers yellow. *Distr.* USA (California).

Chaenomeles (From the Greek *chainein*, to gape, and *melon*, apple, alluding to the erroneous belief that the fruit splits.) JAPONICA, FLOWERING QUINCE. A genus of 3 species of deciduous, typically thorny shrubs with simple, alternate leaves and numerous saucer-shaped flowers that are followed by yellow, apple-like fruits. Several species and a number of cultivars are grown as ornamentals. The fruit is edible when cooked in much the same way as the COMMON QUINCE, *Cydonia oblonga*. *Distr.* E Asia. *Fam.* Rosaceae. *Class* D.

C. japonica JAPANESE QUINCE, JAPONICA, MAULE'S QUINCE. Spreading shrub. Flowers to 4cm across, orange-red to scarlet. *Distr.* Japan.

C. maulei See *C. japonica*

C. sinensis See *Pseudocydonia sinensis*

C. speciosa Spreading shrub. Flowers red. Several cultivars of this species have been raised and it is a popular bonsai subject in Japan. *Distr.* China.

Chaenorrhinum (From the Greek *chaino*, to gape, and *rhis*, snout, alluding to the open-mouthed flowers.) A genus of 20 species of annual and perennial herbs with simple, opposite, or alternate leaves and 2-lipped, tubular, spurred flowers. Several species are grown as ornamentals. *Distr.* Europe, N Africa, W Asia. *Fam.* Scrophulariaceae. *Class* D.

C. glareosum Perennial herb. Flowers violet to pink with yellow markings. *Distr.* S Spain.

C. origanifolium Perennial herb. Flowers violet or white, marked with yellow, borne in loose racemes. *Distr.* SW and Central Europe.

Chaerophyllum (From the Greek *chairo*, to please, and *phyllon*, a leaf, alluding to the fragrant foliage.) A genus of about 35 species of annual, biennial and perennial herbs with taproots or tubers, finely divided leaves and compound umbels of typically white flowers. *Distr.* N temperate regions. *Fam.* Umbelliferae. *Class* D.

C. aureum GOLDEN CHERVIL. Perennial. Foliage lime- to golden-green. Grown as an ornamental. *Distr.* Central and S Europe to W Asia.

C. bulbosum TURNIP-ROOTED CHERVIL, PARSNIP CHERVIL. Erect biennial to 2m high. The root is eaten as a vegetable. *Distr.* Central and E Europe, naturalized North America.

chain of love *Antigonon leptopus*.

chalcedonicus, -a, -um: of Chalcedon (now Kadikoy), near Istanbul, Turkey.

chalk lettuce plant *Dudleya pulverulenta*.

chalksticks, blue *Senecio serpens*.

Chamaebatiaria (From the name of a related genus, *Chamaebatia*.) A genus of 1 species of deciduous subshrub to 1.5m high with pinnately divided, balsam-scented leaves and panicles of small, white flowers. *Distr.* W North America. *Fam.* Rosaceae. *Class* D.

C. millefolium Grown as a rock-garden ornamental.

chamaebuxus: dwarf *Buxus*.

chamaecistus: dwarf *Cistus*.

Chamaecyparis (From the Greek *chamai*, dwarf, and *kuparissos*, cypress.) FALSE CYPRESS. A genus of 7 species of evergreen, coniferous shrubs and trees with minute, scale-like leaves borne on finely branched shoots. Some species are a commercial source of timber; others are grown as ornamentals. *Distr.* E Asia, North America. *Fam.* Cupressaceae. *Class* G.

C. lawsoniana LAWSON'S CYPRESS, OREGON CEDAR. Tree to 60m high. A very important source of timber in W North America and much planted as an ornamental elsewhere. *Distr.* W North America.

C. leylandii See × *Cupressocyparis leylandii*

C. nootkatensis NOOTKA CYPRESS, SITKA CYPRESS, YELLOW CEDAR, YELLOW CYPRESS. Narrow tree to 40m high. A source of timber. Several ornamental cultivars have been raised from this species. *Distr.* W North America.

C. obtusa JAPANESE CYPRESS. *Distr.* Central and S Japan, Taiwan.

C. pisifera SAWARA CYPRESS. *Distr.* Central and S Japan.

chamaecyparissus: dwarf cypress.

Chamaecytisus (From the Greek *chamai*, dwarf, and the genus name *Cytisus*.) A genus of 30 species of shrubs and small trees with alternate, trifoliate leaves and racemes of irregular, pea-like flowers. This genus is sometimes included within the genus *Cytisus*. Several species are grown as ornamentals. *Distr.* Europe, Canary Islands. *Fam.* Leguminosae. *Class* D.

C. albus Shrub. Flowers white or pale yellow. *Distr.* SE and Central Europe.

C. purpureus Densely branched, low shrub. Flowers large, red to pale pink. *Distr.* SE Europe.

Chamaedaphne (From the Greek *chamai*, dwarf, and *daphne*, the LAUREL.) A genus of 1 species of low evergreen shrub with simple leaves and small white flowers. *Distr.* N temperate regions. *Fam.* Ericaceae. *Class* D.

C. calyculata LEATHERLEAF, CASSANDRA. A cultivated ornamental previously used by North American Indians to make tea.

Chamaedorea (From the Greek *chamai*, dwarf, and *dorea*, a gift, alluding to the easily reached fruits.) A genus of 110–140 species of small to medium-sized palms with slender, erect, climbing or creeping stems, pinnate or entire leaves and pendent inflorescences. Some species have edible fruits. *Distr.* North, Central and South America, Europe. *Fam.* Palmae. *Class* M.

C. elegans Very common house-plant in USA. Fruit and immature inflorescence edible. *Distr.* Mexico, Guatemala.

chamaedrifolius, -a, -um: from the genus *Chamaedrys*, and *folius*, leaved.

chamaedrioides: from the genus *Chamaedrys*, with the ending *-oides*, indicating resemblance.

Chamaelirium (From the Greek *chamai*, dwarf, and *lirion*, lily.) A genus of 1 species of tuberous herb with a basal rosette of spoon-shaped leaves and a dense cylindrical spike of small white flowers. The male and female flowers are borne on separate plants, the female plants being somewhat larger than the male. *Distr.* E North America. *Fam.* Melanthiaceae. *Class* M.

C. luteum BLAZING STAR, UNICORN ROOT, FAIRY WAND. The dried tubers are used medicinally.

Chamaemelum (From the Greek *chamai*, dwarf, and *melon*, apple.) CHAMOMILE. A genus of 3 species of annual to perennial herbs with alternate bi- or tripinnate leaves and yellow flowers in daisy-like heads that may or may not bear distinct white rays. *Distr.* Europe, Mediterranean. *Fam.* Compositae. *Class* D.

C. nobile CHAMOMILE. Decumbent, aromatic perennial. An important medicinal herb; Chamomile is also a source of essential oil used to flavour liqueurs. Its dried leaves and flowers are used, to make tea. In the garden it has been cultivated as lawns that require minimum maintenence and have good drought tolerance. Although considered a weed of grass lawns it is still included in some seed mixtures for particularly hard-wearing areas. Several cultivars are available, including a non-flowering form for use in lawns. *Distr.* S and W Europe, Mediterranean.

Chamaenerion See *Epilobium*

C. angustifolium See *Epilobium angustifolium*

Chamaepericlymenum See *Cornus*

C. canadense See *Cornus canadensis*

Chamaerops (From the Greek *chamai*, dwarf, and *rhops*, a bush.) A genus of 1 species of shrubby, clump-forming palm with fan-shaped leaves on spiny stalks and short stiff inflorescences of yellow flowers. *Distr.* N Africa, S Europe. *Fam.* Palmae. *Class* M.

C. excelsa See *Trachycarpus fortunei*

C. humilis DWARF FAN PALM. Grown as an ornamental. The leaves are a source of 'vegetable horse-hair' (Algerian fibre, crin vegetable), used in upholstery, and the young buds are edible.

Chamaespartium See *Genista*

chamomile *Chamaemelum, C. nobile* **dyer's** ~ *Anthemis tinctoria* **false** ~ *Boltonia* **yellow** ~ *Anthemis tinctoria.*

champak *Michelia champaca.*

chaplet flower *Stephanotis floribunda.*

charitopes with a graceful stalk.

charity *Polemonium caeruleum.*

Charlie, creeping *Pilea nummulariifolia.*

Chasmanthe (From the Greek *chasme*, gaping, and *anthos*, flower.) A genus of about 3 species of perennial herbs with corms, sword-shaped leaves and spikes of irregular, red to yellow flowers. *Distr.* S and SW Africa. *Fam.* Iridaceae. *Class* M.

C. aethiopica Leaves narrow. Flowers scarlet. Grown as an ornamental. *Distr.* South Africa (Cape Province).

C. floribunda Leaves broad. Flowers orange or scarlet. Grown as an ornamental. *Distr.* South Africa.

Chasmanthium A genus of 5 species of perennial grasses with flat leaves and large panicles. *Distr.* North America. *Fam.* Gramineae. *Class* M.

C. latifolium NORTH AMERICAN WILD OATS, SPANGLE GRASS, SEA OATS. Grown for the large decorative panicles. *Distr.* E USA, N Mexico.

chaste tree *Vitex agnus-castus.*

chathamicus, -a, -um: of the Chatham Islands, South Pacific.

chayote *Sechium edule.*

checkerberry *Gaultheria procumbens.*

checkerbloom *Sidalcea malviflora.*

cheeses *Malva sylvestris.*

cheese wood *Pittosporum undulatum.*

Cheilanthes (From the Greek *cheilos*, lip, and *anthos*, flower, because the sori are clustered on the edge of the frond.) RESURRECTION FERN, LIP FERN. A genus of about 180 species of small ferns, often with scaly leaves. The foliage shrivels during drought, rehydrating after rains. Some species are grown as ornamentals. *Distr.* Widespread in tropical and warm temperate zones. *Fam.* Adiantaceae. *Class* F.

C. lanosa Leaves tripinnately divided, to 30cm long. *Distr.* North America.

cheilanthifolius, -a, -um: from the genus *Cheilanthes*, and *folius*, leaved.

cheiranthoides: from the genus *Cheiranthus*, with the ending *-oides*, indicating resemblance.

Cheiranthus See *Erysimum*

C. cheiri See *Erysimum cheiri*

C. rupestris See *Erysimum pulchellum*

Cheiropleuriaceae A family of 1 genus and 1 species of terrestrial or lithophytic, medium-sized ferns with broad, 2-lobed vegetative leaves and narrow, strap-shaped fertile leaves. *Distr.* Indochina to Malaysia. *Class* F.

chelidoniifolius, -a, -um: from the genus *Chelidonium*, and *folius*, leaved.

chelidonioides: from the genus *Chelidonium*, with the ending *-oides*, indicating resemblance.

Chelidonium (From the Greek *chelidon*, swallow, alluding to the flowering time which coincided with the appearance of the birds.) A genus of 1 species of biennial or perennial herb with orange latex, pinnately lobed or cut leaves, and regular flowers that bear 2 sepals, 4 petals and numerous stamens. *Distr.* Temperate and subarctic Eurasia, naturalized in E USA. *Fam.* Papaveraceae. *Class* D.

C. majus GREATER CELANDINE, SWALLOWWORT. Occasionally grown as an ornamental. The latex has long been used medicinally in the treatment of eye and skin disorders.

Chelone (From the Greek *chelone*, tortoise, alluding to the flowers which are said to look like a tortoise's head.) SHELLFLOWER. A genus of 5–6 species of perennial herbs with simple, opposite leaves and spike-like racemes of 2-lipped, 5-lobed, tubular flowers that are closed at the mouth. Several species are grown as ornamentals. *Distr.* North American. *Fam.* Scrophulariaceae. *Class* D.

C. barbata See *Penstemon barbatus*

C. glabra BALMONY, TURTLEHEAD, SNAKEHEAD. Erect herb to 60cm high. Flowers pink. Formerly used medicinally as a vermifuge. *Distr.* E North America.

C. obliqua TURTLEHEAD. Erect herb to 60cm. Flowers pink-purple. *Distr.* USA.

chenille plant *Echeveria pulvinata.*

Chenopodiaceae The Goosefoot family. 113 genera and 1300 species of perennial herbs, shrubs and trees with deep penetrating roots and simple or lobed leaves. The flowers are very small and typically bear 5 sepals, no petals, and a superior ovary. The fruit is a small nut. Many species are capable of living in soils with a high concentration of salts. Sugar Beet (*Betas vulgaris*) is the most important crop in this family although *Chenopodium quinoa* is also important as a vegetable and grain crop in South America. *Distr.* Widespread. *Class* D.
See: Atriplex, Beta, Camphorosma, Chenopodium, Rhagodia, Spinacia.

Chenopodium (From the Greek *chen*, a goose, and *podion*, a little foot, alluding to the leaf shape.) GOOSEFOOT. A genus of 150 species of annual and perennial herbs and subshrubs with alternate, lobed leaves and spikes or panicles of numerous small flowers that bear 5 papery sepals, no petals, and 5 stamens. Some species are edible, usually as leaf vegetables, occasionally as grains. *Distr.* Temperate and tropical regions. *Fam.* Chenopodiaceae. *Class* D.

C. album FAT HEN, GOOSEFOOT. An annual herb. Leaf surfaces mealy. Leaves eaten as a vegetable. *Distr.* Europe, occurring as a weed of agricultural land.

C. bonus-henricus GOOD KING HENRY, ALL GOOD, WILD SPINACH, MERCURY. Cultivated since at least medieval times as a vegetable, the leaves being eaten like SPINACH and the young shoots like ASPARAGUS. *Distr.* Europe.

C. botrys FEATHER GERANIUM. Erect annual herb. Leaves aromatic. *Distr.* S Europe to Central Asia.

C. foliosum Once used as a leaf vegetable. *Distr.* Asia, S Europe, and NW Africa.

C. quinoa QUINOA. Grown for its edible leaves and seeds; particularly important in the Peruvian Andes.

cherimoya *Annona cherimolia.*

cherry *Prunus avium* **amarelle** ~ *P. cerasus* **American** ~ *P. serotina* **bird** ~ *P. avium, P. padus* **black** ~ *P. serotina* **bladder** ~ *Physalis alkekengi* **Christmas** ~ *Solanum pseudocapsicum* **cornelian** ~ *Cornus mas* **ground** ~ *Physalis pruinosa* **Japanese cornelian** ~ *Cornus officinalis* **Japanese flowering** ~ *Prunus sato-zakura* **Jerusalem** ~ *Solanum pseudocapsicum* **Madeira winter** ~ *S. pseudocapsicum* **morello** ~ *Prunus cerasus* **plum** ~ *P. serotina* **purple ground** ~ *Physalis peruviana* **rum** ~ *Prunus serotina* **sour** ~ *P. cerasus* **steppe** ~ *P. fruticosa* **Surinam** ~ *Eugenia uniflora* **sweet** ~ *Prunus avium* **wild** ~ *P. avium* **winter** ~ *Solanum pseudocapsicum, Physalis alkekengi.*

cherry pie *Heliotropium arborescens.*

chervil *Anthriscus cerefolium* **golden** ~ *Chaerophyllum aureum* **parsnip** ~ *C. bulbosum* **turnip-rooted** ~ *C. bulbosum* **white** ~ *Cryptotaenia canadensis.*

chess *Bromus.*

chestnut: Australian ~ *Castanospermum australe* **Chinese** ~ *Castanea mollissima* **Chinese water** ~ *Eleocharis dulcis* **earth** ~ *Lathyrus tuberosus* **golden** ~ *Chrysolepis chrysophylla* **horse** ~ *Aesculus, A. hippocastanum*, Hippocastanaceae **Indian horse** ~ *Aesculus indica* **Japanese horse** ~ *A. turbinata* **Moreton Bay** ~ *Castanospermum australe* **red horse** ~ *Aesculus pavia* **Spanish** ~ *Castanea sativa* **sweet** ~ *C. sativa* **water** ~ *Eleocharis dulcis, Trapa, T. natans*

Chiastophyllum (From the Greek *chiastos*, diagonally arranged, and *phyllon*, leaf.) A genus of 1 species of rhizomatous, trailing herb with large, rounded, fleshy leaves and a many-flowered cyme of nodding, somewhat bell-shaped yellow flowers. *Distr.* Caucasus. *Fam.* Crassulaceae. *Class* D.

C. *oppositifolium* Grown as an ornamental.

C. *simplicifolium* See *C. oppositifolium*

chicken gizzard *Iresine herbstii.*

chickweed *Cerastium, Stellaria, Trientalis* **common mouse-ear** ~ *Cerastium fontanum* **mouse-ear** ~ *Cerastium.*

chicory *Cichorium intybus.*

chicot *Gymnocladus dioica.*

chiger flower *Asclepias tuberosa.*

Chile nut *Gevuina avellana.*

Chile nut tree *Araucaria araucana.*

chilensis: of Chile.

Chiliotrichum (From the Greek *chilioi*, thousand, and *thrix*, hair.) A genus of 2 species of erect shrubs with alternate, evergreen leaves and flowers in solitary, daisy-like heads that bear distinct rays. *Distr.* Temperate South America. *Fam.* Compositae. *Class* D.

C. *diffusum* Much branched shrub. Rays white. *Distr.* S South America, Falkland Islands.

chilli *Capsicum annuum.*

chiloensis, -e: of Chile.

chimaera: a monster; an organism made by combining genetically different parts.

Chimaphila (From the Greek *cheima*, winter, and *phileo*, to love, alluding to the evergreen habit.) A genus of 4–5 species of evergreen herbs with a creeping rootstock and simple, leathery leaves. The nodding, white to pink flowers are borne in a flat-topped cluster. Some species have medicinal qualities. *Distr.* Eurasia, North and tropical America. *Fam.* Ericaceae. *Class* D.

C. *maculata* A cultivated ornamental. *Distr.* E North America.

Chimonanthus (From the Greek *cheimon*, winter, and *anthos*, flower.) A genus of 4–6 species of evergreen or deciduous shrubs with simple, opposite leaves and regular, yellow-white flowers that bear numerous fleshy tepals and open in late winter. *Distr.* China, Japan. *Fam.* Calycanthaceae. *Class* D.

C. *fragrans* See *C. praecox*

C. *praecox* WINTERSWEET. Grown as an ornamental for its waxy, fragrant flowers.

Chimonobambusa (From the Greek *cheimon*, winter, and the Latin *bambusa*, bamboo.) BAMBOO. A genus of 20 species of spreading, rhizomatous bamboos often with thorny or swollen nodes. *Distr.* Himalaya, China and Japan. *Fam.* Gramineae. *Class* M.

C. *hookeriana* See *Drepanostachyum hookerianum*

C. *marmorea* KAN CHIKU. Stems to 3m high. New growth tinged pink. Grown as a hardy ornamental. *Distr.* Japan.

C. *quadrangularis* Stem 4-angled. Often grown as a large hardy ornamental. *Distr.* China.

china tree, wild *Sapindus drummondii.*

chincherinchee *Ornithogalum thyrsoides.*

chinensis, -e: of China.

Chinese root *Rheum palmatum.*

chingma *Abutilon theophrasti.*

chinkapin, golden *Chrysolepis chrysophylla.*

chinkapin, water *Nelumbo lutea.*

Chiogenes See *Gaultheria*

C. *hispidula* See *Gaultheria hispida*

Chionanthus (From the Greek *chion*, snow, and *anthos*, flower.) FRINGE FLOWER. A genus of 120 species of deciduous shrubs and trees with simple, opposite leaves and panicles of regular, white flowers that bear 4 slender, fused petals and 2 stamens. Several species are grown as ornamentals. *Distr.* E Asia, E North America. *Fam.* Oleaceae. *Class* D.

C. *retusa* CHINESE FRINGE TREE. Arching shrub. Flowers pure white, borne in large masses. *Distr.* Taiwan.

C. *virginicus* FRINGE TREE. Bushy shrub or small tree. Leaves dark green. Flowers in drooping sprays. *Distr.* E North America.

chionanthus, -a, -um: with snow-white flowers.

Chionochloa (From the Greek *chion*, snow, and *chloe*, grass.) SNOW GRASS. A genus of about 20 species of perennial grasses with robust, narrow, grooved leaves and persistent, showy panicles. An important genus in alpine areas of New Zealand, it also contains a number of ornamental species. *Distr.* Mountains of New Zealand and Australia. *Fam.* Gramineae. *Class* M.

C. conspicua HUNANGAMOHO GRASS, PLUMED TUSSOCK GRASS. Stems densely tufted. Leaves to 1.2m long. *Distr.* New Zealand.

C. rubra RED TUSSOCK GRASS. Leaves tinged red. *Distr.* New Zealand.

Chionodoxa (From the Greek *chion*, snow, and *doxa*, glory, alluding to the flowers which appear as the snow recedes.) GLORY OF THE SNOW. A genus of 6 species of small, bulbiferous herbs with linear leaves and racemes of blue to pink or white flowers which bear 6 fused tepals. All the species are grown as ornamentals for their early, often rich blue flowers. *Distr.* W Turkey, Crete, Cyprus. *Fam.* Hyacinthaceae. *Class* M.

C. forbesii Flowers deep blue with a white centre. Several cultivars are available. *Distr.* W Turkey.

C. gigantea See *C. luciliae*

C. luciliae GLORY OF THE SNOW. Flowers few, upward-pointing, blue with a white centre. *Distr.* W Turkey.

C. sardensis Flowers nodding, deep blue without a white centre. *Distr.* W Turkey.

C. tmolusi See *C. forbesii*

chionophilus, -a, -um: snow-loving.

× *Chionoscilla* (From the names of the parent genera.) A genus of hybrids between members of the genera *Chionodoxa* and *Scilla*. *Fam.* Hyacinthaceae. *Class* M.

×*C. allenii* (After James Allen.) Bulbiferous herb with star-shaped, dark blue flowers. *Distr.* Garden origin.

Chironia (After Chiron in Greek mythology, who taught the uses of plants.) A genus of 15 species of perennial herbs with pairs of sessile, lance-shaped leaves and 5-lobed, salver-shaped flowers that are borne singly or in terminal clusters. Several species are grown as ornamentals. *Distr.* Madagascar, Tropical and S Africa. *Fam.* Gentianaceae. *Class* D.

C. baccifera CHRISTMAS BELLS. Shrubby perennial. Flowers dark pink. Fruit a berry-like, orange-red capsule. *Distr.* S Africa.

chittamwood *Cotinus obovatus*.

chives *Allium schoenoprasum* **Chinese** ~ *A. tuberosum* **garlic** ~ *A. tuberosum*.

Chlidanthus (From the Greek *chlide*, a luxury or costly ornament, and *anthos*, flower.) A genus of 2 species of bulbiferous, perennial herbs with linear leaves that appear after the umbels of funnel-shaped, yellow flowers. *Distr.* S Peru, Europe. *Fam.* Amaryllidaceae. *Class* M.

C. fragrans Grown as a tender ornamental for its scented flowers.

Chloranthaceae A family of 4 genera and 58 species of herbs, shrubs and trees usually with aromatic leaves and small inconspicuous flowers. Some species have medicinal uses; others may be grown as ornamentals. *Distr.* Tropical and subtropical regions, absent from Africa. *Class* D.

chloranthus, -a, -um: green-flowered.

chlorifolium from the genus *Chlora* (now *Blackstonia*), and *folius*, leaved.

chloropetalum: with green petals.

Chlorophytum (From the Greek *chloros*, green, and *phyton*, a plant.) A genus of about 300 species of rhizomatous, evergreen perennials with basal, linear, or lance-shaped leaves and racemes or panicles of small white-green flowers. The flowers are often replaced by small plantlets. A few species are cultivated as house-plants. *Distr.* Tropical America, Africa, Australia and India. *Fam.* Anthericaceae. *Class* M.

C. comosum RIBBON PLANT, SPIDER PLANT, SPIDER IVY. Leaves basal, linear to lance-shaped, often striped white. An extremely common house-plant. *Distr.* South Africa.

chlorops: with a green eye.

chlorosarcus, -a, -um: fleshy green.

chlorosticta: green-spotted.

chocho *Sechium edule*.

chocolate Indian *Geum rivale*.

chocolate plant *Theobroma cacao*.

chocolate root *Geum rivale.*

Choisya (After Jacques D. Choisy (1799–1854), Swiss botanist.) A genus of 7 species of aromatic shrubs with palmate, opposite leaves and clusters of regular flowers that bear 4–5 petals and 8–10 stamens. Several species are grown as ornamentals. *Distr.* SW North America. *Fam.* Rutaceae. *Class* D.

C. arizonica Small shrub to 1m high. Flowers white, around 2cm across. *Distr.* USA (Arizona).

C. ternata MEXICAN ORANGE. Shrub to 3m high. Flowers white, to 3cm across, scented. *Distr.* SW Mexico.

choisyanum: referring to the genus *Choisya.*

chokeberry *Aronia* **red** ~ *A. arbutifolia.*

Chondrodendron (From Greek chondras, cortilage and deadron, true.) A genus of 10 species of large deciduous lianas with large, rounded, simple leaves and clusters of small unisexual flowers. *Distr.* Central and tropical South America. *Fam.* Menispermaceae. *Class* D.

C. tomentosum CURARE, PAREIRA ROOT, PAREIRA BRAVA. The roots are a source of D-tubocurarine, a muscle relaxant used in surgery and a component of the arrow poison curare. *Distr.* Brazil and Peru.

Chordospartium (From the Greek *chorde*, string, and *spartos* BROOM.) A genus of 2 species of shrubs and small trees with pinnate leaves only present in young plants, replaced by slender, green, weeping or erect, branches in adult plants. The irregular, pea-like flowers are borne in crowded, erect, or pendulous racemes. *Distr.* New Zealand (South Island). *Fam.* Leguminosae. *Class* D.

C. muritai Stems erect. Flowers white with violet veins.

C. stevensonii WEEPING TREE BROOM. Branches weeping. Flowers lilac, veined purple.

Chorisia (After L. I. Choris, an early 19th-century artist.) A genus of 5 species of deciduous trees with thick spiny trunks, alternate palmate leaves and large showy flowers that bear 5 fused sepals and 5 free petals. *Distr.* Tropical America. *Fam.* Bombacaceae. *Class* D.

C. insignis Grown as an ornamental. The fibre from the seeds has allegedly been used to make arrow-proof clothing. *Distr.* Argentina, Peru.

Chorizema (From the Greek *koros*, a dance, and *zema*, a drinking vessel; the first European to collect *C. ilicifolia* was close to exhaustion when he came across it next to a water source.) A genus of 18 species of shrubs and twining climbers with alternate, simple or lobed leaves and racemes of irregular, pea-like flowers. Several species are grown as ornamentals. *Distr.* S Australia. *Fam.* Leguminosae. *Class* D.

C. ilicifolium HOLLY FLAME PEA. Erect or spreading subshrub or shrub to 1m high. Leaves holly-like. Flowers orange, red and yellow. *Distr.* W Australia.

chow chow *Sechium edule.*

Christmas bells Tasmanian *Blandfordia punicea*

Christmasberry tree *Schinus terebinthifolius*

Christmas bush, **Victoria** *Prostanthera lasianthos*

Christmas flower *Euphorbia pulcherrima.*

Christmas star *E. pulcherrima* ~ *tree Abiesalba, Picea aboes*

Christmas tree, New Zealand *Metrosideros excelsa.*

christophine *Sechium edule.*

christ plant *Euphorbia milii.*

Christ's thorns *Paliurus spina-christi.*

chrysacanthion: with golden spines, from the Greek *chryseus*, golden, and *acantha*, spine.

Chrysalidocarpus (From the Greek *chrysalis*, chrysalis, and *karpos*, fruit, alluding to the resemblance of the fruit to butterfly pupae.) A genus of about 20 species of medium-sized palms with feather-shaped, short-stalked leaves and pendent inflorescences of small flowers. Several species are grown as tender ornamentals. *Distr.* Madagascar. *Fam.* Palmae. *Class* M.

C. lutescens BAMBOO PALM, CANE PALM, YELLOW PALM, ARECA PALM, BUTTERFLY PALM.

Stems numerous, clustered. Leaves to 3m long. Flowers yellow. Grown as a house-plant.

chrysanthemoides: from the genus *Chrysanthemum*, with the ending *-oides*, indicating resemblance.

Chrysanthemum (From the Greek *chrysos*, gold, and *anthemon*, flower.) A genus of 2 species of annual herbs with alternate leaves and flowers in daisy-like heads that bear distinct rays. This genus is not to be confused with the CHRYSANTHEMUM of gardeners and florists. Several species and numerous cultivars are available for the garden. *Distr.* N Africa, Europe. *Fam.* Compositae. *Class* D.

C. alpinum See *Leucanthemopsis alpina*

C. arcticum See *Arctanthemum arcticum*

C. argenteum See *Tanacetum argenteum*

C. balsamita See *Tanacetum balsamita*

C. cinerariifolium See *Tanacetum cinerariifolium*

C. clusii See *Tanacetum corymbosum*

C. coccineum See *Tanacetum coccineum*

C. coronarium CROWN DAISY. Flowerheads solitary, flowers yellow. Leaves and flowers eaten in China and Japan. *Distr.* Mediterranean.

C. corymbosum See *Tanacetum corymbosum*

C. foeniculaceum See *Argyranthemum frutescens*

C. frutescens See *Argyranthemum frutescens*

C. haradjanii See *Tanacetum haradjanii*

C. leucanthemum See *Leucanthemum vulgare*

C. macrophyllum See *Tanacetum macrophyllum*

C. maximum See *Leucanthemum maximum*

C. nipponicum See *Nipponanthemum nipponicum*

C. parthenium See *Tanacetum parthenium*

C. ptarmiciflorum See *Tanacetum ptarmiciflorum*

C. roseum See *Tanacetum coccineum*

C. rubellum See *Dendranthema zawadskii*

C. segetum CORN MARIGOLD. Leaves clasping the stem, somewhat fleshy. Flowers yellow. Leaves eaten in China. *Distr.* Europe, W Asia, naturalized in North America.

C. uliginosum See *Leucanthemella serotina*

C. welwitschi See *Chrysanthemum segetum*

C. weyrichii See *Dendranthema weyrichii*

C. yezoense See *Dendranthema yezoense*

chrysanthemum *Dendranthema x grandiflorum* **florist's** ~ *D. x grandiflorum*.

chrysanthus, -a, -um: with golden flowers, from the Greek *chryseus*, golden, and *anthos*, flower.

Chrysobalanaceae The Coco Plum family. 17 genera and 460 species of shrubs and trees with simple, alternate leaves and irregular, unisexual or bisexual flowers. The fruits of some species are eaten locally, others are a source of seed oil. *Distr.* Tropical and subtropical regions. *Class* D.

Chrysocoma (From the Greek *chrysos*, gold, and *kome*, hair.) A genus of 20 species of small, annual and perennial herbs with simple or pinnately lobed leaves and yellow flowers in solitary, daisy-like heads that lack distinct rays. *Distr.* South Africa. *Fam.* Compositae. *Class* D.

C. coma-aurea Perennial subshrub grown as a free-flowering, scented annual in gardens. *Distr.* South Africa.

chrysocyathus: from the Greek *chryseus*, golden, and *cyathos*, a cup or ladle.

chrysodoron: long and golden.

Chrysogonum (From the Greek *chrysos*, gold, and *gone*, joint, alluding to the flowers, which are borne at the nodes of the stem.) GOLDEN KNEE. A genus of 4 species of rhizomatous, perennial herbs with simple leaves and yellow flowers in daisy-like heads that bear distinct yellow rays. *Distr.* E USA, Australia. *Fam.* Compositae. *Class* D.

C. virginianum Grown as ground cover. *Distr.* E USA.

chrysographes: with gold lines.

Chrysolepis (From the Greek *chrysos*, gold, and *lepis*, scale, alluding to the scaly leaves.) A genus of 2 species of evergreen shrubs and trees with simple, alternate leaves that are dark green above and coated in golden scales below. The small unisexual flowers are borne in slender catkins and the fruit consists of 1–3 nuts surrounded by a case (cupule) that is coated in branched spines. *Distr.* W North America. *Fam.* Fagaceae. *Class* D.

C. chrysophylla GOLDEN CHESTNUT, GOLDEN CHINKAPIN. Shrub or small tree. Grown as an ornamental. *Distr.* W USA.

chrysoleucus, -a, -um: gold and white coloured.

chrysophyllus, -a, -um: golden-leaved.

Chrysopsis (From the Greek *chrysos*, gold, and *opsis*, appearance.) GOLDEN ASTER. A genus of 10 species of annual and perennial herbs and subshrubs with simple, hairy leaves and yellow flowers borne in daisy-like heads that bear distinct yellow rays. *Distr.* SE USA, especially Florida. *Fam.* Compositae. *Class* D.

C. gossypina Biennial to perennial herb. Grown as a garden ornamental. *Distr.* Florida to Carolina.

chrysosphaera: golden sphere.

Chrysosplenium (From the Greek *chrysos*, gold, and *splen*, spleen.) GOLDEN SAXIFRAGE. A genus of about 55 species of perennial herbs of bogs and damp areas with round to kidney-shaped leaves and small flowers. *Distr.* Europe, N Africa, NE Asia, Temperate America. *Fam.* Saxifragaceae. *Class* D.

C. americanum WATER MAT, WATER CARPET. Flowering stems creeping, flowers yellow-green. *Distr.* North America.

C. oppositifolium GOLDEN SAXIFRAGE. Leaves opposite. Flowers dark yellow-green. An emergency food plant. *Distr.* Europe.

Chrysothamnus (From the Greek *chrysos*, golden, and *thamnos*, a shrub.) RABBIT BUSH. A genus of 13 species of much-branched shrubs with alternate, simple leaves and daisy-like heads of white or yellow flowers that lack distinct rays. *Distr.* SW North America. *Fam.* Compositae. *Class* D.

C. viscidiflorus Small shrub with white bark. *Distr.* W North America.

Chrysothemis (After Chrysothemis, in Greek mythology, the daughter of Agamemnon.) A genus of 7 species of perennial herbs with simple, opposite leaves and cymes of 5-lobed, tubular flowers. *Distr.* Tropical America. *Fam.* Gesneriaceae. *Class* D.

C. friedrichsthaliana Flowers orange-yellow with faint pink stripes. *Distr.* Columbia, Central America.

chufa *Cyperus esculentus.*

chusan *Trachycarpus fortunei.*

Chusquea (From the South American vernacular name for these bamboos) BAMBOO. A genus of 96 species of bamboo with solid stems and prolific branching. *Distr.* Mexico to S Argentina. *Fam.* Gramineae. *Class* M.

C. culeou Loosely tufted with short to long rhizomes. Grown as an ornamental. *Distr.* Most of Andean South America.

cicely, sweet *Myrrhis odorata* **woolly** ~ *Osmorhiza claytonii.*

Cicer (From the classical Latin name for the CHICK PEA.) A genus of about 40 species of annual and perennial herbs with pinnate or trifoliate leaves and racemes of small, pea-like flowers. The fruit is a pod containing 1–4 large seeds. *Distr.* Europe, Central and W Asia to Ethiopia. *Fam.* Leguminosae. *Class* D.

C. arietinum CHICK PEA, GARBANZO BEAN, EGYPTIAN BEAN. Annual herb. Leaves pinnate. Seeds are a very important pulse, eaten fresh, dried, or made into flour. *Distr.* S Europe, SW Asia and N Africa.

Cicerbita A genus of 18 species of perennial, latex-producing herbs with simple to pinnately lobed leaves and flowers in dandelion-like heads. Several species may be grown in wild areas of the garden. *Distr.* N temperate regions, especially mountainous areas. *Fam.* Compositae. *Class* D.

C. alpina ALPINE SOW THISTLE, MOUNTAIN SOW THISTLE. Perennial to 2.5m. Flowers pale blue. *Distr.* Europe.

C. plumieri Perennial. Flowers blue. *Distr.* Pyrenees to Bulgaria.

Cichorium (The classical Latin name for chicory and endive.) A genus of 8 species of annual and perennial, latex-producing herbs with simple or pinnately lobed leaves and blue flowers borne in numerous daisy-like heads. *Distr.* Europe, Mediterranean, Ethiopia. *Fam.* Compositae. *Class* D.

C. endivia ENDIVE. Annual to perennial herbs with glabrous leaves. Leaves eaten in salads. *Distr.* Himalaya to the Mediterranean.

C. intybus CHICORY, WITLOOF, SUCCORY. Perennial herb with long, stout taproot. Leaves edible, usually blanched; also used for skin complaints. Root used as an adulterant of or substitute for coffee. *Distr.* Mediterranean.

C. spinosum Dwarf perennial. Flowers blue. Grown as a garden ornamental. *Distr.* Mediterranean.

cidra *Cucurbita ficifolia.*

cidron *Aloysia triphylla.*

cigar flower *Cuphea ignea.*

cigar tree *Catalpa speciosa.*

ciliaris, -e: edged with hairs.

ciliatus, -a, -um: fringed with hairs.

cilicicus, -a, -um: of Cilicia, Asia Minor.

ciliicalyx: with sepals fringed with hairs.

ciliolatus, -a, -um: fringed with small hairs.

ciliospinosa: with hairy spines.

ciliosus, -a, -um: with many hairs.

Cimicifuga (From the Latin *cimex*, a bug, and *fugo*, to drive away, alluding to the use of *Cimicifuga foetida* as a vermifuge.) BUGBANE, BLACK COHOSH. A genus of 10–15 species of upright perennial herbs with compound, alternate leaves and panicles or spike-like racemes of small white flowers. Several species are grown as ornamentals. *Distr.* N temperate regions. *Fam.* Ranunculaceae. *Class* D.

C. foetida FOETID BUGBANE. Stems to 1m high. Flowers tinged green, borne in branched racemes. *Distr.* Siberia, E Asia.

C. racemosa BLACK COHOSH, BLACK SNAKEROOT. Herb to 2.5m high. Flowers fragrant, borne in narrow racemes to 1m long. The dried rhizomes are used medicinally. *Distr.* E USA.

C. simplex Flowers tiny, slightly fragrant, borne in arching spikes. *Distr.* E Asia (Kamchatka).

Cineraria (From the Latin *cinereus*, alluding to the grey, downy hairs on the leaves of some species.) A genus of about 50 species of perennial herbs or subshrubs with pinnately lobed leaves and yellow flowers in small daisy-like heads that bear distinct yellow rays. Not to be confused with Horist's FLORIST'S CINERARIA in the genus *Pericallis*. *Distr.* Africa, Madagascar. *Fam.* Compositae. *Class* D.

C. albida Small shrub with white, felty, woolly foliage. *Distr.* S Africa.

C. maritima See *Senecio cineraria*

cineraria, florist's *Pericallis × hybrida.*

cinerariifolium: grey-leaved.

cinerascens greyish

cinereus, -a, -um: grey.

cinnabarinus, -a, -um: cinnabar red, a red with a touch of yellow.

cinnamomeus, -a, -um: cinnamon-coloured, pale reddish-brown.

cinnamomifolius, -a, -um: with cinnamon-coloured (pale reddish-brown) leaves.

Cinnamomum (From the Greek name *kinnamomum*.) A genus of about 250 species of evergreen shrubs and trees with aromatic bark, simple or occasionally pinnate, leathery leaves and panicles of small bisexual flowers. The wood and bark yield products for flavouring, scent, and medicine. *Distr.* Topical and subtropical Asia, Australia and the Pacific Islands. *Fam.* Lauraceae. *Class* D.

C. aromaticum BASTARD CINNAMON, CASSIA, CHINESE CINNAMON. Bark used as a spice. *Distr.* Burma.

C. camphora CAMPHOR. A source of timber and the original source of camphor, which is now synthesized artificially. *Distr.* Tropical Asia, Malaysia, Taiwan, Japan.

C. zeylanicum CINNAMON, CEYLON CINNAMON. Bark is the spice cinnamon of commerce although it is often adulterated with the coarser, inferior bark of *C. aromaticum*. *Distr.* S India, Sri Lanka.

cinnamon *Cinnamomum zeylanicum* **bastard** ~ *C. aromaticum* **Ceylon** ~ *C. zeylanicum* **Chinese** ~ *C. aromaticum.*

cinquefoil *Potentilla P. reptans,* **shrubby** ~ *P. fruticosa.*

Cionura A genus of 1 species of small, deciduous shrub with blue-grey, marbled leaves and numerous, white, fragrant flowers. *Distr.* E Mediterranean. *Fam.* Asclepiadaceae. *Class* D.

C. erecta Grown as an ornamental.

Circaea (After Circe, the enchantress in the *Odyssey*.) A genus of 7 species of rhizomatous herbs with simple, opposite leaves and racemes of irregular flowers that bear their parts in 2s. *Distr.* Widespread N hemisphere. *Fam.* Onagraceae. *Class* D.

C. lutetiana ENCHANTER'S NIGHTSHADE. Erect or decumbent perennial. Flowers white or pink. A weed of cultivation. *Distr.* Eurasia, E North America.

Circaeasteraceae A family of 1 genus and 1 species of rhizomatous herb with simple, opposite leaves and small bisexual flowers that bear 2 sepals and 2–6 stamens but no petals. *Distr.* Himalaya to NW China. *Class* D.

circinalis: coiled inwards, like a young fern frond.

circinatus, -a, -um: coiled inwards, like a young fern frond.

cirrhatus, -a, -um: with tendrils.

cirrhiflora with tendrilled flowers.

cirrhosus, -a, -um: with tendrils.

Cirsium (from the Greek *kirsos*, a swollen vein, alluding to the effect when pricked by one of the bristles.) PLUME THISTLE. A genus of about 250 species of biennial and perennial herbs with alternate, pinnately lobed, or simple spiny leaves and flowers in large daisy-like heads that lack distinct rays. A number of species are cultivated as ornamentals although they have a tendency to become weedy. *Distr.* N temperate regions, especially temperate America. *Fam.* Compositae. *Class* D.

C. acaule Stemless perennial. Flowers purple. *Distr.* Eurasia.

C. dissectum Perennial, occasionally biennial. Flowers purple. *Distr.* W Europe.

C. eriophorum Biennial. Flowers purple. *Distr.* W Europe to the Balkans.

C. helenioides Perennial to 1.5m high. Flowers purple. *Distr.* Europe to Russia.

C. heterophyllum See *C. helenioides*.

C. japonicum NO AZAMI. Biennial or perennial to 2m high. Flowers lilac-pink. *Distr.* Japan.

C. palustre SWAMP THISTLE. Tall biennial. Stem pith is eaten by North American Indians. *Distr.* Europe, E North America.

C. rivulare Tall perennial. Flowers purple. *Distr.* SW Europe.

C. vulgare SPEAR THISTLE, BULL THISTLE. Tall biennial. Flowers purple. *Distr.* Europe and Mediterranean, naturalized in North America.

Cissus (From Greek *kissos*, ivy, alluding to the climbing habit of many species.) GRAPE IVY. A genus of about 350 species of herbs, shrubs, and vines, with simple to palmately compound leaves and clusters of inconspicuous flowers followed by dry berries. Several species are grown as ornamentals and houseplants. *Distr.* Tropical and warm regions. *Fam.* Vitaceae. *Class* D.

C. antarctica KANGAROO VINE. Evergreen vine. Fruit small, black. Grown as a houseplant. *Distr.* E Australia.

Cistaceae The Rock Rose family. 7 genera and 175 species of shrubs, subshrubs and occasionally herbs of dry sunny habitats. The leaves are simple and often have oil glands or glandular hairs. The flowers are regular, bisexual, typically with 5 sepals, 5 petals, and numerous stamens. Many species are grown as ornamentals; others are a source of fragrant resins. *Distr.* N temperate regions, NE tropical Africa, tropical South America. *Class* D.

See: Cistus, Halimium, X Halimiocistus, Helianthemum, Tuberaria.

Cistus (From *kistos*, the classical Greek name for these plants.) ROCK ROSE. A genus of 17 species of low, evergreen or semi-evergreen shrubs characteristic of scrub and dry woodland with simple, opposite leaves and regular, delicate, rose-like flowers that bear 3–5 sepals, 4–5 petals, and numerous stamens. Individual flowers often only last one day. All the species and numerous hybrids are grown as ornamentals. *Distr.* Mediterranean, Canary Islands. *Fam.* Cistaceae. *Class* D.

C. albidus WHITE-LEAVED ROCK ROSE. Compact shrub to 2m high. Leaves densely white-hairy. Flowers pink to pale purple, blotched yellow. *Distr.* SW Europe.

C. crispus Small rounded shrub. Petals white with a yellow basal spot. *Distr.* S Europe.

C. formosus See *Halimium lasianthum.*

C. ladanifer GUM CISTUS, LAUDANUM. Shrub to 2.5m high. Stems sticky. Leaves hairy below. Flowers white with a maroon blotch at the base of each petal. The source of the fragrant resin laudanum used in soap and medicinally. Also the parent of many ornamental hybrids. *Distr.* W Mediterranean.

C. lasianthus See *Halimium lasianthum*

C. monspeliensis MONTPELIER ROCK ROSE. A compact shrub. Flowers white. *Distr.* SW Europe.

C. ocymoides See *Halimium ocymoides*

C. sahucii See *X Halimiocistus sahucii*

C. salviifolius Small spreading shrub. Petals white with a yellow basal spot. The leaves have been used as a tea substitute. *Distr.* S Europe.

C. tomentosus See *Helianthemum nummularium*

C. wintonensis See × *Halimiocistus wintonensis*

C. × verguinii Shrub. Flowers while. *Distr.* S France.

cistus, gum *Cistus ladanifer.*

Citharexylum (From the Greek *kithara*, lyre, and *xylon*, wood.) FIDDLEWOOD. A genus of about 70 species of deciduous shrubs and trees with simple, opposite or whorled flowers and racemes or spikes of small, 5-lobed, tubular flowers. Several species provide high quality timber. *Distr.* Tropical America to Argentina. *Fam.* Verbenaceae. *Class* D.

C. quadrangulare See *C. spinosum*

C. spinosum Spiny shrub or small tree to 12m high. Flowers cream. Occasionally grown as a half-hardy ornamental. *Distr.* West Indies, N South America.

citrange × *Citroncirus* × *webberi.*

citratus, -a, -um: resembling the genus *Citrus.*

citrifolius, -a, -um: from the genus *Citrus*, and *folius*, leaved.

citriniflorum: with lemon-yellow flowers.

citrinus, -a, -um: lemon-yellow-coloured.

citriodorus, -a, -um: lemon-scented.

× Citrofortunella (From the names of the parent genera.) A genus of hybrids between members of the genera *Citrus* and *Fortunella. Fam.* Rutaceae. *Class* D.

×C. floridana LIMEQUAT. Evergreen tree. Flowers white. Fruit yellow. *Distr.* Garden origin.

×C. microcarpa CALAMONDIN, PANAMA ORANGE. A hybrid of *C. reticulata* x *F. margarita.* Small evergreen tree. Fruit bright orange, very bitter. A popular house-plant. *Distr.* Garden origin.

×C. mitis See *X. microcarpa*

citron *Citrus, C. medica.*

X Citroncirus (From the names of the parent genera.) A genus of hybrids between members of the genera *Citrus* and *Poncirus. Distr.* Garden origin. *Fam.* Rutaceae. *Class* D.

×C. webberi CITRANGE. A hybrid of *C. sinensis* × *P. trifoliata.* Semi-deciduous shrub or small tree. Flowers white, fragrant. Grown as a more or less hardy citrus fruit although its fruits are rather bitter. *Distr.* Garden origin.

Citronella (The diminutive form of Greek *kitron*, citrius, alluding to the fragrant flowers.) A genus of 21 species of evergreen shrubs and trees with spirally arranged, simple leaves and panicles of regular, bisexual fragrant flowers. *Distr.* Indonesia and Malaysia, Pacific, tropical America. *Fam.* Icacinaceae. *Class* D.

Citrullus From the fruits were thought to resemble those of the genus *Citrus.*) A genus of 3 species of annual and perennial herbs with trailing or climbing stems, pinnately lobed leaves, and branched tendrils. The unisexual, bell-shaped, yellow flowers are borne singly in the axils of the leaves. *Distr.* Tropical and S Africa. *Fam.* Cucurbitaceae. *Class* D.

C. colocynthis VINE OF SODOM, BITTER APPLE. Perennial. Fruit round, to 5cm across. The pulp of the fruit is used medicinally. *Distr.* N Africa, W Asia.

C. lanatus WATER MELON. Much branched trailing annual herb. Fruit large, globose, with a hard rind and red fleshy pulp. Flesh of fruit eaten. *Distr.* Tropical Africa, now widely cultivated throughout tropical and subtropical regions.

Citrus (From the classical Latin name for *Citrus medica*.) CITRON. A genus of about 16 species of evergreen shrubs and small trees with simple leaves and corymbs of small flowers that bear 4–5 petals and 16–20 stamens. The fruit is technically termed a hesperidium and consists of a glandular rind surrounding a number of segments that contain vesicles filled with watery juice. Members of this genus are amongst the most important fruit trees, particularly in warmer countries. Many of the citrus fruits familiar today have been produced through years of breeding and no longer resemble wild species. *Distr.* S and SE Asia, Malay peninsula. *Fam.* Rutaceae. *Class* D.

C. aurantifolia LIME. Fruit small, green tinged yellow when ripe. The bitter-tasting fruit is used in cooking and fruit drinks. Oil from rind and seeds used in soap. *Distr.* Tropical Asia.

C. aurantium SEVILLE ORANGE, SOUR ORANGE, BIGARADE. Tree to 10m high. Flowers large, white, fragrant. Fruit large, bright orange, becoming hollow in the centre. Flowers yield neroli oil for scent and the fruit is used for marmalade and in liqueurs.

C. grandis POMELO. Pulp of fruit eaten fresh. *Distr.* SE Asia, widely cultivated in S Asia.

C. japonica See *Fortunella japonica*

C. limon LEMON. Small tree to 7m high. Flowers white, tinged purple. Fruit yellow when ripe. The fruit is eaten and is a source of essential oil. *Distr.* Origin unknown but could possibly have arisen as a cross between *C. medica* and *C. aurantifolia*, occurring in cultivation around 1000 A.D.

C. maxima POMELO, SHADDOCK, POMPELMOUS. Fruit very large with thick rind. The source of bitter narinjin, used in drinks and sweets. Of little economic note as a crop but the parent of many important fruits. *Distr.* W Indies.

C. medica CITRON. Shrub or small tree. Fruit large, fragrant, yellow when ripe. The first citrus fruit to be introduced into the Mediterranean region, cultivated in Mesopotamia since 4000 BC *Distr.* India.

C. × paradisi GRAPEFRUIT. A complex hybrid involving *C. maxima* and *C. sinensis*. The fruit is popular as a breakfast. *Distr.* Arose in the W Indies in the 18th century.

C. reticulata MANDARIN, TANGARINE, SATSUMA, CLEMENTINE. Shrub or small tree. Fruit orange when ripe, sweet tasting. The fruit is traditionally eaten around Christmas time in Britain. *Distr.* SE Asia.

C. sinensis SWEET ORANGE, ORANGE. Medium-sized tree to 13m high. Flower very fragrant. Fruit large, orange, sweet to the taste. The most important citrus crop. *Distr.* Origin uncertain, probably from numerous hybrid crosses in cultivation.

C. × tangelo UGLI FRUIT, TANGELO. A hybrid of *C. paradisi* × *C. reticulata*. Fruit bright orange with a thin skin, edible. *Distr.* Garden origin.

Cladothamnus See *Elliottia*

Cladrastis (From the Greek *klados*, branch, and *thraustos*, fragile.) YELLOW WOOD. A genus of 6 species of small deciduous trees with pinnate leaves and panicles or racemes of irregular, pea-like flowers. Several species are grown as ornamentals. *Distr.* North America, E Asia. *Fam.* Leguminosae. *Class* D.

C. lutea YELLOW ASH. Rounded tree. Flowers white, borne in drooping panicles. A source of fine wood and yellow dye. *Distr.* SE USA.

clandestinus, -a, -um: concealed.

clarinervium: with clear veins.

Clarkia (After Captain William Clark (1770–1838), American explorer.) A genus of 33 species of annual herbs with simple leaves and spikes or racemes of funnel- or bell-shaped flowers. A number of species and cultivars are grown as ornamentals. *Distr.* W North America, South America. *Fam.* Onagraceae. *Class* D.

C. concinna RED RIBBONS. Flowers pink-red, streaked with white and purple. *Distr.* USA (California).

clary *Salvia sclarea* **meadow** ~ *S. pratensis* **wild** ~ *S. verbenaca*.

clavatus, -a, -um: club-shaped.

Clavinodum See *Arundinaria*

Claytonia (After John Clayton (1686–1773), American botanist.) PURSLANE, SPRING BEAUTY. A genus of 15–25 species of small succulent herbs with simple leaves and racemes of white flowers. Several species are grown as ornamentals. *Distr.* North America, E Asia, Australia, New Zealand. *Fam.* Portulacaceae. *Class* D.

C. australasica WHITE PURSLANE. Creeping perennial. Flowers fragrant. *Distr.* Australia.

C. virginica SPRING BEAUTY. Perennial herb from corms. Flowers tinged pink. *Distr.* E North America.

cleavers *Galium.*

cleftstone plant *Pleiospilos.*

clematiflorus, -a, -um: flowers like clematis

Clematis (A classical Greek name used for a number of different climbing plants and based on *clema*, a tendril.) A genus of 230 species of evergreen and deciduous climbing shrubs and subshrubs with simple or compound, typically opposite leaves, and regular uni- or bisexual flowers that bear 4–8 petal-like sepals and numerous stamens, but lack true petals. The fruits are wind-dispersed and have long, feathery, persistent styles. Many species and cultivars are grown as ornamentals and some have medicinal uses. *Distr.* Temperate regions of both hemispheres and tropical African mountains. *Fam.* Ranunculaceae. *Class* D.

C. afoliata Scrambling shrub. Stems dark green. Leaves small, sparse. Flowers green-white. *Distr.* New Zealand.

C. armandii Evergreen climbing shrub. Leaves leathery, trifoliate. Flowers white or cream. *Distr.* SW China.

C. cirrhosa Evergreen climber. Leaves simple or 3-lobed. Flowers cream, to 7cm across. *Distr.* Mediterranean.

C. florida Deciduous or semi-evergreen climber. Leaves compound. Flowers cream with green markings. *Distr.* E China, Japan.

C. heracleifolia Herbaceous climber. Leaves trifoliate. Flowers tubular, scented, deep blue, borne in small clusters. *Distr.* NE China.

C. integrifolia Perennial herb or subshrub. Flowers bell-shaped, deep blue. *Distr.* Central Europe.

C. montana Vigorous climber. Leaves trifoliate. Flowers white or pink, hairy on the veins outside. *Distr.* Himalaya, W China (Tibet, Yunnan).

C. patens Deciduous climber. Leaves pinnately divided. Flowers cream to bright blue. *Distr.* Japan, China (Shandong, Liaoning).

C. rehderiana Deciduous climber. Leaves large, pinnately divided. Flowers bell-shaped, yellow or pale green. *Distr.* China (Tibet).

C. virginiana WOODBINE, LEATHER FLOWER, VIRGIN'S BOWER, DEVIL'S DARNING NEEDLE. Deciduous semi-woody climber. Leaves trifoliate. Flowers dull-white, borne in many-flowered panicles. *Distr.* E North America.

C. vitalba OLD MAN'S BEARD, TRAVELLER'S JOY. Deciduous, semi-woody climber. Leaves pinnate. Flowers white-green. Lengths of stem have been used as a tobacco substitute and the young shoots are edible. A common plant of British hedgerows. *Distr.* Eurasia, N Africa.

clementine *Citrus reticulata.*

Cleome (A Greek name used by Theophrastus.) A genus of 150 species of annual and perennial herbs with alternate, palmately lobed leaves and racemes of small flowers that bear 4 sepals, 4 narrow petals, and 6 stamens. Several species are grown as ornamentals and some are eaten or have medicinal properties. *Distr.* Tropical and warm regions. *Fam.* Capparaceae. *Class* D.

C. hassleriana SPIDER FLOWER. Annual. Flowers white to purple-pink, borne in rounded heads. Grown as ornamentals and for cut flowers. *Distr.* SE Brazil to Argentina.

Clerodendrum (From the Greek *kleros*, chance, and *dendron*, a tree, alluding to the unpredictable medicinal qualities of these plants.) A genus of about 400 species of shrubs, trees and lianas with simple, opposite or whorled leaves and cymes or panicles of large, fragrant, salver-shaped flowers that bear 4 long stamens. A number of species are grown as ornamentals. *Distr.* Tropical and warm regions, especially E hemisphere. *Fam.* Verbenaceae. *Class* D.

C. paniculatum PAGODA FLOWER. Small erect shrub. Flowers orange-red. *Distr.* SE Asia.

C. philippinum GLORY BOWER. Shrub to 3m high. Flowers white or fragrant. *Distr.* E Asia, naturalized in S USA.

C. thomsoniae BLEEDING HEART VINE, BAG FLOWER. Evergreen twining shrub. Flowers dark red. *Distr.* Tropical W Africa.

Clethra (From *klethra*, the Greek name for ALDER.) WHITE ALDER, SUMMER SWEET. A genus of 64 species of evergreen or deciduous shrubs and small trees with simple leaves and racemes of regular, white flowers that bear 5 sepals, 5 petals, 2 whorls of 5 stamens, and a superior ovary. Several species are grown as ornamentals for their scented flowers. *Distr.* Tropical America, Asia, North America, Madeira. *Fam.* Clethraceae. *Class* D.

C. alnifolia BUSH PEPPER, SWEET PEPPER BUSH, PINK SPIRE, NANA. Deciduous shrub or small tree. Flowers in erect, spike-like clusters. *Distr.* E North America.

C. arborea LILY OF THE VALLEY TREE, FOLHADO. Evergreen shrub or small tree. Flowers small, strongly fragrant, borne in long, nodding clusters. *Distr.* E North America.

C. barbinervis Deciduous shrub. Leaves turn red in autumn. Flowers in more or less erect spikes. *Distr.* E China, Japan.

C. delavayi Deciduous open shrub. Flowers tinged pink, borne in spreading clusters. *Distr.* W China.

Clethraceae A family of 1 genus and 64 species. *Class* D.
See: Clethra.

clethroides from the genus *Clethra*, with the ending *-oides*, indicating resemblance.

Cleyera (After Dr Cleyer, 18th-century Dutch botanist.) A genus of 17 species of evergreen and deciduous shrubs and trees with solitary or clustered, cream or yellow flowers that bear 5 petals and around 25 stamens. Several species are grown as ornamentals. *Distr.* E Asia, Central America. *Fam.* Theaceae. *Class* D.

C. japonica SAKAKI. Evergreen shrub to 4.5m high. Flowers creamy white. *Distr.* Japan, Korea, China.

Clianthus (From the Greek *kleos*, glory, and *anthos*, flower.) A genus of 2 species of procumbent subshrubs and shrubs with alternate, pinnate leaves and panicles or racemes of brightly coloured, pea-like flowers that bear a large, sharply pointed keel. Both species are grown as ornamentals. *Distr.* Arid regions of Australia, NE New Zealand. *Fam.* Leguminosae. *Class* D.

C. formosus DESERT PEA, GLORY PEA, STURT'S DESERT PEA. Annual or perennial herb or subshrub. Flowers scarlet with a boss-like central black blotch. *Distr.* Arid regions of Australia.

C. puniceus GLORY PEA, PARROT'S BEAK, LOBSTER CLAW. Evergreen shrub. Flowers large, borne in drooping racemes, scarlet, pink or white. Very rare in the wild. *Distr.* New Zealand (North Island).

cliff brake *Pellaea* **green** ~ *P. viridis* **purple** ~ *P. atropurpurea.*

cliffbush *Jamesia americana.*

cliff green *Paxistima canbyi.*

Clinopodium (From the Greek *klinein*, a slope, and *pous*, a foot.) A genus of 4 species of erect perennial herbs with simple, opposite leaves and spike-like inflorescences that consist of whorls of 2-lipped flowers. Several species are grown as ornamentals. *Distr.* Temperate. *Fam.* Labiatae. *Class* D.

C. ascendens See *Calamintha sylvatica*

C. calamintha See *Calamintha nepeta*

C. grandiflorum See *Calamintha grandiflora*

C. vulgare WILD BASIL, CUSHION CALAMINT. Herb to 80cm high. Leaves faintly aromatic. Flowers rose-purple. *Distr.* Eurasia.

Clintonia (After De Witt Clinton, Governor of New York State in the early 19th century.) BEAR-TONGUE, WOOD LILY. A genus of 5 species of rhizomatous herbs with simple basal leaves and erect flowering stems that bear umbels or racemes of white, yellow or purple, star- or bell-shaped flowers. Several species are grown as ornamentals. *Distr.* E Asia and North America. *Fam.* Convallariaceae. *Class* M.

C. andrewsiana Leaves to 25cm long. Flowers deep purple, borne in umbels. *Distr.* USA (California).

C. borealis CORN LILY, BLUEBEARD. Flowers in a loose umbel, nodding, yellow-green. Young leaves eaten as a salad vegetable. *Distr.* E North America.

C. umbellulata SPECKLED WOOD LILY. Flowers in umbels, nodding, white, spotted red and green. *Distr.* E North America.

Clivia (After Charlotte Florentina Clive, later (died 1866) Duchess of Northumberland.) KAFFIR LILY. A genus of 4 species of perennial, evergreen, bulbiferous herbs with linear leaves and umbels of short, funnel-shaped flowers. *Distr.* South Africa. *Fam.* Amaryllidaceae. *Class* M.

C. miniata Umbel 10–20 flowered. Flowers red with a yellow throat. Cultivated ornamentally with several cultivars of different flower colours available. *Distr.* South Africa.

clover *Trifolium* **Alpine** ~ *T. alpinum* **bush** ~ *Lespedeza* **cavalry** ~ *Medicago intertexta* **crimson** ~ *Trifolium incarnatum* **Dutch** ~ *T. repens* **elk** ~ *Aralia californica* **Greek** ~ *Trigonella foenum-graecum* **holy** ~ *Onobrychis viciifolia* **Hungarian** ~ *Trifolium pannonicum* **Italian** ~ *T. incarnatum* **lucky** ~ *Oxalis tetraphylla* **musk** ~ *Erodium moschatum* **purple** ~ *Trifolium pratense* **red** ~ *T. pratense* **water** ~ *Marsilea* **white** ~ *Trifolium repens.*

cloveroot *Geum urbanum.*

clubmoss *Lycopodium* **little** ~ **Selaginella spreading** ~ *Selaginella kraussiana.*

Clusia (After Charles de l'Ecluse (Carolus Clusius) (1526–1609), Flemish botanist.) A genus of about 145 species of shrubs, trees and strangling epiphytes with large, leathery, opposite leaves and regular, yellow or pink flowers that bear 4–9 petals and numerous stamens. Several species have medicinal uses. *Distr.* Tropical and warm regions of America. *Fam.* Guttiferae. *Class* D.

C. major BALSAM APPLE, COPEY, SCOTCH ATTORNEY, AUTOGRAPH TREE, PITCH APPLE, FAT PORK TREE. Evergreen shrub or tree to 20m high. Leaves deep green. Flowers cup-shaped, pink. Fruit a green, round, leathery, resinous capsule. Grown as a tender ornamental. *Distr.* Tropical and subtropical America.

Clusiaceae See Guttiferae

Clytostoma (From the Greek *klytos*, glorious, and *stoma*, mouth, alluding to the flowers.) A genus of 9 species of evergreen woody climbers with opposite, pinnate leaves that bear a terminal tendril. The funnel- to bell-shaped flowers typically have 4 stamens and are borne in pairs or panicles. *Distr.* Tropical South America. *Fam.* Bignoniaceae. *Class* D.

C. callistegioides ARGENTINE TRUMPET VINE. Flowers paired, yellow with purple markings. Grown as a tender ornamental. *Distr.* S Brazil, Argentina.

Cneoraceae A family of 1 genus and 3 species. *Class* D.
See: Cneorum.

Cneorum (From the classical Greek name *kneoron*, probably applied to species of the genus *Daphne*.) A genus of 3 species of evergreen shrubs and small trees with simple, leathery leaves and regular flowers that bear 3–4 sepals and 3–4 yellow petals. *Distr.* Canary Islands, W Mediterranean, Cuba. *Fam.* Cneoraceae. *Class* D.

C. tricoccon SPURGE OLIVE. Erect shrub. Flowers small, dark yellow. Fruit red, ripening to black. Grown as an ornamental. The fruits act as a violent purgative. *Distr.* W Mediterranean.

Cnicus (From the Greek name *knikos*, used for a thistle-like plant.) A genus of 1 species of annual herb with leathery, minutely spiny leaves and yellow flowers in daisy-like heads that lack distinct rays. *Distr.* Mediterranean, Europe. *Fam.* Compositae. *Class* D.

C. benedictus BLESSED THISTLE. Previously an important source of the drug cnicin, used in treatment of gout and as a tonic. Occasionally grown as an ornamental.

coarctata: pressed together, confined.

Cobaea (After B. Cobo (1572–1659), a Jesuit priest who travelled in South America.) A genus of about 10 species of shrubby and herbaceous climbers with pinnate leaves that bear a terminal tendril. The solitary bell-shaped flowers bear 5 fused sepals, 5 fused petals, and 5 stamens. The fruit is a somewhat fleshy capsule. *Distr.* Tropical America. *Fam.* Cobaeaceae. *Class* D.

C. scandens CUP AND SAUCER VINE, MEXICAN IVY, MONASTERY BELLS. Perennial herbaceous climber. Flowers cream-green and foetid at first, becoming purple and sweet smelling. The sepals have the appearance of a saucer and the bell-shaped petal tube that of a cup. Grown as an ornamental annual. *Distr.* Mexico.

Cobaeaceae A family of 1 genus and 10 species. *Class* D.
See: Cobaea.

cobnut *Corylus avellana.*

cobra plant *Arisaema nepenthoides.*

coca Erythroxylaceae.

cocaine plant *Erythroxylum coca.*

coccifera: berry-bearing.

coccineus, -a, -um: scarlet.

Cocculus (The diminutive of the Greek *kokkos*, a berry.) MOONSEED, SNAILSEED. A genus of about 7 species of evergreen and deciduous shrubs, climbers and small trees with alternate, entire or lobed leaves, and panicles or racemes of inconspicuous, unisexual flowers that bear their parts in 6s. The fruit is a small black or red drupe. Several species are grown as ornamentals and a few have local medicinal uses. *Distr.* Tropical E Asia, Africa, temperate North America. *Fam.* Menispermaceae. *Class* D.

C. carolinus CAROLINA MOONSEED, RED MOONSEED, SNAILSEED. Deciduous twining climber. Leaves downy below. Flowers green-white. Fruit bright red. *Distr.* SE USA.

cochenillifera: cochineal-bearing.

Cochlearia (From the Latin *cochlea*, spoon, alluding to the leaf shape in some species.) A genus of about 25 species of annual and perennial herbs with small white flowers that are often tinged with yellow or purple. *Distr.* N temperate regions. *Fam.* Cruciferae. *Class* D.

C. officinalis COMMON SCURVY GRASS. Biennial or perennial, occasionally annual with white flowers. Previously eaten as a vegetable rich in vitamin C, particularly by sailors. *Distr.* Coastal NW Europe and Alps.

cochleariifolius, -a, -um: with spoon-shaped leaves.

cochlearispathus: with a spoon-like spathe.

cochlearis: spoon-shaped.

cochleatus, -a, -um: spoon-shaped.

Cochlospermaceae The Rose Imperial family. 2 genera and 15 species of shrubs and trees with alternate, palmately lobed leaves and typically showy yellow flowers that bear 5 sepals, 5 petals, and numerous stamens. Some species are grown as ornamentals in warm countries, others are a source of fibre. *Distr.* Tropical regions. *Class* D.

cockle, white *Silene latifolia.*

cocklebur *Agrimonia.*

cockscomb *Celosia argentea*, Amaranthaceae *Dactylis glomerata.*

cock's foot *Dactylis glomerata.*

cockspur *Echinochloa crusgalli.*

cocoa *Theobroma cacao.*

coconut *Cocos nucifera.*

Cocos (From the Spanish and Portuguese *coco*, a smiling face, alluding to the marks on the fruit.) A genus of 1 species of large palm to 30m high with feather-shaped leaves to 3m long and pendent inflorescences that bear cream flowers in groups of 3, 2 males and 1 female. The fruit consists of a single large seed (coconut) in a fibrous husk. *Distr.* Malaysia, naturalized throughout the tropics. *Fam.* Palmae. *Class* M.

C. nucifera COCONUT. The coconut palm is the source of a great many products which include: dried and fresh flesh from within the nut, along with coconut milk, oil, and a number of fermented drinks; coir, the fibre from the husk, which is used for matting and cordage as well as in soilless composts; timber, known as porcupine wood, which is used for building. The inflorescence axis can be tapped to provide a toddy or a type of sugar (jaggery) and the apical bud of the tree can be eaten (palm hearts).

cocoyam *Colocasia esculenta.*

Codiaeum (From the native Malay name, *kodiho.*)A genus of 6 species of evergreen shrubs and trees with thick leathery leaves and slender racemes of small unisexual flowers. The male flowers bear 5 scale-like petals but the females only bear 5 sepals. *Distr.* Malaya, Pacific Islands. *Fam.* Euphorbiaceae. *Class* D.

C. variegatum CROTON. Small shrub. Leaves variegated yellow or white. Male flowers white. Over 100 ornamental cultivars have been produce from this species. *Distr.* S India to Malaysia.

Codonanthe (From the Greek *kodon*, bell and *anthe*, flower.) A genus of 13 species of scrambling or epiphytic shrubs with equal or unequal pairs of simple leaves and funnel- to bell-shaped 2-lipped flowers. The seeds resemble ant eggs and so are distributed by unwary ants which take them back to their nests. Several species are grown as tender ornamentals. *Distr.* Tropical America. *Fam.* Gesneriaceae. *Class* D.

C. gracilis Epiphytic shrub. Flowers white-yellow, marked with red. *Distr.* S Brazil.

Codonopsis (From the Greek *kodon*, bell, and *opsis*, appearance, alluding to the flowers.) A genus of about 30 species of spreading herbs with tuberous roots, sword-shaped opposite or alternate leaves, and nodding, bell-shaped flowers. A number of species are grown as ornamentals and some have medicinal properties. *Distr.* Central and E Asia to Malaysia. *Fam.* Campanulaceae. *Class* D.

C. clematidea Stems sprawling and twining. Flowers pale blue with orange and black markings. *Distr.* Central Asia.

C. convolvulacea Twining climber. Flowers spreading, bell- to saucer-shaped, purple-blue. *Distr.* SW China.

C. ovata Stems ascending to erect. Flowers pale blue with dark veins. *Distr.* W Himalaya.

C. tangshen Stems twining. Flowers yellow to olive green, marked with purple. Root used as a substitute for GINSENG. *Distr.* W China.

coelestis, -e: sky-blue-coloured.

coelicus, -a, -um: of the sky, heavenly.

coeruleus, -a, -um: blue.

Coffea (From the Arabic name for these plants.) COFFEE. A genus of 40 species of shrubs and small trees with simple, leathery or membranous leaves and cream, funnel-shaped, fragrant flowers that are followed by red or yellow berries. These plants are economically very important as the source of the beverage coffee which is extracted from their seeds (coffee beans). *Distr.* Tropical regions of the Old World, especially Africa. *Fam.* Rubiaceae. *Class* D.

C. arabica ARABIAN COFFEE. The source of what are considered to be the higher quality coffees. *Distr.* Tropical Africa.

C. canephora ROBUSTA COFFEE, RIO NUNEZ COFFEE. The source of a bitter coffee considered to be of lower quality than that produced by *C. arabica* but often used in blends and in the production of instant coffee. *Distr.* W Africa

C. robusta See *C. canephora*

coffee *Coffea*, Rubiaceae **Arabian** ~ *Coffea arabica* **Rio Nunez** ~ *C. canephora* **robusta** ~ *C. canephora* **wild** ~ *Polyscias guilfoylei*, *Triosteum perfoliatum*.

coffee tree *Polyscias guilfoylei* **Kentucky** ~ *Gymnocladus dioica*.

cohosh *Actaea* **black** ~ *Cimicifuga*, *C. racemosa*, *Actaea spicata* **blue** ~ *Caulophyllum*

coigue *Nothofagus dombeyi*.

Coix (From the Greek *koix*, a name for a reed-like plant.) A genus of about 4 species of annual and perennial rhizomatous grasses with broad, flat leaves and small spikelets that are followed by hard, pear-shaped fruits. *Distr.* Tropical E Asia. *Fam.* Gramineae. *Class* M.

C. lacryma-jobi JOB'S TEARS. Annual. Inflorescences arching. Grown as an ornamental curiosity. *Distr.* SE Asia.

cokernut, little *Jubaea chilensis*.

Cola (From the African vernacular name.) A genus of about 125 species of trees with palmately lobed or divided leaves and unisexual flowers that bear 5 sepals and no petals. *Distr.* Tropical Africa. *Fam.* Sterculiaceae. *Class* D.

C. acuminata COLA, ABATA, GOORA NUT. Evergreen tree to 20m high. The caffeine-containing seeds are chewed or used in the manufacture of soft drinks. *Distr.* Tropical Africa.

cola *Cola acuminata*.

Colchicaceae The Autumn Crocus family. 16 genera and about 170 species of herbs and tuberous climbers with simple leaves and flowers which bear 6 free or fused tepals, 6 stamens, and 3 styles. *Distr.* Widespread in the Old World. *Class* M.

See: Bulbocodium, Burchardia, Colchicum, Gloriosa, Littonia, Merendera, Sandersonia, Wurmbea.

colchiciflorus, -a, -um: with flowers resembling those of the genus *Colchicum.*

Colchicum (From the name of Colchis, on the Black Sea.) AUTUMN CROCUS, NAKED LADIES. A genus of 45–65 species of herbs with linear leaves arising from a corm. The stalkless flowers often develop with or before the leaves in the autumn and have 6 fused tepals, 6 stamens, and 3 free styles. Although it appears superficially like the genus *Crocus* it is different enough to be placed in a separate family. Many species and cultivars are grown as ornamentals. *Distr.* E Europe and N Africa to W Asia and N India. *Fam.* Colchicaceae. *Class* M.

C. agrippinum A vigorous sterile hybrid of *C. variegatum* × *C. autumnale*. *Distr.* Garden origin.

C. autumnale AUTUMN CROCUS, MEADOW SAFFRON, NAKED BOYS. Flowers purple, pink or white, appearing in late summer or early autumn. Dried seeds and corms have been used medicinally as a source of pain-killer. *Distr.* W and Central Europe to Byelorussia and Ukraine.

C. bivonae Flowers large, pink, chequered purple, appearing in the autumn. *Distr.* S Europe, W Turkey.

C. bowlesianum See *C. bivonae*

C. byzantinum A robust autumn-flowering cultigen, probably derived from *C. cilicicum*, with pink-purple flowers. *Distr.* Garden origin.

C. cilicicum Flowers cup- or funnel-shaped, pink-purple, appearing in the autumn. *Distr.* Turkey, Syria, Lebanon.

C. luteum Flowers pale to deep yellow appearing in the spring and summer. *Distr.* Central Asia, Afghanistan, N India, SW China (Xizang).

C. sibthorpii See *C. bivonae*

C. speciosum Flowers cup- or bell-shaped, green below, appearing in the autumn. *Distr.* N Turkey, Iran, Caucasus.

C. variegatum Flowers strongly chequered violet-purple, appearing in the autumn. *Distr.* Greece, SW Turkey.

colchicus, -a, -um: of Colchis, on the East coast of the Black Sea.

cole, red *Armoracia rusticana.*

colenso *Dianella nigra.*

Coleonema (From the Greek *koleos*, sheath, and *nema*, thread, alluding to the stamens which are enfolded in channels in the petals.) A genus of 8 species of shrubs with simple, glandular leaves and solitary, white or pink flowers that bear 5 sepals and petals and 5 stamens. Several species are grown as ornamentals. *Distr.* South Africa. *Fam.* Rutaceae. *Class* D.

C. pulchrum Shrub to 1.2m high. Flowers pink, to 2cm across. *Distr.* South Africa (Cape Province).

Coleus See *Plectranthus*

coleus *Solenostemon scutellarioides.*

colic root *Aletris, A. farinosa.*

Colletia (After Philibert Collet (1643–1718), French botanist.) A genus of 17 species of thorny shrubs with flattened green branches and small or absent leaves. The small, bell-shaped or tubular flowers are borne singly or in clusters and lack petals. Several species are grown as ornamentals. *Distr.* S South America. *Fam.* Rhamnaceae. *Class* D.

C. cruciata See *C. paradoxa*

C. paradoxa Shrub to 2m high. Leaves about 6mm long, soon falling. Flowers in small clusters, yellow-white. *Distr.* S Brazil, Uruguay.

collimamol *Luma apiculata.*

Collinsonia (After Peter Collinson (1694–1768), English amateur botanist and gardener.) A genus of 5 species of aromatic perennial herbs with simple opposite leaves and panicles of tubular, 2-lipped flowers that bear 4 almost equal lobes and 1 larger, toothed or fringed lobe. *Distr.* E North America. *Fam.* Labiatae. *Class* D.

C. canadensis HORSE BALM, STONE ROOT. Clump-forming perennial to 80cm high. Flowers yellow, lower lip fringed. Occasionally grown as an ornamental and formerly used medicinally. *Distr.* E USA.

collinus, -a, -um: of hills.

Collomia (From the Greek *kolla*, glue, alluding to the sticky seeds of many species.)

A genus of 15 species of annual and perennial herbs with simple leaves and funnel-shaped, 5-lobed flowers. Several species are grown as ornamentals. *Distr.* W North America to Bolivia and Patagonia. *Fam.* Polemoniaceae. *Class* D.

C. debilis ALPINE COLLOMIA. rhizomatous herb. Flowers borne in dense clusters, cream tinged pink or blue. *Distr.* W North America.

collomia, Alpine *Collomia debilis.*

Colobanthus (From the Greek *kolobos*, mutilated, and *anthos*, flower, alluding to the lack of petals.) A genus of 20 species of tuft-forming perennial herbs with deep tap roots, simple, slightly fleshy leaves, and solitary, small, green flowers that bear 4–5 sepals and 4–5 stamens but no petals. *Distr.* Australia, New Zealand, S Pacific Islands, temperate South America. *Fam.* Caryophyllaceae. *Class* D.

C. muscoides Occasionally grown as an ornamental for its bright green cushions of foliage. *Distr.* New Zealand.

Colocasia (From the Arabic *kolkas*, orginally used for the root of a species of Nelumbo.) ELEPHANT'S EAR. A genus of 7 species of tuberous herbs with conspicuously veined, peltate leaves and small flowers borne on a spadix that is surrounded and exceeded by a hooded spathe. *Distr.* Tropical Asia. *Fam.* Araceae. *Class* M.

C. esculenta COCOYAM, TARO, DASHEEN, IMPERIAL TARO, BLACK CALADIUM. Widely cultivated in warm-temperate and tropical regions for its edible tubers and young leaves and for its ornamental foliage. Several cultivars are available. *Distr.* Tropical E Asia, naturalized throughout the tropics.

colorans: colouring.

coloratus, -a, -um: coloured.

colossus: huge, gigantic.

Colquhounia (After Sir Robert Colquhoun (died 1838), a patron of Calcutta Botanic Garden.) A genus of about 3 species of evergreen, erect or climbing shrubs with simple, opposite leaves and whorls or racemes of 2-lipped flowers. *Distr.* E Himalaya, SW China. *Fam.* Labiatae. *Class* D.

C. coccinea Lax shrub to 3m high. Flowers, red, yellow within, borne in whorls. *Distr.* SW China.

coltsfoot *Tussilago farfara, Galax urceolata* **Alpine** ~ *Homogyne alpina* **sweet** ~ *Petasites.*

columbaria: dove-like or of doves; pale dove-grey.

columbianus, -a, um. of British Columbia, Canada.

columbine *Aquilegia, A. vulgaris* **Canadian** ~ *A. canadensis.*

Columelliaceae A family of 1 genus and 2 species of evergreen trees and shrubs with simple opposite leaves and slightly irregular, bisexual flowers. *Distr.* NW South America. *Class* D.

columnae: columns.

Columnea (After Fabio Colonna (1567–1640), Italian botanist and author of the first botanical book with copperplate illustrations.) A genus of 160 species of shrubby epiphytes and lianas with pendulous stems, simple, opposite leaves and tubular, 5-lobed, 2-lipped flowers. Numerous species and cultivars are grown as ornamentals. *Distr.* Tropical America. *Fam.* Gesneriaceae. *Class* D.

C. scandens Slender creeping shrub. Flowers irregular, red and yellow. Numerous cultivars are available. *Distr.* Lesser Antilles.

columnifera: column-bearing.

Colutea (A classical Greek name used for a tree which grew in the Lipari Islands.) A genus of 28 species of deciduous shrubs or small trees with alternate, pinnate or trifoliate leaves and racemes of irregular, pea-like flowers. Several species are grown as ornamentals. *Distr.* Mediterranean to E Africa and E Asia. *Fam.* Leguminosae. *Class* D.

C. arborescens BLADDER SENNA. Open shrub. Flowers yellow, pods inflated. *Distr.* S Europe.

Coluteocarpus (From the genus name *Colutea*, and the Greek *karpos*, fruit.) A genus of 2 species of dwarf, tufted, perennial herbs with numerous golden-yellow flowers. *Distr.* SW Asian mountains. *Fam.* Cruciferae. *Class* D.

C. vesicaria Decorative seed pods used in arrangements of dried flowers. *Distr.* E Mediterranean.

colza *Brassica napus.*

comans: tufted hairs, or a tuft of leaves.

Comarum See *Potentilla*

Combretaceae A family of 20 genera and 500 species of trees, shrubs and lianas with simple leaves and small, often clustered flowers which may produce large amounts of nectar. Some of the tree species are important locally for their timber; others have medicinal qualities. *Distr.* Tropical and subtropical regions. *Class* D.
See: Quisqualis, Terminalia.

comfrey *Symphytum* **blue** ~ *S.* × *uplandicum* **prickly** ~ *S. asperum* **Russian** ~ *S.* × *uplandicum.*

Commelina (After Johan Commelin (1629–92), and his nephew Caspar Commelin (1667–1731), Dutch botanists.) DAY FLOWER, WIDOW'S TEARS. A genus of 150–200 species of tuberous herbs typically with horizontal rooting stems and simple leaves. The small, irregular flowers are somewhat concealed by large bracts. Some species have edible tubers. *Distr.* Chiefly tropical regions. *Fam.* Commelinaceae. *Class* M.

C. coelestis Tuberous. Flowers blue. Bract streaked dark purple. *Distr.* Central and South America.

C. tuberosa Tuberous. Leaves narrow, lance-shaped. *Distr.* Central and South America.

Commelinaceae The Spiderwort family. 38 genera and 620 species of somewhat succulent, annual or perennial herbs with short or jointed stems and alternate, entire leaves. The flowers are typically regular and bisexual with 3 free sepals and 3 free petals. A number of important garden and house plants occur in this family, including *Tradescantia* and *Commelina. Distr.* Widespread in tropical and subtropical regions, rarer in temperate areas. *Class* M.
See: Callisia, Commelina, Cyanotis, Geogenanthus, Gibasis, Tradescantia, Tripo-gandra, Weldenia.

Commiphora A genus of 185 species of deciduous, resinous shrubs and trees with simple to pinnate leaves and stalked clusters of small regular flowers that bear 4 petals and 8 stamens. *Distr.* Africa, Madagascar, Arabia to W India and South America. *Fam.* Burseraceae. *Class* D.

C. abyssinica MYRRH. Source of the myrrh of the Bible, once used medicinally as a purgative. *Distr.* NE Africa to Arabia.

C. opobalsamum MECCA MYRRH. Resin is a source of scent. *Distr.* Arabia.

commixtus, -a, -um: mixed together.

communis, -e: common.

commutatus, -a, -um: changeable.

comosus, -a, -um: with a tuft of hairs or leaves.

compactus, -a, -um: compact.

compass plant *Silphium laciniatum.*

complanata: flattened.

complexus, -a, -um: embraced, surrounded.

Compositae The Daisy family. 1509 genera and 21000 species of herbs, shrubs, and trees, typically with simple, sometimes compound leaves. The flowers are small and borne in characteristic dense heads that are subtended by a whorl or whorls of bracts. This enormous family provides many food crops and raw materials and is of great ecological importance, especially in drier areas. *Distr.* Widespread. *Class* D.
See: Achillea, Actinella, Agathaea, Ageratina, Ajania, Allardia, Anacyclus, Anaphalis, Andryala, Antennaria, Anthemis, Arctanthemum, Arctotis, Argyranthemum, Arnica, Artemisia, Aster, Asteriscus, Baccharis, Balsamita, Bellis, Bellium, Berkheya, Berlandiera, Bidens, Boltonia, Brachyglottis, Brachyscome, Buphthalmum, Calendula, Calocephalus, Calomeria, Carduncellus, Carduus, Carlina, Carthamus, Cassinia, Cassinia x Helichrysum, Catananche, Celmisia, Centaurea, Chaenactis, Chamaemelum, Chevreulia, Chiliotrichum, Chrysanthemopsis, Chrysanthemum, Chrysocoma, Chrysogonum, Chrysopsis, Chrysothamnus, Cicerbita, Cichorium, Cineraria, Cirsium, Cnicus, Coreopsis, Corethrogyne, Cosmos, Cotula, Craspedia, Cremanthodium, Crepis, Cynara, Dahlia, Dasyphyllum, Dendranthema, Dendroseris, Dicoma,

Dimorphotheca, Dolichothrix, Doronicum, Echinacea, Echinops, Emilia, Engelmannia, Erigeron, Eriophyllum, Eumorphia, Eupatorium, Euryops, Ewartia, Farfugium, Felicia, Gaillardia, Galactites, Gamolepis, Gazania, Gerbera, Gnaphalium, Grindelia, Gynura, Haplocarpha, Haplopappus, Helenium, Helianthella, Helianthus, Helichrysum, Helichrysum X Raoulia, Heliopsis, Heteropappus, Hieracium, Hirpicium, Homogyne, Humea, Hymenoxys, Hypochoeris, Inula, Isocoma, Jurinea, Kalimeris, Lactuca, Lagenophora, Lapsana, Leibnitzia, Leontodon, Leontopodium, Leptinella, Leucanthemella, Leucanthemopsis, Leucanthemum, Leucogenes, × *Leucoraoulia, Leuzea, Liatris, Ligularia, Machaeranthera, Matricaria, Microglossa, Mikania, Mutisia, Nardophyllum, Nipponanthemum, Olearia, Onopordum, Osteospermum, Otanthus, Othonna, Ozothamnus, Parthenium, Perezia, Pericallis, Petasites, Picris, Plectostachys, Ptilostemon, Pulicaria, Pyrethropsis, Raoulia, Ratibida, Rudbeckia, Santolina, Saussurea, Scolymus, Scorzonera, Senecio, Seriphidium, Serratula, Silphium, Silybum, Solidago,* × *Solidaster, Sonchus, Sphaeromeria, Spilanthes, Staehelina, Steirodiscus, Stokesia, Syncarpha, Tagetes, Tanacetum, Taraxacum, Telanthophora, Telekia, Tithonia, Tolpis, Tonestus, Townsendia, Tragopogon, Tripleurospermum, Tussilago, Urospermum, Ursinia, Vernonia, Zinnia.*

compositus, -a, -um: compound, united.

compressus, -a, -um: compressed or flattened, usually laterally.

Comptonia (Named for Henry Compton (1632–1713), Bishop of London, gardener and patron of botany.) A genus of 1 species of diffuse shrub with pinnate, fern-like, fragrant leaves and catkins of small unisexual flowers. *Distr.* E North America. *Fam.* Myricaceae. *Class* D.

C. peregrina Grown as a hardy ornamental for its sweet smelling foliage.

concentricus, -a, -um: concentric.

concholoba: with shell-shaped lobes.

concinnus, -a, -um: elegant.

concolor: one-coloured throughout.

coneflower *Echinacea* **coneflower** *Rudbeckia* **long-head** ~ *Ratibida columnifera* **prairie** ~ *R. columnifera* **sweet** ~ *Rudbeckia subtomentosa.*

confertiflorus, -a, -um: with flowers pressed together.

confertus, -a, -um: crowded, densely pressed together.

confluens confluent, running together.

confusus, -a, -um: confused.

congestiflorus, -a, -um: with flowers crowded together, congested.

congestus, -a, -um: crowded together, congested.

conglomeratus, -a, -um: conglomerate; clustered, often spherically.

congolensis: of the Congo basin, West to Central Africa.

conicus, -a, -um: cone-shaped.

coniferus, -a, -um: bearing cones.

Conium (From the Greek *komion*, small cone.) A genus of 2–3 species of biennial herbs with much divided leaves and compound umbels of small white flowers. All parts of these plants are highly poisonous. *Distr.* Temperate Eurasia, naturalized in South Africa. *Fam.* Umbelliferae. *Class* D.

C. maculatum HEMLOCK, POISON HEMLOCK, SPOTTED HEMLOCK. Biennial to 2.5m high. Stem spotted purple towards the base. Contains the toxic principle coniine that causes death by paralysis of the respiratory system. *Distr.* Europe, naturalized in North America.

conjunctus, -a, -um: connected, joined.

conjungens: conjugating, fusing.

conker tree *Aesculus hippocastanum.*

Connaraceae The Zebra Wood family. 12 genera and about 300 species of trees and twining shrubs with alternate trifoliate or pinnate leaves and panicles of more or less regular flowers that bear their parts in 5s. *Distr.* Widespread in tropical regions. *Class* D.

connatus, -a, -um: fused.

connectilis, -e: well connected.

conoideus, -a -um: almost cone-shaped.

conophalloides: with a cone-shaped spadix.

Conopodium (From the Greek *konos*, cone, and *pous*, foot.) A genus of 20 species of perennial, tuberous herbs with divided leaves and compound umbels of unequal white flowers. *Distr.* Eurasia, Mediterranean. *Fam.* Umbelliferae. *Class* D.

C. majus EARTH NUT, PIG NUT. Stems erect to 1m high. Tubers edible. *Distr.* Europe.

conopseus, -a, -um: from the Greek *conops*, a gnat.

Conradina (After Solomon W. Conrad (1753–1814), German botanist.) A genus of 4 species of small shrubs with crowded narrow leaves and clusters of 2-lipped, purple flowers. *Distr.* SE North America. *Fam.* Labiatae. *Class* D.

C. verticillata Much-branched, low shrub. Flowers pale purple with darker marks. Occasionally grown as an ornamental. *Distr.* USA (Tennessee).

Consolida (The medieval name for these plants, from the Latin *consolido*, to make firm, alluding to their medicinal properties.) LARKSPUR. A genus of 40 species of erect annual herbs with palmate leaves and racemes or panicles of irregular, spurred flowers. Several species and a number of cultivars are grown as ornamentals. *Distr.* Mediterranean area to Central Asia. *Fam.* Ranunculaceae. *Class* D.

C. ajacis See *C. ambigua*

C. ambigua Flowers bright blue, occasionally pink. Found in garlands on Egyptian mummies. *Distr.* Mediterranean region.

C. regalis Flowers deep blue to pink, borne in panicles. *Distr.* SE Europe, SW Asia.

conspicuus, -a, -um: conspicuous, distinguished.

consumption weed: *Baccharis halimifolia.*

contaminatus, -a, -um: contaminated.

continentalis, -e: of the continent.

contortus, -a, -um: twisted.

controversus, -a, -um: controversial.

Convallaria (From the Latin *convallis*, a valley.) LILY OF THE VALLEY. A genus of 3 species of creeping, rhizomatous herbs with erect stems and simple, folded leaves. The bell-shaped, highly fragrant, white flowers are pendent on a one-sided raceme. *Distr.* N temperate regions. *Fam.* Convallariaceae. *Class* M.

C. majalis A popular spring flowering ornamental with numerous cultivars available. The flowers are used in the manufacture of scent and the rhizomes are reported to have medicinal properties although they are poisonous. *Distr.* Europe, naturalized in North America.

Convallariaceae The Lily of the Valley family. 25 genera and about 180 species of typically rhizomatous herbs with erect or arching stems and simple, clasping or sheathing leaves. The flowers have 6 tepals, that are often fused into a bell- or funnel-shaped tube, 6 stamens, and a 2–3-lobed stigma. The fruit is a berry. This family is sometimes included within the family Liliaceae. Many genera contain cultivated ornamental species. *Distr.* Predominantly N temperate regions. *Class* M.
See: Aspidistra, Clintonia, Convallaria, Disporum, Liriope, Maianthemum, Medeola, Ophiopogon, Peliosanthes, Polygonatum, Reineckea, Rohdea, Speirantha, Streptopus, Tricyrtis, Uvularia.

convallarioides: resembling LILY of the valley.

convolutus, -a, -um: convolute, rolled up.

Convolvulaceae The Bindweed family. 55 genera and about 1500 species of herbaceous and woody climbers with simple, alternate leaves and showy, bell-shaped or tubular flowers which are usually short-lived and have their parts in 5s. The roots of *Ipomoea batatas* are eaten as the SWEET POTATO. *Distr.* Widespread. *Class* D.
See: Calystegia, Convolvulus, Evolvulus, Ipomoea, Merremia, Stictocardia.

convolvulacea: resembling the genus *Convolvulus.*

convolvuloides: from the genus *Convolvulus*, with the ending *-oides*, indicating resemblance.

Convolvulus (From the Latin *convolvere*, to twine, alluding to the stems.) BINDWEED. A genus of about 250 species of annual and perennial, erect or climbing herbs and shrubs with simple leaves and short-lived, funnel-shaped flowers. Several species are grown as ornamentals. *Distr.* Cosmopolitan, especially temperate regions. *Fam.* Convolvulaceae. *Class* D.

C. althaeoides Perennial. Leaves heart-shaped, silver-grey. Flowers purple-pink. *Distr.* S Europe.

C. arvensis FIELD BINDWEED. Perennial climber. Flowers white or pink, sweet-scented. A persistent weed. *Distr.* Eurasia.

C. boissieri Cushion-forming perennial. Flowers pale-pink. *Distr.* Spain.

C. cneorum SILVERBUSH. Bushy evergreen subshrub. Leaves silky, silver-green. Flowers white with a yellow centre. *Distr.* S Europe, N Africa.

C. nitidus See *C. boissieri*

C. tricolor Annual or short-lived perennial. Flowers blue with a white centre. Numerous cultivars available. *Distr.* S Europe, N Africa.

coolwort *Tiarella cordifolia.*

Cooperia See *Zephyranthes*

copey *Clusia major.*

Coprosma (From the Greek *kopros*, dung, and *osme*, odour, alluding to the malodorous foliage of some species.) A genus of 90 species of evergreen shrubs and small trees with simple opposite or occasionally whorled leaves and inconspicuous, funnel-shaped flowers. Several species are grown as ornamentals and some are a source of dye or are used medicinally. *Distr.* Malaysia, Australia, Pacific. *Fam.* Rubiaceae. *Class* D.

C. repens LOOKING-GLASS PLANT, MIRROR PLANT. Mat-forming shrub or small tree. Leaves lustrous green, leathery. Fruits orange-red. Several cultivars have been raised from this species. *Distr.* New Zealand.

Coptis (From the Greek *kopto*, to cut, alluding to the deeply cut leaves.) GOLD THREAD, MOUTH ROOT. A genus of about 10 species of rhizomatous herbs with divided, leathery basal leaves and small, regular, white or yellow flowers that are borne singly or in clusters. Several species are grown as ornamentals and some have medicinal properties. The rhizomes produce a yellow dye. *Distr.* Temperate North America, Asia. *Fam.* Ranunculaceae. *Class* D.

C. laciniata Leaves divided into 3s. Flowers solitary or in small clusters, white tinged green. *Distr.* E North America.

coquito *Jubaea chilensis.*

coracan *Eleusine coracana.*

coralbells *Heuchera sanguinea.*

coralberry *Symphoricarpos orbiculatus, Ardisia crenata, Symphoricarpos.*

coral gem *Lotus berthelotii.*

corallinus, -a, -um: coral red-coloured.

corallitta *Antigonon leptopus.*

coralloides: coral-like.

Corallospartium (From the Greek *korallion*, coral, and *spartium*, BROOM.) A genus of 1 species of erect shrub with green branches, small rounded leaves (which are rarely produced) and dense clusters of yellow-white flowers. *Distr.* New Zealand (South Island). *Fam.* Leguminosae. *Class* D.

C. crassicaule CORAL BROOM. Grown as an ornamental.

coral plant *Berberidopsis corallina, Russelia equisetiformis.*

coral tree *Erythrina, E. crista-galli* **cockspur** ~ *E. crista-galli* **common** ~ *E. crista-galli.*

corazon de Jesus *Begonia fuchsioides.*

Corchorus (From the Greek name *korchoros*, blue pimpernel.) A genus of about 40 species of annual herbs or subshrubs with simple alternate leaves and small yellow flowers. *Distr.* Tropical regions to S Australia. *Fam.* Tiliaceae. *Class* D.

C. capsularis JUTE. Annual erect herb. Source of the fibre JUTE used in rope-making. *Distr.* China; now widely cultivated.

cordatus, -a, -um: heart-shaped.

cordifolius, -a, -um: with heart-shaped leaves.

cordiformis, -e: heart-shaped.

Cordyline (From the Greek *kordyle*, a club, alluding to the shape of the root.) CABBAGE TREE. A genus of 15–20 species of tufted or tree-like herbs with large rosettes of leaves. Grown for their striking appearance. *Distr.* SE Australasia to Hawaii. *Fam.* Agavaceae. *Class* M.

C. australis PALM LILY. Large tree to 20m high. Leaves a source of fibre and of a syrup with a very high fructose content. *Distr.* New Zealand.

C. fruticosa See *C. terminalis*

C. terminalis Shrub to 4m high. A number of ornamental cultivars of this species are available.

coreanus, -a, -um: of Korea, Eastern Asia.

Coreopsis (From Greek *koris*, bug, and *opsis*, like, alluding to the fruits.) TICKSEED. A genus of 114 species of annual and perennial herbs with simple to tripinnate leaves and flowers in daisy-like heads that bear distinct yellow rays. Several species are grown as ornamentals. *Distr.* New World. *Fam.* Compositae. *Class* D.

C. atkinsoniana See *C. tinctoria*

C. auriculata Stoloniferous perennial. *Distr.* SE North America.

C. grandiflora Perennial or occasionally annual. Flowers yellow and orange. *Distr.* S Central USA.

C. lanceolata Erect perennial to 60cm high. *Distr.* Central and SE USA.

C. rosea Annual or perennial to around 60cm high. Flowers red, rays yellow. *Distr.* E Coast of North America.

C. tinctoria Glabrous annual to 1.5m high. Some local medicinal uses. *Distr.* North America.

C. tripteris Tall perennial. Flowers yellow, turning purple with age. *Distr.* E USA.

C. verticillata Glabrous perennial. Flowers yellow. *Distr.* E USA.

Corethrogyne (From the Greek *korethron*, brush, and *gyne*, female, alluding to the style appendages.) A genus of 3 species of perennial herbs with simple leaves and flowers in daisy-like heads with distinct rays. Some species are grown as tender ornamentals. *Distr.* USA (California). *Fam.* Compositae. *Class* D.

C. californica Stem decumbent. Rays violet-purple. *Distr.* California.

coriaceus, -a, -um: leathery.

coriander *Coriandrum sativum*
Roman ~ *Nigella sativa.*

Coriandrum (A name used by Pliny, from Greek *koris*, bug, alluding to the aromatic leaves.) A genus of 2 species of aromatic annual herbs with pinnately divided leaves and compound umbels of pink to purple flowers. *Distr.* W Mediterranean. *Fam.* Umbelliferae. *Class* D.

C. sativum CORIANDER, CHINESE PARSLEY. Stems to 50cm high. Lower leaves rounded, broad. Upper leaves narrow. The fruits are dried, ground, and used as a spice in curries and the leaves are used fresh as a herb. *Distr.* E Mediterranean, introduced Asia.

Coriaria (From the Latin *corium*, leather, alluding to the use of some species in tanning.) A genus of 5 species of herbs, shrubs and small trees with simple, whorled, or opposite leaves and racemes of small green flowers. Several species are grown as ornamentals. *Distr.* Central and South America, E Asia, New Zealand and the Pacific Islands, S Europe. *Fam.* Coriariaceae. *Class* D.

C. japonica Erect subshrub. Flowers bright red, turning blue black in fruit. *Distr.* Japan.

C. myrtifolia Shrub to 3m high. Fruit shiny black, surrounded by dark red, persistent petals. The leaves and bark are used for tanning and the poisonous, hallucinogenic fruit is used as a fly killer. *Distr.* W Mediterranean.

C. terminalis Deciduous arching shrub. Flowers minute, green. Fruit black. *Distr.* W China.

Coriariaceae A family of 1 genus and 5 species. *Class* D.
See: Coriaria.

Coris (A classical Greek name for an unknown plant.) A genus of 2 species of biennial or perennial herbs with simple, alternate leaves and spike-like racemes of irregular flowers that bear their parts in 5s. *Distr.* Mediterranean, Somalia. *Fam.* Primulaceae. *Class* D.

C. monspeliensis Erect perennial. Flowers purple. Grown as an alpine ornamental. *Distr.* Mediterranean.

corkscrew *Euphorbia mammillaris.*

corkwood, Florida Leitneriaceae.

corn *Zea mays* **Guinea** ~ *Sorghum* **sweet** ~ *Zea mays.*

Cornaceae The Dogwood family. 8 genera and 70 species of trees, shrubs and a few herbs with simple, occasionally evergreen leaves. The flowers are small, bisexual or unisexual with 4–5 sepals and 4–5 petals; they are often borne in clusters that are subtended by colourful bracts. The family contains a number of popular ornamentals. *Distr.* N temperate regions to tropical Asia, Africa and South America. *Class* D.
See: Chamaepericlymenum, Cornus, Corokia, Davidia, Nyssa.

corn cob *Euphorbia mammillaris.*

corn cockle *Agrostemma, A. githago.*

cornel *Cornus* **Bentham's** ~ *C. capitata* **Japanese** ~ *C. officinalis* **western** ~ *C. glabrata.*

cornflower *Centaurea cyanus* **perennial** ~ *C. montana.*

corniculatus, -a, -um: with small horns.

corn plant *Dracaena fragrans.*

corn salad *Valerianella, V. locusta* **common** ~ *V. locusta.*

corn spurrey *Spergula arvensis.*

Cornus (The classical Latin name for *Cornus mas.*) DOGWOOD, CORNEL. A genus of 45 species of rhizomatous herbs, subshrubs, shrubs and small trees typically with deciduous, opposite, simple leaves. The small bisexual flowers are borne in heads or clusters that are often subtended by 1–4 showy bracts. A number of species and hybrids are grown as ornamentals for their inflorescences and coloured bark in winter. *Distr.* N temperate regions, Europe, rarer in South America, Africa. *Fam.* Cornaceae. *Class* D.

C. alba TARTARIAN DOGWOOD, RED-BARKED DOGWOOD. Deciduous shrub. Stems bright red in autumn and winter. *Distr.* Central and E Asia.

C. canadensis BUNCHBERRY, CRACK-ERBERRY, CREEPING DOGWOOD. Rhizomatous herb. Flowers purple-tinged. Bracts 4, white. Used as ground cover. *Distr.* E Asia, North America.

C. candidissima See *C. racemosa*

C. capitata BENTHAM'S CORNEL. Evergreen tree. Flowers cream, very small. Bracts cream-yellow. Fruit large, strawberry-like. *Distr.* China.

C. controversa GIANT DOGWOOD. Deciduous tree to 16m high. Leaves turning purple in autumn. *Distr.* China, Japan, Himalaya.

C. florida EASTERN FLOWERING DOGWOOD, COMMON WHITE DOGWOOD. Deciduous shrub or small tree. Leaves white-woolly below. Flowers green. Bracts pink-white. *Distr.* E North America.

C. glabrata WESTERN CORNEL, BROWN DOGWOOD. Deciduous shrub. Young stems brown-red. *Distr.* W North America.

C. macrophylla Deciduous spreading tree. Flowers profuse, yellow-white. *Distr.* China, Japan, Himalaya.

C. mas CORNELIAN CHERRY, SORBET. Deciduous spreading shrub or small tree. Flowers small, bright yellow, produced before the leaves. The bright red fruit is used in jam and as the basis of the alcoholic drink Vin de Cornoulle. *Distr.* Central and S Europe, SW Asia.

C. obliqua SILKY DOGWOOD. Deciduous, spreading shrub. Leaves silky hairy below. *Distr.* E North America.

C. officinalis JAPANESE CORNEL, JAPANESE CORNELIAN CHERRY. Deciduous shrub or tree. Flowers yellow, appearing before the leaves. Fruit red. *Distr.* China, Japan.

C. racemosa PANICLED DOGWOOD. Deciduous shrub. Flowers in arching panicles. *Distr.* E North America.

C. sanguinea COMMON DOGWOOD, BLOOD-TWIG, PEGWOOD. Deciduous shrub. Flowers profuse, white, heavily fragrant. Wood used for skewers and bobbins. *Distr.* Europe.

cornutus, -a, -um: horned.

Corokia (From the Maori name for these plants, *korokia-taranga.*) A genus of about 4 species of evergreen shrubs and trees with small, simple, alternate or clustered leaves and star-shaped yellow flowers that bear their parts in 5s. Several species and hybrids are grown as ornamentals. *Distr.* New Zealand. *Fam.* Cornaceae. *Class* D.

C. cotoneaster WIRE-NETTING BUSH. Branches tortuous, tangled. Flowers solitary. *Distr.* New Zealand.

coronarius, -a, -um: like a crown.

coronatus, -a, -um: crowned.

Coronilla (From the Latin *coronilla*, a little crown, alluding to the inflorescence.) A genus of 20 species of herbs and shrubs with blue-green, pinnate leaves that are sometimes reduced to a single leaflet, and radiating umbels of pea-like flowers. *Distr.* Atlantic islands, Mediterranean, Europe. *Fam.* Leguminosae. *Class* D.

C. emerus SCORPION SENNA. Evergreen shrub. Flowers fragrant, yellow. *Distr.* Europe.

C. valentina BASTARD SENNA. Spreading evergreen shrub. Flowers golden yellow, strongly fragrant. Used in erosion control. *Distr.* Mediterranean.

C. varia CROWN VETCH. Perennial herb. Flowers white, pink or purple. *Distr.* Central and S Europe.

Correa (After Jose Francesco Correa da Serra (1750–1823), Portuguese botanist.) AUSTRALIAN FUCHSIA. A genus of 7 species of evergreen shrubs and small trees with simple, leathery, opposite leaves and solitary or clustered, tubular, 4-lobed flowers that bear 8 stamens. Several species are grown as ornamentals. *Distr.* Temperate regions of Australia. *Fam.* Rutaceae. *Class* D.

C. reflexa COMMON CORREA. Erect or prostrate shrub. Flowers pendent, white, pink or red. *Distr.* Australia.

C. speciosa See *C. reflexa*

correa, common *Correa reflexa.*

corrugatus, -a, -um: wrinkled.

Corsiaceae A family of 2 genera and 26 species of rhizomatous and tuberous, saprophytic herbs with scale-like, spirally arranged leaves and no chlorophyll. The flowers are solitary and irregular with 1 large and 5 small tepals, 6 stamens and an inferior ovary. *Distr.* New Guinea, Australasia, Pacific, Chile. *Class* M.

Corsican curse *Soleirolia soleirolii.*

corsicus, -a, -um: of Corsica.

Cortaderia (From *cortadera*, the local name for these plants in Argentina.) PAMPAS GRASS. A genus of 24 species of tussock-forming perennial grasses to 3m high with rough narrow leaves and tall, plumed, silver-white inflorescences. Several species are grown as ornamentals. *Distr.* Tropical and temperate South America, New Zealand. *Fam.* Gramineae. *Class* M.

C. argentea See *C. selloana*

C. selloana PAMPAS GRASS. Cultivated as a ornamental specimen plant and commercially for the dried inflorescences. *Distr.* South America.

Cortusa (After J. A. Cortusi, 16th-century Italian botanist.) A genus of about 8 species of perennial herbs with long-stalked basal leaves and umbels of funnel-shaped flowers that bear their parts in 5s. *Distr.* Mountains of Central Europe to N Asia. *Fam.* Primulaceae. *Class* D.

C. matthioli Herb to 40cm high. Flowers purple. Grown as a rock garden ornamental. *Distr.* W Europe.

cortusifolius, -a, -um: from the genus *Cortusa*, and *folius*, leaved.

cortusoides: from the genus *Cortusa*, with the ending *-oides*, indicating resemblance.

coruscus, -a, -um: trembling.

Corydalis (From the Greek *korydalis*, the crested lark, alluding to flowers said to resemble the shape of a lark's head.) A genus of about 300 species of rhizomatous and tuberous, annual and perennial herbs with compound leaves and racemes of tubular, spurred flowers. Numerous species and cultivars are grown as ornamentals. *Distr.* N temperate regions and mountainous areas of tropical Africa. *Fam.* Papaveraceae. *Class* D.

C. bulbosa See *C. solida*

C. cashmeriana Tuberous perennial. Flowers bright blue. *Distr.* Himalaya.

C. cava Tuberous perennial. Tubers hollow. Flowers violet or white. *Distr.* Central Europe.

C. lutea Rhizomatous perennial. Flowers golden yellow. *Distr.* Europe, naturalized in British Isles.

C. parnassica See *C. bulbosa*

C. solida FUMEWORT. Tuberous perennial. Flowers red, borne in erect racemes. *Distr.* N Europe, Asia.

Corylaceae The Hazel family. 4 genera and 57 species of shrubs and trees typically with deciduous, alternate, simple leaves that are carried in 2 ranks and tiny unisexual flowers that are borne in catkins. *Distr.* N temperate regions. *Class* D.
See: Carpinus, Corylus, Ostrya, Ostryopsis.

Corylopsis (From the Greek *korylos*, HAZEL, and *opsis*, appearance, alluding to the similarity of the leaves to those of the genus *Corylus*.) WINTER HAZEL. A genus of 7 species of deciduous shrubs and small trees with simple, stalked leaves and pendent racemes of small, fragrant, yellow flowers that bear their parts in 5s. The flowers usually appear before the leaves in spring. Several species are grown as ornamentals. *Distr.* E Asia. *Fam.* Hamamelidaceae. *Class* D.

C. glabrescens Shrub or small tree. Leaves blue-green below, bristle-toothed. Flowers bell-shaped. *Distr.* Korea, Japan.

C. pauciflora BUTTERCUP WITCH HAZEL. Spreading shrub. Leaves bronze when young. Flowers large, cup-shaped. *Distr.* Taiwan, Japan.

C. sinensis Loosely spreading shrub. Leaves downy below. Flowers pale yellow. *Distr.* China.

C. spicata SPIKE WITCH HAZEL. Spreading shrub. Flowers pale-yellow. *Distr.* Japan.

Corylus (The classical name for these plants.) HAZEL, FILBERT. A genus of 10 species of deciduous shrubs and small trees with alternate, simple, double-toothed leaves that are borne in 2 ranks. The tiny unisexual flowers are borne in separate catkins and appear in winter. The fruit is a small nut. *Distr.* Temperate Asia, Europe, and North America. *Fam.* Corylaceae. *Class* D.

C. avellana HAZEL, COBNUT. Shrub or rarely a small tree. Much grown for coppice, especially for hurdles and garden poles. The seeds can be used as a source of oil for cooking. A number of ornamental cultivars are also available. *Distr.* SE Europe, SW Asia, N Africa.

C. colurna TURKISH HAZEL. Stout tree to 30m high. Cultivated for nuts and timber. *Distr.* SE Europe, W Asia.

C. maxima FILBERT. Shrub or small tree. Leaves slightly lobed. A number of ornamental cultivars are available. *Distr.* SE Europe, SW Asia.

C. sieboldiana JAPANESE HAZEL. Large shrub. Leaves often marked red. *Distr.* N China, Japan and Korea.

corymbifera: bearing flowers in a corymb.

corymbiflora: bearing flowers in a corymb.

corymbosus, -a, -um: like a corymb.

Corynephorus (From Greek *koryne*, club, and *phoreo*, to bear, alluding to the shape of the own.) HAIR GRASS. A genus of 6 species of tufted perennial grasses with rough, grey-green leaves and spreading panicles. *Distr.* Europe, Mediterranean. *Fam.* Gramineae. *Class* M.

C. canescens GREY HAIR GRASS. A coastal plant of sand dunes. *Distr.* Europe, naturalized in North America.

Corynocarpaceae A family of 1 genus and 4 species. *Class* D.
See: Corynocarpus.

Corynocarpus (From Greek, *koryne*, club, and *carpos*, fruit, alluding to its shape.) A genus of 4 species of shrubs and trees with alternate, simple leaves and racemes of flowers that bear their parts in 5s. The fruit is a drupe. *Distr.* New Guinea, NE Australia, New Zealand. *Fam.* Corynocarpaceae. *Class* D.

cosh, blue *Caulophyllum thalictroides.*

Cosmos (From the Greek *kosmos*, ornament.) A genus of 26 species of annual and perennial herbs with simple to tripinnate leaves and flowers in daisy-like heads which usually bear distinct rays. Several species and numerous cultivars are grown as tender ornamentals. *Distr.* Tropical and warm America, especially Mexico. *Fam.* Compositae. *Class* D.

C. atrosanguineus Perennial. Flowers vanilla- or chocolate-scented, rays dark velvety red. Several cultivars of this species are available. *Distr.* Mexico.

Costaceae A family of 4 genera and 110 species of rhizomatous herbs with spirally

arranged leaves and irregular flowers that bear 3 fused petals, 1 fertile stamen and a large petal-like, sterile stamen. The ovary is inferior and the fruit a capsule. This family is sometimes included within Zingiberaceae. *Distr.* Widespread in tropical regions. *Class* M.
See: Costus, Tapeinochilos.

costaricana: of Costa Rica, Central America.

costaricensis: of Costa Rica, Central America.

costatus, -a, -um: with a midrib.

costmary *Tanacetum balsamita.*

Costus (A classical Latin name for these plants, probably derived from the Arabic plant name *koost*.) SPIRAL FLAG, SPIRAL GINGER. A genus of about 90 species of clump-forming, rhizomatous herbs with spirally arranged, somewhat fleshy leaves, and pyramidal inflorescences of showy, irregular, 1-lipped flowers. Some species have local medicinal uses; others are grown as ornamentals. *Distr.* Widespread in tropical regions. *Fam.* Costaceae. *Class* M.

C. cuspidatus FIERY COSTUS. Stem flushed red. Leaves red below. Flowers orange. *Distr.* Brazil.

C. igneus See *C. cuspidatus*

C. speciosus WILD GINGER, CREPE GINGER, MALAY GINGER. Inflorescence ovoid, bracts spiny. Flowers white-pink with orange markings. Cultivated throughout the tropics. *Distr.* SE Asia to New Guinea.

costus, fiery *Costus cuspidatus.*

Cotinus (From the Greek *kotinos*, wild olive.) SMOKEWOOD. A genus of 2 species of shrubs and trees with orange wood, deciduous simple leaves, and clusters of small yellow flowers. *Distr.* SE North America, S Europe to China. *Fam.* Anacardiaceae. *Class* D.

C. americanus See *C. obovatus*

C. coggygria SMOKE TREE, SMOKE BUSH, WIG TREE, VENETIAN SUMAC, TYROLEAN SUMAC, TURKISH SUMAC, INDIAN SUMAC, HUNGARIAN SUMAC. Shrub with very finely divided panicles of very small flowers. Leaves turning red in autumn. Leaves used for tanning, wood gives a yellow dye. Numerous cultivated ornamentals are available. *Distr.* S Europe to China.

C. obovatus CHITTAMWOOD, AMERICAN SMOKEWOOD. A source of orange dye. *Distr.* SE USA.

Cotoneaster (From the Latin *cotoneum*, quince, and *-aster*, incomplete resemblance.) A genus of about 50 species of evergreen and deciduous shrubs and small trees with simple, alternate leaves and solitary or clustered, small, white flowers that are followed by red or black fruits. Numerous species and cultivars are grown as ornamentals, particularly for their brightly coloured fruits which usually last the entire winter. *Distr.* Temperate regions of the Old World. *Fam.* Rosaceae. *Class* D.

C. conspicuus Prostrate, evergreen or semi-evergreen shrub. Fruit shiny, scarlet. *Distr.* SE Tibet.

C. dammeri Graceful evergreen shrub to 5m high. Fruit typically bright red. This species has given rise to numerous garden cultivars. *Distr.* China.

C. horizontalis Flat deciduous shrub. Branches forming a herring-bone pattern, spreading. Flowers red. Fruit orange or red. *Distr.* W China.

C. humifusus See *C. dammeri*

cotton *Gossypium* **lavender** ~ *Santolina, S. chamaecyparissus* **Levant** ~ *Gossypium herbaceum* **Sea Island** ~ *G. barbadense* **tree** ~ *G. arboreum.*

cotton plant, common *Eriophorum angustifolium.*

cotton seed tree *Baccharis halimifolia.*

cotton weed *Otanthus maritimus.*

cottonwood *Populus trichocarpa, Populus deltoides.*

Cotula (From the Greek *kotyla*, a small cup, alluding to the leaf bases.) BRASS BUTTONS. A genus of about 80 species of annual to perennial herbs, typically with pinnate leaves and yellow flowers in solitary, daisy-like heads which do not bear distinct rays. *Distr.* Widespread, particularly in the S Hemisphere. *Fam.* Compositae. *Class* D.

C. atrata See *Leptinella atrata*

C. coronopifolia BRASS BUTTONS. Somewhat fleshy annual or perennial. *Distr.* S Africa.

C. goyenii See *Leptinella goyenii*

C. hispida Tuft-forming perennial. *Distr.* S Africa.

C. lineariloba Tuft- or mat-forming perennial. Flower red-yellow. *Distr.* S Africa.

C. pectinata See *Leptinella pectinata*

C. perpusilla Mat-forming perennial. *Distr.* New Zealand.

C. potentilloides See *Leptinella potentillina*

C. pyrethrifolia See *Leptinella pyrethrifolia*

C. reptans See *Leptinella reptans*

C. rotundata See *Leptinella rotundata*

C. scariosa See *Leptinella scariosa*

C. sericea See *Leptinella albida*

C. squalida See *Leptinella squalida*

Cotyledon (From Greek *kotyle*, cavity, alluding to the cupped leaves of some species.) A genus of about 35 species of subshrubs and shrubs with simple, fleshy leaves that are borne in opposite pairs or whorls and many panicles of numerous, nodding, bell-shaped flowers that bear their parts in 5s. A number of species are grown as ornamentals. *Distr.* S and E Africa, Arabia. *Fam.* Crassulaceae. *Class* D.

C. chrysantha See *Rosularia pallida*

C. oppositifolia See *Chiastophyllum oppositifolium*

C. orbiculata Much branched shrub. Leaf margins crimped. Flowers orange to deep red. Leaf juice is toxic but said to be beneficial in the treatment of epilepsy. *Distr.* South Africa (Cape Province), SW Africa.

C. simplicifolia See *Chiastophyllum oppositifolium*

courgette *Cucurbita pepo.*

Cowania (After James Cowan (died 1823), an English merchant who introduced of plants into cultivation from Central and South America.) CLIFF ROSE. A genus of 5 species of evergreen shrubs and small trees with pinnately lobed, leathery leaves and solitary, rose-like flowers that are followed by feathery seed heads. Several species are grown as ornamentals. *Distr.* SW North America. *Fam.* Rosaceae. *Class* D.

C. mexicana CLIFF ROSE. Upright shrub. Flowers pale-yellow to white, fragrant. *Distr.* SW North America.

C. stansburyana See *C. mexicana*

cowberry *Vaccinium vitis-idaea.*

cow cockle *Vaccaria hispanica.*

cow itch tree *Lagunaria patersonii.*

cowslip *Primula veris* **American** ~ *Dodecatheon D. meadia*, **Cape** ~ *Lachenalia.*

crackerberry *Cornus canadensis.*

Crambe (From the Greek for CABBAGE.) A genus of 20 species of annual to perennial herbs with large leaves and numerous fragrant, white to yellow flowers. *Distr.* Eurasia, Mediterranean, Macaronesia, tropical African mountains. *Fam.* Cruciferae. *Class* D.

C. cordifolia Perennial with white flowers. *Distr.* Caucasus.

C. koktebelica Tall perennial with yellow sepals and white petals. *Distr.* Coast of the Black Sea.

C. maritima SEA KALE. Grown for its edible succulent shoots. *Distr.* Coast of W Europe to SW Asia.

crampbark *Viburnum opulus.*

cranberry *Vaccinium oxycoccos*, *V. vitis-idaea*, *V. macrocarpon*, *Viburnum trilobum* **American** ~ *Vaccinium macrocarpon* **European** ~ *V. oxycoccos* **highbush** ~ *Viburnum trilobum* **hog** ~ *Arctostaphylos uva-ursi* **large** ~ *Vaccinium macrocarpon* **mountain** ~ *V. vitis-idaea* **small** ~ *V. oxycoccos* **southern mountain** ~ *V. erythrocarpum.*

cranberry bush, American *Viburnum trilobum* ~ **European** *V. opulus.*

crane flower *Strelitzia reginae.*

cranesbill *Geranium* **bloody** ~ *G. sanguineum* **dusky** ~ *G. phaeum* **meadow** ~ *G. pratense* **shining** ~ *G. lucidum* **wood** ~ *G. sylvaticum.*

Craspedia (From the Greek *kraspedon*, edging, alluding to the furry margins of some species.) BILLY BUTTONS, BACHELOR'S BUTTONS. A genus of 6 species of annual and perennial herbs with basal rosettes of entire leaves and flowers in daisy-like heads which lack distinct rays. *Distr.* Temperate Australia, New Zealand. *Fam.* Compositae. *Class* D.

C. glauca COMMON BILLY BUTTONS, BACHELOR'S BUTTONS. Tufted perennial. Flowers yellow-cream. *Distr.* S Australia.

C. globoides DRUMSTICKS. Perennial. Leaves woolly. Flowers yellow. *Distr.* SE Australia.

C. incana Perennial. Leaves woolly. Flowers yellow. *Distr.* New Zealand.

C. richea See *C. glauca*

crassicaulis, -e: with a thick stem.

crassifolius, -a, -um: thick-leaved.

crassipes: with a thick stalk.

Crassula (The diminutive of the Latin *crassus*, thick, alluding to their succulent texture.) A genus of about 300 species of annual, biennial, and perennial succulent herbs and small shrubs with simple, variable leaves and small tubular to star-shaped flowers that may be borne singly or in large branched clusters. Many species are grown as ornamentals. *Distr.* Widespread but chiefly S Africa. *Fam.* Crassulaceae. *Class* D.

C. arborescens SILVER JADE PLANT. Shrubby perennial. Leaves grey-green with a red-purple margin. Flowers cream, rarely produced in cultivation. Much grown as a houseplant. *Distr.* S Africa.

C. falcata AEROPLANE PROPELLER PLANT. Bushy perennial. Leaves blue-grey, twisted, clasping stem. *Distr.* South Africa.

C. sarcocaulis Half-hardy evergreen shrub. Flowers red, tiny, borne in clusters. *Distr.* S Africa.

C. tillaea MOSSY STONECROP. Small tufted annual. Leaves tinged red. Flowers solitary, white. *Distr.* S and W Europe.

Crassulaceae The Stonecrop family. 37 genera and about 1500 species of succulent herbs and small shrubs with more or less fleshy, simple leaves and showy corymbs or panicles of regular, bisexual flowers that typically bear their parts in 5s. Many species are valued as ornamental rock-garden and greenhouse plants. *Distr.* Widespread, except Australia. *Class* D.

See: Aeonium, Aichryson, Chiastophyllum, Cotyledon, Crassula, Dudleya, Echeveria, Graptopetalum, Greenovia, Hylotelephium, Jovibarba, Kalanchoe, Orostachys, Pachyphytum, × Pachyveria, Rhodiola, Rosularia, Sedum, Sempervivum, Umbilicus, Villadia.

crassulifolius: with slightly thick leaves.

crassulus, -a, -um: rather thick.

crassus, -a, -um: thick.

crataegifolius: from the genus *Crataegus*, and *folius*, leaved.

Crataegus (From the Greek *kratos*, strength, alluding to the tough wood.) HAWTHORN, THORN APPLE. A genus of about 280 species of deciduous, typically spiny shrubs and small trees with simple or lobed leaves and corymbs of regular, white flowers. Several species and numerous cultivars are grown as ornamentals and as hedging plants. *Distr.* N temperate regions. *Fam.* Rosaceae. *Class* D.

C. crus-galli COCKSPUR THORN. Tree to 12m high. Thorns long, curved. *Distr.* E North America.

C. laevigata ENGLISH HAWTHORN, QUICKSET THORN, WHITE THORN, MAY. Shrub or small tree to 10m high. The buds and young leaves of this species are edible (being called 'bread and cheese' in some parts of Britain) and the wood is used in a similar way to that of BOX. It is also grown as an ornamental with a number of cultivars having been raised. *Distr.* N and Central Europe.

C. monogyna COMMON HAWTHORN, ENGLISH HAWTHORN. Shrub or small tree to 8m high. A number of ornamental cultivars have been raised from this species. *Distr.* Europe.

C. oxyacantha See *C. laevigata*

C. phaenopyrum WASHINGTON THORN. Round-headed tree to 10m high. *Distr.* North America.

crazyweed *Oxytropis.*

creambush *Holodiscus discolor.*

cream cups *Platystemon californicus.*

creashak *Arctostaphylos uva-ursi.*

creeper: Australian bluebell ~ *Sollya heterophylla* **bleubell** ~ *S. heterophylla* **canary** ~ *Tropaeolum peregrinum* **Japanese** ~ *Parthenocissus tricuspidata* **Mexican** ~ *Antigonon leptopus* **Rangoon** ~ *Quisqualis indica* **Virginia** ~ *Parthenocissus, P. quinquefolia, P. tricuspidata.*

Cremanthodium (From the Greek *kremao*, to hang up, and *anthodium*, flower-head, alluding to the nodding inflorescences.) A genus of 50 species of perennial herbs with kidney-shaped leaves and flowers in large, daisy-like, nodding heads that bear distinct rays. *Distr*. Himalaya. *Fam*. Compositae. *Class* D.

C. nobile Perennial. Rays golden-yellow. *Distr*. China and Tibet.

crenatus, -a, -um: with shallow, rounded teeth.

crenulatus, -a, -um: with small, shallow, rounded tooth.

crepe flower *Lagerstroemia indica.*

Crepis (From the Greek *krepis*, a classical Greek plant name that also meant sandal.) HAWK'S BEARD. A genus of about 200 species of annual and perennial herbs with simple to pinnate leaves and flowers in dandelion-like heads. *Distr*. N hemisphere. *Fam*. Compositae. *Class* D.

C. aurea Perennial. Flowers yellow-orange. Occasionally grown as an ornamental. *Distr*. Alps to Balkans.

C. incana Small perennial. Flowers magenta-pink. *Distr*. Greece.

cress: American ~ *Barbarea verna* **Belle Isle** ~ *B. verna* **bitter** ~ *B. vulgaris* **common garden** ~ *sativum sativum* **early winter** ~ *Barbarea verna* **garden** ~ *sativum sativum* **Indian** ~ *Tropaeolum, T. majus* **Italian** ~ *Eruca* **meadow** ~ *Cardamine pratensis* **Normandy** ~ *Barbarea verna* **rocket** ~ *B. vulgaris* **stone** ~ *Aethionema* **wall** ~ *Arabis*.

cretaceus, -a, -um: chalky white in colour.

cretensis: of Crete.

creticus, -a, -um: of Crete.

Crinitaria: See *Aster*

crinitus, -a, -um: long-haired.

Crinodendron (From the Greek *krinon*, lily, and *dendron*, tree.) A genus of 2 species of evergreen shrubs and small trees with simple leathery leaves and pendent, tubular to urn-shaped flowers that bear their parts in 5s. *Distr*. Chile and Argentina. *Fam*. Elaeocarpaceae. *Class* D.

C. hookerianum Flowers numerous, red. *Distr*. Argentina, Chile.

C. patagua Flowers white. *Distr*. Chile.

Crinum (From the Greek *krinon*, a lily.) CAPE LILY. A genus of 125 species of bulbiferous, perennial herbs with large linear leaves and umbels of regular or irregular, more or less funnel-shaped flowers. Several species are grown as tender ornamentals. *Distr*. Tropical and warm temperate regions. *Fam*. Amaryllidaceae. *Class* M.

C. amoenum Flowers fragrant, white tinged red. *Distr*. Burma.

C. bulbispermum Flowers fragrant, curved, white with red stripes. *Distr*. South Africa.

C. capense See *C. bulbispermum*

C. moorei Flowers pink or white. Has some local medicinal uses. *Distr*. South Africa (Natal, Cape Province).

C. × powellii Leaves grooved. Flowers pink, fragrant. *Distr*. South Africa.

crispifolius, -a, -um: with wavy leaves.

crispula: irregularly wavy.

crispus, -a, -um: irregularly wavy, usually referring to the edges of the leaves.

crista-galli: cock's comb.

cristatus, -a, -um: crested.

Crithmum (From the Greek *krithe*, barley, alluding to the shape of the seeds.) A genus of 1 species of perennial herb with somewhat succulent, divided leaves and compound, flat-topped umbels of very small, yellow-white flowers. *Distr*. Atlantic coastal regions or Europe. *Fam*. Umbelliferae. *Class* D.

C. maritimum SEA SAMPHIRE, SAMPHIRE. Leaves gathered and used in salads and pickles.

croatica: of Croatia.

crocatus, -a, -um: saffron yellow in colour.

crocodile jaws *Aloe humilis*.

Crocosmia (From the Greek *krokos*, saffron, and *osme*, a smell; the dried flowers smell of saffron when wetted.) MONTBRETIA. A genus of about 10 species of perennial herbs with corms, 2 ranks of linear leaves and spikes of funnel-shaped, orange or yellow flowers. Several species and hybrids are grown as ornamentals. *Distr.* S Africa. *Fam.* Iridaceae. *Class* M.

C.* × *crocosmiiflora A hybrid of *C. aurea* x *C. pottsii*. Forms the basis of a wide range of garden cultivars. *Distr.* Garden origin.

C. masonorum Leaves pleated. Flowers orange-red, borne in 2 rows. *Distr.* South Africa (Transkei).

C. paniculata Flowers orange, borne on a branched spike. *Distr.* South Africa.

C. rosea See *Tritonia rubrolucens*

Crocus (From the Greek *krokos*, saffron, derived from Hebrew *karkon*.) A genus of 80 species of perennial herbs with corms, linear to lance-shaped leaves, and funnel-shaped flowers that are borne at ground level and bear 3 stamens. Numerous species and cultivars are grown as autumn- or spring-flowering ornamentals. *Distr.* Europe and W and Central Asia. *Fam.* Iridaceae. *Class* M.

C. cartwrightianus WILD SAFFRON. Flowers white, lilac or mauve with dark veins, appearing at the same time as the leaves in early winter. *Distr.* S Greece.

C. chrysanthus Flowers creamy-yellow, sometimes striped maroon, appearing at the same time as the narrow leaves in late winter. A number of cultivars have been raised from this species. *Distr.* Balkans.

C. flavus Flowers golden-yellow, scented, appearing at the same time as the leaves in the spring. *Distr.* Balkans to NW Turkey.

C. nudiflorus Flowers white, flushed mauve, appearing before the leaves in autumn. Formerly cultivated in the British Isles by the Knights of St John as a saffron substitute. *Distr.* SW France, NE Spain, naturalized in British Isles.

C. sativus SAFFRON CROCUS, SAFFRON. Flowers white to lilac or mauve with darker veins. The source of the spice and dye Saffron which consists of the red stigmas. Saffron was first used as a dye in the Mediterranean basin over 3500 years ago and was cultivated in Britain until the end of the last century. It is still widely grown in Spain, India, and China. *Distr.* A sterile cultigen probably derived from *C. cartwrightianus*, unknown as a wild plant but occasionally persisting as a relic of cultivation.

C. speciosus Flowers white and mauve with purple veins, appearing before the leaves in autumn. *Distr.* Crimea, Caucasus, N Iran, N and Central Turkey.

C. vernus Flowers mauve to white, appearing with leaves in spring or early summer. A vigorous species with numerous cultivars available. *Distr.* Central and SE Europe.

crocus: autumn ~ Colchicaceae, *Colchicum*, *C. autumnale* **Chilean** ~ *Tecophilaea cyanocrocus* **Indian** ~ *Pleione* saffron ~ *Crocus sativus*.

Crossandra (From the Greek *krossoi*, fringe, and *aner*, male, alluding to the fringed stamens.) A genus of about 50 species of evergreen subshrubs and shrubs with whorls of simple leaves and 4-sided spikes of funnel-shaped flowers subtended by showy bracts. *Distr.* Tropical Africa, Arabia, Madagascar. *Fam.* Acanthaceae. *Class* D.

C. infundibuliformis FIRECRACKER FLOWERS. Spikes to 10cm high. Bracts downy. Flowers bright orange or pink. Grown as a tender ornamental, allegedly an aphrodisiac. *Distr.* S India.

C. nilotica Spikes dense. Flowers orange-red, funnel-shaped with a long thin tube. Grown as an ornamental. *Distr.* Tropical Africa.

Crossosomataceae A family of 3 genera and 8 species of shrubs with simple, alternate leaves and regular, bisexual flowers that bear 5 sepals, 5 petals and numerous stamens. *Distr.* SW USA, Mexico. *Class* D.

crosswort *Crucianella*.

Crotalaria (From the Greek *krotalon*, a castanet, alluding to the seeds which rattle in the mature pod.) A genus of about 600 species of annual and perennial herbs and shrubs with simple or palmate leaves and typically racemes of irregular, pea-like flowers. Several species are grown as ornamentals and some are used as fodder. *Distr.* Tropical and subtropical regions. *Fam.* Leguminosae. *Class* D.

C. agatiflora CANARY BIRD BUSH. Spreading evergreen shrub. Flowers yellow-green with a large standard. *Distr.* Highland areas of E Africa.

C. capensis YELLOW PEA, CAPE LABURNUM. Shrub or small tree. Leaves trifoliate. Flowers yellow. *Distr.* S Africa.

C. grevyi Shrub. Flowers yellow with red or purple markings. *Distr.* Madagascar.

croton *Codiaeum variegatum.*

crowberry Empetraceae, *Empetrum nigrum* **black** ~ *E. nigrum.*

Crowea (After James Crow (1750–1807), English botanist.) A genus of 3 species of perennial herbs and shrubs with simple, glandular, alternate leaves and solitary, regular flowers that bear 5 petals and 10 stamens. *Distr.* S Australia. *Fam.* Rutaceae. *Class* D.

C. exalata Shrub to 1m high. Flowers mauve to pink. Grown as a half-hardy ornamental. *Distr.* Australia (New South Wales, Victoria).

crowfoot *Ranunculus* **water** ~ *R. aquatilis.*

crown imperial *Fritillaria imperialis.*

crow root *Aletris farinosa.*

Crucianella (A diminutive of Latin *crux*, cross.) CROSSWORT. A genus of about 30 species of annual and perennial herbs and subshrubs with 4-angled stems, whorls of 4 or 6 simple leaves and spikes or clusters of tubular flowers. Several species are grown as ornamentals. *Distr.* Europe, Central Asia. *Fam.* Rubiaceae. *Class* D.

C. maritima Perennial herb. Foliage blue-green. Flowers cream-yellow, borne in spikes. *Distr.* W Mediterranean.

C. stylosa See *Phuopsis stylosa*

cruciatus, -a, -um: cross-shaped.

Cruciferae The Mustard family. 381 genera and about 3000 species of annual to perennial herbs and occasionally shrubs or even aquatics with simple, lobed, or divided leaves. The flowers are very characteristic for the family, having 4 free sepals, 4 free petals, and 6 stamens (4 of which are long and 2 short). The fruit is a 2-chambered capsule. There are many important vegetable, fodder, and oilseed crops in this family as well as a number of ornamental species. *Distr.* Widespread, but mainly in temperate and warm-temperate regions. *Class* D.
See: Aethionema, Alliaria, Alyssoides, Alyssum, Arabis, Armoracia, Aubrieta, Barbarea, Biscutella, Bornmuellera, Brassica, Braya, Cardamine, Cheiranthus, Cochleria, Coluteocarpus, Crambe, Degenia, Dentaria, Draba, Eruca, Erysimum, Farsetia, Fibigia, Heliophila, Hesperis, Hugueninia, Hutchinsia, Iberis, Isatis, Lepidium, Lesquerella, Lunaria, Malcolmia, Matthiola, Megacarpaea, Morisia, Murbeckiella, Nasturtium, Pachyphragma, Parrya, Petrocallis, Phoenicaulis, Physaria, Raphanus, Rorippa, Schivereckia, Sisymbrium, Thlaspi, Wasabia

cruel plant *Araujia sericofera.*

cruentus, -a, -um: blood-red.

crustatus, -a, -um: with a hardened crust.

cry-baby tree *Erythrina crista-galli.*

Cryptanthus (From the Greek *krypto*, to hide, and *anthos*, a flower, alluding to the way in which the flowers are obscured by the bracts.) EARTH STAR. A genus of 20 species of stemless or short-stemmed herbs typically with flattened rosettes of triangular, often variegated leaves, and clusters of small flowers borne in the centre of each rosette. Several species and numerous hybrids are grown for their ornamental foliage. Inter-generic hybrids have also been produced with plants of the genus *Billbergia*. *Distr.* E Brazil. *Fam.* Bromeliaceae. *Class* M.

C. acaulis GREEN EARTH STAR, STARFISH PLANT. Leaves green. Flowers white, fragrant.

C. bromelioides RAINBOW STAR. Leaves variegated, olive green. Flowers ivory. *Distr.* Brazil.

C. zonatus ZEBRA PLANT. Leaves dark-green with silver banding.

× *Cryptbergia* (From the names of the parent genera.) A genus of hybrids between members of the genera *Cryptanthus* and *Billbergia*. *Distr.* Garden origin. *Fam.* Bromeliaceae. *Class* M.

Crypteroniaceae A family of 3 genera and 9 species of shrubs and trees with 4-angled stems, simple, opposite leaves, and racemes or panicles of very small flowers that typically bear their parts in 5s but may lack petals. *Distr.* Tropical Asia. *Class* D.

Cryptogramma (From the Greek *kryptos*, hidden, and *gramme*, a line, alluding to the lines of sori that are hidden before maturity.) ROCK BRAKE, PARSLEY FERN. A genus of

4 species of dwarf deciduous ferns. Some species are cultivated as ornamentals. *Distr.* Alpine and N regions. *Fam.* Adiantaceae. *Class* F.

C. crispa PARSLEY FERN. Leaves bright green, somewhat resembling parsley. *Distr.* Europe and SW Asia.

Cryptomeria (From the Greek *krypto*, to hide, and *meris*, a part.) A genus of 1 species of evergreen tree to 40m high with linear leaves and solitary, globose fruiting cones. *Distr.* China, Japan. *Fam.* Taxodiaceae. *Class* G.

C. fortunei See *C. japonica*

C. japonica JAPANESE CEDAR. Grown as an ornamental and for timber in Japan.

cryptomerioides from the genus *Cryptomeria*, with the ending *-oides*, indicating resemblance.

Cryptotaenia (From the Greek *kryptos*, hidden, and *tainia*, bands.) A genus of 4 species of annual and perennial herbs with divided leaves and compound umbels of white flowers. *Distr.* N temperate regions and the mountains of tropical Africa. *Fam.* Umbelliferae. *Class* D.

C. canadensis HONEYWORT, WHITE CHERVIL, MITSUBA. Perennial herb to 1m high. This species is grown in Japan as a salad crop and for its roots which are eaten as a fried vegetable. *Distr.* E Asia, North America.

crystallinus, -a, -um: crystalline, glistening.

Ctenanthe (From the Greek *kteis*, a comb, and *anthos*, flower, alluding to the appearance of the bracts on the inflorescence.) A genus of about 20 species of rhizomatous herbs with aerial roots, rosettes of long-stalked, often patterned leaves, and spikes of small, 3-petalled flowers. Several species are grown for their ornamental foliage. *Distr.* Costa Rica, Brazil. *Fam.* Marantaceae. *Class* M.

C. lubbersiana Leaves long, pointed, deep green, marked with pale green. *Distr.* Brazil.

C. oppenheimiana NEVER-NEVER PLANT. Leaves lance-shaped, green with a silver or cream pattern above, tinged red below. *Distr.* E Brazil.

Ctenolophonaceae A family of 1 genus and 1 species of tree with opposite, simple, leathery leaves and panicles of regular, bisexual flowers that bear 5 sepals, 5 petals, 10 stamens, and a superior ovary. It is sometimes used as a source of timber. *Distr.* W tropical Africa, Indonesia and Malaysia. *Class* D.

cuckold *Bidens*.

cuckoo bread *Oxalis acetosella*.

cuckoo flower *Cardamine pratensis*.

cuckoo pint *Arum*.

cuckoo plant *A. maculatum*.

cucullatus, -a, -um: hooded.

cucumber *Cucumis sativus* **squirting** ~ *Ecballium elaterium*.

cucumber root *Medeola virginica* **Indian** ~ *M. virginica*.

cucumber tree *Magnolia acuminata*.

Cucumis (The classical Latin name for CUCUMBER.) A genus of about 30 species of scambling or climbing, annual and perennial herbs with simple tendrils and simple to deeply lobed leaves. The bell-shaped yellow flowers are unisexual, the sexes often being carried on separate plants. The fruit is a fleshy, smooth, or spiny berry. *Distr.* Tropical regions of the Old World. *Fam.* Cucurbitaceae. *Class* D.

C. melo MELON. Soft-hairy, annual, trailing vine. Fruit large, with a hard rind and soft pulp, variable according to cultivar. The pulp of the fruit is eaten as a dessert and there are numerous cultivars available. *Distr.* Tropical Africa; now cultivated widely.

C. sativus CUCUMBER, GHERKIN. Trailing or climbing annual with stiff bristly hairs. The fruits are picked before they are fully mature and eaten fresh or pickled. *Distr.* India; cultivated since ancient times.

Cucurbita (The classical Latin name of a type of gourd.) A genus of 27 species of annual and perennial, prostrate or climbing herbs with simple tendrils and large simple leaves. The bell-shaped yellow flowers are unisexual, both sexes being carried on the same plant. The fruit is a large berry, often with a hard beige rind. Several species have edible fruits. *Distr.* Warm and tropical America. *Fam.* Cucurbitaceae. *Class* D.

C. ficifolia CIDRA, SIDRA, MALABAR GOURD. Perennial. Fruits to 35cm long, striped white.

Fruit pulp made into a kind of sweet preserve. *Distr.* Mexico to Chile.

C. foetidissima MISSOURI GOURD, FOETID WILD PUMPKIN, BUFFALO GOURD. Perennial. Fruit to 7cm long, striped green. The seeds are a source of oil and protein. *Distr.* S USA, Mexico.

C. maxima PUMPKIN, WINTER SQUASH. Long, climbing or scrambling annual. Fruits eaten as a vegetable, larger varieties fed to livestock. *Distr.* South America; first cultivated in Peru.

C. pepo MARROW, PUMPKIN, SUMMER SQUASH, COURGETTE, ZUCCHINI. Slightly bushy or trailing annual. Cultivars of various habits and fruit types are available. Fruits eaten as a vegetable or fed to livestock. *Distr.* Central America; cultivated in Mexico and S USA for over 8,000 years.

Cucurbitaceae The Gourd family. 120 genera and about 700 species of climbing and trailing herbs with palmately veined leaves and spiralling tendrils. The yellow, funnel-shaped flowers are usually unisexual and bear their parts in 5s. The fruit is a large fleshy berry. There are 3 economically important genera, *Cucurbita* (which provides the PUMPKIN, SQUASH, MARROW, and COURGETTE), *Cucumis* (CUCUMBER) and *Citrullus* (the WATER MELON). *Distr.* Widespread but centred in tropical regions. *Class* D.
See: Bryonia, Citrullus, Cucumis, Cucurbita, Ecballium, Luffa, Sechium.

cudweed *Artemisia ludoviciana, Gnaphalium.*

Cuitlauzina (After King Cuitlahuatzin, who brought rare plants to the gardens of W Mexico.) A genus of 1 species of epiphytic or terrestrial orchid with leathery leaves and long racemes of lemon-scented flowers. *Distr.* Mexico. *Fam.* Orchidaceae. *Class* M.

C. pendula This species was formerly included in the genus *Odontoglossum*. Grown as a tender ornamental.

culinaris: edible, of use in cooking.

cultriformis, -e: resembling the blade of a knife.

culver's root *Veronicastrum virginicum.*

cumin *Cuminum cyminum* **black ~** *Nigella sativa.*

Cuminum (From the Greek name for these plants, *kyminon*.) A genus of 2 species of annual herbs with divided leaves and compound umbels of white or pink flowers. *Distr.* Mediterranean to E Africa and Central Asia. *Fam.* Umbelliferae. *Class* D.

C. cyminum CUMIN. The fruits (cumin seeds) are used as a flavouring, particularly in curry powder, although it is often replaced by CARAWAY. *Distr.* Mediterranean.

cuneatus, -a, -um: pointed, wedge-shaped.

cuneifolius, -a, -um: with leaves narrowing to a wedge shape at the base.

cuneiformis, -a, -um: wedge-shaped, narrowing to a point.

Cunila A genus of 15 species of aromatic herbs and shrubs with simple, opposite, sometimes purple-spotted leaves and 2-lipped flowers. *Distr.* E North America to Uruguay. *Fam.* Labiatae. *Class* D.

C. origanoides OIL OF DITTANY PLANT. Straggling perennial herb. Flowers in loose cymes, purple or occasionally white. Sometimes grown as an ornamental and the source of a medicinal oil. *Distr.* E North America.

Cunninghamia (After James Cunningham (died 1709), who collected in China.) A genus of 2 species of evergreen coniferous tree with stiff, lance-shaped leaves and globose fruiting cones. *Distr.* China and Taiwan. *Fam.* Taxodiaceae. *Class* G.

C. lanceolata Tree to 50m high. Used in reafforestation projects. *Distr.* Central and SE China.

Cunoniaceae A family of 23 genera and 340 species of shrubs and trees typically with trifoliate or pinnate, leathery leaves and small regular flowers that bear numerous stamens. The family is a source of light timber. *Distr.* S Hemisphere, especially Australia and New Caledonia. *Class* D.
See: Weinmannia.

cup and saucer *Campanula medium.*

cup and saucer plant *Silphium perfoliatum.*

cupflower *Nierembergia.*

Cuphea (From the Greek *kyphos*, curved, alluding to the curved fruit capsule.) A genus of 250 species of annual and perennial herbs and subshrubs with sticky stems, simple leaves, and racemes or panicles of irregular, 6-lobed, tubular flowers. Several species are grown as ornamentals. *Distr.* North to South America. *Fam.* Lythraceae. *Class* D.

C. cyanea Rounded evergreen subshrub. Flowers orange-red with a yellow-green tip. *Distr.* Mexico.

C. hyssopifolia FALSE FEATHER, ELFIN HERB. Dense evergreen subshrub. Flowers with spreading petals, pink or white. *Distr.* Mexico, Guatemala.

C. ignea CIGAR FLOWER, FIRECRACKER PLANT, RED-WHITE-AND-BLUE FLOWER. Evergreen bushy subshrub. Flowers orange-red with a white ring at the mouth. *Distr.* Mexico, Jamaica.

Cupid flower *Ipomoea quamoclit.*

Cupid's darts *Catananche caerulea.*

Cupressaceae The Cypress family. 18 genera and 133 species of coniferous shrubs and trees. The leaves are needle-like in seedlings but usually become scale-like and appressed in mature plants. The female cones are made up of only a few scales which typically open to release the seeds. Sometimes the scales become fleshy, notably in the genus *Juniperus*. Members of this family are important for their timber, resin, and fruit, and as ornamentals. *Distr.* Widespread. *Class* G. *See: Austrocedrus, Callitris, Calocedrus, Chamaecyparis, X Cupressocyparis, Cupressus, Diselma, Fitzroya, Fokienia, Juniperus, Libocedrus, Microbiota, Pilgerodendron, Thuja, Thujopsis, Widdringtonia.*

× *Cupressocyparis* (From the names of the parent genera.) A genus of 3 species of hybrids between species of *Cupressus* and *Chamaecyparis*. *Distr.* Garden origin. *Fam.* Cupressaceae. *Class* G.

×*C. leylandii* A hybrid of *Cupressus macrocarpa* × *Chamaecyparis nootkatensis*. Evergreen coniferous tree to 35m high with minute, scale-like leaves on much-branched, fan-like shoots. Numerous cultivars have been raised of this hybrid which are very widely planted as fast-growing shelter belts and ornamentals. *Distr.* Garden origin.

cupressoides: resembling the genus *Cupressus*.

Cupressus (From *kuparissos*, a classical Greek name for *Cupressus sempervirens*.) CYPRESS. A genus of about 15 species of evergreen coniferous trees with needle-like juvenile leaves and scale-like adult leaves. Several species and numerous hybrids are grown as ornamentals; some are a source of timber. *Distr.* North America, N Africa, Asia. *Fam.* Cupressaceae. *Class* G.

C. arizonica ARIZONA CYPRESS. Tree to 25m high. Several ornamental cultivars have been raised from this species. *Distr.* SW North America.

C. funebris CHINESE WEEPING CYPRESS. Conical or cylindrical tree to 25m high. *Distr.* China.

C. glabra See *C. arizonica*

C. lusitanica MEXICAN CYPRESS, CEDAR OF GOA. Tree to 45m high. *Distr.* Mexico and Guatemala; long naturalized in Europe.

C. macrocarpa MONTEREY CYPRESS. Tree to 25m high. Numerous ornamental cultivars of this species have been raised. *Distr.* USA (California).

curare *Chondrodendron tomentosum.*

curassavicus, -a, -um: Of Curacao, in the Caribbean.

Curcuma (From the Arabic *kurkum*, yellow, alluding to the rhizome of *Curcuma longa*.) A genus of 40 species of herbs with reed-like stems, thick rhizomes, lance-shaped leaves, and terminal inflorescences of irregular flowers. *Distr.* Tropical Asia. *Fam.* Zingiberaceae. *Class* M.

C. longa TURMERIC. The rhizome is used as the source of a yellow dye and a spice used in curry powder. *Distr.* India.

currant Saxifragaceae, *Ribes* **Alpine** ~ *R. alpinum* **American** ~ *R. sanguineum* **American black** ~ *R. americanum* **black** ~ *R. nigrum* **buffalo** ~ *R. aureum, R. odoratum* **clove** ~ *R. odoratum* **European black** ~ *R. nigrum* **fetid** ~ *R. glandulosum* **flowering** ~ *R. sanguineum* **golden** ~ *R. aureum* **Indian** ~ *Symphoricarpos orbiculatus* **Missouri** ~ *Ribes odoratum* **mountain** ~ *R. alpinum* **skunk** ~ *R. glandulosum* **winter** ~ *R. sanguineum.*

curtipendulus, -a, -um: short and pendulous.

Curtonus See *Crocosmia*

curtus, -a, -um: short.

curvibracteatus, -a, -um: with curved bracts.

curviflorus, -a, -um: with curved flowers.

curvispinus, -a, -um: with curved spines.

curvulus, -a, -um: slightly curved.

cuscutiformis: from the genus *Cuscuta*, and *formis*, in the shape of.

cush cush *Dioscorea trifida*.

cuspidatus, -a, -um: with a short sharp point.

cyanandrus: with blue flowers.

Cyananthus (From the Greek *kyanos*, blue, and *anthos*, flower.) TRAILING BELL-FLOWER. A genus of about 30 species of perennial, tufted, alpine herbs with alternate leaves and solitary, bell-shaped, blue, yellow, or white flowers. A number of species are grown as ornamentals. *Distr*. China, Himalaya. *Fam*. Campanulaceae. *Class* D.

C. integer Leaves white and hairy. Flowers violet-blue. *Distr*. W Himalaya.

C. lobatus Leaves somewhat fleshy, deeply lobed. Flowers bright purple to blue. *Distr*. W Himalaya.

C. microphyllus Mat-forming. Leaves numerous, small. Flowers violet-blue with tufted white hairs within. *Distr*. Nepal, N India.

cyanaster: a blue star.

Cyanastraceae A family of 1 genus and 6 species of tuberous herbs with sheathing basal leaves. The flowers bear 2 whorls of 3 tepals, 2 whorls of 3 stamens, and a half-inferior ovary. *Distr*. Tropical Africa. *Class* M.

Cyanella (From the diminutive of the Greek *kyanos*, blue.) A genus of 8 species of perennial herbs with leaves in flat rosettes and racemes of showy flowers. *Distr*. Africa, especially SW Cape. *Fam*. Tecophilaeaceae. *Class* M.

C. orchidiformis Flowers pale mauve with darker markings. *Distr*. S Africa.

cyaneus, -a, -um: blue.

cyanocarpum: with blue fruit.

cyanocrocus: like a blue *Crocus*.

Cyanotis (From the Greek *kyanos*, blue, and *anthos*, a flower, alluding to the colour of the flower.) A genus of 30 species of tuberous herbs with narrow downy leaves and regular, small, blue flowers that bear 3 sepals, 3 petals, and 6 stamens. Several species are grown as ornamentals. *Distr*. Old World. *Fam*. Commelinaceae. *Class* M.

C. somaliensis Leaves succulent. Flowers very small, blue-mauve. *Distr*. Origin obscure but thought to be E Africa.

Cyatheaceae A family of 1 genus and 600 species of tree-ferns with scaly new growth and tripinnate leaves in a massive rosette. The sori are borne on the veins on the underside of the leaves and may lack an indusium. *Distr*. Tropical to warm-temperate regions. *Class* F.

cyatheoides: resembling the genus *Cyathea*.

cyathistipula: with cup-like stipules.

Cyathodes (From the Greek *kyathos*, cup, and *odous*, tooth, alluding to the nectar disc.) A genus of about 15 species of evergreen, heath-like shrubs with small overlapping leaves and small tubular flowers borne singly or in terminal spikes. Several species are grown as ornamentals. *Distr*. Australasia. *Fam*. Epacridaceae. *Class* D.

C. colensoi Low shrub. Flowers white. Berries red or white. *Distr*. New Zealand.

C. empetrifolia Low shrub. Flowers fragrant, cream. *Distr*. New Zealand.

cyathophorum: with a cup-shaped tube formed by united floral parts.

Cybistetes (From the Greek *kybistetes*, one who tumbles, alluding to the means of seed dispersal whereby the complete umbel is detached and blow along the ground.) A genus of 1 species of bulbiferous, perennial herb with 2 ranks of curved leaves which appear after the umbels of large, fragrant, pink and white flowers. *Distr*. South Africa. *Fam*. Amaryllidaceae. *Class* M.

C. longifolia Cultivated as a tender ornamental.

cycad *Cycas* **prickly** ~ *Encephalartos altensteinii.*

Cycadaceae A family of 1 genus and 20 species. *Class* G.
See: Cycas.

cycadina: cycad-like.

Cycas (The classical Greek name for a kind of palm.) CYCAD, FALSE SAGO. A genus of about 20 species of palm-like coniferous plants with tall trunks clothed in old leaf bases. The leaves are pinnate with each segment bearing a single vein. The female sporophylls do not form a true cone but are arranged in a terminal mass. The pith and seeds of some species are eaten after preparation but contain carcinogens and alkaloids that may cause nervous disorders. *Distr.* Tropical regions of Africa, Asia and Australasia. *Fam.* Cycadaceae. *Class* G.

C. circinalis FALSE SAGO, SAGO PALM. Palm to 5m high. Leaves feathery. *Distr.* Tropical Asia.

Cyclamen (From the Greek *kyklos*, circular.) PERSIAN VIOLET, ALPINE VIOLET. A genus of about 17 species of perennial herbs with large, rounded tubers, heart- or kidney-shaped, patterned leaves, and nodding flowers that bear 5 strongly reflexed petals. All the species and many cultivars are grown as ornamentals. *Distr.* Mediterranean to Iran. *Fam.* Primulaceae. *Class* D.

C. hederifolium SOWBREAD. Leaves variable, lobed, patterned silver-green. Flowers deep pink. The tuber was formerly considered a cure for baldness when used as snuff. *Distr.* S Europe to SW Asia, naturalized in British Isles.

C. latifolium See *C. persicum*

C. neapolitanum See *C. hederifolium*

C. persicum Leaves marked pale green or silver. Flowers pale pink or white, fragrant. Numerous cultivars have been raised from this species. *Distr.* SE Europe, N Africa.

cyclamineus: resembling the genus *Cyclamen.*

Cyclanthaceae A family of 12 genera and 190 species of rhizomatous herbs and woody climbers with palm-like, deeply divided leaves and small flowers crowded into a spadix. This family is closely related to the palms although much more highly developed. The leaves are a source of fibre. *Distr.* Tropical regions of the New World. *Class* M.
See: Carludovica.

Cyclocheilaceae A family of 2 genera and 4 species of shrubs. *Distr.* E and NE tropical Africa. *Class* D.

cycloglossa: with a round tongue.

cyclophyllus: with round leaves.

cyclosecta: divided into round sections.

Cydonia (After the town of Cydon on the island of Crete.) A genus of 1 species of deciduous shrub or small tree with simple, alternate, dark green leaves, and white, 5-petalled flowers that are followed by aromatic, fleshy, yellow fruits. *Distr.* W Asia, widely naturalized in S Europe. *Fam.* Rosaceae. *Class* D.

C. japonica See *Chaenomeles japonica*

C. oblonga QUINCE, COMMON QUINCE, PORTUGUESE QUINCE. The tart-tasting fruit is used in the preparation of preserves and in cooking in combination with other fruits. It has long been associated with love and fertility and was perhaps the fruit Paris gave to Aphrodite in Greek mythology.

C. seibosa See *C. oblonga*

cylindraceus, -a, -um: resembling a cylinder.

cylindricus, -a, -um: cylindrical.

Cymbalaria (From the Greek *kymbalon*, cymbal, alluding to the shape of the leaves of some species.) A genus of about 10 species of creeping, perennial herbs with round or kidney-shaped leaves and solitary, 2-lipped, 5-lobed flowers that are closed at the mouth by a hairy palate. Several species are grown as ornamentals. *Distr.* W Europe, Mediterranean. *Fam.* Scrophulariaceae. *Class* D.

C. muralis KENILWORTH IVY, IVY-LEAVED TOADFLAX, MOTHER OF THOUSANDS, PENNYWORT, WANDERING SAILOR. Stems trailing and rooting at nodes. Flowers blue-violet with a yellow palate. Frequently grown in rock gardens and on walls. *Distr.* Europe, naturalized in Great Britain, North America.

Cymbidium (From the Greek *kymbe*, a boat, alluding to the shape of the lip.) A genus of 40–50 species of epiphytic or terrestrial orchids with evergreen glossy leaves. The flowers are waxy to 14cm across with similar sepals and petals and a 3-lobed, fleshy lip; the lateral lobes are held vertically and the midlobe is somewhat reflexed. This is the most popular cultivated orchid genus both as pot plants and for cut flowers. There are very many hybrids available. *Distr.* Asia to Australia. *Fam.* Orchidaceae. *Class* M.

C. atropurpureum Flowers scented of coconut, sepals and petals maroon or tinged maroon, lip white or cream. *Distr.* Thailand to the Philippines.

C. bicolor Flowers as many as 25, fragrant, cream with maroon stripes and a white lip. *Distr.* Malaysia to India and China.

C. devonianum Flowers to 3.5cm across, sepals and petals yellow to maroon, lip red. *Distr.* NE India to N Thailand.

C. ensifolium CHIEN LAN ORCHID, FUKIEN ORCHID. Flowers yellow-green to 5cm across. Extract from the flowers is used as an eye treatment in China. *Distr.* India to China and the Philippines.

C. floribundum Flowers crowded, sepals and petals brown or green, lip white, blotched maroon. *Distr.* S China, Taiwan.

C. hookerianum Flowers to 14cm across, scented, sepals and petals green marked red, lip yellow. *Distr.* N India, China.

C. virescens Flowers eaten in hot water or plum vinegar. *Distr.* China.

cymbiformis: boat-shaped.

Cymbopogon (From the Greek *kymbe*, boat, and *pogon*, beard.) A genus of 45–60 species of annual and perennial, rhizomatous grasses to 2m high with dense, compound inflorescences. Many species yield aromatic oils that are used both medicinally and as flavourings. *Distr.* Warm and tropical regions of the Old World. *Fam.* Gramineae. *Class* M.

C. citratus LEMON GRASS, SERI. Strongly lemon-scented perennial. Leaves blue-green. Used as a flavouring, particularly in Asian cookery. *Distr.* S India, Sri Lanka, cultivated in Florida.

Cymodoceaceae A family of 5 genera and 16 species of submerged perennials with creeping herbaceous or woody stems, linear or curved leaves and small solitary flowers. *Distr.* Tropical and subtropical coasts. *Class* M.

cymosus, -a, -um: cyme-like.

Cynara (From the Greek *kyon*, a dog, alluding to the bracts which resemble dogs' teeth.) A genus of 10 species of coarse perennial herbs with spiny, lobed leaves and flowers in daisy-like heads which lack distinct rays. The leaves are eaten as a vegetable. *Distr.* Mediterranean, Canary Islands. *Fam.* Compositae. *Class* D.

C. cardunculus CARDOON. Leaves leathery with yellow spines. Flowers purple. *Distr.* SW Mediterranean, Morocco.

C. hystrix Leaves white-hairy. Flowers purple. *Distr.* Morocco.

C. scolymus GLOBE ARTICHOKE, FRENCH ARTICHOKE. Flower-heads surrounded by fleshy bracts. Young flower-heads and bracts are eaten as a vegetable. *Distr.* N Mediterranean.

cynaroides: from the genus *Cynara*, with the ending *-oides*, indicating resemblance.

Cynoglossum (From the Greek *kyon*, dog, and *glossa*, tongue, alluding to the strap-shaped, lumpy leaves.) HOUND'S TONGUE. A genus of about 55 species of annual, biennial, and perennial herbs with simple, alternate leaves and cymes of funnel- to bell-shaped flowers. Several species are grown as ornamentals. *Distr.* Temperate and warm regions. *Fam.* Boraginaceae. *Class* D.

C. amabile CHINESE FORGET-ME-NOT. Biennial. Stems erect. Flowers numerous, white, pink or blue. *Distr.* E Asia.

C. grande Perennial. Stems erect. Flowers deep blue. *Distr.* W North America.

C. nervosum HIMALAYAN HOUND'S TONGUE. Perennial. Stems erect, branched. Flowers small, blue. *Distr.* W Pakistan, NW India.

C. officinale HOUND'S TONGUE. Biennial. Stems not branched. Flowers small, dark blue, borne in panicles. A medicinal herb occasionally eaten in salads. *Distr.* Europe.

Cynomoriaceae A family of 1 genus and 2 species of red to purple-black, rhizomatous parasites with a club-shaped inflorescence. *Distr.* Mediterranean to N Asia. *Class* D.

Cynosurus (From the Greek *kyon*, dog, and *oura*, tail.) SILKY BENT. A genus of 5 species of annual and perennial, clump-forming grasses with flat leaves and 1-sided panicles. Predominantly grasses of pasture and meadow; some species are occasionally used for hard-wearing lawns or as ornamentals. *Distr.* Mediterranean. *Fam.* Gramineae. *Class* M.

C. cristatus CRESTED DOG'S TAIL. Wiry tufted perennial. *Distr.* Europe, naturalized in North America.

C. echinatus ROUGH DOG'S TAIL GRASS. Erect annual. *Distr.* Mediterranean to SW Asia, introduced into N Europe.

Cypella (From the Greek *kypellon*, a goblet, alluding to the shape of the flowers.) A genus of 20 species of bulbiferous herbs with pleated leaves and short-lived, *Iris*-like flowers on branched stems. Several species are grown as half-hardy ornamentals. *Distr.* Mexico to Argentina. *Fam.* Iridaceae. *Class* M.

C. coelestis Flowers blue, marked yellow-brown, around 8cm across. *Distr.* Brazil, Uruguay, Argentina.

C. herbertii Flowers chrome-yellow with mauve stripes. The most frequently cultivated species. *Distr.* Brazil, Uruguay, Argentina.

Cyperaceae The Sedge family. 102 genera and about 3500 species of mainly perennial, grass-like herbs. The aerial stems are usually solid and 3-angled with leaves in 3 ranks as opposed to hollow and round with leaves in 2 ranks as found in grasses (Gramineae). Uses include paper-making, packaging, fodder, and construction. *Distr.* Widespread, especially in damp, temperate and subarctic regions. *Class* M.
See: Carex, Cymophyllus, Cyperus, Eleocharis, Eriophorum, Schoenus, Scirpus, Trichophorum, Uncinia.

Cyperus (From the Greek *cyperus*, edge, alluding to the sharp-edged leaves of some species.) GALINGALE. A genus of 500–600 species of annual and perennial, rhizomatous herbs with rounded or 3-angled stems, grass-like leaves and very small, typically wind pollinated flowers. The flowers are borne in spikes that are arranged in umbels or clusters which are often subtended by a whorl of leaf-like bracts. *Distr.* Cosmopolitan but more frequent in tropical regions. *Fam.* Cyperaceae. *Class* M.

C. albostriatus Leaves and bracts with white veins. *Distr.* South Africa.

C. alternifolius See *C. involucratus*

C. diffusus See *C. albostriatus*

C. esculentus CHUFA, TIGER NUT, EARTH ALMOND, ZULU NUT. Tuber eaten. *Distr.* Mediterranean, widely naturalized in subtropical and warm temperate regions.

C. involucratus UMBRELLA PLANT. Umbel of spike subtended by a whorl of leaf-like bracts. A common house-plant. *Distr.* Africa, widely naturalized throughout tropical and subtropical regions.

C. longus SWEET GALINGALE. Violet-scented tuber used in the manufacture of scent. *Distr.* Europe, SW and Central Asia, N Africa.

C. papyrus PAPYRUS. The source of Egyptian papyrus, the leaf pith being sliced into thin strips, layered into sheets, and compressed to form a parchment-like material last used in around the 8th century AD *Distr.* S and tropical Africa, extending N to Egypt, naturalized in Sicily.

C. rotundus COCO GRASS. A very pernicious weed. *Distr.* Tropical regions, widely naturalized.

C. sumula One of the very few insect-pollinated members of the Cyperaceae. *Distr.* Australia.

Cyphomandra (From the Greek *kyphos*, tumour, and *aner*, man, alluding to the shape of the anthers.) A genus of 30 species of herbs, shrubs, lianas-and trees with simple or compound leaves and racemes of 5-lobed flowers that are followed by 2-chambered berries. *Distr.* Tropical America. *Fam.* Solanaceae. *Class* D.

C. betacea See *C. crassicaulis*

C. crassicaulis TREE TOMATO, TAMARILLO. Shrub or small tree to 4m high. Leaves simple, somewhat fleshy. Flowers pink. The fruit is edible and used in jellies. *Distr.* South America.

cypress Cupressaceae, *Cupressus* **African** ~ *Widdringtonia* **Arizona** ~ *Cupressus arizonica* **bald** ~ *Taxodium, T. distichum* **Chinese swamp** ~ *Glyptostrobus pensilis* **Chinese weeping** ~ *Cupressus funebris* **false** ~ *Chamaecyparis* **Japanese** ~ *C. obtusa*

Lawson's ~ *C. lawsoniana* **Mexican** ~ *Cupressus lusitanica* **Monterey** ~ *C. macrocarpa* **Nootka** ~ *Chamaecyparis nootkatensis* **Patagonian** ~ *Fitzroya cupressoides* **sawara** ~ *Chamaecyparis pisifera* **Sitka** ~ *C. nootkatensis* **southern** ~ *Taxodium distichum* **swamp** ~ *T. distichum* **yellow** ~ *Chamaecyparis nootkatensis*.

Cypripedium (From the Greek *Kypris*, Venus or Aphrodite, and *pous*, foot.) LADY'S SLIPPER, MOCCASIN FLOWER, SLIPPER ORCHID. A genus of about 35 species of terrestrial orchids with short rhizomes and flowers borne in slender-stalked racemes. The sepals and the upper petals are free, the lateral petals are fused and held below the lip, which is inflated so as to resemble a slipper. There are 2 fertile anthers. Some of the species are grown as cultivated ornamentals of the garden or alpine house. Numerous hybrids are available. *Distr.* Widely distributed throughout N hemisphere. *Fam.* Orchidaceae. *Class* M.

C. acaule Leaves 2, broad. Flowers solitary, lip red to white. *Distr.* NE North America.

C. calceolus Flowers to 4, maroon-green with a yellow lip. One of the rarest plants in Great Britain due to over-collection, now probably represented by only a single plant. *Distr.* Europe and North America.

C. debile Flowers small, solitary, pale yellow-green marked red. *Distr.* Japan, W China.

C. guttatum Flowers solitary, white-yellow spotted red and violet. *Distr.* Europe, temperate Asia to Alaska and NW Canada.

C. japonicum Flowers solitary, green, spotted maroon, lip rose-pink, spotted mauve. *Distr.* Japan, Taiwan, S and Central China.

C. macranthum A complex species with a white-lined or tinged red lip. *Distr.* China and Central Asia.

cyprius, -a, -um: of Cyprus.

Cyrilla (After Dominico Cyrillo (1734–1799), professor of botany at Naples.) A genus of 1 species of deciduous or evergreen shrub, with simple, alternate leaves and slender spike-like racemes of small white flowers. *Distr.* SE North America to N South America. *Fam.* Cyrillaceae. *Class* D.

C. racemiflora LEATHERWOOD. Grown as an ornamental, chiefly for its flowers.

Cyrillaceae A family of 3 genera and 14 species of deciduous or evergreen shrubs and small trees with simple, alternate leaves and racemes of regular flowers with their parts in 5s. *Distr.* Tropical to warm-temperate America. *Class* D.
See: Cyrilla.

Cyrtanthus (From the Greek *kyrtos*, arched, and *anthos*, a flower, alluding to the curved flower-tube.) FIRE LILY. A genus of 47 species of bulbiferous, perennial herbs with a rosette of linear, fleshy leaves and umbels of funnel-shaped flowers. *Distr.* S Africa. *Fam.* Amaryllidaceae. *Class* M.

C. elatus GEORGE LILY, SCARBOROUGH LILY. Flowers bright scarlet or deep pink, tinged green in the throat. *Distr.* S Africa.

C. falcatus Flowers pendulous, red. *Distr.* South Africa (Natal).

C. mackenii IFAFA LILY. Flowers white, fragrant. *Distr.* South Africa.

C. purpureus See *C. elatus*

C. speciosus See *C. elatus*

Cyrtomium (From the Greek *kyrtos*, arched, alluding to the arching leaves.) A genus of about 20 species of terrestrial ferns with pinnately divided leaves. Several species are grown as ornamentals. *Distr.* Hawaii, E Asia, South Africa, Central and South America. *Fam.* Dryopteridaceae. *Class* F.

C. falcatum JAPANESE HOLLY FERN. *Distr.* E Africa to China and Hawaii.

Cyrtosperma (From the Greek *kyrtos*, curved, and *sperma*, seed, alluding to the shape of the seeds of several of the species.) A genus of 11 species of rhizomatous or tuberous, aquatic and marginal herbs that bear large simple leaves with long, warty or prickly leaf-stalks. The small flowers are borne on a short spadix that is subtended by and sometimes surrounded by a purple-brown spathe. Several species are grown as ornamentals and some have edible tubers. *Distr.* India to Malaysia. *Fam.* Araceae. *Class* M.

Cystopteris (From the Greek *kystis*, bladder, and *pteris*, fern, alluding to the shape of the indusium.) BLADDER FERN. A genus of about 18 species of small ferns of rocky areas.

Several species are grown as ornamentals. *Distr.* Temperate and warm regions especially in the N hemisphere. *Fam.* Woodsiaceae. *Class* F.

C. fragilis BLADDER FERN, BRITTLE BLADDER FERN, FRAGILE FERN. To 45cm high. Leaves delicate, pinnately divided. *Distr.* Widespread in temperate regions.

Cytisus (From the Greek *kytisos*, a kind of clover.) BROOM. A genus of 33 species of evergreen or deciduous shrubs and small trees with ribbed green stems, trifoliate leaves and white or yellow, irregular, pea-like flowers. Several species and numerous cultivars are grown as ornamentals. *Distr.* Europe, N Africa, Canary Islands, W Asia. *Fam.* Leguminosae. *Class* D.

C. ardoinoi Deciduous prostrate shrub. Flowers deep yellow, solitary or in small groups. *Distr.* SW Alps, S France.

C. nigrescens See *C. nigricans*

C. nigricans Erect shrub. Leaves dark green. Flowers pale yellow in slender racemes. *Distr.* Europe.

C. scoparius COMMON BROOM, SCOTCH BROOM. Erect shrub to 2m high. Leaves trifoliate but simple on young shoots. Flowers axillary, solitary or in pairs, deep yellow. Over 80 ornamental cultivars of this species have been produced. *Distr.* Europe.

D

Daboecia (After the Irish saint, Dabeoc.) A genus of 2 species of evergreen shrubs with simple, silver-green leaves and loose racemes of small, urn-shaped flowers. *Distr.* Ireland to Spain and the Azores. *Fam.* Ericaceae. *Class* D.

D. cantabrica ST DABEOC'S HEATH, CONNEMARA HEATH. Grown as an ornamental. *Distr.* W Europe, fossils have been found as far N as Shetland.

dacryberry *Dacrycarpus* **New Zealand** ~ *D. dacrydioides.*

Dacrycarpus (A combination of the names of two closely related genera, *Dacrydium* and *Podocarpus.*) DACRYBERRY. A genus of 9 species of evergreen coniferous shrubs and trees with linear, flat juvenile leaves and scale-like adult leaves. The 2 types of leaves are often found together even on mature trees. The female cones bear red, fleshy receptacles. Several species are grown as ornamentals and some are a source of timber. *Distr.* Burma to New Zealand. *Fam.* Podocarpaceae. *Class* G.

D. dacrydioides NEW ZEALAND DACRYBERRY. Tree to 65m high. Fruit eaten by Maoris. *Distr.* New Zealand.

dacrydioides: like a tear, pear-shaped.

Dacrydium (From the Greek *dakrydion*, a small tear, alluding to the small drops of resin that these trees occasionally exude.) A genus of 25 species of evergreen coniferous shrubs and trees with linear juvenile leaves and scale-like adult leaves. The 2 types of leaf are occasionally found together on mature plants. Several species are grown as ornamentals. *Distr.* Pacific. *Fam.* Podocarpaceae. *Class* G.

D. bidwillii See *Halocarpus bidwillii*

D. cupressinum RIMU, RED PINE. Tree to 40m high. *Distr.* New Zealand.

D. franklinii See *Lagarostrobos franklinii*

D. laxifolium See *Lepidothamnus laxifolium*

dactylifera: finger-bearing.

Dactylis (From the Greek *daktylos*, a finger, alluding to the shape of the inflorescence.) A genus of 1 species of perennial, tufted grass with flat leaves and an erect, open panicle. *Distr.* Europe, N Africa and temperate Asia. *Fam.* Gramineae. *Class* M.

D. glomerata COCK'S FOOT, ORCHARD GRASS. A grass of meadows and waste places.

Dactylorhiza (From the Greek *daktylos*, finger, and *rhiza*, root, in allusion to the finger-like tubers.) SALEP. A genus of 30 species of deciduous terrestrial orchids with flat tubers and leaves that are often spotted purple. The purple, yellow, or white flowers are borne in spikes. The upper sepal forms a hood with the petals and the lip is spurred. The tubers of some species are eaten in W Asia (Persian salep). *Distr.* Europe, N Africa, Asia, and North America. *Fam.* Orchidaceae. *Class* M.

D. elata ROBUST MARSH ORCHID. Flowers maroon to pink in dense or loose spikes. *Distr.* SW Europe.

D. foliosa MADEIRAN ORCHID. Flowers pink to pale maroon in a dense spike. *Distr.* Madeira.

D. fuchsii COMMON SPOTTED ORCHID. Leaves spotted purple. Flower colour variable from white to purple with mauve markings. *Distr.* Europe and W Asia, typically on calcareous soils.

D. incarnata EARLY MARSH ORCHID. Flowers pink, cream or purple in a cylindrical spike. *Distr.* Mediterranean.

D. maculata Flowers in a dense spike, mauve to red or white. *Distr.* Europe, N Africa and W Asia.

D. majalis BROAD-LEAVED MARSH ORCHID. Flowers in dense spikes, sepals magenta-lilac, lip white and purple. *Distr.* Europe, Turkey, and the Baltic.

D. mascula See *Orchis mascula*

D. praetermissa SOUTHERN MARSH ORCHID. Flowers pale garnet in a dense spike. *Distr.* NW Europe.

D. purpurella NORTHERN MARSH ORCHID. Flowers claret to maroon. *Distr.* N Europe.

daffodil Amaryllidaceae, *Narcissus* **autumn** ~ *Sternbergia* **hoop petticoat** ~ *Narcissus bulbocodium* **Peruvian** ~ *Hymenocallis, H. narcissiflora*, **sea** ~ *Pancratium maritimum* **white hoop petticoat** ~ *Narcissus cantabricus* **wild** ~ *N. pseudonarcissus.*

dagger plant *Yucca aloifolia.*

Dahlia (After Andreas Dahl (1751–89), Swedish botanist and pupil of Linnaeus.) A genus of 28 species of perennial, typically tuberous herbs with simple to pinnate leaves and flowers in daisy-like heads that bear distinct rays. Many of the cultivated forms have highly modified flower-heads which are usually double. Originally grown in Central America for the edible tubers which were reported to have medicinal qualities; they are now important ornamental and cut flowers. Very many cultivars are available that no longer resemble the wild species. *Distr.* Mexico to Colombia, introduced into Europe. *Fam.* Compositae. *Class* D.

D. coccinea Herb to 3m high. Flower-heads in clusters of 2–3, flowers and rays yellow. *Distr.* Mexico to Guatemala.

D. imperialis TREE DAHLIA. Large subshrub to 9m high. Flower-heads numerous, often nodding. *Distr.* Guatemala to Columbia.

D. merckii Herb to 2m. Flower-heads numerous, rays white to purple. *Distr.* Mexico.

D. pinnata Herb to 2m high. Flower-heads in groups of 2–4, rays pale purple. *Distr.* Mexico.

dahlia, tree *Dahlia imperialis.*

dahoon *Ilex cassine* **myrtle** ~ *I. myrtifolia.*

dahuricus, -a, -um: of Dahuria, SE Siberia.

Daiswa (From *dai swa*, the Nepalese name for these plants.) A genus of 15 species of rhizomatous herbs with solitary stems, a whorl of simple leaves and solitary star-shaped flowers that bear 4–8 petals and up to 20 stamens. Most of the species are grown as ornamentals. *Distr.* Himalaya to SE Asia. *Fam.* Trilliaceae. *Class* M.

D. polyphylla Herb to 1m high. Petals thread-like, yellow-green. *Distr.* Himalaya.

daisy Compositae, *Bellis*, *perennis* **African** ~ *Arctotis* **Barberton** ~ *Gerbera, G. jamesonii* **blue** ~ *Felicia, F. amelloides* **blue-eyed African** ~ *Arctotis venusta* **common** ~ *Bellis perennis* **crown** ~ *Chrysanthemum coronarium* **dog** ~ *Leucanthemum vulgare* **eastern** ~ *Townsendia exscapa* **globe** ~ *Globularia* **Hilton** ~ *Gerbera aurantiaca* **Kingfisher** ~ *Felicia, F. bergeriana* **Livingstone** ~ *Dorotheanthus bellidiformis* **Michaelmas** ~ *Aster tripolium, A. novi-belgii* **moon** ~ *Leucanthemum vulgare* **orange** ~ *Erigeron aurantiacus* **ox-eye** ~ *Buphthalmum, Leucanthemum vulgare* **painted** ~ *Tanacetum coccineum* **seaside** ~ *Erigeron glaucus* **Shasta** ~ *Leucanthemum* × *superbum* **Transvaal** ~ *Gerbera, G. aurantiaca, G. jamesonii* **turfing** ~ *Tripleurospermum* **yellow paper** ~ *Helichrysum bracteatum.*

daisy bush *Olearia.*

daisy-fleabane *Erigeron annuus.*

dalmaticus, -a, -um: of Dalmatia, on the Adriatic Coast.

damascenus, -a, -um: of Damascus, Syria.

damson *Prunus domestica.*

Danae (After Danae, daughter of King Acrisius of Argos in Greek mythology.) A genus of 1 species of clump-forming subshrub with very small leaves that are replaced by stiff, leaf-like branches (cladophylls). The cream flowers are borne in short terminal racemes and followed by orange-red berries. *Distr.* SW Asia. *Fam.* Ruscaceae. *Class* M.

D. racemosa ALEXANDRIAN LAUREL. Grown as an ornamental.

dandelion *Taraxacum* **common** ~ *T. officinale.*

danewort *Sambucus ebulus.*

Daphne (After Daphne, who was changed into a bay laurel by Apollo, in Greek mythology.) A genus of about 50 species of evergreen and deciduous shrubs with simple, often leathery leaves and clusters or racemes of tubular flowers that lack petals. Many species and cultivars are grown as ornamentals. *Distr.* Eurasia. *Fam.* Thymelaeaceae. *Class* D.

D. laureola SPURGE LAUREL. Much-branched evergreen shrub to 1.5m high. Flowers fragrant, yellow-green. *Distr.* Eurasia, naturalized in Great Britain.

D. mezereum MEZEREON, FEBRUARY DAPHNE. Short-lived deciduous shrub. Flowers violet-pink, fragrant. Numerous ornamental cultivars have been raised from this species. The fruit was formerly used as a pepper substitute but is poisonous. *Distr.* Eurasia.

daphne Thymelaeaceae **February** ~ *Daphne mezereum.*

Daphniphyllaceae A family of 1 genus and 10 species. *Class* D.
See: Daphniphyllum.

Daphniphyllum (From the genus name *Daphne*, and the Greek *phyllum*, leaf.) A genus of 10 species of evergreen shrubs and trees with simple, spirally arranged leaves and racemes of small unisexual flowers that lack petals. *Distr.* China to tropical Australia. *Fam.* Daphniphyllaceae. *Class* D.

D. humile Leaves used like tobacco. *Distr.* N Japan, Korea.

D. macropodum Leaves dark green with red veins. Flowers pungent-smelling. Grown as a hardy ornamental. *Distr.* China, Korea, Japan.

Darlingtonia (After Dr William Darlington (1782–1863), American botanist.) A genus of 1 species of rhizomatous, carnivorous herb with rosettes of pitcher-like leaves that trap and digest insects. The yellow-green flowers are borne on a stalk up to 1m above the rosette of leaves. *Distr.* USA (N California to SW Oregon). *Fam.* Sarraceniaceae. *Class* D.

D. californica COBRA LILY, CALIFORNIA PITCHER PLANT. Occasionally grown as an ornamental.

Darmera A genus of 1 species of large, rhizomatous, perennial herb with round, peltate leaves and cymes of small, white to pink flowers. *Distr.* USA (California, Oregon). *Fam.* Saxifragaceae. *Class* D.

D. peltata UMBRELLA PLANT, INDIAN RHUBARB.

dasheen *Colocasia esculenta.*

dasyanthus, -a, -um: with very hairy flowers.

Dasylirion (From the Greek *dasys*, thick, and *lirion*, lily, in allusion to the thick stems.) SOTOL, BEAR GRASS. A genus of about 15 species of subshrubs with thick, stout trunks and linear leaves that are clustered towards the tips of the branches. The small bell-shaped flowers are borne in large panicles in the axils of papery bracts. The spiny-edged leaves of some species are used for thatching and basketry and their sap is used as a drink (SOTOL). *Distr.* S USA to Mexico. *Fam.* Dracaenaceae. *Class* M.

D. acrotrichum Stem stout, to 1.5m high. Leaves to 1m long. *Distr.* Mexico.

D. gracile See *D. acrotrichum*

D. longissimum Stem to 3.5m high. Leaves long, 4-angled. *Distr.* Mexico.

D. texanum TEXAS SOTOL. Leaves narrow with yellow spines on the margin. *Distr.* USA (Texas), N Mexico.

dasypetalus, -a, -um: with very hairy petals.

dasyphyllus, -a, -um: with very hairy leaves.

dasypogon: with a shaggy beard.

Dasypogonaceae A family of 2 genera and 3 species of herbs and trees with spirally arranged, often spiny leaves. The small regular flowers have 6 tepals, 6 stamens and a superior ovary. *Distr.* Australia. *Class* M.
See: Kingia.

dasystemon: with hairy stamens.

date *Phoenix* **Indian** ~ *P. sylvestris*
Trebizond ~ *Elaeagnus angustifolia*
wild ~ *Phoenix sylvestris.*

date plum *Diospyros lotus.*

Datisca A genus of 2 species of perennial herbs with alternate, pinnate, or trifoliate leaves and clusters of small unisexual flowers that lack petals. *Distr.* W North America, S Asia. *Fam.* Datiscaceae. *Class* D.

D. cannabina Clump-forming perennial. Stems to 2m high. Leaves pinnate. Flowers yellow. Grown as an ornamental and as a source of yellow dye for silk. *Distr.* Turkey to India.

Datiscaceae A family of 3 genera and 4 species of perennial herbs and trees with pinnate or simple leaves and small flowers that bear 3–9 sepals and 8 to many stamens but no petals. *Distr.* E Mediterranean to Australia, SW USA and Mexico. *Class* D. *See: Datisca.*

Datura (From an Indian vernacular name.) A genus of 8 species of annual or short-lived perennial herbs and subshrubs with simple, alternate leaves and large, 5-lobed, funnel-shaped flowers that are followed by spiny capsules. All species contain toxic alkaloids. *Distr.* S North America, widely naturalized. *Fam.* Solanaceae. *Class* D.

D. arborea See *Brugmansia arborea*

D. cornigera See *Brugmansia arborea*

D. inoxia DOWNY THORN APPLE, INDIAN APPLE, ANGEL'S TRUMPET. Annual to 1m high. Flowers white tinged pink. Formerly a sacred hallucinogen in SW North America. *Distr.* S North America.

D. metel HORN OF PLENTY, DOWNY THORN APPLE. Glabrous annual to 1m high. Flowers white or purple, frequently double. *Distr.* SW USA, widely naturalized throughout the tropics.

D. meteloides See *D. inoxia*

D. sanguinea See *Brugmansia sanguinea*

D. stramonium THORN APPLE, JAMESTOWN WEED, JIMSON WEED. Annual to 2m high. Flowers white or purple, to 20cm long. The source of stramonium, a drug formerly used to treat asthma, and a hallucinogenic drug used in rituals by the Algonquin Indians of E USA. *Distr.* North America, but widely naturalized, including Great Britain.

D. suaveolens See *Brugmansia suaveolens*

D. tatula See *D. stramonium*

D. versicolor See *Brugmansia versicolor*

daucoides: like a carrot, from the genus *Daucus*, with the ending *-oides*, indicating resemblance.

Daucus (From the classical Greek name for these plants, *daukos*.) A genus of about 22 species of annual and biennial herbs with finely divided leaves and compound umbels of white or yellow flowers. The umbels are usually subtended by a whorl of finely divided bracts. *Distr.* Widespread. *Fam.* Umbelliferae. *Class* D.

D. carota CARROT, WILD CARROT. Biennial herb. Taproot substantial. The edible carrot we know today has been derived from this species through many years of cultivation and is usually considered a distinct subspecies. *Distr.* Eurasia.

Davallia (After E. Davall (1763–1798), Swiss botanist.) HARE'S FOOT FERN. A genus of about 35 species of deciduous or semi-evergreen, terrestrial and epiphytic ferns. Several species are grown as ornamentals. *Distr.* Tropical and warm temperate regions of the Old World. *Fam.* Davalliaceae. *Class* F.

Davalliaceae A family of 10 genera and about 200 species of typically epiphytic, occasionally terrestrial ferns with creeping fleshy stems and pinnate leaves borne in 2 alternate rows. The sori are borne at the ends of the veins or dorsally. *Distr.* Tropical and temperate regions. *Class* F. *See: Davallia, Humata, Leucostegia, Rumohra, Scyphularia.*

Davidia (After Armand David (1826–1900), French missionary who brought it into cultivation.) A genus of 1 species of deciduous tree to 25m high with broad ovate leaves that are silky-hairy below. The small unisexual flowers are borne in dense rounded heads that are subtended by 2 white, unequal bracts. The larger bract resembles a handkerchief. The fruit is a rounded, purple-brown drupe. *Distr.* SW China. *Fam.* Cornaceae. *Class* D.

D. involucrata DOVE TREE, GHOST TREE, HANDKERCHIEF TREE. Grown as an ornamental.

David's harp *Polygonatum multi-florum.*

Davidsonia A genus of 1 species of small tree with spirally arranged, pinnate leaves that reach 1m long and panicles of small tubular flowers that lack petals. The fruit is plum-like and fleshy with a velvety surface. *Distr.* NE Australia. *Fam.* Davidsoniaceae. *Class* D.

D. pruriens DAVIDSON'S PLUM. The fruit is edible and often made into preserves.

Davidsoniaceae A family of 1 genus and 1 species. *Class* D. *See: Davidsonia.*

day flower *Commelina.*

dealbatus, -a, -um: with a white powdery covering.

death camas *Zigadenus*.

debilis, -e: weak.

Decaisnea (After Joseph Decaisne (1807–1882), director of the Jardin des Plantes, Paris.) A genus of 2 species of erect deciduous shrubs with pinnate leaves and drooping inflorescences of small, male and bisexual flowers. *Distr.* E Asia. *Fam.* Lardizabalaceae. *Class* D.

D. fargesii Leaves with 11–25 leaflets. Flowers yellow-green. Fruit pod-like, blue. Grown as an ornamental. The fruits are eaten in China. *Distr.* W China.

decandrus, -a, -um: with ten stamens, from the Greek *decem-*, ten, and *andros*, male.

decapetalus, -a, -um: ten-petalled.

deciduus, -a, -um: deciduous.

decipiens: deceptive, resembling another species.

decolorans: discoloured, faded.

decompositus, -a, -um: much divided.

decoratus, -a, -um: decorative.

decorus, -a, -um: beautiful.

Decumaria (From the Latin *decimus*, tenth, alluding to the number of sepals, petals, and stamens.) A genus of 2–3 species of deciduous or semi-evergreen climbing shrubs with adhesive, aerial roots, simple, opposite leaves and umbels of small white flowers. *Distr.* E Asia, E North America. *Fam.* Hydrangeaceae. *Class* D.

D. barbara CLIMBING HYDRANGEA, WOOD VAMP. Climbing to 10m. Flowers sweetly fragrant. *Distr.* SE USA.

D. sinensis Climbing to 4m. Flowers musky fragrant. *Distr.* China.

decumbens: prostrate, low-growing.

decurrens: decurrent, with the base of the leaf gradually merging with the stem.

decussatus, -a, -um: with the leaves in pairs, one pair at right angles to the next.

deflexus, -a, -um: deflected downwards abruptly.

deformis, -e: deformed.

Degeneriaceae A family of 1 genus and 1 species of tree with simple, spirally arranged leaves and regular flowers that bear 3 sepals, 3–5 whorls of petals, numerous broad stamens and a single, slightly split carpel. A primitive family. *Distr.* Fiji. *Class* D.

Degenia (After Dr Arpad von Degen (1866–1934), director of the Budapest seed-testing station.) A genus of 1 species of perennial herb with rosettes of simple leaves and yellow flowers. *Distr.* SE Europe. *Fam.* Cruciferae. *Class* D.

D. velebitica Cultivated for silver-grey foliage. *Distr.* E Alps.

Deinanthe (From the Greek *deinos*, strange, and *anthos*, flower, alluding to the size of the flowers which is unusual for this family.) A genus of 2 species of rhizomatous herbs with simple, toothed or lobed leaves and clusters of large, fertile and small, sterile, nodding flowers. Both species are grown as rock garden ornamentals. *Distr.* China and Japan. *Fam.* Hydrangeaceae. *Class* D.

D. bifida Leaves crinkly. Flowers white. *Distr.* Japan.

D. caerulea Leaves smooth. Flowers pale purple-blue. *Distr.* China.

dejectus, -a, -um: fallen or low.

delicatus, -a, -um: delicate.

deliciosus, -a, -um: tasting delicious.

Delonix (From the Greek *delos*, conspicuous, and *onyx*, claw, alluding to the petals.) A genus of 10 species of evergreen and deciduous trees with pinnate leaves and branched racemes of irregular, white or red flowers that bear 5 petals and 10 stamens. *Distr.* Tropical Africa, Madagascar, India. *Fam.* Leguminosae. *Class* D.

D. regia FLAMBOYANT, PEACOCK FLOWER, FLAME TREE. Deciduous tree. Flowers large, scarlet to orange. Grown as a half-hardy ornamental. *Distr.* Madagascar.

Delosperma (From the Greek *delos*, conspicuous, and *sperma*, seed, alluding to the open seed capsules.) A genus of 140 species of annual to perennial herbs and shrubs typically with mat-forming stems, succulent leaves, and many-petalled flowers. Several species are grown as ornamentals. *Distr.* S and E Africa, Arabia. *Fam.* Aizoaceae. *Class* D.

D. cooperi Stems prostrate. Leaves cylindrical. Flowers purple-red. *Distr.* South Africa (Orange Free State, Transvaal).

D. tradescantioides Small creeping. Leaves somewhat furrowed. Flowers white. *Distr.* South Africa (E Cape, Natal, Transvaal).

delphinantha: with flowers like the genus *Delphinium*.

Delphinium (From the Greek name for these plants, *delphinion*, derived from *delphis*, a dolphin, alluding to the shape of the flowers.) LARKSPUR. A genus of about 250 species of annual, biennial and perennial herbs with palmately lobed leaves and spike-like racemes of irregular, spurred flowers that bear 5 petal-like sepals and 2 pairs of petals. Many species and cultivars are grown ornamentally, all are poisonous. *Distr.* N temperate regions, mountainous areas of E Africa. *Fam.* Ranunculaceae. Class D.

D. brunonianum Erect perennial. Flowers pale blue. *Distr.* Afghanistan to China.

D. cardinale Short-lived erect perennial to 2m high. Flowers scarlet with a yellow centre. *Distr.* California.

D. chinense See *D. grandiflorum*

D. elatum Erect perennial. Flowers blue, borne in dense racemes. An important parent of many garden hybrids. *Distr.* Central Europe to Siberia.

D. grandiflorum Erect perennial, often grown as an annual. Flowers violet, blue or white, borne in loose racemes. *Distr.* Siberia, China.

D. nudicaule Short-lived erect perennial. Flowers red, occasionally yellow. *Distr.* W USA.

D. staphisagria Biennial. Flowers purple-blue tinged with green, borne in dense spikes. The seeds were formerly used as an insecticide. *Distr.* Mediterranean area, SW Asia.

D. tatsienense Erect short-lived perennial. Flowers brilliant blue, borne in loose clusters. *Distr.* W China.

deltoideus, -a, -um: deltoid, i.e. triangular; shaped like the Greek capital letter delta.

demersus, -a, -um: submerged, underwater.

demissus, -a, -um: drooping, lowly, humble

Dendranthema (From the Greek *dendron*, tree, and *anthemon*, flower.) A genus of 18 species of perennial herbs with alternate, pinnately lobed or entire leaves and flowers in daisy-like heads which bear distinct white, yellow or purple rays. A number of species are cultivated as ornamentals, the most important of which is *D.* × *grandiflorum*. *Distr.* Europe, Central and E Asia. *Fam.* Compositae. *Class* D.

D.* × *grandiflorum FLORIST'S CHRYSANTHEMUM, CHRYSANTHEMUM. A highly variable perennial herb or subshrub to 1.5m high. Grown as a cultivated ornamental particularly for cut flowers. Numerous cultivars are available. *Distr.* Probably derived in cultivation from *D. indicum* in China.

D. indicum Stoloniferous perennial. Rays yellow. *Distr.* Japan.

D. weyrichii Mat-forming. Rays white or pink. *Distr.* Japan.

D. yezoense Stoloniferous. Rays white. *Distr.* Japan.

D. zawadaskii Rhizomatous. Rays white to purple. *Distr.* N and Central Russia.

Dendriopoterium See *Sanguisorba*

Dendrobenthamia See *Cornus*

Dendrocalamus (From the Greek *dendron*, tree, and *kalamos*, reed.) GIANT BAMBOO. A genus of about 30 species of very large, clump-forming bamboos similar to the genus *Bambusa* but of much greater stature. *Distr.* SE Asia. *Fam.* Gramineae. *Class* M.

D. asper Grown for edible young shoots. *Distr.* SE Asia.

D. giganteus GIANT BAMBOO. The largest known bamboo. Stems used for manufacture of buckets, rafts and split for chop sticks. *Distr.* Burma.

D. strictus MALE BAMBOO, CALCUTTA BAMBOO. Solid stems used in the construction industry and as a source of paper pulp. *Distr.* India.

Dendromecon (From the Greek *dendron*, tree and *mekon*, poppy.) A genus of 1 species of evergreen shrub or small tree with simple, alternate leaves and solitary, fragrant flowers that bear 4 yellow petals and numerous stamens. *Distr.* SW North America. *Fam.* Papaveraceae. *Class* D.

D. rigida Grown as a half-hardy ornamental.

Dennstaedtiaceae A family of 18 genera and about 400 species of terrestrial, occasionally epiphytic ferns with creeping rhizomes and 2–3-pinnate or rarely simple leaves. The sori are borne marginally or submarginally. *Distr.* Widespread. *Class* F.
See: Histiopteris, Hypolepis, Microlepia, Paesia, Pteridium, Sphenomeris.

dens-canis: dog's tooth.

densiflorus, -a, -um: densely flowered.

densifolius, -a, -um: densely leaved.

densispinus, -a, -um: densely spiny.

densus, -a, -um: dense.

Dentaria See *Cardamine*
D. californica See *Cardamine californica*
D. digitata See *Cardamine pentaphyllos*
D. pinnata See *Cardamine heptaphylla*

dentatus, -a, -um: toothed.

denticulatus, -a, -um: slightly toothed.

denudatus, -a, -um: naked, the flowers opening before the leaves emerge.

depauperatus, -a, -um: depauperate, lacking something.

dependens: suspended, hanging down.

depressus: flattened; low-growing, for example.

Dermatobotrys (From the Greek *derma*, skin or bark, and *botrys*, cluster.) A genus of 1 species of epiphytic deciduous shrub with simple, fleshy leaves and whorls of tubular, red and yellow flowers that are followed by egg-shaped berries. *Distr.* South Africa. *Fam.* Scrophulariaceae. *Class* D.

D. saundersii Grown as an ornamental.

Deschampsia (After Louis August Deschamps (1765–1842), French botanist.) HAIR GRASS. A genus of 40–50 species of tussock-forming, perennial grasses with slender stems, thread-like leaves, and tall panicles. *Distr.* Temperate and cold regions, and mountainous areas in the tropics. *Fam.* Gramineae. *Class* M.

D. caespitosa TUSSOCK GRASS, TUFTED HAIR GRASS. Grown as a fodder crop and as an ornamental with various cultivars available. *Distr.* Temperate regions and the mountainous regions of tropical Africa and Asia.

D. flexuosa WAVY HAIR GRASS, CRINKLED HAIR GRASS. A grass of acid heaths and moors sometimes grown as an ornamental. *Distr.* Europe, Asia, North and South America, mountainous regions of the Old World tropics.

desert candle *Eremurus.*

Desfontainia (After René Louiche Desfontaines (1752–1833), French botanist.) A genus of 1 species of evergreen bushy shrub with opposite, simple, leathery leaves and solitary, tubular, red-orange flowers that bear their parts in 5s. *Distr.* Andes (Costa Rica to Cape Horn). *Fam.* Loganiaceae. *Class* D.

D. spinosa Grown as an ornamental. A source of yellow dye and an allegedly hallucinogenic tea.

Desmazeria (After J. B. Desmazières (1796–1862), French botanist.) FERN GRASS. A genus of 3 species of annual grasses with narrow leaves and rigid panicles. *Distr.* W Europe to SW Asia. *Fam.* Gramineae. *Class* M.

D. marina SEA FERN GRASS. A coastal grass of sand, shingle and rock. *Distr.* W Europe, N Africa.

Desmodium (From the Greek *desmos*, a chain, alluding to the shape of the pods.) A genus of 300 species of annual and perennial herbs and deciduous shrubs and trees with alternate, pinnate or trifoliate leaves and racemes or panicles of irregular, pink, purple or white flowers. Several species are grown as ornamentals and some are used locally as fodder or as a green manure. *Distr.* Tropical and subtropical regions. *Fam.* Leguminosae. *Class* D.

D. elegans Deciduous shrub. Leaves trifoliate. Flowers palelilac to deep pink. *Distr.* China.

D. tiliifolium See *D. elegans*

Deutzia (After John van de Deutz (1743–88), friend and patron of C. P. Thunberg who described the genus.) A genus of about 40 species of deciduous and occasionally evergreen shrubs with pithy stems, simple, opposite leaves, and regular flowers that bear 5 petals and 10 stamens. Many species and numerous cultivars are grown as ornamentals. *Distr.* Temperate Asia, mountains of Central America. *Fam.* Hydrangeaceae. *Class* D.

D. gracilis Upright or spreading shrub. Flowers clustered, pure white. *Distr.* Japan.

D. longifolia Deciduous. Flowers deep pink. *Distr.* China.

D. × magnifica A hybrid of uncertain parentage that has given rise to a number of cultivars. *Distr.* Garden origin.

D. monbeigii Small arching shrub. Flowers small, white. *Distr.* SW China.

D. pulchra Deciduous. Flowers white tinged with pink, borne in pendent clusters. *Distr.* Taiwan, Philippines.

D. scabra Deciduous shrub. Leaves dark green. Flowers pure white . *Distr.* Japan.

D. setchuenensis Upright, deciduous. Flowers white, narrow-petalled. *Distr.* W China.

D. × wellsii See *D. scabra*

devil flower *Tacca chantrieri*.

devil's, claw *Physoplexis comosa*.

devil's darning needle *Clematis virginiana*

devil's paintbrush *Hieracium aurantiacum*.

devil's tongue *Amorphophallus, A. rivierei*.

devil's walking-stick *Aralia spinosa*.

dewberry *Rubus caesius* **European** ~ *R. caesius*.

dewdrop, golden *Duranta erecta*.

diacanthus, -a, -um: two-spined.

Dialypetalanthaceae A family of 1 genus and 1 species of tree with simple, opposite leaves and clusters of flowers that bear 4 sepals, 4 petals, around 18 stamens and an inferior ovary. *Distr.* Brazil. *Class* D.

Dianella (Diminutive of Diana, Roman goddess of the chase.) FLAX LILY. A genus of about 25 species of perennial herbs with slender stems, grass-like leaves and loose panicles of regular, blue or white flowers. Several species are grown as ornamentals and the berries are sometimes used as a source of blue dye. *Distr.* Tropical Asia, Australia, New Zealand and Polynesia. *Fam.* Phormiaceae. *Class* M.

D. caerulea Tuft-forming. Flowers blue. Berries blue. *Distr.* E and S Australia.

D. nigra COLENSO, NEW ZEALAND BLUEBERRY. Leaves to 1.5m long. Flowers white tinged blue or green. Berries blue. *Distr.* New Zealand.

D. tasmanica Herb to 1.5m high. Flowers with reflexed blue-purple tepals. Berries dark blue. *Distr.* SE Australia, Tasmania.

dianthiflora: with flowers resembling those of the genus *Dianthus*.

Dianthus (From the Greek *dios*, of Zeus or Jove, and *anthos*, flower.) PINK, CARNATION. A genus of about 300 species of annual and perennial herbs with opposite, linear, often grey-green leaves and frequently fragrant flowers. The flowers are subtended by a whorl of bracts and bear 5 erect sepals and 5 long-clawed petals that have large, toothed limbs. Most of the cultivated forms are doubles and bear many petals. Several of the species have given rise to some of the most important garden and cut flowers. There are more than 30,000 cultivars recorded in the International Dianthus Register and its supplements. The term Carnation is usually applied to cultivars that are grown for cut flowers or the summer border; Pink is loosely applied to any member of the genus. *Distr.* Eurasia, North America, mountains of Africa. *Fam.* Caryophyllaceae. *Class* D.

D. armeria DEPTFORD PINK. Tufted perennial. Flowers pink to white, large solitary. *Distr.* Europe, W Asia.

D. barbatus SWEET WILLIAM. Short-lived perennial often grown as a biennial. Flowers in large dense heads. *Distr.* S Europe, long-cultivated and widely naturalized.

D. caesius See *D. gratianopolitanus*

D. caryophyllus CARNATION, CLOVE PINK. A source of oil for soap and scent. Of very great importance as a parent of the Carnation cultivars. *Distr.* Origin uncertain, thought to be around the Mediterranean.

D. chinensis CHINESE PINK. Biennial or short-lived perennial. Flowers large, borne in loose clusters. *Distr.* C and E China.

D. deltoides MAIDEN PINK, MEADOW PINK. Mat-forming or loosely tufted perennial. Flowers usually solitary. *Distr.* Europe, temperate Asia.

D. gratianopolitanus CHEDDAR PINK. Mat-forming. Flowers large, solitary, strongly fragrant. Rare and protected in the British Isles. *Distr.* W and Central Europe.

D. monspessulanus Loosely tufted perennial. Flowers usually clustered. Petals deeply divided, pink or white. *Distr.* Mountains of S and Central Europe.

D. myrtinervius Dense mat-forming. Flowers solitary, bright pink. *Distr.* Balkans.

D. plumarius CLOVE PINK. Loosely tufted perennial. Of importance in the breeding of Carnations. *Distr.* E Central Europe; cultivated since the 17th century, naturalized throughout Europe.

D. superbus Robust perennial. Flowers fragrant, solitary. Petals deeply divided. *Distr.* Europe.

Diapensia (From an early Greek name.) A genus of 4 species of tuft-forming, evergreen subshrubs with simple leaves and solitary, regular flowers that bear their parts in 5s. *Distr.* N Temperate regions. *Fam.* Diapensiaceae. *Class* D.

D. lapponica Leaves deep green, narrow. Flowers white. Not discovered in Scotland until 1951 and now a protected species in Great Britain. It is grown as an alpine ornamental. *Distr.* N cold temperate regions.

Diapensiaceae A family of 7 genera and 28 species of small shrubs and stemless herbs with simple leaves and regular, bisexual flowers that bear their parts in 5s. *Distr.* N temperate regions and Himalaya. *Class* D.
See: Diapensia, Galax, Schizocodon, Shortia.

Diascia (From the Greek *dis*, two, and *askos*, sac, alluding to the 2-spurred flowers.) A genus of about 50 species of annual and perennial herbs with simple, often glandular hairy leaves and terminal racemes of 5-lobed, 2-spurred flowers. Several species are grown as ornamentals. *Distr.* South Africa. *Fam.* Scrophulariaceae. *Class* D.

D. elegans See *D. vigilis*

D. fetcaniensis Compact perennial herb. Flowers cup-shaped, rose-pink. *Distr.* South Africa.

D. rigescens Trailing perennial. Flowers in dense racemes, salmon pink, flat-faced. *Distr.* South Africa.

D. vigilis Prostrate perennial. Flowers in loose racemes, upward-facing, pink. *Distr.* South Africa.

Diascia × Linaria See *Nemesia*

Dicentra (From the Greek *dis*, two, and *kentron*, spurred, alluding to the shape of the flowers.) A genus of 19 species of annual and perennial herbs with compound leaves and panicles or racemes of heart-shaped, 2-spurred flowers. A number of species are grown ornamentally and some have local medicinal uses. *Distr.* Asia, North America. *Fam.* Papaveraceae. *Class* D.

D. eximia See *D. formosa*

D. formosa WILD BLEEDING HEART. Spreading perennial. Flowers purple or red, pendent. *Distr.* W North America.

D. spectabilis BLEEDING HEARTS, DUTCHMAN'S BREECHES. Hummock-forming perennial. Flowers distinctly heart-shaped, pink-red, pendent from arching stems. *Distr.* Japan.

Dichapetalaceae A family of 3 genera and 180 species of shrubs, climbers and small trees with simple alternate leaves and cymes of small flowers that typically bear their parts in 5s. *Distr.* Tropical and warm-temperate regions. *Class* D.

Dichelostemma (From the Greek *dis*, two, *chelos*, pronged, and *stemma*, garland.) A genus of 6–7 species of perennial herbs with linear keeled leaves and umbels of small flowers. Several species are cultivated as ornamentals. *Distr.* W North America. *Fam.* Alliaceae. *Class* M.

D. congestum OOKOW. Flowers violet-blue, tubular. *Distr.* W USA.

D. ida-maia FIRECRACKER FLOWER. Flowers red, sometimes yellow. *Distr.* W USA.

D. multiflorum WILD HYACINTH. Flowers violet, rarely white. *Distr.* W USA.

D. pulchellum BLUE DICKS, WILD HYACINTH. Flowers white to violet. *Distr.* W USA.

D. volubile SNAKE LILY. Flowering stem flexuous, twining; flowers pendent, rose-pink. *Distr.* USA (California).

dichotomus, -a, -um: dichotomously branching, i.e. dividing always into two equal branches.

dicks, blue *Dichelostemma pulchellum.*

Dicksonia (After James Dickson (1738–1822), British botanist.) A genus of 25 species of very large tree-ferns with 4-pinnately divided fronds. Several species are grown ornamentally. *Distr.* Tropical America, Australasia. *Fam.* Dicksoniaceae. *Class* F.

D. antarctica To 15m high. Leaves to 2m long. The pith has been used as a source of starch by Aborigines. *Distr.* Australia.

Dicksoniaceae A family of 7 genera and about 50 species of tree-ferns with creeping to erect trunks and leaves that are pinnately divided. The sori are terminal on the veins and have a 2–valved indusium. *Distr.* Montane tropical and temperate regions. *Class* F.
See: Dicksonia, Thyrsopteris.

Dicliptera (From the Greek *diklis*, double folded, and *pteron*, wing, alluding to the two-winged fruit.) A genus of about 150 species of annual and perennial herbs, shrubs and lianas with simple opposite leaves and clusters of 2-lipped tubular flowers. *Distr.* Tropical and warm temperate regions. *Fam.* Acanthaceae. *Class* D.

D. suberecta Grown as a tender ornamental for its downy-grey foliage and rust-red flowers. *Distr.* Uruguay.

Dicranostigma (From the Greek *dicranos*, two-headed, and *stigma*, stigma.) A genus of 2–3 species of annual and perennial herbs with basal rosettes of pinnate leaves and regular flowers that bear 2 sepals, 4 petals, and numerous stamens. *Distr.* Himalaya, W China. *Fam.* Papaveraceae. *Class* D.

D. lactucoides Perennial. Flowers orange, to 5cm across. Grown as an ornamental. *Distr.* Himalaya.

Dictamnus (From *diktamnon*, the classical Greek name for DITTANY.) A genus of 1 species of perennial subshrub with alternate pinnate leaves and large racemes of showy, white to red or lilac flowers that bear 5 sepals, 5 petals and 10 stamens. The foliage of this plant releases volatile oils that have been known to catch light spontaneously. They may also produce an allergic blistering of the skin. *Distr.* Central and S Europe to N China. *Fam.* Rutaceae. *Class* D.

D. albus BURNING BUSH, DITTANY. Grown as an ornamental.

D. fraxinella See *D. albus*

Dictyolimon (From Greek dictyon, net, and *leimon*, meadow.) A genus of 4 species of perennial herbs with rosettes of simple basal leaves and spike-like racemes of small flowers that bear their parts in 5s. *Distr.* Afghanistan to India. *Fam.* Plumbaginaceae. *Class* D.

D. macrorrhabdos Sometimes grown as an ornamental.

Dictyosperma (From the Greek *dictyon*, net, and *sperma*, seed, alluding to the fibrous seed-coats.) A genus of 1 species of large palms with solitary stems, feather-shaped leaves and long, ascending inflorescences of yellow to maroon flowers. *Distr.* Mascarenes. *Fam.* Palmae. *Class* M.

D. album PRINCESS PALM. Grown as a tender ornamental. Now rare in the wild.

didder *Briza media.*

Didiereaceae A family of 4 genera and 11 species of typically columnar cactus-like plants with small simple leaves and unisexual flowers, the sexes being borne on separate plants. *Distr.* Dry areas of Madagascar. *Class* D.

Didymelaceae A family of 1 genus and 2 species of evergreen trees with simple, spirally arranged leaves and small unisexual flowers that lack sepals and petals. *Distr.* Madagascar. *Class* D.

Didymochlaena (From the Greek *didymos*, double, and *chlaina*, cloak, alluding to the indusia.) A genus of 1 species of terrestrial fern with bipinnate leathery leaves. *Distr.* Tropical and South Africa. *Fam.* Dryopteridaceae. *Class* F.

D. lunulata See *D. truncatula*

D. truncatula Occasionally grown as a tender ornamental.

Didymosperma See *Arenga*

didymus, -a, -um: in pairs.

Dieffenbachia (After J. F. Dieffenbach, a gardener at Schönbrunn in Austria during the early 19th century.) DUMB CANE, LEOPARD LILY. A genus of about 25 species of perennial herbs with stout fleshy stems and simple leaves. The small flowers are borne on a long spadix that is surrounded by a green spathe. Several species and numerous cultivars are grown as house plants chiefly for their variegated foliage. *Distr*. Tropical America. *Fam*. Araceae. *Class* M.

D. maculata This species, along with *D. seguine*, has given rise to the majority of the numerous cultivars available.

D. seguine DUMB CANE. The stem of this plant is poisonous, causing loss of speech and death if chewed; it was previously used to torture slaves.

Diegodendraceae A family of 1 genus and 1 species of shrub or small tree. *Distr*. Madagascar. *Class* D.

Dierama (From the Greek *dierama*, a funnel, alluding to the shape of the flowers.) WANDFLOWER, ANGEL'S FISHING ROD. A genus of 10–20 species of clump-forming perennial herbs with linear leaves and pendulous bell-shaped flowers that hang from an arching inflorescence. A number of the species are grown as ornamentals. *Distr*. S and E topical Africa. *Fam*. Iridaceae. *Class* M.

D. dracomontanum Flowers rose-pink to mauve. *Distr*. South Africa.

D. pendulum ANGEL'S FISHING ROD. Leaves arching. Flowers magenta, occasionally white. The most frequently cultivated species with numerous cultivars available. *Distr*. South Africa.

D. pulcherrimum Flowers deep purple, marked white. *Distr*. South Africa.

D. pumilum See *D. dracomontanum*

Diervilla (After M. Dierville, a French physician who introduced *D. lonicera* into cultivation at the end of the 17th century.) BUSH HONEYSUCKLE. A genus of 2–3 species of low-growing deciduous shrubs with simple opposite leaves and clusters of tubular, 2-lipped flowers. Several species are grown as ornamentals. *Distr*. North America. *Fam*. Caprifoliaceae. *Class* D.

D. lonicera BUSH HONEYSUCKLE. Spreading shrub. Flowers green-yellow. *Distr*. E North America.

D. middendorffiana See *Weigela middendorffiana*

D. rivularis Branches densely hairy. Flowers yellow, turning red. *Distr*. SE North America.

D. sessilifolia Branches angled. Flowers in pairs, yellow. *Distr*. SE North America.

Dietes A genus of 6 species of rhizomatous herbs with linear leaves and iris-like flowers. Some species are grown as ornamentals. *Distr*. Tropical and S Africa. *Fam*. Iridaceae. *Class* M.

D. iridioides Flowers white, 8cm across, last only one day. *Distr*. S Africa to Kenya.

diffusiflorus, -a, -um: with diffuse spreading flowers.

diffusus, -a, -um: spreading, loose, diffuse.

Digitalis (From the Latin *digitus*, finger, alluding to the shape of the flowers.) FOXGLOVE. A genus of about 19 species of biennial and perennial herbs with simple leaves and racemes of showy, 5-lobed, tubular flowers. Several species and numerous cultivars are grown ornamentally. *Distr*. Mediterranean, Europe to Central Asia. *Fam*. Scrophulariaceae. *Class* D.

D. ambigua See *D. grandiflora*

D. grandiflora LARGE YELLOW FOXGLOVE. Biennial or perennial to 1m high. Flowers yellow with brown veins. *Distr*. Eurasia.

D. lanata AUSTRIAN DIGITALIS, GRECIAN FOXGLOVE. Biennial or perennial. Flowers off-white. A source of the cardiac stimulant digitalin. *Distr*. Central and SE Europe, naturalized North America.

D. lutea STRAW FOXGLOVE. Perennial to 1m high. Flowers white to pale yellow. *Distr*. W and SW Europe, N Africa.

D. parviflora Perennial to 60cm high. Flowers small, red-brown. *Distr*. N Spain.

D. purpurea COMMON FOXGLOVE. Biennial or perennial herb to 1.5m high. Flowers purple, pink or occasionally white. Numerous ornamental cultivars of this species have been raised. A source of the cardiac stimulant digitalin. *Distr*. Europe, particularly in the W.

digitalis, Austrian *Digitalis lanata*.

digitatus, -a, -um: digitate, finger-like, lobed like a hand.

digynus, -a, -um: with two ovaries.

dilatatus, -a, -um: dilated, widened, expanded.

dill *Anethum graveolens*.

Dilleniaceae A family of 11 genera and 300 species of shrubs, trees and climbers with simple, alternate, typically deciduous leaves and regular, bisexual, white or yellow flowers that bear 5 petals and numerous stamens. *Distr*. Tropical and subtropical to Tasmania. *Class* D.
See: Hibbertia.

dimorphophyllus, -a, -um: with two types of leaves.

Dimorphotheca (From the Greek *dis*, two, *morphe*, form, and *theke*, case, alluding to the 2 different shapes of fruit produced.) SUN MARIGOLD. A genus of 7 species of annual and perennial herbs and shrubs with simple or pinnate leaves and flowers in daisy-like heads that bear golden rays. Several species are grown as ornamentals. *Distr*. S and tropical Africa. *Fam*. Compositae. *Class* D.

D. pluvialis WEATHER PROPHET. Annual, to 40cm high. Flowers white, rays white with a purple base. *Distr*. South Africa, Namibia.

dioicus, -a, -um: dioecious, with separate male and female plants.

Dionaea (After Dione, in Greek mythology the mother of Aphrodite.) A genus of 1 species of carnivorous perennial herb with leaves in basal rosettes and erect stems that bear coiled racemes of small white flowers. The leaves have a large winged stalk and a blade that is hinged about the midrib. When an insect touches trigger hairs on the upper surface of the leaf, the 2 lobes of the trap snap shut. *Distr*. SE USA. *Fam*. Droseraceae. *Class* D.

D. muscipula VENUS' FLY TRAP. Sometimes grown as a curiosity.

Dioncophyllaceae A family of 3 genera and 3 species of shrubs and lianas with spirally arranged simple leaves, the tips of which are often extended into hooks [illegible] tendrils. The regular bisexual flowers bear 5 sepals, 5 petals, and 10–30 stamens. *Distr*. Tropical Africa. *Class* D.

Dionysia (After Dionysos, the Greek god of wine.) A genus of 42 species of cushion-forming subshrubs with rosettes of simple, sometimes scale-like leaves, and umbels of salver-shaped, 5-lobed flowers. A number of species are grown as ornamentals. *Distr*. Central Asia, Iraq, Iran, Afghanistan. *Fam*. Primulaceae. *Class* D.

D. aretioides Dense cushion-forming subshrub. Flowers yellow. *Distr*. N Iran.

D. involucrata Tuft-forming subshrub. Flowers lilac with a white centre. *Distr*. Central Asia.

D. tapetodes Cushion-forming perennial. Flowers yellow. *Distr*. NE Iran to W Pakistan.

Dioon (From the Greek *dis*, two, and *oon*, an egg, alluding to the paired seeds.) A genus of 4 species of tree-like cycads with simple erect stems and a whorl of pinnate leathery leaves. *Distr*. Central America. *Fam*. Zamiaceae. *Class* G.

D. edule Stem to 1.5m high. Leaves to 1.5m long, bearing over 100 leaflets. Seeds edible if cooked. *Distr*. Mexico.

Dioscorea (After Dioscorides, 1st-century Greek physician and herbalist.) YAM. A genus of about 600 species of tuberous perennial herbs with annual twining stems, simple or palmately lobed leaves and inconspicuous yellow-green flowers. Several species are grown as ornamentals but a number of more important species are grown for their edible tubers which form the staple source of starch for many of the world's people. Most species have poisonous tubers and only those that have been domesticated should be eaten. A few species are grown as a source of diosgenin, a precursor in the manufacture of progesterone for use in contraceptive pills. *Distr*. Tropical and subtropical regions. *Fam*. Dioscoreaceae. *Class* M.

D. alata WHITE YAM, WATER YAM, GUY-ANA ARROWROOT. The most widely grown species of edible YAM, with numerous cultivars available. *Distr*. Tropical Asia.

D. batatas CHINESE YAM, CINNAMON YAM. Stem bears small axillary tubers. *Distr*.

Temperate areas of E Asia, naturalized North America.

D. bulbifera AIR POTATO. Stem bears axillary tubers that are edible in some cultivars. *Distr.* Tropical Africa and Asia.

D. discolor ORNAMENTAL YAM. Leaves heart-shaped, light and dark green above, red beneath. Grown as a house plant. *Distr.* Tropical South America.

D. elephantipes ELEPHANT'S FOOT, HOTTENTOT BREAD. Tuber large, with a fissured covering of cork, protruding above the soil surface, producing green stems only in the wet season. An emergency food source, also grown as an ornamental. *Distr.* S Africa.

D. esculenta POTATO YAM, YAM POTATO. A much cultivated edible YAM. *Distr.* E Asia.

D. trifida CUSH CUSH, YAMPEE. Produces numerous small tubers. Leaves deeply lobed. *Distr.* South America, West Indies.

Dioscoreaceae The Yam family. 8 genera and 630 species of herbaceous or woody climbers with well-developed tubers or rhizomes. The inconspicuous flowers have 6 fused tepals, 2 whorls of 3 stamens and an inferior ovary. *Distr.* Widespread in tropical and subtropical regions, rarely in temperate areas. *Class* M.
See: Dioscorea, Tamus.

Diosma (From the Greek *dios*, divine, and *osme*, smell, alluding to the fragrant foliage.) A genus of 28 species of upright or spreading shrubs with simple glandular leaves and white to cream flowers that bear their parts in 5s. Several species are grown as ornamentals. *Distr.* South Africa (Cape Province). *Fam.* Rutaceae. *Class* D.

D. ericoides BREATH OF HEAVEN. Evergreen rounded shrub. Leaves needle-like. Flowers numerous, white, fragrant. *Distr.* SW Cape Province.

D. asperuloides See *Trachelium asperuloides*

Diospyros (From the Greek *dios*, of Zeus or Jove, and *pyros*, grain, alluding to the edible fruits.) EBONY. A genus of about 475 species of evergreen and deciduous shrubs and trees with simple alternate leaves and bell-shaped, typically unisexual flowers. The fruit is a fleshy berry. Several species are grown for their timber and fruits. *Distr.* Tropical regions. *Fam.* Ebenaceae. *Class* D.

D. ebenum EBONY. Principal source of the wood EBONY. *Distr.* India and Sri Lanka.

D. kaki JAPANESE PERSIMMON, KAKI. Much grown in Japan and China for dried fruit. *Distr.* E Asia.

D. lotus DATE PLUM. Fruit eaten fresh or dried. *Distr.* Asia.

D. virginiana PERSIMMON, AMERICAN PERSIMMON, POSSUMWOOD. Fruits eaten only after they have been slightly frosted. *Distr.* SE USA.

Dipelta (From the Greek *dis*, two, and *pelte*, a shield, alluding to the floral bracts.) A genus of 4 species of deciduous shrubs with simple long-pointed leaves and large tubular to bell-shaped flowers. After flowering the bracts enlarge so as to conceal the fruit. Several species are grown as ornamentals. *Distr.* China. *Fam.* Caprifoliaceae. *Class* D.

D. floribunda Flowers numerous, fragrant, pink, marked yellow in the throat. *Distr.* W and Central China.

D. yunnanensis Flowers cream-white, marked orange in the throat. *Distr.* Yunnan.

Dipentodontaceae A family of 1 genus and 1 species of small deciduous tree with simple, spirally arranged leaves and umbels of flowers that bear 5–7 sepals, petals, and stamens. *Distr.* W China, Burma. *Class* D.

dipetalus, -a, -um: with two petals.

Diphylleia (From the Greek *dis*, two, and *phyllon*, a leaf.) A genus of 3 species of rhizomatous herbs with large, plate-like, 2-lobed, long-stalked leaves and cymes of small white flowers that bear 12 tepals and 6 stamens. *Distr.* E North America and Japan. *Fam.* Berberidaceae. *Class* D.

D. cymosa UMBRELLA LEAF. To 1m high. Leaves 60cm across. Grown as a hardy ornamental. *Distr.* E North America.

diphyllus, -a, -um: two-leaved.

Diplacus See *Mimulus*

Dipladenia See *Mandevilla*

dipladenia, white *Mandevilla boliviensis*.

Diplarrhena (From the Greek *diplos*, double, and *arren*, male, alluding to the flowers which bear 2 fertile stamens.) A genus of 1 species of rhizomatous perennial herb with fans of sword-shaped leaves and clusters of showy white flowers that bear 3 large and 3 small tepals. *Distr.* Australia and Tasmania. *Fam.* Iridaceae. *Class* M.

D. moraea Grown as a half-hardy garden ornamental.

Dipsacaceae The Teasel family. 7 genera and 250 species of annual to perennial herbs and subshrubs with opposite or whorled leaves and dense heads of irregular flowers. Several genera contain ornamental species. *Distr.* Eurasia and Africa. *Class* D.
See: Cephalaria, Dipsacus, Knautia, Pterocephalus, Scabiosa, Succisa.

Dipsacus (The classical Greek name for these plants.) TEASEL. A genus of 15 species of biennial or short-lived perennial herbs with erect stems and pairs of simple leaves that are often united at the base so as to form a cup around the stem. The stems, leaf veins and leaf margins are often spiny. The small tubular flowers are borne in cone-like heads amongst numerous sharp bracts. Several species are grown ornamentally and some have medicinal properties. *Distr.* Eurasia, N Africa. *Fam.* Dipsacaceae. *Class* D.

D. fullonum COMMON TEASEL. Biennial to 2m high. Flowers pale purple. Fruit heads much used in dried flower arrangements. *Distr.* Eurasia, naturalized North America.

D. pilosus SHEPHERD'S ROD, SMALL TEASEL. Biennial to 1m high. Flowers white. *Distr.* Eurasia.

D. sativus FULLER'S TEASEL. Heads previously used to nap cloth. *Distr.* SW and Central Europe.

Dipteracanthus See *Ruellia*

Dipteridaceae A family of 1 genus and 8 species of terrestrial or epiphytic ferns with 2-lobed leaves and small, round, scattered sori. *Distr.* India and China to Fiji. *Class* F.

Dipterocarpaceae A family of 16 genera and 530 species of small to large trees with buttressed trunks and simple alternate leaves. The flowers are regular and bisexual with 5 sepals and 5 petals. These plants are one of the world's major sources of hardwood timber and are the dominant trees in much of the lowland forest of SE Asia and parts of Africa. *Distr.* Tropical regions of the Old World, especially Malaysia and Indonesia; one genus occurs in the Guyana Highlands of South America. *Class* D.

dipterocarpus, -a, -um: with a two-winged fruit.

Dipteronia (From the Greek *di*, double, and *pteron*, wing, alluding to the fruit.) A genus of 2 species of deciduous trees with pinnate leaves that bear 7–15 leaflets. The flowers are borne in erect panicles at the ends of the branches. *Distr.* Central and S China. *Fam.* Aceraceae. *Class* D.

D. sinensis Large shrub or tree. Flowers pale green. Fruit turning red in autumn. Grown as an ornamental. *Distr.* Central China.

dipterus, -a, -um: two-winged (e.g. the fruit).

Disanthus (From the Greek *dis*, twice, and *anthos*, flower, alluding to the arrangement of the flowers.) A genus of 1 species of deciduous shrub with simple stalked leaves and small, inconspicuous, spider-like flowers that are borne in pairs and bear their parts in 5s. *Distr.* Japan. *Fam.* Hamamelidaceae. *Class* D.

D. cercidifolius Grown as an ornamental for its spectacular autumn foliage.

Discaria (From the Greek *diskos*, disc, alluding to the nectiferous disc that occurs in the flowers.) A genus of about 15 species of thorny shrubs and small trees with small leaves and dense clusters of small flowers that often lack petals. Several species are grown as ornamentals. *Distr.* South America, Australia, New Zealand. *Fam.* Rhamnaceae. *Class* D.

D. toumatou WILD IRISHMAN. Shrub or small tree to 5m high. Flowers solitary or in small clusters. Petals absent. *Distr.* New Zealand.

Discocactus (From the Greek *diskos*, a disc, and the old genus name *Cactus*, alluding to the flattened stems.) A genus of about 5 species of low-growing cacti with almost spherical, ribbed stems and prominent spines. The tubular or salver-shaped white flowers are borne at the apex of the stem and open at night. Most of the species are grown as ornamentals. Some horticulturalists consider that there are over 30 species in this genus. *Distr.* E South America. *Fam.* Cactaceae. *Class* D.

D. placentiformis Stems somewhat squat, broader than high. Spines to 4cm long. *Distr.* E Brazil.

discolor: two-coloured.

Diselma (From the Greek *dis*, two, and *selma*, upper, alluding to the fertile scales in the cone.) A genus of 1 species of evergreen coniferous shrub with very small scale-like leaves. *Distr.* Tasmania. *Fam.* Cupressaceae. *Class* G.

D. archeri Occasionally grown as an ornamental.

dispar: unlike other closely related species.

Disporum (From the Greek *dis*, two, and *spora* seed, alluding to the 2-seeded fruits of some species.) FAIRY BELLS. A genus of 15–30 species of rhizomatous herbs with erect or spreading stems and simple alternate leaves. The flowers are borne singly or in umbels and bear 6 free, white or yellow-green tepals. Several species are grown as ornamentals. *Distr.* E and tropical Asia, W North America. *Fam.* Convallariaceae. *Class* M.

D. hookeri Flowers clustered, green-white, drooping. Berries orange. *Distr.* NW USA.

D. sessile Flowers solitary or in clusters of 2–3, white with green tips. Berries black. *Distr.* Japan.

dissectus, -a, -um: dissected, deeply divided.

distichus, -a, -um: in two ranks, referring to the leaves for example.

Distictis (From the Greek *di-*, twice, and *stiktos*, spotted, alluding to the 2 rows of seeds in the capsules.) A genus of 9 species of evergreen woody climbers with leaves consisting of 2 leaflets and a 3-branched tendril. The large flowers are tubular to funnel-shaped and borne in small racemes or panicles. Several of the species are grown as tender ornamentals. *Distr.* Mexico, West Indies. *Fam.* Bignoniaceae. *Class* D.

D. buccinatoria MEXICAN BLOOD FLOWER. Flowers red with a yellow tube. *Distr.* Mexico.

Distylium (From the Greek *dis*, twice and *stylos*, style.) A genus of 12 species of evergreen shrubs and small trees with simple, leathery, alternate leaves and racemes of small flowers that bear 3–5 sepals and no petals. The fruit is a small capsule to which the 2 styles remain attached. *Distr.* Tropical Asia, Central America. *Fam.* Hamamelidaceae. *Class* D.

D. racemosum ISU TREE. Spreading shrub. Stamens red, conspicuous. Grown as an ornamental and for the fine-grained wood that is used for furniture and art. *Distr.* Japan.

distylus, -a, -um: with two styles.

ditchmoss *Elodea*.

dittany *Dictamnus albus* **Cretan** ~ *Origanum dictamnus*.

Diuranthera See *Chlorophytum*

divaricatus, -a, -um: spreading.

divergens: divergent, going in different directions.

diversifolius, -a, -um: diverse-leaved.

dock *Rumex* **prairie** ~ *Silphium*, *Parthenium integrifolium* **sour** ~ *Rumex acetosa*.

dockmackie *Viburnum acerifolium*.

Dodecatheon (From the Greek for twelve gods, alluding to the habit of the flowers.) AMERICAN COWSLIP, SHOOTING STAR. A genus of 14 species of rhizomatous herbs with rosettes of simple leaves and umbels of nodding, *Cyclamen*-like flowers that bear 5 reflexed petals. Several species are grown as ornamentals. *Distr.* North America, E Siberia. *Fam.* Primulaceae. *Class* D.

D. hendersonii SAILORS CAP, MOSQUITO BILLS. Leaves rather fleshy. Flowers violet. *Distr.* California.

D. meadia SHOOTING STAR, AMERICAN COWSLIP. Flowers purple with a white centre. *Distr.* E USA.

Dodonaea (After Rembert Dodoens (1516–85), Dutch herbalist.) A genus of about 50 species of shrubs and trees with simple or pinnately divided leaves and cymes of inconspicuous flowers that lack petals. The foliage is often covered in numerous yellow glands. Several species are used medicinally and some are used as fodder. *Distr.* Tropical and warm regions, especially Australia. *Fam.* Sapindaceae. *Class* D.

D. viscosa NATIVE HOPS. Spreading shrub to 2m high. Grown as a half-hardy ornamental. *Distr.* S Africa, Australia.

dog-hobble *Leucothoe fontanesiana.*

dog's mercury *Mercurialis perennis.*

dog's tail, crested *Cynosurus cristatus.*

dogwood *Cornus*, Cornaceae **brown** ~ *Cornus glabrata* **common** ~ *C. sanguinea* **common white** ~ *C. florida* **creeping** ~ *C. canadensis* **eastern flowering** ~ *C. florida* **giant** ~ *C. controversa* **panicled** ~ *C. racemosa* **red-barked** ~ *C. alba* **silky** ~ *C. obliqua* **Tartarian** ~ *C. alba* **Victoria** ~ *Prostanthera lasianthos.*

dolabratus, -a, -um: axe- or hatchet-shaped.

dolabriformis, -e: resembling the shape of a hatchet or axe.

Dolichothrix (From the Greek *dolichos*, long, and *thrix*, hair, alluding to the hairs on the fruit.) A genus of 1 species of heath-like shrub with small leaves and flowers in daisy-like heads which lack distinct rays but which have somewhat showy bracts. *Distr.* South Africa (Cape Province). *Fam.* Compositae. *Class* D.

D. ericoides Grown as a garden ornamental. Several cultivars are available.

doll's eyes *Actaea alba.*

dolomiticus, -a, -um: of the Dolomite Mountains.

domesticus, -a, -um: cultivated.

donax: Greek name for a kind of reed.

donkey's tail *Sedum morganianum.*

Doodia (After Samuel Doody (1656–1706), apothecary and keeper of the Chelsea Physic Garden.) A genus of 11 species of small terrestrial ferns with pinnate leaves. Several species are grown as ornamentals. *Distr.* Sri Lanka to Australia. *Fam.* Blechnaceae. *Class* F.

D. caudata SMALL RASP FERN. *Distr.* Australia to Polynesia.

D. media COMMON RASP FERN. *Distr.* Australia, New Zealand, New Caledonia.

Doronicum (From the Arabic name for these plants, *doronigi.*) LEOPARD'S BANE. A genus of about 35 species of rhizomatous or tuberous, perennial herbs with simple leaves and yellow flowers in daisy-like heads that bear distinct yellow rays. Several species are grown as hardy ornamentals and some species are reported to have medicinal qualities. *Distr.* Eurasia, Mediterranean. *Fam.* Compositae. *Class* D.

D. austriacum Stems to 1.2m high. Flower-heads several. *Distr.* Central and S Europe.

D. caucasicum See *D. orientale*

D. columnae Stems erect, to 60cm high. *Distr.* S Europe to W Asia.

D. cordatum A name erroneously applied in horticulture to plants of *D. columnae* or *D. orientale*.

D. orientale Tuft-forming. Flower-heads solitary. *Distr.* Caucasus to SE Europe.

D. pardalianches GREAT LEOPARD'S BANE. Tuft-forming. Flower-heads 2–6. *Distr.* W Europe.

D. plantagineum Tuft-forming. Flower-heads solitary. *Distr.* W Europe.

Dorotheanthus (After Dorothea, mother of G.F. Schwantes, the botanist who described the genus, and the Greek *anthos*, flower.) A genus of 14 species of succulent annual herbs with linear leaves and numerous, large, many-petalled flowers. The leaves and stems are covered in crystal-like papillae. Several species are grown as ornamentals. *Distr.* South Africa. *Fam.* Aizoaceae. *Class* D.

D. bellidiformis LIVINGSTONE DAISY. Leaves mostly basal, rough. Flowers white or red. *Distr.* South Africa.

D. gramineus Stems branching from base. Flowers red with a dark centre. Several cultivars are available. *Distr.* South Africa (Cape Province).

Doryanthaceae A family of 1 genus and 2 species. *Class* M.
See: Doryanthes.

Doryanthes (From the Greek *dory*, spear, and *anthos*, flower, in allusion to the tall flowering stem.) SPEAR LILY. A genus of 3 species of perennial herbs with strap-shaped pointed leaves and heads of large red flowers borne on tall flowering stems. All the species are grown as tender ornamentals. *Distr.* E Australia. *Fam.* Doryanthaceae. *Class* M.

D. palmeri Leaves ribbed, arching. Flowers red. *Distr.* Australia (Queensland).

Dorycnium See *Lotus*

Doryopteris (From the Greek *dory*, spear, and *pteris*, fern, alluding to the frond shape in some species.) A genus of 35 species of small to medium-sized, tufted ferns. *Distr.* Tropical and warm regions. *Fam.* Adiantaceae. *Class* F.

Douglasia See *Androsace*
D. laevigata See *Androsace laevigata*
D. montana See *Androsace montana*
D. vitaliana See *Vitaliana primuliflora*

dove flower *Peristeria elata.*

dove tree *Davidia involucrata.*

down tree *Ochroma lagopus.*

Doxantha See *Macfadyena*

Draba (The classical Greek name for *Lepidium draba.*) A genus of about 300 species of annual and perennial, typically cushion-forming, herbs with small flowers that bear 4 petals. *Distr.* N temperate regions and the mountains of South America. *Fam.* Cruciferae. *Class* D.

D. acaulis Cushion-forming perennial with yellow flowers. *Distr.* Turkey.

D. aizoides YELLOW WHITLOW GRASS. Tuft-forming perennial. Flowers yellow. *Distr.* Central and SE Europe, Balkans. This species also occurs in S Wales although it is thought that this may be an introduction.

D. arabisans Somewhat diffuse perennial with white flowers. *Distr.* E North America.

D. bruniifolia Cushion- or mat-forming perennial with woolly leaves and golden yellow flowers. *Distr.* Mediterranean.

D. bryoides See *D. rigida*

D. cinerea Densely white-hairy, robust perennial. *Distr.* Arctic Europe.

D. compacta Minute tufted perennial with pale yellow flowers. *Distr.* N Balkan Peninsula, E Carpathians.

D. dedeana Cushion-forming perennial with white flowers. *Distr.* Pyrenees.

D. haynaldii Very small perennial with deep yellow flowers. *Distr.* S Carpathians.

D. hoppeana Very small perennial with pale yellow flowers. *Distr.* Alps.

D. incana Biennial or perennial with white flowers. *Distr.* Europe.

D. mollissima Cushion-forming perennial with yellow flowers. *Distr.* Caucasus.

D. norvegica Tufted perennial with white flowers. *Distr.* NW Europe, Greenland.

D. rigida Tufted perennial with yellow flowers. *Distr.* Turkey and Armenia.

D. rupestris See *D. norvegica*

D. ventosa Tufted perennial with yellow flowers. *Distr.* N America.

Dracaena (From the Greek *drakaina*, a female dragon.) A genus of 40–60 species of subshrubs and trees with smooth bark, lance-shaped leathery leaves, and panicles of small flowers that are fragrant at night. The resin from the stems of some species is a source of dragon's blood used in varnishes and photo-engraving. *Distr.* Mostly Africa. *Fam.* Dracaenaceae. *Class* M.

D. concinna MADAGASCAR DRAGON TREE. Small shrub. Leaves with red margins. *Distr.* Madagascar.

D. deremensis Shrub to 4.5m high. Leaves often variegated. Flowers red, white within, bad-smelling. *Distr.* E tropical Africa.

D. draco DRAGON TREE. Tall trees. Trunk silver-grey. Leaves in large terminal rosettes. An endangered species in the wild owing to over-exploitation. *Distr.* Canary Islands.

D. fragrans CORN PLANT. Sparsely branched shrub, to 15m tall. Flowers yellow, very fragrant. Used as live fencing in tropical regions. *Distr.* Tropical Africa.

D. marginata A name often misapplied to *D. concinna* in cultivation.

D. sanderiana RIBBON PLANT. Stems cane-like. Leaves with yellow margins. *Distr.* Cameroon.

Dracaenaceae A family of 6 genera and 156 species of rhizomatous herbs to trees with tough leathery leaves. The flowers have 6 fused tepals, 6 stamens and a superior ovary. This family is sometimes included within the family Liliaceae or in the family Agaveaceae.

Distr. Tropical and subtropical regions, to SW USA. *Class* M.
See: Beaucarnea, Dasylirion, Dracaena, Nolina, Sansevieria.

Dracocephalum (From the Greek *drakon*, dragon, and *kephale*, head, alluding to the shape of the flowers.) DRAGON HEAD. A genus of about 45 species of annual or perennial herbs and shrubs with simple opposite leaves and spike-like inflorescences consisting of whorls of 2-lipped flowers. Several species are grown ornamentally. *Distr.* Eurasia, North America. *Fam.* Labiatae. *Class* D.

D. forrestii Erect perennial to 50cm high. Flowers deep purple-blue. *Distr.* W China.

D. grandiflorum Erect perennial to 35cm high. Flowers hooded, deep-blue, in long spikes. *Distr.* Siberia.

D. mairei See *D. renatii*

D. moldavicum Aromatic annual. Flowers violet or white, borne in whorls. *Distr.* E Europe, Central Asia, Siberia.

D. renatii Low perennial herb. Flowers 2cm long, cream with red marks. *Distr.* Morocco.

D. ruyschianum Erect perennial herb. Flowers violet-blue. *Distr.* Central Europe to Siberia.

D. virginicum See *Physostegia virginiana*

Dracophyllum (From the Greek *drakon*, dragon, and *phyllon*, leaf, alluding to the similarity of the leaves to those of the DRAGON TREE, *Dracaena draco*.) A genus of about 48 species of shrubs and small trees with grass-like or linear leaves and small tubular or bell-shaped flowers that bear their parts in 5s. Several species are grown ornamentally. *Distr.* New Zealand, Australia, New Caledonia. *Fam.* Epacridaceae. *Class* D.

D. capitatum Small erect shrub. Flowers in dense panicles, white. *Distr.* W Australia.

Dracula (From the Latin *dracula*, little dragon, alluding to the appearance of the flowers.) A genus of 60 species of tufted lithophytic or epiphytic orchids with pendulous flowers. The lips of the flowers are distinctly divided into claw and blade. Members of this genus were formerly included in the genus *Masdevallia*. Some species are grown as tender ornamentals. *Distr.* Central and South America. *Fam.* Orchidaceae. *Class* M.

D. bella Flower nodding, solitary, green and white with red spots. *Distr.* Colombia.

D. chimaera Flowers 1–6, green with red markings. *Distr.* Colombia.

Dracunculus (From the diminutive of the Latin *draco*, dragon.) A genus of 3 species of tuberous herbs with large divided leaves and small flowers that are borne on a spadix surrounded by a spathe and which have a similar pollination mechanism to that found in *Arum*. The inflorescences are often very malodorous. Members of this genus are sometimes grown as ornamentals and curiosities. *Distr.* Mediterranean region. *Fam.* Araceae. *Class* M.

D. muscivorus HAIRY ARUM, DRAGON'S MOUTH. *Distr.* W Mediterranean Islands.

D. vulgaris DRAGON ARUM. *Distr.* Central and E Mediterranean.

dragon, water *Calla palustris.*

dragon head *Dracocephalum* **false** ~ *Physostegia.*

dragon mouth *Horminum pyrenaicum.*

dragon's mouth *Dracunculus muscivorus.*

dragon tree *Dracaena draco*
Madagascar ~ *D. concinna.*

Drapetes (From the Greek *drapetes*, fugitive.) A genus of 1–4 species of small, tuft-forming shrubs with simple alternate leaves and heads of small, tubular, 4-lobed flowers that lack petals. *Distr.* Borneo and New Guinea to South America. *Fam.* Thymelaeaceae. *Class* D.

D. tasmanica Very low shrub. Flowers white, silky-hairy. Grown as a rock garden ornamental. *Distr.* Tasmania.

Dregea A genus of 3 species of evergreen woody climbers with simple leaves and umbels of tubular flowers that bear 5 sepals and 5 basally fused petals. *Distr.* Warm regions of the Old World. *Fam.* Asclepiadaceae. *Class* D.

D. sinensis Grown as a hardy ornamental. *Distr.* China.

Drepanostachyum (From the Greek *drepanon*, sickle, and *stachys*, spike, alluding to the sickle-shaped inflorescence.) A genus of 15 species of clump-forming bamboos with many branches and open panicles. *Distr.* Subtropical regions. *Fam.* Gramineae. *Class* M.

D. hookerianum Grown as a half-hardy ornamental for its striped stems. *Distr.* India, Sikkim, Bhutan.

Drimys (From the Greek *drimys*, acrid, alluding to the taste of the bark.) A genus of about 300 species of evergreen shrubs and trees with simple, alternate, aromatic leaves and regular flowers that bear 2 or more petals and numerous stamens. Several species are grown as ornamentals. *Distr.* S hemisphere. *Fam.* Winteraceae. *Class* D.

D. winteri WINTER'S BARK. Shrub or tree to 20m high. Used medicinally. *Distr.* Chile, Argentina.

dropwort *Filipendula vulgaris*
hemlock water ~ *Oenanthe crocata.*

Drosanthemum (From the Greek *drosos*, dew, and *anthos*, flower, alluding to the papillae of the stems and leaves.) A genus of 90–100 species of herbs and shrubs with erect or prostrate stems, succulent, flat to cylindrical leaves, and many-petalled flowers. The leaves and stems are covered in numerous, small, often transparent papillae. *Distr.* S and SW Africa. *Fam.* Aizoaceae. *Class* D.

D. floribundum Cushion-forming. Flowers numerous, pale pink. *Distr.* South Africa (Cape Province).

D. hispidum Stems arching and rooting, rough white-hairy. Flowers deep purple-red. *Distr.* SW Africa.

Drosera (From the Greek *droseros*, dewy, alluding to the appearance of the glandular leaves.) SUNDEW. A genus of about 80 species of carnivorous perennial herbs with basal rosettes of leaves and erect stems that bear coiled racemes of small white flowers. The leaves are covered in numerous, sticky, glandular hairs that trap and digest small insects. Many species are grown as ornamentals and curiosities. *Distr.* Cosmopolitan, especially in the S hemisphere. *Fam.* Droseraceae. *Class* D.

D. anglica GREAT SUNDEW, ENGLISH SUNDEW. Leaves more or less erect, around 3cm long, tapering to a long stalk. Flowers white. *Distr.* N Europe, N Asia, N North America.

D. capensis CAPE SUNDEW. Leaves narrow. Flowers purple. *Distr.* S Africa.

D. rotundifolia ROUND-LEAVED SUNDEW. Leaves round, on a long stalk. Flowers white or pink.

D. spathulata SPOON-LEAF SUNDEW. Leaves spoon-shaped. Flowers pink or white. *Distr.* E China to New Zealand.

Droseraceae The Sundew family. 4 genera and 85 species of carnivorous annual and perennial herbs with leaves that are covered in sessile or stalked glands which secrete digestive enzymes. Insects are trapped either by sticky hairs or rapid leaf movements. The flowers are regular, typically have their parts in 5s, and are borne in coiled racemes. Some species are grown as curiosities. *Distr.* Widespread. *Class* D.
See: Dionaea, Drosera.

drumsticks *Craspedia globoides*
narrow-leaved ~ *Isopogon anethifolius.*

drupaceus, -a, -um: with fleshy fruit, a drupe.

Dryadella (After Dryas, a wood nymph in Greek mythology.) A genus of 25 species of epiphytic orchids with short stems, thick leaves and small solitary flowers. Several species are grown ornamentally. *Distr.* Central and South America. *Fam.* Orchidaceae. *Class* M.

D. edwallii Flowers golden yellow, marked with red. *Distr.* Brazil.

D. simula Flowers white, marked with purple. *Distr.* Central America.

Dryandra (After Jonas Dryander (1748–1810), Swedish botanist.) DRYANDRA. A genus of about 55 species of shrubs and small trees with woody tubers, leathery leaves and yellow flowers borne in dense heads that are surrounded by a whorl of bracts. Numerous species are grown as ornamentals. *Distr.* W Australia. *Fam.* Proteaceae. *Class* D.

D. formosa SHOWY DRYANDRA. Shrub to 2m high. Flower-heads terminal, golden-orange. *Distr.* W Australia.

D. praemorsa CUT-LEAVED DRYANDRA. Upright shrub to 3.5m high. Leaves wavy, white beneath. *Distr.* W Australia.

D. pteridifolia TANGLED HONEYPOT. Tuft-forming shrub. Flower-heads large, cream. *Distr.* W Australia.

dryandra *Dryandra* **cut-leaved** ~ *D. praemorsa* **showy** ~ *D. formosa*.

dryandroides from the genus *Dryandra*, with the ending *-oides*, indicating resemblance.

Dryas (After Dryas, a wood nymph in Greek mythology for whom the oak was sacred, alluding to the oak-shaped leaves of *D. octopetala*.) A genus of 2 species of mat-forming, evergreen shrubs with alternate leathery leaves and solitary, white to yellow flowers. *Distr*. Cool regions of the N hemisphere. *Fam*. Rosaceae. *Class* D.

D. octopetala MOUNTAIN AVENS. Mat-forming shrub to 50cm high. Flowers erect, white. Grown as a rock garden ornamental. *Distr*. Mountainous regions of the N hemisphere.

drynarioides: from the fern genus *Drynaria*, with the ending *-oides*, indicating resemblance.

dryophyllus, -a, -um: with leaves resembling oak leaves.

Dryopteridaceae The Male Fern family. 47 genera and about 1000 species of terrestrial, occasionally epiphytic, ferns with pinnate, sometimes simple leaves and round sori that are borne on the underside of the leaf. *Distr*. Widespread. *Class* F.
See: Arachniodes, Cyrtomium, Didymochlaena, Dryopteris, Lastreopsis, Phanerophlebia, Polystichum, Tectaria.

Dryopteris (From the Greek *dryas*, oak, and *pteris*, fern; several species are associated with oak woodland.) SHIELD FERN, BUCKLER FERN. A genus of about 150 species of medium-sized terrestrial ferns with erect or spreading, pinnate to tripinnate leaves. Several species are grown as ornamentals and some were previously used medicinally to remove tapeworms, a use that has stopped because of the dangers to the patient. *Distr*. Cosmopolitan. *Fam*. Dryopteridaceae. *Class* F.

D. filix-mas MALE FERN. Leaves deciduous, to 30cm long. A common woodland fern with a number of cultivars available. *Distr*. N temperate regions.

D. pedata HAND FERN. Cultivated ornamental. *Distr*. Tropical America.

Drypis (From the Greek *drypto*, to tear or scratch, alluding to the spiny leaves.) A genus of 1 species of perennial herb with simple, opposite, spiny leaves and clusters of small flowers which bear 5 pink or white petals. *Distr*. S Europe. *Fam*. Caryophyllaceae. *Class* D.

D. spinosa Grown as a rock garden ornamental.

dubius, -a, -um: doubtful.

Duchesnea (After Antoine Nicolas Duchesne (1747–1827), French horticulturalist.) A genus of 2 species of stoloniferous perennial herbs with palmately divided leaves and small yellow flowers that are followed by fleshy red fruits. *Distr*. S and E Asia, naturalized North America. *Fam*. Rosaceae. *Class* D.

D. indica INDIAN STRAWBERRY, MOCK STRAWBERRY. Grown as ornamental ground cover for the rock garden although it may become an invasive weed. *Distr*. India to Japan, naturalized in North America.

Duckeodendraceae A family of 1 genus and 1 species of tree with spirally arranged, simple leaves and small cymes of funnel-shaped flowers that have their parts in 5s. *Distr*. Brazil. *Class* D.

duckmeat *Lemna*.

duck plant *Sutherlandia frutescens*.

duckweed *Lemna*, Lemnaceae, *Spirodela* **common** ~ *Lemna minor* **great** ~ *Spirodela polyrrhiza* **lesser** ~ *Lemna minor* **star** ~ *L. trisulca* **tropical** ~ *Pistia stratiotes*.

Dudleya A genus of about 40 species of succulent perennial herbs with rosettes of simple leaves and upright flowering stems that bear panicles of tubular or star-shaped flowers. Several species are grown ornamentally. *Distr*. SW North America. *Fam*. Crassulaceae. *Class* D.

D. pulverulenta CHALK LETTUCE PLANT. Leaves strap-shaped, pointed, silver-grey. Flowers numerous, star-shaped, red. *Distr*. California.

D. virens ALABASTER PLANT. Leaves grey-green, in stalked rosettes. Flowers white. *Distr*. Coastal S California.

dulcis, -e: sweet, pleasant.

dumb cane *Dieffenbachia*, *D. seguine*.

dumetorum: of hedges.

dumosus, -a, -um: bushy, shrubby.

dumulosus, -a, -um: like a small shrub.

Duranta (After Castor Durante (1523–90) Italian physician and botanist.) A genus of about 30 species of shrubs and trees with simple, opposite or whorled leaves and racemes or panicles of small, salver-shaped, 5-lobed flowers. *Distr.* Caribbean to South America. *Fam.* Verbenaceae. *Class* D.

D. erecta PIGEON BERRY, GOLDEN DEWDROP. Shrub or small tree. Flowers white to purple-blue. Fruit a yellow berry. *Distr.* Tropical America.

D. repens See *D. erecta*

durian Bombacaceae, *Durio zibethinus*.

durifolius, -a, -um: hard-leaved.

Durio (From the Malay name *duryon*.) A genus of 27 species of large evergreen trees with 2 ranks of simple leathery leaves and clusters of cup-shaped flowers that are often borne directly on the trunk. *Distr.* Burma to W Malaysia. *Fam.* Bombacaceae. *Class* D.

D. zibethinus DURIAN. The large, spiny, bad-smelling fruit is highly regarded for its sweet-tasting flesh, the seeds are also eaten roasted. *Distr.* W Malaysia to Indonesia.

durior: harder

durum *Triticum durum*.

dusty miller *Artemisia stelleriana*, *Tanacetum ptarmiciflorum*, *Lychnis coronaria*, *Senecio cineraria*.

Dutchman's breeches *Dicentra spectabilis*.

Dutchman's pipe *Aristolochia*, *A. macrophylla*.

Dutch mice *Lathyrus tuberosus*.

dysentericus, -a, -um: a cause of or cure for dysentery.

E

earth star *Cryptanthus*.

Easter ledger *Polygonum bistorta*.

Ebenaceae The Ebony family. 3 genera and about 500 species of small trees with a pointed crown and sprays of flattened foliage. The tubular flowers are typically unisexual with the sexes being borne on separate plants. This family is a source of hard dark timber, as well as having fruits that are eaten locally. *Distr*. Widespread in tropical regions, particularly in the lowland rainforest of E Asia. *Class* D.
See: Diospyros.

Ebenus (From *ebenos*, a classical Greek name for a plant in the Pea family.) A genus of 18 species of perennial herbs and subshrubs with pinnate or trifoliate alternate leaves and racemes or heads of irregular pea-like flowers. *Distr*. E Mediterranean to Central Asia. *Fam*. Leguminosae. *Class* D.

E. cretica Evergreen subshrub. Flowers borne in dense heads, bright pink. Grown as a half-hardy ornamental. *Distr*. Crete.

ebenus, -a, -um: ebony, black.

ebony Ebenaceae, *Diospyros*, *D. ebenum*.

ebracteatus, -a, -um: without bracts.

eburneus, -a, -um: ivory-white.

ecalcaratus, -a, -um: without a spur.

Ecballium (From the Greek *ekballein*, to throw out, alluding to the fruits that burst open.) A genus of 1 species of bushy or trailing, perennial herb with palmately lobed leaves and yellow bell-shaped flowers. The fruit is a 3-5cm long, cylindrical berry filled with a mixture of mucilage and seeds under pressure. At maturity the stalk breaks off and the berry is propelled, rocket fashion, a distance of up to 2m. *Distr*. Dry regions around the Mediterranean and Black Sea. *Fam*. Cucurbitaceae. *Class* D.

E. elaterium SQUIRTING CUCUMBER. Occasionally planted as a curiosity; the fruit is used as a purgative.

Eccremocarpus (From the Greek *ekkremus*, hanging, and *karpos*, a fruit, alluding to the pods.) A genus of 5 species of annual climbers and woody evergreen vines with bipinnate opposite leaves that bear terminal tendrils. The red to yellow, tubular flowers are borne in terminal racemes. The mouths of the floral tubes are somewhat closed. Several species are grown as tender ornamentals. *Distr*. Peru and Chile. *Fam*. Bignoniaceae. *Class* D.

E. scaber GLORY FLOWER. Climbing subshrub. Flowers orange. Several cultivars are available with yellow and red flowers. *Distr*. Chile.

Ecdeiocoleaceae A family of 1 genus and 1 species of rhizomatous herb with small, regular, wind-pollinated flowers. This family is sometimes included within the family Restionaceae. *Distr*. W Australia. *Class* M.

Echeveria (After Atanasio Echeverria Godoy, 19th-century botanical artist.) A genus of about 150 species of succulent herbs and shrubs with simple fleshy leaves borne in rosettes or scattered along the stems. The small tubular flowers are borne in tall, branched, often 1–sided, inflorescences. Many species are grown as ornamentals. *Distr*. Warm America, especially Mexico. *Fam*. Crassulaceae. *Class* D.

E. agavoides Leaves light green, pointed, borne in basal rosettes. Flowers red. *Distr*. Mexico.

E. derenbergii (After J. Derenberg (1873–1928).) Rosettes clump-forming. Flowers yellow with red markings. *Distr*. Mexico.

E. elegans MEXICAN SNOWBALL, WHITE MEXICAN ROSE, MEXICAN GEM. Rosettes clump-forming. Leaves very fleshy, pale silver-blue. Flowers pink outside, yellow within, borne on a red stem. *Distr*. Mexico.

E. harmsii (After Dr Hermann Harms (1870–1942), German botanist.) Bushy perennial. Flowers red, yellow within. *Distr.* Mexico.

E. pulvinata PLUSH PLANT, CHENILLE PLANT. Bushy perennial. Leaves densely hairy. Flowers yellow, marked red. *Distr.* S Mexico.

echidnus, -a, -um: a snake.

Echinacea (From Greek *echinos*, hedgehog, alluding to the prickly receptacle scales.) CONE FLOWER. A genus of 9 species of rhizomatous perennial herbs with simple leaves and flowers in conical daisy-like heads. The rhizomes are used in Indian medicine. *Distr.* E USA. *Fam.* Compositae. *Class* D.

E. angustifolia Rays purple-pink.

E. pallida Rays purple.

E. purpurea Flowers orange, rays red-purple.

echinaceus, -a, -um: spiny.

echinatus, -a, -um: covered with prickles.

Echinochloa (From the Greek *echinos*, hedgehog, and *chloe*, grass, alluding to the spiky appearance of the heads of some species.) A genus of 20–25 species of annual and perennial grasses with flat leaves and inflorescences of flattened racemes arranged along a central axis. *Distr.* Widespread in tropical and warm temperate regions. *Fam.* Gramineae. *Class* M.

E. crus-galli COCKSPUR, BARNYARD GRASS, BARNYARD MILLET. Sometimes grown for its inflorescences which are used in dried flower arrangements. *Distr.* Cosmopolitan.

Echinodorus (From the Greek *echinos*, hedgehog, and *doros*, bag, alluding to the clustered spiny fruits.) BURHEAD, SWORD PLANT. A genus of 47 species of annual and perennial aquatic herbs with both submerged and aerial leaves and regular white or pink flowers. Several species are grown as ornamental aquarium plants. *Distr.* Tropical and subtropical regions of Africa and the Americas. *Fam.* Alismataceae. *Class* M.

E. cordifolius TEXAS MUD BABY. *Distr.* S USA, Mexico.

E. magdalenensis DWARF AMAZON SWORD PLANT. *Distr.* Colombia.

E. paniculatus AMAZON SWORD PLANT. *Distr.* Brazil and Venezuela to Paraguay.

Echinops (From the Greek *echinos*, hedgehog and, *opsis*, appearance.) GLOBE THISTLE. A genus of about 120 species of robust, perennial, rarely annual herbs with spherical heads comprising numerous 1-flowered flower-heads each with a whorl of bracts. Several species are cultivated as ornamentals. *Distr.* Europe to Central Asia and tropical African mountains. *Fam.* Compositae. *Class* D.

E. bannaticus Perennial, to 1.2m high. *Distr.* SE Europe, W Asia.

E. exaltatus Perennial, to 2m high. Flowers white-grey. *Distr.* E Europe to Russia.

E. giganteus Shrub to 5m high. *Distr.* Ethiopia.

E. horridum See *Genista horrida*

E. ritro (From a local vernacular name.) Perennial, to 60cm high. Flowers blue. *Distr.* E Europe, W Asia.

E. sphaerocephalus Perennial, to 2m high. Flowers grey-white. *Distr.* Europe, W Asia.

echinus: sea urchin or hedgehog.

echioides: from the genus *Echium*, with the ending *-oides*, indicating resemblance.

Echium (From *echion*, the Greek name for these plants.) A genus of 40 species of annual to perennial herbs and shrubs with simple leaves and coiled cymes of funnel-shaped, somewhat irregular flowers. The green parts are usually stiffly hairy. Some species are grown as ornamentals and several were formerly used medicinally. *Distr.* Canary Islands, Europe, W Asia, N and S Africa. *Fam.* Boraginaceae. *Class* D.

E. bouraeanum See *E. wildpretii*

E. vulgare VIPER'S BUGLOSS. Biennial, to 1m high. Flowers blue, occasionally white. Dwarf cultivars are popular. *Distr.* Europe, Asia, naturalized North America.

E. wildpretii Biennial. Leaves silver hairy. Flowers red, in tall spike-like inflorescences. *Distr.* Canary Islands.

ecristatus, -a, -um: without a crest.

edelweiss *Leontopodium*, *L. alpinum* **New Zealand** ~ *Leucogenes* **North Island** ~ *L. leontopodium* **South Island** ~ *L. grandiceps*.

Edgeworthia (After Michael Pakenham Edgeworth (1812–81), plant collector with the East India Co.) PAPER BUSH. A genus of 3 species of semi-evergreen to deciduous shrubs with simple, somewhat leathery leaves and a profusion of fragrant, tubular, 4-lobed flowers that lack petals. The bark is sometimes used in the manufacture of high quality paper. *Distr.* China, Japan. *Fam.* Thymelaeaceae. *Class* D.

E. papyrifera PAPER BUSH. Flowers white. Grown as an ornamental. *Distr.* China, long cultivated in Japan.

Edraianthus (From the Greek *hedraios*, sitting, and *anthos*, a flower, alluding to the sessile flowers.) GRASSY BELLS. A genus of 24 species of short-lived perennial herbs with narrow linear leaves and solitary or clustered, delicate, bell-shaped flowers surrounded by a whorl of bracts. Several species are grown ornamentally. *Distr.* SE Europe, Caucasus. *Fam.* Campanulaceae. *Class* D.

E. dalmaticus Stems to 10cm high. Flowers blue-violet. *Distr.* SE Europe.

E. graminifolius Leaves in rosettes. Flowers in rounded clusters. *Distr.* SE Europe.

E. pumilio Low, tuft-forming. Flowers appear from amongst foliage, deep blue. *Distr.* SE Europe.

E. serpyllifolius Foliage forming tight mats. Flowers erect, relatively large, deep purple. *Distr.* SE Europe.

edulis, -e: edible.

effusus, -a, -um: loose, spreading, usually referring to the growth habit of the plant.

Egeria (After Egeria, a water nymph in Roman mythology.) A genus of 2 species of aquatic perennial herbs with bright green linear leaves and small unisexual flowers borne above the surface of the water. The male and female flowers are borne on separate plants and both have 3 green sepals and 3 white or yellow petals. *Distr.* Subtropical South America. *Fam.* Hydrocharitaceae. *Class* M.

E. densa Cultivated as an oxygenator of cold and tropical aquaria. *Distr.* South America, naturalized elsewhere.

E. naias Occasionally grown as an aquarium plant.

egg plant *Solanum melongena.*

eggs and bacon *Lotus corniculatus.*

Ehretia (After Georg Dionysius Ehret (1708–1770), German botanical artist who lived in London.) A genus of 50 species of deciduous and evergreen shrubs and trees with alternate simple leaves and panicles of tube- to bell-shaped, white, blue, or yellow flowers. *Distr.* Tropical and warm regions. *Fam.* Boraginaceae. *Class* D.

E. acuminata KODA WOOD, KODO WOOD. Deciduous tree, to 10m high. Exploited as a source of timber. *Distr.* Tropical Asia to Japan and Australia.

E. dicksonii Deciduous tree. Flowers small, white, fragrant. Grown as an ornamental. *Distr.* China and Taiwan.

E. thyrsifolia See *E. acuminata*

Ehrharta (After Jakob Friedrich Ehrhart (1742–95), pupil of Linnaeus.) A genus of 25–30 species of annual and perennial, spreading grasses with flat or in-rolled leaves and narrow panicles. Some species are grown as pasture grasses. *Distr.* South Africa, few naturalized in North America and Australia. *Fam.* Gramineae. *Class* M.

E. erecta A cultivated ornamental. *Distr.* S Africa.

Eichhornia (After J. A. F. Eichhorn (1779–1856), Prussian politician.) WATER HYACINTH, WATER ORCHID. A genus of 7 species of annual and perennial, free-floating or rooting aquatic herbs with rosettes of more or less round leaves on inflated stalks. The tubular blue flowers are borne on spikes or in panicles. *Distr.* Tropical South America, widely naturalized. *Fam.* Pontederiaceae. *Class* M.

E. crassipes WATER HYACINTH, WATER ORCHID. Although an attractive ornamental this plant is a serious weed of waterways in many parts of the tropics. Because of its very fast growth rate it has been suggested that it could be used as a source of biomass for the production of methane. When grown on sewage it can yield 800 kg dry matter/ha/day. It is used as pig fodder in Indonesia. *Distr.* Brazil, now widely naturalized throughout the tropics.

ejow *Arenga pinnata.*

Elaeagnaceae The Oleaster family. 3 genera and 45 species of much branched, often thorny shrubs with silver, brown, or golden hairs. The leaves are simple and leathery and

the flowers are regular with 2–8 fused sepals and no petals. Some species are grown as ornamentals. *Distr.* Temperate N regions to tropical Asia and Australia. *Class* D.
See: Elaeagnus, Hippophae, Shepherdia.

elaeagnifolius, -a, -um: from the genus *Elaeagnus*, and *folius*, leaved.

elaeagnos: Greek name of a willow.

Elaeagnus (From Greek *helodes*, growing in marshes, and *hagnos*, pure; the name was originally applied to a WILLOW with massed white fruit.) OLEASTER. A genus of about 40 species of evergreen or deciduous shrubs and trees frequently with spiny stems, simple alternate leaves, and small regular flowers that lack petals. Some species are grown as ornamentals and some are important for their fruit. *Distr.* Europe, Asia, North America. *Fam.* Elaeagnaceae. *Class* D.

E. angustifolia RUSSIAN OLIVE, TREBIZOND DATE, OLEASTER, WILD OLIVE, SILVER BERRY. Deciduous shrub or small tree. Leaves narrow, silver-scaly beneath. Flowers yellow, fragrant. *Distr.* SE Europe, W Asia.

E. argentea See *E. commutata*

E. commutata SILVER BERRY. Deciduous shrub. Leaves broad, silver on both surfaces. Flowers silver and yellow, fragrant. *Distr.* North America.

E. × ebbingei (After J. W. E. Ebbinge of Boskoop.) A hybrid of *E. macrophylla* × *E. pungens*. Evergreen shrub. Flowers cream-white. Several cultivars are available. *Distr.* Garden origin.

E. macrophylla Bushy evergreen shrub. Leaves broad oval, silver beneath. Flowers cream-yellow, fragrant. Fruit red. *Distr.* Korea, Japan.

E. multiflora Deciduous shrub. Flowers solitary, fragrant. Fruit dark-red, edible. *Distr.* Japan, China.

Elaeis (From Greek *elaia*, olive tree, alluding to the oily fruits.) A genus of 2 species of small to medium-sized palms with feather-shaped, spiny-stalked leaves and much-branched, unisexual inflorescences. *Distr.* South America, Tropical Africa. *Fam.* Palmae. *Class* M.

E. guineensis OIL PALM, AFRICAN OIL PALM, MACAW FAT. This species is the main source of palm oil which is extracted from the fruits and used in the manufacture of a variety of products from margarine and soap to candles and diesel substitutes. *Distr.* W Africa, now widely planted throughout the tropics.

Elaeocarpaceae A family of 9 genera and about 200 species of shrubs and trees with simple, opposite, or alternate leaves. The flowers have 4–5 sepals, 4–5 or no petals, and numerous stamens. *Distr.* Tropical, subtropical and occasionally temperate regions but not Africa. *Class* D.
See: Aristotelia, Crinodendron, Elaeocarpus, Vallea.

Elaeocarpus (From Greek *elaia*, olive, and *karpos*, fruit.) A genus of about 60 species of evergreen trees and shrubs typically with simple alternate leaves and racemes of small fragrant flowers that bear 3-5 fringed petals and numerous stamens. Several species are grown as tender ornamentals and several have edible fruits. *Distr.* Tropical and warm regions of the Old World except Africa. *Fam* Elaeocarpaceae. *Class* D.

E. serratus CEYLON OLIVE. Fruit used in cooking. *Distr.* India, Malaysia.

E. sphaericus BREAD OF INDIA TREE. Small tree. Flowers white, fragrant. Fruit round, purple. *Distr.* India, Malaysia.

Elaphoglossum (From Greek *elaphos*, stag, and *glossa*, tongue, alluding to the shape of the leaves.) A genus of about 400 species of epiphytic ferns with simple leaves. Several species are grown as ornamentals. *Distr.* Tropical and subtropical regions. *Fam.* Lomariopsidaceae. *Class* F.

E. crinitum ELEPHANT'S EAR FERN. *Distr.* West Indies, Central America.

elasticus, -a, -um: elastic, flexible.

Elatinaceae A family of 2 genera and 32 species of annual to perennial herbs and occasionally subshrubs typically with opposite simple leaves and regular inconspicuous flowers. Some species occur as weeds of irrigation ditches and rice fields. *Distr.* Temperate and tropical regions. *Class* D.

elatior: taller.

elatus, -a, um: tall.

elder *Sambucus, S. nigra* **American** ~ *S. pubens* **American red** ~ *S.*

canadensis **black** ~ *S. nigra* **blue** ~ *S. caerulea* **box** ~ *Acer negundo* **common** ~ *Sambucus nigra* **Dane's** ~ *S. ebulus* **dwarf** ~ *S. ebulus* **European** ~ *S. nigra* **european red** ~ *S. racemosa* **ground** ~ *Aegopodium podagraria* **red-berried** ~ *Sambucus pubens*, *S. racemosa* **stinking** ~ *S. pubens* **sweet** ~ *S. canadensis* **yellow** ~ *Tecoma stans*.

elderberry *Sambucus*, *S. nigra* **blue** ~ *S. caerulea*.

elecampane *Inula helenium*.

elegans: elegant.

elegantissimus, -a, -um: very elegant.

elegantulus, -a, -um: elegant.

Eleocharis (From Greek *helodes*, growing in marshes, and *charis*, grace.) SPIKE RUSH. A genus of about 150 species of aquatic and marginal, tuberous, rush-like herbs with much reduced leaves and very small, wind-pollinated flowers borne in solitary spikelets. Many species are used for matting and for women's skirts in New Guinea; some species have edible tubers. *Distr.* Cosmopolitan. *Fam.* Cyperaceae. *Class* M.

E. acicularis HAIR GRASS, NEEDLE SPIKE RUSH, SLENDER SPIKE RUSH. Rootstock very thin, mat-forming. Stems tufted. *Distr.* Subarctic and temperate regions of the N hemisphere.

E. dulcis WATER CHESTNUT, CHINESE WATER CHESTNUT. Grown in a similar way to lowland rice, in flooded fields, but harvested for its crunchy edible tubers that are a major vegetable of Chinese cooking. *Distr.* Asia.

E. palustris CREEPING SPIKE RUSH. Roots horizontal, creeping. Stems stout. *Distr.* North America, Europe, Asia.

elephantipes: resembling an elephant's foot.

elephant's ear *Colocasia*, *Bergenia*.

elephant's ear plant *Alocasia*.

elephant's ears *Caladium*.

elephant's foot *Dioscorea elephantipes*.

elephant's foot tree *Beaucarnea recurvata*.

elephant wood *Bolusanthus speciosus*.

Elettaria (From a vernacular name used in Malabar.) A genus of about 6 species of rhizomatous herbs with 2 ranks of leaves and inflorescences on long prostrate stems. *Distr.* Tropical Asia. *Fam.* Zingiberaceae. *Class* M.

E. cardamomum CARDAMOM, CEYLON CARDAMOM, MALABAR CARDAMOM. Widely cultivated in Asia for spicy seeds used in medicine and cooking, imported to Europe since the Roman period. *Distr.* India.

Eleusine (From Eleusis, site of a temple to Ceres, goddess of fertility.) A genus of 9 species of annual and perennial stoloniferous grasses with folded or flat leaves and palmately arranged racemes of spikelets. *Distr.* E and NE Africa, South America. *Fam.* Gramineae. *Class* M.

E. coracana FINGER MILLET, CORACAN, AFRICAN MILLET. Staple cereal crop of parts of Africa and India, used in gruels, fermenting, and for flour. *Distr.* Africa, cultivated since antiquity.

Eleutherococcus (From the Greek *eleutheros*, free, and *kokkos*, pip, alluding to the fruits.) A genus of 30 species of prickly shrubs and trees with palmately compound leaves and clusters of small, pale green flowers. Several species are grown as ornamentals and some are a source of timber. *Distr.* E and SE Asia, Himalaya. *Fam.* Araliaceae. *Class* D.

E. senticosus SIBERIAN GINSENG. Source of a tonic. *Distr.* NE Asia.

E. sieboldianus Deciduous shrub. Leaves glossy green. Variegated cultivars are available. *Distr.* E China, introduced Japan.

eleven o'clock plant *Portulaca grandiflora*.

elfin herb *Cuphea hyssopifolia*.

Elliottia (After Stephen Elliott (1771–1830), American botanist.) A genus of 1 species of deciduous shrub or rarely small tree with alternate, simple, thin leaves and sweet-scented flowers that bear 4, rarely 5, free white petals. *Distr.* North America. *Fam.* Ericaceae. *Class* D.

E. racemosa Grown as an ornamental.

ellipsoidalis, -e: ellipsoidal.

ellipticus, -a, -um: elliptic.

elm *Ulmus*, Ulmaceae **Caucasian**

~ *Zelkova carpinifolia* **Chinese** ~ *Ulmus parvifolia* **Dutch** ~ *U. x hollandica* **English** ~ *U. procera* **water** ~ *Planera aquatica* **wych** ~ *Ulmus glabra*.

Elmera (After A.D.E. Elmer (1870–1942), American botanist.) A genus of 1 species of hairy perennial herb with basal kidney-shaped leaves and loose racemes of small white-green flowers. *Distr.* USA (Washington state). *Fam.* Saxifragaceae. *Class* D.

E. racemosa A cultivated ornamental.

Elodea (From Greek *helodes*, growing in marshes.) WATERWEED, PONDWEED, DITCHMOSS. A genus of about 12 species of submerged aquatic herbs with long rooting stems or rhizomes and whorls of 3 linear-oblong leaves. The unisexual flowers have 3 sepals, 3 petals and 3 whorls of 3 stamens. The male and female flowers are usually borne on separate plants. *Distr.* North and subtropical South America, widely naturalized elsewhere. *Fam.* Hydrocharitaceae. *Class* M.

E. callitrichoides *Distr.* South America, naturalized British Isles.

E. canadensis CANADIAN POND-WEED, WATERWEED. Male flowers break from plant whilst still in bud and float to the surface where they open. Female flowers extend ovary to reach surface of water for fertilization. A widely used oxygenator for aquaria. *Distr.* North America, introduced British Isles 1836, initially spread widely, now scarce.

E. crispa See *Lagarosiphon major*

E. nuttallii (After Thomas Nuttall (1786–1859) who collected plants in North America.) *Distr.* North America, introduced British Isles 1966.

elongatus, -a, -um: lengthened.

Elsholtzia (After Johann Sigismund Elsholtz (1623–88), German physician.) A genus of about 35 species of annual and perennial herbs and shrubs with simple opposite leaves and dense whorls of 2-lipped flowers in which the upper lip often forms a hood. Several species are grown ornamentally. *Distr.* Temperate areas of the Old World and mountainous regions of the tropics. *Fam.* Labiatae. *Class* D.

E. fruticosa Much branched, aromatic subshrub. Flowers pink or white. *Distr.* Nepal, China.

E. stauntonii (After Sir George Staunton (1737–1801) who travelled in China.) Subshrub. Flowers in a somewhat 1-sided inflorescence, small, dark pink. *Distr.* N China.

Elymus (From *elymos*, the Greek name for millet.) WILD RYE, LYME GRASS. A genus of about 100 species of rhizomatous perennial grasses typically with flat leaves and spike-like inflorescences. Several species are cultivated as ornamentals for their blue-tinged foliage. *Distr.* N temperate regions. *Fam.* Gramineae. *Class* M.

E. arenarius See *Leymus arenarius*

E. canadensis CANADIAN WILD RYE. Leaves tinged blue-green. Inflorescence pendent. Grown as an ornamental. *Distr.* Canada and USA.

emarginatus, -a, -um: shallowly notched.

Emblingiaceae A family of 1 genus and 1 species of prostrate subshrub. This family is sometimes included within the family Polygalaceae. *Distr.* W Australia. *Class* D.

Embothrium (From Greek *en*, in, and *bothrion*, a small pit; the anthers are borne in small pits.) A genus of 8 species of evergreen shrubs and trees with simple leathery leaves and clusters or racemes of bisexual flowers. *Distr.* Temperate South America. *Fam.* Proteaceae. *Class* D.

E. coccineum FIRE BUSH, CHILEAN FIRE BUSH, CHILEAN FLAME FLOWER. Variable shrub or small tree with bright red flowers. Grown as a half-hardy ornamental. *Distr.* Chile, W Argentina.

emersus, -a, -um: raised above, standing out.

Emilia (Derivation obscure but presumably commemorative.) A genus of 24 species of annual herbs with simple or lobed leaves and flowers in daisy-like heads that lack distinct rays. Several species are grown as ornamental. *Distr.* Tropical regions of the Old World, 3 species are pantropical weeds. *Fam.* Compositae. *Class* D.

E. coccinea TASSEL FLOWER. Flowers scarlet. *Distr.* Tropical Africa.

E. javanica In horticulture this name is often misapplied to plants of *E. coccinea* although it is strictly a synonym of *E. sonchifolia*.

E. sonchifolia Flowers purple-red. *Distr.* Tropical Asia and Africa.

Eminium (Greek name used by Dioscorided.) A genus of 7 species of tuberous herbs that bear their flowers and fruit on a spadix at ground level. *Distr.* E Mediterranean to Central Asia. *Fam.* Araceae. *Class* M.

Emmenopterys (From Greek *emmenes*, enduring, and *pteryx*, a wing, alluding to the flowers which sometimes bear a single enlarged sepal.) A genus of 2 species of deciduous trees with simple, opposite, leathery leaves and panicles of bell- or funnel-shaped, 5-lobed flowers. *Distr.* SE Asia, China. *Fam.* Rubiaceae. *Class* D.

E. henryi (After Augustine Henry (1857–1930), who collected plant in China.) Small tree. Flowers large, white. Grown as an ornamental but rarely flowers. *Distr.* Central and SW China, Burma, Thailand.

emodi: of the Himalaya.

Empetraceae The Crowberry family. 3 genera and 5 species of heath-like evergreen shrubs with small overlapping leaves and inconspicuous flowers that bear 4–6 tepals in 2 similar whorls. *Distr.* N temperate regions plus the Andes and S Atlantic Islands. *Class* D.
See: Empetrum.

empetrifolius, -a, -um: from the genus name *Empetrum*, and *folius*, leaved.

empetriformis, -e: resembling the genus *Empetrum*.

Empetrum (From Greek *en*, on, and *petros*, a rock, alluding to its habitat.) A genus of 2 species of evergreen heath-like shrubs with small whorled leaves and small unisexual flowers that bear 3 sepals and 3 stamens but no petals. The fruit is a round fleshy drupe. *Distr.* N temperate and Arctic regions, S Andes, Falkland Islands, Tristan da Cunha. *Fam.* Empetraceae. *Class* D.

E. nigrum CROWBERRY, BLACK CROWBERRY, CRAKE BERRY, CURLEW BERRY, MONOX. Fruit glossy black, edible. *Distr.* N temperate regions.

emu bush Myoporaceae.

Encephalartos (From Greek *en–*, within, *kephale*, the head, and *artos*, bread.) A genus of about 30 species of cycads with subterranean or erect stems and pinnate leaves to 3m long. *Distr.* Tropical and S Africa. *Fam.* Zamiaceae. *Class* G.

E. altensteinii PRICKLY CYCAD. To 7m high. Leaves spiny. Grown as an ornamental. *Distr.* South Africa.

Encyclia (From Greek *enchos*, to encircle, alluding to the lip which encircles the column.) A genus of about 150 species of epiphytic and lithophytic orchids with fleshy or leathery leaves from fat or thin pseudobulbs. The showy flowers are borne in typically pendulous inflorescences and have similar sepals and petals which are spreading or reflexed; the lip is simple or 3-lobed and sometimes encloses the column. A number of species and a great many hybrids are cultivated as ornamentals. *Distr.* Central and South America. *Fam.* Orchidaceae. *Class* M.

E. adenocaula Inflorescence a many-flowered panicle to 1m long. Flowers pink to pale purple with dark markings. *Distr.* W Mexico.

E. alata Inflorescence to 1.5m long, many-flowered. Flowers yellow-green, marked purple. *Distr.* S Mexico to Costa Rica.

E. baculus Inflorescence 2–3 flowered. Flowers large, cream-white, fragrant. *Distr.* S Mexico to Brazil.

E. brassavolae Flowers large, green-yellow, lip rose-purple. *Distr.* S Mexico and W Panama.

E. citrina Flowers 1–2, yellow-gold. *Distr.* Mexico.

E. cochleata COCKLE ORCHID, CLAMSHELL ORCHID. Flowers yellow-green, marked purple, with a black-purple lip. *Distr.* Florida to S and W Mexico, Colombia and Venezuela.

E. cordigera Flowers purple-brown or purple-green, lip cream, flushed red. *Distr.* S Mexico to Colombia and Venezuela.

E. fragrans Flowers to 2cm across, white-green striped maroon, fragrant. *Distr.* Greater Antilles, S Mexico to Brazil.

E. mariae Inflorescence short, few-flowered. Flowers large, green with a white lip. *Distr.* E Mexico.

E. polybulbon Flowers solitary, pale-yellow, flushed brown. *Distr.* S Mexico, Guatemala, Honduras, Cuba and Jamaica.

E. prismatocarpa Flowers numerous, fragrant, yellow-green spotted black-purple. *Distr.* Costa Rica, Panama.

E. radiata Flowers cream, striped maroon, fragrant. *Distr.* Central America.

E. selligera Inflorescence to 1m, many-flowered. Flowers fragrant, green, tinged red, lip white. *Distr.* Mexico and Guatemala.

E. tampensis Inflorescence to 75cm, many-flowered. Flowers yellow-green, lip white, marked maroon. *Distr.* USA (Florida).

E. vitellina Flowers showy, vermilion to scarlet. *Distr.* S Mexico and Guatemala.

endive *Cichorium endivia.*

Endymion See *Hyacinthoides*

Engelmannia (After George Engelmann (1809–84), American physician and botanist.) A genus of 1 species of erect perennial herb with pinnately lobed leaves and flowers in daisy-like heads that bear golden yellow rays. *Distr.* North America. *Fam.* Compositae. *Class* D.

E. pinnatifida Grown as a hardy ornamental.

Enkianthus (From Greek *enkyos*, pregnant, and *anthos*, a flower; in *E. quinqueflorus* each flower appears to bear another inside it.) A genus of 10 species of deciduous, pagoda-shaped shrubs with small drooping flowers. *Distr.* Himalaya to Japan. *Fam.* Ericaceae. *Class* D.

E. campanulatus Shrub with cream to pink flowers. *Distr.* Japan.

E. cernuus Deciduous shrub with white flowers. *Distr.* Japan.

E. chinensis Shrub with cream and rose flowers. *Distr.* W China and Upper Burma.

E. deflexus Shrub or small tree with yellow-red flowers. *Distr.* Himalaya and W China.

E. perulatus Deciduous shrub with white flowers. *Distr.* Japan.

enneaphyllus, -a, -um: with nine leaves.

ensatus, -a, -um: sword-like.

Ensete (From the Amharic name *anset.*) A genus of about 7 species of very large tree-like herbs similar to BANANA (*Musa*) but lacking leaf stalks and flowering only once before dying. *Distr.* Tropical Africa and Asia. *Fam.* Musaceae. *Class* M.

E. ventricosum ABYSSINIAN BANANA, ETHIOPIAN BANANA. Grown as a staple food crop in parts of Ethiopia, the stems and corms being boiled or fermented. Fibre from the stems is used for sacking and cordage. *Distr.* Ethiopia to Angola.

ensete: a local African name for *Musa*.

ensifolius, -a, -um: with sword-shaped leaves.

ensiformis, -e: sword-shaped.

Entelea (From Greek *enteles*, perfect, alluding to the fact that all the stamens are fertile.) A genus of 1 species of evergreen shrub or small tree with simple alternate leaves and cymes of small white flowers that bear 4–5 petals and numerous stamens. *Distr.* New Zealand. *Fam.* Tiliaceae. *Class* D.

E. arborescens The source of a very lightweight timber.

enulae *Inula helenium.*

Eomecon (From Greek *heoros*, eastern, and *mekon*, poppy, alluding to its distribution.) A genus of 1 species of rhizomatous perennial herb with tuft-forming, kidney-shaped leaves and poppy-like, nodding, white flowers that bear 2 sepals, 4 petals, and numerous stamens. *Distr.* E China. *Fam.* Papaveraceae. *Class* D.

E. chionanthum SNOW POPPY. Grown as an ornamental.

Epacridaceae The Australian Heath family. 31 genera and 400 species of shrubs and small trees with simple narrow leaves and small flowers. This family is related to Ericaceae and occupies a similar ecological position in Australia and parts of Asia. *Distr.* Mostly Australasia but also tropical Asia, the Pacific Islands and South America. *Class* D. *See: Cyathodes, Dracophyllum, Leucopogon, Richea, Sphenotoma, Styphelia, Trochocarpa.*

epaulette tree *Pterostyrax*
fragrant ~ *P. hispida.*

Ephedra (From a Greek name used for a similar plant, *hippuris.*) JOINT FIR. A genus of about 40 species of coniferous shrubs, climbers, and small trees with whorls of 3 scale-like leaves and numerous green branches. The female cones bear of 1–3 flower-like structures consisting of an ovule surrounded by an integument and subtended by 2–4 bracts. Used medicinally as a source of ephedrine. *Distr.* N temperate regions, temperate South America and tropical Asia. *Fam.* Ephedraceae. *Class* G.

E. distachya Shrub to 1m high. *Distr.* S Europe to Siberia.

E. gerardiana Creeping shrub to 5cm high. *Distr.* China, Himalaya.

Ephedraceae A family of 1 genus and 40 species. *Class* G.
See: Ephedra.

ephedroides: from the genus *Ephedra*, with the ending *-oides*, indicating resemblance.

ephemerus, -a, -um: ephemeral.

Epidendrum (From Greek *epi*, upon, and *dendron*, a tree, alluding to their epiphytic habit.) A genus of about 500 species of terrestrial and epiphytic orchids with cane-like stems or pseudobulbs. The flowers have similar sepals and petals and a simple or 3-lobed lip that often encloses the column. A number of species and hybrids are cultivated as tender ornamentals. *Distr.* Tropical America. *Fam.* Orchidaceae. *Class* M.

E. ciliare Pseudobulbs tufted. Flowers large, white-green to pale yellow. *Distr.* Central America, N South America, W Indies.

E. ibaguense Stems to 2m, scrambling. Flowers showy, in a pyramidal raceme, orange-yellow to red or white. *Distr.* Central America, N South America, W Indies.

E. medusae See *Nanodes medusae*

E. nocturnum Stems to 1m. Flowers few, large, showy. *Distr.* USA (Florida), Central America, N South America, W Indies.

E. parkinsonianum (After John Parkinson (1772–1847), Consul-General in Mexico, who sent plants to the Royal Botanic Gardens, Kew.) Stems short. Leaves flesh. Flowers showy and fragrant. *Distr.* Central America.

E. stamfordianum Pseudobulbs 2-leaved. Flowers fragrant, yellow-green to brown, marked red-brown. *Distr.* Central America to Venezuela.

Epigaea (From Greek *epi*, on and *gaia*, the earth, referring to their creeping habit.) A genus of 3 species of small, creeping, evergreen shrubs with scented flowers. *Distr.* Japan, E USA, Caucasus and E Turkey. *Fam.* Ericaceae. *Class* D.

E. asiatica Flowers rose-coloured. *Distr.* Japan.

E. repens TRAILING ARBUTUS, MAY-FLOWER. Flowers white to pink. Several cultivars available. *Distr.* North America.

Epigeneium (From Greek *epi*, upon, and *geneion*, chin, alluding to the position of the lateral sepals and petals in relation to the column foot.) A genus of 35 species of epiphytic orchids with 2-leaved pseudobulbs and showy waxy flowers. The sepals and petals are similar and borne in the same vertical plane, the lip is 3-lobed, broad and motile. Several species and hybrids are cultivated as tender ornamentals. *Distr.* E Asia. *Fam.* Orchidaceae. *Class* M.

E. lyonii Flowers to 12cm across, sepals and petals triangular, yellow-white, flushed red, lip red. *Distr.* Philippines.

Epilobium (From Greek *epi*, upon, and *lobos*, a pod; the petals are borne on a pod-like ovary.) WILLOW-HERB. A genus of about 200 species of perennial herbs and subshrubs with simple, spirally arranged, opposite or whorled leaves, and red to magenta or yellow flowers that bear 4 sepals, 4 fused petals, and 8 stamens. Several species and cultivars are grown ornamentally. *Distr.* Temperate regions, especially W North America. *Fam.* Onagraceae. *Class* D.

E. angustifolium ROSE BAY WILLOWHERB, GREAT WILLOWHERB, FRENCH WILLOW, FIREWEED, WICKUP. Erect perennial. Flowers profuse, purple-pink, rarely white. This plant has increased greatly in the British Isles in the latter half of the 20th century; this is associated with its colonizing the bomb-sites of World War II although it may also be due to introduced forms from North America. It is considered a weed of some agricultural crops but the leaves were formerly eaten as a vegetable by North American Indians and made into tea by the Russians. *Distr.* N temperate regions.

E. canum HUMMINGBIRD'S TRUMPET, CALIFORNIA FUCHSIA. Small shrub to 60cm high. Flowers funnel-shaped, red. Grown as an ornamental with a number of cultivars having been raised. *Distr.* SW USA, Mexico.

Epimedium (From *epimedion*, a Greek name for a different plant.) BARRENWORT, BISHOP'S HAT, BISHOP'S MITRE. A genus of about 25 species of rhizomatous herbs with 3-lobed leaves and clusters of pendulous regular flowers which bear 8 outer tepals and 4 smaller, nectar-producing inner tepals. A number of species and hybrids are grown as ornamentals. *Distr.* S Europe, NW Africa, SW Asia, China and Japan. *Fam.* Berberidaceae. *Class* D.

E. alpinum BARRENWORT. Leaves evergreen. Outer tepals red, inner tepals slipper-shaped, yellow. *Distr.* S Europe, naturalized in N Europe.

E. × cantabrigiense (Of Cambridge, where it was found by W.T. Stearn and R.Thoday in 1950.) A hybrid of *E. alpinum* × *E. pubigerum*. Flowers numerous, on red-hairy stalks. *Distr.* Garden origin.

E. grandiflorum Leaves bronze at first. Flowers with long spurs, deep-lilac or red. *Distr.* Japan, N China, N Korea.

E. macranthum See *E. grandiflorum*

E. × perralchicum A hybrid of *E. perralderianum* × *E. pinnatum*. Flowers yellow, on short spikes. *Distr.* Garden origin

E. perralderianum (After Henri René le Tourneux de la Perraudière (1831–61), French naturalist.) Leaves bronze when young. Flowers green with a yellow centre. *Distr.* Algeria.

E. pinnatum Leaves dark green, hairy at first. Flowers bright yellow. *Distr.* N Iran, Caucasus.

E. pubigerum Flowers cream-white or pink. *Distr.* SE Europe, Turkey, Black Sea Coast.

E. × rubrum A hybrid of *E. alpinum* × *E. grandiflorum*. Leaves marked brown-red when young. Flowers red. *Distr.* Garden origin.

E. × warleyense A hybrid of *E.alpinum* × *E. pinnatum*. Leaves tinged purple-red. Flowers orange and yellow, in erect racemes. *Distr.* Garden origin.

E. × youngianum A hybrid of *E. diphyllum* × *E. grandiflorum*. Flowers white. *Distr.* Japan.

Epipactis (A Greek name used for this genus.) HELLEBORINE. A genus of about 20 species of terrestrial rhizomatous orchids typically with 2-ranked leaves and spikes of flowers. The sepals and petals are spreading or curved forward and the lip is spurless and constricted at the base. Some species and cultivars are grown as hardy ornamentals. *Distr.* N temperate regions to tropical Africa, Thailand and Mexico. *Fam.* Orchidaceae. *Class* M.

E. atrorubens BROAD-LEAVED HELLEBORINE, DARK RED HELLEBORINE. To 1m high. Flowers purple-red, nodding, vanilla-scented. *Distr.* Europe to Central Asia.

E. gigantea To 90cm. Flowers pink-green with a red and yellow lip. *Distr.* NW America, Himalaya.

E. helleborine BROAD-LEAVED HELLEBORINE. To 1m high. Flowers in a one-sided spike, green-purple. *Distr.* Europe, N Africa, SW Asia to Himalaya.

E. palustris MARSH HELLEBORINE. To 50cm. Sepals maroon-green, petals cream-maroon, lip white, lined pink within. *Distr.* Eurasia, N Africa.

E. phyllanthes To 60cm high. Flowers yellow-green. *Distr.* NW and Central Europe.

Epipremnum (From Greek *epi*, upon, and *premnum*, tree-trunk, referring to the epiphytic habit.) A genus of about 10 species of evergreen lianas with leathery, long-stalked leaves and small flowers that are borne on a short spadix subtended by a yellow to purple spathe. Some species are grown as ornamentals while some are reported to have medicinal properties. *Distr.* Tropical SE Asia and the Pacific Islands. *Fam.* Araceae. *Class* M.

E. aureum DEVIL'S IVY, GOLDEN POTHOS, HUNTER'S ROBE. Leaves irregularly variegated yellow or white. *Distr.* Solomon Islands.

E. mirabile TONGA PLANT. Leaves pinnately cut. *Distr.* Polynesia to Australia and Malaysia.

E. pinnatum Leaves perforated. *Distr.* Malaysia, Indonesia, and New Guinea.

Episcia (From Greek *episkios*, shaded; these plants are shade-loving.) A genus of 6 species of perennial, sometimes epiphytic herbs with creeping stolons and pairs of simple leaves. The 5-lobed, funnel-shaped flowers are borne singly or in clusters. Several species and many hybrids are grown ornamentally. *Distr.* Tropical America. *Fam.* Gesneriaceae. *Class* D.

E. cupreata FLAME VIOLET. Creeping perennial. Leaves downy. Flowers scarlet, yellow within. *Distr.* Brazil, Venezuela, Columbia.

E. dianthiflora LACE FLOWER. Creeping perennial, Leaves thick, velvety. Flowers white, fringed. *Distr.* Costa Rica, Mexico.

E. lilacina Creeping perennial. Leaves woolly. Flowers scarlet. *Distr.* Costa Rica.

equestris: equestrian, of horses.

Equisetaceae A family of 1 genus and 29 species. *Class* F.
See: Equisetum.

equisetiformis, -e: resembling the genus *Equisetum*, horsetail.

Equisetum (From Latin *equus*, horse, and *saeta*, bristle, hair.) HORSETAIL, SCOURING RUSH. A genus of 25–30 species of rhizomatous perennial herbs with hollow jointed stems and whorls of branches. The leaves take the form of a toothed sheath that occurs at the nodes of the stem. The sporangia are aggregated into terminal strobili. Several species are grown ornamentally. These plants have an affinity for gold in solution and concentrate it more than any other plant, to a maximum 0.25g of gold per kg of plant tissue. *Distr.* Cosmopolitan except for Australia and New Zealand. *Fam.* Equisetaceae. *Class* F.

E. arvense COMMON HORSETAIL, FIELD HORSETAIL. To 60cm high. A common plant of ditches and damp fields. The shoots were formerly eaten like ASPARAGUS and used to make tea. *Distr.* Temperate regions.

E. hyemale SCOURING RUSH, ROUGH HORSETAIL. Stems to 1.5m high, rough. Formerly used as a scouring pad. *Distr.* N temperate Eurasia.

Eragrostis (From Greek *eros*, love, and *agrostis*, a grass.) LOVE GRASS. A genus of about 300 species of annual and perennial grasses with narrow leaves and loose or dense panicles. *Distr.* Temperate and tropical regions. *Fam.* Gramineae. *Class* M.

E. chloromelas Used to bind soil and prevent erosion. *Distr.* S Africa.

E. curvula WEEPING LOVE GRASS, AFRICAN LOVE GRASS. Grown as a pasture and fodder grass. *Distr.* South Africa.

E. mexicana MEXICAN LOVE GRASS. Annual. Inflorescence tinged purple. *Distr.* USA and Mexico.

E. tef TEFF, T'EF. Grown for edible seeds in parts of Africa. *Distr.* NE Africa.

E. trichodes Grown as an ornamental for its showy panicles. *Distr.* USA.

Eranthis (From Greek *er*, spring, and *anthos*, a flower.) WINTER ACONITE. A genus of 6–8 species of rhizomatous herbs with palmately lobed, basal leaves and solitary, regular, yellow or white flowers that bear 2 whorls of petal-like tepals and are subtended by a whorl of 3 leaves. Several species are grown as ornamentals. *Distr.* Eurasia. *Fam.* Ranunculaceae. *Class* D.

E. hyemalis WINTER ACONITE. Perennial. Flowers yellow. *Distr.* Europe, SW Asia, naturalized North America.

Ercilla (After Don Alonso de Ercilla (1533–95), Spanish botanist.) A genus of 2 species of evergreen ivy-like climbers with thick leathery leaves and dense spikes of small flowers. *Distr.* Chile and Peru. *Fam.* Phytolaccaceae. *Class* D.

E. spicata Climbing shrub to 6m high. Flowers purple-green. Occasionally grown as an ornamental. *Distr.* Chile and Peru.

E. volubilis See *E. spicata*

erectus, -a, -um: erect.

Eremolepidaceae A family of 3 genera and 13 species of semiparasitic epiphytic shrubs with spirally arranged, simple leaves and spikes or catkins of small unisexual flowers. *Distr.* Tropical America. *Class* D.

Eremosynaceae A family of 1 genus and 1 species of annual herb. This family is sometimes included within the family Saxifragaceae. *Distr.* W Australia. *Class* D.

Eremurus (From Greek *eremos*, desert, and *oura*, tail, alluding to their habitat and the shape of the inflorescence.) FOXTAIL LILY, GIANT ASPHODEL, DESERT CANDLE. A genus of about 35 species of rhizomatous herbs with rosettes of narrow basal leaves and tall column-like racemes of yellow, pink, or white flowers. These plants are characteristic of dry, high altitude grasslands of W Asia. Several species are cultivated as ornamentals. *Distr.* W and Central Asia, Europe. *Fam.* Asphodelaceae. *Class* M.

E. bungei See *E. stenophyllus*

E. himalaicus Flowering stem to 2.5m high. Flowers white. *Distr.* Afghanistan, NW Himalaya.

E. × isabellinus A hybrid of *E. stenophyllus* × *E. olgae* Numerous garden cultivars have been produced from this cross. *Distr.* Garden origin.

E. olgae Flowers white. *Distr.* N Iran, N Afghanistan, Soviet Central Asia.

E. robustus Flowering stem to 3m high. Flowers pink. *Distr.* Afghanistan, Soviet Central Asia.

E. stenophyllus Flowering stems to 1.5m high. Flowers yellow. *Distr.* Soviet Central Asia, Afghanistan, Iran, W Pakistan.

Eria (From Greek *erion*, wool, alluding to the covering on the perianth.) A genus of about 400 species of epiphytic and terrestrial orchids with small flowers. The lateral petals are small and the lip 3-lobed or entire. Some species and hybrids are cultivated as tender ornamentals. *Distr.* India, SE Asia and the Pacific Islands. *Fam.* Orchidaceae. *Class* M.

E. coronaria Flowers to 2.5cm across, waxy, slightly fragrant. *Distr.* E Himalaya to Thailand and Malay Peninsula.

E. javanica Flowers numerous, star-like, 4cm across, yellow-green, tinged maroon. *Distr.* Himalaya, Thailand, Malaysia, Indonesia and Philippines.

E. spicata Flowers very small, slightly fragrant, yellow. *Distr.* E Himalaya and Burma.

eriacanthus, -a, -um: with woolly spines.

erianthus, -a, -um: with woolly flowers.

Erica (From the Greek name *ereike* for HEATHER, derived from *ereiko* (to break); an infusion of the leaves was said to break bladder stones.) HEATH, HEATHER. A genus of about 665 species of evergreen shrubs and small trees with simple, often needle-like leaves and bell-shaped pendulous flowers. Many species are grown as cultivated ornamentals. *Distr.* South Africa (650 species are found S of the Limpopo of which 580 species are in South Cape Province, where 520 are endemic), tropical African mountains, Mediterranean, Europe. *Fam.* Ericaceae. *Class* D.

E. arborea TREE HEATH. Tree-like shrub with white fragrant flowers. Woody nodules at ground level traditionally used to make briar pipes. *Distr.* Mediterranean to tropical African mountains.

E. australis SPANISH HEATH. Erect shrub with red-pink flowers. *Distr.* Spain and Portugal.

E. baccans Erect, much-branched shrub with rose-pink flowers. *Distr.* South Africa.

E. bauera BRIDAL HEATH. Open shrub with small, white or pink flowers. *Distr.* South Africa (SW Cape).

E. carnea WINTER HEATH. Dwarf procumbent shrub with purple-pink flowers. *Distr.* E Europe.

E. ciliaris DORSET HEATH. Spreading shrub with red-pink flowers. Many garden cultivars available. *Distr.* W Europe.

E. cinerea BELL HEATHER. Low shrub with flowers varying from red through pink to white. Numerous garden cultivars are available. *Distr.* W Europe, covering large areas of drier moorland.

E. curviflora Erect, much branched shrub with red, orange, or yellow flowers. *Distr.* South Africa (S and SW Cape).

E. × darleyensis (Of Darley Dale, Derbyshire, where it was raised.) A hybrid of *E. erigena* × *E. herbacea*. *Distr.* Garden origin.

E. erigena IRISH HEATH. Erect bushy shrub with white to rose-pink flowers. *Distr.* W Europe.

E. glandulosa Erect to sprawling shrub with spreading, pink-orange flowers. *Distr.* South Africa (SW Cape).

E. herbacea See *E. carnea*

E. hibernica See *E. erigena*

E. × hyemalis FRENCH HEATHER, WHITE WINTER HEATHER. Small shrub with white-pink flowers. *Distr.* Origin unknown.

E. lusitanica PORTUGUESE HEATH. Somewhat tree-like shrub. Flowers white. *Distr.* W Iberian Peninsular to SW France, naturalized in SW England.

E. mediterranea See *E. erigena*

E. perspicua PRINCE OF WALES HEATH. Much-branched shrub with finely hairy flowers. *Distr.* South Africa (SW Cape).

E. scoparia BESOM HEATH. Slender erect shrub with green and red flowers. *Distr.* SW France, Spain, N Africa and the Canary Islands.

E. sessiliflora Erect, much-branched shrub with pale green flowers. *Distr.* South Africa (SW Cape).

E. terminalis CORSICAN HEATH. Erect shrub with rose-pink flowers. *Distr.* SW Europe.

E. tetralix CROSS-LEAVED HEATH. Dwarf shrub with pale pink flowers Numerous cultivars are available for the garden. A source of yellow dye. *Distr.* N and W Europe, covering great areas of wetter moorland.

E. umbellata DWARF SPANISH HEATH. Dwarf shrubs with rose to pink-purple flowers. *Distr.* W Iberian Peninsula.

E. vagans CORNISH HEATH. Erect shrub with lilac to white flowers. *Distr.* SW Europe.

E. × watsonii (After H. C. Watson* who discovered it near Truro.) A hybrid of *E. ciliaris* × *E. tetralix*. *Distr.* NW Europe.

E. × williamsii (After P. D. Williams, who introduced it to cultivation in 1910.) A hybrid of *E. vagans* × *E. tetralix*. *Distr.* Cornwall.

Ericaceae The Heath family. 116 genera and about 3000 species of shrubs and climbers plus a few herbs, trees and even saprophytes. The leaves are always simple and often evergreen. The flowers are usually regular and bisexual with 4–5 fused sepals and an equal number of fused petals. Several genera are important as ornamentals, notably *Rhododendron* and *Erica*. *Distr.* Widespread, typically on acidic soils, less common in Australia. *Class* D.
See: Agapetes, Agarista, Andromeda, Arbutus, Arctostaphylos, Botryostege, Bruckenthalia, Bryanthus, Calluna, Cassiope, Chamaedaphne, Chimaphila, Chiogenes, Craibiodendron, Daboecia, Elliottia, Enkianthus, Epigaea, Erica, Gaultheria, Gaylussacia, Harrimanella, Kalmia, Kalmiopsis, X Kalmiothamnus, X Ledodendron, Leiophyllum, Leucothoe, Loiseleuria, Lyonia, Menziesia, Oxydendrum, X Phylliopsis, Phyllodoce, X Phyllothamnus, Pieris, Pyrola, Rhododendron, Rhodothamnus, Tripetaleia, Tsusiophyllum, Vaccinium, Zenobia.

ericifolius, -a, -um: with leaves like the genus *Erica*.

ericoides: from the genus *Erica*, with the ending *-oides*, indicating resemblance.

erigenus, -a, -um: of Ireland.

Erigeron (From Greek *eri*, early, and *geron*, an old man, alluding to the fluffy white seed-heads.) FLEABANE. A genus of about 200 species of annual and perennial herbs with simple or lobed leaves and flowers in daisy-like heads that bear distinct, usually white rays. *Distr.* Cosmopolitan, especially North America. *Fam.* Compositae. *Class* D.

E. acer BLUE FLEABANE. Small, annual or biennial herb. Rays erect, blue-purple. *Distr.* Temperate N hemisphere.

E. alpinus Perennial. Flowers lilac. *Distr.* Mountains of S and Central Europe.

E. annuus DAISY FLEABANE. Erect perennial. Rays white. *Distr.* North America, naturalized in Europe.

E. atticus Erect perennial. Rays purple. *Distr.* Mountains of Europe.

E. aurantiacus ORANGE DAISY. Mat-forming perennial. Leaves velvety. Flowers orange-yellow. *Distr.* Turkestan.

E. aureus Small perennial herb. Rays yellow. *Distr.* NW USA.

E. borealis ALPINE FLEABANE. Perennial to 30cm. Flowers lilac. *Distr.* NW Europe.

E. caespitosus Perennial. Rays white to blue. *Distr.* North America.

E. delicatus Rhizomatous perennial herb. Rays blue. *Distr.* California.

E. glaucus BEACH ASTER, SEASIDE DAISY. Tufted, succulent, sprawling perennial. Rays purple-violet. *Distr.* W USA.

E. karvinskianus (After Wilhelm Friedrich Karwinski von Karwin (1780–1855), who collected in South America.) Subshrub. Rays purple-red. *Distr.* Mexico to Venezuela.

E. linearis Perennial herb to 25mm. Rays yellow. *Distr.* Central and SW USA.

E. mucronatus See *E. karvinskianus*

E. multiradiatus HIMALAYAN FLEABANE. Perennial. Rays long, purple-pink. *Distr.* Pakistan to Bhutan.

E. nanus Tufted perennial. Rays purple-white. *Distr.* W Central USA.

E. philadelphicus Biennial or short-lived perennial. Rays mauve. *Distr.* N USA and Canada.

E. pinnatisectus Tufted perennial. Rays purple-blue. *Distr.* Central USA.

E. pumilis Perennial, to 50cm high. Rays usually white. *Distr.* W USA.

E. rotundifolius See *Bellis rotundifolia*

E. speciosus Erect perennial. Rays blue. Numerous cultivars have been raised from this species. *Distr.* W North America.

E. uniflorus Small perennial. Rays white or pale purple. *Distr.* Arctic.

E. vagus Perennial, to 30cm high. Rays white-pink. *Distr.* W USA.

Erinacea (From Latin *erinaceus*, a hedgehog, alluding to its spines.) A genus of 1 species of spiny, clump-forming, evergreen subshrub with simple or trifoliate leaves and clustered, pea-like, purple flowers. *Distr.* W Mediterranean region. *Fam.* Leguminosae. *Class* D.

E. anthyllis HEDGEHOG BROOM, BRANCH THORN. Grown as an ornamental rock garden plant.

E. pungens See *E. anthyllis*

erinaceus, -a, -um: resembling a hedgehog.

eringoe *Eryngium.*

Erinus (From *erinos*, the classical Greek name for a kind of BASIL.) A genus of 2 species of perennial prostrate herbs with simple alternate leaves and racemes of small, almost regular, pink flowers that bear 5 fused petals, each of which is slightly notched. *Distr.* N Africa, N Spain, Pyrenees, Alps. *Fam.* Scrophulariaceae. *Class* D.

E. alpinus FAIRY FOXGLOVE, ALPINE BALSAM, LIVER BALSAM. Flowers pink-purple. Grown as ornamental ground cover for the rock garden. *Distr.* N Spain, Pyrenees, Alps.

Eriobotrya (From Greek *erion*, wool, and *botrys*, a bunch of grapes, alluding to the woolly inflorescences.) A genus of about 27 species of evergreen shrubs and small trees with simple leathery leaves and panicles of regular, small, white flowers that are followed by fleshy, few-seeded fruits. Several species are grown as ornamentals or for their edible fruits. *Distr.* Himalaya to E Asia and Indonesia. *Fam.* Rosaceae. *Class* D.

E. japonica LOQUAT, JAPANESE LOQUAT, JAPANESE MEDLAR, NISPERO. Grown throughout the tropical and warm temperate regions for its edible apricot-like fruits. In temperate regions it is grown as an ornamental and for its fragrant flowers, as the fruit is usually spoilt by frost. *Distr.* China, Japan.

eriocarpus, -a, -um: with woolly fruits.

Eriocaulaceae The Pipewort family. 9 genera and about 1200 species of perennial, occasionally annual, herbs with grass-like leaves and flowers borne in dense heads. Dried flower-heads are sold for decoration. *Distr.* Widespread but predominantly in New World tropical and subtropical regions. *Class* M.

Eriochloa (From Greek *erion*, wool, and *chloe*, grass.) CUP GRASS. A genus of about 30 species of annual and perennial clump-forming grasses with small bead-like structures in their inflorescences. Some species are used as fodder. *Distr.* Tropical and warm regions. *Fam.* Gramineae. *Class* M.

E. villosa HAIRY CUP GRASS. Annual. Grown as an ornamental. *Distr.* Asia.

Eriogonum (From Greek *erion*, wool, and *gonia*, joint.) WILD BUCKWHEAT. A genus of about 150 species of annual and perennial herbs and subshrubs with simple, opposite, alternate, or whorled leaves and heads of small flowers that bear 6 petal-like tepals and 9 stamens. Several species are grown as ornamentals; others have edible roots and leaves. *Distr.* W and S North America. *Fam.* Polygonaceae. *Class* D.

E. allenii UMBRELLA PLANT. Perennial herb to 1m high. Flowers bright yellow, borne in flat-topped clusters. *Distr.* USA (Virginia).

E. caespitosum Mat-forming perennial herb. Flowers yellow, tinged red. *Distr.* W USA.

E. umbellatum SULPHUR FLOWER. Perennial herb or subshrub to 1m high. Flowers borne in a loose umbel, sulphur yellow. *Distr.* W North America.

Eriophorum (From Greek *erion*, wool, and *phoreo*, to bear, alluding to the hairy fruits.) COTTON GRASS. A genus of 20 species of tufted rhizomatous herbs with grass-like leaves. The very small wind-pollinated flowers are borne in many-flowered spikelets that are arranged in umbels. The female flowers have tepals represented by woolly hairs that lengthen in fruit to aid dispersal. The hairy fruits are sometimes used as a stuffing for pillows. *Distr.* N temperate and Arctic regions, 1 species in South Africa. *Fam.* Cyperaceae. *Class* M.

E. angustifolium COMMON COTTON GRASS. Leaves grooved. A common plant of wet moorland. *Distr.* N Europe, North America, Siberia.

E. latifolium BROAD-LEAVED COTTON GRASS. Leaves to 8mm wide. *Distr.* Europe, Turkey, Siberia, North America.

Eriophyllum (From Greek *erion*, wool, and *phyllon*, leaf.) WOOLLY SUNFLOWER. A genus of 11 species of annual and perennial herbs and shrubs with simple or pinnately lobed leaves and flowers in daisy-like heads that bear a few distinct rays. *Distr.* W North America. *Fam.* Compositae. *Class* D.

E. lanatum Perennial. Rays yellow. *Distr.* W North America.

eriophyllus, -a, -um: with woolly leaves.

Eriopsis (From Greek *opsis*, resemblance, and the genus name *Eria*.) A genus of 6 species of epiphytic orchids with 2–3-leaved pseudobulbs and small flowers. The sepals are fleshy, the petals are narrow, and the lip 3-lobed. *Distr.* S Central America and tropical South America. *Fam.* Orchidaceae. *Class* M.

E. biloba Raceme to 1m long. Flowers yellow-cream. *Distr.* Tropical America from Costa Rica to Peru and Bolivia.

Eriospermaceae A family of 1 genus and 80 species of tuberous herbs with linear to lance-shaped, basal leaves and tall racemes of yellow flowers which bear 6 free tepals, 6 stamens, and a superior ovary. The fruit is a heart-shaped capsule. This family is sometimes included in the Liliaceae. *Distr.* Tropical and S Africa. *Class* M.

eriostachyus: with a woolly spike.

eriostemon: with woolly stamens.

Eritrichium: (From Greek *erion*, wool, and *trichos*, hair; green parts of the plants are woolly.) AMERICAN FORGET-ME-NOT. A genus of about 30 species of tuft-forming perennial herbs with simple leaves and cymes of funnel-shaped flowers. Several species are grown as ornamentals. *Distr.* N temperate regions. *Fam.* Boraginaceae. *Class* D.

E. canum Flowers to 7mm across, blue. *Distr.* W Himalaya.

E. elongatum Mat-forming. Flowers blue with a yellow centre. *Distr.* W North America.

E. nanum FAIRY FORGET-ME-NOT, KING OF THE ALPS. Densely tufted. Flowers blue with a yellow centre. *Distr.* Alps.

E. rupestre See *E. canum*

E. strictum See *E. canum*

erodioides: from the genus *Erodium*, with the ending *-oides*, indicating resemblance.

Erodium (From Greek *erodios*, a heron, alluding to the shape of the fruits which resemble the head and bill of a heron.) HERON'S BILL, STORK'S BILL. A genus of about 60 species of annual and perennial herbs and shrubs with pinnately lobed or divided leaves and regular or irregular flowers that bear their parts in 5s and may be borne singly or in umbels. This genus is closely related to the genus *Geranium* but has 5 rather than 10 stamens. Several species are grown as ornamentals, some are weedy and some edible. *Distr.* Europe, Central Asia, temperate Australia, tropical South America. *Fam.* Geraniaceae. *Class* D.

E. chamaedrioides See *E. reichardii*

E. chrysanthum Tufted perennial. Leaves finely divided, silver-hairy. Flowers cup-shaped, white to yellow. *Distr.* Greece.

E. corsicum Clump-forming perennial. Leaves grey-green. Flowers regular, flat, pink, with darker veins. *Distr.* Corsica, Sardinia.

E. manescavi (After Manescau (died 1875), Italian merchant and naturalist.) Tufted perennial. Leaves finely divided, blue-green. Flowers deep pink with an irregular, darker central patch. *Distr.* Pyrenees.

E. moschatum WHITE-STEMMED FILAREE, MUSK CLOVER. Annual to biennial. Leaves divided, slightly hairy. Flowers regular, pink. *Distr.* Europe, naturalized in the Americas.

E. reichardii ALPINE GERANIUM. Mound-forming perennial. Leaves very small, lobed. Flowers regular, white, with pink veins. *Distr.* Balearic Islands.

erosus, -a, -um: with an irregularly toothed or gnawed margin.

Erpetion See *Viola*

erubescens: blushing, becoming red.

Eruca (The Latin name of an edible species.) ROCKET SALAD, ROQUETTE, ITALIAN CRESS. A genus of 5 species of annual to perennial herbs with white, yellow, or violet flowers. *Distr.* Mediterranean. *Fam.* Cruciferae. *Class* D.

E. vesicaria SALAD ROCKET, JAMBA, GARDEN ROCKET. Annual herb. Eaten as salad greens. Oilseed substitute for RAPE in India. *Distr.* Mediterranean.

erumpens: breaking through, erupting.

Eryngium (From *eyringion*, the Greek name for *E. campestre*.) ERYNGO, ERINGOE, SEA HOLLY. A genus of 230 species of annual, biennial, and perennial herbs with basal rosettes of simple or divided, leathery, spiny leaves, and heads of small flowers. The heads are usually subtended by a whorl of spiny bracts and the foliage and bracts are often mottled white. Many species are grown ornamentally. *Distr.* Widespread. *Fam.* Umbelliferae. *Class* D.

E. agavifolium Clump-forming perennial. *Distr.* Argentina.

E. alpinum Erect perennial. Flower-heads and bracts suffused purple. *Distr.* Europe.

E. amethystinum Flower-heads and bracts blue. *Distr.* Europe.

E. bourgatii (After M. Bourgat who collected in the Pyrenees.) Clump-forming perennial. Flower-heads blue-green, borne on stalks above the foliage. *Distr.* Mediterranean region.

E. campestre ERYNGO. Perennial to 60cm high. Candied roots formerly used as a confectionery and medicinally. *Distr.* Europe, SW Asia.

E. eburneum Arching perennial. Flowering heads rounded, borne on long stems. *Distr.* South America.

E. giganteum MISS WILLMOTT'S GHOST. Clump-forming biennial or short-lived perennial. Flowering heads large, rounded, blue. *Distr.* Caucasus.

E. maritimum SEA HOLLY, SEA HOLM, SEA ERYNGIUM. Blue-grey perennial. Roots candied and formerly valued as tonic. *Distr.* Coastal region of Europe, naturalized in North America.

eryngium, sea *Eryngium maritimum.*

eryngo *Eryngium*, *E. campestre.*

Erysimum (From Greek *eryo*, to draw out, alluding to the blistering properties of some species.) A genus of about 80 species of annual, biennial, and perennial herbs with small, typically yellow flowers. *Distr.* Mediterranean, Asia. *Fam.* Cruciferae. *Class* D.

E. arkansanum See *E. asperum*

E. asperum WESTERN WALLFLOWER. Biennial or short-lived perennial with bright, yellow-orange flowers. *Distr.* North America.

E. capitatum COASTAL WALLFLOWER. Biennial. Flowers yellow-cream to brown-maroon. *Distr.* W North America.

E. cheiri WALLFLOWER. Perennial herb with yellow-orange, red-striped flowers. *Distr.* S Europe.

E. concinnum See *E. suffrutescens.*

E. perofskianum (After V. A. Perofsky (1794–*c.*1857).) Biennial or perennial cultivated as an annual with red-orange flowers. *Distr.* Afghanistan, Pakistan.

E. pulchellum Tufted perennial with golden yellow flowers. *Distr.* Greece, W Asia.

E. rupestre See *E. pulchellum*

E. semperflorens Subshrub with white flowers. *Distr.* Morocco.

E. suffrutescens Woody perennial with yellow flowers. *Distr.* Coastal California.

E. torulosum Biennial with yellow flowers. *Distr.* W North America.

Erythraea See *Centaurium*

Erythrina (From Greek *erythros*, red, alluding to the colour of the flowers.) CORAL TREE. A genus of 108 species of deciduous or evergreen shrubs and trees with alternate, pinnate, or trifoliate leaves and typically red, pea-like flowers. Several species are grown ornamentally or as shade trees and the flowers of some are cooked and eaten. *Distr.* Warm regions. *Fam.* Leguminosae. *Class* D.

E. crista-galli (Latin for cock's comb.) CORAL TREE, COMMON CORAL TREE, COCKSPUR CORAL TREE, CRY-BABY TREE. Shrub or small tree. Flowers dark red, with copious nectar; bird-pollinated. *Distr.* South America.

erythrocarpus, -a, -um: with red fruit.

Erythronium (From Greek *erythros*, red, the colour of the European species.) DOG'S TOOTH VIOLET, ADDER'S TONGUE, TROUT LILY, FAWN LILY. A genus of about 20 species of perennial herbs with tooth-like bulbs, simple, stalked, mottled leaves, and nodding flowers which bear 6 reflexed tepals and 6 stamens. Most of the species are grown as ornamentals. *Distr.* Temperate North America, 1 species in Eurasia. *Fam.* Liliaceae. *Class* M.

E. albidum WHITE DOG'S TOOTH VIOLET. Flowers solitary, pink-blue outside, white and yellow within. *Distr.* E Central North America.

E. americanum YELLOW ADDER'S TONGUE, TROUT LILY, AMBERBELL. Flowers solitary, marked yellow-brown. Medicinal. *Distr.* E North America.

E. dens-canis DOG'S TOOTH VIOLET. Flowers solitary, pink. Numerous ornamental cultivars available. The bulbs are a source of starch and are eaten with reindeer milk in Siberia and Mongolia. *Distr.* Eurasia.

E. grandiflorum AVALANCHE LILY. Leaves not mottled. Flowers bright yellow. *Distr.* NW North America.

E. mesochoreum See *E. albidum*

E. revolutum Leaf margins crisped. Flowers pink with a yellow centre. Several cultivars available. *Distr.* W North America.

erythropodus, -a, -um: red-stalked.

erythrosepala: with red sepals.

erythrospermus, -a, -um: with red seeds.

Erythroxylaceae The Coca family. 4 genera and 260 species of shrubs and trees with spirally arranged, simple, ovate leaves and small bell-shaped flowers that bear their parts in 5s. Some species are a source of timber, others of drugs. *Distr.* Tropical and subtropical regions. *Class* D.
See: Erythroxylum.

Erythroxylum (From Greek *erythros*, red, and *xylon*, wood.) A genus of about 250 species of shrubs with spirally arranged, simple leaves, and small bell-shaped flowers. *Distr.* Tropical regions, especially South America and Madagascar.

E. coca COCAINE PLANT. Grown as a hedging plant and for the leaves which are chewed locally as a mild narcotic; they are also the source of cocaine (the narcotic drug and anaesthetic). *Distr.* E Andes.

Escallonia (After Señor Escallon, 18th-century Spanish traveller in South America.) A genus of 39 species of evergreen and deciduous shrubs and trees typically with alternate, sessile, toothed leaves and panicles or racemes of funnel-shaped flowers that bear their parts in 5s. Many species and cultivars are grown ornamentally. *Distr.* S America, especially the Andes. *Fam.* Escalloniaceae. *Class* D.

E. leucantha Upright evergreen shrub. Leaves dark green. Flowers white, cup-shaped, in long racemes. *Distr.* Chile, Argentina.

E. macrantha See *E. rubra*

E. punctata See *E. rubra*

E. rubra Evergreen shrub. Flowers pink-red, borne in lax panicles. Used as a hedging plant. *Distr.* Chile, Argentina.

E. virgata Deciduous spreading shrub. Flowers white, cup-shaped, borne in racemes. *Distr.* Chile, Argentina.

Escalloniaceae A family of 20 genera and about 160 species of evergreen and deciduous trees and shrubs with simple leaves and small bisexual flowers that typically have their parts in 5s. This family is sometimes included within the families Saxifragaceae or Grossulariaceae. *Distr.* S hemisphere to the tropics, especially S America, Australasia. *Class* D.
See: Escallonia, Itea, Quintinia.

esculentus, -a, -um: edible.

Esmeralda A genus of 2 species of scrambling epiphytic orchids with strap-shaped leaves and flowers in racemes. The sepals and petals are similar and spreading, the lip is ridged and mobile. This genus is closely related to *Vanda*. *Distr.* Himalaya and SE Asia. *Fam.* Orchidaceae. *Class* M.

E. cathcartii Stem to 2m, pendent. Flowers yellow with brown stripes, lip white, striped red. *Distr.* E Himalaya.

esparto *Stipa tenacissima.*

estragon *Artemisia dracunculus.*

Etlingera (After A. E. Etlinger, an 18th-century botanist who described several species of *Salvia*.) A genus of 57 species of rhizomatous herbs with cane-like stems, simple leaves and irregular flowers. *Distr.* Sri Lanka to New Guinea. *Fam.* Zingiberaceae. *Class* M.

E. elatior TORCH GINGER, PHILIPPINE WAXFLOWER. Grown as an ornamental. *Distr.* Malaysia.

etruscus, -a, -um: of Tuscany.

Eucalyptus (From Greek *eu*, well, and *kalypto*, to cover, alluding to the sepals which form a lid over the flowers in bud.) A genus of about 450 species of aromatic woody plants ranging from dwarf shrubs to some of the tallest trees known at 100m high. Most species show a distinct progression of foliage types. Juvenile foliage leaves are typically broad and borne in opposite pairs that are often fused together. Adult foliage leaves are typically stalked, lance-shaped, and borne alternately. The flowers are small and bear a cap (consisting of the fused sepals and petals) that is shed to reveal the numerous stamens. This genus is of great ecological importance in Australia, often forming the dominant vegetation. A number of species are planted as fast-growing plantation trees and have become part of the landscape in many warm regions of the N hemisphere. As well as providing timber they are also a source of oil and fodder and are planted as ornamentals. *Distr.* Australia with a few extending to Malaysia. *Fam.* Myrtaceae. *Class* D.

E. citriodora LEMON-SCENTED GUM. Tree to 50m high. A source of a lemon-scented oil and structural timber. *Distr.* Queensland.

E. coccifera MOUNT WELLINGTON PEPPERMINT, TASMANIAN SNOW GUM. Shrub or tree to 10m high. Bark blue-grey and white, peeling. *Distr.* Tasmania.

E. dalrympleana BROAD-LEAVED KINDLING BARK, MOUNTAIN GUM. Tree to 25m high. Bark white, turning pink and peeling. *Distr.* Tasmania, SE Australia.

E. delegatensis ALPINE ASH. Tree to 40m. Bark fibrous. An important source of timber. *Distr.* SE Australia.

E. divaricata See *E. gunnii*

E. ficifolia RED FLOWERING GUM. Straggling tree to 10m high. Widely cultivated for showy panicles of red flowers. *Distr.* W Australia.

E. glaucescens TINGIRINGI GUM. Spreading tree to 12m high. Leaves silver-blue. *Distr.* SE Australia.

E. globulus TASMANIAN BLUE GUM, FEVER TREE, BLUE GUM. Tree to 50m. Most widely cultivated species and a major source of eucalyptus oil, firewood and pulp. *Distr.* Tasmania, SE Australia, naturalized California.

E. gunnii (After Ronald Campbell Gunn (1808–81), a magistrate and botanist in Tasmania.) CIDER GUM. Tree to 25m. Bark brown, peeling. *Distr.* Tasmania.

E. muelleriana YELLOW STRINGBARK. Tree to 40m high. A source of rot-resistant timber. *Distr.* S and E Australia.

E. niphophila See *E. pauciflora*

E. pauciflora CABBAGE GUM, WHITE SALLY, SNOW GUM, GHOST GUM. Spreading tree to 20m high. Bark white, peeling. *Distr.* Tasmania, SE Australia.

E. perriniana (After George Samuel Perrin (1849–1900), a forester in SE Australia.) SPINNING GUM, ROUND-LEAVED GUM. Small spreading tree or shrub to 6m high. *Distr.* SE Australia.

E. viminalis RIBBON GUM, WHITE GUM, MANNA GUM. Spreading tree to 50m high. A local source of timber. *Distr.* Tasmania, SE Australia.

eucalyptus Myrtaceae.

Eucharis (From Greek *eu*, good, and *charis*, attraction, alluding to the flowers.) A genus of about 20 species of bulbiferous perennial herbs with ovate to lance-shaped, stalked leaves and umbels of funnel-shaped, white-green flowers. Some species are grown as tender ornamentals. *Distr.* Tropical South America. *Fam.* Amaryllidaceae. *Class* M.

E. amazonica Flowers white, tepal lobes spreading. Emetic tea prepared from whole plant including bulbs. *Distr.* Peru, Ecuador.

E. candida Perianth lobes spreading or recurved. *Distr.* Colombia.

E. × grandiflora AMAZON LILY, EUCHARIST LILY, STAR OF BETHLEHEM. Flowers large, scented, white. *Distr.* Andes of Colombia and Peru.

E. sanderi Flowers funnel-shaped. *Distr.* Colombia.

Eucodonia (From Greek *eu*, good, and *kodon*, bell.) A genus of 2 species of small rhizomatous herbs with simple whorled leaves and solitary, bell- or funnel-shaped flowers. *Distr.* Central America. *Fam.* Gesneriaceae. *Class* D.

E. andrieuxii Flowers violet with a white throat. Grown as an ornamental. *Distr.* S Mexico.

Eucomis (From Greek *eu*, good, and *kome*, hair, alluding to the bracts at the top of the inflorescence.) PINEAPPLE LILY. A genus of about 10 species of bulbiferous herbs with strap-shaped leaves and flowers in terminal racemes topped with cluster of sterile leafy bracts. Several species are grown as ornamentals. *Distr.* S Africa. *Fam.* Hyacinthaceae. *Class* M.

E. autumnalis Leaf margins wavy. Flowers white. *Distr.* South Africa, Rhodesia, Malawi, Swaziland and Botswana.

E. comosa Leaves purple-spotted. *Distr.* South Africa.

E. punctata See *E. comosa*

E. undulata See *E. autumnalis*

Eucommia (From Greek *eu*, good, and *kommi*, gum, alluding to the fact that it is the only hardy tree that can produce rubber.) A genus of 1 species of small deciduous tree with simple leaves and clusters of unisexual flowers that lack sepals and petals. Latex is produced in the leaves; if they are broken and pulled slowly apart thin strands can be seen. *Distr.* Central China. *Fam.* Eucommiaceae. *Class* D.

E. ulmoides GUTTA PERCHA TREE. The bark is used as a tonic and medicinally for arthritis. The latex is the source of gutta percha.

Eucommiaceae A family of 1 genus and 1 species. *Class* D.
See: Eucommia.

Eucryphia (From Greek *eu*, well and *kryphios*, covered; the sepals form a protective cap which falls off as the flower opens.) A genus of 5 species of shrubs and trees with simple or pinnate, opposite leaves and regular, large, white flowers that bear 4 petals and numerous stamens. Several species and hybrids are grown ornamentally. *Distr.* SE Australia, Chile. *Fam.* Eucryphiaceae. *Class* D.

E. cordifolia ULMO, ROBLE DE CHILE. Evergreen tree to 15m. Leaves simple. A commercial source of timber and tannin. *Distr.* Chile.

E. glutinosa NIRRHE. Large deciduous shrub or tree. Leaves simple. Flowers fragrant. Several cultivars are available. *Distr.* Chile.

E. × hillieri: A hybrid of *E. lucida* × *E. moorei* . *Distr.* Garden origin.

E. × intermedia A hybrid of *E. glutinosa* × *E. lucida*. *Distr.* Garden origin.

E. lucida LEATHERWOOD, PINKWOOD. Small evergreen tree. Leaves simple, flowers fragrant. Grown for timber as well as ornamentally. *Distr.* Tasmania.

E. milliganii (After Joseph Milligan (1807–83), a Scottish surgeon who collected in Tasmania.) Similar to *E. lucida* but slower-growing. *Distr.* Tasmania.

E. moorei STINKWOOD, PINKWOOD, PLUM TREE. Evergreen tree to 15m. Leaves pinnate. *Distr.* SE Australia.

E. × nymansensis (Of Nymans, Sussex, where it originated.) A hybrid of *E. cordifolia* × *E. glutinosa*. Erect tree. Leaves pinnate and simple. Flowers numerous, appearing in late summer. *Distr.* Garden origin.

Eucryphiaceae A family of 1 genus and 5 species. *Class* D.
See: Eucryphia.

Eugenia (After Prince Eugene of Savoy (1663–1736), a patron of botany.) STOPPER. A genus of about 1000 species of evergreen shrubs and trees with simple opposite leaves and small regular flowers which bear 4–5 white petals and numerous stamens. The fruit is a juicy berry. Many species have edible fruits and a few are grown as ornamentals. *Distr.* Tropical regions, especially America. *Fam.* Myrtaceae. *Class* D.

E. uniflora SURINAM CHERRY. Shrub or small tree to 10m high. Widely cultivated in the tropical regions as a hedge-plant, for edible fruit used in jams, and the leaves which act as an insect-repellent when crushed. *Distr.* Tropical America.

eugenioides from the genus *Eugenia*, with the ending *-oides*, indicating resemblance.

eulalia *Miscanthus sinensis*.

Eulophia (From Greek *eulophos*, well-plumed, alluding to the prominent crests on the lips of many species.) INDIAN SALEP. A genus of about 250 species of terrestrial, occasionally epiphytic orchids with leaves that appear after flowering. The flowers are small or large with dissimilar sepals and petals and a 3-lobed lip. The tubers of some species provide Indian salep. *Distr.* Tropical regions, especially Africa. *Fam.* Orchidaceae. *Class* M.

E. guineensis Flowers green-purple with a lilac lip. *Distr.* Tropical Africa, from Gambia to Angola and Uganda.

E. horsfallii Flowers numerous, sepals green, petals pink, lip purple. *Distr.* Tropical Africa from Sierra Leone to Mozambique.

E. streptopetala Flowers numerous, sepals green, blotched brown, petals and lip yellow. *Distr.* S Africa, Zimbabwe.

Eulophiella (Diminutive of the genus name *Eulophia*.) A genus of 3–4 species of stout epiphytic orchids with long pseudobulbs and racemes of large flowers. *Distr.* Madagascar. *Fam.* Orchidaceae. *Class* M

E. elizabethae Flowers white with a large yellow blotch on the lip. *Distr.* Madagascar.

E. roempleriana Flowers pink with a more or less round lip. *Distr.* Madagascar.

Eumorphia (From Greek *eumorphos*, attractive.) A genus of 4 species of heath-like shrubs with small leaves and flowers in daisy-like heads that bear distinct rays. *Distr.* S Africa. *Fam.* Compositae. *Class* D.

E. sericea Much-branched shrub. Flowers yellow, rays white. *Distr.* South Africa.

Eunomia See *Aethionema*

Euodia See *Tetradium*

Euonymus (The classical Greek name for these plants.) A genus of 177 species of deciduous and evergreen, erect or scrambling shrubs and trees with simple, typically opposite leaves. The flowers are small and inconspicuous and bear 4–5 sepals and petals. The fruit is a capsule that splits to reveal a brightly coloured interior containing seeds enclosed in colourful arils. Several species and numerous cultivars are grown as ornamentals. *Distr.* N temperate regions, Australia. *Fam.* Celastraceae. *Class* D.

E. alatus WINGED SPINDLE TREE. Deciduous shrub. Stems winged. Arils orange-vermilion. *Distr.* China, Japan.

E. americanus STRAWBERRY BUSH. Erect deciduous shrub. Fruit pink, warty. Arils bright red. *Distr.* E USA.

E. europaeus SPINDLE TREE, COMMON SPINDLE TREE. Deciduous shrub. Flowers yellow-green. Inside of fruit pink-red. Arils orange. The timber has been used to manufacture spindles, violin bows, and gunpowder charcoal. The seeds are a source of yellow dye formerly used to colour butter. *Distr.* Europe, W Asia.

E. fortunei (After Robert Fortune (1812–80), who collected in China and introduced the TEA plant from China to India.) Evergreen shrub. Numerous ornamental cultivars available including variegated forms. *Distr.* E Asia.

E. hamiltonianus (After Francis Buchanan-Hamilton (1762–1829, director of the Calcuthon Botanic Garden.) Deciduous shrub or small tree. Used as BOX. substitute. *Distr.* Himalaya, E Asia.

E. japonicus Evergreen shrub or small tree. Pod pink. Seeds white. Aril orange. Numerous cultivars are available including variegated forms and house plants. *Distr.* Japan.

E. latifolius Deciduous erect shrub or small tree. Fruit winged, red. Seeds white. Aril orange. *Distr.* S Europe, Turkey.

E. radicans See *E. fortunei*

E. yedoensis See *E. hamiltonianus*

Eupatorium (After Eupator, King of Pontus, who used one of these plants as an antidote for poison.) A genus of 38 species of perennial herbs, subshrubs, and shrubs with simple leaves and flowers in daisy-like heads which lack distinct rays. This genus formerly included some 1200 species from the tropics, but these are now segregated into a number of other small genera. Several species are cultivated as ornamentals. *Distr.* E USA, Eurasia. *Fam.* Compositae. *Class* D.

E. ageratoides See *Ageratina altissima*

E. altissimum TALL THOROUGHWORT. Downy herb. Flowers white. *Distr.* USA.

E. aromaticum See *Ageratina aromatica*

E. cannabinum HEMP AGRIMONY. Perennial, to 2m high. Flowers white to mauve. Formerly important medicinally. *Distr.* Europe, N Africa to Central Asia.

E. fraseri See *Ageratina altissima*

E. ligustrinum See *Ageratina ligustrina*

E. maculatum JOE-PYE WEED. Perennial herb. Flower-heads in a flat topped cyme. *Distr.* E North America.

E. micranthum See *Ageratina ligustrina*

E. perfoliatum BONESET, THOROUGHWORT. Perennial herb. Flower-heads very large. *Distr.* SE USA.

E. purpureum TRUMPET WEED, GRAVEL ROOT. Tall perennial. Leaves vanilla-scented when crushed. Formerly used medicinally for urinary problems. *Distr.* E North America.

E. rugosum See *Ageratina altissima*

E. weinmannianum See *Ageratina ligustrina*

Euphorbia (After Euphorbus, physician of the 1st century AD, who used latex from some species for medicinal purposes.) SPURGE. A genus of about 1600 species of annual and perennial herbs, succulents, shrubs and trees with milky latex. The flowers are often highly reduced and in an inflorescence surrounded by a whorl of showy bracts. The succulent species occupy the same ecological position in Africa as the Cactaceae in America and may be very specialized. The latex of all species is toxic and can bring out skin allergies. Many species are cultivated as curiosities and ornamentals. *Distr.* Cosmopolitan, especially in warm regions. *Fam.* Euphorbiaceae. *Class* D.

E. amygdaloides WOOD SPURGE. Perennial herb. Stems clump-forming, erect. Flowers lime-green in loose clusters. *Distr.* Europe, SW Asia.

E. candelabrum Tree to 20m high. Branches ascending, fleshy, green, angled. Leaves small, soon falling. *Distr.* S and E Africa.

E. characias Hardy evergreen shrub. Leaves linear. Flowers green in large, rounded heads. *Distr.* Mediterranean region.

E. cyparissias CYPRESS SPURGE. Perennial herb. Leaves grey-green. Flowers lime-green, borne in loose umbels. Grown as ground ornamental cover. Formerly used as a cosmetic in the Ukraine. *Distr.* Europe, naturalized in North America.

E. fulgens SCARLET PLUME. Evergreen arching shrub. Leaves lance-shaped. Flowers with bright red bracts, borne in spike-like plumes. *Distr.* Mexico.

E. horrida AFRICAN MILK BARREL. Succulent spiny shrub. *Distr.* South Africa (Cape Province).

E. hyberna IRISH SPURGE. Rhizomatous perennial herb. Fruit warty, pink-brown. *Distr.* W Europe.

E. lathyris CAPER SPURGE, MOLE PLANT, MYRTLE SPURGE. Biennial herb with blue-green leaves. Seeds toxic but fruit used as a poor substitute for CAPER. Plant incorrectly believed to be mole-repellent. *Distr.* Europe, naturalized North America.

E. mammillaris CORNCOB CACTUS, CORN COB, CORKSCREW. Dwarf spiny succulent. *Distr.* S Africa.

E. milii (After M. Millius, Governor of the Isle of Bourbon where it was grown.) CROWN OF THORNS, CHRIST THORN, CHRIST PLANT. Spiny shrub with bright red or yellow bracts. Often grown as a house plant. *Distr.* Madagascar.

E. obesa GINGHAM GOLF BALL, LIVING BASEBALL. Highly reduced succulent. Stems spherical, ridged, striped purple. *Distr.* S Africa.

E. portlandica PORTLAND SPURGE. Perennial herb with bright green leaves. *Distr.* W Europe.

E. pulcherrima POINSETTIA, CHRISTMAS STAR, CHRISTMAS FLOWER, PAINTED LEAF, LOBSTER PLANT, MEXICAN FLAMELEAF. Shrub to 3m high. Bracts bright red, leaf-like. Frequently given as a pot-plant at Christmas. In order to produce plants of a suitable size and colour it is necessary to grow them in artificial day length and use growth retardants. *Distr.* Central America, Mexico.

E. robbiae See *E. amygdaloides*

Euphorbiaceae The Spurge family. 331 genera and about 8000 species of herbs, shrubs and trees typically with alternate simple or occasionally palmate leaves. The flowers are regular and unisexual, typically with 5 tepal, segments, although these may be missing or there may be an additional whorl of petals. Some of the species are highly developed succulents. The latex present in many of the species is of particular importance in the genus *Hevea* and a number of other rubber-producing genera. The family also contains several important food plants. *Distr.* Widespread. *Class* D.
See: Acalypha, Breynia, Codiaeum, Euphorbia, Hevea, Mercurialis, Monadenium, Pedilanthus, Ricinus, Sapium.

Euphroniaceae A family of 1 genus and 3 species of shrubs and trees. This family is sometimes included within the family Vochysiaceae or the Trigoniaceae. *Distr.* Colombia, Venezuela, N Brazil. *Class* D.

Eupomatiaceae A family of 1 genus and 2 species of shrubs and small trees with simple, spirally arranged leaves and showy flowers. The flowers do not bear distinct sepals and petals but have numerous stamens and a numerous unfused carpels. *Distr.* New Guinea and E Australia. *Class* D.

Euptelea (From Greek *eu*, well, and *ptelea*, elm.) A genus of 2 species of shrubs and small trees with simple, spirally arranged leaves and small flowers that lack sepals and petals but have 7–30 stamens and 6–18 primitive carpels. *Distr.* Himalaya, China, and Japan. *Fam.* Eupteleaceae. *Class* D.

E. franchetii See *E. pleiosperma*

E. pleiosperma Tree to 10m high. Leaves turn red in autumn. Grown as a hardy ornamental. *Distr.* E Himalaya, China.

Eupteleaceae A family of 1 genus and 2 species. *Class* D.
See: Euptelea.

europaeus, -a, -um: European.

Eurya (From the Greek *eurys*, broad, possibly alluding to the leaves of some species.) A genus of about 70 species of evergreen shrubs and trees with simple alternate leaves and clusters of small unisexual flowers that bear 5 petals. Several species are grown as ornamentals. *Distr.* Tropical and warm regions of Asia. *Fam.* Theaceae. *Class* D.

E. japonica Evergreen shrub or tree to around 10m high. Leaves dark, glossy green. Flowers white. *Distr.* Himalaya, Japan, SE Asia.

Eurychone (From Greek *eurys*, broad, and *chone*, funnel.) A genus of 2 species of epiphytic orchids with short stems and racemes of large scented flowers that bear funnel-shaped lips. *Distr.* Tropical Africa. *Fam.* Orchidaceae. *Class* M.

E. rothschildiana Flowers white, tinged green. *Distr.* Tropical Africa, from Sierra Leone to Uganda.

Euryops (From Greek *euryops*, with large eyes, alluding to the large flowers.) A genus of 97 species of annual and perennial herbs and shrubs with simple overlapping leaves and flowers in large daisy-like heads which may or may not bear distinct rays. Some species are cultivated as ornamentals. *Distr.* Africa to Arabia and Socotra. *Fam.* Compositae. *Class* D.

E. abrotanifolius Erect shrub. Rays yellow. *Distr.* South Africa.

E. acraeus Dense rounded shrub. Rays yellow. Cultivated ornamental. *Distr.* Drakensberg Mountains, South Africa.

E. chrysanthemoides Erect shrub, to 2m high. Rays long, narrow, yellow. *Distr.* South Africa.

E. evansii Subshrub or shrub, to 2m high. This name is frequently misapplied to *E. acraeus* in cultivation. *Distr.* South Africa.

E. pectinatus Fast-growing shrub, to 2m high. *Distr.* South Africa.

E. sericea See *Ursinia sericea*

E. virgineus Compact shrub, to 4m high. *Distr.* South Africa.

eurysiphon: with a broad tube.

Eustephia (From Greek *eu*, well, and *stephos*, crown, alluding to the circle of stamens.) A genus of 6 species of bulbiferous perennial herbs with erect linear leaves and an umbel of pendulous, red or green, funnel-shaped flowers. Several species are cultivated as ornamentals. *Distr.* Peru, Argentina. *Fam.* Amaryllidaceae. *Class* M.

E. coccinea Flowers warm red, green at base. *Distr.* Peruvian Andes.

E. jujuyensis (After the province in which it occurs, Jujuy.) Flowers bright orange-red. *Distr.* N Argentina.

Eustoma (From Greek *eu*, good, and *stoma*, mouth, alluding to the throat of the flower.) A genus of 3 species of annual, biennial and perennial herbs with opposite, simple, sessile leaves and 5- to 6-lobed, funnel- or bell-shaped flowers. *Distr.* S North America to N South America. *Fam.* Gentianaceae. *Class* D.

E. grandiflorum PRAIRIE GENTIAN. Annual or biennial. Flowers purple, blue or white, darker in the centre. Grown as an ornamental. *Distr.* North America.

E. russellianum See *E. grandiflorum*

evergreen, Chinese *Aglaonema.*

everlasting *Antennaria*, *Helichrysum*, *H. bracteatum*, *Gnaphalium* **Alpine** ~ *Helichrysum acuminatum* **button** ~ *H. scorpiodes* **golden** ~ *H. bracteatum* **green** ~ *Gnaphalium californicum* **orange** ~ *Helichrysum acuminatum* **pearly** ~ *Anaphalis*, *A. margaritacea* **pink** ~ *Gnaphalium ramosissimum.*

everlasting flower *Helichrysum* **yellow** ~ *H. arenarium.*

evodia See *Tetradium*

Evolvulus (From Latin *evolvere*, to untwist, alluding to the non-climbing habit compared with other members of the same family.) A genus of 98 species of annual and perennial herbs and subshrubs with simple leaves and small bell-shaped flowers. Several species are grown as ornamentals. *Distr.* Warm and tropical America, 2 species extending to the Old World. *Fam.* Convolvulaceae. *Class* D.

E. glomeratus Prostrate or decumbent herb. Leaves narrow. Flowers bright blue. *Distr.* Brazil.

Exacum (From Latin *ex*, out, and *agere*, to drive, alluding to the purgative properties of some of these plants.) A genus of about 25 species of annual and biennial herbs with simple opposite leaves and salver- to cup-shaped flowers. *Distr.* Tropical regions of the Old World. *Fam.* Gentianaceae. *Class* D.

E. affine PERSIAN VIOLET, ARABIAN VIOLET, GERMAN VIOLET. Annual or short-lived perennial. Flowers cup-shaped, violet-purple with bright yellow stamens. A popular pot-plant. *Distr.* Socotra.

exaltatus, -a, -um: very tall.

exasperatus, -a, -um: roughened, covered in short hard points.

excavatus, -a, -um: excavated, hollowed out in a curve.

excelsior: higher, taller.

excelsus, -a, -um: high, tall.

excisus, -a, -um: excised, cut out.

excorticatus, -a, -um: with peeling bark or lacking bark.

exiguus, -a, -um: weak, feeble, little.

eximeus, -a, -um: distinguished.

Exochorda (From Greek *exo*, outside, and *chorda*, cord, alluding to the position of fibres in the fruit.) PEARLBUSH. A genus of 4 species of deciduous shrubs with simple alternate leaves and racemes of regular, abundant, white flowers that are followed by star-shaped fruits. Several species are grown ornamentally. *Distr.* Central Asia, China. *Fam.* Rosaceae. *Class* D.

E. giraldii (After Giuseppe Giraldi (1848–1901), an Italian missionary in China who introduced it into cultivation.) Arching shrub. Flowers large, borne in erect racemes. *Distr.* NW China.

E. × macrantha A hybrid of *E. korolkowii* × *E. racemosa*. A popular hybrid from which several cultivars have been raised, the most striking of which is 'The Bride'. *Distr.* Garden origin.

exoletus, -a, -um: mature.

exsertus, -a, -um: sticking out, protruding.

exsuccus, -a, -um: juiceless.

exsul: banished, exiled.

exul: see *exsul*.

F

Fabaceae See Leguminosae

fabaceus, -a, -um: resembling the species *Vicia faba*, the broad bean.

Fabiana (After Archbishop Francisco Fabian y Fuero (1719–1801), Spanish botanist.) A genus of 25 species of evergreen shrubs with small, densely packed, alternate leaves and white, tubular or bell-shaped, 5-lobed flowers. *Distr.* Warm temperate South America. *Fam.* Solanaceae. *Class* D.

F. imbricata PICHI. Grown as an ornamental and reported to have some medicinal properties. *Distr.* Chile.

faeroensis: of the Faeroe Islands.

Fagaceae The Beech family. 8 genera and about 1000 species of deciduous or evergreen trees and a few shrubs. The leaves are simple and entire or pinnately lobed. The flowers are small and unisexual, both sexes being borne on the same tree. The nut-like fruits are surrounded by a characteristic 'cupule' which may be spiny and all-enclosing as in *Castanea* (SWEET CHESTNUT) or just a cup as in *Quercus* (OAK). Members of this family are a valuable source of timber, others provide edible fruits or are useful ornamentals. *Distr.* Widespread, except tropical Africa. *Class* D. *See: Castanea, Castanopsis, Chrysolepis, Fagus, Lithocarpus, Nothofagus, Quercus.*

Fagopyrum (From Greek *fogo*, to eat, and *pyros*, wheat.) A genus of about 6 species of annual herbs with simple alternate leaves and racemes of small, white or pink flowers. *Distr.* Temperate Eurasia. *Fam.* Polygonaceae. *Class* D.

F. esculentum BUCKWHEAT. Seeds milled to produce flour or roasted and then boiled to make a gruel. *Distr.* E Siberia, brought through cultivation to W Asia and Europe.

Fagus (The classical Greek and Latin name for these trees.) BEECH. A genus of about 10 species of deciduous trees with smooth bark, simple, alternate, 2-ranked leaves and small unisexual flowers. The male flowers are borne in pendent stalked clusters and the females are in groups of 2–3 on short stalks. The fruit is a stiffly hairy case (cupule) surrounding 2 or occasionally 3 nuts. Several species are important for their timber and as ornamentals. *Distr.* N temperate regions. *Fam.* Fagaceae. *Class* D.

F. crenata JAPANESE BEECH. To 40m high. An important forest tree. *Distr.* Japan.

F. engleriana CHINESE BEECH. To 20m high. Branching almost from base. *Distr.* Central China.

F. sylvatica COMMON BEECH, EUROPEAN BEECH, COPPER BEECH, WEEPING BEECH. To 50m high. Leaves with soft hairs. An important timber tree in Europe occurring in natural and artificial pure stands. Casts dense shade and produces growth inhibitors in old leaves to prevent the generation of undergrowth. Several ornamental cultivars are available including forma *purpurea*, the COPPER BEECH, with purple leaves, and forma *pendula*, the WEEPING BEECH. *Distr.* Europe, Turkey.

fair maids of France *Saxifraga granulata.*

fairies' thimbles *Campanula cochleariifolia.*

fairy duster *Calliandra eriophylla.*

fairy wand *Chamaelirium luteum* ~

fairy washboard *Haworthia limifolia.*

falcatus, -a, -um: falcate, scythe-shaped.

falcifolius, -a, -um: from Latin *falcatus*, scythe-shaped, and *folius*, leaved.

falcinellus, -a, -um: slightly scythe-shaped.

falklandicus, -a, -um: of the Falkland Islands.

fallax: deceptive, fallacious.

falling stars *Campanula isophylla.*

fameflower *Talinum.*

Farfugium (Name used by Pliny.) A genus of 2 species of tuft-forming, evergreen, perennial herbs with simple leaves and flowers in daisy-like heads that bear distinct rays. *Distr.* E Asia. *Fam.* Compositae. *Class* D.

F. japonicum Leaves woolly. Flowers yellow. *Distr.* Japan.

farinaceus, -a, -um: starchy.

farinosus, -a, -um: mealy, covered with a floury bloom.

Farsetia (After Philip Farseti, Venetian botanist.) A genus of 20 species of perennial herbs and subshrubs with white or yellow flowers. *Distr.* Morocco to NW India and tropical African mountains. *Fam.* Cruciferae. *Class* D.

F. aegyptica Shrubby perennial with white, suffused yellow flowers. *Distr.* N Sudan.

F. clypeata See *Fibigia clypeata*

fasciatus, -a, -um: banded.

Fascicularia (From Latin *fasciculus*, a small bundle, alluding to the way the leaves grow.) A genus of 5 species of terrestrial or occasionally epiphytic herbs with rosettes of spiny leaves. The small blue flowers are borne in dense clusters in the centre of each rosette. Several species are grown as ornamentals. *Distr.* Chile, 1 species naturalized in Europe. *Fam.* Bromeliaceae. *Class* M.

F. andina See *F. bicolor*

F. bicolor Leaves to 50cm long. Inner leaves red at flowering. *Distr.* Central Chile.

F. pitcairniifolia Leaves to 1m long. Inner leaves bright red at flowering. *Distr.* S Chile, naturalized locally in Europe.

fasciculatus, -a, -um: clustered, growing in bundles.

fasciculiflorus, -a, -um: with clustered flowers.

fastigiatus, -a, -um: upright, clustered branches.

fastuosus, -a, -um: haughty, proud.

fat hen *Chenopodium album.*

fat pork tree *Clusia major.*

× *Fatshedera* (From the names of the parent genera.) A genus of 1 hybrid between the genera *Fatsia* and *Hedera*. *Distr.* Garden origin (France, 1910). *Fam.* Araliaceae. *Class* D.

×*F. lizei* (After Lize Frères of Nantes who raised it in 1910.) A hybrid between *Fatsia japonica* × *Hedera helix*. Large shrub with palmate leaves. Grown as an ornamental. *Distr.* Garden origin.

Fatsia (From a Japanese name, *fatsi*.) A genus of 1 species of evergreen shrub with large, leathery, palmately lobed leaves and large compound inflorescences that bear numerous small flowers. *Distr.* Japan. *Fam.* Araliaceae. *Class* D.

F. japonica FALSE CASTOR OIL PLANT, GLOSSY-LEAVED PAPER PLANT. Grown as an ornamental for its bold foliage. *Distr.* Japan.

F. papyrifera See *Tetrapanax papyriferus*

Faucaria (From the Latin *fauces*, jaw, alluding to the paired, toothed leaves that resemble open jaws.) TIGER'S JAWS. A genus of about 35 species of clump-forming succulent herbs with 4-ranked, toothed leaves and large many-petalled flowers. Several species are grown as ornamentals. *Distr.* South Africa (Cape Province). *Fam.* Aizoaceae. *Class* D.

F. tigrina TIGER JAWS. Leaves grey-green with stout hooked teeth. Flowers golden yellow. *Distr.* South Africa (Cape Province).

feaberry *Ribes uva-crispa.*

feather, false *Cuphea hyssopifolia.*

Feathertop, Abyssinian *Pennisetum villosum.*

febrifugus, -a, -um: a cure for fever.

Felicia (After Herr Felix (died 1846), German official.) BLUE MARGUERITE, BLUE DAISY, KINGFISHER DAISY. A genus of 83 species of annual and perennial herbs, subshrubs, and shrubs with simple leaves and flowers in daisy-like heads that bear distinct blue to mauve or white rays. Some species are cultivated as ornamentals. *Distr.* South Africa, with few extending to tropical Africa and Arabia. *Fam.* Compositae. *Class* D.

F. amelloides BLUE MARGUERITE, BLUE DAISY. Densely hairy subshrub. Rays white. *Distr.* South Africa.

F. amoena Annual to perennial, downy herb. Rays pale blue. *Distr.* South Africa.

F. bergeriana (After A. Berger (1871–1931), German horticulturalist.) KINGFISHER DAISY. Mat-forming annual. Rays bright blue. *Distr.* South Africa.

F. capensis See *F. amelloides*

F. echinata Subshrub. Rays white to lilac. *Distr.* South Africa.

F. natalensis See *F. rosulata*

F. pappei See *F. amoena*

F. petiolata Prostrate perennial. Rays violet to white. *Distr.* South Africa.

F. rosulata Rhizomatous perennial. Rays blue. *Distr.* South Africa.

F. uliginosa Procumbent perennial. Rays lilac. *Distr.* South Africa.

felt bush *Kalanchoe beharensis.*

fenestralis, -e: windowed, with openings.

fenestrellatus, -a, -um: with little windows, small openings.

fennel *Foeniculum vulgare* **dog** ~ *Anthemis* **Florence** ~ *Foeniculum vulgare* **giant** ~ *Ferula, F. communis.*

fenugreek *Trigonella foenum-graecum.*

fern: adder's ~ *Polypodium vulgare* **American sword** ~ *Polystichum munitum* **asparagus** ~ *Asparagus setaceus* **bear's foot** ~ *Humata tyermannii* **bear's paw** ~ *Aglaomorpha meyeniana* **beech** ~ *Phegopteris connectilis* **bird's nest** ~ *Asplenium nidus* **black caterpillar** ~ *Scyphularia pentaphylla* **bladder** ~ *Cystopteris, C. fragilis,* **Boston** ~ *Nephrolepis N. exaltata,* **bristly shield** ~ *Lastreopsis hispida* **brittle bladder** ~ *Cystopteris fragilis* **buckler** ~ *Dryopteris* **Button** ~ *Tectaria cicutaria, Pellaea rotundifolia* **Byfield** ~ *Bowenia serrulata* **chain** ~ *Woodwardia* **Christmas** ~ *Polystichum acrostichoides* **cinnamon** ~ *Osmunda cinnamomea* **climbing** ~ *Lygodium, Stenochlaena palustris* **climbing maidenhair** ~ *Lygodium microphyllum* **common lady** ~ *Athyrium filix-femina* **common maidenhair** ~ *Adiantum capillus-veneris* **common rasp** ~ *Doodia media* **common stag's horn** ~ *Platycerium bifurcatum* **crape** ~ *Todea barbara* **crêpe** ~ *T. barbara* **deer** ~ *Blechnum spicant* **delta maidenhair** ~ *Adiantum raddianum* **dish** ~ *Pteris* **eagle** ~ *Pteridium aquilinum* **elephant's ear** ~ *Elaphoglossum crinitum* **elk's horn** ~ *Platycerium, P. bifurcatum,* **felt** ~ *Pyrrosia* **filmy** ~ *Hymenophyllum,* Hymenophyllaceae **five-fingered maidenhair** ~ *Adiantum pedatum* **floating** ~ *Salvinia auriculata* **flowering** ~ *Osmunda regalis, Anemia* **fork** ~ *Psilotum* **fragile** ~ *Cystopteris fragilis* **golden** ~ *Pityrogramma chrysophylla* **golden maidenhair** ~ *Polypodium vulgare* **hand** ~ *Dryopteris pedata* **hard** ~ *Blechnum, B. spicant* **hard shield** ~ *Polystichum aculeatum* **hare's foot** ~ *Davallia* **hart's tongue** ~ *Asplenium scolopendrium* **hen and chick** ~ *A. bulbiferum* **holly** ~ *Polystichum acrostichoides* **hound's tongue** ~ *Microsorum diversifolium* **interrupted** ~ *Osmunda claytoniana* **iron** ~ *Rumohra adiantiformis* **Japanese felt** ~ *Pyrrosia lingua* **Japanese holly** ~ *Cyrtomium falcatum* **kangaroo** ~ *Microsorum diversifolium* **king** ~ *Marattia fraxinea, Todea barbara* **lace** ~ *Paesia scaberula* **ladder** ~ *Nephrolepis* **lady** ~ *Athyrium* **leather** ~ *Rumohra adiantiformis* **leatherleaf** ~ *R. adiantiformis* **licorice** ~ *Polypodium glycyrrhiza* **limestone** ~ *Gymnocarpium robertianum* **lip** ~ *Cheilanthes* **long beech** ~ *Phegopteris connectilis* **maidenhair** ~Adiantaceae, *Adiantum* **male** ~*Dryopteris filix-mas,* Dryopteridaceae **marsh** ~ *Thelypteris palustris* **meadow** ~ *T. palustris* **miniature tree** ~ *Blechnum gibbum* **mosquito** ~ *Azolla* **North American maidenhair** ~ *Adiantum pedatum* **northern holly** ~ *Polystichum lonchitis* **northern oak** ~ *Gymnocarpium robertianum* **oak** ~ *G. dryopteris* **ostrich feather** ~ *Matteuccia struthiopteris* **para** ~ *Marattia fraxinea* **parsley** ~ *Cryptogramma, C. crispa* **Prince of Wales's** ~ *Leptopteris superba* **rabbit's foot** ~ *Phlebodium aureum* **resurrection** ~ *Cheilanthes* **royal** ~ *Osmunda, O. regalis,* **scented** ~ *Paesia scaberula* **sensitive** ~ *Onoclea sensibilis* **shield** ~ *Dryopteris* **shuttlecock** ~ *Matteuccia struthiopteris*

silver ~ *Pityrogramma calomelanos* **skeleton fork** ~ *Psilotum nudum* **small rasp** ~ *Doodia caudata* **snake** ~ *Lygodium microphyllum* **snuff box** ~ *Thelypteris palustris* **soft shield** ~ *Polystichum setiferum* **southern maidenhair** ~ *Adiantum capillus-veneris* **spider** ~ *Pteris multifida* **stag's horn** ~ *Platycerium* **sword** ~ *Nephrolepis* **table** ~ *Pteris* **trim shield** ~ *Lastreopsis decomposita* **water** ~ *Azolla* **whisk** ~ *Psilotum*.

fern rhapis *Rhapis excelsa*.

fern tree *Jacaranda*.

ferox: wild, fierce; often used for very thorny plants.

Ferraria (After Giovanni Battista Ferrari (1584–1655), Italian botanist.) A genus of about 10 species of small perennial herbs with 2-ranked linear leaves and cymes of yellow, red, green, or purple flowers which bear 3 large outer tepals and 3 small inner tepals. Several species are cultivated as ornamentals despite their malodorous flowers. *Distr.* S and tropical Africa. *Fam.* Iridaceae. *Class* M.

F. crispa Flowers mottled green, brown, and purple, margins wavy. *Distr.* South Africa (Cape Province).

F. undulata See *F. crispa*

ferrugineus, -a, -um: rusty red-brown.

Ferula (The classical Latin name for these plants.) GIANT FENNEL. A genus of 172 species of perennial herbs with a substantial rootstock, much-divided leaves, and a compound umbel of yellow flowers. Some of the outer umbels of the inflorescence may be entirely sterile and showy. Several species are grown as large ornamentals and some are a source of medicinal resins. *Distr.* Mediterranean to Central Asia. *Fam.* Umbelliferae. *Class* D.

F. assa-foetida ASAFOETIDA. Herb to 2m high. The source of a resin that is sometimes used in veterinary medicine. *Distr.* W Iran.

F. communis GIANT FENNEL. Herb to 3m high. Several ornamental cultivars are available with bronze or purple foliage. *Distr.* S Europe, W Asia.

ferulifolius, -a, -um: from the genus *Ferula*, and *folius*, leaved.

fescue *Festuca* **blue** ~ *F. glauca* **giant** ~ *F. gigantea* **grey** ~ *F. glauca* **red** ~ *F. rubra* **sheep's** ~ *F. ovina*.

Festuca (From Latin *festuca*, stalk or stem.) FESCUE, BLUEGRASS. A genus of about 300 species of rhizomatous or tufted grasses with leaves that are flat or rolled in dry weather and dense or loose panicles that sometimes bear small plantlets. *Distr.* Temperate and subtropical regions and mountainous areas of the tropics. *Fam.* Gramineae. *Class* M.

F. gigantea GIANT FESCUE. To 1.5m high. Panicles nodding. Grown as an ornamental. *Distr.* Europe, temperate Asia, N Africa.

F. glauca BLUE FESCUE, GREY FESCUE. A cultivated ornamental with numerous cultivars available. *Distr.* SE and Central Europe.

F. longifolia An important lawn grass. *Distr.* W Europe.

F. ovina SHEEP'S FESCUE. An important pasture grass. *Distr.* Temperate Old World.

F. rubra RED FESCUE. A lawn grass. *Distr.* N temperate.

F. tenuifolia An important lawn grass. *Distr.* Europe.

F. vivipara Flowers replaced by numerous small plantlets. A grass of moors and mountains. *Distr.* N Europe.

festucoides from the genus *Festuca*, with the ending *-oides*, indicating resemblance.

fetter bush *Pieris floribunda*.

feverbark *Ilex verticillata*.

fever bush *Garrya elliptica*, *G. fremontii*, *Lindera*.

feverfew: *Tanacetum parthenium* **American** ~ *Parthenium integrifolium*.

fever root *Triosteum perfoliatum*.

fever tree *Eucalyptus globulus*.

feverwort *Triosteum*.

Fibigia (After Johann Fibig, died 1792, professor at Mainz.) A genus of 14 species of perennial herbs with yellow, sometimes violet flowers. *Distr.* E Mediterranean to Afghanistan. *Fam.* Cruciferae. *Class* D.

F. clypeata Grown for the white-woolly foliage and yellow flowers. *Distr.* Europe to Iran.

fibrosus, -a, -um: fibrous.

ficaria: relating to *Ficus*, the fig.

ficifolius, -a, -um: from the genus *Ficus*, and *folius*, leaved.

ficoideus, -a, -um: resembling the genus *Ficus*, fig.

fictolacteum: false *Rhododendron lacteum*.

Ficus (The classical Latin name for the edible fig.) FIG. A genus of about 8003 species of deciduous and evergreen shrubs, trees, and climbers with simple or palmately lobed leaves and minute unisexual flowers that are borne within a fleshy, flask-shaped receptacle (the fig). Pollination is carried out by small specialized wasps that lay their eggs in the fig. Most species contain latex and some grow as large epiphytes that slowly overwhelm and strangle their hosts. These plants are the source of a large number of natural products ranging from edible fruits and medicines through timber and fibres to latex for rubber, and they are also of some religious importance. *Distr.* Cosmopolitan in tropical and subtropical regions, especially Asia. *Fam.* Moraceae. *Class* D.

F. australis See *F. rubiginosa*

F. benghalensis BANYAN TREE, EAST INDIAN FIG TREE, INDIAN BANYAN. Very large, initially epiphytic tree, spreading by aerial roots that drop down to become accessory trunks so that a single plant may spread to a diameter of 200m. A source of timber and fibre and sacred to the Hindus. Possibly the tree under which Buddha achieved enlightenment. *Distr.* Pakistan, India, Bangladesh.

F. benjamina (From *benjan*, the Indian name.) WEEPING FIG, JAVA FIG, BENJAMIN TREE, TROPIC LAUREL, JAVA TREE, SMALL-LEAVED RUBBER PLANT. Weeping tree, often starting as a strangling epiphyte. A popular house plant. *Distr.* Tropical Asia.

F. carica COMMON FIG, EDIBLE FIG. Deciduous tree. Figs pear-shaped. The fleshy, pear-shaped figs are eaten fresh or dried and are the source of the laxative syrup of figs. The latex is used against warts. *Distr.* SW Asia, widely naturalized through over 6000 years of cultivation.

F. deltoidea MISTLETOE FIG, MISTLETOE RUBBER PLANT. Slow growing, bushy shrub. *Distr.* Malaysia.

F. diversifolia See *F. deltoidea*

F. elastica RUBBER PLANT, INDIA RUBBER TREE, INDIA RUBBER FIG, SNAKE TREE, ASSAM RUBBER. Large tree with buttresses, usually starting as an epiphyte. Juvenile forms make popular house plants. Formerly an important source of rubber but now of little economic importance as it has been replaced by *Hevea brasiliensis*. *Distr.* Tropical Asia.

F. lyrata FIDDLE LEAF FIG, FIDDLE-LEAF, BANJO FIG. Robust tree to 12m high. Leaves lobed. Juvenile forms cultivated as house plants. *Distr.* Tropical W Africa.

F. pumila CREEPING FIG, CLIMBING FIG, CREEPING RUBBER PLANT. Liana with aerial roots which secrete a gummy exudate, becoming a large tree. Fruit used for jelly in China (okgue). Juvenile forms are much cultivated as conservatory plants. *Distr.* N Vietnam to China, Taiwan, Japan and Ryukyu Islands.

F. rubiginosa RUSTY FIG, PORT JACKSON FIG, LITTLE-LEAF FIG, BOTANY BAY FIG. Dense-headed tree with buttressed trunk. Leaves rusty-hairy below. *Distr.* E Australia.

ficus-indica: Indian fig; see *Opuntia*.

fiddlehead *Osmunda cinnamomea*.

fiddle-leaf *Ficus lyrata*.

fiddleneck *Phacelia tanacetifolia*.

fiddlewood *Citharexylum*.

fig *Ficus* **banjo** ~ *F. lyrata* **Barbary** ~ *Opuntia ficus-indica* **Botany Bay** ~ *Ficus rubiginosa* **climbing** ~ *F. pumila* **common** ~ *F. carica* **creeping** ~ *F. pumila* **devil's** ~ *Argemone mexicana* **edible** ~ *Ficus carica* **fiddle-leaf** ~ *F. lyrata* **Hottentot** ~ *Carpobrotus edulis* **Indian** ~ *Opuntia ficus-indica* **Java** ~ *Ficus benjamina* **little-leaf** ~ *F. rubiginosa* **mistletoe** ~ *F. deltoidea* **Port Jackson** ~ *F. rubiginosa* **rusty** ~ *F. rubiginosa* **silk** ~ *Musa x paradisiaca* **weeping** ~*Ficus benjamina*.

fig tree, East Indian *Ficus benghalensis*.

figwort *Scrophularia* **common** ~ *S. nodosa* **water** ~ *S. auriculata*.

filamentosus, -a, -um: with filaments.

filbert *Corylus, C. maxima*.

filicifolius, -a, -um: with fern-like leaves.

filifera: thread-bearing.

filifolius, -a, -um: with thread-like leaves.

filiformis, -e: thread-like.

Filipendula (From Latin *filum*, thread, and *pendulus*, hanging, alluding to the threads that connect the root tubers.) A genus of 10 species of rhizomatous herbs with pinnately divided or palmately lobed leaves and panicles of small regular flowers that bear 4–5 round petals and numerous long stamens. Several species are grown as ornamentals. *Distr.* N temperate regions. *Fam.* Rosaceae. *Class* D.

F. alnifolia See *F. ulmaria*

F. hexapetala See *F. vulgaris*

F. kamtschatica Clump-forming perennial to 1.5m high. Leaves palmately lobed. Flowers fragrant, pale pink to white. *Distr.* Japan, Manchuria, Kamchatka.

F. purpurea Erect perennial to 1m high. Leaves deeply palmately divided. Flowers purple-red. *Distr.* Japan.

F. rubra QUEEN OF THE PRAIRIE. Erect perennial. Flowers borne in large plumes, soft pink-red. *Distr.* E North America.

F. ulmaria MEADOW-SWEET, QUEEN OF THE MEADOWS. Erect perennial to 2m high. Leaves pinnately divided, fragrant if crushed. Flowers cream-white. Formerly used medicinally and as a source of fragrant oil. *Distr.* Europe, Asia, naturalized in North America.

F. vulgaris DROPWORT. Erect perennial to 80cm high. Leaves pinnately divided, not fragrant when crushed. Flowers white, tinged purple. Formerly used medicinally for kidney disorders. *Distr.* Europe, N Africa, W Asia, Siberia.

filipendulina: resembling the genus *Filipendula*.

filipes: slender.

fimbriatus, -a, -um: fimbriate, with a fringed margin.

finger-nail plant *Neoregelia spectabilis*.

finger plant, painted *Neoregelia spectabilis*.

fingers: five *Syngonium auritum*, *Pseudopanax arboreus* **lady's** ~ *Anthyllis vulneraria*, *Abelmoschus esculentus* **seven** ~ *Schefflera digitata*.

fir *Abies* **balsam** ~ *A. balsamea* **Colorado** ~ *A. concolor* **Douglas** ~ *Pseudotsuga*, *P. menziesii*, **giant** ~ *Abies grandis* **hedgehog** ~ *A. pinsapo* **joint** ~ *Ephedra* **noble** ~ *Abies procera* **silver** ~ *Abies*, *A. alba* **Spanish** ~ *A. pinsapo* **white** ~ *A. concolor*.

fire bush *Embothrium coccineum* *Streptosolen jamesonii*. **Chilean** ~ *E. coccineum*.

firecracker: Brazilian ~ *Manettia luteo-rubra*.

firecracker flower *Dichelostemma ida-maia*.

firecracker flowers *Crossandra infundibuliformis*

firecracker plant *Cuphea ignea*. *Russelia equisetiformis*.

fireweed *Epilobium angustifolium*.

Firmiana (After Karl Josef von Firmian (died 1782), governor of Lombardy and patron of the Padua botanic garden.) A genus of 9 species of shrubs and trees with simple or palmately lobed leaves and panicles or racemes of bell-shaped flowers that lack petals and are followed by leaf-like fruits. Several species are grown as ornamentals and some are a source of timber. *Distr.* Tropical regions of the Old World. *Fam.* Sterculiaceae. *Class* D.

F. simplex CHINESE PARASOL TREE, CHINESE BOTTLE TREE, JAPANESE VARNISH TREE, PHOENIX TREE. Deciduous tree to 20m high. Leaves large. Grown as an ornamental. *Distr.* E Asia.

firmus, -a, -um: firm.

fissistipulus, -a, -um: with split stipules.

fissus, -a, -um: split, divided.

fistulosus, -a, -um: hollow.

Fittonia (After Elizabeth and Sarah Mary Fitton, authors of *Conversations on Botany* (1823).) A genus of 2 species of evergreen perennial herbs with creeping rooting stems and simple leaves with colourful veins. The small inconspicuous flowers are borne in slender terminal racemes. *Distr.* NE South America. *Fam.* Acanthaceae. *Class* D.

F. verschaffeltii (After M. Verschaffelt, a 19th-century Belgian nurseryman.) MOSAIC PLANT, NERVE PLANT, SILVER NET PLANT, SILVER FITTONIA, SILVER NERVE, SILVER THREADS, SNAKE-SKIN PLANT. Grown as an ornamental for its colourful foliage. Numerous cultivars are available. *Distr.* Peru.

fittonia, silver *Fittonia verschaffeltii.*

Fitzroya (After Captain Robert Fitzroy (1805–65), commander of the *Beagle* during Darwin's voyage.) A genus of 1 species of evergreen coniferous shrub or tree with peeling bark and scale-like leaves that are borne in whorls of 3. *Distr.* Chile, Argentina. *Fam.* Cupressaceae. *Class* G.

F. cupressoides ALERCE, PATAGONIAN CYPRESS. A source of timber for shingles. Occasionally grown as a slow-growing ornamental.

fivecorner, green *Styphelia viridis.*

five spot *Nemophila maculata.*

flabellatus, -a, -um: fan-shaped.

flaccidus, -a, -um: flaccid.

Flacourtiaceae A family of 79 genera and about 800 species of shrubs and trees with simple leaves and regular unisexual or bisexual flowers that bear 6–16 tepals, numerous stamens and a superior ovary. *Distr.* Tropical and occasionally warm-temperate regions. *Class* D.
See: Azara, Berberidopsis, Casearia, Idesia, Poliothyrsis.

flag *Iris* **myrtle** ~ *Acorus calamus* **soft** ~ *Typha angustifolia* **spiral** ~ *Costus* **sweet** ~ *Acorus, A. calamus.*

Flagellariaceae A family of 1 genus and 4 species of erect or climbing herbs. The leaves are long and often end in a tendril. The flowers are regular, bisexual, and arranged in branching racemes. The stems of some species are used in basket-making. *Distr.* Tropical regions of the Old World. *Class* M.

flagellaris, -e: flagelliform.

flagelliflorus, -a, -um: flagellate flowers.

flagroot *Acorus calamus.*

flamboyant *Delonix regia.*

flamboyant tree *Caesalpinia pulcherrima.*

flame flower *Pyrostegia venusta, Tropaeolum* **Chilean** ~ *Embothrium coccineum.*

flame of the forest *Butea monosperma*

flame plant *Anthurium scherzerianum*

flame tree *Delonix regia.*

flaming Katy *Kalanchoe blossfeldiana.*

flamingo flower *Anthurium, A. andreanum, A. scherzerianum.*

flamingo plant *Justicia carnea, Hypoestes phyllostachya.*

flaming sword *Vriesea splendens.*

flammeus, -a, -um: scarlet, flame-coloured.

flannel bush *Fremontodendron*

flannel flower *Actinotus helianthi*

flannel plant *Verbascum thapsus.*

flavens: yellow.

flavescens: becoming pale yellow, yellowish.

flavidus, -a, -um: yellowish.

flavus, -a, -um: pale yellow.

flax *Linum, L. usitatissimum,* Linaceae **blue** ~ *L. narbonense* **bush** ~ *Astelia nervosa* **golden** ~ *Linum flavum* **holy** ~ *Santolina, S. rosmarinifolia* **mountain** ~ *Phormium colensoi* **New Zealand** ~ *P. tenax* **pale** ~ *Linum bienne* **perennial** ~ *L. perenne* **sticky** ~ *L. viscosum* **tree** ~ *L. arboreum* **yellow** ~ *L. flavum, Reinwardtia indica.*

flaxseed, water *Spirodela polyrrhiza.*

fleabane *Pulicaria, Erigeron* **Alpine** ~ *E. borealis* **blue** ~ *E. acer* **Himalayan** ~ *E. multiradiatus.*

flexicaulis, -e: flexible-stemmed.

flexilis flexible.

flexuosus, -a, -um: bending alternately in each direction, zig-zag.

floccosus, -a, -um: with tufted soft hairs, appearing somewhat matted.

floradora *Stephanotis floribunda.*

florentinus, -a, -um: of Florence, Italy.

floribundus, -a, -um: profusely flowering.

floridanus, -a, -um: of Florida, USA.

floridus, -a, -um: profusely flowering.

floriferum, -a, -um: bearing many flowers.

flos-cuculi: cuckoo flower.

flos-jovis: flower of Jupiter.

flower of the western wind *Zephyranthes candida.*

fluctuans: fluctuating.

fluitans: floating, swimming.

fluminensis: of Rio de Janeiro, Brazil.

fluviatilis, -e: of rivers.

foam, meadow *Limnanthes.*

foam flower *Tiarella cordifolia* **Japanese** ~ *Tanakaea radicans.*

foemineus, -a, -um: female.

foeminus, -a, -um: feminine.

Foeniculum (The classical Latin name for the culinary herb FENNEL.) A genus of 1 species of biennial or perennial herb to 2m high with a hollow stem, finely divided leaves and compound umbels of yellow flowers. *Distr.* Europe, Mediterranean. *Fam.* Umbelliferae. *Class* D.

F. vulgare FENNEL, FLORENCE FENNEL. This plant has been used since antiquity as a flavouring. Today it is grown either for its swollen stem bases (Florence fennel) or for its fruits which are crushed for oil.

foetens: fetid, stinking.

foetidissima: very fetid, extremely smelly.

foetidus, -a, -um: stinking, fetid.

Fokienia (After Fujien Province (originally transliterated Fukien), China, where it grows.) A genus of 1 species of evergreen coniferous shrub with minute scale-like leaves that are borne in whorls of 4. *Distr.* SW and SE China. *Fam.* Cupressaceae. *Class* G.

F. hodginsii (After Captain A. Hodgins the first European to discover it.) Occasionally grown as an ornamental. The roots are used as a source of oil for scent-making.

folhado *Clethra arborea.*

foliaceus, -a, -um: leafy.

foliolosus, -a, -um: with leaflets.

foliosissimus, -a, -um: very leafy.

foliosus, -a, -um: leafy.

follicularis: bearing follicles, dry one-chambered fruits.

forget-me-not Boraginaceae, *Myosotis* **Alpine** ~ *M. alpestris* **American** ~ *Eritrichium* **Chatham Island** ~ *Myosotidium hortensia* **Chinese** ~ *Cynoglossum amabile* **creeping** ~ *Omphalodes verna* **fairy** ~ *Eritrichium nanum* **garden** ~ *Myosotis sylvatica* **water** ~ *M. scorpioides.*

formosanus, -a, -um: of Taiwan (Formosa).

formosissimus, -a, -um: very beautiful.

formosus, -a, -um: beautiful, good looking.

forrestii: after George Forrest (1873–1932), Scottish plant collector who made several trips to W China and collected many new garden plants.

Forsythia (After William Forsyth (1737–1804), Scottish gardener who became superintendent of the royal garden, Kensington Palace.) A genus of 7 species of deciduous shrubs with opposite, typically simple leaves and yellow flowers that appear before the leaves in spring and bear 4 narrow fused petals and 2 stamens. Several species and numerous cultivars are grown as ornamentals. *Distr.* SE Europe, E Asia. *Fam.* Oleaceae. *Class* D.

F. × intermedia A hybrid of *F. suspensa* × *F. viridissima* that has given rise to numerous cultivars. *Distr.* China, long cultivated Japan.

F. ovata KOREAN FORSYTHIA. Bushy shrub. Leaves toothed, dark green. Flowers small, bright yellow. *Distr.* Korea.

F. suspensa GOLDEN BELL. Arching shrub. Leaves often trifoliate. Flowers bright yellow, nodding. The fruit wall is used medicinally in China. *Distr.* China.

F. viridissima Erect shrub. Leaves maroon in autumn. Flowers often tinged green, appearing in late spring. *Distr.* China.

forsythia: Korean ~ *Forsythia ovata* **white** ~ *Abeliophyllum distichum.*

Fortunella (After Robert Fortune (1812–80), who collected in China and introduced the TEA plant from China to India.) KUMQUAT. A genus of 4–5 species of spiny shrubs and small trees with simple leaves and regular white flowers that bear 5 petals and 16–20 stamens. The fruit is fleshy, aromatic, similar to that of the genus *Citrus* and may be eaten whole. *Distr.* S China. *Fam.* Rutaceae. *Class* D.

F. japonica KUMQUAT, ROUND KUMQUAT, MARUMI KUMQUAT. Fruit round, with sweet peel and large seeds. *Distr.* S China.

F. margarita OVAL KUMQUAT, NAGAMI KUMQUAT. Fruit oblong, peel somewhat spicy. *Distr.* SE China, possibly arose through cultivation of other species.

Fosterella (After M. B. Foster, American bromeliad specialist.) A genus of about 13 species of stemless terrestrial herbs with rosettes of soft leaves and panicles of small white flowers. Several species are grown as ornamentals. *Distr.* Central America and W South America. *Fam.* Bromeliaceae. *Class* M.

F. penduliflora Leaves tapering to apex. Inflorescence loose. *Distr.* Central Peru and NW Argentina.

Fothergilla (After Dr John Fothergill (1712–80), English physician and gardener.) A genus of 2 species of small deciduous shrubs with simple, toothed, alternate leaves and brush-like cylindrical spikes of small fragrant flowers that bear very long white stamens. Both species are grown as ornamentals for their bottle-brush-like inflorescences in early spring and their spectacular autumn colour. *Distr.* SE North America. *Fam.* Hamamelidaceae. *Class* D.

F. gardenii (After Dr Alexander Garden (1730–91), a Scottish physician and botanist who lived in South Carolina.) Small spreading shrub. *Distr.* N Carolina to S Alabama.

F. major Erect shrub to 3m high. *Distr.* Allegheny Mountains.

fountain plant *Russelia equisetiformis.*

Fouquieriaceae A family of 1 genus and 11 species of spiny shrubs and trees with small succulent leaves and showy flowers which bear 5 sepals, 5 petals, and 10 stamens. *Distr.* SW North America to Mexico. *Class* D.

four o'clock plant *Mirabilis jalapa.*

foxberry *Vaccinium vitis-idaea.*

foxglove Scrophulariaceae, *Digitalis* **common** ~ *D. purpurea* **fairy** ~ *Erinus alpinus* **Grecian** ~ *Digitalis lanata* **large yellow** ~ *D. grandiflora* **straw** ~ *D. lutea.*

foxglove tree *Paulownia tomentosa.*

fox's brush *Centranthus ruber.*

foxtail *Setaria italica, Alopecurus* **meadow** ~ *A. pratensis.*

fractiflexus, -a, -um: zig-zag.

Fragaria (From *fraga*, the classical Latin name for STRAWBERRY, derived from *fragrans*, fragrant, alluding to the fruit.) STRAWBERRY. A genus of about 12 species of stoloniferous herbs with compound leaves and white flowers that bear 5–8 petals. After flowering the receptacle swells and becomes red and fleshy, forming the characteristic strawberry 'fruit'. It should be noted that this is not a true fruit in the strict botanical sense. The true fruit is represented by the pips on the surface of the receptacle. See *Malus*, APPLE. Several species have edible fruits and some are grown as ornamentals. *Distr.* N temperate regions and S South America. *Fam.* Rosaceae. *Class* D.

F. alpina See *F. vesca*

F. × ananassa (From the genus name *Ananas*, because it was originally called the pineapple or pin strawberry.) GARDEN STRAWBERRY, CULTIVATED STRAWBERRY. A complex group of hybrids with large fruits from crosses between *F. chiloensis* × *F. virginiana*. These hybrids provide the all the commercial cultivars of strawberries. *Distr.* Garden origin.

F. chiloensis BEACH STRAWBERRY. Leaves numerous. Fruit red with white flesh. *Distr.* South America.

F. indica See *Duchesnea indica*

F. moschata HAUTBOIS STRAWBERRY, PLYMOUTH STRAWBERRY. Leaves bright green. Flowers large. Fruit red. Eaten in a similar way to *F. vesca*. *Distr.* Central Europe.

F. vesca ALPINE STRAWBERRY, WILD STRAWBERRY. Leaves long-stalked. Fruits small. The fruit has a more intense flavour than those of the commercial varieties. They are gathered in the wild and several cultivars have been raised with yellow fruits. *Distr.* N Temperate regions.

F. virginiana SCARLET STRAWBERRY. Leaves blue-green, somewhat leathery. Flowers often unisexual. Fruit deep red. *Distr.* North America.

fragarioides: from the genus *Fragaria*, with the ending *-oides*, indicating resemblance.

fragiformis, -e: strawberry-like.

fragilis, -e: fragile.

fragrans: fragrant.

fragrantissima: very fragrant.

Francoa (After Francisco Franco, a 16th-century Spanish physician.) A genus of 5 species of perennial evergreen herbs with pinnately lobed basal leaves and spikes of small pink-white flowers. *Distr.* Chile. *Fam.* Saxifragaceae. *Class* D.

F. appendiculata See *F. sonchifolia*

F. ramosa BRIDAL WREATH. Flowers white.

F. sonchifolia BRIDAL WREATH, WEDDING FLOWER. Flowers pink with dark spots. *Distr.* Chile.

frangipani *Plumeria rubra.*

Frankenia (After Johan Frankenius (1590–1661), professor of botany at Uppsala.) A genus of about 25 species of herbs and subshrubs with wiry branches and typically rolled, hairy leaves. The small bisexual flowers bear 4–6 sepals, petals, and stamens and the fruit is a 3-angled capsule. Some species are used medicinally or as poisons, others are cultivated as rock-plants. *Distr.* Temperate and subtropical regions, particularly in salty habitats. *Fam.* Frankeniaceae. *Class* D.

F. laevis SEA HEATH. Small evergreen shrub. Flowers solitary, pink. Grown as a rock garden plant. *Distr.* Europe.

F. thymifolia Tuft-forming evergreen shrub. Flowers clustered, pink. Grown as a rock garden plant. *Distr.* N Africa.

Frankeniaceae A family of 5 genera and about 40 species of salt-tolerant herbs and shrubs with small scale-like leaves and regular bisexual flowers that bear a fused calyx and 4–7 free petals. *Distr.* Warm-temperate and subtropical regions. *Class* D.
See: Frankenia.

frankincense Burseraceae, *Boswellia sacra.*

Franklinia (After Benjamin Franklin (1706–90), American statesman.) A genus of 70 species of deciduous and evergreen trees and shrubs with simple alternate leaves and solitary regular flowers that bear 5 petals and numerous stamens. Several species are grown as ornamentals. *Distr.* SE USA, SE Asia. *Fam.* Theaceae. *Class* D.

F. alatamaha (After the Altamaha River, Georgia, near which it grew.) Deciduous shrub or tree to 10m high. This plant is now extinct in the wild and was last seen in 1803. It has survived in cultivation and is now quite widely grown as an ornamental. *Distr.* SE USA.

F. axillaris Evergreen shrub or tree to 10m high. Flowers yellow-white. *Distr.* China, Taiwan.

F. lasianthus LOBLOLLY BAY. Evergreen shrub or tree to 20m high. Flowers pure white. *Distr.* SE USA.

Frasera (After John Fraser (1750–1811), Chelsea nurseryman who collected North American plants.) A genus of 15 species of biennial and perennial herbs with thick taproots, simple basal leaves, and crowded inflorescences of 4-lobed, cream-white flowers. Some species are used locally as medicines and tonics. Several species are grown as ornamental. *Distr.* North America. *Fam.* Gentianaceae. *Class* D.

F. speciosa GREEN GENTIAN. Perennial to 2m high. Flowers flushed green and spotted purple. *Distr.* Rocky Mountains.

fraternus, -a, -um: brotherly, closely allied.

fraxinellus, -a, -um: like a small ash.

fraxineus, -a, -um: resembling the genus *Fraxinus*, ash.

fraxinifolius, -a, -um: from the genus *Fraxinus*, and *folius*, leaved.

fraxinoides: from the genus *Fraxinus*, with the ending *-oides*, indicating resemblance.

Fraxinus (The classical Latin name for ash.) ASH. A genus of 65 species of deciduous trees with opposite pinnate leaves and panicles or racemes of inconspicuous flowers that bear 4 petals and 2 stamens. The fruit is characteristic, consisting of a single seed with a long flattened wing. The flowers may lack sepals and petals and be wind-pollinated or bear showy petals and be insect-pollinated. Several species and cultivars are grown as ornamentals and some are important as a source of timber. *Distr.* N temperate regions, a few species extending into the tropics. *Fam.* Oleaceae. *Class* D.

F. americana AMERICAN ASH, CANADIAN ASH, WHITE ASH. Spreading tree to 30m high. Used in a similar way to *F. excelsior*. *Distr.* E North America.

F. chinensis Tree to 25m high. Leaves bear 4 leaflets. Insects feeding on this tree exuded a wax that is used, in China, in the manufacture of candles and high-quality paper and for polishing jade and soapstone. *Distr.* China.

F. excelsior COMMON ASH, EUROPEAN ASH. Tree to 40m high. Leaves bear up to 11 leaflets. Flowers borne in dense clusters. An important source of high-quality, elastic timber. The bark was formerly used medicinally and young shoots are sometimes fed to livestock. *Distr.* Europe, Caucasus, SW Asia.

F. ornus FLOWERING ASH, MANNA ASH. Domed tree to 8m high. Flowers white, fragrant, insect-pollinated. Cultivated in Sicily for manna sugar or syrup which is exuded from damaged branches and is a mild laxative. *Distr.* S Europe, W Asia.

F. pennsylvanica AMERICAN ASH, CANADIAN ASH, RED ASH, WHITE ASH, GREEN ASH. Tree to 18m high. Young branches and underside of leaves velvety. *Distr.* E North America.

freckle face *Hypoestes phyllostachya*.

Freesia (After Friedrich Heinrich Theodor Freese (died 1876), German physician.) A genus of 10–20 species of perennial herbs with conical corms, fans of linear leaves, and 1-sided spikes of somewhat irregular flowers. A number of species and very many hybrids are grown as ornamentals and for the cut-flower market. *Distr.* South Africa (Cape Province). *Fam.* Iridaceae. *Class* M.

F. corymbosa Flowers ivory to pale pink with a cream centre. Very many garden cultivars of this species are available. *Distr.* South Africa (Cape Province).

F. × hybrida FLORIST'S FREESIA. A complex group of hybrids and cultivars based on crosses between 4 main species *F. alba, F. corymbosa, F. refracta,* and *F. leichtlinii*. *Distr.* Garden origin.

F. refracta Flowers greenish-yellow, very fragrant. *Distr.* South Africa (Cape Province).

freesia, florist's *Freesia × hybrida*.

Fremontodendron (After Major-General John Charles Fremont (1813–90), who collected *F. californicum*, and Greek *dendron*, tree.) FLANNEL BUSH, CALIFORNIAN BEAUTY. A genus of 2 species of evergreen shrubs and trees with simple or palmately lobed leaves and solitary flowers that bear petal-like sepals and fused stamens but no petals. Both species and a hybrid are grown as ornamentals. *Distr.* SW North America. *Fam.* Sterculiaceae. *Class* D.

F. californicum Shrub to 7m high. Flowers showy. *Distr.* California, W Arizona, N Baja California.

F. mexicanum Shrub to 6m high. Flowers somewhat concealed by the leaves. *Distr.* S California, Mexico.

French tree *Tamarix gallica*.

French weed *Thlaspi arvense*.

Freylinia (After Count L. Freylino, a 19th-century Italian gardener.) A genus of 4 species of evergreen shrubs with simple, whorled, or scattered leaves and panicles or racemes of 5-lobed tubular flowers. *Distr.* Tropical and South Africa. *Fam.* Scrophulariaceae. *Class* D.

F. cestroides See *F. lanceolata*

F. lanceolata HONEY BELLS. Dense shrub to 3m high. Flowers cream with a yellow interior, fragrant, borne in panicles. Grown as a half-hardy ornamental. *Distr.* S Africa.

friar's cap *Aconitum napellus.*

friars cowl *Arisarum vulgare.*

friendship plant *Pilea involucrata, Billbergia nutans.*

frigidus, -a, -um: cold.

fringe: mountain ~ *Adlumia fungosa* **water** ~ *Nymphoides peltata.*

fringebell *Shortia soldanelloides.*

fringe cups *Tellima grandiflora.*

fringed galax *Shortia soldanelloides.*

fringe flower *Chionanthus.*

fringe tree *Chionathus virginicus* **Chinese** ~ *C. retusa.*

Fritillaria (From the Latin *fritillus*, dice-box, alluding to the chequered pattern on the flowers of many species.) FRITILLARY. A genus of about 100 species of bulbiferous herbs with linear or lance-shaped leaves and tubular or bell-shaped flowers that bear 2 whorls of 3 free tepals and 6 stamens. Many of the species and numerous cultivars are grown as ornamentals. *Distr.* Temperate regions of the N hemisphere. *Fam.* Liliaceae. *Class* M.

F. camschatcensis BLACK SARANA. Stems to 75cm high. Flowers clustered, purple-black. *Distr.* Japan, E Siberia, North America.

F. cirrhosa Leaves with tendril-like tips. Flowers green, tinged purple or yellow. The bulbs are used in the treatment of chest ailments. *Distr.* Himalaya, China.

F. imperialis CROWN IMPERIAL. Whole plant smells of fox. Flowers orange-red, borne in a terminal whorl. A number of cultivars are available. The bulbs were formerly used as a source of starch and are reported to have medicinal properties. *Distr.* S Turkey to NW India.

F. meleagris SNAKE'S HEAD, GUINEA FLOWER, SNAKE'S HEAD FRITILLARY, GUINEA-HEN FLOWER, CHEQUERED LILY, LEPER FLOWER, LEPER LILY. To 30cm high. Flowers purple, chequered shades of red and dark purple. Numerous ornamental cultivars are available. The common names LEPER LILY and LEPER FLOWER come from the resemblance of the flowers to the bells that lepers were required to carry. *Distr.* Europe, naturalized in British Isles.

F. roylei (After John Forbes Royle (1798–1858), surgeon with the East India Co., who collected in India and the Himalaya.) Stem to 60cm. Flowers green, tinged yellow. This species was used medicinally in China in a similar way to *F. cirrhosa* in the W. *Distr.* Himalaya.

fritillary *Fritillaria* **snake's head** ~ *F. meleagris.*

frogbit *Hydrocharis, H. morsus-ranae.*

frondosus, -a, um: leafy.

frutescens: becoming shrubby.

frutex: shrub, bush.

fruticans: becoming shrubby.

fruticosus, -a, -um: shrubby, bushy.

fruticulosus, -a, -um: small shrub or bush.

fucatus, -a, -um: painted, coloured, stained.

Fuchsia (After Leonhart Fuchs (1501–66), German physician and herbalist.) LADY'S EARDROPS. A genus of 100 species of deciduous and evergreen shrubs and trees with simple opposite or whorled leaves and typically pendulous, bird-pollinated, tubular flowers that are followerd by edible berries. This is a very popular ornamental genus with over 8000 cultivars having been raised, 2000 of which are still in cultivation; they are most familiar in the British garden as subjects for hanging baskets or grown as standards. *Distr.* Central and S America to New Zealand. *Fam.* Onagraceae. *Class* D.

F. magellanica Erect or climbing shrub. Flowers paired or solitary, purple-pink. *Distr.* S South America, more or less naturalized in S Ireland.

F. procumbens TRAILING FUCHSIA. Prostrate shrub. Leaves tiny. Flowers small, erect. *Distr.* New Zealand.

fuchsia: Australian ~ *Correa* **Californian** *Ribes speciosum, Epilobium* *canum* **trailing** ~ *Fuchsia procumbens* **tree** ~ *Schotia brachypetala.*

fuchsioides: from the genus *Fuchsia*, with the ending *-oides*, indicating resemblance.

fugax: ephemeral, transient, the flowers not lasting long for example.

fulgens: brightly coloured, shining.

fulgidus, -a, -um: shining.

fuliginosus, -a, -um: sooty, dirty black-brown.

fuller's herb *Saponaria officinalis.*

fulvescens: becoming a tawny brown.

fulvus, -a, -um: tawny, dull yellow-brown.

Fumaria (From Latin *fumus terrae*, smoke of the earth, alluding to the diffuse foliage of some species.) FUMITORY. A genus of 55 species of scrambling and climbing annual herbs with pinnately divided leaves and racemes of irregular, tubular, spurred flowers. *Distr.* Europe to Central Asia and Himalayas, 1 species tropical E African highlands. *Fam.* Papaveraceae. *Class* D.

F. officinalis COMMON FUMITORY. Suberect or climbing herb. Flowers pink with black tips. *Distr.* Europe, SW Asia.

Fumariaceae See Papaveraceae

fumewort *Corydalis solida.*

fumitory *Fumaria* **climbing** ~ *Adlumia fungosa* **common** ~ *Fumaria officinalis.*

Funkiaceae See Hostaceae

furcatus, -a, -um: forked.

Furcraea: (After Antoine François de Fourcroy (1755–1809), French chemist.) A genus of about 20 species of thick-stemmed succulent herbs with rosettes of large leaves and very large inflorescences. Some species are a source of fibres (fique) and some are cultivated as ornamentals. *Distr.* Tropical and subtropical America. *Fam.* Agavaceae. *Class* M.

F. foetida: MAURITIUS HEMP, GREEN ALOE. Stemless or short-stemmed herb with strongly scented green-yellow flowers. Often grown as a conservatory plant in temperate regions. *Distr.* South America, naturalized and introduced elsewhere.

furfuraceus, -a, -um: scurfy, covered in brown powdery scales.

furze *Ulex, U. europaeus* **needle** ~ *Genista anglica.*

fuscus, -a, -um: dark brown.

fusiformis, -e: swollen in the middle, tapering towards each end.

fustuq *Pistacia vera.*

G

gage *Prunus domestica*.

Gagea (After Sir Thomas Gage (1761–1820) of Suffolk, an amateur botanist.) A genus of 50–70 species of small bulbiferous herbs with linear to lance-shaped leaves and umbels or racemes of bell- or star-shaped, yellow or occasionally white flowers. Several species are grown as ornamentals. *Distr*. Mainly found in Central Asia and temperate Eurasia (23 species in Europe) (2 species in the British Isles). *Fam*. Liliaceae. *Class* M.

G. *lutea* YELLOW STAR OF BETHLEHEM. To 30cm high. Flowers yellow, tinged green. *Distr*. Europe, Caucasus.

G. *peduncularis* Flowers star-shaped, yellow with green stripes, on long stalks. *Distr*. S Europe, N Africa and Turkey.

Gaillardia (After M. Gaillard de Charentoneau, 18th-century French magistrate and patron of botany.) BLANKET FLOWER, INDIAN BLANKET. A genus of 28 species of annual to perennial herbs with simple to pinnate leaves and purple flowers in daisy-like heads which bear distinct, yellow to red rays. Several species are cultivated as ornamentals. *Distr*. North and temperate South America. *Fam*. Compositae. *Class* D.

G. *aristata* Hairy perennial. Ray florets yellow, occasionally with a red base. *Distr*. North America (Rocky Mountains).

G. × *grandiflora* A hybrid of *G. aristata* × *G. pulchella*. Hairy perennial. Flowerheads large. *Distr*. Garden origin (Belgium *c*.1857), naturalized in SW USA.

galactinus, -a, -um: referring to milk.

Galactites (From Greek *gala*, milk, alluding to the white leaf veins.) A genus of 3 species of annual herbs with pinnately lobed and toothed leaves and flowers in daisy-like heads that lack distinct rays. *Distr*. Canaries and Mediterranean (2 species in Europe). *Fam*. Compositae. *Class* D.

G. *tomentosa* Leaves white-veined. Grown as an ornamental. *Distr*. Mediterranean.

galangal *Alpinia officinarum*.

galantheus, -a, -um: with milk-white flowers.

Galanthus (From Greek *gala*, milk, and *anthos*, flower, referring to the colour of the flowers, 'Snowdrop' is derived from German *Schneetropfen*, pendants or ear-rings fashionable in the 16th and 17th centuries.) SNOWDROP. A genus of 12 species of small, bulbiferous, perennial herbs with linear to elliptic leaves and solitary, nodding, white flowers. The outer 3 tepals are longer than, and conceal, the inner 3 tepals which usually bear green markings. All the species and many cultivars and hybrids are grown as garden ornamentals. *Distr*. Europe, Turkey, Iran and the Caucasus. *Fam*. Amaryllidaceae. *Class* M.

G. *allenii* (After James Allen, a snowdrop grower who collected it.) Leaves broad. Flowers almond-scented. *Distr*. Caucasus.

G. *alpinus* Leaves broad. Flowers scented of bitter almonds. *Distr*. Caucasus.

G. *bortkewitschianus* Leaves narrow, strongly hooded at tip. *Distr*. Caucasus.

G. *byzantinus* See *G. plicatus*

G. *cabardensis* See *G. lagodechianus*

G. *caucasicus* Flowers often large. Many of the plants grown under this name in cultivation do not resemble those in the wild. *Distr*. Caucasus.

G. *corcyrensis* See *G. reginaeolgae*

G. *elwesii* (After H. J. Elwes (1846–1922), English naturalist, who introduced it into cultivation.) Leaves hooded at apex. Flowers honey-scented. *Distr*. N Greece, Bulgaria, W Turkey.

G. *fosteri* (After Sir Michael Foster F.R.S. (1836–1907) English physician and gardener.) Leaves broad, recurved. *Distr*. Turkey, Lebanon.

G. *gracilis* Leaves narrow and twisted. Flowers sometimes violet-scented. *Distr*. Bulgaria, Greece, Turkey.

G. *graecus* See *G. gracilis*

G. ikariae A variable species with leaves typically widest above the middle. *Distr.* Aegean Islands, N Turkey, N Iran, Caucasus.

G. kemulariae See *G. lagodechianus*

G. lagodechianus Leaves linear, with a recurved margin. *Distr.* Caucasus.

G. latifolius See *G. ikariae*

G. nivalis COMMON SNOWDROP. Leaves narrow, linear. Spring-flowering. A variable and very widely cultivated species with many different forms available. *Distr.* Europe, SW Turkey, naturalized in Great Britain.

G. platyphyllus See *G. ikariae*

G. plicatus Leaves folded, widest at their middle. *Distr.* Turkey, E Romania, Crimea.

G. reginae-olgae Similar to *G. nivalis* but flowering in the autumn. *Distr.* W Turkey, Balkans, Sicily.

G. rizehensis Leaves linear, deep, dull green. *Distr.* N Turkey, N Iran.

galaticus, -a: of central Asia Minor.

Galax (From Greek *gala*, milk, alluding to the white flowers.) A genus of 1 species of tuft-forming, evergreen, perennial herb with round to heart-shaped leaves and narrow spike-like racemes of very small white flowers that bear their parts in 5s. The leaves turn red-bronze in autumn and winter. *Distr.* SE North America. *Fam.* Diapensiaceae. *Class* D.

G. aphylla See *G. urceolata*

G. urceolata WAND FLOWER, WAND PLANT, BEETLE WEED, GALAXY, COLTSFOOT. Grown as an ornamental for ground cover and its inflorescences.

Galaxia (From Greek *galaxaios*, milky, alluding to the sap.) A genus of 14 species of perennial herbs with rounded corms, linear leaves and solitary funnel-shaped flowers. Several species are grown as ornamentals. *Distr.* SW South Africa. *Fam.* Iridaceae. *Class* M.

G. ovata Flowers yellow. *Distr.* South Africa (SW Cape).

G. versicolor Flowers mauve to pink, with a yellow throat. *Distr.* South Africa (SW Cape).

galaxy *Galax urceolata.*

gale, sweet *Myrica gale.*

Galeandra (From Latin *galea*, helmet-shaped, describing the anther cap of some species.) A genus of about 8 species of epiphytic and terrestrial orchids with cane-like pseudobulbs and few-flowered racemes of large flowers. Some species are grown as ornamentals. *Distr.* Central and South America. *Fam.* Orchidaceae. *Class* M.

G. batemanii See *G. baueri*

G. baueri Flowers ochre to chocolate brown, lip pale purple to white. *Distr.* Mexico to Panama and Surinam.

G. devoniana Flowers green or pale brown with red-brown lines, lip white. *Distr.* Tropical South America, from Venezuela to N Brazil.

galeatus, -a, -um: provided with a helmet.

Galega (From Greek *gala*, milk; *G. officinalis* was thought to improve the flow of goats' milk.) A genus of 6 species of perennial herbs with pinnate leaves and racemes of white or blue, pea-like flowers. *Distr.* Eurasia, E African mountains. *Fam.* Leguminosae. *Class* D.

G. officinalis GOAT'S RUE. Stems erect, to 1.5m high. Flowers lavender to white. Grown for fodder and as an ornamental, previously used medicinally. *Distr.* Europe, W Asia.

Galeobdolon See *Lamium*

G. luteum See *Lamium galeobdolon*

gale from an old English vernacular name. See *Myrica.*

galingale *Cyperus* **sweet** ~ *C. longus.*

galioides: from the genus *Galium*, with the ending *-oides*, indicating resemblance.

Galium (From Greek *gala*, milk, alluding to the use of *G. verum* to curdle milk.) BEDSTRAW, CLEAVERS. A genus of about 400 species of slender, annual and perennial herbs typically with square stems, whorled simple leaves and 4-lobed tubular flowers. Several species are grown as ornamentals. *Distr.* Cosmopolitan. *Fam.* Rubiaceae. *Class* D.

G. mollugo HEDGE BEDSTRAW, WHITE BEDSTRAW, WHITE MADDER, FALSE BABY'S BREATH. Perennial herb to 1.5m high. Flowers white, marked with purple, borne in large panicles. *Distr.* Europe, naturalized in North America.

G. odoratum SWEET WOODRUFF, WOODRUFF. Perennial herb to 45cm high. Leaves fragrant on drying. Flowers white. The dried foliage of this plant has been used as a flavouring, in pot-pouri, and medicinally since the Middle Ages. *Distr.* Europe, N Africa, W Asia.

G. verum LADY'S BEDSTRAW, YELLOW BEDSTRAW, OUR LADY'S BEDSTRAW. Perennial herb to 1m high. Flowers yellow, smelling of urine, attractive to flies. Formerly used in the manufacture of cheese and as a source of red dye. *Distr.* Europe, W Asia, naturalized North America.

gallberry *Ilex glabra.*

gallicus, -a, -um: of France.

Galtonia (After Sir Frances Galton (1812–1911), who travelled in South Africa.) A genus of 3 species of bulbiferous herbs with fleshy leaves and racemes of nodding, white or green flowers. Several species are grown as ornamentals. *Distr.* SE Africa. *Fam.* Hyacinthaceae. *Class* M.

G. candicans SUMMER HYACINTH, BERG LILY. Leaves white. Flowers white, fragrant. *Distr.* Orange Free State, Natal, Lesotho.

G. viridiflora Flowers pale green. *Distr.* South Africa (Orange Free State, Natal, Lesotho).

Gamolepis See *Steirodiscus*

gand flower *Polygala vulgaris.*

ganja *Cannabis sativa.*

Gardenia (After Dr Alexander Garden (1730–91), Scottish physician and botanist who lived in South Carolina.) A genus of about 200 species of evergreen shrubs and trees with leathery, opposite or whorled leaves and large, bell- or funnel-shaped, scented flowers. Several species and numerous cultivars are grown ornamentally and were previously an important source of cut flowers. Locally these plants are used medicinally and as a source of timber and dye. *Distr.* Tropical and warm regions of the Old World. *Fam.* Rubiaceae. *Class* D.

G. angusta CAPE JASMINE. Evergreen shrub or tree to 12m high. Flowers ivory-white, intensely fragrant. Numerous cultivars of this species have been raised. *Distr.* E Asia.

G. florida See *G. angusta*

G. grandiflora See *G. angusta*

G. jasminoides See *G. angusta*

G. thunbergia WHITE GARDENIA. Shrub or tree. Flowers cream to white. *Distr.* S Africa.

gardenia, white *Gardenia thunbergia.*

garganicus, -a, -um: of Monte Gargano, South Italy.

garland flower *Hedychium coronarium.*

garlic *Allium sativum* **Canada** ~ *A. canadense* **crow** ~ *A. vineale* **daffodil** ~ *A. neapolitanum* **false** ~ *A. vineale, Nothoscordum* **flowering** ~ *Allium neapolitanum* **fragrant-flowered** ~ *A. ramosum* **German** ~ *A. senescens* **giant** ~ *A. scorodoprasum* **grace** ~ *Nothoscordum* **hedge** ~ *Alliaria petiolata* **keeled** ~ *Allium carinatum* **Levant** ~ *A. ampeloprasum* **mouse** ~ *A. angulosum* **Naples** ~ *A. neapolitanum* **Oriental** ~ *A. tuberosum* **rosy** ~ *A. roseum* **round-headed** ~ *A. sphaerocephalon* **Sicilian honey** ~ *Nectaroscordum siculum* **society** ~ *Tulbaghia* **Spanish** ~ *Allium scorodoprasum* **stag's** ~ *A. vineale* **sweet** ~ *Tulbaghia fragrans* **wild** ~ *Allium ursinum, Tulbaghia* **wood** ~ *Allium ursinum.*

Garrya (After Nicholas Garry (died 1830) of the Hudson Bay Co.) SILK TASSEL, TASSEL TREE. A genus of 13 species of evergreen shrubs and small trees with opposite simple leaves and catkin-like inflorescences of small unisexual flowers. Several species are grown as ornamentals. *Distr.* W USA to Panama and the West Indies. *Fam.* Garryaceae. *Class* D.

G. elliptica FEVERBUSH. Shrub or small tree. Leaves elliptic. Male catkins 20cm long, grey with maroon-purple flowers. Female catkins similar, much shorter. *Distr.* W USA.

G. fremontii FEVERBUSH, SKUNK BUSH, QUININE BUSH. Shrub. Male catkins yellow, to 20cm long. Formerly used medicinally. *Distr.* California, Oregon.

garrya Garryaceae.

Garryaceae A family of 1 genus and 13 species. *Class* D.
See: Garrya.

Gasteria (From Greek *gaster*, belly, alluding to the swollen base of the corolla tube.) A genus of about 13 species of stemless perennial succulents with rosettes of leaves and racemes of typically pendulous, tubular flowers. Several species and numerous varieties are cultivated as ornamentals. *Distr.* South Africa. *Fam.* Aloaceae. *Class* M.

G. acinacifolia Leaf rosettes solitary or clustered. Leaves densely white-spotted. *Distr.* South Africa.

G. angulata See *G. carinata*

G. bicolor Leaves dark green, spotted with white. *Distr.* E Cape Province.

G. caespitosa See *G. bicolor*

G. candicans See *G. acinacifolia*

G. carinata Leaf rosettes form small dense clusters. Leaves covered in small warts. *Distr.* South Africa.

G. disticha Leaves 2-ranked. *Distr.* South Africa.

G. excavata See *G. carinata*

G. liliputiana See *G. carinata*

G. maculata See *G. bicolor*

G. marmorata See *G. bicolor*

G. nigricans See *G. disticha*

G. nitida Leaf rosettes solitary or with a few offsets. Flowers bright red pinks. *Distr.* South Africa.

G. picta See *G. bicolor*

G. subcarinata See *G. carinata*

G. verrucosa See *G. carinata*

Gastrochilus (From Greek *gaster*, belly, and *cheilos*, lip, alluding to the swollen lip of the flowers.) A genus of 15–20 species of epiphytic orchids with short stems and 2 ranks of leathery leaves. The flowers are usually large, resembling those of *Vanda*, and are borne in axillary racemes. Several species are cultivated as ornamentals. *Distr.* Himalaya and E Asia. *Fam.* Orchidaceae. *Class* M.

G. acutifolius Flowers to 20cm across, yellow or pale green. *Distr.* Nepal, NE India.

G. bellinus Flowers to 4cm across, fragrant, yellow, blotched red. *Distr.* Burma, Thailand.

G. dasypogon Flower bright yellow, spotted maroon, to 2.5cm across. *Distr.* India, Thailand and Sumatra.

Gaultheria (After Dr Gaultier (*c.*1708–58), Canadian botanist and physician.) WINTERGREEN. A genus of 150 species of evergreen shrubs with simple serrated leaves and 5-lobed, bell- or urn-shaped flowers. Some species are a source of the medicinal wintergreen, others are cultivated as ornamental shrubs. *Distr.* E Asia to New Zealand and North to South America. *Fam.* Ericaceae. *Class* D.

G. adenothrix Dwarf shrub with solitary white flowers. *Distr.* Japan.

G. antipoda Procumbent diminutive to large erect shrub with solitary white flowers. *Distr.* New Zealand.

G. cumingiana Small shrub with white flowers. *Distr.* S China to the Philippines.

G. cuneata Dwarf shrub with racemes of white flowers. *Distr.* W China.

G. depressa Dwarf cushion-forming shrub with pink to white flowers. *Distr.* New Zealand.

G. eriophylla Small shrub with red-pink, hairy flowers. *Distr.* SE Brazil.

G. forrestii Rounded bushy shrub with fragrant flowers. *Distr.* Yunnan Province, China.

G. fragrantissima Large shrub or small tree with yellow, very fragrant flowers. *Distr.* The mountains of India.

G. furiens See *G. insana*

G. hispida SNOW BERRY, WAXBERRY. Small shrub with compact racemes of small flowers and globose white fruit. *Distr.* Australia and Tasmania.

G. hookeri Shrub with pink flowers and violet fruit. *Distr.* E Himalaya, Sikkim.

G. humifusa ALPINE WINTERGREEN. Dwarf shrub with solitary white-pink flowers and red fruit. *Distr.* W North America.

G. insana Erect shrub with racemes of white flowers. *Distr.* S Chile and Argentina.

G. itoana Dwarf shrub with racemes of white flowers. *Distr.* China and Taiwan.

G. macrostigma Prostrate shrub with solitary white flowers and pink fruit. *Distr.* New Zealand.

G. miqueliana Small diffuse shrub with racemes of white flowers. *Distr.* Japan.

G. mucronata Robustly branched shrub with solitary white-pink flowers. Numerous garden hybrids are available. *Distr.* Chile and Argentina.

G. myrsinoides Prostrate or creeping shrubs with solitary white flowers and blue-black fruit. *Distr.* Costa Rica to Central Chile.

G. nana See *G. parvula*

G. nummularioides Prostrate small shrubs with solitary white to red flowers and blue-black fruit. *Distr.* Himalaya and W China.

G. ovalifolia See *G. fragrantissima*

G. parvula Dwarf, decumbent or erect shrub with solitary flowers and flat fruit. *Distr.* New Zealand.

G. phillyreifolia Medium-sized shrub with racemes of white flowers. *Distr.* Argentina.

G. procumbens CREEPING WINTERGREEN, PARTRIDGE BERRY, WINTERGREEN, CHECKERBERRY, TEABERRY, MOUNTAIN TEA. Small creeping shrub with white to pale pink flowers and aromatic red fruit. Wintergreen was originally extracted from this species; it is now extracted from *Betula lenta*. *Distr.* E North America.

G. prostrata See *G. myrsinoides*

G. pumila Diffuse prostrate shrub with solitary white flowers. *Distr.* Falkland Islands, Patagonia and Tierra del Fuego.

G. pyroloides Small stoloniferous shrub with short racemes of white-pink flowers. *Distr.* Himalaya.

G. rupestris Erect to procumbent shrub with racemes of small white flowers. *Distr.* New Zealand.

G. semi-infera Medium-sized shrub with racemes of white flowers. *Distr.* Sikkim to China.

G. shallon (From the local vernacular name.) SABAL, SHALLON, SALAL. Stoloniferous shrub with white-pink flowers and red fruit. *Distr.* W North America, naturalized in Great Britain.

G. sinensis Procumbent to compact shrub with solitary white flowers. *Distr.* Upper Burma, Yunnan and Tibet.

G. tasmanica Mat-forming shrub with solitary white flowers and red fruit. *Distr.* Tasmania.

G. tetramera Procumbent to erect shrub with green-white flowers. *Distr.* Tibet and W China.

G. thymifolia Dwarf shrub with solitary white-pink flowers. *Distr.* Upper Burma.

G. trichophylla Carpet- or cushion-forming shrub with solitary red to white flowers. *Distr.* W China and Himalaya.

G. willisiana See *G. eriophylla*

G. × wisleyensis (From Wisley, the site of the garden of the Royal Hoticultural Society.) A hybrid of *G. shallon* × *G. mucronata*.

G. yunnanensis Openly branching shrub with white-green flowers and black fruit. *Distr.* Yunnan Province, China.

Gaura (From Greek *gauros*, superb, alluding to the flowers.) A genus of 21 species of annual, biennial, and perennial herbs with large, pinnately lobed basal leaves and spikes of irregular or regular, tubular flowers that bear their parts in 4s. *Distr.* North America. *Fam.* Onagraceae. *Class* D.

G. lindheimeri Robust perennial. Flowers irregular, white, flushed pink, only lasting 1 day. Grown as an ornamental. *Distr.* Texas, Louisiana.

gay feather *Liatris* **Kansas** ~ *L. spicata*.

Gaylussacia (After J. L. Gay-Lussac (1778–1850), French chemist.) HUCKLEBERRY. A genus of 48 species of deciduous or evergreen shrubs with alternate simple leaves and racemes of tubular or bell-shaped, 5-lobed flowers. The fruit is edible, especially in pies, notably *G. baccata*. *Distr.* North and South America. *Fam.* Ericaceae. *Class* D.

G. baccata BLACK HUCKLEBERRY. Deciduous shrub to 1m high with dull red flowers and black fruit. The fruit is edible and often baked in pies. *Distr.* E North America.

G. brachycera BOX HUCKLEBERRY. Creeping evergreen shrubs with red-striped flowers. *Distr.* E USA.

Gazania (After Theodore of Gaza (1398–1478), who translated the botanical works of Theophrastus into Latin.) TREASURE FLOWER. A genus of 16 species of annual to perennial herbs with a milky latex, variable leaves and orange-brown flowers in large daisy-like heads that bear distinct rays. Several species are cultivated as half-hardy ornamentals. *Distr.* Tropical and S Africa. *Fam.* Compositae. *Class* D.

G. krebsiana Stemless perennial. Rays pale yellow, marked deep yellow at base. *Distr.* South Africa.

G. rigens TREASURE FLOWER. Decumbent perennial. Rays orange with a black spot at the base. *Distr.* South Africa.

G. splendens See *G. rigens*
G. uniflora See *G. rigens*

Geissolomataceae A family of 1 genus and 1 species of shrub of dry regions with simple sessile leaves arranged in 4 ranks along the stems. The flowers have 6 persistent bracts, and 4 petal-like sepals but no true petals. *Distr.* South Africa. *Class* D.

Geissorhiza (From Greek *geisson*, coping, and *rhiza*, root, alluding to the corms which appear to be covered in tiles.) A genus of 60–80 species of perennial herbs with hard corms, typically linear leaves and spikes of more or less regular, funnel-shaped flowers. *Distr.* S Africa. *Fam.* Iridaceae. *Class* M.

G. aspera Flowers star-like, white or blue.
G. crosa See *G. inflexa*
G. inflexa Flowers pink, red or purple, somewhat star-like.
G. ovata Flowers white to pink with a dark red-purple centre.

Gelasine (From Greek *gelasinos*, one who laughs.) A genus of 1–4 species of perennial herbs with flat corms, folded leaves and clusters of cup-shaped flowers. *Distr.* Subtropical South America. *Fam.* Iridaceae. *Class* M.

G. azurea See *G. elongata*
G. elongata Stems to 60cm high. Flowers bright blue with a white centre. Grown as an ornamental. *Distr.* Uruguay, S Brazil.

Gelidocalamus See *Indocalamus*

Gelsemium (From Italian *gelsomino*, jasmine.) A genus of 2 species of evergreen twining shrubs with simple opposite leaves and clusters of fragrant tubular flowers that bear their parts in 5s. *Distr.* S USA to Guatemala, 1 species SE Asia. *Fam.* Loganiaceae. *Class* D.

G. sempervirens YELLOW JASMINE, FALSE JASMINE, EVENING TRUMPET FLOWER, CAROLINA JASMINE. Grown as an ornamental and used medicinally for neuralgia and migraine. *Distr.* S USA, Central America.

geminiflorus, -a, -um: with twin flowers.

geminispinus, -a, -um: with twin spines.

gemmifera: bearing buds.

genevensis, -e: of Geneva, Switzerland.

geniculatus, -a, -um: bent, with a knee-like bend.

Genista (Classical Latin name for broom.) BROOM, WOADWAXEN. A genus of 87 species of shrubs and small trees with green stems, simple or trifoliate, occasionally absent leaves and racemes or heads of yellow pea-like flowers. The fruit is an explosive capsule. Numerous species are grown as ornamentals. *Distr.* Europe, Canary Islands, Mediterranean, W Asia. *Fam.* Leguminosae. *Class* D.

G. aetnensis MOUNT ETNA BROOM. Rounded tree. Leaves few or absent. Flowers yellow, borne in great profusion. *Distr.* Sicily, Sardinia.
G. anglica PETTY WHIN, NEEDLE FURZE. Procumbent to erect, deciduous, spiny shrub. *Distr.* W Europe.
G. cinerea Arching deciduous shrub. Flowers fragrant, abundant. *Distr.* SW Europe, N Africa.
G. hispanica SPANISH GORSE. Spiny, bushy, deciduous shrub. Flowers crowded into racemes. *Distr.* SW Europe.
G. horrida Cushion-forming spiny shrub. *Distr.* Pyrenees.
G. tinctoria DYER'S GREENWEED. Erect or ascending, non-spiny shrub. Flowers yield a yellow dye. *Distr.* Europe to Siberia, naturalized in North America.

genistifolius, -a, -um: from the genus *Genista*, and *folius*, leaved.

gentian Gentianaceae, *Gentiana* **bottle** ~ *G. andrewsii* **closed** ~ *G. andrewsii* **fringed** ~ *Gentianopsis* **great yellow** ~ *Gentiana lutea* **green** ~ *Frasera speciosa* **horse** ~ *Triosteum* **marsh** ~ *Gentiana pneumonanthe* **prairie** ~ *Eustoma grandiflorum* **spring** ~ *Gentiana verna* **trumpet** ~ *G. acaulis* **willow** ~ *G. asclepiadea* **yellow** ~ *G. lutea*.

Gentiana (After Gentius, king of Illyria in the 2nd century BC, who is said to have discovered the medicinal properties of *G. lutea* roots.) GENTIAN. A genus of about 300 species of typically perennial, occasionally annual or biennial herbs with whorls or pairs of simple leaves. The 4–7-lobed, funnel-shaped flowers are typically a stunning blue-purple but may also be yellow, white, or red. Very many species and cultivars are grown as ornamentals; the roots of a few are said to have medicinal qualities. *Distr.* Temperate, Arctic and mountainous regions, absent from Africa. *Fam.* Gentianaceae. *Class* D.

G. acaulis TRUMPET GENTIAN. Tuft-forming perennial. Flowers dark blue to 5cm tall. *Distr.* Alps and Pyrenees.

G. andrewsii CLOSED GENTIAN, BOTTLE GENTIAN. Erect perennial herb to 60cm high. Flowers tubular, closed at mouth, blue with a white tip. *Distr.* NE North America.

G. asclepiadea WILLOW GENTIAN. Perennial to 60cm high. Leaves narrow. Flowers clustered, erect, deep blue. *Distr.* Europe, W Asia.

G. clusii (After Charles de l'Ecluse (Carolus Clusius) (1526–1609), Flemish botanist.) Tufted perennial. Flowers solitary, trumpet-shaped, azure-blue with green spots. *Distr.* Alps.

G. crinita See *Gentianopsis crinita*

G. × hascombensis See *G. septemfida*

G. kochiana See *G. acaulis*

G. lutea YELLOW GENTIAN, GREAT YELLOW GENTIAN. Erect unbranched perennial to 1.4m high. Flowers yellow, borne in whorls about the stem. Commercial source of gentian root used as a tonic and in flavouring liquors. *Distr.* Europe.

G. pneumonanthe MARSH GENTIAN. Perennial. Stems slender. Flowers deep purple with lines of green dots. A source of a blue dye. *Distr.* Europe, Caucasus, N Asia.

G. saxosa Small hummock-forming perennial. Leaves leathery. Flowers white. *Distr.* New Zealand.

G. septemfida Erect or ascending perennial. Flowers in clusters, deep-blue with pale spots. *Distr.* W Asia.

G. sino-ornata Prostrate spreading perennial. Leaves narrow. Flowers blue. *Distr.* W China, Tibet.

G. verna SPRING GENTIAN. Small tuft-forming perennial. Flowers bright blue with a white throat. *Distr.* Europe to mountains of Asia.

Gentianaceae The Gentian family. 76 genera and 1200 species of perennial herbs and occasionally shrubs with opposite simple leaves and bell-shaped, typically purple-blue flowers. Many species are grown as ornamentals and some have medicinal properties. *Distr.* Wide-spread. *Class* D.
See: Blackstonia, Centaurium, Chironia, Crawfurdia, Eustoma, Exacum, Frasera, Gentiana, Gentianopsis, Orphium.

gentianoides from the genus *Gentiana*, with the ending *-oides*, indicating resemblance.

Gentianopsis (From the genus name *Gentiana*, and Greek *opsis*, resemblance.) FRINGED GENTIAN. A genus of 16–25 species of annual and biennial herbs with simple opposite leaves and blue or white, 4-lobed, funnel-shaped, fringed flowers. Several species are grown as ornamentals. *Distr.* N temperate Asia and America. *Fam.* Gentianaceae. *Class* D.

G. crinita Annual to biennial herb to 1m high. Flowers 5cm long, bright blue. *Distr.* E North America.

Geogenanthus (From Greek *ge*, earth, *gen-*, producing, and *anthos*, flower.) A genus of 2–3 species of herbs with creeping stems and a few broad leaves. The flowers are borne in sparse cymes and have 3 sepals, 3 petals and 5 unequal stamens. *Distr.* Amazonian Brazil, Colombia, Ecuador and Peru. *Fam.* Commelinaceae. *Class* M.

G. poeppigii SEERSUCKER PLANT. Leaves nearly round, dark green with silver stripes. Grown as a tropical bedding plant. *Distr.* Brazil, Peru.

geometrizans: regularly marked.

georgianus, -a, -um: of Georgia.

Geraniaceae The Geranium family. 11 genera and about 700 species of annual or perennial herbs and rarely shrubs with simple or compound, often glandular hairy leaves. The flowers are showy, regular or slightly irregular, with their parts in 5s. The genus *Pelargonium* is particularly important as the source of the cultivated GERANIUM. *Distr.* Widespread but mainly in temperate and subtropical regions. *Class* D.
See: Erodium, Geranium, Monsonia, Pelargonium.

geraniifolius, -a, -um: with geranium-like leaves.

geranioides: geranium-like.

Geranium (From Greek *geranos*, crane, alluding to the beak-like fruits.) CRANESBILL. A genus of 300 species of annual or perennial herbs and a few thick-stemmed shrublets with simple or more typically palmately divided leaves. The flowers bear 5 sepals, 5 petals, and 10 stamens and are borne in diffuse or umbel-like inflorescences. This genus should not be confused with the closely related genus *Pelargonium*, which bears the common name GERANIUM. Many species and cultivars are grown as ornamentals, especially for groundcover. *Distr.* Temperate and tropical mountain regions. *Fam.* Geraniaceae. *Class* D.

G. lucidum SHINING CRANESBILL. Erect annual. Stems shiny red, somewhat succulent. Leaves glossy. Flowers small, deep pink. Weed. *Distr.* Europe and Mediterranean to Himalaya.

G. phaeum MOURNING WIDOW, DUSKY CRANESBILL, BLACK WIDOW. Clump-forming perennial. Leaves palmately lobed. Flowers purple-maroon in lax inflorescences. *Distr.* Europe.

G. pratense MEADOW CRANESBILL. Clump-forming perennial. Leaves deeply divided. Flowers in dense inflorescences, violet-blue to white. *Distr.* Europe, Central Asia, Himalaya.

G. robertianum HERB ROBERT. Annual or biennial. Leaves palmate with each section pinnately lobed. Flowers small, dark pink. Formerly used medicinally. *Distr.* N temperate regions.

G. sanguineum BLOODY CRANESBILL. Hummock-forming perennial. Leaves dark green, deeply divided. Flowers striking magenta-pink. *Distr.* Europe, Caucasus.

G. sylvaticum WOOD CRANESBILL. Erect, clump-forming perennial. Flowers purple-violet to blue, in dense inflorescences. *Distr.* Europe, N Asia.

G. wallichianum (After Nathaniel Wallich (1786–1854), Danish surgeon and botanist with the East India Company, who introduced it into cultivation in 1819.) Perennial with leaves in distinct rosettes. Petals notched, pink-purple. The roots have been used in tanning and dying. *Distr.* Afghanistan, Himalaya.

geranium Geraniaceae, *Pelargonium* **almond** ~ *P. quercifolium* **Alpine** ~ *Erodium reichardii* **bedding** ~ *Pelargonium x hortorum* **beefsteak** ~ *Begonia rex* **feather** ~ *Chenopodium botrys* **gouty** ~ *Pelargonium gibbosum* **hanging** ~ *P. peltatum* **ivy** ~ *P. peltatum* **knotted** ~ *P. gibbosum* **mint** ~ *Tanacetum balsamita* **oak-leaved** ~ *Pelargonium quercifolium* **peppermint** ~ *P. tomentosum* **rock** ~ *Heuchera americana* **rose** ~ *Pelargonium graveolens* **strawberry** ~ *Saxifraga stolonifera* **sweet-scented** ~ *Pelargonium graveolens* **village oak** ~ *P. quercifolium*.

Gerbera (After Traugott Gerber (died 1743), German naturalist who travelled in Russia.) TRANSVAAL DAISY, BARBERTON DAISY. A genus of 35 species of perennial, often stemless herbs usually with simple leaves and flowers in daisy-like heads that bear distinct rays. Several species and numerous cultivars are grown as tender ornamentals and for cut flowers. *Distr.* Tropical and South Africa to Bali. *Fam.* Compositae. *Class* D.

G. aurantiaca TRANSVAAL DAISY, HILTON DAISY. Leaves simple. Flowers pink or red. *Distr.* South Africa (Natal).

G. jamesonii BARBERTON DAISY, TRANSVAAL DAISY. The parent of many hybrids and cultivars that produce long-lasting cut flowers in a great variety of colours. *Distr.* Transvaal.

germander *Teucrium, T. chamaedrys* **American** ~ *T. canadense* **shrubby** ~ *T. fruticans* **tree** ~ *T. fruticans* **wall** ~ *T. chamaedrys* **wood** ~ *T. scorodonia*.

germanicus, -a, -um: of Germany.

Gesneria (After Conrad Gessner (1516–65), Swiss naturalist.) A genus of 47 species of perennial herbs, shrubs and small trees with simple, opposite, or alternate leaves and tubular or bell-shaped, 5-lobed flowers. Some species are cultivated as ornamentals. *Distr.* Tropical America. *Fam.* Gesneriaceae. *Class* D.

G. acaulis Subshrub. Leaves to 24cm long, red-green-hairy below. Flowers numerous, red-orange with a yellow interior. *Distr.* Jamaica.

G. cardinalis See *Sinningia cardinalis*

Gesneriaceae The African violet family. 133 genera and about 2500 species of herbs,

shrubs and rarely trees with simple leaves and bisexual, typically irregular flowers. A number of species are grown as ornamentals. *Distr.* Tropical and subtropical, occasionally warm-temperate regions. *Class* D.
See: Achimenes, × *Achimenantha, Aeschynanthus, Alsobia, Asteranthera,* × *Brigandra, Briggsia, Chrysothemis, Codonanthe,* × *Codonatanthus, Columnea, Episcia, Eucodonia, Gesneria, Gloxinia, Haberlea,* × *Heppimenes, Koellikeria, Kohleria, Mitraria, Nematanthus,* × *Nephimenes, Niphaea, Petrocosmea, Ramonda, Rhabdothamnus, Saintpaulia, Sinningia, Smithiantha,* × *Smithicodonia,* × *Streptocarpella, Streptocarpus.*

gesneriiflorus, -a, -um: from the genus *Gesneria*, and *flos*, flower.

Geum (The classical Latin name for these plants.) AVENS. A genus of 65 species of rhizomatous herbs and subshrubs with simple or pinnately lobed leaves and regular flowers that bear 5 petals and numerous stamens. A number of species and cultivars are grown as ornamentals. *Distr.* Temperate and cold regions. *Fam.* Rosaceae. *Class* D.

G.* × *borisii A hybrid of *G. bulgaricum* × *G. reptans* that has given rise to numerous garden cultivars. *Distr.* Occurs naturally in Bulgaria but has been repeated many times in cultivation.

G. chiloense Leaves lobed or cut. Flowers scarlet. Several garden cultivars have been raised from this species. *Distr.* Chile.

G. coccineum See *G. chiloense*

G. macrophyllum YELLOW AVENS. Stems to 1m high. Leaves lobed. Flowers small, yellow. *Distr.* E Asia, North America.

G. montanum ALPINE AVENS. Clump-forming perennial. Flowers golden-yellow. *Distr.* S Europe.

G. reptans CREEPING AVENS. Clump-forming perennial. Flowers bright yellow. *Distr.* Europe.

G. rivale WATER AVENS, INDIAN CHOCOLATE, PURPLE AVENS, CHOCOLATE ROOT. Rhizome thick. Flowers bell-shaped, nodding. Sepals purple-brown, petals yellow-white. Fruit prickly. *Distr.* Europe, Asia, North America.

G. triflorum PURPLE AVENS, LION'S BEARD, OLD MAN'S WHISKERS, GRANDFATHER'S BEARD, PRAIRIE SMOKE. Flowers pink-yellow. Fruit hairy. *Distr.* North America.

G. urbanum WOOD AVENS, HERB BENNET, CLOVEROOT. Stems to 60cm high. Flowers erect, pale-yellow. Fruit hooked. Formerly used medicinally. *Distr.* Europe, W Asia, Mediterranean.

Gevuina (From the local vernacular name.) A genus of 1 species of evergreen shrub or tree with pinnately divided, leathery leaves and racemes of tubular cream flowers that are followed by red-black fruits. *Distr.* S South America. *Fam.* Proteaceae. *Class* D.

G. avellana CHILEAN HAZELNUT, CHILE NUT. Grown as a half-hardy ornamental. The seeds are edible and taste a little like hazelnuts.

gherkin *Cucumis sativus.*

ghost plant *Graptopetalum paraguayense.*

ghost tree *Davidia involucrata.*

Gibasis (From Latin *gibbus*, swollen, and Greek *basis*, base, alluding to the sack-like base of the sepals.) A genus of about 11 species of annual to perennial tuberous herbs with simple pointed leaves and numerous regular flowers that bear 3 sepals, 3 petals, and 6 stamens. *Distr.* Tropical America. *Fam.* Commelinaceae. *Class* M.

G. pellucida BRIDAL VEIL. Sparsely hairy perennial with white flowers. Often grown in hanging baskets. *Distr.* Mexico.

gibberosus, -a, -um: swollen in a lopsided way.

gibbiflorus, -a, -um: with flowers swollen on one side.

gibbosus, -a, -um: swollen, often in a lopsided way.

gibbus, -a, -um: swollen in a lopsided way.

gibraltaricus, -a, -um: of Gibraltar.

giganteus, -a, -um: gigantic, very large.

gigas: giant.

Gilia (After Felipe Gil, 18th-century Spanish botanist.) A genus of about 25 species of annual and perennial herbs with pinnately lobed, alternate leaves and clusters of tubular to funnel-shaped flowers that bear their parts in 5s. Some species are cultivated as ornamentals. *Distr.* NW North America. *Fam.* Polemoniaceae. *Class* D.

G. aggregata See *Ipomopsis aggregata*

G. tricolor BIRD'S EYES. Annual herb. Flowers violet-blue with a yellow-orange tube. *Distr.* California.

gilia, scarlet *Ipomopsis aggregata.*

Gillenia (After Arnold Gille (Gillenius), 17th-century German botanist.) A genus of 2 species of rhizomatous herbs with alternate trifoliate leaves and loose panicles of white or pink flowers that bear 5 petals and numerous stamens. Both species are grown as ornamentals and their roots were formerly used medicinally. *Distr.* E North America. *Fam.* Rosaceae. *Class* D.

G. stipulata AMERICAN IPECACUANHA. Stems to 1.2m high. Flowers white.

G. trifoliata INDIAN PHYSIC, ROMAN'S ROOT. Stem to 80cm high. Flowers tinged purple.

Gill over the ground *Glechoma hederacea.*

gin, black *Kingia australis.*

gingelly *Sesamum indicum.*

ginger Zingiberaceae, *Zingiber officinale* **Bengal** ~ *Z. purpureum* **Canton** ~ *Z. officinale* **cassumar** ~ *Z. purpureum* **common** ~ *Z. officinale* **crepe** ~ *Costus speciosus* **Japanese** ~ *Zingiber mioga* **kahili** ~ *Hedychium gardnerianum* **Malay** ~ *Costus speciosus* **mioga** ~ *Zingiber mioga* **red** ~ *Alpinia purpurata* **shell** ~ *A. zerumbet* **spiral** ~ *Costus* **stem** ~ *Zingiber officinale* **torch** ~ *Etlingera elatior* **variegated** ~ *Alpinia vittata* **white** ~ *Hedychium coronarium* **wild** ~ *Asarum*, *A. canadense*, *Costus speciosus* **yellow** ~ *Hedychium flavescens.*

gingham golf ball *Euphorbia obesa.*

Ginkgo (From the Chinese *yin-kuo*, silver apricot.) A genus of 1 species of deciduous, coniferous tree with characteristic fan-shaped leaves in which the veins always branch into 2 equal parts. The fruiting 'cone' is fleshy and smells of rancid butter. 200-million-year-old fossils of nearly identical plants have been found, prompting this plant to be labelled 'a living fossil'. *Distr.* Cultivated in China since antiquity but not known in the wild. *Fam.* Ginkgoaceae. *Class* G.

G. biloba MAIDENHAIR TREE. A slow-growing but very attractive ornamental and street tree.

ginkgo Ginkgoaceae.

Ginkgoaceae A family of 1 genus and 1 species. *Class* G.
See: Ginkgo.

ginseng *Panax ginseng* **American** ~ *P. quinquefolius* **Siberian** ~ *Eleutherococcus senticosus.*

gipsywort *Lycopus europaeus.*

Gisekiaceae A family of 1 genus and 2 species of weedy herbs with simple hairy leaves and numerous small flowers that lack petals. *Distr.* Africa, Arabia and tropical Asia. *Class* D.

glabellum, -a, -um: rather smooth, hairless.

glaber: smooth, hairless.

glaberrimus: very smooth.

glabra: smooth, hairless.

glabratus, -a, -um: rather smooth, hairless.

glabrifolius, -a, -um: with smooth hairless leaves.

glabriusculus, -a, -um: rather smooth, hairless.

glacialis, -e: growing in icy places.

gladdon *Iris foetidissima.*

Gladiolus (From Latin *gladiolus*, the diminutive of sword, alluding to the shape of the leaves.) A genus of about 180 species of perennial herbs with rounded or rhizome-like corms, sword-shaped leaves and spikes of irregular, funnel-shaped flowers. Many species and numerous cultivars are grown as ornamentals. *Distr.* Africa, Madagascar and the Mediterranean Region, extending to N Europe *Fam.* Iridaceae. *Class* M.

G. callianthus Flowers white with purple markings, highly scented. *Distr.* E Africa.

G. cardinalis WATERFALL GLADIOLUS, NEW YEAR LILY. Inflorescence arching. Flowers few, large, bright red. *Distr.* South Africa (SW Cape Province).

G. × hortulanus FLORIST'S GLADIOLUS. A complex group of hybrids involving numerous parent species. *Distr.* Garden origin.

G. illyricus Inflorescence 2-sided. Flowers magenta with white marks. *Distr.* SW Europe, naturalized North America and British Isles.

G. italicus FIELD GLADIOLUS. Inflorescence a loose spike of 10–20 pink-purple flowers. *Distr.* S Europe, NW Africa to Afghanistan.

G. segetum See *Gladiolus italicus*

gladiolus: field ~ *Gladiolus italicus* **florist's** ~ *G. x hortulanus* **water** ~ *Butomus umbellatus* **waterfall** ~ *Gladiolus cardinalis.*

gladwyn *Iris foetidissima* **stinking** ~ *I. foetidissima.*

glanduliferus, -a, -um: bearing glands.

glanduliflorus, -a, -um: with glandular flowers.

glandulosissimus, -a, -um: with very many glands.

glandulosus, -a, -um: glandular.

glareosus, -a, -um: pertaining to gravel.

glaucescens: rather blue-green, downy.

Glaucidiaceae A family of 1 genus and 1 species. *Class* D.
See: Glaucidium.

Glaucidium (From the name of the genus *Glaucium*, with which it shares similar flowers.) A genus of 1 species of erect rhizomatous herb with round or kidney-shaped, lobed leaves and poppy-like bright yellow flowers that bear 4 petal-like sepals and no true petals. *Distr.* Japan. *Fam.* Glaucidiaceae. *Class* D.

G. palmatum Grown as a hardy ornamental.

glaucifolius, -a, -um: with glaucous, blue-green leaves.

glaucinus, -a, -um: slightly glaucous, blue-green.

Glaucium (From Greek *glaukos*, grey-green, alluding to the colour of the foliage.) HORNED POPPY. A genus of about 25 species of annual, biennial, and perennial herbs with pinnately lobed leaves and panicles of cup-shaped, yellow to red flowers that bear 2 sepals, 4 petals, and numerous stamens. The fruit is a large horn-like capsule. Several species are grown as ornamentals. *Distr.* Europe, SW and Central Asia. *Fam.* Papaveraceae. *Class* D.

G. corniculatum RED HORNED POPPY. Biennial to 45cm high. Flowers red, occasionally orange-yellow. *Distr.* Europe, W Asia.

G. flavum YELLOW HORNED POPPY, SEA POPPY, HORNED POPPY. Biennial or perennial to 1m high. Flowers golden-yellow. The seeds were formerly a source of lamp oil. *Distr.* Europe, N Africa, W Asia, naturalized North America.

G. phoenicium See *G. corniculatum*

glaucoalbus, -a, -um: glaucous white.

glaucocarpus, -a, -um: with blue-green, glaucous fruit.

glaucophyllus, -a, -um: with blue-green, glaucous leaves.

glaucus, -a, -um: glaucous.

Glaux (A classical Greek plant name.) A genus of 1 species of perennial succulent herb with a creeping stem, simple leaves, and small, pink or white, 5-lobed flowers that lack petals. *Distr.* N temperate regions. *Fam.* Primulaceae. *Class* D.

G. maritima SEA MILKWORT, BLACK SALTWORT. A plant of coastal salt marshes and cliffs.

Glechoma (From Greek *glechon*, a kind of mint.) A genus of about 10 species of perennial, creeping, stoloniferous herbs with simple, opposite, almost round leaves and 1-sided inflorescences of 2-lipped flowers. *Distr.* Temperate Eurasia. *Fam.* Labiatae. *Class* D.

G. hederacea GROUND IVY, ALEHOOF, FIELD BALM, GILL OVER THE GROUND, RUN AWAY ROBIN. Mat-forming perennial. Several variegated cultivars are widely grown, particularly in hanging baskets. The leaves were formerly used to make a medicinal tea and added to beer as a preservative on long voyages. *Distr.* Europe, Asia, naturalized in North America.

glechoma resembling the genus *Glechoma*.

Gleditsia (After Johann Gottlieb Gleditsch (1714–86), director of the Berlin botanic garden.) HONEY LOCUST. A genus of 14 species of deciduous thorny trees with alternate pinnate leaves and racemes of small green-white flowers that bear 6–10 prominent stamens. Several species are grown as ornamentals and others are used as a source of timber. *Distr.* Asia, North and South America. *Fam.* Leguminosae. *Class* D.

G. caspica CASPIAN LOCUST. To 15m high. Trunk armed with large branched spines. Flowers densely packed. *Distr.* N Iran, Transcaucasus.

G. triacanthos HONEY LOCUST. To 45m high. Trunk armed with simple or branched spines. Flowers 3mm across, green. Numerous cultivars have been produced from this tree. *Distr.* North America.

Gleicheniaceae A family of 4 genera and 140 species of terrestrial, small to very large ferns with at least partially creeping stems and pinnately divided leaves. The sori are round, borne on the underside of the leaf, and lack an indusium. *Distr.* Widespread in tropical regions. *Class* F.

Globba (From *galoba*, the local vernacular name in Amboina, Indonesia.) A genus of about 70 species of rhizomatous herbs with reed-like stems, linear leaves, and pendulous racemes of irregular, long-tubed flowers subtended by showy bracts. *Distr.* E and tropical Asia. *Fam.* Zingiberaceae. *Class* M.

G. winitii Flowers yellow. Bracts pink to purple. *Distr.* Thailand.

globeflower *Trollius*, *T. europaeus*.

globifera: bearing globes or spheres.

globoides: resembling a globe or sphere.

globosus, -a, -um: spherical.

Globularia (From Latin *globulus*, small ball, alluding to the flower heads.) GLOBE DAISY. A genus of 22 species of perennial hummock- or mat-forming herbs and shrubs with somewhat leathery, simple leaves and rounded heads of small, 2-lipped, tubular flowers. Several species are cultivated as ornamentals especially as rock-garden plants. *Distr.* Europe, Turkey, Canary Islands. *Fam.* Globulariaceae. *Class* D.

G. bellidifolia See *G. meridionalis*

G. cordifolia Dwarf, creeping, mat-forming shrub. Leaves small. Flower-heads short-stalked, lavender-blue. *Distr.* Alps.

G. meridionalis Dwarf, robust, creeping shrub. Flower-heads lavender-blue, short-stalked. *Distr.* Alps to SE Europe.

G. nana See *G. repens*

G. nudicaulis Tufted herb. Leaves long. Flower-heads of long leafless stalks. *Distr.* Alps, Pyrenees.

G. repens Creeping dwarf shrub. Leaves folded along the midrib. Flower-heads blue, sessile. *Distr.* SW Europe.

Globulariaceae A family of 2 genera and 24 species of perennial herbs and subshrubs with spirally arranged, simple leaves and showy heads of small 2-lipped flowers. *Distr.* Macaronesia to Turkey and NE tropical Africa. *Class* D.
See: Globularia.

globulariifolius, -a, -um: from the genus *Globularia*, and *folius*, leaved. Globulifera.

globulus, -a, -um: spherical, globular.

glochidiatus, -a, -um: with barbs or stiff bristles.

glomeratus, -a, -u,: clustered closely together in spherical form.

glomerulatus, -a, -um: clustered together in small spheres.

Gloriosa (From Latin *gloriosus*, glorious.) A genus of 1 species of tuberous climbing herb with thin stems and simple leaves, the tips of which are developed into tendrils. The large, nodding, yellow to red or purple flowers are borne on long stalks from the axils of the leaves and have 6 reflexed tepals and 6 stamens. *Distr.* Tropical Africa, tropical Asia. *Fam.* Colchicaceae. *Class* M.

G. caramii See *G. superba*

G. carsonii See *G. superba*

G. lutea See *G. superba*

G. rothschildiana See *G. superba*

G. superba GLORY LILY, CLIMBING LILY, CREEPING LILY, FLAME LILY. A variable species with many cultivars available, some of which have been given species names. Tubers are poisonous.

gloriosus, -a, -um: glorious.

glory bower *Clerodendrum philippinum.*

glory bush *Tibouchina urvilleana.*

glory flower *Eccremocarpus scaber.*

glory of the snow *Chionodoxa, C. luciliae.*

glory of the sun *Leucocoryne ixioides.*

glory plant, purple *Sutera grandiflora.*

Gloxinia (After Benjamin Peter Gloxin (fl. 1785), German botanist.) A genus of 15 species of rhizomatous herbs with broad simple leaves and solitary or paired, 5-lobed, 2-lipped, tubular flowers. This genus must not be confused with the related species *Sinningia speciosa*, FLORIST'S GLOXINIA. Several species and cultivars are grown ornamentally. *Distr.* Tropical America. *Fam.* Gesneriaceae. *Class* D.

G. perennis CANTERBURY BELLS. Leaves mid-green, heart-shaped. Flowers bell-shaped, pale purple. *Distr.* Colombia to Peru.

gloxinia: creeping ~ *Asarina erubescens* **florist's** ~ *Sinningia speciosa.*

glumaceus, -a, -um: like husks or glumes (floral parts) of grasses.

glutinicaulis, -e: with sticky stems.

glutinosus, -a, -um: sticky.

gluttatus, -a, -um: spotted.

Glyceria (From Greek *glykys*, sweet, alluding to the edible seeds of some species.) SWEET GRASS, MANNA GRASS. A genus of about 16 species of rhizomatous, aquatic or marginal grasses with reed-like stems, folded leaves and delta-shaped panicles of spikelets. Nutritious pasture grasses for cattle; the seeds of some species were formerly eaten by North American Indians. *Distr.* Cosmopolitan. *Fam.* Gramineae. *Class* M.

G. aquatica See *G. maxima*

G. fluitans SWEET GRASS, MANNA GRASS. *Distr.* N temperate regions.

G. maxima REED GRASS, REED SWEET GRASS, REED MEADOW GRASS. Grown as a bog-garden ornamental. *Distr.* Temperate Europe and Asia.

G. spectabilis See *G. maxima*

Glycine (From Greek *glykys*, sweet; the leaves and roots of some species taste sweet.) A genus of 9 species of hairy, erect, or sprawling herbs typically with trifoliate leaves and racemes of purple-pink, pea-like flowers. *Distr.* Asia to Australia. *Fam.* Leguminosae. *Class* D.

G. max SOYA BEAN, SOJA BEAN, SOYBEAN. An erect bushy annual to 2m high. The most highly protein-rich vegetable crop known. Seeds eaten in many forms, crushed for oil, milled for flour, or fermented. Also grown as a forage crop. *Distr.* Asia to Australia, widely naturalized elsewhere.

glycyphyllos: with sweet-smelling leaves.

Glycyrrhiza (From Greek *glykys*, sweet, and *rhiza*, root.) A genus of 20 species of sticky perennial herbs with trifoliate or pinnate leaves and pea-like flowers. Local sources of liquorice. *Distr.* Eurasia, Australia, North America, temperate South America. *Fam.* Legu-minosae. *Class* D.

G. glabra LIQUORICE, SWEETWOOD. Grown for the rhizomes, which are the source of licorice of commerce. *Distr.* Mediterranean to Central Asia.

glycyrrhiza: with sweet roots.

glyptostroboides: from the genus *Glyptostrobus*, with the ending *-oides*, indicating resemblance.

Glyptostrobus (From Greek *glypto*, to carve, and *strobilus*, cone, alluding to the pitted cone scales.) A genus of 1 species of deciduous, coniferous tree with scale-like and linear leaves and small pear-shaped fruiting cones. *Distr.* SE China. *Fam.* Taxodiaceae. *Class* G.

G. lineatus See *G. pensilis*

G. pensilis CHINESE SWAMP CYPRESS. Occasionally grown as an ornamental.

Gnaphalium (The Greek name for these plants.) CUDWEED, EVERLASTING. A genus of about 150 species of annual to perennial, often aromatic herbs with simple leaves and flowers in small daisy-like heads that lack distinct rays but have showy papery bracts. Several species are cultivated as ornamentals. *Distr.* Cosmopolitan. *Fam.* Compositae. *Class* D.

G. californicum GREEN EVERLASTING. Erect biennial herb. Bracts silver-white. *Distr.* W North America.

G. ramosissimum PINK EVERLASTING. Erect biennial. Bracts pink-white. *Distr.* California.

Gnetaceae A family of 1 genus and 28 species. *Class* G.
See: Gnetum.

Gnetum A genus of 28 species of deciduous vines, shrubs and trees with simple entire leaves. The inflorescence consists of a spike with a number of collars, each containing small, simple, male and female flowers. These plants have been considered to represent an evolutionary link between the true flowering plants and the conifers. They are a source of fibre and edible seeds. *Distr.* Widespread in tropical regions. *Fam.* Gnetaceae. *Class* G.

G. gnemon Tree to 18m high. Seed ground for flour. *Distr.* Tropical Asia.

gnidioides: from the genus *Gnidia*, with the ending *oides*, indicating resemblance.

goat nut *Simmondsia chinensis.*

goat root *Ononis natrix.*

goat's beard *Tragopogon, T. pratensis, Aruncus, A. dioicus false* ~ *Astilbe biternata.*

goat's rue *Galega officinalis.*

godetia See *Clarkia*

god's eye *Veronica chamaedrys.*

Goetzeaceae A family of 4 genera and 7 species of shrubs and small trees. This family is sometimes included within the Solanaceae. *Distr.* West Indies. *Class* D.

gold and silver flower *Lonicera japonica.*

golden chain *Laburnum anagyroides* **Alpine** ~ *L. alpinum.*

golden club *Orontium aquaticum.*

golden drop *Onosma tauricum.*

golden knee *Chrysogonum.*

golden pothos *Epipremnum aureum.*

golden rod *Solidago.*

golden seal *Hydrastis canadensis.*

golden shower *Pyrostegia venusta.*

golden shower tree *Cassia fistula.*

golden star *Hypoxis hygrometrica.*

golden top *Lamarckia aurea.*

golden wonder *Senna didymobotrya.*

gold guinea plant *Hibbertia scandens.*

goldilocks *Aster linosyris.*

gold thread *Coptis.*

Gomesa (After Dr Bernadino Antonio Gomes, 19th-century Portuguese botanist and physician.) A genus of about 8 species of epiphytic orchids with short pseudobulbs and racemes of small fragrant flowers. *Distr.* South America, mostly Brazil. *Fam.* Orchidaceae. *Class* M.

G. crispa Flowers 2cm across, yellow, tinged green. *Distr.* Brazil.

G. planifolia Flowers yellow-green, highly fragrant, borne in an arching raceme. *Distr.* Brazil, Paraguay, Argentina.

Gomortegaceae A family of 1 genus and 1 species of tree with simple opposite leaves, that are dotted with aromatic oil glands, and racemes of primitive regular flowers. *Distr.* Chile. *Class* D.

Gomphocarpus physocarpus See *Asclepias physocarpa*

gomuti *Arenga pinnata.*

Gongora (After Don Antonio Caballero y Gongora, Bishop of Cordoba in the late 18th century.) A genus of about 25 species of epiphytic orchids with pseudobulbs and large leaves. The flowers are typically dull-coloured, highly scented and borne in arching or pendent racemes. *Distr.* Tropical America. *Fam.* Orchidaceae. *Class* M.

G. galeata Flowers numerous, to 5cm across, yellow-brown. *Distr.* Mexico.

G. quinquenervis Flowers pale yellow, spotted and striped red-brown. *Distr.* Central and South America.

goniocalyx: with an angled whorl of sepals.

Goniolimon (From Greek *gonio*, angled, alluding to the branches, and the genus name

Limonium.) A genus of 20 species of perennial herbs and subshrubs with basal rosettes of leathery or somewhat fleshy leaves and panicles or spikes of small flowers. Some species are cultivated as ornamentals. *Distr.* Europe to Mongolia, NW Africa. *Fam.* Plumbaginaceae. *Class* D.

G. tataricum STATICE, TARTARIAN STATICE. Perennial herb to 30cm high. Flowers maroon, borne in dense spikes. *Distr.* S Europe, N Africa, Caucasus, Russia.

Goniophlebium (From Greek *gonion*, angle, and *phlebion*, vein.) A genus of about 20 species of epiphytic ferns with pinnate leaves. Several species are grown as ornamentals. *Distr.* Tropical Asia to the Pacific region. *Fam.* Polypodiaceae. *Class* F.

G. subauriculatum Fronds to 1m long, pendent. *Distr.* NE India and SW China to Australia.

Goodenia (After Samuel Goodenough (1743–1827), Bishop of Carlirle, botanist, and vice-president of the Royal Society of London.) A genus of 170 species of herbs and shrubs with simple, basal or alternate leaves and racemes or cymes of 5-lobed tubular flowers. Several species are grown as tender ornamentals. *Distr.* Australasia, 1 species SE Asia. *Fam.* Goodeniaceae. *Class* D.

G. humilis Small herb with hairy yellow flowers. *Distr.* S Australia.

Goodeniaceae A family of 12 genera and about 450 species of herbs and some shrubs with entire to pinnately cut leaves and irregular, bisexual, 2-lipped flowers that typically bear their parts in 5s. *Distr.* Australasia, tropical and E Asia, South America, and the shores of the Indian Ocean. *Class* D.
See: Goodenia, Scaevola, Selliera.

Good King Henry *Chenopodium bonus-henricus.*

good luck leaf *Oxalis tetraphylla.*

good luck plant *O. tetraphylla.*

Goodyera (After John Goodyer (1592–1664), English botanist.) A genus of about 40 species of terrestrial, rarely epiphytic orchids with fleshy leaves and spikes of small flowers. Some species and hybrids are grown as ornamentals. *Distr.* N temperate regions, to SE Asia and Australasia. *Fam.* Orchidaceae. *Class* M.

G. oblongifolia Spike 1-sided, flowers pale green, upper sepal and petals forming a hood. *Distr.* W USA.

G. pubescens Spike dense, cylindrical, many-flowered. Flowers white. *Distr.* E USA.

gooseberry *Ribes, R. grossularioides, R. uva-crispa* **Barbados** ~ *Pereskia aculeata* **Cape** ~ *Physalis peruviana* **Chinese** ~ *Actinidia deliciosa* **dwarf Cape** ~ *Physalis pruinosa* **fuchsia-flowered** ~ *Ribes speciosum.*

goosefoot *Chenopodium, C. album,* Chenopodiaceae.

gorse *Ulex, U. europaeus.* **common** ~ *U. europaeus* **dwarf** ~ *U. minor* **Irish** ~ *U. europaeus* **Spanish** ~ *Genista hispanica.*

gossypinus, -a, -um: cottony.

Gossypium (From Latin *gossypion*, cotton plant.) COTTON. A genus of 39 species of annual and perennial herbs, shrubs, and trees with palmately lobed or divided leaves and typically solitary, regular, bisexual flowers that bear 5 petals and 10 fused stamens. The fruit is a glandular capsule containing densely hairy seed. The seed hairs of these plants are the world's most important fibre crop and have been used in Asia and America since prehistoric times, becoming important in Europe only in the last few hundred years. These plants are also a source of oils, dyes, and numerous other natural products. Many commercial cultivars have been produced from a combination of Old and New World species. *Distr.* Warm temperate and tropical regions. *Fam.* Malvaceae. *Class* D.

G. arboreum TREE COTTON. Shrub to 5m high. Flowers yellow to deep red-purple. *Distr.* Tropical and subtropical Asia.

G. barbadense SEA ISLAND COTTON. Shrub to 3m high. Flowers solitary, yellow with a dark central spot. One of the major sources of genetic material for modern commercial crop plants. *Distr.* Tropical South America.

G. herbaceum LEVANT COTTON. Annual or perennial herb. Flowers yellow with a purple centre. *Distr.* Africa, India.

Goupiaceae A family of 1 genus and 3 species of shrubs and trees. This family is sometimes included in the family Celastraceae. *Distr.* N South America. *Class* D.

gourd Cucurbitaceae **buffalo** ~ *Cucurbita foetidissima* **Malabar** ~ *C. ficifolia* **Missouri** ~ *C. foetidissima.*

gracifolius, -a, -um: with slender leaves.

graciliformis, -e: thin, slender-shaped.

gracilipes: slender-stalked.

gracilis, -e: thin, slender, graceful.

gracilistylus, -a, -um: with slender styles.

gracillimus, -a, -um: somewhat slender.

graecum, -a, -us: of Greece.

gramagrass, blue *Bouteloua gracilis.*

Gramineae The Grass family. 657 genera and about 7000 species of annual and perennial herbs with leaves borne in two rows, usually on a short stem at ground level. Upright stems can be cane-like or even woody. The leaves are composed of two parts, the sheath and the blade. The sheath is wrapped tightly around the stem, helping to protect and support it. The blade is joined to the top of the sheath and is typically long and narrow. The blade has a growing point at its base so that it can continue to grow even when it has been grazed by an animal, thus giving this family an enormous ecological advantage over others. The flowers are represented by florets and lack distinct sepals and petals but are contained within a spikelet. Spikelets are usually arranged into larger inflorescences. One of the most important families for humans, providing all the world's cereals and most of the sugar, as well as grazing for domestic animals. *Distr.* Cosmopolitan. *Class* M.
See: Agropyron, Agrostis, Alopecurus, Ampelodesmus, Andropogon, Antho-xanthum, Arrhenatherum, Arundinaria, Arundo, Avena, Avenula, Bambusa, Beckmannia, Bothriochloa, Bouteloua, Brachypodium, Brachystachyum, Briza, Bromus, Calamagrostis, Catapodium, Chasmanthium, Chimonobambusa, Chionochloa, Chomonobambusa, Chondrosum, Chrysopogon, Chusquea, Clavinodum, Coix, Cortaderia, Corynephorus, Cymbopogon, Cynosurus, Dactylis, Dasypyrum, Dendrocalamus, Deschampsia, Desmazeria, Diarrhena, Drepanostachyum, Echinaria, Echinochloa, Ehrharta, Eleusine, Elymus, Eragrostis, Eriochloa, Fargesia, Festuca, Fingerhuthia, Gaudinia, Glyceria, Gynerium, Hakonechloa, Helictotrichon, × *Hibanobambusa, Hierochloe, Himalayacalamnus, Holcus, Hordeum, Hystrix, Imperata, Indocalamus, Koeleria, Lagurus, Lamarckia, Leymus, Melica, Melinis, Mibora, Milium, Miscanthus, Molinia, Muhlenbergia, Nassella, Oplismenus, Oryza, Oryzopsis, Otatea, Panicum, Paspalum, Pennisetum, Pentaschistis, Phaenosperma, Phalaris, Phleum, Phragmites, Phyllostachys, Piptatherum, Pleioblastus, Poa, Pogonatherum, Polypogon, Pseudosasa, Rhynchelytrum, Rostraria, Saccharum, Sasa, Sasaella, Sasamorpha, Semiarundinaria, Sesleria, Setaria, Shibataea, Sinarundinaria, Sinobambusa, Sorghastrum, Sorghum, Spartina, Spodiopogon, Stenotaphrum, Stipa, Thamnocalamus, Triticum, Uniola, Yushania, Zea, Zizania.*

gramineus, -a, -um: grass-like.

graminifolius, -a, -um: from *Gramineae*, the grasses, and *folius*, leaved.

graminoides grass-like.

Grammangis: (From Greek *gramma*, marking or letter, and *angos*, vessel, alluding to the markings on the flowers.) A genus of 1–2 species of epiphytic orchids with fleshy leaves and racemes of large flowers. The upper sepal is free, the laterals being fused to the lateral petals to form a column foot. Some have local medicinal uses or are considered an aphrodisiac. *Distr.* Indonesia and Malaysia to Polynesia. *Fam.* Orchidaceae. *Class* M.

G. ellisii Flowers yellow, marked red and red-brown. *Distr.* Madagascar.

Grammatophyllum (From Greek *gramma*, marking or letter, and *phyllon*, leaf, alluding to the conspicuous markings of the sepals and petals.) A genus of about 10 species of large terrestrial or epiphytic orchids with pseudobulbs in tight cane-like clusters and many-flowered racemes of small-lipped flowers. *Distr.* SE Asia. *Fam.* Orchidaceae. *Class* M.

G. scriptum Inflorescence pendent, many-flowered. Flowers to 4.5cm across, yellow-green, marked brown. *Distr.* Borneo, Philippines, Indonesia (Moluccas).

G. speciosum With inflorescences to 3m high of up to 100 flowers around 15cm across. Spectacular cultivated ornamental epiphyte. *Distr.* Malaysia, Indonesia, Philippines.

Grammitidaceae A family of 14 genera and about 500 species of epiphytic, occasionally terrestrial ferns with hairy, simple to tripinnate leaves. The sori are borne on the underside of the leaves and lack an indusium. *Distr.* Widespread in tropical and S temperate areas. *Class* F.

granadensis: of Granada, Colombia or Southern Spain.

Granadilla See *Passiflora*

granadilla *Passiflora edulis* **giant** ~ *P. quadrangularis* **purple** ~ *P. edulis* **red** ~ *P. coccinea* **sweet** ~ *P. ligularis* **yellow** ~ *P. laurifolia.*

granatensis: of Granada, Colombia or southern Spain.

grandfather's beard *Geum triflorum.*

grandiceps: large-headed.

grandiflorus, -a, -um: large-flowered.

grandifolius, -a, -um: large-leaved.

grandis,-e: large.

granny bonnets *A. vulgaris.*

granny's bonnet *Aquilegia.*

granulatus, -a, -um: granular, with granules.

grape *Vitis vinifera*, Vitaceae **bear's** ~ *Arctostaphylos uva-ursi* **Javan** ~ *Tetrastigma* **Oregon** ~ *Mahonia aquifolium.*

grapefruit *Citrus × paradisi.*

grape hyacinth *Muscari* **common** ~ *M. neglectum.*

Graptopetalum (From Greek *graptos*, written upon, and *petalon*, petal.) A genus of 10 species of succulent herbs and shrubs with rosettes of simple fleshy leaves and cymes of star-shaped flowers. Several species are grown as ornamentals. *Distr.* S W North America. *Fam.* Crassulaceae. *Class* D.

G. amethystinium JEWEL LEAF PLANT. Clump-forming subshrub. Leaves rounded, grey-green, tinged red. Flowers yellow-green with red markings. *Distr.* Mexico.

G. paraguayense GHOST PLANT, MOTHER OF PEARL PLANT. Sprawling, perennial subshrub. Leaves grey-green, tinged pink. Flowers yellow, spotted red. *Distr.* Mexico.

grass Gramineae **African fountain** ~ *Pennisetum villosum* **African love** ~ *Eragrostis curvula* **Aleppo** ~ *Sorghum halepense* **Algerian** ~ *Stipa tenacissima* **Alkali** ~ *Zigadenus, Z. elegans* **Amur silver** ~ *Miscanthus sacchariflorus* **annual beard** ~ *Polypogon monspeliensis* **annual meadow** ~ *Poa annua* **barnyard** ~ *Echinochloa crus-galli* **basket** ~ *Oplismenus hirtellus* **bear** ~ *Camassia, Dasylirion, Xerophyllum tenax, Yucca Y. smalliana* **beard** ~ *Andropogon* **blue** ~ *Poa* **blue-eyed** ~ *Sisyrinchium, S. angustifolium* **blue oat** ~ *Helictotrichon sempervirens* **bottle-brush** ~ *Hystrix patula* **broad-leaved cotton** ~ *Eriophorum latifolium* **brome** ~ *Bromus* **buffalo** ~ *Stenotaphrum secundatum* **Californian blue-eyed** ~ *Sisyrinchium bellum* **canary** ~ *Phalaris canariensis* **cat tail** ~ *Rostraria cristata, Phleum* **China** ~ *Boehmeria, B. nivea* **Chinese fountain** ~ *Pennisetum alopecuroides* **cloud** ~ *Agrostis nebulosa* **coco** ~ *Cyperus rotundus* **common cord** ~ *Spartina anglica* **common meadow** ~ *Poa pratensis* **common quaking** ~ *Briza media* **common scurvy** ~ *Cochleria officinalis* **common viper's** ~ *Scorzonera hispanica* **cord** ~ *Spartina* **cotton** ~ *Imperata cylindrica, Eriophorum* **crab** ~ *Panicum* **creeping soft** ~ *Holcus mollis, H. lanatus* **crested hair** ~ *Koeleria macrantha* **crinkled hair** ~ *Deschampsia flexuosa* **cup** ~ *Eriochloa* **diss** ~ *Ampelodesmus mauritanicus* **dog** ~ *Agropyron* **early sand** ~ *Mibora mimima* **eel** ~ *Vallisneria, V. spiralis*, Zosteraceae **elk** ~ *Xerophyllum tenax* **esparto** ~ *Stipa tenacissima* **European dune** ~ *Leymus* *arenarius* **European feather** ~ *Stipa pennata* **feather** ~ *Stipa* **fern** ~ *Desmazeria* **fountain** ~ *Pennisetum setaceum* **freshwater cord** ~ *Spartina pectinata* **golden-eyed** ~ *Sisyrinchium californicum* **golden weather** ~ *Hypoxis hygrometrica* **goose** ~ *Potentilla anserina* **grama** ~ *Bouteloua* **great quaking** ~ *Briza maxima* **green bottle** ~ *Setaria viridis* **grey hair** ~ *Corynephorus canescens* **hair** ~ *Eleocharis acicularis*, *Corynephorus*, *Koeleria*, *Deschampsia* **hairy cup** ~ *Eriochloa villosa* **holy** ~ *Hierochloe odorata* **hunangamoho** ~ *Chionochloa conspicua* **Indian** ~ *Sorghastrum nutans* **Indian basket** ~ *Xerophyllum tenax* **Johnson** ~ *Sorghum halepense* **Kentucky blue** ~ *Poa pratensis* **lemon** ~ *Cymbopogon citratus* **lesser quaking** ~ *Briza minor* **love** ~ *Eragrostis* **lyme** ~ *Leymus arenarius*, *Elymus* **manna** ~ *Glyceria*, *G. fluitans* **marsh** ~ *Spartina* **meadow** ~ *Alopecurus*, *Poa* **means** ~ *Sorghum halepense* **Mexican love** ~ *Eragrostis mexicana* **millet** ~ *Milium effusum* **mosquito** ~ *Bouteloua gracilis* **needle** ~ *Stipa* **New Zealand wind** ~ *S. arundinacea* **oat** ~ *Arrhenatherum* **old witch** ~ *Panicum capillare* **orchard** ~ *Dactylis glomerata* **pampas** ~ *Cortaderia*, *C. selloana* **panic** ~ *Panicum* **pearl** ~ *Briza maxima* **pepper** ~ *Lepidium sativum* **pheasant's tail** ~ *Stipa arundinacea* **plumed tussock** ~ *Chionochloa conspicua* **prairie cord** ~ *Spartina pectinata* **purple beard** ~ *Bothriochloa caucasica* **purple-eyed** ~ *Olsynium douglasii* **purple moor** ~ *Molinia caerulea*, *Sesleria albicans* **purple viper's** ~ *Scorzonera purpurea* **quaking** ~ *Briza* **rabbit's foot** ~ *Polypogon monspeliensis* **red tussock** ~ *Chionochloa rubra* **reed** ~ *Calamagrostis*, *Glyceria maxima* **reed canary** ~ *Phalaris arundinacea* **reed meadow** ~ *Glyceria maxima* **reed sweet** ~ *G. maxima* **rough dog's tail** ~ *Cynosurus echinatus* **scorpion** ~ *Myosotis* **sea fern** ~ *Desmazeria marina* **sea lyme** ~ *Leymus arenarius* **shaking** ~ *Briza* **side oats** ~ *Bouteloua curtipendula* **silk** ~ *Oryzopsis hymenoides* **slough** ~ *Spartina pectinata* **small cord** ~ *S. maritima* **smilo** ~ *Oryzopsis miliacea* **smooth cord** ~ *Spartina alterniflora* **snow** ~ *Chionochloa* **spangle** ~ *Uniola*, *Chasmanthium latifolium* **spear** ~ *Poa*, *Stipa* **spike** ~ *Uniola* **spring** ~ *Anthoxanthum odoratum* **squaw** ~ *Xerophyllum tenax* **squirrel-tail** ~ *Hordeum jubatum* **St Augustine** ~ *Stenotaphrum secundatum* **star** ~ *Hypoxis* **swamp foxtail** ~ *Pennisetum alopecuroides* **sweet** ~ *Glyceria*, *G. fluitans* **sweet vernal** ~ *Anthoxanthum odoratum* **switch** ~ *Panicum virgatum* **tape** ~ *Vallisneria spiralis* **Texas blue** ~ *Poa arachnifera* **tor** ~ *Brachypodium pinnatum* **tufted hair** ~ *Deschampsia caespitosa* **tussock** ~ *D. caespitosa* **uva** ~ *Gynerium sagittatum* **vernal** ~ *Anthoxanthum* **water star** ~ *Heteranthera dubia* **wavy hair** ~ *Deschampsia flexuosa* **weeping love** ~ *Eragrostis curvula* **white star** ~ *Hypoxis capensis* **willow** ~ *Polygonum amphibium* **witch** ~ *Panicum capillare* **wood** ~ *Sorghastrum nutans* **worm** ~ *Spigelia marilandica* **yellow whitlow** ~ *Draba aizoides*.

grassnut *Triteleia laxa*.

grass of Parnassus *Parnassia*, *P. palustris*.

grass tree Xanthorrhoeaceae.

grass widow *Olsynium douglasii*.

Gratiola (From Latin *gratia*, grace or thanks, alluding to the medicinal properties of *G. officinalis*.) HEDGE HYSSOP. A genus of about 20 species of rhizomatous herbs with small opposite leaves and solitary, tubular, 2-lipped, 5-lobed flowers. *Distr.* Temperate regions. *Fam.* Scrophulariaceae. *Class* D.

G. officinalis HEDGE HYSSOP, GRATIOLE. Stems erect, to 60cm high. Flowers yellow-white. Formerly used as a medicinal herb. *Distr.* N temperate regions.

gratiole *Gratiola officinalis*.

gratissimus, -a, -um: very pleasing.

gratus, -a, -um: pleasing, agreeable.

gravel root *Eupatorium purpureum.*

graveolens: strong-smelling.

Grecian urn plant *Quesnelia marmorata.*

Greek hay *Trigonella foenum-graecum.*

green earth star *Cryptanthus acaulis.*

greenheart *Rubiaceae scabra.*

greenhood *Pterostylis.*

Greenovia (After George Bellas Greenough (1778–1855), English geologist.) A genus of 4 species of succulent tuft-forming herbs with dense rosettes of fleshy leaves and 1-sided inflorescences of yellow star-shaped flowers. *Distr.* Canary Islands. *Fam.* Crassulaceae. *Class* D.

G. aurea Grown as a tender ornamental.

greenweed, dyer's *Genista tinctoria.*

gregarius, -a, -um: gregarious, growing in groups.

Grevillea (After Charles Francis Greville (1749–1809), a founder of the Royal Horticultural Society.) SPIDER FLOWER. A genus of about 250 species of evergreen shrubs and small trees with simple or lobed, opposite leaves and racemes or panicles of tubular flowers. A number of species and hybrids are grown as half-hardy and tender ornamentals and some are exploited for their timber. *Distr.* Australia, a few species in New Guinea and Indonesia. *Fam.* Proteaceae. *Class* D.

G. alpina Rounded wiry shrub. Flowers red, borne in small clusters. *Distr.* Australia (Victoria, New South Wales).

G. robusta SILKY OAK. Conical tree. Leaves pinnately lobed, silky-hairy beneath. Flowers more or less bell-shaped, yellow-orange. Often grown as a house plant while a seedling. A source of timber, sometimes being used as a shade crop for COFFEE. *Distr.* Australia (Queensland).

G. rosmarinifolia Rounded shrub. Leaves narrow, silky-hairy beneath. Flowers red. *Distr.* Australia (New South Wales).

Grewia (After Nehemiah Grew (1641–1712), English botanist.) A genus of 150 species of shrubs, trees, and climbers with simple alternate leaves and small flowers that bear 5 petals and numerous stamens. The seeds of the African species are dispersed by elephants. Twigs of the Australian species are used as paint-brushes; other species yield edible timber and fruit. *Distr.* Warm regions of the Old World. *Fam.* Tiliaceae. *Class* D.

G. biloba Deciduous shrub to 3m high. Flowers yellow. Fruit red. Grown as a hardy ornamental. *Distr.* E Asia.

G. parviflora See *G. biloba*

grey goddess *Brahea armata.*

Greyia (After Sir George Grey (1812–98), Governor-General of Cape Colony.) NATAL BOTTLEBRUSH. A genus of 3 species of deciduous or evergreen shrubs and small trees with simple alternate leaves and spike-like racemes of small red flowers that bear 10 long, exerted stamens. *Distr.* South Africa. *Fam.* Greyiaceae. *Class* D.

G. sutherlandii CAPE BOTTLEBRUSH. Deciduous rounded shrub. Leaves leathery, deeply toothed. Inflorescence a cylindrical red spike. Grown as a tender ornamental.

Greyiaceae A family of 1 genus and 3 species. *Class* D.
See: Greyia.

Griffinia (After William Griffin (died 1827), who collected several species.) A genus of 6–7 species of bulbiferous perennial herbs with broad stalked leaves and umbels of numerous, funnel-shaped, small flowers. Some species are cultivated as tender ornamentals. *Distr.* Brazil. *Fam.* Amaryllidaceae. *Class* M.

G. hyacinthina Perianth segments white, blue at the tips. *Distr.* Brazil.

Grindelia (After David H. Grindel (1776–1836), German botanist.) GUM PLANT, TARWEED, ROSIN WEED. A genus of about 60 species of annual and perennial herbs and shrubs with leaves that bear resinous glands and flowers in large daisy-like heads that bear distinct rays. Some species are cultivated as ornamentals, others have local medicinal uses. *Distr.* W North and South America. *Fam.* Compositae. *Class* D.

G. chiloensis Subshrub to 1m high. *Distr.* Argentina, Chile.

G. integrifolia Perennial herb to 80cm high. *Distr.* W North America.

G. robusta Perennial herb to 1.2m high. *Distr.* California, Baja California.

G. squarrosa Biennial to perennial herb, to 1m high. Used medicinally to treat skin complaints. *Distr.* NW North America, naturalized in Great Britain and Australia.

G. stricta Perennial herb, to 1m. *Distr.* W North America.

Grindelia, Pacific *Grindelia stricta.*

Griselinia (After Francesco Griselini (1717–83), Venetian naturalist.) A genus of 6 species of evergreen shrubs or trees with alternate leathery leaves and panicles or racemes of minute unisexual flowers that bear their parts in 5s. Several species are grown ornamentally and some are a source of timber. *Distr.* New Zealand, Chile. *Fam.* Griseliniaceae. *Class* D.

G. littoralis Shrub or small tree. Timber used for railway-sleepers. *Distr.* New Zealand.

G. lucida Erect shrub. Leaves large, flowers green. Variegated cultivars are available.

Griseliniaceae A family of 1 genus and 6 species. *Class* D.
See: Griselinia.

griseoargenteus, -a, -um: silver-grey.

griseus, -a, -um: grey.

groenlandicus, -a, -um: of Greenland.

gromwell *Lithospermum officinale, Mertensia maritima* **corn** ~ *Lithospermum arvense.*

gromwell's puccoon *Lithospermum.*

Gronophyllum (After Jan Fredrik Gronovius (1686–1762), Dutch botanist.) A genus of 14 species of medium to large palms with feather-shaped leaves and more or less pendent, much-branched inflorescences of cream or white flowers. *Distr.* E Malaysia, N Australia. *Fam.* Palmae. *Class* M.

G. ramsayi NORTHERN KENTIA PALM. To 35m high. Leaves grey-green. Flowers white. Fruit red. Grown as an ornamental. *Distr.* N Australia.

Grossulariaceae A family of 1 genus and about 180 species. *Class* D.
See: Anopterus, Ribes.

grossulariifolius, -a, -um: from the genus *Grossularia*, and *folius*, leaved.

grossularioides: from the genus *Grossularia*, with the ending *-oides*, indicating resemblance.

grossus, -a, -um: coarse, rough, thick.

groundnut *Arachis hypogaea.*

groundsel *Senecio vulgaris* **bush** ~ *Baccharis halimifolia* **giant** ~ *Ligularia wilsoniana* **tree** ~ *Baccharis halimifolia.*

groundwort *Stachys.*

Grubbiaceae A family of 1 genus and 3 species of heath-like shrubs with leaves in 4 rows and small flowers in cone-like inflorescences. *Distr.* South Africa. *Class* D.

grummel *Lithospermum officinale.*

guatemalensis, -e: of Guatemala, Central America.

guava *Psidium, P. guajava* **Chilean** ~ *Ugni molinae.*

Guettarda (After Jean Etienne Guettard (1715–86), French botanist.) A genus of about 60 species of shrubs and small trees with simple, opposite, or whorled leaves and salver-shaped, 4- to 9-lobed, fragrant flowers. Several species are grown ornamentally and some have edible fruit. *Distr.* Tropical America. *Fam.* Rubiaceae. *Class* D.

G. scabra VELVET BERRY, GREENHEART. Small tree to 10m high. Flowers white, occasionally tinged pink. Grown as a tender ornamental. *Distr.* West Indies, E South America, USA (Florida).

Guichenotia (After Antoine Guichenot, 19th-century French horticulturalist.) A genus of 5 species of evergreen shrubs with simple opposite leaves and racemes of pink or purple flowers that bear their parts in 5s. *Distr.* SW Australia. *Fam.* Sterculiaceae. *Class* D.

G. ledifolia Shrub to 2m high. Leaves white-hairy. Flowers mauve. Grown as a tender ornamental. *Distr.* SW Australia.

guinea flower *Fritillaria meleagris.*

guinea-hen flower *F. meleagris.*

guineensis: of Guinea, West Africa.

gum: black ~ *Nyssa sylvatica* **blue** ~ *Eucalyptus globulus* **cabbage** ~ *E. pauciflora* **cider** ~ *E. gunnii* **cotton** ~ *Nyssa aquatica, N. sylvatica* **Formosan** ~ *Liquidambar formosana* **ghost** ~ *Eucalyptus pauciflora* **lemon-scented** ~ *E. citriodora* **manna** ~ *E. viminalis* **mountain** ~ *E. dalrympleana* **red** ~ *Liquidambar styraciflua* **red flowering** ~ *Eucalyptus ficifolia* **ribbon** ~ *E. viminalis* **round-leaved** ~ *E. perriniana* **snow** ~ *E. pauciflora* **spinning** ~ *E. perriniana* **sweet** ~ *Liquidambar* **Tasmanian blue** ~ *Eucalyptus globulus* **Tasmanian snow** ~ *E. coccifera* **tingiringi** ~ *E. glaucescens* **white** *E. viminalis.*

gumbo *Abelmoschus esculentus.*

gummiferus, -a, -um: gum-bearing.

gum olibanum *Boswellia sacra.*

gum plant *Grindelia.*

Gunnera (After Ernst Gunnerus (1718–73), Norwegian bishop and botanist.) A genus of about 40 species of rhizomatous herbs with rounded leaves and spikes or panicles of minute flowers. Some species are small and form dense mats while others are enormous, form large tufts, and have some of the largest leaves known. Several species are grown as ornamentals. *Distr.* S warm and tropical regions. *Fam.* Gunneraceae. *Class* D.

G. chilensis See *G. tinctoria*

G. magellanica Mat-forming perennial. Leaves to 9cm across, turning bronze in winter. *Distr.* S South America, Falkland Islands.

G. manicata GIANT RHUBARB. Waterside clump-forming perennial. Leaves to 2m across on 2m-long stalks, with spiny veins below. A most spectacular garden plant. *Distr.* Colombia.

G. scabra See *G. tinctoria*

G. tinctoria Compact clump-forming perennial. Leaves to 150cm across. Peeled young leaf-stalks eaten in Chile. *Distr.* Chile.

Gunneraceae A family of 1 genus and 40 species. *Class* D.
See: Gunnera.

gunpowder plant *Pilea microphylla.*

gutta: a droplet.

gutta percha tree *Eucommia ulmoides, Palaquium gutta.*

guttatus, -a, -um: spotted.

Guttiferae The St John's Wort family. 47 genera and 1350 species of shrubs and trees with simple opposite leaves. The flowers are regular, bisexual or unisexual, with 2 whorls of 4–5 tepals and numerous stamens in 2 whorls of 5 bundles. The family is a source of timber, fruit and essential oils. *Distr.* Widespread. *Class* D.
See: Clusia, Hypericum.

Guzmania (After Anastasio Guzman (died 1802), Spanish apothecary and botanist.) A genus of about 130 species of typically epiphytic, stemless or short-stemmed herbs with dense rosettes of maroon- or purple-marked leaves. The small yellow-white flowers are borne in panicles with brightly coloured bracts. Several species are grown as tender ornamentals. *Distr.* South America. *Fam.* Bromeliaceae. *Class* M.

G. lingulata Leaves strap-shaped, arching. Flowers surrounded by red bracts. *Distr.* West Indies to Bolivia and Brazil.

G. monostachia Leaves pale-yellow. Inflorescence a spike. *Distr.* USA (Florida), West Indies and Nicaragua to N Brazil and Peru.

G. sanguinea Leaf rosette cup-shaped. Flowers borne in a cluster in the centre of each mature rosette. *Distr.* Costa Rica, Colombia, Ecuador, Trinidad and Tobago.

Gymnadenia (From Greek *gymnos*, naked, and *aden*, gland; the pollinia are uncovered.) A genus of about 10 species of terrestrial orchids with palmately lobed tubers and a dense cylindrical spike of flowers. The upper sepal and the petals are fused into a hood over the column. *Distr.* Eurasia, E North America. *Fam.* Orchidaceae. *Class* M.

G. conopsea SCENTED ORCHID, FRAGRANT ORCHID. To 70cm. Flowers fragrant, pink, lilac or red, rarely purple or white, spur to 2cm. *Distr.* Eurasia to Japan.

G. odoratissima SHORT-SPURRED FRAGRANT ORCHID. To 30cm. Flowers spur to 15mm. *Distr.* Europe to the Urals.

Gymnocarpium (From Greek *gymnos*, naked, and *karpos*, fruit; the sori are not covered by an indusium.) A genus of about 6 species of deciduous terrestrial herbs with 2 ranks of bipinnate leaves that are triangular in outline. Some species are cultivated as ornamental ground cover. *Distr.* N temperate regions of the Old World. *Fam.* Woodsiaceae. *Class* F.

G. dryopteris OAK FERN. Leaves bright yellow green, 40cm long. *Distr.* North America, Eurasia.

G. robertianum LIMESTONE POLYPODY, LIMESTONE FERN, NORTHERN OAK FERN. Leaves to 50cm long, fragrant when bruised. *Distr.* N America, Europe.

gymnocarpus, -a, -um: naked fruit.

Gymnocladus (From Greek *gymnos*, naked, and *klados*, branch, alluding to the deciduous leaves.) A genus of 5 species of deciduous trees with large bipinnate leaves and racemes or panicles of flowers that bear 5 regular petals and 10 stamens. *Distr.* E North America, E Asia. *Fam.* Leguminosae. *Class* D.

G. dioica KENTUCKY COFFEE TREE, CHICOT. Grown as an ornamental as well as being a source of timber and a coffee substitute. *Distr.* E North America.

Gymnopteris (From Greek *gymnos*, naked, and *pteris*, fern, alluding to the absence of indusia and hence naked sori.) A genus of about 5 species of small to medium sized ferns with divided leaves. *Distr.* Warm America, Asia. *Fam.* Adiantaceae. *Class* F.

G. rufa A small fern with brown-hairy leaves. *Distr.* Tropical America.

Gymnospermium (From Greek *gymnos*, naked, and *sperma*, seed, probably alluding to the capsules that open at the end revealing the seeds.) A genus of about 6 species of rhizomatous herbs with leaves that are divided into 3 pinnately divided leaflets. The yellow flowers bear 6 petal-like sepals, 6 very small petals and 6 stamens and are borne in racemes. *Distr.* Balkan Peninsula, Central Asia, China. *Fam.* Berberidaceae. *Class* D.

G. albertii Grown as a hardy ornamental. *Distr.* Central Asia.

Gynandriris (From Greek *gyne*, female, *andros*, male, and the name of the genus *Iris*.) A genus of 9 species of perennial herbs with corms, narrow, curled leaves, and cymes of short-lived, *Iris*-like flowers. *Distr.* S Africa, Mediterranean, SW Asia. *Fam.* Iridaceae. *Class* M.

G. sisyrinchium Flowers violet to lavender. Grown as an ornamental. *Distr.* Portugal to SW Asia and Pakistan.

Gynerium (From Greek *gyne*, woman, and *erion*, wool, alluding to the hairy female inflorescences.) A genus of 1 species of aquatic perennial grass with reed-like stems to 10m high, fan-like arrangements of leaves, and feathery inflorescences. *Distr.* Tropical America. *Fam.* Gramineae. *Class* M.

G. argenteum See *Cortaderia selloana*

G. sagittatum UVA GRASS. Young shoots are used as a kind of shampoo.

Gynura (From Greek *gyne*, female, and *oura*, tail, alluding to the long rough stigma.) VELVET PLANT. A genus of about 50 species of perennial herbs and shrubs with simple or pinnately lobed leaves and bad-smelling flowers in daisy-like heads that lack distinct rays. Several species are cultivated as tender ornamentals; others have medicinal uses or are edible. *Distr.* Tropical regions of the Old World. *Fam.* Compositae. *Class* D.

G. aurantiaca VELVET PLANT, PURPLE PASSION VINE, PURPLE VELVET PLANT, ROYAL VELVET PLANT. Ornamental foliage plant with velvety purple leaves. *Distr.* Java, Sulawesi, widely naturalized.

G. procumbens Climbing or trailing perennial. Leaves almost glabrous. *Distr.* W Africa to China.

G. sarmentosa See *G. procumbens*

gypsicola: growing in lime soils.

Gypsophila (From Greek *gypsos*, gypsum or chalk, and *philos*, loving; some species grow on alkaline soils.) A genus of about 125 species of annual and perennial herbs with simple, linear, opposite leaves. The flowers bear 5 sepals, 5 petals, and 10 stamens and are small, numerous, and borne in spreading panicles or occasionally larger and solitary. Several species are grown as ornamentals and some are reported to have medicinal qualities. *Distr.* Temperate Eurasia. *Fam.* Caryophyllaceae. *Class* D.

G. cerastioides Creeping perennial. Leaves somewhat fleshy. Flowers saucer-shaped, pink. *Distr.* Himalaya.

G. dubia See *G. repens*

G. elegans Bushy annual. Flowers small, white, in much branched, loose heads. *Distr.* SW Asia.

G. paniculata BABY'S BREATH. Robust perennial. Flowers in large spreading panicles. Used as a cut flower, particularly in wedding bouquets. *Distr.* Europe to Central Asia.

G. repens Mat-forming. Flowers numerous, white or pink. *Distr.* Mountains of Central Europe.

gypsyweed *Veronica officinalis*.

Gyrostemonaceae A family of 5 genera and 14 species of shrubs and trees with simple, spirally arranged leaves and small unisexual flowers without true sepals or petals but with 6 to many stamens and a superior ovary. *Distr.* Australia. *Class* D.

H

Habenaria (From the Latin *habena*, thong or strap, alluding to the long slender spur, petal, and lip lobes of many species.) A genus of about 500 species of terrestrial, occasionally epiphytic orchids with fleshy roots or tubers and terminal racemes of typically green-white flowers. *Distr.* Tropical and subtropical regions. *Fam.* Orchidaceae. *Class* M.

H. macrandra Flowers green-white, around 5cm across. Cultivated as an ornamental. *Distr.* Tropical Africa.

H. radiata See *Pecteilis radiata*

Haberlea (After Carl Constantin Haberle (1764–1832), professor of botany at Budapest.) A genus of 1 species of perennial herb with rosettes of simple toothed leaves and umbels of 5-lobed, 2 lipped, pale purple flowers. *Distr.* Balkans. *Fam.* Gesneriaceae. *Class* D.

H. ferdinandi-coburgii See *H. rhodopenis*

H. rhodopenis Grown as an ornamental with several cultivars available.

Habranthus (From Greek *habros*, graceful, and *anthos*, flower.) A genus of 10–20 species of bulbiferous perennial herbs with grass-like leaves and solitary, irregular, funnel-shaped flowers. Some species are cultivated as tender ornamentals. *Distr.* Temperate South America. *Fam.* Amaryllidaceae. *Class* M.

H. andersonii See *H. tubispathus*

H. brachyandrus Flowers pink, dark red at base. *Distr.* S Brazil, Paraguay.

H. gracifolius Flowers pink or white, green at base. *Distr.* Uruguay, Argentina.

H. robustus Leaves fleshy. Flowers pink, green at base. *Distr.* S Brazil.

H. texanus See *H. tubispathus*

H. tubispathus Leaves narrow. Flowers orange-yellow. *Distr.* S Brazil, Uruguay, E Argentina and S Chile.

habrotrichus, -a, -um: with beautiful hairs.

hackberry *Celtis*, *C. occidentalis*
Mediterranean ~ *C. australis*.

Hacquetia (After Balthasar Hacquet (1740–1815), Austrian botanist.) A genus of 1 species of clump-forming perennial herb with palmately lobed, basal leaves and simple umbels of yellow flowers. The umbels are subtended by a whorl of bright green, petal-like bracts. *Distr.* Central Europe. *Fam.* Umbelliferae. *Class* D.

H. epipactis Grown as a rock garden ornamental.

Haemanthus (From Greek *haima*, blood, and *anthos*, flower, alluding to their colour.) BLOOD LILY, RED CAPE TULIP. A genus of 21 species of bulbiferous perennial herbs with 2 ranks of linear to lance-shaped leaves and dense or lax umbels of regular flowers. The umbels are often so tightly packed with flowers that they resemble a paint or shaving brush. Some species are cultivated as tender ornamentals. *Distr.* S Africa. *Fam.* Amaryllidaceae. *Class* M.

H. albiflos WHITE PAINT BRUSH. Umbel dense, 50-flowered, to around 7cm across. *Distr.* South Africa.

H. coccineus CAPE TULIP. Leaves flat, often bearing maroon markings. Umbel compact, to 10cm across, very-many-flowered. *Distr.* S and SW Africa.

H. humilis Flowers very numerous, borne in a dense head with exerted yellow anthers. *Distr.* S Africa.

H. kalbreyeri See *Scadoxus multiflorus*

H. katherinae See *Scadoxus multiflorus*

H. natalensis See *Scadoxus puniceus*

H. pubescens Umbel dense, to around 6cm across, stamens exerted. *Distr.* S and SW Africa.

H. sanguineus Umbel to 8cm across, 25-flowered. Flowers red-pink. *Distr.* South Africa.

haemanthus, -a, -um: with blood-red flowers.

haematocalyx: with a blood-red whorl of sepals.

haematocarpus, -a, -um: with blood-red fruit.

haematodes: blood-red coloured.

Haemodoraceae The Bloodwort family. 14 genera and about 100 species of herbs with linear leaves and hairy inflorescences. The flowers have 1 or 2 whorls of 3 petals. Some species are cultivated as ornamentals. *Distr.* South Africa, New Guinea, Australia, and tropical regions of the New World. *Class* M.
See: Anigozanthos, Macropidia, Wachendorfia.

hagberry *Prunus padus.*

Hakea (After Baron Christian Ludwig von Hake (1745–1818), German patron of botany.) PINCUSHION TREE. A genus of about 125 species of shrubs and small trees with alternate simple leaves and racemes or clusters of small flowers. Several species are grown as ornamentals although some are considered invasive weeds. A few species are used as a source of timber and in the reclamation of arid lands. *Distr.* Australia. *Fam.* Proteaceae. *Class* D.

H. lissosperma NEEDLE BUSH. Erect evergreen shrub. Leaves needle-like. Flowers white, spidery. *Distr.* Australia (Victoria, New South Wales, Tasmania).

Hakonechloa (After Hakone, a region of Japan, and Greek *chloa*, a grass.) A genus of 1 species of rhizomatous grass with soft stems, smooth leaves and nodding inflorescences. *Distr.* Japan. *Fam.* Gramineae. *Class* M.

H. macra Grown as an ornamental cultivar with a number of variegated varieties available that turn red in autumn.

halepensis, -e: of Aleppo, Syria.

Halesia (After Dr Stephen Hales (1677–1761), English botanist.) SILVER BELL, BELL TREES, SNOWDROP TREE, SILVERBELL TREE. A genus of 5 species of deciduous shrubs and small trees with simple alternate leaves and pendulous clusters of bell-shaped, 4-lobed, white flowers that are followed by winged fruits. Several species are grown as ornamentals. *Distr.* SE USA, E China. *Fam.* Styracaceae. *Class* D.

H. carolina See *H. tetraptera*

H. diptera Large shrub or small tree to 8m high. Flowers white. *Distr.* USA.

H. monticola MOUNTAIN SNOWDROP TREE. Tree to 25m high. Flowers white. *Distr.* USA.

H. tetraptera SNOWDROP TREE, SILVER BELL, OPOSSUM WOOD. Shrub or small tree to 6m high. Flowers white. *Distr.* E USA.

halimiifolius, -a, -um: with leaves resembling those of the species *Atriplex halimus.*

× ***Halimiocistus*** (From the names of the parent genera.) A genus of hybrids between members of the genera *Cistus* and *Halimium*, some of which occur in the wild and some of which are of garden origin. *Fam.* Cistaceae. *Class* D.

×*H. algarvensis* See *Halimium ocymoides*

×*H. revolii* A hybrid of *H. alyssoides* x *C. salviifolius*. Densely branched low shrub with flowers that bear 5 sepals and 5 yellow to white petals. *Distr.* S France.

×*H. sahucii* (After M. Sahuc, a member of the party that discovered it.) A hybrid of *H. umbellatum* x *C. salviifolius*. Compact shrub to 1m high with flowers that bear 3–5 sepals and 5 white petals. *Distr.* S France.

×*H. wintonensis* (Of Winchester, where it was raised by Hillier's.) A hybrid of *H. ocymoides* x *C. salviifolius*. Spreading shrub to 50m high with flowers that bear 5 white petals that have yellow and maroon markings at the base. *Distr.* Garden origin.

Halimium (From Greek *halimos*, belonging to the sea.) A genus of 9 species of evergreen shrubs with simple opposite leaves and racemes or cymes of regular flowers that bear 3–5 sepals, 5 yellow or white petals, and numerous stamens. This genus is closely related to the genus *Helianthemum* and to *Cistus*, with which it forms hybrids. Most of the species and several hybrids are grown as ornamentals. *Distr.* Mediterranean. *Fam.* Cistaceae. *Class* D.

H. formosum See *H. lasianthum*

H. lasianthum Erect bushy shrub. Leaves grey-green. Petals golden yellow with a maroon basal spot. *Distr.* S Portugal, S Spain.

H. ocymoides Erect or prostrate shrub. Leaves narrow. Petals bright, golden yellow with a purple-black basal spot. *Distr.* Spain, Portugal.

H. umbellatum Upright shrub. Leaves narrow. Flowers white. *Distr.* Mediterranean.

H. wintonensis See × *Halimiocistus wintonensis*

Halimodendron (From Greek *halimum*, maritime, and *dendron*, tree.) A genus of 1 species of spiny shrub or small tree with pinnate leaves (that bear deciduous leaflets and a persistent spiny stalk) and racemes of pale purple flowers. *Distr.* Europe to Central Asia. *Fam.* Leguminosae. *Class* D.

H. halodendron SALT TREE. Grown as an ornamental for its silvery foliage.

Halocarpus (From Greek *halos*, halo, and *karpos*, fruit.) A genus of 3 species of evergreen coniferous shrubs and trees with linear juvenile leaves and scale-like adult leaves. *Distr.* New Zealand. *Fam.* Podocarpaceae. *Class* G.

H. bidwillii TARWOOD, MOUNTAIN PINE. Shrub to 6m high. Occasionally grown as an ornamental. *Distr.* New Zealand.

halodendron salt tree, from the Greek *halo-*, salt, and *dendron*, tree.

Halophytaceae A family of 1 genus and 1 species of succulent annual herb with small flowers. This family is sometimes included within the family Chenopodiaceae. *Distr.* Argentina. *Class* D.

Haloragaceae A family of 9 genera and 120 species of delicate aquatic herbs and terrestrial shrubs or small trees with small regular flowers. *Distr.* Wide-spread, especially Australia. *Class* D.
See: Myriophyllum.

Hamamelidaceae The Witch Hazel family. 29 genera and about 100 species of shrubs and trees with alternate, simple, or palmate leaves and variable small flowers with distinct or absent petals. *Distr.* Mostly subtropical and temperate regions. *Class* D.
See: Corylopsis, Disanthus, Distylium, Fothergilla, Hamamelis, Liquidambar, Loropetalum, Parrotia, Parrotiopsis, Sinowilsonia, × *Sycoparrotia, Sycopsis.*

Hamamelis (From a Greek name for an unidentified pear-shaped fruit.) WITCH HAZEL. A genus of 5–6 species of deciduous shrubs and small trees with simple alternate leaves and clusters of small, spider-like, fragrant flowers that bear 4, typically yellow, narrow petals. Several species are grown as ornamentals. *Distr.* E North America, E Asia. *Fam.* Hamamelidaceae. *Class* D.

H. × intermedia A hybrid of *H. japonica* × *H. mollis*. The bark and leaves are a source of medicinal witch hazel for bruises, haemorrhoids, varicose veins, and eye-lotions. The common name comes from the Old English *wych*, a pliable stick, alluding to its use as a divining rod. *Distr.* Garden origin.

H. japonica JAPANESE WITCH HAZEL. Shrub or tree. Flowers yellow, appearing in spring. *Distr.* Japan.

H. mollis CHINESE WITCH HAZEL. Shrub or small tree. Flowers yellow, tinged with red, appearing in late winter. *Distr.* W China.

H. vernalis Small erect shrub. Flowers red or yellow, appearing in the winter. *Distr.* SE North America.

H. virginiana VIRGINIAN WITCH HAZEL. Shrub or small tree. Flowers yellow, appearing in the late autumn. *Distr.* E North America.

handkerchief tree *Davidia involucrata.*

Hanguanaceae A family of 1 genus and 1 species of tuberous herbs of wet areas with rolled leaves and panicles of small green flowers that bear 6 tepals, 3 large and 3 small stamens, and a superior ovary. Male and female flowers are borne on separate plants. *Distr.* Tropical Asia. *Class* M.

Haplocarpha (From Greek *haplos*, single, and *karphos*, stalk.) A genus of 10 species of mat-forming perennial herbs with simple to pinnately lobed leaves and flowers in solitary daisy-like heads which bear distinct rays. *Distr.* Africa. *Fam.* Compositae. *Class* D.

H. rueppellii Leaves variable. Flowerheads to 2cm across. *Distr.* Mountains of E Africa.

Haplopappus (From Greek *haplos*, single, and *pappos*, down, alluding to the absence of an outer pappus.) A genus of about 160 species of annual to perennial herbs and shrubs with simple to pinnate leaves and yellow flowers in daisy-like heads that bear distinct yellow rays. Some species are used to indicate the present of selenium; others are cultivated as ornamentals. A popular genus for geneticists to work on because of the small number of chromosomes (2n = 4). *Distr.* SW North America. *Fam.* Compositae. *Class* D.

H. brandegei See *Erigeron aureus*

H. coronopifolius See *H. glutinosus*

H. glutinosus Cushion-forming perennial. *Distr.* Chile, Argentina.

Hardenbergia (After Countess von Hardenberg, sister of Baron von Hügel, who collected plants in W Australia in 1833.) A genus of 3 species of subshrubs and climbers with alternate pinnate leaves, that are sometimes reduced to a single leaflet, and long racemes of irregular pea-like flowers. Cultivated as ornamentals. *Distr.* Australia. *Fam.* Leguminosae. *Class* D.

H. comptoniana (After Compton, the family name of Lady Northampton, who grew it *c.*1810.) WESTERN AUSTRALIA CORAL PEA. Twinning climber. Leaves with 3–5 leaflets. Flowers purple-blue, crowded. *Distr.* W Australia.

H. violacea VINE LILAC, PURPLE CORAL PEA, FALSE SARSPARILLA. Twinning climber. Leaves reduced to a single leaflet. Flowers purple to white. *Distr.* E Australia, Tasmania.

hardhack, golden *Potentilla fruticosa.*

hardheads *Centaurea nigra.*

harebell *Camparella rotundifolia.*

hare's ear, shrubby *Bupleurum fruticosum.*

hare's ear, tail *Lagurus ovatus.*

harlequin flower *Sparaxis tricolor.*

harmal *Peganum harmala.*

harvest lice *Agrimonia.*

hashish *Cannabis sativa.*

hastatus, -a, -um: with equal, more or less triangular, lobes directed outwards.

hastilis, -e: spear-shaped.

haw, black *Viburnum prunifolium.*

hawkbit *Leontodon* **autumn** ~ *L. autumnalis* **rough** ~ *L. hispidus.*

hawk's beard *Crepis.*

hawkweed *Hieracium.*

Haworthia (After Adrian Hardy Haworth (1768–1833), English plantsman.) A genus of about 70 species of small stemless or short-stemmed rosette-plants with succulent warty leaves and small simple flowers borne in a simple raceme. The genus contains many important succulent ornamentals. *Distr.* Dry areas of S Africa. *Fam.* Aloaceae. *Class* M.

H. angustifolia Rosettes stemless, clustered, dull green. *Distr.* South Africa.

H. arachnoidea COBWEB ALOE. Rosettes stemless. Leaf margins with long white teeth. *Distr.* South Africa.

H. armstrongii Leaves spirally arranged on short stems. *Distr.* South Africa.

H. attenuata Rosettes stemless or short-stemmed. Leaves erect. *Distr.* South Africa.

H. batesiana Stemless rosettes aggregated into clumps. *Distr.* South Africa.

H. coarctata Leaves spirally arranged on stems to 20cm long. *Distr.* South Africa.

H. cuspidata Probably a hybrid of *H. cymbiformis* × *H. retusa*. *Distr.* Cape Province.

H. cymbiformis A variable species with stemless rosettes to 10cm across. *Distr.* South Africa.

H. fasciata ZEBRA HAWORTHIA. Rosettes stemless or short-stemmed. Leaves erect. *Distr.* South Africa.

H. glabrata Rosettes stemless or short-stemmed. Leaves fleshy. Commonly cultivated. *Distr.* Not known in the wild.

H. glauca Leaves spirally arranged on short stems. *Distr.* South Africa.

H. herbacea See *H. arachnoidea*

H. limifolia FAIRY WASHBOARD. Rosettes stemless with 20 green-yellow leaves. *Distr.* South Africa.

H. longiana Leaves long, ascending, clumped. *Distr.* South Africa.

H. maughanii (After Dr R. Maughan Brown.) A highly specialized plant of dry habitats. The leaves are borne below ground, only exposing their translucent tips at soil level. Light is channelled from these 'windows', through the internal tissues of the leaves, to the photosynthetic structures. *Distr.* South Africa.

H. minima Rosettes large, stemless, covered in tubicles. *Distr.* South Africa.

H. mirabilis Rosettes stemless, small. *Distr.* South Africa.

H. radula Rosettes stemless or short-stemmed. Leaves covered in small white tubicles. *Distr.* South Africa.

H. reinwardtii (After Caspar Reinwardt (1773–1854), founder of Bogor Botanic Garden, Java.) Leaves spirally arranged around thin stems. *Distr.* South Africa.

H. reticulata Rosettes stemless, clustered, pale green. *Distr.* South Africa.

H. retusa Rosettes stemless, to 14cm across. *Distr.* South Africa.

H. translucens Rosettes stemless. Leaf margins with long white teeth. *Distr.* South Africa.

H. truncata Leaves arranged in 2 series. *Distr.* South Africa.

H. turgida Rosettes stemless. Leaves fleshy, soft, pale green. *Distr.* South Africa.

H. variegata Rosettes stemless. Leaves narrow. *Distr.* South Africa.

H. viscosa Leaves in 3 spiralled series on stems to 20cm long. *Distr.* South Africa.

haworthia, zebra *Haworthia fasciata.*

hawthorn *Crataegus* **common** ~ *C. monogyna* **English** ~ *C. laevigata*, *C. monogyna* **Indian** ~ *Rhaphiolepis indica* **water** ~ Aponogetonaceae, *Aponogeton distachyos.*

hazel *Corylus*, *C. avellana*, Corylaceae **Japanese** ~ *Corylus sieboldiana* **Turkish** ~ *Corylus colurna* **winter** ~ *Corylopsis.*

hazelnut, Chilean *Gevuina avellana.*

heal all *Prunella.*

heartnut *Juglans ailantifolia.*

heartsease *Viola tricolor.*

heartseed *Cardiospermum grandiflorum.*

hearts entangled *Ceropegia linearis.*

heath *Erica*, Ericaceae **Australian** ~ Epacridaceae **besom** ~ *Erica scoparia* **bridal** ~ *E. bauera* **Connemara** ~ *Daboecia cantabrica* **Cornish** ~ *Erica vagans* **Corsican** ~ *E. terminalis* **cross-leaved** ~ *E. tetralix* **Dorset** ~ *E. ciliaris* **dwarf Spanish** ~ *E. umbellata* **Irish** ~ *E. erigena* **Portuguese** ~ *E. lusitanica* **Prince of Wales** ~ *E. perspicua* **sea** ~ *Frankenia laevis* **Spanish** ~ *Erica australis* **spike** ~ *Bruckenthalia spiculifolia* **St Dabeoc's** ~ *Daboecia cantabrica* **tree** ~ *Erica arborea* **winter** ~ *E. carnea.*

heather *Erica* **bell** ~ *E. cinerea* **Brewer's mountain** ~ *Phyllodoce breweri* **French** ~ *Erica x hyemalis* **pink mountain** ~ *Phyllodoce empetriformis* **purple** ~ *P. breweri* **Scottish** ~ *Calluna vulgaris* **white** ~ *Cassiope mertensiana* **white winter** ~ *Erica x hyemalis* **yellow mountain** ~ *Phyllodoce glanduliflora.*

Hebe (After Hebe, the Greek goddess of youth.) A genus of 75 species of evergreen shrubs and small trees with simple opposite leaves that are sometimes reduced and scale-like and sometimes fleshy. The tubular, 4-lobed, white or blue flowers bear 2 stamens and are borne in racemes or heads. Many species, hybrids and cultivars are grown ornamentally. *Distr.* Australasia, temperate South America. *Fam.* Scrophulariaceae. *Class* D.

H. albicans Dense, mound-forming shrub. Foliage blue-grey. Flowers white. *Distr.* New Zealand.

H. brachysiphon Bushy dense shrub. Flowers white, borne in dense spikes. *Distr.* New Zealand.

H. buchananii Hummock-forming shrub. Leaves blue-green. Flowers white. *Distr.* New Zealand (Southern Alps).

H. canterburiensis Spreading shrub. Leaves dark green. Flowers white, produced in profusion. *Distr.* New Zealand.

H. carnosula Small, tuft-forming shrub. Leaves fleshy, overlapping. Flowers white, borne in rounded heads. *Distr.* New Zealand (South Island).

H. catarractae See *Parahebe catarractae*

H. cupressoides Bushy shrub. Foliage conifer-like, fragrant. Leaves very small. Flowers tiny, white. *Distr.* New Zealand (South Island).

H. hectoris Variable bush. Branches slender, covered in scale-like leaves. Flowers pink-white, borne in congested heads. *Distr.* New Zealand (South Island).

H. hulkeana (After T. H. Hulke, who is said to have discovered it.) NEW ZEALAND

LILAC. Sprawling shrub. Leaves leathery, spreading, often flushed red. Flowers lilac-lavender to white, borne in large heads. *Distr.* New Zealand.

H. lyallii See *Parahebe lyallii*

H. ochracea Dense, bushy shrub. Branches slender, covered in scale-like leaves. Flowers white, borne on short spikes. *Distr.* New Zealand (South Island).

H. perfoliata See *Parahebe perfoliata*

H. salicifolia Variable erect shrub. Leaves narrow, pointed. Flowers white-lilac, scented. Perhaps the most hardy cultivated species. *Distr.* New Zealand, Chile.

hebecarpus, -a, um: with soft downy fruit.

Hebenstretia (After Johan Ernst Hebenstreit (1703–57), professor of botany at Leipzig.) A genus of 25 species of annual and perennial herbs and shrubs with simple alternate leaves and dense terminal spikes of white or yellow, 4-lobed, tubular flowers. Several species are cultivated as ornamentals. *Distr.* Tropical and S Africa. *Fam.* Scrophulariaceae. *Class* D.

H. dentata Annual herb to 60cm high. Flowers yellow or white, marked orange-yellow. *Distr.* South Africa.

hebepetalus, -a, -um: with soft downy petals.

Hectorellaceae A family of 2 genera and 2 species of tuft-forming perennial herbs. This family is sometimes included within the family Portulacaceae. *Distr.* New Zealand, Kerguelen. *Class* D.

Hedera (Classical Latin name.) IVY. A genus of 11 species of evergreen climbers with palmately lobed juvenile leaves on climbing rooting stems and simple mature leaves or erect flowering stems. The flowers are small, have their parts in 5s, and are borne in umbels. Most of the species are grown as ornamentals. *Distr.* Europe, Mediterranean to W Asia. *Fam.* Araliaceae. *Class* D.

H. chinensis See *H. nepalensis*

H. colchica PERSIAN IVY, COLCHIS IVY. Leaves very large, smelling of celery when crushed, red-hairy beneath. *Distr.* Caucasus, Turkey to N Iran.

H. cristata See *H. helix*

H. helix COMMON IVY, ENGLISH IVY. Leaves leathery, hairless. Flowers white. Fruit orange-yellow to black. Grown as an ornamental with numerous cultivars available for the garden and the house. It is reported have medicinal properties against the effects of alcohol and the fruits are an important source of late winter food for birds. The wood has been used as a substitute for box wood. *Distr.* Europe, Mediterranean and W Asia.

H. hibernica IRISH IVY, ATLANTIC IVY. *Distr.* W Europe.

H. maroccana MOROCCAN IVY. *Distr.* Morocco, naturalized in S Spain and Canary Islands.

H. nepalensis NEPAL IVY. *Distr.* Himalaya.

H. rhombea JAPANESE IVY. *Distr.* Japan.

hederaceus, -a, -um: resembling the genus *Hedera*, IVY.

hederifolius, -a, -um: with leaves like the genus *Hedera*, IVY.

Hedychium (From Greek *hedys*, sweet, and *chion*, snow, alluding to the fragrant white flowers of *H. coronarium*) GINGER LILY, GARLAND LILY. A genus of 40–50 species of robust rhizomatous herbs with cane-like stems and dense spikes of showy, tubular, fragrant flowers. Many species are cultivated as ornamentals. *Distr.* Tropical Asia, Madagascar, Himalaya. *Fam.* Zingiberaceae. *Class* M.

H. coccineum SCARLET GINGER LILY, RED GINGER LILY. Flowers red-orange. *Distr.* Himalaya.

H. coronarium BUTTERFLY LILY, GARLAND FLOWER, WHITE GINGER. Flowers white, very fragrant. *Distr.* SE Asia, widely naturalized in tropical America.

H. densiflorum Flowers fragrant, orange or yellow. *Distr.* Himalaya, Assam.

H. flavescens YELLOW GINGER. Flowers yellow with red lips. *Distr.* E Himalaya.

H. gardnerianum (After Edward Gardner (born 1784), political resident in Nepal.) KAHILI GINGER. Flowers lemon-yellow and red. *Distr.* Himalaya, Assam.

H. michauxii (After André Michaux (1746–1803), French traveller and plant collector.) CREEPING BLUETS. Vigorous perennial. Flowers star-shaped, blue, with a yellow centre. *Distr.* SE USA.

Hedysarum (From Greek *hedys*, sweet, alluding to the fragrant flowers of *H. coronarium*.) A genus of 100 species of perennial herbs and shrubs with pinnate leaves and racemes of irregular pea-like flowers. Some species are cultivated ornamentally. *Distr.* N temperate regions. *Fam.* Leguminosae. *Class* D.

H. coronarium FRENCH HONEYSUCKLE. Perennial herb to 1m high. Flowers fragrant, red to purple. Sometimes grown as a fodder crop. *Distr.* Europe.

H. multijugum Deciduous shrub. Flowers magenta. *Distr.* W Mongolia, E Tibet.

Hedyscepe (From Greek *hedys*, sweet, and *skepe*, a covering, alluding to the dense clusters of flowers.) A genus of 1 species of medium-sized palm to 10m high with solitary stems, feather-like leaves and few-branched inflorescences of small yellow flowers. *Distr.* Lord Howe Island, off Australia. *Fam.* Palmae. *Class* M.

H. canterburyana UMBRELLA PALM. Grown as an ornamental.

Heimerliodendron See *Pisonia*

Heimia (After Dr Heim of Berlin (died 1834).) A genus of 3 species of perennial herbs and shrubs with 4-angled stems, small simple leaves, and tubular flowers that are borne singly or in small clusters. Several species are grown as ornamentals, as well as being used medicinally as a contraceptive and a mild hallucinogen. *Distr.* S USA to Argentina. *Fam.* Lythraceae. *Class* D.

H. salicifolia Deciduous shrub. Flowers solitary, yellow. *Distr.* S USA, Central and South America.

Helenium (From *helenion*, the Greek name for another plant (perhaps *Inula helenium*), said to have been named after Helen of Troy.) SNEEZEWEED. A genus of 40 species of annual to perennial herbs with simple to pinnately lobed leaves and yellow flowers in daisy-like heads that usually have distinct yellow rays. *Distr.* America. *Fam.* Compositae. *Class* D.

H. amarum BITTERWEED, BITTER SNEEZEWEED. Slender annual, to 70cm high. *Distr.* SE USA.

H. autumnale SNEEZEWEED. Perennial herbs to 1.5m high. *Distr.* North America.

H. bigelovii Perennial herb, to 90cm high. *Distr.* USA (California, Oregon).

H. hoopesii SNEEZEWEED. A weedy perennial, to 1m high. Causes 'spewing sickness' in live stock. *Distr.* W USA.

Heliamphora (From Greek *helios*, the sun, and *amphoreus*, jug or pitcher.) SUN PITCHER. A genus of 5 species of perennial herbs and subshrubs with rosettes of pitcher-shaped leaves that trap and digest insects. The flowers are borne in a raceme and bear 4 petal-like sepals and 10–20 stamens but no petals. Several species are grown as ornamentals. *Distr.* Guyana Highland of South America. *Fam.* Sarraceniaceae. *Class* D.

H. nutans Pitchers tinged red. Flowers white, ageing red. *Distr.* Guyana.

Helianthella (Diminutive of the genus name *Helianthus*.) A genus of 8 species of perennial herbs with simple leaves and yellow flowers in large, usually solitary, daisy-like heads that bear distinct yellow rays. *Distr.* W North America. *Fam.* Compositae. *Class* D.

H. quinquenervis Tall herb, to 5m high. Flower-heads nodding. *Distr.* Central USA, Mexico.

Helianthemum (From Greek *helios*, sun, and *anthemon*, flower; the flowers tend to open only in bright sunshine.) SUN ROSE, ROCK ROSE. A genus of about 110 species of dwarf evergreen shrubs with simple opposite leaves and 1-sided racemes of regular, saucer-shaped flowers that bear 5 sepals, 5 fragile petals, and numerous stamens. Individual flowers often only last a single day. Several species and very many hybrids are grown as ornamentals. *Distr.* Europe, NE Africa, C Asia, North and South America. *Fam.* Cistaceae. *Class* D.

H. alpestre See *H. nummularium*

H. apenninum Much-branched spreading shrub. Flowers pure white. *Distr.* SW Europe, N Africa.

H. canum Procumbent shrub. Flowers yellow. *Distr.* Europe, Mediterranean.

H. chamaecistus See *H. nummularium*

H. globulariifolium See *Tuberaria globulariifolia*

H. nummularium Procumbent or ascending shrub. Flowers usually yellow, occasionally pink, white or orange. *Distr.* Europe, W Asia, Caucasus.

H. ovatum See *H. nummularium*

H. serpyllifolium See *H. nummularium*

H. umbellatum *See Halimium umbellatum*

helianthoides: from the genus *Helianthus*, with the ending *-oides*, indicating resemblance.

Helianthus (From Greek *helios*, sun, and *anthos*, flower.) SUNFLOWER. A genus of 67 species of annual and perennial herbs with simple alternate leaves and flowers in large, showy, daisy-like heads that bear distinct yellow or red rays. Several species are grown as ornamentals. *H. tuberosus* and *H. annuus* are both of commercial importance. *Distr.* America. *Fam.* Compositae. *Class* D.

H. annuus SUNFLOWER. Tall annual herb with a deep taproot and large flower-heads. Seeds are eaten whole or crushed for oil, which is widely used in cooking; the remaining cake is fed to livestock. *Distr.* W North America.

H. atrorubens DARK-EYED SUNFLOWER. Tall perennial. Flowers maroon, rays yellow-orange. *Distr.* SE USA.

H. decapetalus THIN-LEAF SUNFLOWER. Tall bushy perennial. Rays pale yellow. *Distr.* E North America.

H. × laetiflorus A hybrid of *H. pauciflorus* x *H. tuberosus*. *Distr.* Central USA.

H. orgyalis See *H. salicifolius*

H. pauciflorus Rough-hairy perennial, to 2m high. *Distr.* USA.

H. quinquenervis See *Helianthella quinquenervis*

H. rigidus See *H. pauciflorus*

H. salicifolius Perennial, to 3m high. Flower-heads numerous. *Distr.* S Central USA.

H. scaberrimus See *H. x laetiflorus*

H. tuberosus JERUSALEM ARTICHOKE, GLOBE ARTICHOKE. Tuberous perennial to 3m high. The tubers are edible and are a good carbohydrate food for diabetics; they are also fed to livestock. *Distr.* E North America.

Helichrysum (From Greek *helios*, sun, and *chryson*, golden.) EVERLASTING, EVERLASTING FLOWER. A genus of about 500 species of annual to perennial herbs and shrubs with simple, often glandular leaves and flowers in daisy-like heads which are surrounded by showy coloured bracts. Many species and cultivars are grown as ornamentals; the flower-heads are often dried for use as winter decorations. *Distr.* Warm Old World especially South Africa and Australia. *Fam.* Compositae. *Class* D.

H. acuminatum ORANGE EVERLASTING, ALPINE EVERLASTING. Perennial herb, to 30cm high. Bracts orange-gold. *Distr.* SE Australia.

H. alveolatum See *H. splendidum*

H. ambiguum Woody perennial. Bracts yellow. *Distr.* Spain (Balearic Islands).

H. angustifolium See *H. italicum*

H. arenarium YELLOW EVERLASTING FLOWER. Tuft-forming perennial. Bracts yellow, tinged red. *Distr.* Europe.

H. argyrophyllum Mat-forming shrub. Bracts lemon-yellow. *Distr.* South Africa (E Cape Province, W Transkei).

H. arwae Tufted subshrub. *Distr.* Yemen.

H. asperum Subshrub. Bracts, erect, yellow-brown. *Distr.* South Africa.

H. bellidioides Prostrate perennial herb. Bracts papery, white. *Distr.* New Zealand.

H. bracteatum EVERLASTING, STRAWFLOWER, GOLDEN EVERLASTING, YELLOW PAPER DAISY. Annual to perennial herb to 1.5m high. Bracts yellow-gold. The flower-heads are popular as dried flowers. Numerous cultivars are available. *Distr.* Australia.

H. chionophilum Tufted, hairy perennial. Bracts yellow. *Distr.* Turkey, Iraq, Iran.

H. coralloides See *Ozothamnus coralloides*

H. depressum See *Ozothamnus depressum*

H. doerfleri Densely tufted perennial. *Distr.* Crete.

H. ericifolium See *Ozothamnus ericifolius*

H. ericoides See *Dolichothrix ericoides*

H. foetidum Biennial herb. Flower-heads numerous. *Distr.* South Africa.

H. frigidum Mat-forming subshrub. Flower-heads solitary. *Distr.* Corsica, Sardinia.

H. gunnii See *Ozothamnus gunnii*

H. hookeri See *Ozothamnus hookeri*

H. italicum Tufted aromatic herb. Oil said to have anti-viral activity. *Distr.* Mediterranean.

H. lanatum See *H. thianschanicum*

H. ledifolium See *Ozothamnus ledifolius*

H. marginatum See *H. milfordiae*

H. microphyllum See *Plectostachys serpyllifolia*

H. milfordiae (After Mrs Helen A. Milford (died 1940) who collected in South

Africa.) Cushion-forming subshrub. Bracts glossy white with red tips. *Distr.* South Africa.

H. orientale Erect subshrub. *Distr.* SE Europe, W Asia.

H. petiolare LIQUORICE PLANT. Spreading shrub. Ornamental foliage. *Distr.* South Africa, naturalized in Portugal.

H. petiolatum See *H. petiolare*

H. plicatum Perennial herb with a woody stock. Plants under this name in cultivation are often actually *H. stoechas*. *Distr.* SE Europe.

H. plumeum Small stout shrub. *Distr.* New Zealand.

H. praecurrens Mat-forming perennial. *Distr.* South Africa (Natal), Lesotho.

H. rosmarinifolium See *Ozothamnus rosmarinifolius*

H. scorpiodes BUTTON EVERLASTING. Tuft-forming perennial herb. *Distr.* SE Australia, Tasmania.

H. serotinum See *H. italicum*

H. serpyllifolium See *Plectostachys serpyllifolia*

H. sibthorpii (After John Sibthorp (1758–96) professor of botany at Oxford.) Cushion-forming herb. *Distr.* Greece.

H. siculum See *H. stoechas*

H. splendidum Shrub to 1.5m high. Bracts bright yellow. *Distr.* South Africa.

H. stoechas Perennial herb to subshrub. Bracts white-glandular. *Distr.* SW Europe.

H. thianschanicum Perennial hairy herb. Bracts orange-yellow. *Distr.* Turkestan.

H. thyrsoideum See *Ozothamnus thyrsoideus*

H. trilineatum See *H. splendidum*

H. tumidum See *Ozothamnus selago*

H. virgineum See *H. sibthorpii*

H. woodii Small shrub. Leaves grey-white hairy. *Distr.* South Africa (Natal).

Heliconia (From Mount Helicon in Greece, the home of the Muses, reflecting the closeness of this genus to *Musa*.) LOBSTER CLAW, FALSE BIRD OF PARADISE, WILD PLANTAIN. A genus of about 100 species of substantial rhizomatous herbs with large leaves that sometimes form pseudo-stems. The irregular flowers are borne in spectacular, pendent, or erect inflorescences that bear large brightly coloured bracts. Many species and cultivars are grown as tender ornamentals, especially in North America. *Distr.* Tropical America and SW Pacific. *Fam.* Heliconiaceae. *Class* M.

H. mariae BEEFSTEAK HELICONIA. Inflorescence pendent, to 80cm. Bracts thick, blood-red, turning black with age. *Distr.* Central and N South America.

H. rostrata Inflorescence pendent. Bracts red, yellow at tips and green on margins. *Distr.* Peru to Argentina.

heliconia, beefsteak *Heliconia mariae*.

Heliconiaceae A family of 1 genus and 100 species. *Class* M

See: Heliconia.

Helictotrichon (From Greek *heliktos*, twisted, and *trichos*, hair, alluding to the shape of the awn.) OAT GRASS. A genus of about 60 species of perennial, tussock-forming grasses with ribbed leaves and erect or nodding inflorescences. *Distr.* N temperate regions plus South Africa and mountainous regions of the Old World tropics. *Fam.* Gramineae. *Class* M.

H. sempervirens BLUE OAT GRASS. Leaves rigid, silver-blue. Inflorescence straw-coloured, open. Grown as an ornamental. *Distr.* SW Europe.

heliolepis: with shining scales; from Greek *helios*, sun.

Heliophila (From Greek *helios*, sun and *philos*, loving.) A genus of 71 species of annual to perennial herbs and subshrubs with white to blue or pink flowers. *Distr.* South Africa. *Fam.* Cruciferae. *Class* D.

H. longifolia Herb with blue flowers reminiscent of flax.

H. scandens Perennial woody climber.

Heliopsis (From Greek *helios*, sun, and *-opsis*, alluding to the showy flower-heads.) OX-EYE. A genus of 13 species of perennial herbs with toothed leaves and flowers in daisy-like heads that bear distinct yellow rays. *Distr.* Upland tropical America. *Fam.* Compositae. *Class* D.

H. helianthoides To 1.5m high. Flower-heads to 8cm across, numerous. Cultivated ornamentally. *Distr.* North America.

heliotrope *Heliotropium*, *H. arborescens* **garden** ~ *Valeriana officinalis* **winter** ~ *Petasites fragrans*.

Heliotropium (From Greek *helios*, sun, and *trope*, to turn, alluding to the belief that the flower-heads turned with the sun.) HELIOTROPE, TURNSOLE. A genus of about 250 species of annual and perennial herbs, subshrubs, and shrubs with simple alternate leaves and curved racemes of cylindrical or funnel-shaped flowers. Several species are grown as ornamentals and some are important locally as medically herbs. *Distr.* Tropical and temperate regions. *Fam.* Boraginaceae. *Class* D.

H. arborescens HELIOTROPE, CHERRY PIE. Large shrub. Flowers purple or white, fragrant. Cultivated ornamental. Flowers used in scent making. *Distr.* Peru.

H. europaeum Annual herb. Flowers white or blue. Grown as an ornamental. *Distr.* Europe to Central Asia.

H. peruvianum See *H. arborescens*

helix: winding around, helical.

hellebore *Helleborus* **black** ~ *H. niger* **false** ~ *Veratrum* **green** ~ *Helleborus viridis* **stinking** ~ *H. foetidus* **white** ~ *Veratrum album.*

helleborifolius, -a, -um: with leaves like the genus *Helleborus.*

helleborine *Epipactus* **broad-leaved** ~ *E. atrorubens*, *E. helleborine* **dark red** ~ *E. atrorubens* **marsh** ~ *E. palustris.*

Helleborus (From *helleboros*, the classical Greek name for *H. orientale.*) HELLEBORE. A genus of 15–20 species of rhizomatous herbs with compound, typically leathery leaves, and regular bisexual flowers that bear 5 showy sepals, many funnel-shaped nectaries and numerous stamens. Many species, hybrids, and cultivars are grown as ornamentals for their early spring flowers. All contain poisonous substances. *Distr.* Europe, Mediterranean region, W Asia. *Fam.* Ranunculaceae. *Class* D.

H. atrorubens Clump-forming perennial. Flowers cup-shaped, purple. *Distr.* E Alps.

H. colchicus See *H. orientalis*

H. cyclophyllus Clump-forming perennial. Flowers saucer-shaped, yellow-green. *Distr.* Balkans.

H. foetidus STINKING HELLEBORE, BEAR'S FOOT, SETTERWORT, STINKWORT. Clump-forming perennial. Flowers green, tinged red at the tips. Formerly used as a dangerous heart drug and veterinary medicine. *Distr.* W and S Europe.

H. niger CHRISTMAS ROSE, BLACK HELLEBORE. Flowers white to pink, becoming green after fertilization. *Distr.* Alps and Apennines.

H. orientalis LENTEN ROSE. Flowers cream. *Distr.* SE Europe, W Asia.

H. viridis GREEN HELLEBORE. Slightly hairy perennial. Flowers green. *Distr.* W and Central Europe, naturalized North America.

helmet flower *Scutellaria*, *Aconitum napellus.*

helodoxa: glory of the marsh.

Helonias (From Greek *helos*, swamp, alluding to the habitat of the plant.) A genus of 1 species of rhizomatous herb with a basal rosette of stalked leaves and racemes of small, star-shaped, pink, fragrant flowers. *Distr.* E North America. *Fam.* Melanthiaceae. *Class* M.

H. bullata SWAMP PINK. Occasionally grown as an ornamental.

helonioides: from the genus *Helonias*, with the ending *-oides*, indicating resemblance.

Heloniopsis (From the genus name *Helonias* and Greek *opsis*, appearance.) A genus of 4 species of rhizomatous herbs with a basal rosette of stalked leaves and solitary or clustered, star-shaped, pink or violet flowers. *Distr.* Japan, Korea, and Taiwan. *Fam.* Melanthiaceae. *Class* M.

H. japonica See *H. orientalis*

H. orientalis Flowers in clusters of 2–10, pink-violet. Cultivated as an ornamental. *Distr.* Japan, Korea.

helveticus, -a, -um: of Switzerland.

Helwingia (After Georg Helwing (1666–1748), German botanist.) A genus of 3 species of deciduous shrubs with alternate simple leaves and clusters of small unisexual flowers borne on the leaf surface close to the midrib. *Distr.* E Asia. *Fam.* Helwingiaceae. *Class* D.

H. japonica Occasionally grown as an ornamental curiosity. *Distr.* China, Japan.

Helwingiaceae A family of 1 genus and 3 species. *Class* D.
See: Helwingia.

helxine *Soleirolia soleirolii.*

Hemerocallidaceae A family of 1 genus and about 15 species. *Class* M.
See: Hemerocallis.

Hemerocallis (From Greek *hemera*, day, and *kallos*, beauty; the flowers last for only one day.) DAY LILY, SPIDER LILY. A genus of about 15 species of perennial, clump-forming, rhizomatous herbs with 2-ranked leaves and funnel-shaped, yellow to orange flowers. Several species and many hybrids are cultivated as ornamentals. *Distr.* Central Europe to China and Japan. *Fam.* Hemerocallidaceae. *Class* M.

H. aurantiaca Flowers densely clustered, orange, flushed purple. *Distr.* China.

H. citrina Flowers yellow, opening at night, fragrant. *Distr.* China.

H. dumortieri (After B. C. Dumortier (1797–1828).) Vigorous clump-former with broad, almost flat flowers. *Distr.* Japan, E USSR, Korea.

H. flava See *H. lilio-asphodelus*

H. forrestii Flowers non-fragrant, yellow. *Distr.* SW China.

H. fulva Flowers orange, scentless. Dried flowers used as food-flavouring in China and Japan. *Distr.* Origin uncertain, perhaps China or Japan, naturalized Asia to Europe and North America.

H. lilio-asphodelus Flowers yellow, fragrant. *Distr.* E Siberia to Japan.

H. middendorffii (After Alexander Theodor von Middendorf (1815–94), Russian traveller and plant collector.) Flowers fragrant, yellow. *Distr.* Japan, E USSR, Korea, N China.

H. minor Dense clump-former with lemon-yellow flowers. *Distr.* Japan, China, E USSR.

H. multiflora Flowers yellow inside, red-brown outside. *Distr.* China.

H. thunbergii (After Carl Peter Thunberg (1743–1828), professor of botany at Uppsala University, who travelled in E Asia.) Flowers fragrant, short, yellow. *Distr.* China, Korea, Japan.

H. vespertina See *H. thunbergii*

Hemionitis (From Greek *hemionos*, mule, alluding to the fact that the plants were thought to be sterile.) A genus of 6–7 species of small to medium-sized ferns. Leaves lobed but not compound. Several species are cultivated as ornamentals. *Distr.* Tropical Asia and America. *Fam.* Adiantaceae. *Class* F.

H. arifolia Leaves irregularly lobed when young, arrowhead-shaped later. *Distr.* E Asia.

H. palmata Leaves palmately lobed when young, simple later. *Distr.* Tropical America, West Indies.

hemisphaericus, -a, -um: hemispherical, half a sphere.

hemitrichotus, -a, -um: half-hairy.

hemlock *Conium maculatum*, *Tsuga* **Canada** ~ *T. canadensis* **eastern** ~ *T. canadensis* **poison** ~ *Conium maculatum* **spotted** ~ *C. maculatum* **western** ~ *Tsuga heterophylla.*

hemp *Cannabis sativa*, Cannabaceae **African** ~ *Sparmannia africana* **African bowstring** ~ *Sansevieria hyacinthoides* **Bahama** ~ *Agave sisalana* **bowstring** ~ *Sansevieria* **Chinese** ~ *Abutilon theophrasti* **Manila** ~ *Musa textilis* **Mauritius** ~ *Furcraea foetida.*

hempweed, climbing *Mikania scandens.*

henbane *Hyoscyamus niger* **black** ~ *H. niger.*

henna *Lawsonia inerma*, Lythraceae.

Hepatica (From Greek *hepar*, liver, alluding to the shape and colour of the leaves, said to cure diseases of the liver.) A genus of 6–10 species of perennial herbs with 3– to 5-lobed, basal leaves and solitary, regular, white, pink, or purple flowers that bear 5–12 petal-like sepals and numerous stamens but no true petals. The leaves are often mottled purple-brown. Several species are grown ornamentally. *Distr.* N temperate. *Fam.* Ranunculaceae. *Class* D.

H. nobilis Rhizomatous perennial. Flowers white, tinged purple or pink. Formerly used medicinally. *Distr.* Eurasia.

H. triloba See *H. nobilis*

hepaticus, -a, -um: deep brown-red, liver-coloured.

heptaphyllus, -a, -um: seven-leaved.

Heptapleurum See *Schefflera*

heracleifolius, -a, -um: from the genus name *Heracleum*, and *folius*, leaved.

Heracleum (After the Greek god Heracles or Hercules.) A genus of 60 species of biennial and perennial herbs with simple or divided leaves and compound umbels of white, occasionally yellow or pink flowers. Several species are grown ornamentally. *Distr*. N temperate regions and tropical mountains. *Fam*. Umbelliferae. *Class* D.

H. mantegazzianum (After Paolo Mantegazzi (1831–1910).) GIANT HOGWEED, CARTWHEEL FLOWER. To 3m high. Umbels to 1m across. The sap causes photodermatitis, sensitizing skin to ultraviolet light and causing blistering. Occasionally grown as an ornamental. *Distr*. Caucasus, naturalized Europe, USA.

H. sphondylium HOGWEED. Biennial or perennial to 2.5m high. *Distr*. Europe, North America, Central Asia.

herbaceus, -a, -um: herbaceous.

herb Bennet *Geum urbanum*.

herb Christopher *Actaea spicata*.

Herbertia (After William Herbert (1778–1847), Dean of Manchester and an authority on bulbiferous plants.) A genus of about 6 species of perennial herbs with lance-shaped, folded leaves and short-lived flowers that bear 3 large outer tepals and 3 small inner tepals. Several species are grown as ornamentals. *Distr*. Temperate South America. *Fam*. Iridaceae. *Class* M.

H. amatorum Flowers violet with brown and white markings and yellow nectaries. *Distr*. S South America.

H. pulchella Flowers violet and mauve with white markings. *Distr*. S South America.

herb Gerard *Aegopodium podagraria*.

herb of grace *Ruta graveolens*.

herb Paris *Paris quadrifolia*.

herb Robert *Geranium robertianum*.

hercules club *Aralia spinosa*.

Hermannia (After Paul Hermann (died 1695), Dutch botanist.) HONEYBELLS. A genus of about 100 species of herbs and subshrubs with simple alternate leaves and bell-shaped flowers that bear their parts in 5s. Several species are grown as ornamentals. *Distr*. Tropical and warm regions, especially South Africa. *Fam*. Sterculiaceae. *Class* D.

H. verticillata HONEYBELLS. Perennial herb to 30cm high. Flowers fragrant, yellow. *Distr*. South Africa.

hermaphroditus, -a, -um: with hermaphrodite flowers.

Hermodactylus (From *Hermes*, Greek name for the god Mercury, and *daktylos*, finger, alluding to the finger-like tubers.) A genus of 1 species of tuberous herbs with linear, 4-angled leaves, and solitary, *Iris*-like, fragrant flowers. *Distr*. E and Central Mediterranean region. *Fam*. Iridaceae. *Class* M.

H. tuberosus SNAKE'S HEAD IRIS, WIDOW IRIS. A hardy ornamental. The tubers were formerly used medicinally.

Hernandiaceae A family of 4 genera and 68 species of shrubs, trees and some lianas with large, simple, or palmate, leaves and small regular flowers that bear 4–10 tepals. The wood of some of the species is used locally for packing and construction. *Distr*. Tropical regions, mainly in coastal areas. *Class* D.

Herniaria (From Latin *hernia*, rupture, alluding to supposed medicinal properties.) RUPTUREWORT. A genus of 15–20 species of annual and perennial, often prostrate herbs with small, sessile, opposite leaves and dense cymes of minute flowers which bear their parts in 5s. *Distr*. Europe, Africa, W Asia. *Fam*. Illecebraceae. *Class* D.

H. glabra RUPTUREWORT, HERNIARY. Annual or short-lived perennial. Flowers white. Grown as ornamental ground cover. *Distr*. Europe, N Africa, W Asia.

herniary *Herniaria glabra*.

heron's bill *Erodium*.

Herpolirion (From Greek *herpo*, to creep, and *lirion*, lily.) A genus of 1 species of rhizomatous perennial herb with 2-ranked, blue-green leaves and solitary blue to cream flowers. *Distr*. SE Australia and New Zealand. *Fam*. Anthericaceae. *Class* M.

H. novae-zelandiae A delicate cultivated alpine. *Distr.* SE Australia and New Zealand.

Herreriaceae A family of 2 genera and 24 species of climbing subshrubs with simple leaves and flowers which bear their parts in threes. This family is sometimes included in the Liliaceae. *Distr.* Madagascar, South America. *Class* M.

Hertia See *Othonna*

Hesperaloe (From Greek *hesperos*, western, and the genus name *Aloe*.) A genus of 3 species of clump-forming, grass-like, stemless herbs with spikes of bell-shaped flowers. Several species are grown as ornamentals. *Distr.* SW North America. *Fam.* Agavaceae. *Class* M.

H. funifera Leaves to 180cm, usually straight. *Distr.* N Mexico.

H. parviflora Leaves to 130cm long, usually arching. *Distr.* USA (Texas).

Hesperantha (From Greek *hespera*, evening, and *anthos*, flower.) A genus of about 55 species of perennial herbs with hard corms, linear leaves, and spikes of tubular or cup-shaped flowers. Several species are cultivated as ornamentals. *Distr.* Sub-Saharan Africa. *Fam.* Iridaceae. *Class* M.

H. buhrii See *H. cucullata*

H. cucullata Flowers white, with outer tepals tinged red, scented, opening in the evening. *Distr.* South Africa (Cape Province).

H. falcata Flowers cream to yellow, tinged orange, opening at night. White-flowered forms fragrant. *Distr.* SW South Africa.

Hesperis (From Greek *hespera*, evening; the flowers are particularly fragrant in the evening.) A genus of 30 species of biennial or short-lived perennial herbs with yellow, white, or purple flowers. *Distr.* Europe, Mediterranean to Iran, Central Asia, W China. *Fam.* Cruciferae. *Class* D.

H. lutea See *Sisymbrium luteum*

H. matronalis SWEET ROCKET, DAME'S VIOLET, DAME'S ROCKET, ROCKET, DAMASK VIOLET. Old-fashioned garden plant with fragrant, white-lilac flowers. Seeds crushed to give an oil. *Distr.* Europe, W and Central Asia, naturalized in North America.

Heteranthera (From Greek *heteros*, different, and Latin *anther*, anthers.) MUD PLANTAIN. A genus of about 10 species of annual and perennial herbs with either linear submerged leaves or round floating leaves or both. The cup-shaped blue or white flowers have 6 tepals, 1 large and 2 smaller stamens. Several species are grown as ornamentals for the aquarium. *Distr.* Tropical and subtropical America and Africa. *Fam.* Pontederiaceae. *Class* M.

H. dubia WATER STAR GRASS. Leaves linear. Flowers borne flat on the surface of the water, pale yellow. *Distr.* North and South America.

heteranthus, -a, -um: with uneven flowers.

Heterocentron (From Greek *heteros*, different, and *kentron*, spur, alluding to the spurs of the anthers.) A genus of 27 species of perennial herbs and subshrubs with simple leaves and white, pink, or mauve flowers that are borne singly or in small clusters and bear 4 petals and 8 unequal stamens. Several species are grown as ornamentals. *Distr.* Mexico, Central America. *Fam.* Melastomataceae. *Class* D.

H. elegans SPANISH SHAWL. Mat-forming subshrub. Flowers solitary, magenta to mauve. *Distr.* Mexico, Central America.

heterocladus, -a, -um: with uneven shoots or branches.

heterodontus, -a, -um: with uneven teeth.

Heteromeles See *Photinia*

Heteropappus (From Greek *heteros*, different, and *pappus*, the fruit appendage in Compositae, alluding to the two different types of fruit.) A genus of 5 species of biennial and perennial herbs with simple alternate leaves and flowers in solitary daisy-like heads that bear distinct rays. *Distr.* Temperate E Asia. *Fam.* Compositae. *Class* D.

H. altaicus Erect herbs, to 40cm high. Rays purple-white. *Distr.* Iran to Himalaya.

heterophyllus, -a, -um: with leaves of very different shapes.

Heuchera (After Johann Heinrich Heucher (1677–1747), professor of medicine at Wittenberg.) ALUM ROOT, CORAL BELLS. A genus of

55 species of perennial herbs with basal, tuft-forming leaves and small flowers that often lack petals. Species are frequently used as evergreen ground cover in the garden. *Distr.* North America. *Fam.* Saxifragaceae. *Class* D.

H. americana ROCK GERANIUM. Leaves hairy with a white mottling. *Distr.* North America.

H. × brizoides A large range of ornamental hybrids involving *H. sanguinea* as one of the parents. *Distr.* Garden origin.

H. chlorantha Leaves double-toothed. Flowers bright green. *Distr.* NW North America.

H. cylindrica Leaves double-toothed. Flowers cream. *Distr.* NW North America.

H. glauca See *H. americana*

H. grossulariifolia Flowers white. *Distr.* USA (Oregon, Washington).

H. micrantha Flowers numerous, white. *Distr.* British Columbia to Sierra Nevada.

H. parvifolia Leaves spreading. Flowers green-white. *Distr.* W North America.

H. pilosissima Flowers dull red. *Distr.* USA (Coastal California).

H. pubescens Lobes of leaves pointed. Flowers mauve to white. *Distr.* USA (Pennsylvania to Virginia).

H. pulchella Leaves deeply lobed, bright green. Flowers pink-purple. *Distr.* S USA.

H. richardsonii Flowers green. *Distr.* W North America.

H. rubescens Flowers somewhat irregular, white-pink. *Distr.* USA (California, Nevada, Utah).

H. sanguinea CORAL BELLS. Flowers bright red. A number of cultivars are available, including white-flowered forms. *Distr.* SW USA, Mexico.

H. versicolor Leaves nearly round. Flowers rose-red. *Distr.* USA (New Mexico).

H. villosa Leaves glandular, pubescent. Flowers green. *Distr.* USA (Appalachians, Arkansas, Tennessee, Kentucky).

× Heucherella (From the names of the parent genera.) A genus of hybrids between members of the genera *Heuchera* and *Tiarella*. *Distr.* Garden origin. *Fam.* Saxifragaceae. *Class* D.

×H. alba A clump-forming perennial herb derived from *Heuchera* × *brizoides* × *Tiarella wherryi*. *Distr.* Garden origin.

×H. tiarelloides Stoloniferous perennial herb with pink flowers, derived from *Heuchera* × *brizoides* × *Tiarella cordifolia*. *Distr.* Garden origin.

heucherifolius, -a, -um: with leaves like the genus *Heuchera*.

Hevea (From a Brazilian vernacular name.) A genus of 9 species of trees with trifoliate, spirally arranged leaves and panicles of small, fragrant, pale yellow flowers. The fruit is an explosive capsule. *Distr.* Amazon Basin. *Fam.* Euphorbiaceae. *Class* D.

H. brasiliensis PARA RUBBER TREE, CAOUTCHOUC TREE. Deciduous tree. Leaves bronze at first. Bark leathery. Flowers yellow-white. This tree is the major source of natural rubber, latex being collected by scoring or tapping the trunks of living trees. Old trees are a source of timber for plywood and chipboard. *Distr.* Basins of the Amazon and Orinoco rivers, cultivated throughout the tropics.

hexandrus, -a, -um: with six stamens.

hexapetalus, -a, -um: with six petals.

hexaphyllus, -a, -um: six-leaved.

hians: gaping, open-mouthed.

Hibbertia (After George Hibbert (1757–1837), English merchant who had a botanic garden at Clapham.) BUTTON FLOWER, GUINEA GOLD VINE. A genus of 122 species of evergreen shrubs and climbers, typically with small simple leaves and regular yellow flowers that bear 5 sepals, 5 petals, and a few to many stamens. Some species have leaf-like stems and reduced leaves. Several species are grown as ornamentals. *Distr.* Madagascar, SE Asia, Australia, New Caledonia, Fiji. *Fam.* Dilleniaceae. *Class* D.

H. dentata Trailing or twining shrub. Leaves to 5cm long. Flowers deep yellow, 5cm across. *Distr.* SE Australia.

H. scandens GUINEA GOLD VINE, SNAKE VINE, GOLD GUINEA PLANT. Climber. Leaves narrow, deep green. Flowers saucer-shaped, bright yellow. *Distr.* Australia.

H. volubilis See *H. scandens*

hibernica: Irish.

Hibiscus (Greek name for a Mallow.) ROSE MALLOW, GIANT MALLOW. A genus of about

200 species of annual and perennial herbs, shrubs, and trees with simple or palmately lobed leaves and regular bisexual flowers that bear 5 petals and numerous stamens that are fused into a column. The fruit is a capsule and the seeds are often hairy. Many species and cultivars are grown as ornamentals and some are used medicinally or as a source of fibre. *Distr.* Warm temperate and tropical regions. *Fam.* Malvaceae. *Class* D.

H. huegelii See *Alyogyne huegelii*

H. militaris SOLDIER ROSE MALLOW, HALBERD-LEAVED MARSH MALLOW, HALBERD-LEAVED ROSE MALLOW. Tall subshrub. Flowers solitary, white, with a red centre. *Distr.* USA.

H. moscheutos COMMON ROSE MALLOW, SWAMP ROSE MALLOW. Hairy subshrub to 2.5m high. Flowers pink or red. This species has given rise to numerous ornamental cultivars. *Distr.* SE USA.

H. rosa-sinensis ROSE OF CHINA, CHINA ROSE, SHOE FLOWER, CHINESE HIBISCUS, HAWAIIAN HIBISCUS, SHOE BLACK. Evergreen shrub or small tree to 5m high. Flowers very variable, usually deep red. Numerous ornamental cultivars of this species have been produced. The flowers are used for shining shoes in India. *Distr.* Origin uncertain, probably tropical Asia; widely cultivated in the tropics.

H. schizopetalus JAPANESE HIBISCUS, JAPANESE LANTERN. Evergreen, spreading shrub to 3m high. Flowers pink or red. Petals deeply cut. *Distr.* Tropical E Africa.

H. syriacus Shrub or small tree. Flowers white, red, or blue. The source of several ornamental cultivars. *Distr.* China, Taiwan.

hibiscus: Chinese ~ *Hibiscus rosa-sinensis* **Hawaiian** ~ *H. rosasinensis* **Japanese** ~ *H. schizopetalus* **lilac** ~ *Alyogyne huegelii* **Norfolk Island** ~ *Lagunaria patersonii.*

hickory *Carya* **broom** ~ *C. glabra* **shagbark** ~ *C. ovata* **small-fruited** ~ *C. glabra* **swamp** ~ *C. cordiformis* **water** ~ *C. aquatica.*

hiemalis, -e: of winter.

hieraciifolius, -a, -um: with leaves like the genus *Hieracium.*

Hieracium (Classical Greek name used for similar plants.) HAWKWEED. A genus of 250–260 species of perennial hairy herbs with milky sap, simple leaves, and flowers in dandelion-like heads. A number of species are cultivated as ornamentals. *Distr.* Temperate regions (except Australia), tropical mountains. *Fam.* Compositae. *Class* D.

H. alpinum Rhizomatous. To 35cm high. *Distr.* Europe, N Asia, E North America.

H. aurantiacum DEVIL'S PAINTBRUSH. Flowers orange-red. *Distr.* Europe.

H. lanatum Densely hairy. Flowers pale yellow. *Distr.* Central Europe.

H. maculatum To 80cm high. Stems often tinged purple. *Distr.* Europe.

H. murorum Rhizomatous. Flowers pale yellow. *Distr.* Europe, Central and N Asia.

H. pannosum Rhizomatous. To 60cm. *Distr.* Balkans, Turkey.

H. villosum Stems wavy. To 40cm high. *Distr.* Europe.

H. waldsteinii (After Count Franz Adam Waldstein-Wartenburg (1759–1823), Austrian botanist.) To 30cm. Leaves mainly basal. *Distr.* Yugoslavia.

H. welwitschii See *H. lanatum*

Hierochloe A genus of about 15 species of annual and perennial grasses to around 75cm high with aromatic leaves and spike-like compressed inflorescences. *Distr.* Temperate and arctic regions, tropical mountains. *Fam.* Gramineae. *Class* M.

H. odorata HOLY GRASS. Burned as incense and spread on church floors in New Mexico. *Distr.* North America, Asia and Europe.

hieroglyphicus, -a, -um: with hieroglyphic markings.

himalaicus, -a, -um: of the Himalaya.

himalayanus, -a, -um: of the Himalaya.

himalayensis, -e: of the Himalaya.

Himantandraceae A family of 1 genus and 1 species of aromatic tree with simple, spirally arranged leaves covered in numerous, minute, downy scales. The large regular flowers bear 2 sepals, 7–9 petals and numerous stamens. *Distr.* E Malaysia to NE Australia. *Class* D.

Hippeastrum (From Greek *hippos*, horse; the inflorescence of *H. puniceum* is said to resemble a horse's head.) AMARYLLIS, KNIGHT'S STAR LILY. A genus of about 80 species of bulbiferous perennial herbs with linear or strap-shaped leaves and umbels of several trumpet-shaped, white, red, or orange flowers. A number of species and numerous named hybrids are grown as ornamentals. *Distr.* Central and South America, 1 species in W Africa. *Fam.* Amaryllidaceae. *Class* M.

H. advenum See *Rhodophiala advena*

H. aulicum LILY OF THE PALACE. Flowers usually 2, crimson with a green throat. *Distr.* Central Brazil, Paraguay.

H. bifidum See *Rhodophiala bifida*

H. elegans Leaves broad. Flowers pale green. *Distr.* South America.

H. hybrida See *Amaryllis belladonna*

H. leopoldii Dwarf. Flowers red with green-white markings. *Distr.* Bolivia.

H. papilio Flowers pale-green, tinged dark red. *Distr.* S Brazil.

H. pardinum Flowers constricted at throat, yellow-green with red spots. *Distr.* Bolivia.

H. phycelloides Flowers brilliant red, yellow in the centre. *Distr.* The Andes of Chile.

H. pratense. *Distr.* Chile.

H. procerum BLUE AMARYLLIS. With lilac flowers and bulb-neck to 1.5 meters Cultivated ornamentally. *Distr.* S Brazil (Organ mountains).

H. psittacinum Flowers white-green, marked crimson-red. *Distr.* S Brazil.

H. puniceum BARBADOS LILY. Flowers red to pink with green-white markings in the throat. *Distr.* Central and N South America, Caribbean.

H. reginae MEXICAN LILY. Leaves appearing after flowers. Flowers drooping, red, marked white-green in throat. *Distr.* Central and N South America.

H. reticulatum Flowers 3–6, mauve with red markings. *Distr.* S Brazil.

H. roseum See *Rhodophiala roseum*

H. rutilum See *H. striatum*

H. solandriflorum See *H. elegans*

H. striatum Flowers crimson and green. *Distr.* Brazil.

H. vittatum Leaves appear after flowers. Flowers red- and white-striped. *Distr.* Peru, Brazil.

Hippocastanaceae The Horse Chestnut family. 2 genera and 15 species of evergreen and deciduous trees with sticky winter buds and palmate leaves. The flowers are irregular with 5 sepals and 4–5 free petals. The seeds are large and lack endosperm. Some trees are grown as ornamentals, others have been used for the production of charcoal or have medicinal qualities. *Distr.* S Europe, tropical and E Asia, North and South America. *Class* D.
See: Aesculus.

hippocastanum: from the Greek *hippos*, horse, and *kastana*, chestnut; also the classical Latin name for horse-chestnut.

Hippocrepis (From Greek *hippos*, horse, and *krepis*, shoe, alluding to the horseshoe-shaped segments of the pods.) A genus of 21 species of annual and perennial herbs and subshrubs with alternate pinnate leaves and head-like racemes of yellow pea-like flowers. *Distr.* Europe, W Asia, Mediterranean. *Fam.* Leguminosae. *Class* D.

H. comosa HORSESHOE VETCH. Cultivated ornamentally and as a fodder crop. *Distr.* Europe.

Hippophae (From *hippophaes*, a Greek name for a kind of spurge.) SEA BUCKTHORN. A genus of 3 species of deciduous shrubs or small trees with alternate, linear, hairy leaves and clusters of small unisexual flowers that lack petals and appear in early spring. The fruit is a round red drupe. *Distr.* Temperate Eurasia. *Fam.* Elaeagnaceae. *Class* D.

H. rhamnoides SEA BUCKTHORN, SALLOW THORN. Shrub or small tree growing typically on sandy coasts and shingle-banks. The fruit is edible and used in sauces for meat and fish in Europe or eaten with milk or cheese in Asia. The wood is suitable for turning. *Distr.* Europe to N China.

hippophaeoides: resembling the genus *Hippophae*.

Hippuridaceae A family of 1 genus and 1 species. *Class* D.
See: Hippuris.

Hippuris (From Greek *hippos*, horse, and *oura*, tail, alluding to the appearance of the aerial stems.) A genus of 1 species of aquatic perennial herb of shallow water with pale green, linear, submerged leaves and darker, broader, emergent leaves. The inconspicuous flowers are borne in the axils of the emergent leaves; they have no sepals or petals but bear a single large stamen. *Distr.* Widespread. *Fam.* Hippuridaceae. *Class* D.

H. vulgaris MARE'S TAIL. Young leaves eaten by Eskimo.

hirsutissimus, -a, -um: very hairy.

hirsutus, -a, -um: hairy.

hirtellus ,-a,-um: rather hairy.

hirtipes hairy-footed.

hirtus, -a, -um: hairy.

hispanicus, -a, -um: of Spain.

hispidulus, -a, -um: rather bristly.

hispidus, -a, -um: bristly, covered in coarse, erect, stubbly hairs.

Histiopteris (From Greek *histion*, sheet or sail, and *pteris*, fern, alluding to the shape of the fronds.) A genus of 1 species of large terrestrial fern with pinnately divided leaves. *Distr.* Tropical and temperate regions. *Fam.* Dennstaedtiaceae. *Class* F.

H. incisa Grown as an ornamental.

hobble bush *Viburnum lantanoides.*

hog brake *Pteridium aquilinum.*

hognut *Carya glabra.*

hogweed *Heracleum sphondylium* **giant** ~ *H. mantegazzianum.*

Hoheria (From *houhere*, the Maori name for *H. populnea.*) LACEBARK. A genus of 5 species of deciduous and evergreen shrubs and small trees with simple or lobed, alternate leaves and regular white flowers that bear 5 notched petals and numerous fused stamens. *Distr.* New Zealand. *Fam.* Malvaceae. *Class* D.

H. angustifolia Tall, columnar tree.

H. lyallii (After David Lyall (1817–95), naval surgeon and naturalist, who collected the type specimen.) LACEBARK. Small spreading tree. Leaves deeply toothed.

H. microphylla See *H. angustifolia*

H. populnea LACEBARK. Spreading, medium-sized tree. The fibrous bark is used for cordage and the wood for cabinet-making.

H. sexstylosa RIBBON WOOD. Small tree or upright shrub.

Holboellia (After Frederik Ludvig Holboell (1765–1829), superintendent of the Copenhagen Botanic Garden.) A genus of about 5 species of evergreen twining shrubs with palmate leaves and corymbs or racemes of regular unisexual flowers in which the petals have been replaced by nectaries and the sepals become petal-like. Several species are grown as ornamentals. *Distr.* Himalaya and SE Asia. *Fam.* Lardizabalaceae. *Class* D.

H. coriacea Leaves with 3 leaflets. Male flowers white. Female flowers larger than males, tinged green or purple. Fruit purple. *Distr.* Central China.

H. latifolia Leaves with 3–7 leaflets. Male flowers green-white. Female flowers purple. Fruit purple. *Distr.* Himalaya.

Holcus (From Greek *holkos*, a kind of cereal.) A genus of about 8 species of annual and perennial, tufted grasses with flat or folded leaves and panicles of compressed spikelets. *Distr.* Europe, temperate Asia, N and S Africa. *Fam.* Gramineae. *Class* M.

H. lanatus CREEPING SOFT GRASS, YORKSHIRE FOG. A perennial, soft-hairy pasture-grass with several ornamental varieties available. *Distr.* Eurasia, naturalized North America.

H. mollis CREEPING SOFT GRASS. A perennial, soft-hairy pasture-grass with several ornamental varieties available. *Distr.* Europe.

holly Aquifoliaceae, *Ilex, I. aquifolium* **American** ~ *I. opaca* **blue** ~ *I. x meserveae* **box** ~ *Ruscus aculeatus* **box-leaved** ~ *Ilex crenata* **Chinese** ~ *I. cornuta, Osmanthus heterophyllus* **common** ~ *Ilex aquifolium* **English** ~ *I. aquifolium* **European** ~ *I. aquifolium* **false** ~ *Osmanthus heterophyllus* **Himalayan** ~ *Ilex dipyrena* **horned** ~ *I. cornuta* **Japanese** ~ *I. crenata* **miniature** ~ *Malpighia coccigera* **myrtle** ~ *Ilex myrtifolia* **sea** ~ *Eryngium, E. maritimum* **Singapore** ~ *Malpighia coccigera* **West Indian** ~ *Leea coccinea.*

holly grape *Mahonia repens.*

hollyhock *Alcea, A. rosea* **Antwerp** ~ *A. ficifolia* **fig-leaved** ~ *A. ficifolia.*

holm, sea *Eryngium maritimum.*

holocarpus, -a, -um: with an entire unlobed fruit.

holochrysus, -a, -um: entirely golden.

Holodiscus (From Greek *holos*, entire, and *diskos*, disc, alluding to the unlobed disc of the flowers.) A genus of 8 species of deciduous shrubs with alternate, simple or pinnately lobed leaves, and large panicles of numerous small flowers that bear 5 petals and 20 stamens. Several species are grown as ornamentals. *Distr.* W North, Central and South America. *Fam.* Rosaceae. *Class* D.

H. discolor OCEAN SPRAY, CREAMBUSH. Large shrub to 5m high. Flowers cream. *Distr.* W North America.

H. dumosus ROCK SPIRAEA. Spreading shrub to 4m high. Flowers white-cream. *Distr.* W North America.

holodontus, -a, -um: with entire teeth.

holophyllus, -a, -um: with entire leaves.

holosericeus, -a, -um: completely silky.

Holy Ghost flower *Peristeria elata.*

home of apple *Arcinos arvencis.*

Homeria (From Greek *homereo*, to meet together, alluding to the filaments united in a sheath around the style.) A genus of 31 species of perennial herbs with linear basal leaves and spikes of funnel- to cup-shaped flowers. *Distr.* S Africa. *Fam.* Iridaceae. *Class* M.

H. breyniana See *H. collina*

H. collina Stems to 40cm high. Flowers peach-pink to yellow, fragrant. *Distr.* South Africa (SW Cape).

H. ochroleuca Stems to 80cm. Flowers pale yellow, marked orange, musky-scented. *Distr.* South Africa (SW Cape).

Homogyne (From Greek *homos*, similar, and *gyne*, female, alluding to the similarity of the male and female flowers.) A genus of 2 species of small perennial herbs with simple basal leaves and flowers in solitary daisy-like heads that lack distinct rays. *Distr.* European mountains. *Fam.* Compositae. *Class* D.

H. alpina ALPINE COLTSFOOT. Leaves kidney-shaped, leathery. Flowers red-purple. *Distr.* Mountains of Europe.

homolepis with scales which are all alike.

honesty *Lunaria annua* **perennial** ~ *L. rediviva.*

honeybells *Hermannia, H. verticillata.*

honey flower *Melissa officinalis, Melianthus major.*

honey palm *Jubaea chilensis.*

honey plant *Hoya carnosa.*

honeypot, tangled *Dryandra pteridifolia.*

honeysuckle Caprifoliaceae, *Aquilegia canadensis, Lonicera* **Australian** ~ *Banksia* **bush** ~ *Diervilla, D. lonicera.* **Cape** ~ *Tecoma capensis* **common** ~ *Lonicera periclymenum* **coral** ~ *L. sempervirens* **fly** ~ *L. xylosteum* **French** ~ *Hedysarum coronarium* **giant** ~ *Lonicera hildebrandiana* **giant Burmese** ~ *L. hildebrandiana* **hairy** ~ *L. hirsuta* **Himalaya** ~ *Leycesteria formosa* **Italian** ~ *Lonicera caprifolium* **Jamaica** ~ *Passiflora laurifolia* **Japanese** ~ *Lonicera japonica* **Minorca** ~ *L. implexa* **New Zealand** ~ *Knightia excelsa* **perfoliate** ~ *Lonicera caprifolium* **scarlet trumpet** ~ *L. x brownii* **swamp** ~ *Rhododendron viscosum* **trumpet** ~ *Lonicera sempervirens.*

honeywort *Cryptotaenia canadensis.*

hongkongensis, -e: of Hong Kong.

hookeri after either Sir William Jackson Hooker (1785–1865) or his son Sir Joseph Dalton Hooker (1817–1911), both of whom were directors of the Royal Botanic Gardens at Kew.

hop *Humulus, H. lupulus* **common** ~ *H. lupulus* **European** ~ *H. lupulus* **false** ~ *Justicia brandegeana* **Japanese** ~ *Humulus japonicus* **wild** ~ *Bryonia dioica.*

Hoplestigmataceae A family of 1 genus and 2 species of trees with large, spirally arranged, simple leaves and cymes of regular bisexual flowers that bear 11–14 petals and numerous stamens. *Distr.* Tropical Africa. *Class* D.

hops, native *Dodonaea viscosa.*

hop tree *Ptelea trifoliata.*

Hordeum (From Latin *horridus*, bearded with bristles, alluding to the awns.) BARLEY. A genus of about 25 species of annual and perennial grasses to 1.5m high with flat or inrolled leaves and dense, cylindrical, spike-like inflorescences. A genus of ancient cereals cultivated since antiquity but now chiefly grown to be malted for the brewing industry, although still eaten as a whole grain and used in the manufacture of soft drinks. Some are cultivated as ornamentals. *Distr.* Temperate regions of the N hemisphere and South America. *Fam.* Gramineae. *Class* M.

H. distichum TWO-ROWED BARLEY. Seeds are produced in 2 ranks on the ear. One of the major cultivated species.

H. jubatum SQUIRREL-TAIL GRASS, SQUIRREL-TAIL BARLEY, FOX-TAIL BARLEY. A cultivated ornamental grown for its feathery inflorescence. *Distr.* N temperate regions.

H. polystichum SIX-ROWED BARLEY, FOUR-ROW BARLEY. Seeds are produced in 4 or 6 ranks on the ear. One of the major cultivated species.

horehound: black ~ *Ballota nigra* **common** ~ *Marrubium vulgare* **water** ~ *Lycopus europaeus* **white** ~ *Marrubium vulgare.*

horizontalis, -e: horizontal, with low growing habit.

Horminum (From Greek *hormao*, I excite, possibly alluding to the use of the plant as an aphrodisiac.) A genus of 1 species of rhizomatous herb with basal rosettes of simple leaves and whorls of violet-blue, 2-lipped flowers borne in a spike-like inflorescence on an erect 4-angled stem. *Distr.* Mountainous regions of S Europe. *Fam.* Labiatae. *Class* D.

H. pyrenaicum PYRENEAN DEAD NETTLE, DRAGON MOUTH. Grown as an ornamental; several cultivars are available.

hornbeam *Carpinus* **American** ~ *C. caroliniana* **easter hop** ~ *ostrya virgiana* **European** ~ *Carpinus betulus* **hop** ~ *Ostrya, O. carpinifolia.*

horn nut *Trapa natans.*

horn of plenty *Datura metel.*

hornwort *Ceratophyllum.*

horridulus, -a, -um: somewhat spiny or bristly.

horridus, -a, -um: bristly or rough.

horsefly weed *Baptisia tinctoria.*

horseradish *Armoracia rusticana* **Japanese** ~ *Wasabia japonica.*

horsetail *Equisetum* **common** ~ *E. arvense* **field** ~ *E. arvense* **rough** ~ *E. hyemale.*

horsetail tree *Casuarina equisetifolia.*

hortensis, -e: of gardens.

Hosta (After Nicolas Thomas Host (1761–1834), Austrian physician.) PLANTAIN LILY. A genus of about 40 species of rhizomatous clump-forming herbs with large, simple, basal leaves that are often variegated, and racemes of pale purple or white flowers. Most of the species and very many cultivars are grown as ornamentals. *Distr.* Japan, Korea, China. *Fam.* Hostaceae. *Class* M.

H. fortunei (After Robert Fortune (1812–80), who collected in China and introduced the TEA plant from China to India.) The source of a group of vigorous garden varieties and cultivars. *Distr.* Origin uncertain.

H. glauca See *H. sieboldiana*

H. plataginea AUGUST LILY, MARUBA, YUSAN. Leaves in open clumps, yellow-green. Flowers white, fragrant at night. *Distr.* China.

H. sieboldiana Leaves in dense clumps, grey-blue, puckered. Flowers lilac. *Distr.* Japan.

Hostaceae A family of 1 genus and 40 species. *Class* M.
See: Hosta.

Hottentot bread *Dioscorea elephantipes.*

Hottonia (After Peter Hotton (1648–1709), Dutch physician and botanist.) A genus of 2 species of aquatic perennial herbs with finely divided leaves and racemes of small 5-lobed flowers that are borne above the surface of the water. *Distr.* W North America, Europe, W Asia. *Fam.* Primulaceae. *Class* D.

H. palustris WATER VIOLET. Creeping perennial. Flowers violet with a yellow throat.

Cultivated as an aquarium ornamental. *Distr.* Europe, Asia.

hot-water plant *Achimenes.*

hound's tongue *Cynoglossum, C. officinale* **Himalayan** ~ *C. nervosum.*

houseleek *Sempervivum* **cobweb** ~ *S. arachnoideum* **common** ~ *S. tectorum* **hen and chickens** ~ *Jovibarba sobolifera* **roof** ~ *Sempervivum tectorum.*

Houstonia (After Dr William Houston (1695–1733), British botanist.) A genus of about 50 species of perennial herbs and subshrubs with simple opposite leaves and salver- or funnel-shaped flowers. Several species are grown as ornamentals. *Distr.* North America, Mexico. *Fam.* Rubiaceae. *Class* D.

H. caerulea BLUETS, QUAKER LADIES, INNOCENCE. Erect perennial herb. Flowers pale blue with a yellow centre. *Distr.* North America.

Houttuynia (After Martin Houttuyn (1720–94), Dutch naturalist.) A genus of 1 species of aromatic perennial herb to 60cm high with palmately lobed, alternate leaves and spikes of small flowers that lack sepals and petals. *Distr.* E Himalaya, Taiwan, Japan and Java. *Fam.* Saururaceae. *Class* D.

H. cordata Grown as an ornamental. The shoots are eaten as a vegetable in China.

Hovenia (After David Hoven, Dutch politician.) A genus of 2 species of deciduous shrubs and small trees with simple alternate leaves and branched inflorescences of small flowers that bear their parts in 5s and are followed by berry-like fruits. *Distr.* E Asia. *Fam.* Rhamnaceae. *Class* D.

H. dulcis CHINESE RAISIN TREE, RAISIN TREE, JAPANESE RAISIN TREE. Tree to 20m high. Flowers yellow-green. Fruit fleshy, red. The fruit is used medicinally, especially for hangovers, and the timber is of high quality. *Distr.* China, Korea, Japan.

Howea (After Lord Howe (1726–99), and the island named after him, where they grow.) SENTRY PALM. A genus of 2 species of medium-sized palms to 15m high with feather-shaped leaves and 2m-long spikes of small, fawn and green flowers. *Distr.* Lord Howe Island, off Australia. *Fam.* Palmae. *Class* M.

H. belmoreana (After de Belmore, a governor of New South Wales.) CURLY SENTRY PALM, BELMORE SENTRY PALM, CURLY PALM. Leaves to 2m long, curved. Grown as a tender ornamental. *Distr.* Lord Howe Island, off Australia.

H. forsteriana (After William Forster, a senator of New South Wales.) PARADISE PALM, KENTIA, SENTRY PALM, FORSTER SENTRY PALM, KENTIA PALM, THATCH LEAF PALM. To 18m high. Leaves flat. Often grown as a stemless house plant. *Distr.* Lord Howe Island, off Australia.

Hoya (After Thomas Hoy (*c.*1750– 1822), gardener at Syon House, London.) WAX FLOWER, WAX PLANT, PORCELAIN FLOWER. A genus of 100–200 species of evergreen climbers and sprawling shrubs with opposite, fleshy or leathery leaves, and umbels or cymes of fragrant star-shaped flowers. Many species are grown as ornamentals. *Distr.* Tropical Asia, Pacific. *Fam.* Asclepiadaceae. *Class* D.

H. australis Twining root-climber. Flowers purple-tinged. *Distr.* Australia.

H. bella See *H. lanceolata*

H. carnosa HONEY PLANT, WAX PLANT. Succulent climber. Flowers white, turning pink. Several ornamental cultivars available. *Distr.* SE Asia to Australia.

H. imperialis Robust climber. Leaves oval, leathery. Flowers to 7cm across, purple-brown with a cream centre. *Distr.* Borneo.

H. lanceolata MINIATURE WAX FLOWER. Small shrub. Flowers sweet-scented, waxy, white. *Distr.* Himalaya.

H. longifolia STRING BEAN PLANT. Epiphyte. Leaves long, narrow, thick. *Distr.* Himalaya to S Thailand and Malaysia.

H. pubicalyx RED BUTTONS. Scrambling shrub. Flowers red-brown. *Distr.* Philippines.

Huaceae A family of 2 genera and 3 species of shrubs and trees with simple, spirally arranged leaves that smell of garlic, and small bisexual flowers. *Distr.* Tropical Africa. *Class* D.

huckleberry *Gaylussacia* **black** ~ *G. baccata* **blue** ~ *Vaccinium membranaceum* **box** ~ *Gaylussacia brachycera* **California** ~ *Vaccinium ovatum* **evergreen** ~ *V. ovatum* **fool's** ~ *Menziesia ferruginea* **he** ~ *Lyonia ligustrina* **red** ~ *Vaccinium parvifolium* **shot** ~ *V. ovatum.*

Huguenina (After M. Huguenin, French botanist.) A genus of 1 species of perennial herb with small yellow flowers. *Distr.* Mountainous regions of S Europe. *Fam.* Cruciferae. *Class* D.

H. tanacetifolia TANSY-LEAVED ROCKET. *Distr.* S Europe.

Humata (From Latin *humus*, soil, alluding to the plants' creeping habit.) A genus of about 50 species of epiphytic or terrestrial ferns with pinnately divided leaves. Several species are grown as ornamentals. *Distr.* Tropical and warm regions of the Old World. *Fam.* Davalliaceae. *Class* F.

H. tyermannii BEAR'S FOOT FERN. *Distr.* Himalaya, China.

humble plant *Mimosa pudica.*

Humea elegans See *Calomeria amaranthoides*

humifusus, -a, -um: procumbent, spreading.

humilis, -e: low-growing, dwarf.

Humiriaceae A family of 8 genera and 50 species of evergreen shrubs and trees with simple, spirally arranged leaves and cymes of regular bisexual flowers. *Distr.* Tropical America, 1 species tropical Africa. *Class* D.

Humulus (From Latin *humus*, soil, alluding to its occasionally prostrate habit.) HOP. A genus of 2–3 species of perennial twining herbs with angled stems and palmately lobed leaves. The small, wind-pollinated, unisexual flowers are borne on separate plants, the males in loose panicles, the females in small cone-like structures made up of a number of bracts. *Distr.* Temperate Eurasia. *Fam.* Cannabaceae. *Class* D.

H. japonicus JAPANESE HOP. Grown as an ornamental. *Distr.* Temperate E Asia.

H. lupulus HOP, COMMON HOP, EUROPEAN HOP, BINE. The female inflorescences are used as the bitter ingredient in beer and may contribute to its keeping qualities by suppressing bacterial growth. The young shoots can be eaten as a vegetable. *Distr.* N temperate regions.

hungaricus, -a, -um: of Hungary.

hunter's robe *Epipremnum aureum.*

huntsman's cup *Sarracenia purpurea.*

huntman's horn *S. purpurea.*

hupehensis, -e: of Hupeh (now Hubei), China.

Hutchinsia See *Thlaspi*

hyacinth *Hyacinthus orientalis* **common** ~ *H. orientalis* **Dutch** ~ *H. orientalis* **Japanese** ~ *Ophiopogon* **Roman** ~ *Bellevalia romana* **summer** ~ *Galtonia candicans* **tassel** ~ *Muscari comosum* **water** ~ *Eichhornia, E. crassipes.* **wild** ~ *Camassia scilloides, Dichelostemma multiflorum, D. pulchellum, Lachenalia contaminata.*

Hyacinthaceae A family of 45 genera and about 500 species of bulbiferous herbs with linear to oblong leaves and flowers that bear 6 free or fused tepals and 6 stamens. This family is sometimes included within the family Liliaceae. *Distr.* Eurasia, Africa, North America. *Class* M.
See: Albuca, Bellevalia, Bowiea, Brimeura, Camassia, Chionodoxa, × Chionoscilla, Eucomis, Galtonia, Hyacinthella, Hyacinthoides, Hyacinthus, Lachenalia, Ledebouria, Leopoldia, Massonia, Muscari, Muscarimia, Ornithogalum, Polyxena, Pseudomuscari, Puschkinia, Schoenolirion, Scilla, Urginea, Veltheimia.

Hyacinthella (Diminutive of the genus name *Hyacinthus*.) A genus of 16 species of small bulbiferous herbs with narrow leaves and racemes of blue to pink, bell-shaped flowers. Some species are cultivated as ornamentals. *Distr.* E and SE Europe to W Asia. *Fam.* Hyacinthaceae. *Class* M.

H. leucophaea Raceme spike-like. Flowers very pale blue. *Distr.* E Europe.

hyacinthifolius, -a, -um: from the genus name *Hyacinthus*, and *folius*, leaved.

hyacinthinus, -a, -um: hyacinth-coloured, violet-blue.

Hyacinthoides (From the genus name *Hyacinthus* and Greek *-oides*, indicating resemblance.) BLUEBELL. A genus of 3–5 species of bulbiferous herbs with linear leaves and racemes of blue flowers which bear 6 fused tepals and are subtended by 2 purple bracts. As well as being important ornamentals these plants have been used as a source of a starch-based glue used in bookbinding. *Distr.* W Europe (3 species) and N Africa. *Fam.* Hyacinthaceae. *Class* M.

H. hispanica SPANISH BLUEBELL. Raceme erect at tip. Flowers not fragrant. Forms hybrids when grown with *H. non-scripta*. *Distr.* SW Europe, N Africa.

H. non-scripta ENGLISH BLUEBELL, BLUEBELL. Raceme drooping at the tip. Flowers fragrant. A very common woodland plant in the British Isles, often becoming the dominant understorey plant in acid woodlands. *Distr.* W Europe.

hyacinthoides: from the genus name *Hyacinthus*, with the ending *-oides*, indicating resemblance.

Hyacinthus (From a Greek name used by Homer, the flowers being said to have sprung from the blood of Hyakinthos, a youth accidentally killed by Apollo.) A genus of 3–4 species of bulbiferous herbs with linear leaves and a short cylindrical raceme of blue flowers. All the species are grown as ornamentals to a greater or lesser extent. *Distr.* Mediterranean region to W and Central Asia. *Fam.* Hyacinthaceae. *Class* M.

H. amethystinus See *Brimeura amethystina*

H. orientalis DUTCH HYACINTH, HYACINTH, COMMON HYACINTH. A much cultivated plant with many cultivars available. The highly fragrant flowers are used in the scent industry. *Distr.* S and Central Turkey, NW Syria, Lebanon, naturalized S Europe.

hybridus, -a, -um: hybrid.

Hydatellaceae A family of 2 genera and 7 species of small, sub-aquatic, annual and perennial herbs with slender leaves and small dense heads of minute flowers. *Distr.* Australia, New Zealand. *Class* M.

Hydnoraceae A family of 2 genera and 17 species of root parasites similar to those in the family Rafflesiaceae, with large, solitary, bisexual flowers. *Distr.* Africa and South America. *Class* D.

Hydrangea (From the Greek *hydor*, water, and *aggos*, jar, alluding to the cup-shaped fruits.) A genus of 23 species of deciduous shrubs and evergreen or deciduous climbers with simple, opposite or whorled leaves and large heads of flowers. Some of the flowers in each head are sterile and much larger than their fertile neighbours. The colour of flowers is dependent on their ability to absorb aluminium ions which may be inhibited on alkaline soils, resulting in 'blue' forms going pink. Some species have medicinal qualities and many species and cultivars are grown as ornamentals. *Distr.* Himalaya to Japan and Philippines, America. *Fam.* Hydrangeaceae. *Class* D.

H. anomala Deciduous climber. Flowers white, in flat heads. *Distr.* Himalaya, China.

H. arborescens Tree-like shrub. Flowers cream, in flat heads. *Distr.* E USA.

H. aspera Upright shrub. Flower-heads rounded. Central fertile flowers blue or purple. Outer flowers white. *Distr.* Himalaya, E Asia.

H. cinerea See *H. arborescens*

H. heteromalla Upright shrub. Leaves narrow. Flowers blue. *Distr.* Himalaya, China.

H. involucrata Spreading shrub. Leaves heart-shaped. Flowers blue. *Distr.* Japan, Taiwan.

H. macrophylla Spreading shrub. Flowers pink or blue. Numerous cultivars have been produced, some of which have only sterile flowers. *Distr.* Japan (cultivated).

H. petiolaris See *H. anomala*

H. quercifolia OAK-LEAVED HYDRANGEA. Mound-forming shrub. Leaves oak-shaped. Flowers white. *Distr.* SE USA.

H. sargentiana See *H. aspera*

H. tiliifolia See *H. anomala*

H. villosa See *H. aspera*

hydrangea, climbing *Decumaria barbara* **oak-leaved** ~ *Hydrangea quercifolia*.

Hydrangeaceae A family of 16 genera and about 160 species of herbs, shrubs, lianas, and trees with simple leaves and large inflorescences consisting of regular bisexual flowers near often irregular, sterile flowers that attract pollinators. *Distr.* N temperate regions to SE Asia. *Class* D.

See: Carpenteria, Decumaria, Deinanthe, Deutzia, Dichroa, Hydrangea, Jamesia, Kirengeshoma, Philadelphus, Schizophragma.

lemon *Citrus limon* **water** ~ *Passiflora laurifolia*.

Lennoaceae A family of 3 genera and 6 species of fleshy, herbaceous root parasites with scale-like leaves and dense heads of small flowers. *Distr.* SW USA to Venezuela. *Class* D.

Lens (Classical Latin name for these plants.) A genus of about 6 species of annual herbs with pinnate leaves and pea-like flowers. *Distr.* W Asia and the Mediterranean region. *Fam.* Leguminosae. *Class* D.

L. culinaris LENTIL, MASUR. The seeds are a very important protein-rich food, sometimes the only source of protein in a poor diet. *Distr.* Cultivated since antiquity in the Mediterranean and W Asia.

Lentibulariaceae The Butterwort family. 3 genera and about 250 species of carnivorous herbs of wet habitats with simple entire leaves which bear sticky hairs or small bladders. The flowers are irregular and 2-lipped. *Distr.* Widespread. *Class* D.
See: Pinguicula, Utricularia.

lentil *Lens culinaris*.

lentisco *Pistacia lentiscus*.

lentus, -am, -um: tough but flexible.

Leonotis (From Greek *leon*, lion, and *otis*, ear, alluding to the hairy flowers.) LION'S EAR. A genus of 15 species of aromatic annual and perennial herbs and subshrubs with 4-angled stems, simple opposite leaves and dense whorls of 2-lipped flowers in which the upper lip is large and fringed with orange hairs. Several species are grown as tender ornamentals. *Distr.* Tropical Africa, 1 species extending to America and Asia. *Fam.* Labiatae. *Class* D.

L. leonurus LION'S EAR. Shrub. Flowers orange-red to scarlet. *Distr.* S Africa.

L. ocymifolia Subshrub to 70cm high. Flowers orange. *Distr.* S Africa.

Leontice (From Greek *leon*, lion, alluding to the apparent pattern of lions' footprints on the leaves.) A genus of 3–5 species of rhizomatous perennial herbs with pinnately divided leaves and racemes of yellow flowers that bear 6–9 petal-like sepals and 6 small petals. Some species have been used medicinally. *Distr.* N Africa, SE Europe, Asia. *Fam.* Berberidaceae. *Class* D.

L. alberti See *Gymnospermium albertii*

L. leontopetalum Grown as an ornamental. *Distr.* N Africa to Iran.

Leontodon (From Greek *leon*, lion, and *odos*, tooth, alluding to the toothed leaves.) HAWKBIT. A genus of about 40 species of perennial herbs with rosettes of simple or pinnately lobed leaves and flowers in daisy-like heads that lack distinct rays. Often occurring as weeds of lawns. *Distr.* Temperate Eurasia. *Fam.* Compositae. *Class* D.

L. autumnalis AUTUMN HAWKBIT. Flowers yellow with red stripes, appearing in late summer. *Distr.* Eurasia, N Africa, introduced to North America.

L. hispidus ROUGH HAWKBIT. Rough-hairy. Flowers pure yellow. *Distr.* Europe to N Iran.

Leontopodium (Classical name for another plant, from Greek *leon*, lion, and *podion*, foot, alluding to the shape of the flower-heads.) EDELWEISS. A genus of 35 species of perennial herbs with simple leaves and flowers in a cluster of daisy-like heads that are surrounded by showy white-woolly leaves. Many are cultivated as ornamentals. *Distr.* Eurasia, especially mountains. *Fam.* Compositae. *Class* D.

L. alpinum EDELWEISS. To 20cm high. Inflorescence leaves forming a star to 10cm across. A popular alpine ornamental. *Distr.* Mountains of Europe.

L. discolor Inflorescence leaves forming a distinct white-hairy star. *Distr.* Japan, SE China to Korea.

L. himalayanum Inflorescence leaves silver-woolly. *Distr.* Himalaya of China, Tibet, and Kashmir.

L. hyachinense To 20cm high. Inflorescence leaves narrow, pointed. *Distr.* Japan (Mount Hayachine).

L. leontopodioides To 50cm high. Inflorescence leaves not forming a distinct star. *Distr.* S Siberia, Mongolia, NE China.

L. sibiricum See *L. leontopodioides*

L. tataricum See *L. discolor*

leontopodium: resembling a lion's foot.

Leonurus (From Greek *leon*, lion, and *oura*, tail, alluding to the hairy flowers.) A genus of 4 species of perennial herbs with simple or palmately divided, opposite leaves and whorls of 2-lipped flowers. *Distr.* Temperate Eurasia. *Fam.* Labiatae. *Class* D.

L. cardiaca MOTHERWORT. Grown as an ornamental but formerly important as a source of green dye and a medicine. *Distr.* Europe, Mediterranean.

leonurus, -a, -um: like a lion's tail.

leopard flower *Belamcanda*

leopard plant *Ligularia.*

leopard's bane *Doronicum* **great** ~ *D. pardalianches.*

Lepechinia (After Frau Lepechin (1737–1802).) A genus of 30 species of perennial herbs and shrubs with simple opposite leaves and 2-lipped flowers. Several species are grown as ornamentals. *Distr.* Warm America, 1 species in Hawaii. *Fam.* Labiatae. *Class* D.

L. calycina PITCHER SAGE. Shrub to 3m high. Flowers pink, broadly bell-shaped, borne in racemes. *Distr.* California.

leper flower *Fritillaria meleagris.*

Lepidium (From Greek *lepis*, scale, alluding to the shape of the fruits.) A genus of about 150 species of annual, biennial, and perennial herbs. *Distr.* Cosmopolitan. *Fam.* Cruciferae. *Class* D.

L. sativum GARDEN CRESS, COMMON GARDEN CRESS, PEPPER GRASS, PEPPERWORT. Annual herb with white-lilac flowers. Eaten as a salad vegetable, as a seedling in 'Mustard and Cress' or nearer to maturity as a salad or soup vegetable. *Distr.* Cultivated throughout the world, probably originated in the Near East and N Africa.

Lepidobotryaceae A family of 1 genus and 1 species of shrub with compound leaves and regular bisexual flowers that bear their parts in fives. This family is sometimes included within the family Oxalidaceae. *Distr.* Tropical Africa. *Class* D.

lepidophyllus, -a, -um: with scaly leaves.

lepidostylus, -a, -um: with a scaly style.

Lepidothamnus (From Greek *lepis*, scale, and *thamnos*, shrub.) A genus of 3 species of evergreen coniferous shrubs and trees with linear juvenile leaves and scale-like adult leaves. *Distr.* S Chile, New Zealand. *Fam.* Podocarpaceae. *Class* G.

L. laxifolium MOUNTAIN RIMU. Prostrate shrub. Occasionally grown as an ornamental. *Distr.* New Zealand.

lepidotus, -a, -um: scaly.

Lepidozamia (From Greek *lepis*, scale, and the genus name *Zamia*, referring to the scale-like leaf bases which clothe the stem.) A genus of 1–2 species of erect cycads with pinnate leaves to 3m long. *Distr.* E Australia. *Fam.* Zamiaceae. *Class* G.

L. peroffskyana Grown as an ornamental. *Distr.* E Australia (Queensland to New South Wales).

lepidus, -a, -um: pleasant.

Lepismium (From Greek *lepisma*, scale, alluding to the scale-like leaf remnants that persist in some species.) A genus of 14 species of epiphytic or terrestrial, shrub-like cacti with ribbed cylindrical stems that may or may not bear spines and small bell-shaped flowers that are followed by a berry-like fruit. Several species are grown as ornamentals. *Distr.* Brazil, Bolivia, Argentina. *Fam.* Cactaceae. *Class* D.

L. ianthothele Spreading or pendent, stems to 60cm. Flowers pale yellow. *Distr.* Argentina.

Leptinella A genus of about 30 species of annual to perennial, creeping, tufted herbs with simple to pinnately lobed leaves and flowers in daisy-like heads which lack distinct rays. Some species are grown as ornamental ground cover. *Distr.* Australasia, temperate South America. *Fam.* Compositae. *Class* D.

L. albida Creeping perennial, forming mats to 1m across. *Distr.* New Zealand.

L. atrata Tufted perennial. Flowers yellow to dark red. *Distr.* New Zealand.

L. dendyi Tufted perennial, to 40cm high. Flowers yellow. *Distr.* New Zealand.

L. goyenii Mat-forming perennial. Leaves covered with long silver hairs. *Distr.* New Zealand.

L. pectinata Creeping tufted perennial, to 15cm high. Flowers usually white. *Distr.* New Zealand.

L. perpusilla See *Cotula perpusilla*

L. potentillina Creeping perennial. Leaves yellow-green. *Distr.* New Zealand.

L. pyrethrifolia Creeping mat-forming perennial. Flowers white. *Distr.* New Zealand.

L. reptans Creeping perennial, to 25cm high. *Distr.* S and SE Australia, Tasmania.

L. rotundata Creeping perennial forming loose turfs. *Distr.* New Zealand (North Island).

L. scariosa Creeping perennial. Leaves leathery. Flowers yellow-green. *Distr.* Temperate South America.

L. squalida Creeping perennial. Leaves bright green. *Distr.* New Zealand.

leptocarpus, -a, -um: with slender fruit.

leptocaulis, -e: with slender stem.

leptolepis: with thin scales.

leptophyllus, -a, -um: with slender leaves.

Leptopteris (From Greek *leptos*, slender, and *pteris*, fern, alluding to the slender leaves.) A genus of 7 species of terrestrial ferns with erect woody rhizomes and tripinnate leaves. Several species are grown as ornamentals. *Distr.* New Zealand, Australia, New Guinea, Polynesia. *Fam.* Osmundaceae. *Class* F.

L. superba PRINCE OF WALES'S FERN, PRINCE OF WALES'S PLUME. Leaves feather-like to 1.2m. *Distr.* New Zealand.

leptopus: with a slender stalk.

leptorrhizus, -a, -um: with a slender root.

leptosepalus, -a, -um: with slender sepals.

Leptospermum (From Greek *leptos*, slender, and *sperma*, seed, alluding to the narrow seeds.) TI TREE, TEA TREE. A genus of 30 species of evergreen shrubs and trees with simple, alternate, aromatic leaves and small white or pink flowers with their parts in fives. Several species are cultivated as ornamentals, the leaves of some are used for tea, and a few are a source of essential oil. *Distr.* Australia, New Zealand, E Asia. *Fam.* Myrtaceae. *Class* D.

L. citratum See *L. petersonii*

L. flavescens See *L. polygalifolium*

L. humifusum See *L. rupestre*

L. petersonii Grown commercially in Kenya and Guatemala as a source of lemon-scented oil. *Distr.* E Australia.

L. polygalifolium Arching shrub. Leaves small, glossy green. Flowers abundant, white tinged with pink. *Distr.* E Australia, Lord Howe Island.

L. rupestre Spreading shrub. Shoots red. Flowers white. *Distr.* Tasmania.

L. scoparium MANUKA, TEA TREE. Shrub to 2m high. The most commonly cultivated species. A source of timber, tea, and essential oil. *Distr.* Australia, New Zealand.

leptotes: delicate.

Lespedeza (After Vincente Manuel de Cespedes, Spanish governor of Eastern Florida around 1790.) BUSH CLOVER. A genus of 40 species of annual and perennial herbs and shrubs with trifoliate leaves and racemes or panicles of small irregular flowers. Cultivated for forage and as green manure or ornamental plants. *Distr.* Temperate North America, E Asia, Australia. *Fam.* Leguminosae. *Class* D.

L. bicolor Deciduous bush to 3m high. Used as a cover crop and for fuel. *Distr.* N China, Japan.

Lesquerella (After Leo Lesquereux, late 19th-century American palaeobotanist.) BLADDER POD. A genus of 40 species of annual to perennial, densely hairy herbs with small flowers. *Distr.* North America. *Fam.* Cruciferae. *Class* D.

L. alpina Tufted perennial with small yellow flowers. *Distr.* Rocky Mountains.

L. grandiflora Annual with yellow flowers. *Distr.* Texas.

lettuce *Lactuca*, *L. sativa* **blue** ~ *L. perennis* **lamb's** ~ *Valerianella locusta* **miner's** ~ *Montia perfoliata* **prickly** ~ *Lactuca serriola* **water** ~ *Pistia stratiotes*.

Leucadendron (From Greek *leukos*, white, and *dendron*, tree, alluding to the silvery foliage of *L. argenteum*.) A genus of about 90 species of shrubs and trees with simple alternate leaves and small unisexual flowers that are borne in clusters subtended by large colourful leaves. Several species are grown as ornamentals and some are a source of timber. *Distr.* South Africa. *Fam.* Proteaceae. *Class* D.

L. argenteum SILVER TREE. Tree to 10m high. Leaves silver-hairy. *Distr.* South Africa (Cape Province, Cape peninsula).

Leucaena (From Greek *leukos*, white, alluding to the flower colour.) A genus of 40 species of evergreen shrubs and trees with bipinnate leaves and fluffy ball-shaped heads of white flowers. *Distr.* Tropical America. *Fam.* Leguminosae. *Class* D.

L. latisiliqua See *L. leucocephala*

L. leucocephala LEAD TREE. Shrub. Flower-heads tinged yellow. Grown as a source of charcoal for fuel. *Distr.* S USA, Mexico.

Leucanthemella (Diminutive of Greek *leukos*, white, and *anthemon*, flower.) A genus of 2 species of perennial herbs with simple leaves and yellow flowers in daisy-like heads that bear distinct rays. *Distr.* SE Europe, E Asia. *Fam.* Compositae. *Class* D.

L. serotina To 1.5m high. Ray florets red or white. *Distr.* SE Europe.

Leucanthemopsis (From the genus name *Leucanthemum*, and the Greek *opsis*, appearance, alluding to the resemblance of the 2 genera.) A genus of 6 species of small tufted perennials with pinnately lobed leaves and flowers in daisy-like heads which have distinct yellow or white rays. *Distr.* Mountains of Europe and N Africa. *Fam.* Compositae. *Class* D.

L. alpina Mat- or tuft-forming. Flowers yellow-orange, rays white. *Distr.* European mountains.

L. pallidum White-hairy. Rays yellow, sometimes white at tips. *Distr.* SE Spain.

Leucanthemum (From Greek *leukos*, white, and *anthemon*, flower.) A genus of about 25 species of annual to perennial herbs with simple or pinnately lobed leaves and yellow flowers in daisy-like heads which usually bear distinct white rays. Several species are cultivated as ornamentals. *Distr.* Mountains of Europe and N Asia. *Fam.* Compositae. *Class* D.

L. hosmariense See *Pyrethropsis hosmariensis*

L. maximum Herb to 1m high. Leaves entire. *Distr.* Pyrenees.

L. nipponicum See *Nipponanthemum nipponicum*

L. × superbum SHASTA DAISY. A hybrid of *L. maximum* × *L. lacustre*. *Distr.* Garden origin.

L. vulgare MOON DAISY, MARGUERITE, DOG DAISY, OX-EYE DAISY. Perennial herb to 1m high. Flowers yellow, rays white, occasionally absent. A popular ornamental, used as a home remedy for catarrh and more or less edible. *Distr.* Eurasia, naturalized North America.

leucanthus, -a, -um: with white flowers.

leucaspis: with a white shield.

leucocarpus, -a, -um: white-fruited.

leucocephalus, -a, -um: with a white head.

leucochilus, -a, -um: white-lipped.

Leucocoryne (From Greek *leukos*, white, and *coryne*, club, alluding to the sterile anthers.) A genus of about 12 species of bulbous spring-flowering herbs with linear leaves and loose umbels of fragrant blue flowers. Several species are grown as ornamentals. *Distr.* Chile. *Fam.* Alliaceae. *Class* M.

L. ixioides GLORY OF THE SUN. Leaves grass-like. Tepals white with a blue border. *Distr.* Chile.

L. purpurea Flowers tinged purple. *Distr.* Chile.

Leucocrinum (From Greek *leukos*, white, and *krinon*, a type of lily.) SAND LILY, STAR LILY, MOUNTAIN LILY. A genus of 1 species of rhizomatous perennial herb with tufts of grass-like leaves and clusters of fragrant white flowers. *Distr.* SW USA. *Fam.* Anthericaceae. *Class* M.

L. montanum SAND LILY, STAR LILY, MOUNTAIN LILY. Cultivated as an ornamental.

leucodermis: with a white skin.

Leucogenes (From Greek *leukos*, white, and *genea*, race.) NEW ZEALAND EDELWEISS. A genus of 2–3 species of silvery-hairy perennial herbs resembling the genus *Leontopodium*, with dense overlapping leaves and flowers in daisy-like heads that lack distinct rays but are subtended by a whorl of showy leaves. 2 species are cultivated as ornamentals. *Distr.* New Zealand. *Fam.* Compositae. *Class* D.

L. grandiceps SOUTH ISLAND EDELWEISS. Leaves to 10cm long, hairy. *Distr.* New Zealand (South Island).

L. leontopodium NORTH ISLAND EDELWEISS. Leaves to 2cm long, silver-hairy. *Distr.* New Zealand (North and South Island).

Leucojum (From Greek *leukos*, white, and *ion*, violet.) SNOWFLAKE. A genus of about 10 species of bulbiferous perennial herbs with linear leaves and umbels of white pendulous flowers that bear 6 equal fused tepals. Several species and a number of hybrids are cultivated as ornamentals. *Distr.* Europe to Morocco and Iran. *Fam.* Amaryllidaceae. *Class* M.

L. aestivum SUMMER SNOWFLAKE, LODDON LILY. To 75cm tall. Flowers 2–7, green at tips of tepals. Seed dispersed by water. *Distr.* Europe, E to Turkey and the Caucasus, naturalized or possibly native in valleys of the Shannon and Thames.

L. autumnale AUTUMN SNOWFLAKE. To 15cm tall. Flowers flushed pink. *Distr.* W Mediterranean.

L. longifolium To 25cm tall. Flowers 1–3. *Distr.* Corsica.

L. nicaeense To 15cm tall. Flowers 1–3, tepals spreading. *Distr.* Coastal region of SE France from Nice E, but almost extinct in the wild.

L. roseum To 15cm. Flowers usually solitary, tinged pink. *Distr.* Corsica, Sardinia.

L. trichophyllum To 25cm tall. Leaves very narrow. Flowers flushed pink-purple. *Distr.* Spain, Portugal, and Morocco.

L. valentinum Leaves narrow. Flowers milk-white. *Distr.* Central Spain, Greece.

L. vernum SPRING SNOWFLAKE. Flowers solitary, tepals marked green-yellow near tip. *Distr.* Europe, naturalized in Great Britain.

leuconeurus, -a, -um: white-veined.

leucophaeus, -a, -um: ash-grey-coloured.

leucophyllus, -a, -um: white-leaved.

Leucopogon (From Greek *leukos*, white, and *pogon*, beard, alluding to the white-hairy flowers.) A genus of 150 species of evergreen, heath-like shrubs and small trees with funnel- to bell-shaped flowers that bear their parts in fives and have hairy throats. Several species are grown as ornamentals. *Distr.* Australia, New Caledonia. *Fam.* Epacridaceae. *Class* D.

L. ericoides Small shrub to 1m high. Flowers in short dense clusters. *Distr.* S and W Australia, Tasmania.

Leucospermum (From Greek *leukos*, white, and *sperma*, seed.) A genus of 46 species of shrubs and small trees with alternate leathery leaves and clusters of tubular flowers. Some species are cultivated as ornamentals. *Distr.* South Africa. Zimbabwe. *Fam.* Proteaceae. *Class* D.

L. cordifolium Shrub to 2m high. Flowers yellow to crimson. *Distr.* South Africa (Cape Province).

leucostachys: with a white spike.

Leucostegia (From Greek *leukos*, white, and *stege*, covering.) A genus of about 2 species of terrestrial or epiphytic ferns with fine pinnately divided leaves. Several species are grown as ornamentals. *Distr.* Tropical Asia. *Fam.* Davalliaceae. *Class* F.

L. immersa Rhizome long, creeping. Leaves to 40cm long. *Distr.* Tropical SE Asia to New Guinea.

Leucothoe (After Leucothoe, daughter of Orchamus, the king of Babylon, and Eurynome in Greek mythology, who is said to have been changed into a shrub by her lover, Apollo.) A genus of 44 species of evergreen and deciduous shrubs with white-pink, tubular or urn-shaped flowers. Cultivated as ornamental shrubs. *Distr.* America and E Asia. *Fam.* Ericaceae. *Class* D.

L. axillaris Evergreen shrub. *Distr.* SE North America.

L. davisiae (After Miss N. J. Davis.) SIERRA LAUREL. Evergreen shrub. *Distr.* California to Oregon.

L. fontanesiana (After René Louiche Desfontaines (1752–1833), French botanist.) DOG-HOBBLE, DROOPING LAUREL, SWITCH IVY. Fast-growing evergreen shrub. *Distr.* SE North America.

L. grayana (After Asa Gray (1810–88) American botanist.) Deciduous to more or less evergreen, slow-growing shrub. *Distr.* Japan.

L. keiskei (After Keisuke Ito (1803–1901), Japanese physician and botanist.) Evergreen shrub. *Distr.* Japan.

L. populifolia See *Agarista neriifolia*

leucotrichus, -a, -um: with white hairs.

leucoxylon: with white wood.

Leuzea (After J. P. F. Deleuze (1753–1835), a friend of the French botanist A. P. de Candolle.) A genus of 3 species of biennial to perennial herbs with simple or pinnately lobed leaves and flowers in daisy-like heads that lack distinct rays. *Distr.* Mediterranean. *Fam.* Compositae. *Class* D.

L. centauroides Perennial, to 1m high. Flowers purple. *Distr.* Pyrenees.

L. conifera Perennial to 30cm high. Flowers white-purple. *Distr.* W Mediterranean, Portugal.

lever wood *Ostrya virginiana.*

Levisticum (Probably derived from *Ligusticum*, a closely related genus.) A genus of 1 species of aromatic perennial herb to 2m high with divided leaves and umbels of bisexual yellow-green flowers. *Distr.* E Mediterranean. *Fam.* Umbelliferae. *Class* D.

L. officinale LOVAGE, BLADDERSEED. Grown as a culinary herb since antiquity for its leaves, stems, and roots. The dried roots were once ground as a condiment.

Lewisia (After Captain Meriwether Lewis (1774–1809) who led an expedition across America in 1806–7.) A genus of about 20 species of perennial succulent herbs with simple leaves and large showy flowers that bear 5–19 white, purple, or yellow petals. A number of species are grown as ornamentals. The thick starchy roots of some species are edible. *Distr.* W North America. *Fam.* Portulacaceae. *Class* D.

L. rediviva BITTER ROOT. Deciduous perennial to 5cm high. A very robust rock-garden ornamental. *Distr.* W North America.

Leycesteria (After William Leycester, Chief Justice in Bengal about 1820.) A genus of 6 species of deciduous or semi-evergreen shrubs with cane-like stems and slender, simple, stalked leaves. The funnel-shaped flowers are borne in pendent or arching racemes. Several species are grown as ornamentals. *Distr.* W Himalaya to SW China. *Fam.* Caprifoliaceae. *Class* D.

L. crocothyrsos Flowers yellow. *Distr.* Himalaya, N Burma.

L. formosa HIMALAYA HONEYSUCKLE. Thicket-forming. Flowers purple. *Distr.* Himalaya, W China, E Tibet, naturalized in Europe.

Leymus (From Greek *elymos*, a kind of millet.) A genus of about 10 species of rhizomatous, perennial grasses with stiff, flat or rolled leaves and spike-like inflorescences. *Distr.* Temperate regions of the Old World. *Fam.* Gramineae. *Class* M.

L. arenarius LYME GRASS, SEA LYME GRASS, EUROPEAN DUNE GRASS. Often used to stabilize sand dunes with its invasive rhizomes. *Distr.* Eurasia.

Liatris GAY FEATHER, BUTTON SNAKE ROOT, BLAZING STAR, SNAKE ROOT. A genus of 34 species of perennial herbs with swollen rootstocks, simple, glandular leaves, and purple, occasionally white flowers in daisy-like heads that lack distinct rays. Several species are cultivated as ornamentals. *Distr.* E North America. *Fam.* Compositae. *Class* D.

L. aspera To 1.2m high. Flower-heads numerous, borne in a loose spike. *Distr.* E North America.

L. pycnostachya BUTTON SNAKE ROOT. Flower-heads crowded into a dense spike. *Distr.* SE USA.

L. scariosa To 1m high. Flower-heads few to many. *Distr.* SE USA.

L. spicata KANSAS GAY FEATHER, BUTTON SNAKEWORT. Stems robust. Flower-heads crowded into a spike around 70cm long. *Distr.* E USA.

libani: of Lebanon.

libanoticus, -a, -um: of Lebanon.

Libertia (After Marie A. Libert (1782–1863), Belgian botanist.) A genus of about 12 species of rhizomatous, perennial herbs with tufted basal leaves and spikes of small, saucer-shaped flowers. Some species are grown as ornamentals. *Distr.* Australasia and temperate South America. *Fam.* Iridaceae. *Class* M.

L. grandiflora Flowers white, tinged olive-green, clustered. *Distr.* New Zealand.

L. ixioides Flowers clustered, white, tinged brown. *Distr.* New Zealand, including Chatham Islands.

Libocedrus (From the Greek *libanos*, incense, and the genus name *Cedrus*, alluding to the scented wood.) A genus of 8 species of evergreen coniferous trees with small scale-like leaves. Some species are grown as ornamentals and some are a source of timber. *Distr.* Indonesia to New Zealand and South America. *Fam.* Cupressaceae. *Class* G.

L. chilensis See *Austrocedrus chilensis*

L. uvifera Tree to 20m high. *Distr.* S South America.

Libonia See *Justicia*

L. floribunda See *Justicia rizzinii*

liburnicus, -a, -um: of Liburnia, part of Croatia, formerly Yugoslavia.

licorice *Glycyrrhiza glabra*.

Licuala (From the Moluccan vernacular name, *leko wala*.) PENANG LAWYERS, PALAS, LOYAK, LOYAR. A genus of about 108 species of short-stemmed or shrubby palms with fan-shaped leaves and branched or simple spikes of small flowers. A number of species are grown as ornamentals in the tropics and their leaves can be used to improve opium burning. *Distr.* SE Asia to Australia and the New Hebrides. *Fam.* Palmae. *Class* M.

L. grandis To 3m high. Inflorescences pendent. A commonly grown ornamental. *Distr.* New Hebrides.

Life of man *Aralia racemosa*.

lignosus, -a, -um: woody.

Ligularia (From Latin *ligula*, a strap, alluding to the strap-like rays.) LEOPARD PLANT. A genus of 180 species of perennial herbs with rounded or kidney-shaped basal leaves and yellow flowers in daisy-like heads which have distinct yellow-orange rays. *Distr.* Temperate Eurasia. *Fam.* Compositae. *Class* D.

L. clivorum See *L. dentata*

L. dentata To 1m high. Flower-heads to 12cm across, rays bright orange. Cultivated ornamental. *Distr.* China, Japan.

L. × hessei A hybrid of *L. dentata* × *L. wilsoniana*. *Distr.* Garden origin.

L. hodgsonii (After C. P. Hodgson, consul in Japan around 1840, who collected it.) Stems succulent. Leaves papery. Rays orange to bright yellow. *Distr.* Japan.

L. japonica Leaves palmately lobed. Flower-heads around 10cm across, rays yellow-orange. *Distr.* China, Korea, Japan.

L. macrophylla To almost 2m. Flower-heads in dense panicles, rays yellow. *Distr.* Altai Mountains of Central Asia.

L. × palmatiloba A hybrid of *L. dentata* × *L. japonica*. *Distr.* Garden origin.

L. przewalskii (After Nicolai M. Przewalski (1839–88), Russian explorer, who collected it.) Stems dark purple. Flower-heads numerous, small, rays yellow. *Distr.* N China.

L. smithii See *Senecio smithii*

L. stenocephala Stems dark purple. Flower-heads numerous in slender racemes. *Distr.* E Asia.

L. tussilaginea See *Farfugium japonicum*

L. veitchiana (After the Veitch family of English nurserymen.) To 2m. Flower-heads numerous, rays bright yellow. *Distr.* China.

L. wilsoniana (After E. H. Wilson (1876–1930), who collected plants in China. GIANT GROUNDSEL. To 2m high. Flower-heads numerous, borne in a long raceme. *Distr.* China.

ligularis: with ligules.

ligusticifolius, -a, -um: from the genus name, *Ligusticum*, LOVAGE and *folius*, leaved.

Ligusticum: (From Greek *ligystikos*, alluding to the abundance of this plant in the province of Liguria, Italy.) A genus of 25 species of perennial herbs with much divided leaves and compound umbels of white or occasionally pale green or purple flowers. *Distr.* N temperate regions. *Fam.* Umbelliferae. *Class* D.

L. scoticum SCOTS LOVAGE. Eaten as a culinary herb. *Distr.* E North America, Europe.

ligusticus, -a, -um: of Liguria, Italy.

ligustrifolius, -a, -um: from the genus *Ligustrum*, and Latin *folius*, leaved.

ligustrinus, -a, -um: resembling the genus *Ligustrum*.

Ligustrum (Classical Latin name for the European species.) PRIVET. A genus of about 50 species of evergreen and semi-evergreen shrubs and trees with simple, opposite, thick leaves and panicles of small, tubular, 4-lobed, white flowers. Several species are commonly grown as hedges; they are all tolerant of city pollution. *Distr.* Europe, N Africa, E and SE Asia to Australia. *Fam.* Oleaceae. *Class* D.

L. japonicum JAPANESE PRIVET. Evergreen bushy shrub. Leaves very dark. Inflorescence conical. *Distr.* N China, Korea, Japan.

L. lucidum GLOSSY PRIVET, CHINESE PRIVET, WHITE WAX TREE. Erect evergreen shrub or tree. Commonly planted as a street tree in S Europe. *Distr.* China, Korea.

L. ovalifolium CALIFORNIAN PRIVET. Evergreen or semi-evergreen dense shrub. Commonly planted as hedges. *Distr.* Japan.

L. sinense CHINESE PRIVET. Deciduous or semi-evergreen bushy shrub. Flowers in large showy panicles, white. *Distr.* China.

L. vulgare EUROPEAN PRIVET, COMMON PRIVET. Deciduous or semi-evergreen shrub. Formerly a common hedging plant but now frequently replaced by *L. ovalifolium*. *Distr.* Europe, N Africa, SW Asia, naturalized in North America.

likiangensis: of Likiang, Yunnan.

lilac *Syringa* **Californian** ~ *Ceanothus* **common** ~ *Syringa vulgaris* **dwarf** ~ *S. meyeri* **Himalayan** ~ *S. emodi* **Japanese** ~ *S. reticulata* **Japanese tree** ~ *S. reticulata* **New Zealand** ~ *Hebe hulkeana* **Persian** ~ *Melia azedarach* **summer** ~ *Buddleja davidii* **vine** ~ *Hardenbergia violacea*.

lilaciflorus, -a, -um: with lilacflowers.

lilacinus, a-, -um: lilac-coloured.

Lilaeaceae A family of 1 genus and 1 species of tufted grass-like annual herb of marshes. *Distr.* British Columbia and Idaho to Mexico and the high Andes. *Class* M.

Liliaceae The Lily family. 11 genera and about 500 species of bulbiferous herbs with linear to heart-shaped leaves and flowers that bear 2 whorls of 3 tepals that are sometimes differentiated into true sepals and petals, 6 stamens, and a superior ovary. *Distr.* Widespread, but mostly N temperate. *Class* M. *See: Amana, Calochortus, Cardiocrinum, Disporopsis, Erythronium, Fritillaria, Gagea, Lilium, Lloydia, Nomocharis, Notholirion, Tulipa.*

liliaceus, -a, -um: resembling a lily.

liliastrum: lily-like.

liliiflorus, -a, -um: with lily-like flowers.

liliifolius, -a, -um: with lily-like leaves.

lilliputiana: of Lilliput, i.e. very small.

Lilium (Latin form of the Greek name *leirion*, used for the MADONNA LILY (*Lilium candidum*).) LILY. A genus of about 100 species of bulbiferous herbs with linear to lance-shaped, whorled leaves and racemes or umbels of cup- to funnel-shaped flowers that bear 6 free tepals, 6 stamens, and a superior ovary. Many of the species and numerous hybrids are grown as ornamentals. *Distr.* Temperate N hemisphere to Philippines. *Fam.* Liliaceae. *Class* M.

L. auratum GOLD-RAYED LILY, MOUNTAIN LILY. Flowers strongly scented, white or yellow marked with crimson to 30cm across. Bulbs eaten in Japan. *Distr.* Japan.

L. bulbiferum ORANGE LILY, FIRE LILY. Bulbils produced in the axils of the leaves. Flowers upward-facing, orange-red. The emblem of the Orange Lodge in Northern Ireland. Bulbs edible. *Distr.* Europe.

L. canadense CANADA LILY, MEADOW LILY, WILD MEADOW LILY. Flowers in umbels, pendulous, yellow. Numerous cultivars available. *Distr.* E North America.

L. candidum MADONNA LILY, BOURBON LILY, WHITE LILY. Flowers white, yellow in the centre, pollen bright yellow. Flowers are used in scent-making. Cultivated since at least 1500 BC. Figured in Cretan frescoes 5000 years old, possibly the Rose of Sharon of the Bible and more recently associated with the Virgin Mary in many religious pictures. *Distr.* Balkans, E Mediterranean areas.

L. lancifolium TIGER LILY, DEVIL LILY, KENTAN. Flowers in racemes, pendulous with reflexed tepals, orange. Bulbs eaten in Japan. *Distr.* Japan, Korea, E China, naturalized in North America.

L. martagon (Turkish word used for both a type of turban and for this plant.) TURK'S CAP LILY, MARTAGON LILY. Flowers in racemes, fragrant, pendulous, tepals dull-pink, spotted maroon, recurved. The most widely spread species. *Distr.* Europe, Asia, Mongolia, naturalized in Great Britain.

L. tigrinum See *L. lancifolium*

lily Liliaceae, *Lilium* **African** ~ *Agapanthus, A. africanus* **African corn** ~ *Ixia* **Amazon** ~ *Eucharis* × *grandiflora* **arum** ~ *Zantedeschia, Z. aethiopica* **atamasco** ~ *Zephyranthes atamasco* **August** ~ *Hosta plataginea* **Australian sword** ~ *Anigozanthos* **avalanche** ~ *Erythronium grandiflorum* **Aztec** ~ *Sprekelia formosissima* **Barbados** ~ *Hippeastrum puniceum* **belladona** ~ *Amaryllis belladonna* **berg** ~ *Galtonia candicans* **blackberry** ~ *Belamcanda, B. chinensis* **blood** ~ *Haemanthus* **blue African** ~ *Agapanthus africanus* **boat** ~ *Tradescantia spathacea* **Bourbon** ~ *Lilium candidum* **butterfly** ~ *Hedychium coronarium* **calla** ~ *Zantedeschia* **Canada** ~ *Lilium canadense* **canna** ~ *Canna* × *generalis* **Cape** ~ *Crinum* **Cayman islands spider** ~ *Hymenocallis latifolia* **chequered** ~ *Fritillaria meleagris* **Chinese lantern** ~ *Sandersonia aurantiaca* **Chinese sacred** ~ *Narcissus tazetta* **chrysolite** ~ *Hymenocallis latifolia* **climbing** ~ *Gloriosa superba* **cobra** ~ *Darlingtonia californica* **corn** ~ *Ixia, Clintonia borealis* **cradle** ~ *Tradescantia spathacea* **crane** ~ *Strelitzia reginae* **creeping** ~ *Gloriosa superba* **Cuban** ~ *Scilla peruviana* **day** ~ *Hemerocallis* **devil** ~ *Lilium lancifolium* **Eucharist** ~ *Eucharis* × *grandiflora* **fairy** ~ *Zephyranthes* **fawn** ~ *Erythronium* **fire** ~ *Cyrtanthus, Lilium bulbiferum* **flame** ~ *Gloriosa superba* **flax** ~ *Phormium, Dianella* **foxtail** ~ *Eremurus* **funeral** ~ *Zantedeschia aethiopica* **garland** ~ *Hedychium* **George** ~ *Cyrtanthus elatus* **giant** ~ *Cardiocrinum giganteum* **ginger** ~ *Hedychium, Alpinia* **glory** ~ *Gloriosa superba* **gold-rayed** ~ *Lilium auratum* **golden hurricane** ~ *Lycoris aurea* **golden spider** ~ *L. aurea* **Guernsey** ~ *Nerine sarniensis* **ifafa** ~ *Cyrtanthus mackenii* **Jacobean** ~ *Sprekelia formosissima* **Japanese toad** ~ *Tricyrtis hirta* **Jersey** ~ *Amaryllis belladonna* **kaffir** ~ *Clivia, Schizostylis coccinea* **knight's star** ~ *Hippeastrum* **lavender globe** ~ *Allium tanguticum* **Lent** ~ *Narcissus pseudonarcissus* **leopard** ~ *Belamcanda chinensis, Dieffenbachia* **leper** ~ *Fritillaria meleagris* **loddon** ~ *Leucojum aestivum* **lotus white** ~ *Nymphaea lotus* **Madonna** ~ *Lilium candidum* **magic** ~ *Lycoris squamigera* **Mariposa** ~ *Calochortus* **martagon** ~ *Lilium martagon* **May** ~ *Maianthemum, M. bifolium* **meadow** ~ *Lilium canadense* **Mexican** ~ *Hippeastrum reginae* **mountain** ~ *Leucocrinum, L. montanum, Lilium auratum,* **New Year** ~ *Gladiolus cardinalis* **orange** ~ *Lilium bulbiferum* **painted wood** ~ *Trillium undulatum* **palm** ~ *Cordyline australis, Yucca gloriosa* **paradise** ~ *Paradisea liliastrum* **peace** ~ *Spathiphyllum, S. wallisii* **peacock** ~ *Moraea* **Peruvian** ~ *Alstroemeria* **pineapple** ~ *Eucomis* **pink porcelain** ~ *Alpinia zerumbet* **plantain** ~ *Hosta* **poor knights** ~ *Xeronema callistemon* **queen** ~ *Phaedranassa* **rain** ~ *Zephyranthes* **red ginger** ~ *Hedychium coccineum* **red spider** ~ *Lycoris radiata* **resurrection** ~ *L. squamigera* **sand** ~ *Leucocrinum L. montanum,* **Scarborough** ~ *Cyrtanthus elatus* **scarlet ginger** ~ *Hedychium coccineum* **sea** ~ *Pancratium maritimum* **snake** ~ *Dichelostemma volubile* **Snowdon** ~ *Lloydia serotina* **spear** ~ Doryanthaceae, *Doryanthes* **speckled wood** ~ *Clintonia umbellulata* **spider** ~ *Hemerocallis, Hymenocallis, Tradescantia, Lycoris radiata* **St Bernard's** ~ *Anthericum, A. liliago* **St Bruno's** ~ *Paradisea liliastrum* **St James** ~ *Sprekelia formosissima* **star** ~ *Leucocrinum, L. montanum, Zigadenus fremontii* **swamp** ~ *Saururus cernuus* **sword** ~ *Iris* **tiger** ~ *Lilium lancifolium* **toad** ~ *Tricyrtis, T. hirta* **torch** ~ *Kniphofia* **triplet** ~ *Triteleia laxa* **trout** ~ *Erythronium, E. americanum* **Turk's cap** ~ *Lilium martagon* **voodoo** ~ *Sauromatum venosum* **water** ~ Nymphaeaceae **white** ~ *Lilium candidum* **wild meadow** ~ *L. canadense* **wood**

~ *Trillium*, *Clintonia* **yellow arum** ~ *Zantedeschia elliottiana* **zephyr** ~ *Zephyranthes atamasco*.

lily of China *Rohdea japonica*

lily of the Incas *Alstroemeria*

lily of the Nile *Agapanthus africanus*

lily of the palace *Hippeastrum aulicum*.

lily of the valley Convallariaceae, *Convallaria* **false** ~ *Maianthemum bifolium*.

lily bush *Pieris japonica* ~ **tree** *Clethra arborea*

lily tree *Magnolia denudata*

lily turf *Liriope*.

lime Tiliaceae, *Tilia*, *T.* × *vulgaris*, *citrus aurantifolia* **American** ~ *Tilia americana* **broad-leaved** ~ *T. platyphyllos* **common** ~ *T.* × *vulgaris* **European** ~ *T.* × *vulgaris* **European white** ~ *T. tomentosa* **house** ~ *Sparmannia africana* **Japanese** ~ *Tilia japonica* **large-leaved** ~ *T. platyphyllos* **silver** ~ *T. tomentosa* **small-leaved** ~ *T. cordata*.

limequat × *Citrofortunella floridana*.

limifolius, -a, -um: with file-like leaves.

Limnanthaceae A family of 2 genera and 8 species of succulent annual herbs with pinnately divided leaves and long-stalked flowers. *Class* D.
See: Limnanthes.

Limnanthes (From Greek *limne*, marsh, and *anthos*, flower.) MEADOW FOAM. A genus of 7 species of fragile annual herbs with pinnate leaves and solitary, regular, white flowers that bear their parts in fives. These plants have been considered a potential oil-seed crop. *Distr.* W North America, especially California. *Fam.* Limnanthaceae. *Class* D.

L. douglasii (After David Douglas (1798–1834), who introduced many plants to European gardens from W North America.) POACHED EGG FLOWER. Flowers fragrant, white with yellow marking. Grown as an ornamental. *Distr.* USA (California to Oregon).

Limnobium (From Greek *limne*, marsh, and *bios*, life.) A genus of 2 species of perennial herbs with rosettes of leaves that form floating mats and unisexual flowers that are held above the water and bear 3 white petals. *Distr.* Warm America, introduced and naturalized elsewhere. *Fam.* Hydrocharitaceae. *Class* M.

L. laevigatum Occasionally grown in tropical aquaria. *Distr.* West Indies, Central and tropical South America.

Limnocharis (From Greek *limne*, a marsh, and *charis*, delight.) A genus of 2 species of annual and perennial herbs with rosettes of stalked leaves and umbels of yellow flowers that bear 3 sepals, 3 petals, and numerous stamens. *Distr.* Warmer regions of America, naturalized elsewhere. *Fam.* Limnocharitaceae. *Class* M.

L. flava Grown as a tender ornamental. Leaves eaten like spinach or used as animal fodder. *Distr.* West Indies, Tropical South America, naturalized in SE Asia.

Limnocharitaceae The Water Poppy family. 3 genera and 12 species of annual and perennial aquatic herbs with latex ducts and showy regular flowers that bear 3 sepals, 3 petals, 3 to numerous stamens and 3 to many free carpels. *Distr.* Tropical and subtropical regions. *Class* M.
See: Hydrocleys, Limnocharis.

limonetto *Aloysia triphylla*.

limonifolius, -a, -um: from the genus *Limonium*, and *folius*, leaved.

Limonium (From Greek *leimon*, meadow, alluding to the habitat of some of the species.) SEA LAVENDER, MARSH ROSEMARY, STATICE. A genus of about 150 species of annual to perennial herbs and subshrubs with basal rosettes of simple or pinnately lobed leaves and clusters of small trumpet-shaped flowers. Numerous species are grown as ornamentals and some are used as cut flowers. *Distr.* Cosmopolitan, especially in coastal areas of the N hemisphere. *Fam.* Plumbaginaceae. *Class* D.

L. bellidifolium Cushion-forming perennial. Leaves dark green. Flowers blue. *Distr.* Europe, Asia.

L. dumosum See *Goniolimon tataricum*

L. tartaricum See *Goniolimon tataricum*

Linaceae The Flax family. 14 genera and 300 species of herbs and shrubs with alternate, simple leaves and showy, regular flowers that typically bear 5 sepals and 5 petals. *Distr.* Widespread but mainly in temperate regions. *Class* D.
See: Linum, Reinwardtia.

Linanthastrum See *Linanthus*

Linanthus (From Greek *linon*, flax, and *anthos*, flower, alluding to the similarity of the flowers to those of the genus *Linum*.) A genus of 35 species of annual and perennial herbs with simple to palmately lobed leaves and heads of funnel-shaped flowers that bear their parts in 5s. Several species are grown as ornamentals. *Distr.* W North America, Chile. *Fam.* Polemoniaceae. *Class* D.

L. nuttallii Bushy perennial herb. Flowers yellow and white. *Distr.* USA (California to Washington).

Linaria (From Greek *linon*, flax, probably alluding to the leaves which resemble those of the genus *Linum*.) TOADFLAX, SPURRED SNAPDRAGON. A genus of about 100 species of annual and perennial herbs with simple narrow leaves and spike-like racemes of spurred, 5-lobed, 2-lipped flowers. The lower lip of the flowers bears a boss that closes the mouth of the petal tube. Some species have local medicinal uses, especially in the treatment of haemorrhoids; many grown as ornamentals. *Distr.* N temperate regions, especially around the Mediterranean. *Fam.* Scrophulariaceae. *Class* D.

L. alpina ALPINE TOADFLAX. Tuft-forming annual to short-lived perennial. Flowers purple-violet with a yellow centre. *Distr.* Alps.

L. cymbalaria See *Cymbalaria muralis*

L. genistifolia Perennial herb to 1m high. Flowers lemon-yellow, boss-bearded, orange. *Distr.* E Europe.

L. origanifolia See *Chaenor-rhinum origanifolium*

L. purpurea PURPLE TOADFLAX. Erect perennial. Flowers purple-blue with white throats. *Distr.* Italy.

L. repens STRIPED TOADFLAX. Creeping rhizomatous perennial. Flowers white to pale lilac with violet veins. *Distr.* W Europe, naturalized in Central Europe.

L. triornithophora (From Greek *tri-*, three, *ornis*, bird, and *phorea*, to bear; the flowers are in threes and resemble small birds with long tails.) THREE BIRDS FLYING, THREE BIRDS TOADFLAX. Erect perennial herb. Flowers purple with a yellow boss. *Distr.* Spain, Portugal.

L. tristis DULL-COLOURED LINARIA, SAD-COLOURED LINARIA. Perennial herb. Flowers yellow tinged purple-brown. *Distr.* S Spain, S Portugal, NW Africa, Canary Islands.

L. vulgaris COMMON TOADFLAX, BUTTER AND EGGS, WILD SNAPDRAGON. Perennial to 80cm high. Flowers yellow, boss orange. *Distr.* Eurasia, naturalized in North America.

linaria: dull-coloured ~ *Linaria tristis* **sad-coloured** ~ *L. tristis.*

linariifolius, -a, -um: from the genus *Linaria*, and *folius*, leaved.

linarioides: resembling the genus *Linaria*.

Lindelofia (After Friedrich von Lindelof of Darmstadt, 19th-century botanist.) A genus of about 12 species of perennial herbs with simple, basal leaves and funnel- to bell-shaped flowers. *Distr.* Central Asia, Himalaya. *Fam.* Boraginaceae. *Class* D.

L. anchusoides To 1m high. Leaves strap-shaped. Flowers blue, pink or purple. Grown as a hardy ornamental. *Distr.* Afghanistan to W Himalaya.

linden *Tilia* **little leaf** ~ *T. cordata.*

Lindera (After Johann Linder (1676–1723), Swedish botanist.) FEVER BUSH, WILD ALLSPICE. A genus of about 80 species of evergreen or deciduous aromatic shrubs or trees with alternate, simple or 3-lobed leaves and umbel-like inflorescences of small, unisexual, yellow flowers. Several species are grown as ornamentals. *Distr.* E Asia, E North America. *Fam.* Lauraceae. *Class* D.

L. benzoin SPICE BUSH, BENJAMIN BUSH. Deciduous rounded shrub. The leaves are used to make a tea and the fruits as a substitute for allspice. *Distr.* SE USA.

L. obtusiloba Deciduous shrub or small tree. Leaves 3-lobed. *Distr.* Korea, China, and Japan.

L. umbellata Deciduous shrub. Leaves simple. *Distr.* Japan.

linearifolius, -a, -um: with linear leaves.

linearis, -e: linear.

lineatus, -a, -um: lined or striped.

ling *Calluna vulgaris.*

lingberry *Vaccinium vitis-idaea.*

lingua: tongue-like.

linguiformis, -e: tongue-shaped.

lingulatus, -a, -um: tongue-shaped.

linifolius, -a, -um: from the genus *Linum*, FLAX, and *folius*, leaved.

Linnaea (Named after Carl Linnaeus (1707–78), Swedish naturalist, at his request.) A genus of 1 species of small, creeping, evergreen subshrub with simple leaves and pairs of nodding, bell-shaped, pink, fragrant flowers. *Distr.* Circumpolar. *Fam.* Caprifoliaceae. *Class* D.

L. borealis TWIN FLOWER. Grown as a rock-garden ornamental.

linnaeoides: resembling the genus *Linnaea.*

Linum (Classical Latin name FLAX.) FLAX. A genus of about 200 species of annual and perennial herbs and shrubs with simple alternate leaves and racemes of regular bisexual flowers that bear their parts in fives. A number of species are grown as ornamentals and for their stem fibres. *Distr.* Temperate, subtropical, and Mediterranean regions. *Fam.* Linaceae. *Class* D.

L. arboreum TREE FLAX. Compact shrub to 1m high. Flowers yellow. *Distr.* Greece, Turkey.

L. bienne PALE FLAX. Annual or perennial herb. Flowers pale blue. *Distr.* W Europe, Mediterranean.

L. flavum GOLDEN FLAX, YELLOW FLAX. Erect perennial herb. Flowers golden-yellow, borne in dense panicles. *Distr.* Central and SE Europe.

L. narbonense BLUE FLAX. Clump-forming perennial herb. Flowers pale blue. *Distr.* SW Europe.

L. perenne PERENNIAL FLAX. Slender, erect perennial herb. Flowers blue. *Distr.* Europe, North America.

L. sibiricum See *L. perenne*

L. usitatissimum FLAX. Erect annual herb. Flowers blue. The stems are the source of the fibre used in linen, canvas, and many other products. The seeds are the source of linseed oil and cattle feed. *Distr.* Asia, widely naturalized.

L. viscosum STICKY FLAX. Glandular-hairy perennial herb. Flowers pink-blue. *Distr.* S Europe.

lion's beard *Geum triflorum.*

lion's ear *Leonotis*, *L. leonurus.*

Lippia (After Auguste Lippi (born 1678), Italian naturalist.) A genus of about 200 species of herbs, shrubs, and small trees with simple, typically aromatic leaves and spikes of small bell-shaped flowers. Several species are grown as ornamentals; the leaves are used to make tea and as a culinary herb in parts of Central and South America. *Distr.* Tropical Africa and America. *Fam.* Verbenaceae. *Class* D.

L. citriodora See *Aloysia triphylla*

L. dulcis YERBA DULCE. Highly aromatic perennial herb to 60cm high. Flowers white. Used medicinally for bronchitis. *Distr.* Mexico to Panama.

lipstick tree *Bixa orellana.*

Liquidambar (From Latin *liquidas*, liquid, and *ambar*, amber, alluding to the fragrant resin obtained from the bark.) SWEET GUM. A genus of 4 species of large deciduous trees with palmately lobed, stalked leaves and clusters of small unisexual flowers. The genus is a valuable source of timber and aromatic balsam (storax) used medicinally and in scent; species are also cultivated as ornamentals for their spectacular autumn colours. *Distr.* North America, Asia. *Fam.* Hamamelidaceae. *Class* D.

L. formosana FORMOSAN GUM. Straight-trunked tree to 40m high. Silkworms fed on it yield 'Marvello hair'. *Distr.* SE Asia.

L. orientalis ORIENTAL SWEET GUM, BALM OF GILEAD. Small bushy tree to 30m. Source of Levant storax, the balm (of Gilead) of the Bible. *Distr.* Turkey.

L. styraciflua SWEET GUM, AMERICAN SWEET GUM, RED GUM, SATIN WALNUT TREE. Tall, straight-trunked tree to 45m high. A source of timber used for furniture and veneers and of storax used to treat skin disease. *Distr.* E USA, Mexico, Guatemala.

liquorice *Glycyrrhiza glabra* **wild ~** *Astragalus glycyphyllos*.

liquorice plant *Helichrysum petiolare*.

Liriodendron (From Greek *leiron*, lily, and *dendron*, tree.) A genus of 2 species of large deciduous trees with alternate, simple leaves and solitary, regular flowers that bear 3 sepal-like tepals, 6 petal-like tepals, and 2 whorls of 3 stamens. The leaves turn yellow in autumn. Both trees are grown as ornamentals and are a source of timber. *Distr.* E Asia, E North America. *Fam.* Magnoliaceae. *Class* D.

L. chinense CHINESE TULIP TREE. Tree to 16m high. Flowers yellow-green to 8cm across. *Distr.* E Asia.

L. tulipifera TULIP TREE, TULIP POPLAR, YELLOW POPLAR, CANARY WHITEWOOD. Tree to 60m high. Flowers to 12cm across, pale green marked orange. Used for cabinetwork, shingles, clapboards etc., formerly for Indian canoes, and for cigar- and bible-boxes for white settlers. *Distr.* E North America.

Liriope (After Liriope, a fountain nymph and the mother of Narcissus in Greek mythology.) LILY TURF. A genus of about 4 species of rhizomatous, stemless, tufted herbs with grass-like leaves and spikes of rounded, grape-like flowers. All the species are cultivated as ornamentals for ground-cover; some are reported to have medicinal and aphrodisiac properties. *Distr.* Japan, China, Korea, Vietnam. *Fam.* Convallariaceae. *Class* M.

L. graminifolia See *L. spicata*

L. hyacinthifolia See *Reineckea carnea*

L. muscari Rhizome thick, dark. Flowers dark mauve, densely clustered. Numerous cultivars of this species are available. *Distr.* China and Japan.

L. spicata Leaves dark green. Flowers pale mauve to white, borne on a mauve stem. *Distr.* China, Vietnam.

Lissocarpaceae A family of 1 genus and 2 species of trees with simple, spirally arranged leaves and cymes of regular, bisexual flowers that bear their parts in fours and have a small corona. *Distr.* Tropical South America. *Class* D.

lissospermus, -a, -um: with smooth seeds.

litchee *Litchi chinensis*.

Litchi (From the Chinese vernacular name.) A genus of 1 species of tree with pinnately divided, spirally arranged leaves and panicles of small white flowers that lack petals. The fruit is a scaly drupe with a single seed that is surrounded by a fleshy white aril. *Distr.* China to W Malaysia. *Fam.* Sapindaceae. *Class* D.

L. chinensis LITCHI, LYCHEE, LITCHEE. The small scaly fruit is eaten fresh or canned.

litchi *Litchi chinensis*.

Lithocarpus (From Greek *lithos*, stone, and *karpos*, fruit, alluding to the hard fruit.) TANBARK OAK. A genus of about 300 species of evergreen trees with simple, leathery leaves and small, wind-pollinated, unisexual flowers. The male flowers are borne on an erect branched spike and the females at the base of the male spike or on a small spike of their own. The fruit is a hard nut surrounded by a scaly case (the cupule). *Distr.* SE Asia; 1 species in North America. *Fam.* Fagaceae. *Class* D.

L. densiflorus TANBARK OAK. Tree to 40m high. Bark red-brown. Leaves densely white-hairy below at first. Male flowers malodorous. *Distr.* W North America.

L. edulis Shrub or small tree to 15m high. Leaves narrow. *Distr.* Japan.

L. glaber Shrub or small tree to 7m high. *Distr.* E China, Japan.

L. henryi (After Augustine Henry (1857–1930), Irish doctor who collected in China.) Large tree to 20m high. *Distr.* S and W China, Japan.

Lithodora (From Greek *lithos*, a stone, and *dorea*, a gift.) A genus of 7 species of evergreen subshrubs and shrubs with linear or ovate leaves and funnel- or salver-shaped, typically blue flowers. Several species are grown as ornamentals. *Distr.* W Europe to Turkey. *Fam.* Boraginaceae. *Class* D.

L. diffusa Straggling shrub. Flowers intense blue-purple. Cultivated as an ornamental shrublet. *Distr.* SW Europe.

L. graminifolia See *Moltkia suffruticosa*

L. × intermedia See *Moltkia × intermedia*

L. oleifolia Stems ascending. Leaves broad. Flowers pink-blue. *Distr.* Spain.

L. rosmarinifolia Leaves narrowly linear. Flowers blue to white. *Distr.* S Italy, N Africa.

L. zahnii Stems erect or ascending. Leaves leathery. Flowers pale-blue. *Distr.* S Greece.

Lithophragma (From Greek *lithos*, stone, and *phragma*, fence, referring to its habitat.) WOODLAND STAR. A genus of 9 species of perennial herbs with rosettes of broad leaves and racemes of small flowers. *Distr.* W North America. *Fam.* Saxifragaceae. *Class* D.

L. parviflora Flowers campion-like, white-pink.

Lithops (From Greek *lithos*, stone, and *opsis*, appearance.) LIVING STONES. A genus of 37 species of succulent stemless herbs with 2 fleshy leaves that are fused for much of their length and resemble small stones. The many-petalled flowers arise from the fissure that separates the 2 leaves. In some species the flattened ends of the leaves are translucent, acting as windows to let light into the rest of the plant, which is often held below the soil surface. A number of species and many cultivars are grown as ornamentals. *Distr.* S Africa. *Fam.* Aizoaceae. *Class* D.

L. dorotheae Leaves grey-green with dark green markings. Flowers yellow. *Distr.* South Africa (NW Cape Province).

L. karasmontana Leaves blue-grey with a pink upper surface. Flowers white. *Distr.* Namibia.

L. marmorata Leaves blue-grey. Flowers white. *Distr.* South Africa (NW Cape Province).

L. schwantesii Leaves grey-blue. Flowers yellow. *Distr.* Namibia.

Lithospermum (From Greek *lithos*, stone, and *sperma*, seed, alluding to the hard seeds.) GROMWELL, PUCCOON. A genus of 59 species of annual and perennial herbs and subshrubs with simple, alternate leaves and funnel- or salver-shaped flowers. Source of some dyes and cultivated as ornamentals. *Distr.* Temperate regions, not Australia. *Fam.* Boraginaceae. *Class* D.

L. arvense BASTARD ALKANET, CORN GROMWELL. Annual of arable fields. A source of dye. *Distr.* Europe and Asia.

L. diffusum See *Lithodora diffusa*

L. officinale GROMWELL, GRUMMEL. Perennial. Flowers green to yellow-white. Leaves used as a tea. *Distr.* Eurasia, naturalized in North America.

L. oleifolium See *Lithodora oleifolia*

L. purpureocaeruleum See *Buglossoides purpuro-caerulea*

litoralis, -e: of the shore.

Littonia (After Dr Samuel Litton (1781–1847), professor of botany at Dublin.) A genus of about 8 species of rhizomatous climbing herbs with zigzagging, unbranched stems and lance-shaped leaves the tips of which have developed into tendrils. The nodding, bell-shaped, orange flowers are borne singly in the leaf axils. *Distr.* Tropical and S Africa, Arabia. *Fam.* Colchicaceae. *Class* M.

L. modesta CLIMBING BELLFLOWER. Cultivated as a half-hardy perennial for the showy flowers and seed pods. *Distr.* South Africa (Transvaal, Natal, Orange Free State).

live long *Hylotelephium telephium.*

lividus, -a, -um: lead-coloured.

living baseball *Euphorbia obesa.*

living granite *Pleiospilos*

living rock *Pleiospilos*

living stones *Lithops.*

Livistona (After Patrick Murray, Baron of Livingston (died 1671) whose plant collection helped to found Edinburgh Botanic Garden in 1670.) A genus of about 30 species of shrubby or tree-like palms to 25m high with fanshaped leaves and much-branched inflorescences of cream flowers. *Distr.* NE Africa, Asia, Australasia. *Fam.* Palmae. *Class* M.

L. australis AUSTRALIAN FAN PALM, GIPPSLAND PALM, CABBAGE PALM, AUSTRALIAN PALM. Grown as a tender ornamental. The young buds are eaten locally as palm cabbages. *Distr.* E Australia.

L. chinensis CHINESE FAN PALM. Grown as a tender ornamental. *Distr.* S Japan, China.

lizard plant *Tetrastigma voinierianum.*

lizard's tail *Saururus.*

Lloydia (After Edward Lloyd (1660–1709), naturalist and Museum keeper of the Ashmolean Collection, Oxford.) A genus of about 12 species of rhizomatous, bulbiferous herbs with narrow, linear leaves and 1–2 star-shaped, white or yellow flowers. Several species are grown as ornamentals. *Distr.* Temperate N hemisphere. *Fam.* Liliaceae. *Class* M.

L. serotina SNOWDON LILY. Flowers to 15mm across, white. A protected species in the British Isles. *Distr.* Temperate N hemisphere.

Loasaceae The Rock Nettle family. 15 genera and 260 species of herbs and shrubs with rough, often stinging hairs, simple or divided leaves and regular, bisexual flowers that bear 4–5 typically free petals. Some species are grown as ornamentals. *Distr.* Chiefly tropical and temperate America with several species in Arabia, Somalia, and Namibia. *Class* D.
See: Mentzelia.

lobatus, -a, -um: lobed.

Lobelia (After Matthias de l'Obel (1538–1616), Flemish botanist and physician to James I.) A genus of 365 species of annual and perennial herbs, shrubs, and small trees with simple, alternate leaves and milky sap. The flowers are 2-lipped with 3 large, spreading, lower lobes and 2 smaller, recurved, upper lobes; they are solitary or borne in racemes. Many species, hybrids, and cultivars are grown as ornamentals. Some species, particularly the thick-stemmed 'Giant Lobelias' of East Africa, are of great ecological importance. *Distr.* Tropical and warm regions, especially America, few in temperate regions. *Fam.* Campanulaceae. *Class* D.

L. angulata See *Pratia angulata*

L. cardinalis CARDINAL FLOWER, INDIAN PINK. Clump-forming perennial herb. Flowers scarlet. *Distr.* E USA.

L. dortmanna WATER LOBELIA. Perennial aquatic herb. Leaves in basal rosettes. Flowers mauve in racemes held above the water. *Distr.* North America, W Europe.

L. erinus EDGING LOBELIA. Small perennial herb. Flowers purple with a yellow or white throat. Often used as an annual bedding plant. Numerous cultivars available. *Distr.* S Africa.

L. inflata INDIAN TOBACCO. Cultivated for leaves which are used medicinally, especially in chest conditions. *Distr.* North America.

L. siphilitica (A reference to supposed medicinal properties.) BLUE CARDINAL FLOWER, GREAT LOBELIA. Clump-forming perennial herb. Flowers blue. *Distr.* E USA.

L. splendens Clump-forming slender perennial. Flowers bright scarlet to blood red. *Distr.* Mexico, Texas.

L. tupa Clump-forming perennial herb. Flowers brick-red, borne in large spikes. Used medicinally and as a narcotic. *Distr.* Chile, Peru.

lobelia: edging ~ *Lobelia erinus* **great** ~ *L. siphilitica* **water** ~ *L. dortmanna.*

lobelioides: from the genus *Lobelia*, with the ending *-oides*, indicating resemblance.

lobophyllus, -a, -um: with lobed leaves.

lobster claw *Heliconia, Clianthus puniceus, Vriesea carinata.*

lobster plant *Euphorbia pulcherrima, Justicia brandegeana.*

locoweed *Oxytropis.*

locust *Robinia* **black** ~ *R. pseudoacacia* **bristly** ~ *R. hispida* **Caspian** ~ *Gleditsia caspica* **honey** ~ *Gleditsia G. triacanthos* **moss** ~ *Robinia hispida* **yellow** ~ *R. pseudoacacia.*

loganberry *Rubus longanobaccus.*

Loganiaceae A family of 20 genera and about 500 species of trees, shrubs, and climbers with opposite, entire leaves and regular, tubular, 4–5-lobed flowers. Many members of the family are poisonous, notably those of the genus *Strychnos* which are the source of strychnine and curare; others are grown as ornamentals. *Distr.* Widespread in tropical regions, rarer in temperate areas. *Class* D.
See: Desfontainia, Gelsemium, Spigelia.

loganioides: resembling the genus *Logania.*

Loiseleuria (After Jean Louis Auguste Loiseleur-Deslongchamps (1774–1849), French physician and botanist.) A genus of 1 species of prostrate, mat-forming, evergreen shrub with curled leaves and regular flowers that bear 5 stamens. *Distr.* N circumpolar. *Fam.* Ericaceae. *Class* D.

L. procumbens ALPINE AZALEA, MOUNTAIN AZALEA, MINEZUO. Grown as a creeping ornamental.

loliaceus, -a, -um: weedy.

lollipop plant *Pachystachys lutea.*

Lomandraceae A family of 6 genera and about 50 species of tufted perennial herbs. *Distr.* New Guinea, Australia, New Coledonia. *Class* M.

Lomaria See *Blechnum*

lomariifolius, -a, -um: from the genus *Lomaria*, and *folius*, leaved.

Lomariopsidaceae A family of 7 genera and about 550 species of terrestrial or epiphytic ferns with simple to pinnate, rarely bipinnate vegetative leaves and reduced fertile leaves that are evenly coated in sporangia on the underside. *Distr.* Widespread in tropical to warm-temperate regions. *Class* F. *See: Elaphoglossum.*

Lomatia (From Greek *loma*, border; the seeds are edged with a wing.) A genus of about 12 species of evergreen shrubs with opposite or alternate, pinnately lobed or compound leaves, and loose racemes of tubular flowers. Some species are cultivated as ornamentals. *Distr.* E Australia, Tasmania, South America. *Fam.* Proteaceae. *Class* D.

L. ferruginea Erect shrub. Flowers yellow or red. *Distr.* Chile, Argentina.

L. silaifolia Bushy shrub. Flowers cream-white. *Distr.* Australia (New South Wales, Queensland).

London pride *Saxifraga* × *urbium.*

longesquamatus, -a, -um: with long scales.

longestylus, -a, -um: with a long style.

longianus, -a, -um: long.

longiauritus: long-eared.

longibracteatus, -a, -um: with long bracts.

longicaulis, -e: with a long stem.

longicuspis: with a long point.

longiflorus, -a, -um: with long flowers.

longifolius, -a, -um: with long leaves.

longilobatus, -a, -um: with long lobes.

longipes: long-stalked.

longipetalus, -a, -um: long-petalled.

longiracemosus, -a, -um: with a long raceme.

longiscapus, -a, -um: with a long scape.

longisepalus, -a, -um: with long sepals.

longispathus, -a, -um: with a long spathe.

longissimus, -a, -um: very long.

longistylus, -a, -um: with a long style.

longituba: with a long floral tube.

longus, -a, -um: long.

Lonicera (After Adam Lonitzer (1528–86), German naturalist and herbalist.) HONEYSUCKLE. A genus of 180 species of deciduous and occasionally evergreen shrubs and climbers with opposite, simple leaves. The tubular or bell-shaped flowers are often fragrant and are borne in pairs, sometimes with the 2 flowers sharing a single ovary. Numerous species and cultivars are grown as ornamentals. *Distr.* N hemisphere to Mexico and Philippines. *Fam.* Caprifoliaceae. *Class* D.

L. × brownii SCARLET TRUMPET HONEYSUCKLE. A hybrid of *L. sempervirens* × *L. hirsuta*. Scrambling deciduous shrub. Flowers trumpet-shaped, 2-lipped.

L. caprifolium (From the Latin *capra*, a goat, and *folium*, leaf.) PERFOLIATE HONEYSUCKLE, ITALIAN WOODBINE, ITALIAN HONEYSUCKLE. Scrambling, deciduous shrub. Flowers white or purplish, fragrant; local medicinal uses. *Distr.* Europe, Caucasus, Turkey.

L. ciliosa Twining evergreen shrub. Flowers orange-red to yellow. Used in the manufacture of shampoo by North American Indians. *Distr.* North America.

L. flexuosa See *L. japonica*

L. fragrantissima More or less deciduous shrub. Flowers white, fragrant, borne in winter. *Distr.* China.

L. hildebrandiana (After A. H. Hildebrand (1852–1918), who collected in India.) GIANT BURMESE HONEYSUCKLE, GIANT HONEYSUCKLE. Scrambling evergreen or somewhat deciduous shrub. Flowers cream turning orange, fragrant. *Distr.* Burma, Thailand, China.

L. hirsuta HAIRY HONEYSUCKLE. Deciduous twiner. Leaves hairy. Flowers in spikes, orange-yellow, 2-lipped. *Distr.* E North America.

L. implexa MINORCA HONEYSUCKLE. Evergreen straggling shrub. Flowers yellow tinged pink. *Distr.* Mediterranean.

L. involucrata TWINBERRY. Erect deciduous shrub. Flowers yellow. Fruit black, subtended by 2 spreading bracts. *Distr.* W North America, Mexico.

L. japonica JAPANESE HONEYSUCKLE, GOLD AND SILVER FLOWER. More or less evergreen climber. Flowers white, becoming yellow-orange, very fragrant. Numerous ornamental cultivars available. *Distr.* Japan, China, Korea, naturalized and serious weed in North America.

L. nitida Evergreen shrub. Leaves small. Flowers cream-white. Much used for hedging, with a number of cultivars available. *Distr.* China.

L. periclymenum COMMON HONEYSUCKLE, WOODBINE. Twining deciduous shrub. Flowers fragrant, red and yellow. Numerous cultivars available. *Distr.* Europe, N Africa.

L. pileata Evergreen or semi-deciduous shrub. Flowers yellow and white. *Distr.* China.

L. sempervirens TRUMPET HONEYSUCKLE, CORAL HONEYSUCKLE. Scrambling evergreen shrub. Flowers scarlet-orange. *Distr.* SE North America.

L. xylosteum FLY HONEYSUCKLE. Erect deciduous shrub. Flowers yellow-white. *Distr.* Eurasia, naturalized in Great Britain.

loofah *Luffa aegyptiaca* **angled** ~ *L. acutangula*.

looking-glass plant *Coprosma repens*.

loosestrife *Lysimachia*, *Lythrum* **purple** ~ *L. salicaria*, *Lysimachia salicaria* **spiked** ~ *Lythrum salicaria* **yellow** ~ *Lysimachia vulgaris*.

lophanthus, -a, -um: with crested flowers.

Lophomyrtus (From Greek *lophos*, crest, and the name of the genus *Myrthus*.) A genus of 2 species of shrubs and small trees with pairs of simple, leathery, glandular leaves and solitary white flowers that bear 4 petals and numerous stamens. Both species are grown as ornamentals. *Distr.* New Zealand. *Fam.* Myrtaceae. *Class* D.

L. bullata To 6m high. Leaves to 5cm long, often tinged red.

L. obcordata Shrub to 4m high. Leaves to 2cm long, notched at tip Distr.

Lophopyxidaceae A family of 1 genus and 1 species of liana and small tree. This family is sometimes included within the family Celastraceae. *Distr.* Malaya, W Pacific Islands. *Class* D.

lophospermus, -a, -um: with crested seeds.

loquat *Eriobotrya japonica* **Japanese** ~ *E. japonica*.

Loranthaceae A family of 77 genera and about 1000 species of small, shrubby, green parasites with suckers for roots. The leaves are evergreen and leathery and the flowers are regular with green or petal-like tepals. The fruit is a berry or drupe and the seed is sticky. *Distr.* Widespread. *Class* D.

lords and ladies *Arum*, *A. maculatum*.

Loropetalum (From Greek *loron*, thong, and *petalon*, petal, alluding to the long narrow petals.) A genus of 1–2 species of evergreen shrubs and small trees with simple, alternate leaves and clusters of spider-like, green-white flowers that bear their parts in fours. *Distr.* India, China, Japan. *Fam.* Hamamelidaceae. *Class* D.

L. chinense Grown as an ornamental. *Distr.* India, China, Japan.

lote tree *Celtis australis*.

Lotus (From the classical Greek name *lotos*, used for several different plants.) A genus of 100 species of annual and perennial herbs with palmate or pinnate leaves and umbel-like racemes of irregular, pea-like flowers. Some species are cultivated as ornamentals and some for forage. *Distr.* N temperate regions. *Fam.* Leguminosae. *Class* D.

L. berthelotii CORAL GEM, PARROT'S BEAK, PELICAN'S BEAK. Perennial herb. Flowers scarlet-red with a large keel. *Distr.* Canary Islands.

L. corniculatus BIRD'S FOOT TREFOIL, EGGS AND BACON. Perennial herb. Flowers yellow with red marks. Grown as fodder but some populations can be poisonous to stock. *Distr.* Eurasia.

L. tetragonolobus ASPARAGUS PEA, WINGED PEA. Annual herb. Flowers maroon to red. Young pods eaten as a vegetable; also grown as a forage crop. *Distr.* S Europe.

lotus *Nelumbo* **American** ~ *N. lutea* **blue** ~ *Nymphaea caerulea* **blue Egyptian** ~ *N. caerulea* **Indian** ~ *Nelumbo nucifera* **sacred** ~ *N. nucifera* **white Egyptian** ~ *Nymphaea lotus.*

louisianicus, -a, -um: of Louisiana, USA.

lousewort *Pedicularis.*

lovage *Levisticum officinale* **black** ~ *Smyrnium olusatrum* **Scots** ~ *Ligusticum scoticum.*

love in a mist *Nigella damascena*

love in idleness *Viola tricolor*

love lies bleeding *Amaranthus caudatus.*

love tree *Cercis siliquastrum.*

Lowiaceae A family of 1 genus and 7 species of rhizomatous herbs with 2 ranks of stalked leaves and orchid-like flowers. *Distr.* S China, India, Borneo. *Class* M.

Loxsomataceae A family of 2 genera and 4 species of medium-sized to large terrestrial ferns with bipinnate to tripinnate leaves. The sori are borne on the margins of the leaves at the ends of veins. *Distr.* Central and S America, New Zealand. *Class* F.

loyak *Licuala.*

loyar *Licuala.*

lucens: shiny, polished.

lucerne *Medicago sativa.*

lucidus, -a, -um: clear, transparent, shining.

Luculia (From *lukuli swa*, the local vernacular name for *Luculia gratissima*.) A genus of 5 species of evergreen shrubs with simple, opposite leaves and corymbs or panicles of fragrant, salver-shaped, 5-lobed flowers. All the species are grown as tender ornamentals. *Distr.* Himalaya and China (Yunnan). *Fam.* Rubiaceae. *Class* D.

L. gratissima Shrub to 5m high. Flowers pink. *Distr.* Himalaya, China.

Luetkea (After Count F. P. Luetke, early 19th-century Russian naval explorer.) A genus of 1 species of tufted evergreen shrub with rigid, pointed leaves and racemes of small white flowers. *Distr.* W North America. *Fam.* Rosaceae. *Class* D.

L. pectinata Grown as ornamental ground cover in the rock garden.

Luffa (From Arabic *louff*, the name used for *Luffa cylindrica*.) A genus of 6 species of annual climbing or trailing herbs with branched tendrils and palmately lobed leaves. The yellow-white, bell-shaped flowers are unisexual and sometimes borne on separate plants. The fruit is marrow-like but dry when ripe. *Distr.* Throughout the tropics. *Fam.* Cucurbitaceae. *Class* D.

L. acutangula ANGLED LOOFAH. Fruits eaten while still young. *Distr.* Asia; widely cultivated throughout the tropics.

L. aegyptiaca LOOFAH, VEGETABLE SPONGE. Bleached vascular system of mature fruit used as a sponge, the loofah of bathrooms. *Distr.* Tropical regions of the Old World.

L. cylindrica See *L. aegyptiaca*

Luma (From Chilean vernacular name for these plants.) A genus of 4 species of shrubs and small trees with simple, somewhat leathery leaves and small clusters of regular, bisexual flowers that bear 4 petals and numerous stamens. *Distr.* Chile, Argentina. *Fam.* Myrtaceae. *Class* D.

L. apiculata ARRAYAN, PALO COLORADO, TEMU, COLLIMAMOL. Large shrub or small tree to 10m high. Stems red. Leaves dark green. Grown as an ornamental. *Distr.* Argentina, Chile.

luma *Amomyrtus luma.*

Lunaria (From Latin *luna*, the moon, referring to the shape of the pods.) A genus of 3 species of biennial to perennial herbs with large purple, rarely white flowers and round, white, papery seed pods. Cultivated as ornamentals. *Distr.* Central and SE Europe. *Fam.* Cruciferae. *Class* D.

L. annua HONESTY, PENNY FLOWER, SILVER DOLLAR. A tall annual with purple-red flowers and flat, more or less round pods. Cultivated as an ornamental, the dried seed-heads being used for winter decoration; the young roots are allegedly edible. *Distr.* SE Europe, naturalized in N Europe and North America.

L. biennis See *L. annua*

L. rediviva PERENNIAL HONESTY. Similar to *L. annua* but perennial with fragrant flowers. *Distr.* Europe.

lungwort *Pulmonaria*, *P. officinalis* **sea** ~ *Mertensia maritima*.

lunulatus, -a, -um: crescent-shaped.

lupin *Lupinus* **Carolina** ~ *Thermopsis villosa* **false** ~ *T. macrophylla* **tree** ~ *Lupinus arboreus* **white** ~ *L. albus*.

lupinoides: from the genus *Lupinus*, LUPIN, with the ending *-oides*, indicating resemblance.

Lupinus (From Latin *lupus*, wolf, alluding to the belief that the plants reduce soil fertility.) LUPIN. A genus of 200 species of annual and perennial herbs and shrubs with alternate, palmate leaves and showy racemes or spikes of pea-like flowers. Several species and numerous cultivars are grown as ornamentals and some as fodder crops. The seeds are eaten but not extensively and are currently being investigated as a possible new proteinrich food source, once the toxins can be bred out of them. *Distr.* North and South America, S Europe and N Africa. *Fam.* Leguminosae. *Class* D.

L. albus WHITE LUPIN. The most frequently eaten species although the seeds are toxic and must be cooked first. It is also grown as a fodder crop and as a green manure. *Distr.* S Balkans, Aegean.

L. arboreus TREE LUPIN. Sprawling semi-evergreen shrub. Flowers small, yellow, borne in erect spikes. Sometimes used in land reclamation projects. *Distr.* USA (California).

L. polyphyllus Erect perennial herb. Flowers blue, purple, pink or white. This species has given rise to many of the garden cultivars grown today. *Distr.* W North America.

lupulus, -a, -um: a small wolf.

luridus, -a, -um: brownish-yellow.

Luronium (Classical Greek name for an unknown plant.) A genus of 1 species of perennial, floating, aquatic herb with narrow submerged leaves, broad aerial leaves and white flowers. Grown as a cultivated ornamental. *Distr.* Europe. *Fam.* Alismataceae. *Class* M.

L. natans The only species. *Distr.* Europe.

lusitanicus, -a, -um: of Portugal.

luteiflorus, -a, -um: with yellow flowers.

luteo-alba: yellow-white.

luteolus, -a,-um: pale yellow.

luteoviridis, -e: yellow-green.

lutescens: becoming yellowish.

luteus, -a, -um: deep yellow.

luxurians: luxuriant.

Luzula (From Latin *luciola*, glow-worm, or *lux*, light, alluding to the use of these plants as lamp-wicks.) WOOD-RUSH. A genus of 80 species of annual and perennial herbs with grass-like leaves and panicles of small, brown-green or white flowers. *Distr.* Cosmopolitan, especially temperate Eurasia (31 species in Europe). *Fam.* Juncaceae. *Class* M.

L. campestris FIELD WOODRUSH. Loosely tufted, to 30cm high. Seeds ant-dispersed with juicy outgrowths. *Distr.* Europe.

L. maxima See *L. sylvatica*

L. nivea SNOW RUSH. Flowers in loose panicles, off-white. *Distr.* W and Central Europe.

L. sylvatica Large, tussock-forming. Leaves channelled, silky-hairy. Several cultivars are available. *Distr.* Europe, Caucasus, W Asia.

Luzuriaga (After Don Ignatio M. R. de Luzuriaga, 19th-century Spanish botanist.) A genus of 2–3 species of trailing and climbing, perennial herbs with simple leaves on twisted stalks and large, solitary, white flowers. *Distr.* Peru, Chile, Falkland Islands, New Zealand. *Fam.* Philesiaceae. *Class* M.

L. radicans Stems slender, climbing. Flowers star-shaped, fragrant. Grown as a half-hardy ornamental. *Distr.* Chile, Peru.

lychee *Litchi chinensis*, Sapindaceae.

Lychnis (From Greek *lychnos*, lamp, alluding to the use of the leaves of *Lychnis coronaria* as lamp-wicks.) ROSE CAMPION, CATCHFLY.

A genus of about 35 species of typically perennial herbs with simple, opposite leaves and regular flowers that bear 5 sepals fused into a tube and 5 petals that have narrow claws and blades that are notched or deeply cleft into two. This genus is closely related to the genus *Silene* but differs in having a 5-branched stigma and fruit capsules that open into 10 teeth. Several species are grown as ornamentals. *Distr.* N temperate and arctic regions. *Fam.* Caryophyllaceae. *Class* D.

L. alpina ALPINE CAMPION. Tuft-forming perennial. Flowers deep pink, borne in rounded heads. Petals frilled. *Distr.* Subarctic regions, mountains of the N hemisphere.

L. chalcedonica MALTESE CROSS, JERUSALEM CROSS. Erect, rough-hairy perennial. Flowers vivid scarlet, borne in flat heads. *Distr.* E European.

L. coronaria ROSE CAMPION, MULLEIN PINK, DUSTY MILLER. Leaves densely woolly. Flowers purple. *Distr.* SE Europe, Central Asia, widely naturalized.

L. dioica See *Silene dioica*

L. flos-cuculi RAGGED ROBIN. Flowers large, rose-red. Petals deeply divided and contorted. *Distr.* Europe, Caucasus, Siberia, naturalized in North America.

L. flos-jovis Flowers bright pink in rounded clusters. *Distr.* Central Alps.

L. lagascae See *Petrocoptis glaucifolia*

L. viscaria GERMAN CATCHFLY. Stems sticky. Flowers purple-red. *Distr.* Europe, W Asia.

Lycianthes (From the genus name *Lycium*, and Greek *anthe*, flower.) A genus of 200 species of shrubs and lianas with simple leaves and tubular flowers that bear 5 spreading lobes. *Distr.* Tropical America, E Asia. *Fam.* Solanaceae. *Class* D.

L. rantonnetii BLUE POTATO BUSH. Shrub to 2m high. Flowers dark blue to violet, often tinged yellow within. Grown as a tender ornamental. *Distr.* Argentina, Paraguay.

Lycium (From Greek *lykion*, after Lycia, Turkey, originally applied to a species of *Rhamnus*.) BOXTHORN, MATRIMONY VINE. A genus of 100 species of deciduous, often thorny shrubs with small, narrow, alternate leaves and small, 5-lobed, funnel-shaped flowers that are followed by showy red berries. Several species are grown as ornamentals, sometimes being used for hedging. *Distr.* Widespread in warm temperate and tropical regions, especially America. *Fam.* Solanaceae. *Class* D.

L. barbarum BOX THORN, DUKE OF ARGYLL'S TEA TREE, COMMON MATRIMONY VINE. Upright or spreading, spiny shrub to 4m high. Flowers dull lilac. *Distr.* SE Europe to China.

lycius, -a, -um: of Lycia, Asia Minor.

Lycopersicon (From Greek *lykos*, wolf, and *persikon*, peach, suggesting that the fruit is inferior to a peach.) A genus of 7 species of annual herbs with coarsely toothed, pinnate leaves and cymes of 5-lobed, yellow flowers. *Distr.* W South America, Galapagos. *Fam.* Solanaceae. *Class* D.

L. esculentum TOMATO, LOVE APPLE. Fruit edible. Regarded as a curiosity in Europe until the end of the 19th century; now grown worldwide as a vegetable. Many cultivars are available with differing sizes, colours, and textures of fruit. *Distr.* Derived from a form found in Peru and Ecuador. Introduced to Europe from Mexico in the 16th century.

Lycopodiaceae A family of 4 genera and about 500 species of terrestrial or epiphytic, fern-like herbs or climbers with small, simple leaves typically arranged in whorls or spirals. The sporangia are borne in the axils of the leaves and may be aggregated into strobili. *Distr.* Widespread. *Class* F.
See: Lycopodium.

lycopodiodes: from the genus *Lycopodium*, with the ending *-oides*, indicating resemblance.

Lycopodium (From Greek *lykos*, wolf, and *podus*, foot, alluding to the rhizomes which resemble a wolf's paws.) CLUBMOSS. A genus of about 450 species of terrestrial and epiphytic, fern-like herbs with simple, crowded, scale-like leaves. Several species are grown as ornamentals and some are a source of fibre. The spores (Lycopodium powder) are utilized in sound experiments in physics. *Distr.* Tropical and temperate regions. *Fam.* Lycopodiaceae. *Class* F.

L. clavatum GROUND PINE, RUNNING PINE. The spores are very flammable and were formerly used in fireworks. *Distr.* N temperate regions.

Lycopsis See *Anchusa*

Lycopus (From Greek *lykos*, wolf, and *pous*, foot, alluding to the rhizomes.) A genus of 4 species of rhizomatous herbs with simple, opposite leaves and dense whorls of 2-lipped flowers. *Distr.* N temperate regions, Australia. *Fam.* Labiatae. *Class* D.

L. europaeus GIPSYWORT, WATER HOREHOUND. Glabrous perennial of damp areas. Flowers pink or white. Roots formerly used to stain skin brown. *Distr.* Eurasia, naturalized in North America.

Lycoris (After Lycoris, a beautiful Roman actress and mistress of Mark Antony.) A genus of 11 species of bulbiferous perennial herbs with linear leaves and an umbel of regular or irregular flowers that bear 6 spreading tepals. Several species are cultivated as tender ornamentals. *Distr.* China and Japan to Burma. *Fam.* Amaryllidaceae. *Class* M.

L. albiflora Leaves narrow. Flowers white. *Distr.* Japan.

L. aurea GOLDEN HURRICANE LILY, GOLDEN SPIDER LILY. Leaves fleshy. Flowers golden-yellow. *Distr.* China, Japan.

L. radiata SPIDER LILY, RED SPIDER LILY. Flowers nodding, pink to red. *Distr.* Japan.

L. sanguinea Flowers dull red. *Distr.* China, Japan.

L. sprengeri Flowers many, erect, dull red. *Distr.* Japan.

L. squamigera MAGIC LILY, RESURRECTION LILY. Flowers nodding, fragrant, rose-pink. *Distr.* Japan.

lydius, -a, -um: of Lydia, W Turkey.

Lygodium (From Greek *lygodes*, twining or flexible, alluding to the climbing habit.) CLIMBING FERN. A genus of about 40 species of climbing ferns with clasping fronds. The stems are used for weaving and yarn. *Distr.* Tropical and warm regions. *Fam.* Schizaeaceae. *Class* F.

L. microphyllum CLIMBING MAIDENHAIR FERN, SNAKE FERN. Clasping leaves to 3m long. *Distr.* Tropical and subtropical Africa, Asia and Australia, naturalized in Jamaica.

L. palmatum Leaves deeply palmately lobed. *Distr.* E USA.

Lyonia (After John Lyon (*c.*1765–1814), Scottish gardener who travelled in the SE USA and introduced many plants into cultivation in Great Britain.) A genus of 35 species of deciduous or evergreen shrubs or small trees with densely packed tubular flowers. Some species are grown as ornamentals. *Distr.* E and SE Asia, E North America, Mexico, and the West Indies. *Fam.* Ericaceae. *Class* D.

L. ligustrina MALE BERRY, HE HUCKLEBERRY, MALE BLUEBERRY, BIG BOY. Deciduous shrub. *Distr.* E North America.

Lyonothamnus (After W. S. Lyon (1851–1916), American who collected it in 1884, and Greek *thamnos*, shrub.) A genus of 1 species of evergreen tree with peeling bark, simple or pinnately lobed leaves and panicles of numerous, small, white flowers. *Distr.* Islands off S Carolina. *Fam.* Rosaceae. *Class* D.

L. floribundus CATALINA IRONWOOD. Grown as a half-hardy ornamental.

lyratus, -a, -um: lyre- or fiddle-shaped.

Lysichiton (From Greek *lysis*, releasing, and *chiton*, cloak, alluding to the shedding of the large spathe.) SKUNK CABBAGE. A genus of 2 species of clump-forming, rhizomatous herbs with soft, bright green leaves that smell musky when bruised. The small flowers are borne on a cylindrical spadix that is subtended by a large, yellow or white spathe. Both species are grown as bog garden plants. *Distr.* N Pacific. *Fam.* Araceae. *Class* M.

L. americanus SKUNK CABBAGE. Spathe bright yellow. The leaves are eaten locally and are reported to have medicinal qualities. *Distr.* W North America.

L. camtschatcensis Spathe white. *Distr.* NE Asia.

Lysimachia (From Greek *lysis*, releasing, and *mache*, strife; King Lysimachos of Thrace (*c.*360–281 BC) was said to have pacified a bull with some LOOSESTRIFE.) LOOSESTRIFE. A genus of about 150 species of perennial herbs and shrubs with simple leaves and 5-lobed flowers. Some species are grown as ornamentals, others have local medicinal uses. *Distr.* Temperate and warm regions. *Fam.* Primulaceae. *Class* D.

L. nemorum YELLOW PIMPERNEL. Evergreen perennial. Flowers yellow. Formerly used medicinally. *Distr.* Europe, Caucasus.

L. nummularia CREEPING JENNY, MONEYWORT. Evergreen perennial. Leaves round.

Flowers yellow. *Distr.* Europe, Caucasus, naturalized in E North America.

L. salicaria PURPLE LOOSESTRIFE. Hairy perennial to 1.2m high. Flowers purple, borne in whorls. *Distr.* Eurasia, N Africa, naturalized in North America.

L. vulgaris YELLOW LOOSESTRIFE. Hairy stoloniferous perennial. Flowers yellow. Grown as an ornamental. *Distr.* Europe, Asia.

lysistemon: with loose stamens.

Lythraceae The Henna family. 30 genera and about 500 species of herbs, shrubs and trees with simple, entire leaves and typically regular, bisexual flowers that bear 4 sepals, 4 petals, and 8 stamens. Several species are sources of dye, notably *Lawsonia inermis*, HENNA. *Distr.* Widespread. *Class* D.
See: Cuphea, Heimia, Lagerstroemia, Lawsonia, Lythrum, Punica.

Lythrum (From Greek *lythron*, blood, alluding to the colour of the flowers.) LOOSESTRIFE. A genus of 38 species of annual and perennial herbs and subshrubs with 4-angled stems, simple leaves, and tubular flowers that bear 4–8 petals. *Distr.* Cosmopolitan, except South America. *Fam.* Lythraceae. *Class* D.

L. salicaria PURPLE LOOSESTRIFE, SPIKED LOOSESTRIFE. Erect, hairy, perennial herb. Leaves lance-shaped. Flowers pink-purple, borne in dense spikes. Several ornamental cultivars have been raised from this species. *Distr.* Eurasia, naturalized in North America.

M

Maackia (After Richard Maack (1825–86), Russian naturalist.) A genus of 8 species of deciduous shrubs and trees with pinnate leaves and dense panicles of white, pea-like flowers. *Distr.* E Asia. *Fam.* Leguminosae. *Class* D.

M. amurensis Grown as an ornamental. *Distr.* NE Asia.

Macadamia (After John Macadam (1827–65) secretary of the Philosophical Institute of Victoria, Australia.) A genus of 6 species of shrubs and trees with simple, whorled leaves and racemes of small flowers that are followed by hard, 1- or 2-seeded fruits. Several species are grown as ornamentals or for their edible seeds. *Distr.* E Australia, Sulawesi. *Fam.* Proteaceae. *Class* D.

M. integrifolia MACADAMIA NUT. Tree to 20m high. Flowers creamy. The seeds of this species are the Macadamia nuts of commerce. *Distr.* E Australia.

M. ternifolia MAROOCHIE NUT. Bushy deciduous tree to 20m high. *Distr.* Australia (New South Wales, Queensland).

M. tetraphylla QUEENSLAND NUT. Bushy tree to 16m high. Flowers tinged purple. Seeds eaten locally. *Distr.* E Australia.

macaronesicus, -a, -um: of Macaronesia.

macaw fat *Elaeis guineensis*.

mace *Myristica fragrans* **reed** ~ *Typha* **sweet** ~ *Tagetes lucida*.

macedonicus, -a, -um: of Macedonia, Balkan Peninsula.

Macfadyena (After James Macfadyen (1789–1850), Scottish botanist and author of *Flora Jamaica* (1837).) A genus of 3–4 species of woody climbers with leaves that consist of 2 leaflets and a terminal, 3-branched tendril. The yellow, trumpet-shaped flowers are borne in axillary cymes and are followed by long narrow capsules. *Distr.* Tropical America. *Fam.* Bignoniaceae. *Class* D.

M. unguis-cati CAT'S CLAW. Flowers yellow, trumpet-shaped. Fruit pods to 30cm long. Grown as a half-hardy ornamental. *Distr.* Mexico to Argentina.

Machaeranthera (From Greek *machaira*, sword, and *anthera*, anther, alluding to the shape of the anthers.) A genus of 26 species of annual to perennial herbs with alternate, often spiny leaves and flowers in daisy-like heads which typically bear distinct rays. *Distr.* W North America. *Fam.* Compositae. *Class* D.

M. bigelovii (After J. M. Bigelow (1804–78), American botanist.) Biennial to perennial herb to 35cm high. *Distr.* W USA.

M. pattersonii See *M. bigelovii*

Machairophyllum (From Greek *machaira*, sword, and *phyllon*, leaf.) A genus of 9 species of cushion-forming herbs with linear, pointed leaves and solitary flowers that bear numerous petals. *Distr.* South Africa (Cape Province). *Fam.* Aizoaceae. *Class* D.

M. acuminatum Flowers golden yellow, opening at night. Grown as an ornamental.

Machilus See *Persea*

Mackaya (After James Thomas Mackay (1775–1862), Scottish botanist and gardener.) A genus of 1 species of evergreen shrub with simple, short-stalked leaves and spikes of pale violet, funnel-shaped flowers. *Distr.* South Africa. *Fam.* Acanthaceae. *Class* D.

M. bella Grown as a tender ornamental. *Distr.* South Africa.

Macleaya (After Alexander Macleay (1767–1848), British civil servant in Australia.) A genus of 2 species of large rhizomatous herbs with palmately lobed leaves and spreading racemes of numerous small flowers that bear many stamens but lack petals. *Distr.* Temperate E Asia. *Fam.* Papaveraceae. *Class* D.

M. cordata PLUME POPPY, TREE CELANDINE. Perennial herb to 2.5m high. Flowers cream, borne in racemes to 1m long. Grown as a hardy ornamental. *Distr.* China, Japan.

Maclura (After William Maclure (1763–1840), American geologist) A genus of 1 species of deciduous tree with simple, entire leaves and heads of small, unisexual, wind-pollinated flowers, the males and females being borne on separate plants. The fruit is of similar form to the mulberry, being a number of fleshy drupes fused into a syncarp, but it is covered with a leathery coat that makes it resemble an orange. It is inedible. *Distr.* E USA. *Fam.* Moraceae. *Class* D.

M. pomifera OSAGE ORANGE. Deciduous tree. Fruit orange. Grown as a living fence and as an ornamental as well as being a source of timber and yellow dye. *Distr.* S and Central USA.

Macodes A genus of 10 species of evergreen terrestrial orchids with prostrate to ascending stems and racemes of numerous small, white flowers. Some species are grown as tender ornamentals. *Distr.* Malaysia to New Guinea. *Fam.* Orchidaceae. *Class* M.

M. petola Leaves in loose rosettes. Inflorescence erect. *Distr.* Malaysia, Philippines to Sumatra.

macqui *Aristotelia chilensis.*

macrantherus, -a, -um: with large anthers.

macranthus, -a, -um: large-flowered.

macrocarpon: with large fruit.

macrocarpus, -a, -um: with large fruit.

macrocephalus, -a, -um: large-headed.

macrodontus, -a, -um: with large teeth.

macroglossus: large-tongued.

macrolepis: with large scales.

macropetalus, -a, -um: with large petals.

macrophyllus, -a, -um: large-leaved.

Macropidia (From *macropus*, the scientific name for the kangaroo). A genus of 1 species of perennial herb to 50cm high with sword-shaped leaves and flowers in branched panicles. *Distr.* W Australia. *Fam.* Haemodoraceae. *Class* M.

M. fuliginosa Occasionally grown as an ornamental.

Macropiper (From Greek *macro*, large or long, and the genus name *Piper*.) A genus of 9 species of shrubs and small trees with simple alternate leaves and spikes of minute unisexual flowers . *Distr.* S Pacific region. *Fam.* Piperaceae. *Class* D.

M. crocatum See *Piper ornatum*

M. excelsum PEPPER TREE. Aromatic shrub or small tree to 6m high. Locally important as a medicine and aphrodisiac. *Distr.* S Pacific region.

macropodus, -a, -um: with a large stalk.

macrorrhizus, -a, -um: with a large root.

macrosepalus, -a, -um: with large sepals.

macrosiphon: with a large tube.

macrostachys: with a large spike.

macrostigmus, -a, -um: with a large stigma.

macrostylus, -a, -um: with a long style.

Macrothelypteris (From Greek *makros*, large, *thylys*, female, and *pteris*, fern.) A genus of 9 species of large terrestrial ferns with tripinnately divided leaves. Several species are grown as ornamentals. *Distr.* SE Asia, Polynesia, Australasia, Tropical America. *Fam.* Thelypteridaceae. *Class* F.

M. torresiana Leaves to 2m long, steeply ascending, finely divided. *Distr.* SE Asia to Japan and Polynesia.

Macrozamia (From Greek *makros*, large, and the genus name *Zamia*.) A genus of 14 species of tree-like cycads with whorls of large pinnate leaves. The seeds are edible if soaked and pounded or baked (Queensland nut). *Distr.* Australia. *Fam.* Zamiaceae. *Class* G.

M. communis BURRAWONG. Grown as an ornamental. *Distr.* SE Australia (New South Wales).

maculatus, -a, -um: spotted.

maculiferus, -a, -um: bearing spots.

maculosus, -a, -um: highly spotted.

madagascariensis, -e: of Madagascar.

madder *Rubia tinctorum* **field** ~ *Sherardia arvensis* **Levant** ~ *Rubia peregrina* **stinking** ~ *Putoria calabrica* **white** ~ *Galium mollugo* **wild** ~ *Rubia peregrina*.

Madeira vine Basellaceae.

maden *Weinmannia trichosperma*.

maderensis, -e: of Madeira.

madrona *Arbutus*, *A. menziesii*.

madrone *A. menziesii*.

madwort *Alyssum*.

magdalenensis, -e: of the Rio Magdalena, Colombia.

magellanicus, -a, -um: of the region of the Magellan Straits.

magellensis, -e: of the Magellan Straits.

magic flower *Cantua buxifolia*.

magnificus, -a, -um: splendid.

magniflorus, -a, -um: with splendid flowers.

Magnolia (After Pierre Magnol (1638–1715), professor of botany at Montpellier.) A genus of 80 species of deciduous and evergreen shrubs and trees with simple, leathery leaves and large, terminal, regular flowers that bear 6–33 petal-like tepals and numerous stamens. Many species are grown as ornamentals for their graceful white, pink or purple flowers and some are a source of timber. *Distr.* Temperate and tropical regions of E Asia and America. *Fam.* Magnoliaceae. *Class* D.

M. acuminata CUCUMBER TREE. Deciduous conical tree. Flowers cup-shaped, green-blue. Timber used for flooring. *Distr.* SE USA.

M. ashei See *Magnolia macrophylla*

M. campbellii (After Dr Archibald Campbell (1805–74), Superintendent of Darjeeling.) Deciduous tree. Flowers large, pink, slightly scented, borne before leaves appear in spring. *Distr.* Himalaya to W China.

M. cordata See *Magnolia acuminata*

M. denudata LILY TREE, YULAN. Deciduous spreading tree. Flowers white, open, bell-shaped, appearing before the leaves in spring. *Distr.* E China.

M. fraseri (After John Fraser (1750–1811), who introduced it to England.) Deciduous spreading tree. Flowers pale-yellow turning milk-white, scented. *Distr.* SE USA.

M. grandiflora BULL BAY, LOB-LOLLY MAGNOLIA. Evergreen, conical or rounded tree. Flowers bowl-shaped, fragrant, cream-white. Often grown against walls in Great Britain. *Distr.* SE USA.

M. heptapeta See *Magnolia denudata*

M. kobus (From *kobushi*, the Japanese name.) Deciduous, broadly conical tree. Flowers numerous, fragrant, white. A local source of timber. *Distr.* Japan.

M. macrophylla Deciduous rounded tree. Leaves very large. Flowers fragrant, white, somewhat hidden by leaves. *Distr.* SE USA.

M. officinalis Deciduous tree. Flowers strongly fragrant, creamy white. The bark and flower buds are used medicinally in China. *Distr.* China.

M. stellata STAR MAGNOLIA. Deciduous shrub or small tree. Flowers profuse, star-shaped, white, borne before the leaves in spring. *Distr.* Japan.

M. tripetala UMBRELLA TREE. Deciduous spreading tree. Flowers cream-white, smelling unpleasant. *Distr.* E North America.

M. virginiana SWEET BAY. Semi-evergreen or deciduous shrub. Flowers cup-shaped, very fragrant, cream-white. A source of narrow timber for tool handles. *Distr.* E USA.

magnolia: loblolly ~ *Magnolia grandiflora* **star** ~ *M. stellata*.

Magnoliaceae A family of 7 genera and 200 species of shrubs and trees with simple alternate leaves and regular flowers that bear 2–3 whorls of petal-like tepals and numerous spirally arranged stamens. This is one of the most primitive of the flowering plant families. Some species are valued as ornamentals. *Distr.* Widespread, tropical and warm-temperate regions, but not in Africa. *Class* D. *See: Liriodendron*, *Magnolia*, *Manglietia*, *Michelia*.

magnoliifolius, -a, -um: from the genus *Magnolia*, and *folius*, leaved.

magnus, -a, -um: great, large.

maguey *Agave*, *A. americana* **Manila** ~ *A. cantala*.

×*Mahoberberis* (From the names of the parent genera.) A genus of species of hybrids between members of the genera *Mahonia* and *Berberis*. *Distr*. Garden origin. *Fam*. Berberidaceae. *Class* D.

×*M. aquisargentii* Evergreen shrub to 2m high. Leaves variable, sharp-spined, tinged red-purple in winter. Flowers rarely occur Distr.

×*M. miethkeana* Evergreen shrub to 2m high. Leaves variable, sharp spined, copper coloured in winter. Flowers pale yellow Distr.

×*M. neubertii* An open evergreen shrub. Leaves simple or compound Distr.

mahogany Meliaceae, *Swietenia mahogoni*.

Mahonia (After Bernard McMahon (born 1816), American nurseryman.) A genus of 70 species of evergreen shrubs with pinnate, thorny leaves and clustered racemes of yellow, scented flowers. Many species and hybrids are grown as ornamentals. *Distr*. Asia, North and Central America. *Fam*. Berberidaceae. *Class* D.

M. acanthifolia See *M. napaulensis*

M. aquifolium OREGON GRAPE. To 2m high. Leaves with 5–13 leaflets, turning red in winter. Flowers profuse, golden-yellow. Berries cooked by North American Indians. *Distr*. W North America.

M. bealei (After T. C. Beale (1774–1856) English botanist who lived in Macao. To 2m high. Leaves with 13–17 leaflets, dull-green above, yellow-green below. Flowers appearing in late winter, strongly-scented. *Distr*. China.

M. japonica To 2m high. Leaves with 9–19 leaflets, red in winter, tinged yellow beneath. Flowers highly fragrant, on pendent racemes. *Distr*. China.

M. lomariifolia To 3m high. Leaves large with 19–37 leaflets. Flowers highly fragrant, in erect racemes. *Distr*. Burma W China.

M.* × *media A hybrid of *M. japonica* × *M. lomariifolia* that has given rise to a number of ornamental cultivars. *Distr*. China, long cultivated in Japan.

M. napaulensis To 3m high. Leaves with up to 11 leaflets. Flowers pale yellow, borne in erect racemes. *Distr*. Nepal.

M. repens HOLLY GRAPE. To 50cm high. Leaves with 5 leaflets, blue-green. Flowers densely clustered, deep yellow. Fruit used to make preserves and drinks. *Distr*. W North America.

Maianthemum (From Greek *maios*, May, and *anthemon*, blossom, alluding to its flowering time.) MAY LILY. A genus of 3 species of creeping, rhizomatous herbs with broad, glossy, clasping leaves and racemes of small white flowers. *Distr*. N temperate region. *Fam*. Convallariaceae. *Class* M.

M. bifolium MAY LILY, FALSE LILY OF THE VALLEY. Grown as ground cover in shaded and woodland gardens. *Distr*. W Europe to Japan.

M. dilatatum See *M. bifolium*

M. kamtschaticum See *M. bifolium*

maidenhair tree *Ginkgo biloba*.

Maihuenia (From local name, *maihuen*.) A genus of 5 species of low, tuft-forming shrubby cacti with segmented stems, small, cylindrical leaves and spines. The yellow, solitary flowers are borne terminally and are followed by fleshy fruits. Several species are grown as ornamentals. *Distr*. Patagonia. *Fam*. Cactaceae. *Class* D.

M. poeppigii Dwarf shrub forming a low mound. Spines white. *Distr*. S Chile.

maikoa *Brugmansia arborea*.

maize *Zea mays* **water** ~ *Victoria amazonica*.

majalis, -e: of the month of May.

major: greater, larger.

majus, -a, -um: larger.

malacoides: soft, mallow-like.

malacophyllus, -a, -um: mallow-leaved.

malanga *Xanthosoma*.

malawiensis, -e: of Malawi.

Malaxis (From Greek *malakis*, softening, alluding to the texture of the leaves.) ADDER'S MOUTH. A genus of 300 species of terrestrial or epiphytic orchids with creeping rhizomes and typically deciduous leaves. The flowers are small, dull-coloured, and occur in dense many-flowered racemes. *Distr*. Temperate and tropical regions. *Fam*. Orchidaceae. *Class* M.

M. discolor Leaves rich purple. *Distr*. Sri Lanka.

M. metallica Leaves metallic red-purple. *Distr*. Borneo.

Malcolmia (After William Malcolm (died 1798) and his son, London nursery-men.) A genus of 35 species of annual to perennial herbs with small, white to purple or red flowers. *Distr*. Greece, Albania. *Fam*. Cruciferae. *Class* D.

M. maritima VIRGINIAN STOCK. Annual with reddish to white flowers. *Distr*. Greece, Albania.

Malephora (From Greek *male*, armhole, and *phorein*, to bear.) A genus of 15 species of erect or creeping shrubs with 4-angled, succulent leaves and yellow or pink flowers that bear numerous petals. Several species are grown as ornamentals. *Distr*. S and SW Africa. *Fam*. Aizoaceae. *Class* D.

M. crocea Leaves crowded, pale green. Flowers yellow inside, tinged red outside. *Distr*. South Africa (Cape Province).

M. lutea Erect shrub. Leaves yellow-green. Flowers yellow. *Distr*. South Africa (Cape Province).

Malesherbiaceae A family of 1 genus and 27 species of herbs and undershrubs with simple, alternate, frequently hairy leaves and regular bisexual flowers that have their parts in fives. *Distr*. Peru, Chile and Argentina. *Class* D.

maliformis, -e: malformed.

mallow Malvaceae, *Malva* **checker** ~ *Sidalcea* **common** ~ *Malva neglecta* **common rose** ~ *Hibiscus moscheutos* **desert** ~ *Sphaeralcea ambigua* **false** ~ *Sphaeralcea* **fringed poppy** ~ *Callirhoe digitata* **giant** ~ *Hibiscus* **globe** ~ *Sphaeralcea* **hairy** ~ *Anisodontea scabrosa* **halberd-leaved marsh** ~ *Hibiscus militaris* **halberd-leaved rose** ~ *H. militaris* **high** ~ *Malva sylvestris* **Indian** ~ *Abutilon theophrasti* **Jew's** ~ *Kerria japonica* **marsh** ~ *Althaea officinalis* **musk** ~ *Abelmoschus moschatus*, *Malva moschata* **poppy** ~ *Callirhoe* **prairie** ~ *Sidalcea* **purple poppy** ~ *Callirhoe involucrata* **rose** ~ *Hibiscus* **soldier rose** ~ *H. militaris* **swamp rose** ~ *H. moscheutos* **tall** ~ *Malva sylvestris* **tree** ~ *Lavatera arborea* **Virginia** ~ *Sida hermaphrodita* **wax** ~ *Malvaviscus arboreus* **white** ~ *Althaea officinalis* **white prairie** ~ *Sidalcea candida*.

Malope (Greek name for a MALLOW.) A genus of 3 species of annual and perennial herbs with simple or palmately lobed leaves and solitary, white, pink, or violet flowers that bear 5 petals and numerous fused stamens. *Distr*. Spain, N Africa. *Fam*. Malvaceae. *Class* D.

M. trifida Annual herb. Flowers deep purple-red. Grown as an ornamental. *Distr*. Spain, N Africa.

Malpighia (After Marcello Malpighi (1628–74), Italian naturalist and professor at Bologna) A genus of 40 species of shrubs and small trees with simple opposite leaves (that sometimes bear stinging hairs) and umbels of irregular flowers that bear their parts in fives. Several species are grown as ornamentals. *Distr*. Tropical and warm regions, especially South America. *Fam*. Malpighiaceae. *Class* D.

M. coccigera MINIATURE HOLLY, SINGAPORE HOLLY. Bushy shrub to 1m high. Leaves leathery with spiny margins. Flowers pale blue or pink. *Distr*. W Indies.

Malpighiaceae A family of 65 genera and about 1000 species of climbers, shrubs, and trees with simple, opposite leaves and racemes of regular or irregular, showy flowers that bear their parts in fives. *Distr*. Tropical and subtropical regions, especially New World. *Class* D.
See: Malpighia.

Maltese cross *Lychnis chalcedonica*.

Malus (Classical Latin name for an apple.) APPLE. A genus of 25 species of deciduous trees with simple, usually toothed or occasionally lobed leaves, and umbels of pink, white, or purple flowers that bear 5 petals and numerous stamens. The fruit is technically termed a pome. The core of the apple represents the true fruit and the fleshy, edible part is a swollen receptacle. See also *Fragaria*, STRAWBERRY. As well as being one of the major fruit crops of temperate regions, apples are also grown as ornamentals with a large number of cultivars having been raised. *Distr.* N temperate regions. *Fam.* Rosaceae. *Class* D.

M. baccata SIBERIAN CRAB APPLE. Round-headed tree to 20m high. Flowers white. Pome small, yellow to red. *Distr.* NE Asia.

M. coronaria Spreading tree to 7m high. Flowers pink-white, fragrant. Pome flattish, globose, yellowish-green, harsh and acidic tasting. Used in garlands. *Distr.* E North America.

M. domestica See *M. pumila*

M. floribunda Spreading tree. Flowers pink. Pome yellow, tart. *Distr.* Japan.

M. hupehensis Spreading tree. Flowers large, fragrant, white. Pomes tinged red, tart. The leaves have been used to make tea. *Distr.* China, India (Assam).

M. pumila APPLE. A very variable group of trees with numerous ornamental and commercial cultivars having been produced. This species is the main source of commercial edible apples. The pomes are eaten, pressed for juice, and fermented for cider and spirits such as Calvados. *Distr.* W Asia.

M. sylvestris EUROPEAN CRAB APPLE. Thorny tree or shrub. Flowers pink-white. Pome yellow flushed with red, tart. Fruit edible when made into a preservc, used in breeding and as a rootstock for *M. pumila*. *Distr.* Europe.

Malva (Classical Latin name for MALLOW, possibly derived from the Greek *malachos*, soothing, alluding to its medicinal properties.) MALLOW. A genus of 30 species of annual and perennial herbs and subshrubs with alternate, rounded leaves and regular, bisexual flowers that bear 5 petals and numerous fused stamens. The fruit is a 1-seeded, disc-shaped capsule. *Distr.* Mediterranean and temperate Europe. *Fam.* Malvaceae. *Class* D.

M. alcea (From Greek *alkaia*, a sort of mallow.) Perennial to 1.5m high. Flowers pale rose-purple. *Distr.* Europe.

M. crispa See *M. verticillata*

M. moschata MUSK MALLOW. Perennial to 1m high. Flowers rose-pink, clustered. *Distr.* Europe, N Africa.

M. neglecta COMMON MALLOW. Annual to 50cm high. Flowers small, pink or white. *Distr.* Europe, N Africa, SW Asia, naturalized in North America.

M. sylvestris TALL MALLOW, HIGH MALLOW, CHEESES. Erect perennial herb to 1m high. Flowers borne in clusters, mauve, pale in the centre. Fruits edible. *Distr.* Europe, N Africa, Asia.

M. verticillata Annual or biennial to 2m high. Flowers in dense clusters, white or purple. *Distr.* Europe, W Asia, naturalized in North America.

Malvaceae The Mallow family. 199 genera and about 2000 species of herbs, shrubs, and trees with alternate, simple or lobed leaves and regular, bisexual flowers that bear 5 sepals, 5 petals, and numerous fused stamens. This family includes COTTON, one of the world's most important fibre crops, as well as many ornamental species. *Distr.* Widespread. *Class* D.
See: Abelmoschus, Abutilon, Alcea, Althaea, Alyogyne, Anisodontea, Anoda, Callirhoe, Gossypium, Hibiscus, Hoheria, Kitaibela, Lagunaria, Lavatera, Malope, Malva, Malvaviscus, Modiolastrum, Nototriche, Pavonia, Phymosia, Plagianthus, Sida, Sidalcea, Sphaeralcea.

malva rosa *Lavatera assurgentiflora.*

Malvaviscus (From Latin *malva*, mallow, and *viscus* glue, alluding to the sticky pulp around the seeds.) A genus of 10–12 species of herbs, shrubs, and climbers with simple, entire to palmately lobed leaves, and red, funnel-shaped flowers that bear 5 petals and numerous fused stamens. The fruit is a berry containing many fleshy-coated seeds. *Distr.* Tropical America. *Fam.* Malvaceae. *Class* D.

M. arboreus WAX MALLOW. Grown as an ornamental. *Distr.* Mexico.

malviflorus, -a, -um: with flowers like *Malva*, MALLOW.

Mammillaria (From Latin *mammilla*, nipple, alluding to the shape of the tubercles.) A genus of 150 species of low, often tuft-forming cacti with globose to cylindrical stems that are covered in tubercles (small mounds) and spines that may be straight or hooked. The small, bell- or funnel-shaped flowers are borne on the sides of the stem and are followed by berry-like fruits that are often bright red. A popular cultivated genus. *Distr.* SW USA to West Indies and N South America. *Fam.* Cactaceae. *Class* D.

M. bocasana POWDER PUFF, SNOWBALL CACTUS. Stems clustered, small, round with long, white, silky hairs and solitary, hooked spines. Flowers pink. *Distr.* Central Mexico.

M. candida Stems clustered, small, round, almost hidden by appressed spines. Flowers pink. *Distr.* NE Mexico.

M. dioica Stems cylindrical, often clustered, woolly and bristly. Flowers cream, unisexual. *Distr.* NW Mexico, USA (California).

M. elongata GOLD LACE CACTUS. Stems erect or prostrate, clustered, densely coated in yellow or golden spines. Flowers white. *Distr.* Central Mexico.

M. geminispina Stems cylindrical to nearly round, clustered, woolly; radial spines short, white, central spines long, black-tipped. Flowers dark red. *Distr.* Central Mexico.

M. hahniana OLD WOMAN CACTUS. Stems solitary or clustered, squat-globose with dense, white, spiny hairs. Flowers small, crimson. *Distr.* Central Mexico.

M. heyderi (After Heyder (1804–84), German cactus-grower.) CORAL CACTUS. Stems round, flattened with bristle-like, white spines and solitary, longer brown spines. Flowers pink. *Distr.* Texas, N Mexico.

M. microhelia Stems cylindrical with numerous yellow or white bristle-like spines and a few longer red-brown spines. Flowers cream. *Distr.* Central Mexico.

M. parkinsonii (After John Parkinson (1772–1847), Consul-General in Mexico in 1838.) Stems round to cylindrical, branching at apex. Flowers yellow. *Distr.* Central Mexico.

M. plumosa FEATHER CACTUS. Stems small, clustered, densely covered in feathery white spines. Flowers white. *Distr.* NE Mexico.

M. prolifera LITTLE CANDLES, SILVER CLUSTER CACTUS. Stems small, round or cylindrical, growing in large clusters, larger spines with bright yellow tips. Flowers pale yellow. *Distr.* West Indies, NE Mexico, SW USA.

M. schiedeana (After Christian Schiede (died 1836), who collected in Mexico.) Stems short, round, covered by short, yellow, appressed spines that age to white. Flowers white. *Distr.* E Central Mexico.

M. sempervivi Stems solitary, squat-rounded with short spines. Flowers purple-pink. *Distr.* E Central Mexico.

M. zielmanniana ROSE PINCUSHION. Stems cylindrical, clump-forming with hooked spines. Flowers purple-pink. *Distr.* Central Mexico.

mammillaris: bearing nipples.

mammulosus, -a, -um bearing nipples.

mandarin *Citrus reticulata* **white** ~ *Streptopus amplexifolius.*

Mandevilla (After Henry John Mandeville (1773–1861), British Minister at Buenos Aires, who introduced *M. suaveolens* to cultivation.) A genus of 114 species of tuberous herbs, subshrubs, and climbers with milky sap, simple leaves, and funnel-shaped flowers. Several species are grown as ornamentals. *Distr.* Tropical America. *Fam.* Apocynaceae. *Class* D.

M. boliviensis WHITE DIPLADENIA. Slender climber. Flowers white with a yellow throat. *Distr.* Ecuador, Bolivia.

M. laxa CHILEAN JASMINE. Climber with wiry stems. Flowers white, highly fragrant. *Distr.* Bolivia, N Argentina.

M. splendens Twining climber. Flowers pink. *Distr.* Brazil.

M. suaveolens See *M. laxa*

Mandragora (Classical Greek name.) MANDRAKE. A genus of 6 species of perennial herbs with stout, fleshy taproots, simple basal leaves and solitary, 5-lobed, bell-shaped flowers. Formerly used medicinally and believed to have magical qualities. *Distr.* Mediterranean region. *Fam.* Solanaceae. *Class* D.

M. officinarum MANDRAKE, DEVIL'S APPLES. Leaves crinkled, borne in a large rosette. Fruit yellow, aromatic, poisonous. Formerly used medicinally as a narcotic and believed to have magical qualities because the root often resembled a human figure. It was believed that should the plant be pulled from the ground it would emit a scream that would deafen humans, so dogs were used to extract them. *Distr.* S Europe.

mandrake *Mandragora, M. officinarum* **American** ~ *Podophyllum peltatum.*

mandschuricus, -a, -um: of Manchuria, China.

mandschuriensis, -e: of Manchuria, China.

Manettia (After Saveria Manetti (1723–85), keeper of the botanic garden in Florence.) A genus of 80 species of evergreen, climbing herbs and subshrubs with simple, opposite leaves and showy, 4-lobed, funnel-shaped flowers. Several species are grown as tender ornamentals. *Distr.* Tropical America. *Fam.* Rubiaceae. *Class* D.

M. luteo-rubra FIRECRACKER VINE, BRAZILIAN FIRECRACKER. Twining climber. Flowers solitary, bright red below, yellow at the tip. *Distr.* N South America.

Manfreda See *Agave*

mangel *Beta vulgaris*.

Manglietia (From the Malayan vernacular name for these plants.) A genus of 25 species of evergreen shrubs and trees with simple leathery leaves and regular bisexual flowers that bear 9 petal-like tepals and numerous stamens. *Distr.* Tropical and subtropical Asia. *Fam.* Magnoliaceae. *Class* D.

M. insignis Tree to 12m high. Flowers scented, white to red. Grown as an ornamental. *Distr.* Central Himalaya to W China.

manicatus, -a, -um: with long sleeves.

manipuliflorus, -a, -um: with bundles of flowers.

manipurensis, -e: of Manipur, India.

manna plant *Tamarix gallica*.

manuka *Leptospermum scoparium*.

manzania *Arbutus*.

manzanita *Arctostaphylos* **dune** ~ *A. pumila* **Eastwood** ~ *A. glandulosa* **Fort Bragg** ~ *A. nummularia* **green** ~ *A. patula* **ione** ~ *A. myrtifolia*.

manzonita *Arbutus*.

maple Aceraceae, *Acer* **common** ~ *Acer campestre* **field** ~ *A. campestre* **flowering** ~ *Abutilon* **great** ~ *Acer pseudoplatanus* **hedge** ~ *A. campestre* **Indian** ~ *Abutilon* **Japanese** ~ *Acer palmatum* **Norway** ~ *A. platanoides* **parlour** ~ *Abutilon* **Scottish** ~ *Acer pseudoplatanus* **striped** ~ *A. saccharum* **sugar** ~ *A. saccharum*.

Maranta (After Bartolommeo Maranti, 16th-century Venetian botanist.) A genus of 30 species of rhizomatous herbs with erect or horizontal stems, obovate, slender-stalked leaves, and narrow spikes of small irregular flowers. The leaves, exhibit sleep movements, adopting different positions during day and night. *Distr.* Tropical America. *Fam.* Marantaceae. *Class* M.

M. arundinacea ARROWROOT, OBEDIENCE PLANT. Rhizome the source of arrowroot, a readily digestible starch used in the treatment of diarrhoea. *Distr.* Tropical America, widely cultivated in the tropics.

M. bicolor Leaves pale green, marked with brown spots. Grown for its showy foliage. *Distr.* Brazil, Guiana.

M. leuconeura PRAYER PLANT, TEN COMMANDMENTS. Leaves dark green with maroon or grey markings. Grown as an ornamental foliage plant. *Distr.* Brazil.

Marantaceae The Arrowroot family. 32 genera and 550 species of perennial herbs, usually with underground rhizomes or tubers and leaves in two rows. The flowers are irregular, borne in spikes and somewhat inconspicuous; they bear 3 sepals, 3 petals, 1 fertile plus 2–4 sterile, petal-like stamens and an inferior ovary. The rhizomes are a source of starch and the leaves of some species are used locally in construction. *Distr.* Tropical, especially New World. *Class* M.
See: Calathea, Ctenanthe, Maranta, Stromanthe, Thalia.

Marattia (After J.F. Marratti (died 1777), Italian botanist.) A genus of 60 species of large ferns with stout fleshy stocks and leaves to 5m long. Several species are grown as ornamentals. *Distr.* Tropical regions of the Old World to Australasia. *Fam.* Marattiaceae. *Class* F.

M. fraxinea KING FERN, PARA FERN. Leaves to 5 x 2m, leathery. The stems of some forms are edible. *Distr.* Tropical Africa, Asia to New Zealand.

Marattiaceae A family of 4 genera and about 100 species of terrestrial ferns with short stout stems and often very large, simple to

several times pinnately divided leaves. The sori are borne on the lateral veins and lack an indusium. *Distr.* Tropical and subtropical regions to Australasia. *Class* F.
See: Marattia.

marble plant *Neoregelia marmorata.*

Marcgraviaceae A family of 5 genera and 108 species of climbing shrubs with simple, leathery leaves and small bisexual flowers that bear their parts in fives. *Distr.* Tropical America. *Class* D.

mare's tall Hippuridaceae, *Hippuris vulgaris.*

marginalis, -e: edged.

marginatus, -a, -um: edged.

marguerite *Leucanthemum vulgare* **blue** ~ *Felicia, F. amelloides.*

Margyricarpus (From Greek *margarites*, pearl, and *karpos*, fruit, alluding to the white fruits.) A genus of 1 species of dwarf evergreen shrub with crowded, spine-tipped, pinnate leaves and small inconspicuous flowers. *Distr.* Andes. *Fam.* Rosaceae. *Class* D.

M. pinnatus PEARL FRUIT. Used in fertility control in Uruguay and grown as a half-hardy ornamental.

M. setosus See *M. pinnatus*

marigold *Tagetes* **African** ~ *T. erecta* **Aztec** ~ *T. erecta* **big** ~ *T. erecta* **bur** ~ *Bidens* **common** ~ *Calendula officinalis* **corn** ~ *Chrysanthemum segetum* **French** ~ *Tagetes patula* **marsh** ~ *Caltha, C. palustris* **pot** ~ *Calendula officinalis* **Scotch** ~ *C. officinalis* **sun** ~ *Dimorphotheca* **sweet-scented** ~ *Tagetes lucida* **sweet-scented Mexican** ~ *T. lucida.*

marihuana *Cannabis sativa.*

marilandicus, -a, -um: of Maryland, USA.

maritimus, -a, -um: of the sea.

marjoram: hop ~ *Origanum dictamnus* **knotted** ~ *O. marjorana* **pot** ~ *O. onites, O. vulgare* **sweet** ~ *O. marjorana* **wild** ~ *O. vulgare.*

Markhamia (After Sir Clements Robert Markham (1830–1916), British explorer and author.) A genus of 13 species of shrubs and trees with opposite, pinnate leaves and panicles of trumpet-shaped, yellow to red-brown flowers that are followed by long, strap-shaped seed pods. Some of the species are exploited for their timber and some are grown as ornamentals. *Distr.* Tropical Africa and Asia. *Fam.* Bignoniaceae. *Class* D.

M. lutea Tree to 15m high. Flowers yellow with red stripes. Fruit over 30cm long. *Distr.* Tropical Africa.

M. platycalyx See *M. lutea*

marmalade bush *Streptosolen jamesonii.*

marmoratus, -a, -um: marbled.

marmoreus, -a, -um: mottled.

maroccanus, -a, -um: of Morocco.

marrow *Cucurbita pepo.*

Marrubium (From Hebrew *marrob*, bitter.) A genus of 30 species of aromatic perennial herbs with opposite, wrinkled, toothed leaves and whorls of 2-lipped, white flowers. Several species are grown as ornamentals. *Distr.* Europe, Mediterranean regions. *Fam.* Labiatae. *Class* D.

M. candidissimum See *M. incanum*

M. incanum Stems much branched, to 50cm high. Leaves densely hairy. *Distr.* Italy, Balkans.

M. vulgare COMMON HOREHOUND, WHITE HOREHOUND. Stems erect to 50cm high. Leaves white-downy below, smelling of THYME. Grown as an ornamental and for the leaves which are used in herbal teas. *Distr.* Eurasia, Mediterranean, Macaronesia.

M. erecta See *Cionura erecta*

Marsilea (After Luigi Ferdinando Marsigli (1658–1730), Italian patron of botany.) PEPPERWORT, WATER CLOVER, NARDOO. A genus of 70 species of small aquatic or amphibious plants with clover-like, 4-lobed leaves. Several species are grown as ornamentals. *Distr.* Tropical regions. *Fam.* Marsileaceae. *Class* F.

M. drummondii COMMON NARDOO. Ground and eaten by Aborigines. *Distr.* Australia.

Marsileaceae A family of 3 genera and 72 species of small to minute ferns of moist to periodically flooded areas. The leaves are pinnate but with several large segments so as to appear palmate. The sori are borne in special hard structures, sporocarps, attached to the leaf-stalk, which split in 2 so as to release the spores. Distr. Widespread. Class F. *See: Marsilea, Pilularia, Regnellidium.*

maruba *Hosta plataginea.*

marvel of Peru *Mirabilis jalapa.*

Mascarena See *Hyophorbe*

mas: male

masculas, -a, -um: male.

Masdevallia (After Jose Masdevall (died 1801), Spanish botanist.) A genus of 300 species of epiphytic, often very small orchids with fleshy leaves and racemes of small or large flowers, the sepals of which are usually extended into long, fine points or tails. Some species and hybrids are cultivated as tender ornamentals. *Distr.* Mexico, Central and South America. *Fam.* Orchidaceae. *Class* M.

M. amabilis Flowers orange-carmine, tube narrow, bent, tails to 5cm, red. *Distr.* Peru.

M. bella See *Dracula bella*

M. caloptera Flowers 3–5, white, striped crimson, tails short, yellow. *Distr.* Colombia, Ecuador, Peru.

M. caudata Flowers slightly fragrant, tails yellow, 5–8cm long. *Distr.* Colombia, Venezuela.

M. chimaera See *Dracula chimaera*

M. coccinea Flowers large, waxy, tails short. *Distr.* Colombia, Peru.

M. coriacea Flowers pale yellow, 8cm across with short, broad tails. Cultivars from magenta to deep crimson and purple. *Distr.* Colombia.

M. infracta Flowers yellow-white, flushed orange. *Distr.* Peru, Brazil.

M. macrura Flowers large, fleshy, sepals thickly studded with numerous black-purple warts within, tails 10–12cm long, yellowish. *Distr.* Colombia, Ecuador.

M. muscosa See *Porroglossum muscosum*

M. peristeria Flowers fleshy, yellow-green with red spots, tails fleshy, to 3cm long. *Distr.* Colombia.

M. rolfeana Flowers chocolate-brown or purple, 8cm long including the yellow tails. *Distr.* Costa Rica.

M. rosea Flowers red and yellow to 10cm across, including tails. *Distr.* Colombia, Ecuador.

M. schlimii Flowers 5–8, yellow with brown-red spots, tails 2–5cm long. *Distr.* Colombia.

M. simula See *Dryadella simula*

M. tovarensis Flowers pure white. *Distr.* Venezuela, Colombia.

M. veitchiana Flowers solitary, large, sepals tawny yellow without, rich orange-scarlet within. *Distr.* Peru.

mask flower *Alonsoa*, *A. warscewiczii.*

Massonia (After Francis Masson (1741–1805), who collected in South Africa.) A genus of 8 species of bulbiferous herbs with 2 somewhat fleshy leaves and white, fragrant flowers borne in an almost stalkless, umbel-like head. Several species are grown as ornamentals. *Distr.* South Africa. *Fam.* Hyacinthaceae. *Class* M.

M. depressa Leaves circular to oblong, acute, hairless or with hairs only on the margins. *Distr.* South Africa (Cape Province).

M. jasminiflora Flowers scented like a ripe pear. *Distr.* South Africa (Cape Province, Orange Free State).

M. pustulata Upper surface and margins of leaves minutely hairy. *Distr.* South Africa (Cape Province).

masterwort *Astrantia* **greater** ~ *A. major.*

mastic *Pistacia lentiscus.*

mastic tree: Peruvian ~ *Schinus molle.*

masur *Lens culinaris.*

matac *Asclepias curassavica.*

matal *Asclepias curassavica.*

Matoniaceae A family of 2 genera and 4 species of medium-sized ferns with creeping hairy stems and fan-like leaves. The sori are round and borne on the underside of the leaf. *Distr.* Indonesia and Malaysia. *Class* F.

Matricaria (From Latin *mater*, mother, alluding to its former use in treating diseases of the uterus.) A genus of 5 species of weedy annual or perennial herbs allied to the genus *Chrysanthemum*, with finely divided leaves and yellow flowers in daisy-like heads that have distinct rays. *Distr.* Eurasia. *Fam.* Compositae. *Class* D.

M. chamomilla See *M. recutita*

M. recutita WILD CAMOMILE, SWEET FALSE CAMOMILE, GERMAN CAMOMILE. Annual sweetly scented herb. Used as a medicinal tea and insecticide. *Distr.* Eurasia, naturalized in North America.

Matteuccia (After Carlo Matteucci (1800–68), Italian physicist.) A genus of about 2 species of medium-sized terrestrial ferns with thick erect rhizomes and pinnately divided leaves. Young fronds are edible if steamed. *Distr.* E Asia, E North America. *Fam.* Woodsiaceae. *Class* F.

M. struthiopteris (From Greek *struthokamelos*, ostrich, and *pteris*, feather or fern.) OSTRICH FEATHER FERN, SHUTTLECOCK FERN. Leaves borne in a whorl, long, tapering to the base. Grown as an ornamental. *Distr.* Eurasia.

Matthiola (After Pierandrea Mattioli (1500–77), Italian botanist and physician.) A genus of 55 species of annual or biennial herbs and subshrubs typically with large, purple, sweet-scented flowers. *Distr.* W Europe, Macaronesia, Mediterranean. *Fam.* Cruciferae. *Class* D.

M. fruticulosa Perennial with a woody base and yellow-purple flowers. *Distr.* S Europe.

M. incana BROMPTON STOCK. A fast-maturing annual used for bedding. Sometimes grown for highly scented cut flowers. *Distr.* S and W European coasts.

M. longipetala NIGHT-SCENTED STOCK. Annual with yellow, pink or red flowers. *Distr.* Greece to SW Asia.

M. sinuata Erect biennial with flowers fragrant at night. *Distr.* Europe.

Matucana See *Oreocereus*

mauritanicus, -a, -um: of Mauritania, North Africa.

mauritianus, -a, -um: of Mauritius.

Maxillaria (From Latin *maxilla*, jaw; in some species the column and lip resemble the jaw of an insect.) A genus of 250 species of epiphytic, occasionally terrestrial orchids with 1-leaved pseudobulbs and solitary, red or yellow to white flowers. A number of species and hybrids are cultivated as tender ornamentals. *Distr.* Tropical and subtropical America. *Fam.* Orchidaceae. *Class* M.

M. alba Flowers white to 4cm across, short-lived. *Distr.* Central America, N South America.

M. coccinea Flowers pink to vermilion, fleshy. *Distr.* W Indies, Colombia, Venezuela.

M. densa Flowers green-yellow-white, tinged red. *Distr.* Central America from Mexico to Honduras.

M. grandiflora Flowers large, fragrant, sepals snow-white, lip 3-lobed, streaked yellow on side lobes, flushed crimson inside. *Distr.* NW South America.

M. luteo-alba Flowers large, creamy-white towards base, tawny-orange upwards, suffused brownishcrimson. *Distr.* Panama, Colombia, Venezuela, Ecuador.

M. marginata Flowers to 3.5cm across, sepals orange-yellow with a dark red margin, petals much smaller than sepals. *Distr.* Colombia, Ecuador, Brazil.

M. nigrescens Flowers port-wine coloured except for the base of the petals which grade to yellow. *Distr.* Colombia, Venezuela.

M. ochroleuca Flowers yellowish-white, strongly scented. *Distr.* Brazil.

M. picta Flowers cream to yellow, streaked and dotted dull purple and chocolate, fragrant. *Distr.* Colombia, Brazil.

M. porphyrostele Flowers yellow, sepals blotched red-brown, column purple. Distr. Brazil.

M. rufescens Flowers cream-brown to dull yellow or orange, not fragrant. A variable species. *Distr.* Central and South America.

M. sanderiana (After H. F. C. Sander (1847–1920), who had orchid nurseries in England and Belgium.) Flowers fleshy to 12–15cm across, sepals broad, slightly hooded, thickly marked blood-red, lip ivory white stained blood-red on rounded side lobes. *Distr.* Ecuador, Peru.

M. sanguinea Flowers dull red-brown and yellow, lip crimson with blackish crest. *Distr.* Costa Rica.

M. sophronitis Flowers cup-shaped, orange-red with a yellow lip. *Distr.* Venezuela.

M. striata Flowers to 15cm across, greenish-yellow striped brown-red, lip white veined purple-red. *Distr.* Peru, Ecuador.

M. tenuifolia Flowers to 3–5cm across, yellow nearly obscured by crimson bars. *Distr.* Mexico to Costa Rica.

M. valenzuelana Flowers to 2.5cm across, yellow-green, lip brown. *Distr.* Central America, N South America, W Indies.

M. variabilis Flowers variable to about 2.5cm across, deep purple to pale yellow. *Distr.* Central America, W Indies, Guyana.

M. venusta Flowers waxy, white, to 12–15cm across, thick and fleshy, very fragrant, nodding. *Distr.* Colombia, Venezuela.

maximus, -a, -um: largest.

medicus, -a, -um: used medicinally

may *Crataegus laevigata.*

Mayacaceae A family of 1 genus and 4 species of small, mat-forming aquatic or amphibious herbs. *Distr.* W and Central tropical Africa, North and South America. *Class* M.

mayblob *Caltha palustris.*

mayflower *Epigaea repens.*

May pops *Passiflora incarnata.*

mayten *Maytenus boaria.*

Maytenus (From *maiten*, a South American vernacular name for *M. boaria*.) A genus of 225 species of evergreen trees with alternate, simple, leaves and clusters of small, white, yellow or red, star-shaped flowers. The fruit is a capsule and the seeds are enclosed in fleshy white or red arils. Several species are grown as ornamentals chiefly for their evergreen foliage. *Distr.* Tropical, warm-temperate regions. *Fam.* Celastraceae. *Class* D.

M. boaria MAYTEN. Small tree with weeping habit. *Distr.* Chile, Argentina.

Mazus (From Greek *mazos*, breast, alluding to the swellings in the throat of the flowers.) A genus of 30 species of annual and perennial, low-growing herbs with simple, toothed leaves and 1-sided racemes of tubular, 2-lipped, 5-lobed flowers that bear 2 distinct mounds in the throat. Several species are grown as ornamental ground cover. *Distr.* Asia to Australia. *Fam.* Scrophulariaceae. *Class* D.

M. pumilio Perennial herb. Stems subterranean, creeping. Flowers white or bluish with a yellow centre. *Distr.* New Zealand, Australia.

M. radicans Perennial herb. Stems creeping, rooting. Flowers white with yellow centre. *Distr.* New Zealand.

M. reptans Perennial herb. Stems slender, prostrate, rooting. Flowers purple-blue, lower lip blotched white, yellow, or red-purple. *Distr.* Himalaya.

mazzard *Prunus avium.*

meadow bright *Caltha palustris.*

meadow-sweet *Filipendula ulmaria.*

measles plant *Hypoestes phyllostachya.*

Meconopsis (From Greek *mecon*, poppy, and *-opsis*, indicating resemblance.) A genus of 42 species of annual, biennial, and perennial herbs typically with basal, pinnately lobed leaves and yellow, blue, or white, poppy-like flowers that bear 2 sepals, 4 petals, and numerous stamens. A number of species and hybrids are grown for their delicate flowers. *Distr.* Himalaya, W China. *Fam.* Papaveraceae. *Class* D.

M. baileyi See *M. betonicifolia*

M. betonicifolia HIMALAYAN BLUE POPPY. Clump-forming perennial. Flowers blue. *Distr.* China.

M. cambrica WELSH POPPY. Spreading perennial. Flowers orange to yellow. *Distr.* W Europe.

M. grandis Erect perennial. Flowers large, blue. *Distr.* Himalaya, W China.

M. napaulensis SATIN POPPY. Clump-forming, short-lived perennial. Flowers red to blue. *Distr.* Himalaya, W China.

M. nudicaule See *Papaver nudicaule*

Medeola (After the sorceress, Medea, of Greek mythology, in allusion to the supposed medicinal properties of this plant.) A genus of 1 species of rhizomatous herb with slender stems, 2 whorls of simple leaves and umbels of small white flowers which are followed by purple berries in the autumn. *Distr.* North America. *Fam.* Convallariaceae. *Class* M.

M. virginica CUCUMBER ROOT, INDIAN CUCUMBER ROOT. Rhizomes crisp and edible like cucumber. *Distr.* E North America.

medic *Medicago.*

Medicago (From *medike*, a classical Greek name for a kind of crop plant.) MEDICK, BURWEED, MEDIC. A genus of 56 species of annual and perennial herbs and small shrubs with trifoliate leaves and short racemes of pea-like flowers. Several species are grown as ornamentals or as fodder crops. *Distr.* Europe, Mediterranean regions, Ethiopia, S Africa. *Fam.* Leguminosae. *Class* D.

M. arborea TREE MEDICAGO, MOON TREFOIL. Evergreen shrub. Flowers yellow. *Distr.* S Europe.

M. echinus See *M. intertexta*

M. intertexta CALVARY CLOVER. Annual herb. Flowers yellow. Fruit twisted and spiny like a crown of thorns. *Distr.* Mediterranean regions.

M. sativa LUCERNE, ALFALFA. Perennial herb. Flowers blue or purple. Grown as fodder and silage and is being developed as an oil-seed crop. *Distr.* SW Asia, naturalized in Europe, North America.

medicago, tree *Medicago arborea.*

medick *Medicago.*

Medinilla (After Jose de Medinilla, Governor of the Mariana Islands in 1820.) A genus of 150 species of evergreen shrubs and climbers with simple leathery or fleshy leaves and panicles or cymes of white or pink flowers that bear 4–6 petals. *Distr.* Philippines. *Fam.* Melasto-mataceae. *Class* D.

M. magnifica Evergreen robust shrub. Flowers pink in pendulous panicles surrounded by pink bracts. Grown as an ornamental. *Distr.* Philippines.

mediterraneus, -a, -um: of the Mediterranean or of inland areas, from *medius*, middle, and *terraneus*, of land, earth.

medius, -a, -um: middle, intermediate.

medlar *Mespilus germanica* **Japanese** ~ *Eriobotrya japonica.*

Medusagynaceae A family of 1 genus and 1 species. *Class* D.
See: Medusagyne.

Medusagyne (After *Medusa* of Greek mythology, and *gynoecium*, the female part of the flower.) A genus of 1 species of shrub with simple opposite leaves and regular bisexual flowers that carry 5 sepals, 5 petals, and numerous stamens. *Distr.* Seychelles. *Fam.* Medusagynaceae. *Class* D.

M. oppositifolia This species was thought to be extinct because it had not been seen since 1903; it was found again in 1976, and has now spread in cultivation.

Medusandraceae A family of 2 genera and 9 species of shrubs and trees with simple, spirally arranged leaves and catkin-like racemes of small regular flowers that bear their parts in fives. *Distr.* Tropical Africa. *Class* D.

Meehania (After Thomas Meehan (1826–1901), London-born Philadelphia nurseryman and writer.) A genus of 2 species of perennial herbs with simple opposite leaves and whorls of 2-lipped flowers. Several species are grown as ornamentals. *Distr.* E Asia, E North America. *Fam.* Labiatae. *Class* D.

M. cordata MEEHAN'S MINT. Stoloniferous, finely hairy, aromatic herb. Flowers blue, marked purple and white. *Distr.* E North America.

M. urticifolia Flowers in more or less 1-sided spikes, purple-blue. *Distr.* Japan, NE China, Korea.

meeting houses *Aquilegia canadensis.*

Megacarpaea (From Greek *megas*, great, and *karpos*, fruit, alluding to the large pods.) A genus of 7 species of perennial herbs with large, white to yellow or violet flowers. *Distr.* Europe to Central Asia, China and Himalaya. *Fam.* Cruciferae. *Class* D.

M. polyandra Perennial with arching, divided foliage. Young leaves edible. *Distr.* Himalaya.

megalanthus, -a, -um: with large flowers.

megalocarpus, -a, -um: with large fruit.

megalophyllus, -a, -um: with large leaves.

megapotamicus, -a, -um: of a large river.

megarhizus, -a, -um: with a large root.

megastigmus, -a, -um: with a large stigma.

Meiracyllium (From Greek *meirakyllion*, a mere lad or little fellow, alluding to the low habit.) A genus of 2 species of small epiphytic, rhizomatous orchids with leathery leaves and relatively large flowers for their stature. *Distr.* Mexico, Guatemala, El Salvador. *Fam.* Orchidaceae. *Class* M.

M. wendlandii Leaves to 5cm long. Flowers purple, yellow at base. *Distr.* Mexico, Guatemala, El Salvador.

Melaleuca (From Greek *melas*, black, and *leukos*, white, alluding to the black trunk and white shoots of many species.) A genus of 150 species of evergreen shrubs with simple, often glandular leaves and spikes or clusters of small flowers with 5 petals and numerous prominent stamens. Several species are grown as ornamentals *Distr.* Indo-malaysia, Australia, Pacific. *Fam.* Myrtaceae. *Class* D.

M. armillaris BRACELET HONEY MYRTLE. Rounded shrub or small tree. Flowers with white stamens in a brush-like whorl around the stem. *Distr.* SE Australia.

M. hypericifolia Rounded shrub. Flowers in brush-like spikes. Stamens red. *Distr.* Australia (New South Wales).

Melandrium See *Vaccaria*

melanocarpus, -a, -um: with black fruit.

melanocentrus, -a, -um: with a black central spine.

melanochrysus, -a, -um: black-gold coloured.

Melanophyllaceae A family of 2 genera and 9 species of shrubs and small trees with simple leaves and racemes of regular, bisexual flowers. This family is sometimes included within the family Cornaceae. *Distr.* Madagascar. *Class* D.

Melanoselinum (From Greek *melas*, black, and the genus name *Selinum*.) A genus of 7 species of biennial or short-lived perennial, shrubby herbs with divided leaves and umbels of white or pale purple flowers. *Distr.* Macaronesia. *Fam.* Umbelliferae. *Class* D.

M. decipiens BLACK PARSLEY. Leaves yellow-green. Leaf-stalk inflated. Flowers fragrant. Grown as an ornamental. *Distr.* Madeira.

melanostictus, -a, -um: with black spots.

melanotrichus, -a, -um: with black hairs.

melanoxylon: with black wood.

Melanthiaceae A family of 24 genera and about 150 species of perennial herbs with simple, stalked or sheathing leaves and racemes or panicles of small flower, that have 6 tepals that may or may not be fused, 6 stamens, and a superior ovary. This family is sometimes included within the family Liliaceae. *Distr.* Widespread, except Africa. *Class* M.
See: Aletris, Chamaelirium, Helonias, Heloniopsis, Melanthium, Narthecium, Stenanthium, Tofieldia, Veratrum, Xerophyllum, Zigadenus.

Melanthium (From Greek *melas*, black, and *anthos*, flower; the persistent flower segments become dark after flowering.) A genus of 5 species of rhizomatous herbs with grass-like leaves and panicles of small, regular flowers. *Distr.* North America. *Fam.* Melanthiaceae. *Class* M.

M. virginicum BUNCHFLOWER. Herb to 1.7m high. Flowers cream becoming black in fruit. Grown as an ornamental. *Distr.* E USA.

Melasphaerula (From Greek *melas*, black, and *sphaerula*, little ball, alluding to the small black bulbs.) A genus of 1 species of perennial herb with a round corm, straggling stem, grass-like leaves, and a spike of small, regular, cream-white flowers. *Distr.* South Africa (SW Cape Province). *Fam.* Iridaceae. *Class* M.

M. graminea See *M. ramosa*

M. ramosa Grown as an ornamental. *Distr.* South Africa.

Melastomataceae A family of 194 genera and about 5000 species of typically shrubs and small trees with simple, opposite leaves and showy flowers that bear 4–5 sepals and petals. *Distr.* Tropical and occasionally temperate regions. *Class* D.
See: Centradenia, Heterocentron, Medinilla, Microlepis, Osbeckia, Tibouchina.

meleagris: spotted.

Melia (Classical Greek name for the ASH, which has similar leaves.) A genus of 3 species of deciduous shrubs and trees with pinnate or bipinnate leaves and panicles of small flowers that bear 5–6 petals and 10–12 stamens. The stamens are fused into a lobed tube. The fruit is a hard drupe. *Distr*. Tropical regions of the Old World. *Fam*. Meliaceae. *Class* D.

M. azedarach BEAD TREE, CHINA BERRY, PERSIAN LILAC. Rounded tree. Flowers lilac, fragrant. Fruit yellow. As well as being grown as an ornamental, this tree is used in reafforestation schemes and as a source of timber. The leaves are used medicinally and the fruits are used as beads. *Distr*. Iran, Himalaya, China.

Meliaceae The Mahogany family. 51 genera and 575 species of shrubs, trees, and rarely herbs with alternate, typically pinnate leaves and regular, usually bisexual flowers that may occur in an inflorescence borne directly on the trunk. This family is a very important source of timber. *Distr*. Widespread in tropical and subtropical regions, rarer in temperate regions. *Class* D.
See: Cedrela, Melia, Quivisia, Swietenia, Toona, Turraea.

Melianthaceae A family of 2 genera and 8 species of shrubs and small trees with alternate, usually pinnate leaves, and racemes of irregular flowers that have their parts in fours or fives. *Distr*. Tropical and S Africa. *Class* D.
See: Melianthus.

Melianthus (From Greek *meli*, honey, and *anthos*, flower, alluding to the abundant nectar.) A genus of 6 species of evergreen shrubs with alternate, pinnate leaves and racemes of irregular flowers that bear 5 petals formed into a hood and a spur. *Distr*. South Africa. *Fam*. Melianthaceae. *Class* D.

M. major HONEY FLOWER. Stems thick, green. Leaves to 50cm long, blue-green, arching. Flowers red-brown, producing large amounts of nectar. Grown as an ornamental. *Distr*. South Africa.

melic *Melica* **Siberian** ~ *M. altissima* **tall** ~ *M. altissima*.

Melica (From Latin *melica*, a kind of vessel, alluding to the swollen stem bases.) MELIC. A genus of 40–50 species of rhizomatous, clump-forming grasses, typically with flat, arching leaves and paniculate inflorescences. Several species are cultivated as ornamentals. *Distr*. Temperate regions. *Fam*. Gramineae. *Class* M.

M. altissima TALL MELIC, SIBERIAN MELIC. Spikelets tawny or red in some ornamental cultivars. *Distr*. Europe, temperate Asia.

M. uniflora WOOD MELICK. A loosely tufted woodland grass. Several ornamental cultivars are available. *Distr*. Europe, SW Asia.

melick, wood *Melica uniflora*.

Melicope (From Greek *meli*, honey, and *kope*, division, alluding to the 4 nectaries at the base of the ovary.) A genus of 20 species of shrubs and trees with simple or trifoliate glandular leaves and cymes or panicles of small flowers that bear 4 petals and 8 stamens. *Distr*. Indomalaysia, Australia, New Zealand. *Fam*. Rutaceae. *Class* D.

M. ternata Shrub or small tree. Leaves trifoliate. Flowers green-white. Grown as a tender ornamental. *Distr*. New Zealand.

Melicytus (From Greek *meli*, honey, and *kytos*, vessel, alluding to the nectaries.) A genus of 4 species of shrubs and trees with simple alternate leaves and clusters of regular flowers that bear their parts in fives and are followed by showy berries. Several species are grown as ornamentals. *Distr*. Solomon Islands, New Zealand, Norfolk Islands, Fiji. *Fam*. Violaceae. *Class* D.

M. ramiflorus Tree to 15m high. Flowers yellow-green. Fruit purple. Formerly a source of timber for charcoal in gunpowder. *Distr*. New Zealand, Norfolk Islands, Fiji, Solomon Islands.

melilot, ribbed *Melilotus officinalis*.

Melilotus (From Greek *meli*, honey, and Lotus.) A genus of 20 species of annual, biennial and short-lived perennial herbs with trifoliate leaves and racemes of small, yellow or white, pea-like flowers. *Distr*. Temperate and subtropical Eurasia, N Africa, Ethiopia. *Fam*. Leguminosae. *Class* D.

M. officinalis RIBBED MELILOT. Biennial. Flowers yellow. Grown as green manure, forage crop and bee plant. *Distr*. Eurasia, naturalized in North America.

Meliosma (From Greek *meli*, honey, and *osme*, scent, alluding to the honey-scented flowers.) A genus of 20–25 species of evergreen and deciduous shrubs and trees with simple or pinnate leaves and large panicles of small fragrant flowers that bear 5 unequal petals and 5 stamens. Several species are grown as ornamentals. *Distr*. Temperate Asia, tropical America. *Fam*. Meliosmaceae. *Class* D.

M. dilleniifolia Deciduous shrub or tree to 15m high. Flowers white. *Distr*. Himalaya, China, Japan.

M. pendens See *M. dilleniifolia*

M. veitchiorum Spreading deciduous tree. Leaves large. Flowers small, white, fragrant. Fruits purple. *Distr*. W China.

Meliosmaceae A family of 2 genera and about 30 species of shrubs and trees with alternate, simple or pinnate leaves and large inflorescences of small, fragrant, irregular flowers that bear their parts in fives. *Distr*. Tropical and E Asia, tropical America. *Class* D.
See: Meliosma.

meliosmifolius, -a, -um: from the genus *Meliosma*, and *folius*, leaved.

Melissa (From Greek *melissa*, honeybee, alluding to the flowers and scented foliage that attract bees.) A genus of 3 species of perennial, deciduous herbs with simple, opposite, short-stalked leaves and whorls of 2-lipped flowers. *Distr*. Europe, W and Central Asia. *Fam*. Labiatae. *Class* D.

M. officinalis HONEY FLOWER, LEMON BALM, BEE BALM. Tuft-forming herb. Leaves broad-toothed, lemon-scented. Flowers pale yellow. Leaves used as a seasoning, medicinally, and in scents. *Distr*. S Europe, naturalized in North America.

melissifolius, -a, -um: from the genus *Melissa*, and *folius*, leaved.

melissophyllus, -a, -um: from the genus *Melissa*, and Greek *phyllon*, leaf.

Melittis (From Greek *melitta*, bee; the plant is attractive to bees.) A genus of 1 species of perennial herb with erect stems, hairy, opposite, honey-scented leaves, and whorls of large 2-lipped flowers that are white with a pink or purple blotch. *Distr*. Europe. *Fam*. Labiatae. *Class* D.

M. melissophyllum BASTARD BALM. Grown as an ornamental. *Distr*. Europe.

mellifera: bearing honey.

Melocactus (From Latin *melo*, melon, and the old genus name *Cactus*.) A genus of 36 species of cacti with large, globose, ribbed stems and small, tubular, red or pink flowers that are borne close to the apex. Several species are grown as ornamentals. *Distr*. E Brazil, Peru, Venezuela, Central America, Caribbean. *Fam*. Cactaceae. *Class* D.

M. broadwayi (After W. R. Broadway, who collected it in 1914.) Stems yellow-green to 20cm high. Flowers purple. *Distr*. West Indies.

M. communis See *M. intortus*

M. intortus Stems to 90cm high. Spines robust. Flowers pink to red. *Distr*. West Indies.

melon *Cucumis melo* **water** ~ *Citrullus lanatus*.

melon shrub *Solanum muricatum*

melon tree *Carica papaya*.

membranaceus, -a, -um: membranous.

Mendoncella (After Luis Mendonca, editor of the orchid journal *Orquidea*.) A genus of 9 species of epiphytic orchids with 1–2-leaved pseudo-bulbs and typically solitary large flowers. Several species are cultivated as ornamentals. *Distr*. Central and tropical South America. *Fam*. Orchidaceae. *Class* M.

M. grandiflora Flowers 2–5, to 8cm across, fragrant, yellow-green, striped brown. *Distr*. Central America, Colombia.

men in a boat *Tradescantia spathacea*.

Menispermaceae The Moonseed family. 76 genera and about 500 species of lianas, herbs, shrubs, and trees with simple alternate leaves and small unisexual flowers that bear 2–3 whorls of 3 tepals. Some species are used as arrow and fish poisons or medicinally. *Distr*. Tropical, subtropical, and occasionally temperate regions. *Class* D.
See: Chondrodendron, Cocculus, Menispermum.

menispermifolius, -a, -um: with leaves like those of the plant family Menispermaceae.

Menispermum (From Greek *mene*, crescent moon, and *sperma*, seed.) MOON SEED. A genus of 2 species of deciduous, twining subshrubs with long-stalked, palmately lobed, more or less peltate leaves and racemes or panicles of small yellow-green flowers. *Distr.* NE Asia, North America. *Fam.* Menispermaceae. *Class* D.

M. canadense MOONSEED, YELLOW PARILLA. Grown as an ornamental. The rhizomes are used medicinally. *Distr.* E North America.

Mentha (Latin name for mint.) MINT. A genus of 25 species of aromatic, rhizomatous herbs with simple, opposite, often hairy leaves, and whorls of small, 2-lipped flowers. Many species and cultivars are grown as ornamentals and as culinary herbs. *Distr.* Temperate regions of the Old World. *Fam.* Labiatae. *Class* D.

M. aquatica WATER MINT. Herb of damp areas. Stems often tinged red. Flowers lilac. *Distr.* Europe, N Africa, Asia.

M. arvensis FIELD MINT, CORN MINT. A variable species. Flowers pink or white. Provides menthol for cigarettes. *Distr.* Eurasia.

M.* × *gracilis GINGER MINT, SCOTCH MINT. A hybrid of *M. spicata* × *M. arvensis*, cultivated for spearmint oil. *Distr.* Garden origin.

M. longifolia HORSE MINT. Leaves linear, strongly scented. Flowers mauve or lilac. *Distr.* Europe, W Asia, S Africa.

M.* × *piperita PEPPERMINT. Naturally occurring hybrid of *M. spicata* × *M. aquatica*. Widely used as a flavouring. *Distr.* Europe.

M. pulegium PENNYROYAL. Used in soap and medicinally. *Distr.* Europe, W Asia.

M. requienii (After Esprit Requien (1788–1851), who studied the flora of Southern France and Corsica) CORSICAN MINT. *Distr.* Italy, Corsica, Sardinia.

M. spicata SPEARMINT, COMMON GREEN MINT. Herb to 1m high. Flowers pink or white, in a terminal spike. Used as a flavouring. *Distr.* Europe.

M. suaveolens ROUND-LEAVED MINT, APPLE MINT, WOOLLY MINT. Herb to 1m high. Leaves densely hairy. Flowers white or pink. Used in mint sauce and jelly. *Distr.* S and W Europe.

M. sylvestris See *M. longifolia*

M. viridis See *M. spicata*

Mentzelia (After Christian Mentzel (1622–1701), German botanist.) A genus of 60 species of annual and perennial herbs and shrubs with simple or pinnate, alternate leaves and cymes or racemes of regular, fragrant flowers. *Distr.* USA (Central California). *Fam.* Loasaceae. *Class* D.

M. lindleyi (After John Lindley (1799–1865), professor of botany at London University.) BLAZING STAR. Bushy annual herb. Flowers cup-shaped, yellow. Grown as an ornamental. *Distr.* USA (Central California).

Menyanthaceae The Bogbean family. 5 genera and 40 species of rhizomatous herbs of bogs and wetlands with simple or trifoliate sheathing leaves and regular, often hairy flowers that typically bear their parts in fives. *Distr.* Widespread. *Class* D.
See: Menyanthes, Nymphoides, Villarsia.

Menyanthes (From Greek *men*, moon or month, and *anthos*, flower.) A genus of 1 species of aquatic or marginal, rhizomatous, perennial herb with trifoliate leaves and erect racemes of regular, white flowers that bear 5 fringed petals. *Distr.* N temperate regions. *Fam.* Menyanthaceae. *Class* D.

M. trifoliata BUCKBEAN, BOGBEAN, MARSH TREFOIL. Grown as a bog garden ornamental; the rhizomes have also been used medicinally and powdered as a kind of flour by the Inuit peoples. The leaves are used as a substitute for hops in beer.

Menziesia (After Archibald Menzies (1754–1842), British naval surgeon and botanist.) A genus of 7 species of deciduous shrubs with simple, entire leaves and drooping bell- or urn-shaped flowers. *Distr.* W North America. *Fam.* Ericaceae. *Class* D.

M. alba See *Daboecia cantabrica*

M. ciliicalyx Erect shrub. *Distr.* Japan.

M. ferruginea RUSTY LEAF, FOOL'S HUCKLEBERRY. Suberect shrub. *Distr.* W North America from Alaska to Oregon.

M. pentandra Softly hairy deciduous shrub. *Distr.* Japan and neighbouring NE Asia.

M. polifolia See *Daboecia cantabrica*

Mercurialis (After the Roman god Mercury.) A genus of 8 species of annual and perennial herbs with simple opposite leaves and small unisexual flowers, the males being arranged along the axillary spikes and the females being borne singly or in small clusters. *Distr.* Eurasia. *Fam.* Euphorbiaceae. *Class* D.

M. perennis DOG'S MERCURY. Rhizomatous perennial, characteristic of disturbed woodland. Used as a source of dye. *Distr.* Eurasia.

mercury *Chenopodium bonus-henricus.*

Merendera (From *quita meriendas*, the Spanish name for *Colchicum.*) A genus of 10 species of herbs with linear leaves arising from a corm. The stalkless flowers usually appear before the leaves and have 6 spreading to erect tepals and 6 stamens. This genus is closely related to *Colchicum* but has less showy flowers. Most of the species are grown as ornamentals. *Distr.* S Europe, N Africa, W Asia, Ethiopia. *Fam.* Colchicaceae. *Class* M.

M. filifolia Flowers appear in spring just before the very narrow leaves. *Distr.* SW Europe, N Africa.

M. montana Flowers appear in autumn just before the strap-shaped leaves. *Distr.* Iberian peninsula, Central Pyrenees.

M. pyrenaica See *M. montana*

M. robusta Flowers appear with leaves in spring. *Distr.* N Afghanistan, Russia, Iran, N India.

meridionalis: southern.

Merinthosorus See *Aglaomorpha*

Merremia (After Blasius Merrem (died 1824), German naturalist.) A genus of 70 species of climbing perennial herbs and subshrubs with simple or compound leaves and bell- or funnel-shaped, white, yellow, or blue flowers. Several species are grown as ornamentals and some are serious weeds of tropical crops. *Distr.* Tropical regions. *Fam.* Convolvulaceae. *Class* D.

M. tuberosa BRAZILIAN JALAP, WOOD ROSE, YELLOW MORNING GLORY, SPANISH WOODBINE. Robust woody climber. Flowers large, bell-shaped, yellow. *Distr.* Tropical America, naturalized throughout the tropics.

merrybells *Uvularia.*

merryhearts *Zigadenus nuttallii.*

Mertensia (After Franz Karl Mertens (1764–1831), professor of botany at Bremen University.) BLUEBELLS. A genus of 50 species of perennial herbs with opposite, linear or heart-shaped leaves and curved cymes of bell- to funnel-shaped flowers. Several species are grown as ornamentals. *Distr.* N temperate regions, S to Mexico. *Fam.* Boraginaceae. *Class* D.

M. echioides Stems clump-forming, erect, to 30cm high. Flowers dark blue. *Distr.* Himalaya.

M. maritima GROMWELL, SEA LUNGWORT, OYSTER LEAF. Leaves fleshy. Flowers pink-blue. Rhizomes eaten by Eskimos. *Distr.* Coastal regions of N Europe and N America.

Meryta (From Greek *merys*, to roll up, the male flowers looking as though rolled together.) A genus of 16 species of small evergreen trees with thick trunks, large simple leaves, and compound inflorescences of small flowers. *Distr.* Norfolk Islands, New Zealand and S Pacific Islands. *Fam.* Araliaceae. *Class* D.

M. sinclairii PUKA. Round-headed tree to 8m high. Grown as an ornamental. *Distr.* New Zealand.

Mesembrianthemum See *Mesembryanthemum*

Mesembryanthemum (From Greek *mesembria*, midday, and *anthemon*, flower. Previously spelled *Mesembrianthemum*, meaning flowering at midday; the current spelling refers to the position of the ovaries.) A genus of 70 species of annual and biennial herbs with fleshy stems and leaves and clustered or solitary, many-petalled flowers. A number of species that were once considered to belong to this genus are now placed in several closely related genera within the Aizoaceae. Several species are grown as ornamentals. *Distr.* S Europe, SW Asia, South Africa, Atlantic islands. *Fam.* Aizoaceae. *Class* D.

M. barklyi Leaves blue-green with prominent veins. Flowers white. *Distr.* SW Africa.

M. bellidiformis See *Dorotheanthus bellidiformis*

M. crystallinum ICE PLANT. Leaves thick, fleshy, covered in glistening papillae. Flowers clustered, white. Grown as a bedding plant. Leaves edible like spinach. *Distr.* SW Africa, naturalized in Mediterranean regions, Canary Islands, Australia, and USA (California).

Mespilus (Classical Latin name for the fruit.) A genus of 1 species of deciduous shrub or tree to 5m high with arching branches, simple, alternate leaves, and white flowers that bear 5 petals and numerous stamens. The fruit resembles a flattened apple and is palatable only when almost rotten. It was formerly widely eaten as a dessert fruit and made into preserves. *Distr.* SE Europe to Central Asia. *Fam.* Rosaceae. *Class* D.

M. germanica MEDLAR. Occasionally grown as an ornamental and for its fruits.

messanensis: of Messina, Italy.

metake *Pseudosasa japonica.*

metallicus, -a, -um: with a metallic sheen.

Metasequoia (From Greek *meta*, changed, and the genus name *Sequoia.*) A genus of 1 species of deciduous coniferous tree to 45m high with bright green linear leaves and small, stalked, fruiting cones. *Distr.* Central China. *Fam.* Taxodiaceae. *Class* G.

M. glyptostroboides DAWN REDWOOD. Grown as an ornamental.

Metaxya A genus of 1 species of terrestrial fern with creeping rhizomes and pinnate leaves. The sori are irregularly arranged on the underside of the leaves. *Distr.* Moist forests of tropical America. *Fam.* Metaxyaceae. *Class* F.

Metaxyaceae A family of 1 genus and 1 species. *Class* F.
See: Metaxya.

Metrosideros (From Greek *metra*, centre, and *sideros*, iron, alluding to the very hard wood.) A genus of 50 species of evergreen shrubs, trees, and climbers with simple, opposite, glandular leaves and cymes of red to white, regular flowers that bear 5 petals and numerous long stamens. Several species are grown as ornamentals. *Distr.* E Malaysia, Pacific regions, S Africa. *Fam.* Myrtaceae. *Class* D.

M. excelsa NEW ZEALAND CHRISTMAS TREE, RATA. Spreading tree to 20m high. Flowers profuse, crimson. Used locally for medicinal purposes. *Distr.* New Zealand.

M. robusta Rounded tree. Flowers red. A source of timber for telegraph poles. *Distr.* New Zealand.

meu *Meum athamanticum.*

Meum (From a Greek name for this plant, *meon.*) A genus of 1 species of aromatic, perennial herb to 60cm high with much-divided leaves and compound umbels of white or pale purple flowers. *Distr.* Eurasia. *Fam.* Umbelliferae. *Class* D.

M. athamanticum BALD MONEY, MEU, SPIGNEL. Grown as an ornamental and used medicinally. The roots are edible.

Mexican flameleaf *Euphorbia pulcherrima.*

Mexican gem *Echeveria elegans*

Mexican hat *Rudbeckia, Ratibida.*

mexicanus, -a, -um: of Mexico.

Meyerophytum (After G. Meyer, a pastor in South Africa in the early 20th century, and the Greek *phyton*, plant.) A genus of 4 species of succulent perennial herbs with pairs of more or less united leaves and solitary, red-purple flowers that bear numerous petals. *Distr.* South Africa (Cape Province). *Fam.* Aizoaceae. *Class* D.

M. meyeri Grown as a mound-forming, tender ornamental.

mezereon *Daphne mezereum.*

Mibora (Name devised by Michel Adanson (1727–1806), French botanist.) A genus of 1 species of small, tufted, annual grass with grey-green leaves and spike-like inflorescences. *Distr.* Mediterranean regions, W Europe. *Fam.* Gramineae. *Class* M.

M. mimima EARLY SAND GRASS, SAND BENT. Grown as an ornamental, sometimes to provide shade for the roots of terrestrial orchids.

micans: shining, slightly metallic.

Michauxia (After André Michaux (1746–1803), French traveller and plant collector.) A genus of 7 species of biennial or short-lived perennial herbs with simple or lobed leaves and dense spikes of white bell-shaped flowers. *Distr.* SW Asia, E Mediterranean. *Fam.* Campanulaceae. *Class* D.

M. campanuloides Grown as an ornamental. *Distr.* E Mediterranean regions.

M. tchihatchewii (After Count Pierre de Tchihatchew (1808–90), Russian traveller and writer.) *Distr.* Turkey.

Michelia (After Pietro Antonio Micheli (1679–1737), Florentine botanist.) A genus of 45 species of evergreen and deciduous shrubs and trees with simple leathery leaves and solitary regular flowers that bear 6–24 tepals and numerous stamens. This genus is closely related to the genus *Magnolia* but bears its flowers in the axils of the leaves rather than terminally. Several species are grown as ornamentals and some are exploited for their timber. *Distr.* Tropical and warm-temperate regions of SE Asia, India, and Sri Lanka. *Fam.* Magnoliaceae. *Class* D.

M. champaca CHAMPAK, SAPU. Evergreen tree to 25m high. Flowers orange to white, highly fragrant. The flowers are used in scent-making, the timber for tea-boxes, and the leaves are fed to silkworms. *Distr.* S China.

M. doltsopa Evergreen rounded tree. Flowers white to pale yellow, strongly scented. A source of good timber. *Distr.* W China, E Himalaya.

M. figo Evergreen tree. Flowers banana-scented, yellow with maroon margins. *Distr.* SE China.

Micranthus (From Greek *mikros*, small, and *anthos*, flower.) A genus of 3 species of perennial herbs with small, globose corms, linear leaves, and spikes of numerous, small, blue flowers. *Distr.* South Africa (SW Cape Province). *Fam.* Iridaceae. *Class* M.

M. plantagineus Flowers pale blue, slightly scented. Grown as an ornamental.

micranthus, -a, -um: with small flowers.

Microbiota (From Greek *mikros*, small, and the old genus name *Biota*.) A genus of 1 species of dwarf, evergreen coniferous shrub to 50cm high with spreading branches and tiny needle-like leaves. *Distr.* SE Siberia. *Fam.* Cupressaceae. *Class* G.

M. decussata Occasionally grown as an ornamental.

Microcachrys (From Greek *mikros*, small, and *kachrys*, cone.) A genus of 1 species of dwarf, evergreen, coniferous shrub with whip-like branches covered in small scale-like leaves and seeds that are surrounded by a fleshy red aril. *Distr.* Australia (Tasmania). *Fam.* Podocarpaceae. *Class* G.

M. tetragona Occasionally grown as an ornamental.

microcarpus, -a, -um: with small fruit.

microcephalus, -a, -um: with a small head.

microchilus, -a, -um: with a small lip.

Microglossa albescens See *Aster albescens*

Microgramma (From Greek *mikros*, small, and *gramma*, letter.) A genus of 20 species of small epiphytic ferns with simple leathery leaves. Several species are grown as ornamentals. *Distr.* Tropical regions of America and Africa. *Fam.* Polypodiaceae. *Class* F.

M. vacciniifolia Rhizomes creeping. Leaves to 5cm long. *Distr.* Tropical Central and South America.

microgynus, -a, -um: with small ovaries.

microhelia: a small sun.

microheliopsis: resembling a small sun.

Microlepia (From Greek *mikros*, small, and *lepis*, scale, alluding to the indusia.) A genus of 45 species of small to medium-sized, terrestrial or occasionally epiphytic ferns. Several species are grown as ornamentals. *Distr.* Pan-tropical, S to New Zealand. *Fam.* Dennstaedtiaceae. *Class* F.

M. strigosa *Distr.* Japan, Polynesia.

Micromeria (From Greek *mikros*, small, and *meris*, part, alluding to the small flowers.) A genus of 70 species of herbs and small shrubs with 4-angled stems, simple opposite leaves, and whorls of 2-lipped flowers. Several species are grown as ornamentals. *Distr.* Canary Islands and Mediterranean regions to Himalaya and SW China. *Fam.* Labiatae. *Class* D.

M. corsica See *Acinos corsicus*

M. croatica Densely hairy, tuft-forming shrub to 20cm high. Flowers violet. *Distr.* Croatia.

M. rupestris See *M. thymifolia*

M. thymifolia Perennial herb with woody stock. Flowers white to violet. *Distr.* Balkans, S Italy.

M. varia Procumbent perennial herb. Flowers purple. *Distr.* Madeira.

micromeris, -e: with small parts.

microphyllus, -a, -um: with small leaves.

micropterus, -a, -um: with small wings.

Microsorum (From Greek *mikros*, small, and *sorus*, the cover of the spore.) A genus of 60 species of epiphytic ferns with simple to pinnately lobed, leathery leaves. Several species are grown as ornamentals. *Distr.* Tropical Asia, Australia and the Pacific. *Fam.* Polypodiaceae. *Class* F.

M. diversifolium KANGAROO FERN, HOUND'S TONGUE FERN. Leaves bright green, arching or pendent. *Distr.* Australia, New Zealand.

microspermus, -a, -um: with small seeds.

Microstrobos (From Greek *mikros*, small, and Latin *strobus*, cone.) A genus of 2 species of dwarf, evergreen, coniferous shrubs with dense, spirally arranged, scale-like leaves. *Distr.* E and SE Australia. *Fam.* Podocarpaceae. *Class* G.

M. fitzgeraldii BLUE MOUNTAIN PINE. Low shrub to 60cm high. Branches pendent. Occasionally grown as an ornamental. *Distr.* Australia (New South Wales).

mignonette Resedaceae, *Reseda* **wild** ~ *R. alba*.

mignonette tree *Lawsonia inerma*.

Mikania (After Joseph G. Mikan (1743–1814), professor of botany in Prague, or perhaps his son J. C. Mikan (1769–1844), who collected in Brazil.) A genus of 300 species of evergreen, herbaceous or woody climbers with simple leaves and flowers in small daisy-like heads that lack distinct rays. *Distr.* Tropical regions. *Fam.* Compositae. *Class* D.

M. dentata Woody climber. Flowers white. *Distr.* Brazil.

M. scandens CLIMBING HEMPWEED. Herbaceous climber. Flower-heads in dense clusters, strongly vanilla-scented. *Distr.* Tropical America.

M. ternata See *M. dentata*

Mila (An anagram of Lima, the capital of Peru.) A genus of 1 species of cactus with clusters of short, cylindrical, ribbed stems and small, funnel- to bell-shaped, yellow flowers. *Distr.* Central Peru. *Fam.* Cactaceae. *Class* D.

M. caespitosa Grown as an ornamental.

mile tree *Casuarina equisetifolia*.

milfoil *Achillea, A. millefolium* **diamond** ~ *Myriophyllum aquaticum* **spiked water** ~ *M. spicatum* **water** ~ *Myriophyllum*.

Milium (Classical Latin name.) MILLET. A genus of 6 species of annual and perennial grasses with stems to 2m high, flat leaves, and loose or contracted, branched inflorescences. *Distr.* N temperate regions. *Fam.* Gramineae. *Class* M.

M. effusum WOOD MILLET, MILLET GRASS. Perennial, cultivated, ornamental grass. *Distr.* Europe, Asia.

milk barrel, African *Euphorbia horrida*.

milkmaids *Burchardia*.

milkvetch, meadow *Oxytropis campestris*.

milkweed Asclepiadaceae, *Asclepias* **swamp** ~ *A. incarnata*.

milkwort Polygalaceae, *Polygala* **chalk** ~ *P. calcarea* **common** ~ *P. vulgaris* **sea** ~ *Glaux maritima*.

Milla (After Juliani Milla, gardener to the Spanish Court in Madrid during the 18th century.) A genus of 6 species of perennial herbs arising from small corms. The leaves are linear and the umbels bear 1–4 tubular flowers. Some species are cultivated as ornamentals. *Distr.* S USA and N Central America. *Fam.* Alliaceae. *Class* M.

M. biflora Flowers white inside, green outside, fragrant. *Distr.* Guatemala, Mexico and S USA.

milleflorus, -a, -um: with many flowers.

millefoliatus, -a, -um: with many leaves.

millefolius, -a, -um: with many leaves.

millet *Panicum, Sorghum, Milium* **African** ~ *Eleusine coracana* **barnyard** ~ *Echinochloa crus-galli* **bulrush** ~ *Pennisetum glaucum* **common** ~ *Panicum miliaceum* **finger** ~ *Eleusine coracana* **German** ~ *Setaria italica* **great** ~ *Sorghum bicolor* **hog** ~ *Panicum miliaceum* **Hungarian** ~ *Setaria italica* **Indian** ~ *Oryzopsis* *hymenoides* **Italian** ~ *Setaria italica* **Japanese** ~ *S. italica* **pearl** ~ *Pennisetum glaucum* **proso** ~ *Panicum miliaceum* **Russian** ~ *P. miliaceum* **spiked** ~ *Pennisetum glaucum* **wood** ~ *Milium effusum.*

Miltonia (After Charles Fitzwilliam, Viscount Milton (1786–1857), horticultural patron.) PANSY ORCHID. A genus of 20 species of epiphytic orchids with 1–2-leaved pseudobulbs and widely spreading flowers with wide petals and a spreading lip. Several species and numerous hybrids are cultivated as tender ornamentals. *Distr.* Central and South America. *Fam.* Orchidaceae. *Class* M.

M. candida Flowers yellow, blotched red-brown, to 8cm across, lip white, clasping column below. *Distr.* E Brazil.

M. clowesii Flowers yellow, blotched chestnut, to 8cm across, lip white. *Distr.* E Brazil.

M. cuneata Flowers brown-tipped, with yellow stripes, lip cream. *Distr.* E Brazil.

M. endresii See *Miltoniopsis warscewiczii*

M. flavescens Flowers fragrant, straw-yellow. *Distr.* Paraguay, N Argentina, S and E Brazil.

M. phalaenopsis See *Miltoniopsis phalaenopsis*

M. regnellii (After Mr Regnell, who introduced it into cultivation.) Flowers cream, suffused red-lilac, lip margins white. *Distr.* S Brazil.

M. roezlii See *Miltoniopsis roezlii*

M. russelliana Flowers to 5cm across, red-brown with yellow-green margins. *Distr.* S Brazil.

M. spectabilis Flowers solitary, white, tinged red. *Distr.* E Brazil Venezuela.

M. vexillaria See *Miltoniopsis vexillaria*

M. warscewiczii See *Miltoniopsis warscewiczii*

Miltoniopsis (From the genus name *Miltonia*, and the Greek ending *-opsis*, indicating resemblance.) A genus of 5 species of epiphytic orchids with 1-leaved pseudobulbs and racemes of showy flattened flowers. Members of this genus were formerly included in the genus *Miltonia*. Several species are grown as tender ornamentals. *Distr.* Central America. *Fam.* Orchidaceae. *Class* M.

M. phalaenopsis Flowers white, lip streaked purple-red. *Distr.* Colombia.

M. roezlii (After Benedict Roezl (1824–85), Czech plant collector.) Flowers white, lip orange-yellow. *Distr.* Colombia.

M. vexillaria Flowers white-pink.

M. warscewiczii (After Josef von Warscewicz, (1812–66), German botanist.) Flowers cream-white blotched red. *Distr.* Costa Rica.

mimimus, -a, -um: very small.

Mimosa (From Greek *mimos*, mimic, alluding to the sensitive leaves.) A genus of 400 species of spiny herbs, shrubs, lianas, and trees with bipinnate leaves and heads of regular flowers. The leaves of many species move rapidly when touched, possibly as a defence against herbivores. Several species are grown as ornamentals. *Distr.* Tropical and warm-temperate regions, especially America. *Fam.* Leguminosae. *Class* D.

M. pudica HUMBLE PLANT, SENSITIVE PLANT, TOUCH ME NOT, ACTION PLANT. Semi-erect subshrub or shrub. Leaves sensitive. Flowers pale pink to violet. *Distr.* Tropical America.

mimosa *Acacia, A. dealbata* **golden** ~ *A. baileyana.*

mimosifolius, -a, -um: from the genus *Mimosa*, and *folius*, leaved.

Mimulus (Diminutive of Latin *mimus*, mimic, alluding to the flowers which resemble a face, or possibly from *mimo*, ape.) MONKEY FLOWER. A genus of 150 species of annual and perennial herbs and shrubs with simple, opposite leaves and spike-like racemes of tubular, 5-lobed, 2-lipped flowers that may be closed at the throat by a raised boss. Several species are grown as ornamentals. *Distr.* South Africa. *Fam.* Scrophulariaceae. *Class* D.

M. aurantiacus BUSH MONKEY FLOWER. Diffuse much-branched shrub. Flowers orange or deep yellow. *Distr.* USA (Oregon, California).

M. × burnetii (After Dr Burnet of Aberdeen, who raised it about 1901.) A hybrid of *M. cupreus* × *M. luteus*. Tuft-forming perennial herb. *Distr.* Garden origin.

M. cardinalis CARDINAL MONKEY FLOWER, SCARLET MONKEY FLOWER. Erect perennial herb. Flowers scarlet with a yellow throat. *Distr.* NW North America.

M. glutinosus See *M. aurantiacus*

M. guttatus COMMON LARGE MONKEY FLOWER. Annual or biennial herb. Flowers yellow with a spotted, hairy throat. *Distr.* W North America.

M. langsdorfii See *M. guttatus*

M. lewisii GREAT PURPLE MONKEY FLOWER. Perennial herb. Flowers magenta. *Distr.* W North America.

M. longiflorus SALMON BUSH MONKEY FLOWER. Sticky shrub. Flowers yellow tinged salmon pink. *Distr.* USA (California).

M. luteus YELLOW MIMULUS, MONKEY MUSK. Perennial herb. Flowers yellow with purple-red spots. *Distr.* Chile.

M. moschatus MUSK PLANT, MUSK FLOWER. Perennial herb. Flowers yellow with brown markings. This plant was previously widely grown for its musky scent but most of the modern cultivars are more or less scentless. *Distr.* N America.

M. ringens ALLEGHENY MONKEY FLOWER. Perennial herb. Flowers violet-blue. *Distr.* North America.

mimulus, yellow *Mimulus luteus.*

Mina See *Ipomoea*

mind your own business *Soleirolia soleirolii.*

minezuo *Loiseleuria procumbens.*

miniatus, -a, -um: flame-red-coloured.

minimiflorus, -a, -um: with the smallest flowers.

minor: smaller.

mint Labiatae, *Mentha* **apple** ~ *M. suaveolens* **common green** ~ *M. spicata* **corn** ~ *M. arvensis* **Corsican** ~ *M. requienii* **field** ~ *M. arvensis* **ginger** ~ *M.* × *gracilis* **horse** ~ *M. longifolia* **lemon** ~ *Monarda citriodora* **Meehan's** ~ *Meehania cordata* **round-leaved** ~ *Mentha suaveolens* **Scotch** ~ *M.* × *gracilis* **water** ~ *M. aquatica* **woolly** ~ *M. suaveolens.*

mint bush, balm ~ *Prostanthera melissifolia* **round-leaved** ~ *P. rotundifolia.*

Minuartia (After Juan Minuart (1693–1768), Spanish botanist.) SANDWORT. A genus of 120 species of perennial and occasionally annual herbs with linear opposite leaves and small flowers that bear 5 sepals and 5 white petals. This genus is closely related to the genus *Arenaria*. Several species are grown as ornamentals. *Distr.* N temperate and arctic regions. *Fam.* Caryophyllaceae. *Class* D.

M. sedoides MOSSY CYPHEL. Cushion-forming plant. Petals very small or absent, tinged yellow. *Distr.* European mountains, Scotland.

M. stricta TEESDALE SANDWORT. A protected species in Great Britain. *Distr.* N Europe, Arctic.

minus: less.

minutiflorus, -a, -um: with very small flowers.

minutissimus, -a, -um: very small, minute.

minutus, -a, -um: very small, minute.

mirabelle *Prunus cerasifera.*

Mirabilis (Latin for wonderful.) A genus of 60 species of annual and perennial tuberous herbs with simple opposite leaves and a cluster of 1 to several tubular flowers subtended by 5 fused bracts. Several species are grown as ornamentals. *Distr.* W North America, Central and South America. *Fam.* Nyctaginaceae. *Class* D.

M. jalapa FOUR O'CLOCK PLANT, MARVEL OF PERU, FALSE JALAP. Erect perennial. Root becoming very large. Flowers fragrant, opening in late afternoon. The flowers are used as a source of food colouring in China and the ground seeds as a cosmetic in Japan. *Distr.* Tropical America.

M. longiflora Annual herb to 1m high. Flowers in clusters of 3 or more, white-pink, fragrant at night. *Distr.* SW USA, Mexico.

M. multiflora The roots are a source of food and contain compounds that may be effective against cancer. *Distr.* SW USA.

mirabilis, -a: wonderful.

mirror plant *Coprosma repens.*

Miscanthus (From Greek *miskos*, stem, and *anthos*, flower, alluding to the stalked spikelets.) A genus of 12 species of perennial, rhizomatous grasses with reed-like stems to 4m high, long, narrow leaves, and pyramidal inflorescences. *Distr.* Tropical regions of the Old World. *Fam.* Gramineae. *Class* M.

M. sacchariflorus AMUR SILVER GRASS. Grown as a perennial ornamental for interesting autumn colour. *Distr.* Asia.

M. sinensis EULALIA. Grown as an ornamental with numerous cultivars available. *Distr.* China, Japan.

Misodendraceae A family of 1 genus and 8 species of semi-parasitic shrubs that are specific to the genus *Nothofagus*. The leaves are deciduous and green or small and scale-like. The minute flowers are unisexual with the sexes borne on separate plants. *Distr.* Temperate South America. *Class* D.

Misopates A genus of 2 species of erect annual herbs with simple leaves and spike-like racemes of 2-lipped, 5-lobed, pink or white flowers. The flowers resemble those of the genus *Antirrhinum*. *Distr.* Mediterranean to Cape Verde Islands, Ethiopia, NW India. *Fam.* Scrophulariaceae. *Class* D.

M. orontium WEASEL'S SNOUT. Grown as an ornamental. *Distr.* Eurasia, naturalized in North America.

missouriensis, -e: of the Missouri River, USA.

Miss Willmott's ghost *Eryngium giganteum*.

mistletoe Viscaceae, *Viscum*, *V. album*.

Mitchella (After Dr John Mitchell (1711–68), Virginian physician and botanist.) A genus of 2 species of mat-forming, evergreen subshrubs with simple, opposite leaves and fragrant, white, funnel-shaped flowers. The fruit is a red or occasionally white berry. *Distr.* North America, Japan. *Fam.* Rubiaceae. *Class* D.

M. repens PARTRIDGE BERRY, TWO-EYED BERRY, RUNNING BOX. Creeping perennial. Flowers borne in pairs and occasionally united at the base so as to produce a double berry. Grown as ornamental ground cover and previously used medicinally. *Distr.* North America.

Mitella (Diminutive of Greek *mitra*, cap or mitre, referring to the fruit.) BISHOP'S CAP. A genus of 20 species of rhizomatous perennial herbs with long-stalked leaves and small, green-yellow or purple, nodding flowers. *Distr.* W North America, E Asia. *Fam.* Saxifragaceae. *Class* D.

M. breweri (After William Henry Brewer (1828–1910), professor of agriculture at Yale University, USA, who worked on the botany of California.) Leaves heart- or kidney-shaped. *Distr.* W North America.

M. caulescens Flowers yellow. *Distr.* W North America.

M. diphylla Petals white, fringed. *Distr.* E North America.

M. stauropetala Flowers white or purple. *Distr.* W North America.

mitis: mild, mellow.

Mitraria (From Greek *mitra*, cap, alluding to the shape of the fruit.) A genus of 1 species of evergreen climbing subshrub with opposite, simple, leathery leaves and solitary, tubular, red flowers. *Distr.* S South America. *Fam.* Gesneriaceae. *Class* D.

M. coccinea Grown as a half-hardy ornamental.

mitrewort, false *Tiarella*.

Mitrophyllum (From Greek *mitra*, mitre, and *phyllon*, leaf, alluding to the way the pairs of leaves are held.) A genus of 25 species of succulent herbs with pairs of leaves, some of which form a cone in the shape of a bishop's hat. The white, pink, or red flowers are borne on short stalks and have numerous petals. *Distr.* South Africa (Cape Province). *Fam.* Aizoaceae. *Class* D.

M. mitratum Grown as a tender ornamental. *Distr.* South Africa (Cape Province).

mitsuba *Cryptotaenia canadensis*.

mixta: mixed.

moccasin flower *Cypripedium*.

mockernut *Carya tomentosa*.

modestus, -a, -um: modest.

Moehringia (After Paul Heinrich Gerard Moehring (1710–92), German physician and botanist.) A genus of 25 species of low-growing perennial herbs with opposite, linear to rounded leaves, and small flowers that bear 4 free, typically white petals. This genus is closely related to the genus *Arenaria*. *Distr.* N temperate and Arctic regions. *Fam.* Caryophyllaceae. *Class* D.

M. muscosa Grown as a rock-garden ornamental for its moss-like foliage and star-like white flowers. *Distr.* Mountains of Central and S Europe.

mohintli *Justicia spicigera.*

mole plant *Euphorbia lathyris.*

Molinia (After Juan Ignacio Molina (1740–1829), Chilean botanist.) A genus of 2 species of perennial, loosely tufted grasses with flat leaves and narrow, loosely branched inflorescences. *Distr.* Eurasia. *Fam.* Gramineae. *Class* M.

M. altissima See *M. caerulea*

M. caerulea PURPLE MOOR GRASS. A commonly grown ornamental with numerous cultivars available. *Distr.* Europe, N and SW Asia.

mollis, -e: soft.

mollissimus, -a, -um: very soft.

Molluginaceae A family of 15 genera and about 130 species of herbs and shrubs with simple leaves and small, regular, bisexual flowers. *Distr.* Chiefly tropical regions. *Class* D.

mollugo: resembling the genus *Mollugo* (Indian chickweed).

Moltkia (After Count Joachim Gadske Moltke (1746–1818), Danish statesman and naturalist.) A genus of 6 species of perennial herbs and shrubs with alternate simple leaves and cymes of funnel-shaped flowers. Several species are grown as rock-garden ornamentals. *Distr.* Europe, SW Asia. *Fam.* Boraginaceae. *Class* D.

M. graminifolia See *M. suffruticosa*

M. × intermedia A hybrid of *M. petraea* × *M. suffruticosa*. *Distr.* Italy, Albania, Greece.

M. petraea Semi-evergreen, bushy shrub. *Distr.* SE Europe.

M. suffruticosa Small deciduous sub-shrubs. *Distr.* Italy.

Moluccella (After the Moluccas, where these plants were first collected.) A genus of 4 species of annual and biennial herbs with 4-angled stems, simple, opposite leaves and whorls of fragrant flowers that bear a 2-lipped petal-tube surrounded by a large, bell-shaped sepal-tube. *Distr.* NW India to the Mediterranean. *Fam.* Labiatae. *Class* D.

M. lavis BELLS OF IRELAND, SHELLFLOWER. Annual herb to 1m high. Grown as an ornamental for the long-lasting, showy sepal-tube. *Distr.* W Asia.

moly *Allium moly.*

monadelphus, -a, -um: with united stamens.

monanthus, -a, -um: one flowered.

Monadenium (From Greek *mono*, one, and *aden*, gland.) A genus of 48 species of succulent herbs and shrubs with simple leaves and poisonous milky latex. The small, highly reduced flowers are borne in irregular inflorescences. Several species are grown as ornamentals. *Distr.* Tropical and S Africa. *Fam.* Euphorbiaceae. *Class* D.

M. lugardae Perennial herb. Stems erect. Leaves spiny on margin. Inflorescence pendent with yellow-orange bracts. *Distr.* S Africa.

monarch of the east *Sauromatum venosum.*

Monarda (After Nicholas Monardes (1493–1588), Spanish botanist and physician.) BERGAMOT. A genus of 12 species of aromatic, annual and perennial herbs with simple opposite leaves, and rounded heads of narrow, tubular, 2-lipped flowers. As well as several species and numerous cultivars being grown as ornamentals, several species have been used medicinally and in herbal teas. *Distr.* North America. *Fam.* Labiatae. *Class* D.

M. citriodora LEMON MINT. Annual to 60cm high. Flowers in rounded heads, pink or white. Used medicinally. *Distr.* S USA, Mexico.

M. didyma OSWEGO TEA, SWEET BERGAMOT, BEE BALM. Perennial to 1.2m high. Flowers in rounded clusters, bright red. A culinary herb and source of Oil of Bergamot. *Distr.* E North America.

M. fistulosa WILD BERGAMOT. Perennial to 1.2m high. Flowers lilac-purple. *Distr.* E North America.

Monardella (Diminutive of the genus name *Monarda*.) A genus of 19 species of aromatic, annual and perennial herbs with small, simple, opposite leaves and rounded, narrow, tubular, 2-lipped flowers. Several species are grown as rock-garden ornamentals and the leaves of some are made into medicinal teas. *Distr.* W North America. *Fam.* Labiatae. *Class* D.

M. macrantha Decumbent shrubby perennial. Flowers red to yellow. *Distr.* S California.

M. odoratissima A variable species. Stems somewhat woody. Flowers typically pink-purple. *Distr.* W North America.

moneywort *Lysimachia nummularia.*

mongolicus, -a, um: of Mongolia.

Monilaria (From Latin *monile*, a string of pearls, alluding to the shape of the stems.) A genus of 11 species of clump-forming, succulent herbs with thick, jointed stems and pairs of more or less fused leaves. The white, yellow or red flowers are borne on long stalks and have numerous petals. Several species are cultivated. *Distr.* South Africa. *Fam.* Aizoaceae. *Class* D.

M. moniliformis Highly succulent. Flowers to 3cm across, yellow sometimes flushed red. Grown as a tender ornamental. *Distr.* South Africa (Cape Province).

Monimiaceae A family of 39 genera and about 500 species of trees and shrubs with opposite, often aromatic leaves and small, unisexual flowers that bear 2-many tepals and 10-numerous stamens. Several species are used locally for timber, edible fruits and in perfumes and medicines. *Distr.* Tropical to warm-temperate regions, mainly in the S hemisphere. *Class* D.
See: Atherosperma, Laurelia.

monkey flower *Mimulus* **Allegheny** ~ *M. ringens* **bush** ~ *M. aurantiacus* **cardinal** ~ *M. cardinalis* **common large** ~ *M. guttatus* **great purple** ~ *M. lewisii* **salmon bush** ~ *M. longiflorus* **scarlet** ~ *M. cardinalis.*

monkey musk *Mimulus luteus.*

monkey plant *Ruellia makoyana.*

monkey puzzle Araucariaceae, *Araucaria.*

monk's hood *Aconitum, A. napellus.*

monocanthus, -a, -um: one-spined.

Monochoria (From Greek *monos*, one, and *chorizo*, to separate, alluding to the stamens.) A genus of 5 species of rhizomatous, aquatic herbs with emergent leaves and elongated racemes of blue, slightly irregular flowers which bear 5 short, yellow stamens and 1 large blue stamen. *Distr.* Africa, Asia, Australia. *Fam.* Pontederiaceae. *Class* M.

M. vaginalis Grown as an ornamental. Roots eaten as a vegetable. *Distr.* S and SE Asia.

monocolor: one-coloured.

Monodora (From Greek *monos*, one, and *dora*, gift, alluding to the solitary flowers.) A genus of 20 species of shrubs and trees with simple, alternate, evergreen leaves and large, fragrant, 6-petalled flowers. *Distr.* Tropical Africa and Madagascar. *Fam.* Annonaceae. *Class* D.

M. myristica CALABASH NUTMEG, JAMAICA NUTMEG. Flowers yellow marked purple. The seeds are used like nutmeg and are reported to have medicinal qualities. *Distr.* W tropical Africa.

monogynus, -a, -um: with one ovary.

monophyllos: one-leaved.

monophyllus, -a, -um: with one leaf.

monosematus, -a, -um: with a single marking.

monospermus, -a, -um: with one seed.

monostachius, -a, -um: with one spike.

monostachyus, -a, -um: with one spike.

monox *Empetrum nigrum.*

Monsonia (After Lady Ann Monson (*c.*1714–76), a correspondent of Linnaeus.) A genus of 25 species of herbs and subshrubs with simple or divided, unequal pairs of leaves and regular, bisexual flowers that bear 5 sepals, 5 petals, and 15 stamens. *Distr.* Africa, Madagascar, SW Asia. *Fam.* Geraniaceae. *Class* D.

M. speciosa Evergreen subshrub. Flowers solitary, red or purple. Grown as a tender ornamental. *Distr.* South Africa (Cape Province).

monspeliensis, -e: of Montpellier, France.

monspessulanus, -a, -um: of Montpellier, France.

Monstera (Derivation obscure, possibly from Latin *monstrum*, a marvel, alluding to the appearance of the leaves.) A genus of 25 species of large epiphytic lianas with simple, perforated or pinnately cut, leathery leaves and corky aerial roots. The small flowers are borne on a spadix which is subtended by a boat-shaped, yellow-white spathe. Several species are grown as ornamentals and some have edible spadices. *Distr.* Tropical regions of the New World. *Fam.* Araceae. *Class* M.

M. acuminata SHINGLE PLANT. Stems flattened against support. Leaves oval with a very lopsided base. *Distr.* Central America.

M. deliciosa SWISS CHEESE PLANT, CERIMAN. Very commonly cultivated as a houseplant, now quite rare in the wild. *Distr.* Mexico, Central America.

monstrosus, -a, -um: monstrous, abnormal.

montanus, -a, -um: of mountains.

montbretia *Crocosmia.*

montevidensis, -e: of Montevideo.

Montia (After Guiseppe Monti (1682–1760), Italian professor of botany.) A genus of 37 species of annual and perennial herbs with somewhat fleshy, stalked, basal leaves and small, white or pink flowers that bear 2–5 petals. Several species are grown as ornamentals. *Distr.* Widespread but not Australia. *Fam.* Portulacaceae. *Class* D.

M. australasica See *Claytonia australasica.*

M. perfoliata WINTER PURSLANE, MINER'S LETTUCE, CUBAN SPINACH. Annual herb. Leaves bright green. Leaves edible. *Distr.* W North America.

monticola: growing on mountains.

Montiniaceae A family of 2 genera and 3 species of shrubs and small trees with simple leaves and regular, bisexual flowers. This family is sometimes included within the family Grossulariaceae. *Distr.* S and E Africa, Madagascar. *Class* D.

Monvillea See *Acanthocereus*

moonflower *Ipomoea alba.*

moonseed Menispermaceae, *Cocculus, Menispermum, M. canadense* **Carolina** ~ *Cocculus carolinus* **red** ~ *C. carolinus.*

moonstones *Pachyphytum oviferum.*

moorberry *Vaccinium uliginosum.*

mooseberry *Viburnum lantanoides.*

moosewood *Viburnum lantanoides.*

Moraceae The Mulberry family. 37 genera and about 1000 species of herbs, shrubs, lianas and trees with simple, spirally arranged or opposite leaves and small, frequently wind pollinated flowers. The leaves and stems frequently contain latex. *Distr.* Tropical to warm-temperate regions. *Class* D.
See: Broussonetia, Ficus, Maclura, Morus.

Moraea (After Robert More (1703–80), English botanist.) PEACOCK LILY. A genus of 111 species of perennial herbs with linear leaves and short-lived *Iris*-like flowers. Several species are grown as ornamentals. *Distr.* S Africa. *Fam.* Iridaceae. *Class* M.

M. gawleri Flowers yellow-marked brick-red. *Distr.* South Africa (Cape Province).

M. iridioides See *Dietes iridioides*

M. spathacea See *M. spathulata*

M. spathulata Leaves solitary. Flowers born in succession, yellow. *Distr.* S Africa.

morel, petty *Aralia racemosa.*

Morina (After Louis Pierre Morin (1636–1715), French botanist.) A genus of 4 species of spiny perennial herbs with tufted, spine-toothed leaves and whorls of 2-lipped, 5-lobed, tubular flowers. *Distr.* Balkans to Central Himalaya. *Fam.* Morinaceae. *Class* D.

M. longifolia WHORL FLOWER. Leaves dark green. Flowers pink and red. Grown as an ornamental. *Distr.* Himalaya.

Morinaceae A family of 12 genera and 4 species. *Class* D.
See: Acanthocalyx, Morina.

Moringaceae A family of 1 genus and 10 species of fast-growing deciduous trees with pinnate or bipinnate leaves and numerous, sweet-scented flowers. *Distr.* Africa, Madagascar, India, naturalized elsewhere. *Class* D.

Morisia (After Guiseppe Giacinto Moris (1796–1869), Italian botanist.) A genus of 1 species of perennial, hairy herb with solitary, golden yellow flowers. *Distr.* Corsica and Sardinia. *Fam.* Cruciferae. *Class* D.

M. hypogaea See *M. monanthos*

M. monanthos A rock-garden ornamental. *Distr.* Corsica, Sardinia.

Mormodes (From Greek *mormo*, goblin, alluding to the strange form of the flowers.) A genus of 50 species of epiphytic orchids with showy flowers, the column twisted to one side of the lip, and the pollinia released explosively. Some species are cultivated as tender ornamentals. *Distr.* Central and South America. *Fam.* Orchidaceae. *Class* M.

M. aromatica Inflorescence ascending, many-flowered. Flowers green-brown, spicy, fragrant. *Distr.* Mexico, El Salvador.

M. buccinator Flowers variable to 6cm across, green to yellow. *Distr.* Venezuela.

M. colossus Flowers to 12cm across, green-yellow, fragrant. *Distr.* Costa Rica.

M. hookeri Flowers to 4cm across, red-brown, fragrant. *Distr.* Panama.

M. maculatum Flowers tawny yellow, spotted red. *Distr.* Mexico.

Mormolyca (From Greek *mormolyka*, hobgoblin, alluding to the form of the flower.) A genus of 5 species of epiphytic orchids with large leathery leaves and fleshy flowers. *Distr.* Central and tropical South America. *Fam.* Orchidaceae. *Class* M.

M. ringens Flowers yellow to lavender with a yellow to maroon lip. *Distr.* Central America, from Mexico to Costa Rica.

morning glory *Ipomoea*, *I. tricolor* **common** ~ *I. purpurea* **red** ~ *I. coccinea* **yellow** ~ *Merremia tuberosa*.

Morus (Classical Latin name for mulberry.) MULBERRY. A genus of 12 species of short-lived, deciduous shrubs and trees with latex-containing shoots, simple or lobed leaves and small, wind pollinated, unisexual flowers. The male flowers are borne in pendulous catkins and the females in erect spikes. The fruit is a syncarp, made up of numerous drupes. *Distr.* Subtropical and warm-temperate regions. *Fam.* Moraceae. *Class* D.

M. alba WHITE MULBERRY. Tree to 16m high. Fruit pink-red, edible but of poor flavour. The leaves of this species are the principle food of the silk worm. *Distr.* Central Asia to China.

M. australis Shrub or small tree. Fruit red, sweet. *Distr.* China, Japan, Taiwan.

M. nigra COMMON MULBERRY, BLACK MULBERRY. Tree to 15m high. Fruit red to deep purple. The best eating mulberry and also a source of timber. *Distr.* W Asia.

M. rubra RED MULBERRY. Tree to 15m high. Fruit orange to purple, sweet with good flavour. A source of timber. *Distr.* E North America.

mosaic plant *Fittonia verschaffeltii*.

moschatel *Adoxa moschatellina*.

moschatus, -a, -um: musk-scented.

Moses in his cradle *Tradescantia spathacea*.

Moses in the bulrushes *Tradescantia spathacea*.

Moses on a raft *Tradescantia spathacea*.

moss: Allegheny ~ *Robinia kelseyi* **ball** ~ *Tillandsia recurvata* **fairy** ~ *Azolla* **Florida** ~ *Tillandsia usneoides*. **Irish** ~ *Soleirolia soleirolii* **mat spike** ~ *Selaginella kraussiana* **Spanish** ~ *Tillandsia usneoides* **spike** ~ *Selaginella* **trailing spike** ~ *S. kraussiana*.

mossy cyphel *Minuartia sedoides*.

mother in law's tongue *Sansevieria*, *S. trifasciata*.

mother of pearl plant *Graptopetalum paraguayense*.

mother of thousands *Cymbalaria muralis*, *Saxifraga stolonifera*, *Soleirolia soleirolii*.

motherwort *Leonurus cardiaca*.

moupinensis, -e: of Moupin, China.

mournful widow *Scabiosa atropurpurea*.

mouse-ear, Alpine *Cerastium alpinum*.

mouse plant *Arisarum proboscideum*.

mouse-tail plant *Arisarum proboscideum*.

mouth root *Coptis*.

mucronatus, -a, -um: mucronate, with a short sharp tip.

mucronifolius, -a, -um: leaves with a short sharp point.

mucronulatus, -a, -um: with a short point.

Muehlenbeckia (After Henri Gustav Muehlenbeck (1789–1845), Swiss botanist and physician.) A genus of 20 species of deciduous, twining shrubs with small, occasionally absent leaves, and minute, 5-lobed flowers. Several species are grown as ornamentals. *Distr.* Australasia, temperate South America. *Fam.* Polygonaceae. *Class* D.

M. adpressa Deciduous twining shrub. Berries edible. *Distr.* Temperate Australia.

M. complexa MAIDENHAIR VINE, WIRE VINE, MATTRESS VINE, NECKLACE VINE. Deciduous twining shrub. Leaves bright green marked with purple. Flowers in small spikes. *Distr.* New Zealand.

mugwort *Artemisia*, *A. vulgaris*. **western** ~ *A. ludoviciana* **white** ~ *A. lactiflora*.

Muilla (Reversal of the genus name *Allium*.) A genus of 5 species of perennial herbs with corms, long thin leaves that are round in cross-section, and umbels of numerous, star-shaped, green-white flowers. *Distr.* SW USA, N Mexcio. *Fam.* Alliaceae. *Class* M.

M. maritima Perennial herb to 50cm high. Flowers to 1cm across, white with purple anthers. Grown as an ornamental. *Distr.* USA (S California), Mexico (Baja California).

Mukdenia (From the name of the Chinese city, Mukden.) A genus of 2 species of rhizomatous herbs with peltate, deeply palmately lobed leaves, and numerous small, white flowers. *Distr.* N China, Manchuria, Korea. *Fam.* Saxifragaceae. *Class* D.

M. rossii Deciduous ground cover for shaded areas of the garden.

mulberry Moraceae, *Morus* **black** ~ *Morus nigra* **common** ~ *M. nigra* **paper** ~ *Broussonetia papyrifera* **red** ~ *Morus rubra* **white** ~ *M. alba*.

mullein *Verbascum* **common** ~ *V. thapsus* **dark** ~ *V. nigrum* **moth** ~ *V. blattaria* **white** ~ *V. lychnitis*.

multibracteatus, -a, -um: with many bracts.

multibulbosus, -a, -um: with many bulbs.

multicaulis, -e: with many stems.

multiceps: with many heads.

multifidus, -a, -um: divided many times.

multiflorus, -a, -um: with many flowers.

multijugus, -a, -um: with many parts joined together.

multiplex: with many folds.

multiradiatus, -a, -um: with many rays.

multiscapoideus, -a, -um: with many scapes.

multisectum: with many divisions.

munitus, -a, -um: armed (e.g. with teeth or spines).

muralis, -e: growing on walls.

muricatus, -a, -um: with a rough, hard-pointed surface.

Murraya (After John Andrew Murray (1740–91), professor of botany and medicine at Göttingen.) A genus of 4 species of aromatic shrubs and trees with alternate, pinnate leaves and large panicles of regular flowers that bear 5 petals and 10 stamens. Several species are grown as tender ornamentals. *Distr.* India and Malaysia to the Pacific. *Fam.* Rutaceae. *Class* D.

M. paniculata CHINESE BOX, SATIN WOOD, ORANGE JASMINE. Shrub or tree. Flowers white, numerous, produced several times a year. The wood is used for cutlery handles and walking sticks. *Distr.* SE Asia, W Malaysia.

Musa (After Antonius Musa, physician to Octavius Augustus, first emperor of Rome (63–14 BC), or possibly from the Arabic *mauz*, banana.) BANANA. A genus of 40 species of large, rhizomatous, tree-like herbs with very large, stalked leaves which have sheathing bases that form a pseudostem. The flowers are borne in the axils of leathery bracts in a large terminal spike. Several species and numerous hybrids are grown for their fruit as well as their beauty. *Distr.* SE Asia. *Fam.* Musaceae. *Class* M.

M. basjoo JAPANESE BANANA. Cultivated in Japan as an ornamental and for fibre. *Distr.* S Japan.

M. ensete See *Ensete ventricosum*

M. × paradisiaca EDIBLE BANANA, FRENCH PLANTAIN, PLANTAIN, SILK FIG, DESSERT BANANA. A complex group of hybrids derived from *M. acuminata* × *M. balbisiana*; most of the edible cultivated bananas are attributed to this group of hybrids. DESSERT BANANAS have a sweet taste when ripe and are eaten raw; the most popular cultivars are the 'Cavendish' cultivars, not least 'Dwarf Cavendish'. PLANTAINS have a high starch content, are somewhat acidic when ripe and are eaten cooked. All edible bananas are sterile and do not have seeds. *Distr.* Tropical regions.

M. textilis MANILA HEMP, ABACA. Grown as a source of fibre. *Distr.* Philippines.

Musaceae The Banana family. 2 genera and 42 species of very large, tree-like herbs with pseudostems formed from leaf sheaths and leaf blades that are large, with numerous veins running from the thick midrib to the margin. The unisexual flowers are borne in clusters in the axils of keeled bracts that are spirally arranged on a large inflorescence. They have 5 reduced and 1 large tepal, 5–6 stamens, and an inferior ovary. The fruit is a fleshy berry. Fruits of some species are eaten; others are a source of fibre. *Distr.* Tropical regions of the Old World. *Class* M.
See: Ensete, Musa.

musaicus, -a, -um: like a mosaic.

Muscari (From Greek *muschos*, musk, alluding to the scent of some species.) GRAPE HYACINTH. A genus of 60 species of bulbiferous herbs with linear, somewhat fleshy leaves and racemes of cylindrical or bell-shaped, blue or white flowers. The upper flowers on the raceme are sometimes sterile and form a tuft or coma. A number of species are grown as ornamentals. *Distr.* Mediterranean, SW Asia. *Fam.* Hyacinthaceae. *Class* M.

M. comosum TASSEL HYACINTH. Raceme loose with brown-tinged flowers and an erect coma. Bulbs edible. *Distr.* S and Central Europe, N Africa, SW Asia.

M. neglectum COMMON GRAPE HYACINTH. Raceme dense. Flowers deep blue, rounded, contracted at mouth. *Distr.* Europe, N Africa, W Asia, Caucasus.

M. paradoxum See *Bellevalia pycnantha*

M. racemosum See *M. neglectum*

muscarioides: from the genus *Muscari*, with the ending *-oides*, indicating resemblance.

muscosus, -a, -um: resembling moss, mossy.

musifolius, -a, -um: from the genus *Musa*, BANANA, and *folius*, leaved.

musk flower *Mimulus moschatus.*

musk plant *M. moschatus.*

muskroot *Adoxa moschatellina.*

muskwood *Olearia argophylla.*

Musschia (After J. H. Mussche (1765–1834), director of the Botanic Garden, Ghent.) A genus of 2 species of small trees with thick trunks, simple leaves, and dull, bell-shaped flowers. *Distr.* Madeira. *Fam.* Campanulaceae. *Class* D.

M. aurea Flowers visited by nectar-seeking lizards. *Distr.* Madeira.

mustard Cruciferae **black** ~ *Brassica nigra* **brown** ~ *B. juncea* **Chinese** ~ *B. juncea* **field** ~ *B. campestris* **garlic** ~ *Alliaria petiolata* **Indian** ~ *Brassica juncea* **Mithridate** ~ *Thlaspi arvense.*

mustard tree *Nicotiana glauca.*

mutabilis, -e: changeable.

mutatus, -a, -um: changed.

Mutisia (After Jose Celestino Mutis (1732–1808), Spanish botanist who studied the South American flora.) A genus of 59 species of evergreen shrubs and lianas with simple or pinnate leaves and flowers in medium to large daisy-like heads which have distinct rays. Some species are cultivated as half-hardy ornamentals. *Distr.* South America. *Fam.* Compositae. *Class* D.

M. clematis Leaves pinnate. Flowerheads pendulous, rays orange-red. *Distr.* Colombia.

M. decurrens Much-branched subshrub. Rays bright orange. *Distr.* Chile, Argentina.

M. felthamii See *Diascia fetcaniensis*

M. ilicifolia Large branched shrub to 3m high. *Distr.* Chile.

M. oligodon Straggling shrub or liana. *Distr.* Chile, Argentina.

M. spinosa Liana to 6m high. Rays pink. *Distr.* Chile.

Myoporaceae The Emu Bush family. 3 genera and about 220 species of shrubs and trees with simple leaves and bisexual, typically 2-lipped, 5-lobed flowers. *Distr.* Mauritius, E Asia to Australasia, S Pacific, West Indies. *Class* D.
See: Myoporum.

Myoporum (From Greek *myo*, to shut, and *poros*, pore, alluding to the transparent spots on the leaves.) A genus of 32 species of evergreen shrubs and trees with simple, alternate, glandular leaves and 5-lobed, tubular to bell-shaped flowers. Several species are grown as ornamentals and some have edible fruits or are used to create shelter belts. *Distr.* E Asia, Australia, New Zealand, Hawaii. *Fam.* Myoporaceae. *Class* D.

M. laetum Rounded shrub or tree. Flowers borne in clusters, bell-shaped, white with purple spots. *Distr.* New Zealand.

M. parvifolium Spreading or prostrate shrub. Leaves somewhat fleshy. Flowers pink, fragrant. *Distr.* Australia.

Myosotidium (From the genus name *Myosotis*, and Greek *-oides*, indicating resemblance.) A genus of 1 species of evergreen perennial herb with large, glossy, ribbed, basal leaves and heads of funnel-shaped, blue flowers. *Distr.* Chatham Islands. *Fam.* Boraginaceae. *Class* D.

M. hortensia CHATHAM ISLAND FORGET-ME-NOT. Grown as a half-hardy perennia.

M. nobile See *M. hortensia*

Myosotis (From the Greek *mus*, mouse, and *otos*, ear, alluding to the shape of the leaves.) FORGET ME NOT, SCORPION GRASS. A genus of 50 species of annual, biennial, and perennial, hairy herbs with simple, alternate leaves and cymes of salver-shaped flowers. The throats of the flowers are almost closed by a ring of scales. Numerous species and cultivars are grown as ornamentals. *Distr.* Europe, Asia. *Fam.* Boraginaceae. *Class* D.

M. alpestris ALPINE FORGET-ME-NOT. Short-lived, clump-forming perennial. Flowers very small, bright blue, yellow in the centre. *Distr.* Europe.

M. australis Short-lived, clump-forming perennial. Flowers open, funnel-shaped, yellow-white. *Distr.* New Zealand.

M. rupicola See *M. alpestris*

M. scorpioides WATER FORGET-ME-NOT. Perennial of damp areas. Leaves narrow. Flowers small, blue. *Distr.* Europe, Asia.

M. sylvatica GARDEN FORGET ME NOT. Biennial or perennial. Flowers blue-purple with a yellow centre. Numerous cultivars are available. *Distr.* Europe, Asia.

Myrceugenella See *Luma*

Myrceugenia A genus of 38 species of aromatic shrubs and trees with simple, opposite leaves and clusters of small flowers that bear 5 petals and numerous stamens. Several species are grown as ornamentals. *Distr.* Juan Fernández Island, Chile, Argentina, SE Brazil. *Fam.* Myrtaceae. *Class* D.

M. exsucca Tree to 14m high. *Distr.* Chile, Argentina.

myrianthus, -a, -um: with numerous flowers.

Myrica (From *myrike*, classical Greek name for Tamarisk.) A genus of 35 species of deciduous and evergreen shrubs and small trees with simple, alternate leaves and small unisexual flowers, the male flowers being borne in catkins and the females in rounded clusters. The fruit is a resinous or waxy drupe. Several species are grown as ornamentals. *Distr.* Widespread, except Australasia. *Fam.* Myricaceae. *Class* D.

M. gale (From an Old English name for this plant, *gagel*.) SWEET GALE, BOG MYRTLE. Deciduous shrub. Fruits yellow-brown, resinous. *Distr.* Maritime Europe and North America.

M. pensylvanica BAYBERRY, CANDLEBERRY, SWAMP CANDLEBERRY. Deciduous or semi-evergreen shrub. Fruits off-white, covered in rough wax. *Distr.* E North America.

Myricaceae A family of 3 genera and 50 species of aromatic shrubs and trees with simple or pinnately cut leaves and unisexual flowers borne in catkin-like spikes, both males and females on the same plant. *Distr.* Widespread. *Class* D.
See: Comptonia, Myrica.

Myricaria (From *myrike*, classical Greek name for a plant of the genus *Tamarix*, to which it is closely related.) FALSE TAMARISK. A genus of 10 species of deciduous subshrubs and shrubs with small, scale-like leaves and narrow racemes of 5-lobed flowers that bear 10 fused stamens. *Distr*. Tropical America. *Fam*. Tamaricaceae. *Class* D.

M. germanica Deciduous shrub to 2m high. Flowers bright red. Grown as a hardy ornamental. *Distr*. Europe to the Himalaya.

myricoides: from the genus *Myrica*, with the ending *-oides*, indicating resemblance.

Myriocarpa (From Greek *myrios*, myriad or very many, and *karpos*, fruit.) A genus of 15 species of shrubs and small trees with simple, alternate leaves and slender, branched inflorescences of small green flowers. *Distr*. Central and South America. *Fam*. Urticaceae. *Class* D.

M. stipitata Shrub or small tree to 15m high. Occasionally grown as a tender ornamental for its foliage. *Distr*. South America.

Myriophyllum (From Greek *myrios*, many, and *phyllon*, leaf, alluding to the finely divided leaves.) WATER MILFOIL. A genus of 40 species of submerged aquatic and terrestrial herbs with pinnate leaves and spikes of small, wind-pollinated flowers. *Distr*. Widespread. *Fam*. Haloragaceae. *Class* D.

M. aquaticum PARROT FEATHER, DIAMOND MILFOIL. Aquatic herb. Flowers solitary in the axils of large feathery bracts borne above the water. Grown in aquaria. *Distr*. South America.

M. brasiliense See *M. aquaticum*

M. proserpinacoides See *M. aquaticum*

M. spicatum SPIKED WATER MILFOIL. Aquatic herb. Emergent inflorescence spike-like. *Distr*. Temperate N Hemisphere.

myriophyllus, -a, -um: with numerous leaves.

Myristica (From Greek *myristikos*, smelling of MYRRH.) A genus of 70–100 species of large evergreen trees with simple, alternate, often waxy leaves and clusters or racemes of small unisexual flowers. The fruit contains a single large seed surrounded by a fleshy aril. *Distr*. Tropical regions of the Old World. *Fam*. Myristicaceae. *Class* D.

M. fragrans NUTMEG, MACE. Tree to 10m high. Flowers yellow. The seed is the NUTMEG of commerce and the red aril, MACE. Nutmeg is used as a flavouring and is a mild hallucinogen. *Distr*. Indonesia (Moluccas), widely cultivated in tropical regions.

Myristicaceae The Nutmeg family. 18 genera and about 400 species of trees with alternate simple leaves often borne on whorled, horizontal branches. The flowers are small and inconspicuous, males and females being borne on separate trees. The fruit contains a single large seed wrapped in a network of brightly coloured tissue, the aril. *Myristica fragrans*, the source of nutmeg and mace, belongs to this family. *Distr*. Tropical regions, mostly Asia and America. *Class* D.
See: Myristica.

Myrothamnaceae A family of 1 genus and 2 species of aromatic shrubs with simple opposite leaves and wind-pollinated flowers. *Distr*. Africa, Madagascar. *Class* D.

myrrh *Commiphora abyssinica* **garden** ~ *Myrrhis odorata* **Mecca** ~ *Commiphora opobalsamum*.

Myrrhis (Name used in Greek.) A genus of 1 species of aromatic, perennial herb to 2m high with hollow stems, divided fern-like leaves and compound umbels of white flowers. *Distr*. Europe. *Fam*. Umbelliferae. *Class* D.

M. odorata SWEET CICELY, GARDEN MYRRH. Grown as an ornamental and as a culinary herb. *Distr*. Europe.

Myrsinaceae A family of 39 genera and 1250 species of shrubs and trees with simple leathery leaves and small regular flowers that bear 4–6 sepals and petals. *Distr*. Widespread in tropical and occasionally temperate regions. *Class* D.
See: Ardisia, Elingamita, Myrsine.

Myrsine (From classical Greek name for Common Myrtle.) A genus of 5 species of evergreen shrubs and trees with alternate, simple, leathery leaves and umbels or clusters of small flowers that bear their parts in fours or fives. Several species are grown as ornamentals. *Distr*. Azores, Africa, Asia. *Fam*. Myrsinaceae. *Class* D.

M. africana CAPE MYRTLE, AFRICAN BOXWOOD. Dense bushy shrub. Leaves small, aromatic. Flowers minute, yellow-brown. *Distr*. E and S Africa, Himalaya, China.

myrsinifolius, -a, -um: from the genus *Myrsine*, and *folius*, leaved.

Myrtaceae The Eucalyptus family. 127 genera and about 4000 species of small shrubs to large trees, typically with opposite, leathery, evergreen leaves which frequently contain oil glands. The flowers are regular, bisexual, and have 4–5 sepals, 4–5 petals, and usually numerous stamens. There are several economically important genera in this family, notably *Eucalyptus*. *Distr.* Tropical, subtropical and S temperate regions. Class D. *See: Acca, Amomyrtus, Baeckea, Beaufortia, Callistemon, Calothamnus, Calytrix, Eucalyptus, Eugenia, Kunzea, Leptospermum, Lophomyrtus, Luma, Melaleuca, Metrosideros, Myrceugenella, Myrceugenia, Myrteola, Myrtus, Pimenta, Psidium, Tristania, Ugni.*

myrtifolius, -a, -um: from the genus *Myrtus*, and Latin *folius*, leaved.

Myrtillocactus (From Latin *myrtillus*, alluding to *Vaccinium myrtillus* which has similar fruits.) A genus of 4 species of shrubby or tree-like cacti with ridged branches and small, star-shaped flowers that open at night. Several species are grown as ornamentals and the fruits are made into jam. *Distr.* Mexico, Guatemala. *Fam.* Cactaceae. *Class* D.

M. geometrizans Tree to 5m high. Stems blue-green. Flowers cream. *Distr.* Mexico.

M. schenkii (After Professor H. Schenk, director of the Darmstadt Botanic Garden.) Tree to 5m high. Stems dark green. Flowers cream. *Distr.* Mexico.

myrtle *Nothofagus cunninghamii* **bog** ~ *Myrica gale* **bracelet honey** ~ *Melaleuca armillaris* **Cape** ~ *Myrsine africana* **common** ~ *Myrtus communis* **crêpe** ~ *Lagerstroemia indica* **Jew's** ~ *Ruscus aculeatus* **Oregon** ~ *Umbellularia californica* **prickly** ~ *Rhaphithamnus spinosus* **sand** ~ *Leiophyllum buxifolium* **snow** ~ *Calytrix alpestris*.

Myrtus (Classical Greek and Latin name for these plants.) A genus of 2 species of aromatic, evergreen shrubs with simple, opposite, glandular leaves and solitary, regular, white flowers that bear 4 spreading petals and numerous stamens. *Distr.* S Europe, N Africa. *Fam.* Myrtaceae. *Class* D.

M. apiculata See *Luma apiculata*

M. bullata See *Lophomyrtus bullata*

M. communis COMMON MYRTLE. Much-branched shrub. Numerous ornamental cultivars have been raised from this species and it has been used medicinally as well as as a source of tannin. *Distr.* SW Europe.

M. lechleriana See *Myrceugenia exsucca*

M. luma See *Luma apiculata*

M. obcordata See *Lophomyrtus obcordata*

N

Nageliella (After Otto Nageli (1871–1938), who collected plants in Mexico.) A genus of 2 species of epiphytic or terrestrial orchids with thick, leathery, tongue-like leaves and racemes of small, pink to red flowers. Both species are cultivated as tender ornamentals. *Distr.* Mexico and Central America. *Fam.* Orchidaceae. *Class* M.

N. angustifolia Flowers nodding, bright magenta. *Distr.* Guatemala.

N. purpurea Flowers red-purple, borne in thin erect racemes. *Distr.* Mexico to Honduras.

nailrod *Typha latifolia.*

naked boys *Colchicum autumnale.*

naked ladies *Colchicum.*

nana *Clethra alnifolia.*

Nancy, early *Wurmbea dioica.*

Nandina (From *nanten*, the Japanese name for this plant.) A genus of 1 species of evergreen or somewhat deciduous shrub with clumps of erect stems and bipinnate leaves that become tinged with red in autumn. The small white flowers are borne in terminal panicles and bear 3–6 petals and 6 stamens. The fruit is a bright red berry. *Distr.* India to Japan. *Fam.* Berberidaceae. *Class* D.

N. domestica HEAVENLY BAMBOO. Grown as an ornamental. *Distr.* India to Japan.

Nandinaceae See Berberidaceae

Nannorrhops (From Greek *nannos*, dwarf, and *rhops*, bush.) A genus of 4 species of bushy palms. *Distr.* Pakistan, Afghanistan, Iran, and Arabia. *Fam.* Palmae. *Class* M.

N. ritchiana Grown as an ornamental; the fibre is used for rope and the leaves for basketry. *Distr.* Pakistan, Afghanistan, Iran and Arabia.

Nanodes (From Greek *nanodes*, pygmy.) A genus of 3 species of small, epiphytic orchids with cane-like stems, 2-ranked leaves, and small waxy flowers. Members of this genus were formerly included in the genus *Epidendrum*. Some species are cultivated as tender ornamentals. *Fam.* Orchidaceae. *Class* M.

N. medusae Flowers yellow-green, tinged red. *Distr.* Ecuador.

nanus, -a, -um: dwarf.

napellus, -a, -um: resembling a small turnip, from the Latin *napus*, a turnip.

narbonensis, -e: of Narbonne, France.

narcissiflorus, -a, -um: with flowers like those of the genus *Narcissus*.

Narcissus (After Narcissus of Greek mythology, who was turned into such a plant after killing himself because he couldn't reach his reflection in a pool.) DAFFODIL. A genus of 50 species of bulbiferous, perennial herbs with linear to very narrow leaves and distinctive yellow or white, sometimes scented flowers. The flowers consist of 6 basally fused tepals and a trumpet-shaped corona, or crown, which originated as a appendage to the stamens but in this genus has become free of them and is often very showy. Several species and numerous cultivars are grown as perennial ornamentals or for cut flowers. *Distr.* S Europe, N Africa, E Mediterranean. *Fam.* Amaryllidaceae. *Class* M.

N. alpestris See *Narcissus pseudonarcissus*

N. assoanus RUSH-LEAVED JONQUIL. Leaves cylindrical. Flowers 2–3, yellow, fragrant. *Distr.* S France, S and E Spain.

N. asturiensis Flowers solitary, drooping, pale yellow. *Distr.* N and Central Spain, N Portugal.

N. aureus See *Narcissus tazetta*

N. bicolor Flowers solitary, tepal lobes cream, corona yellow. *Distr.* Pyrenees.

N. broussoneti Flowers white, corona small. *Distr.* Morocco.

N. bulbocodium (From the Greek *bulbos*, bulb, and *kodion*, wool.) HOOP PETTICOAT DAFFODIL. A very variable species with narrow leaves and a corona that far exceeds the tepals. *Distr.* SW and W France, Spain, Portugal, N Africa.

N. cantabricus WHITE HOOP PETTICOAT DAFFODIL. Flowers white, fragrant. *Distr.* S Spain, Morocco, Algeria.

N. corcyrensis See *Narcissus tazetta*

N. cyclamineus Flowers solitary, drooping, deep yellow. *Distr.* NW Spain, NW Portugal.

N. dubius Flowers white, corona cup-shaped. *Distr.* S France, NE Spain.

N. elegans Tepals white, flat, corona becoming dull orange. *Distr.* W Mediterranean, Italy, Sicily.

N. fernandesii Flowers 1–5, ascending, yellow, corona slightly deeper yellow. *Distr.* S Portugal.

N. gaditanus Flowers 1–3, yellow, cup-shaped. *Distr.* S Portugal, S Spain.

N. gracilis See *Narcissus x tenuior*

N. hedraeanthus Dwarf. Flowers pale yellow. *Distr.* Spain.

N. henriquesii See *Narcissus jonquilla*

N. italicus See *Narcissus tazetta*

N. × incomparabilis A natural hybrid between *N. poeticus* and *N. pseudonarcissus*. *Distr.* S and Central France, naturalized elsewhere in Europe.

N. jonquilla (From Spanish *junquillo*, rush, alluding to the slender leaves.) JONQUIL. Leaves rush-like. Flowers 1–6, yellow, corona cup-shaped. The very fragrant flowers are used in scent-making. *Distr.* Spain, Portugal.

N. juncifolius See *Narcissus assoanus*

N. lobularis See *Narcissus minor*

N. longispathus Spathe long. Flowers yellow. *Distr.* SE Spain.

N. marvieri See *Narcissus rupicola*

N. × medioluteus PEERLESS PRIMROSE. A natural hybrid of *N. poeticus* × *N. tazetta* that typically bears 2 flowers per inflorescence. *Distr.* S France, widely naturalized elsewhere in Europe.

N. minimus See *Narcissus asturiensis*

N. minor Small. Perianth segments yellow, somewhat twisted, corona frilled. *Distr.* Pyrenees, N Spain.

N. nanus See *Narcissus minor*

N. obesus See *Narcissus bulbocodium*

N. obvallaris See *Narcissus pseudonarcissus*

N. × odorus A hybrid of *N. jonquilla* and *N. pseudonarcissus*. *Distr.* Garden origin, naturalized in S Europe.

N. pachybolbus Flowers numerous, small, white. *Distr.* Morocco, Algeria.

N. pallidiflorus See *Narcissus pseudonarcissus*

N. panizzianus See *Narcissus papyraceus*

N. papyraceus PAPER WHITE NARCISSUS. Leaves erect, keeled. Flowers white, fragrant. *Distr.* SW Europe.

N. poeticus POET'S NARCISSUS, PHEASANT'S EYE. Flowers white with a very short, red-tipped corona, fragrant. Used in scent-making. *Distr.* S Europe.

N. polyanthus See *Narcissus papyraceus*

N. pseudonarcissus WILD DAFFODIL, LENT LILY, TRUMPET NARCISSUS. Flowers usually solitary, yellow, or white and yellow. This species is a parent of many of the cultivated hybrids. *Distr.* W Europe.

N. pumilus See *Narcissus minor*

N. radiiflorus See *Narcissus poeticus*

N. requienii See *Narcissus assoanus*

N. romieuxii (After Romieux, of Geneva, who grew it.) Flowers solitary, pale yellow, corona large, flared. *Distr.* Morocco.

N. rupicola Leaves 2-keeled. Flowers solitary, yellow. *Distr.* Central Spain, N Portugal, Morocco.

N. scaberulus Flowers ascending, tepals reflexed, orange-yellow, corona deep yellow. *Distr.* N Portugal.

N. serotinus Flowers ascending, tepal lobes white, twisted, corona minute, dark orange-yellow. *Distr.* Mediterranean regions.

N. tazetta CHINESE SACRED LILY, BUNCH-FLOWERED NARCISSUS, POLYANTHUS NARCISSUS. A very variable species with fragrant flowers that bear white tepals and a yellow corona. *Distr.* W Mediterranean regions.

N. × tenuior A hybrid of *N. jonqilla* × *N. poeticus*. *Distr.* Garden origin.

N. triandrus ANGEL'S TEARS. Flowers nodding, tepal lobes white, reflexed, corona tinged yellow. *Distr.* Portugal, Spain, NW France.

N. viridiflorus Flowers green, malodorous. *Distr.* SW Spain, Morocco.

N. willkommii Flowers solitary, yellow, corona deeply 6-lobed. *Distr.* S Portugal, S W Spain.

narcissus: bunch-flowered ~ *Narcissus tazetta* **paperwhite** ~ *N.*

papyraceus **poet's** ~ *N. poeticus polyanthus* ~ *N. tazetta* **trumpet** ~ *N. pseudonarcissus*.

nardoo *Marsilea* **common** ~ *M. drummondii*.

Narthecium (From Latin *narthex*, rod, previously used for the genus *Ferula*.) BOG ASPHODEL. A genus of 8 species of rhizomatous herbs with rush-like leaves and racemes of star-shaped, yellow flowers. *Distr.* N Temperate regions. *Fam.* Melanthiaceae. *Class* M.

N. ossifragum BOG ASPHODEL. Stems and fruit becoming deep orange after flowering. Has been used as a substitute for saffron and as a hair-dye. Grown as a bog-garden ornamental. *Distr.* W Europe.

Nasturtium See *Rorippa*

N. officinale See *Rorippa nasturtium-aquaticum*

nasturtium Tropaeolaceae, *Tropaeolum*, *T. majus*.

natalensis, -e: of Natal, S. Africa.

natans: floating.

native's comb *Pachycereus pectenaboriginum*.

navel seed *Omphalodes*.

navelwort *Hydrocotyle*, *Umbilicus rupestris*, *Omphalodes* **Venus'** ~ *O. linifolia*.

navicularis: boat-shaped.

neapolensis, -e: of Naples, Italy.

neapolitanus, -a, -um: of Naples, Italy.

nebulosus, -a, -um: clouded, cloudy, with unevenly blended colours.

nectarine *Prunus persica*.

Nectaroscordum (From Greek *nektar*, nectar, and *skordon*, GARLIC.) A genus of 2–3 species of bulbous herbs with linear leaves that smell of ONION when crushed and umbels of small flowers with free tepals. *Distr.* S Europe, Turkey. *Fam.* Alliaceae. *Class* M.

N. siculum SICILIAN HONEY GARLIC. Flowers white, flushed pink. Cultivated as an ornamental. *Distr.* S Europe, Turkey.

needle bush *Hakea lissosperma*.

needles, Spanish *Bidens*.

neglectum: overlooked.

nemoralis, -e: growing in woods.

Neillia (After Patrick Neill (1776–1851), Scottish naturalist.) A genus of 11 species of deciduous subshrubs and shrubs with alternate, typically 3-lobed leaves, and racemes or panicles of white or pink flowers that bear 5 petals. Several species are grown as ornamentals. *Distr.* E Asia. *Fam.* Rosaceae. *Class* D.

N. longiracemosa See *N. tibetica*

N. sinensis Arching shrub to 3m high. Flowers pale pink, borne in racemes. *Distr.* China.

N. tibetica Arching shrub. Flowers pink, borne in slender spikes. *Distr.* China.

Nelumbo (Sinhalese name for *N. nucifera*.) LOTUS. A genus of 2 species of aquatic, rhizomatous, perennial herbs with long-stalked, round, floating leaves and large, solitary, emergent flowers that bear 4–5 sepals, numerous petals, and numerous stamens. Both species are grown as ornamentals. *Distr.* Warm-temperate and tropical Asia and the New World. *Fam.* Nelumbonaceae. *Class* D.

N. lutea AMERICAN LOTUS, WATER CHINQUAPIN. Flowers yellow. The seeds and the rhizomes are eaten. *Distr.* North and Central America to Colombia.

N. nucifera INDIAN LOTUS, SACRED LOTUS. Flowers red. A sacred flower in India, Tibet, and China. Both the flowers and the rhizomes are used in Chinese cooking. *Distr.* Asia to Australia, widely introduced, naturalized in S Europe.

Nelumbonaceae A family of 1 genus and 2 species. *Class* D.
See: Nelumbo.

Nemastylis (From Greek *nema*, thread, and *stylos*, column, alluding to the slender style.) A genus of 7 species of bulbiferous herbs with linear, folded leaves and clusters of a few regular flowers. Some species are cultivated as ornamentals. *Distr.* North America. *Fam.* Iridaceae. *Class* M.

N. tenuis Flowers pale blue, scented. *Distr.* Mexico.

Nematanthus (From Greek *nema*, thread, and *anthos*, flower, alluding to the slender flower stalks of some species.) A genus of 30 species of epiphytic climbing or trailing shrubs with simple opposite leaves and tubular 5-lobed flowers that are inflated in the middle. Several species are grown as house-plants. *Distr.* South America. *Fam.* Gesneriaceae. *Class* D.

N. gregarius Prostrate or trailing shrub. Flowers small, orange-yellow, much inflated. *Distr.* E South America.

N. radicans See *N. gregarius*

N. strigillosa Stems climbing. Flowers orange-yellow, to 2cm long. *Distr.* South America.

Nemesia (From *nemesion*, a Greek name for a similar plant.) A genus of 65 species of annual and perennial herbs and subshrubs with simple, opposite leaves and 2-lipped, 5-lobed flowers. The flowers bear a boss on the lower lip that almost closes the throat. Several species are grown as ornamentals, typically as annual summer-bedding plants. *Distr.* South Africa. *Fam.* Scrophulariaceae. *Class* D.

N. strumosa Bushy annual herb. Flowers yellow, white or purple. Numerous cultivars have been raised from this species. *Distr.* South Africa.

Nemophila (From Greek *nemos* glade, and *phileo*, to love; they grow in shady places.) A genus of 11 species of annual herbs with pinnately lobed leaves and solitary or clustered, bell-shaped, blue or white flowers which bear their parts in 5s. Several species are grown as ornamentals. *Distr.* California. *Fam.* Hydrophyllaceae. *Class* D.

N. maculata FIVE SPOT. Flowers white with a purple spot near the tip of each petal. *Distr.* USA (California).

N. menziesii (After Archibald Menzies (1754–1842), British naval surgeon and botanist.) BABY BLUE EYES. Flowers blue with white centres. *Distr.* USA (California).

nemorosus, -a, -um: of woods.

nemorus, -a, -um: growing in woods.

neocaledonicus, -a, -um: of New Caledonia.

Neofinetia (After Achille Finet (1862–1913), French botanist.) A genus of 1 species of epiphytic orchid to 15cm high with fleshy leaves and racemes of up to 10 spurred, white flowers. *Distr.* Japan. *Fam.* Orchidaceae. *Class* M.

N. falcata Grown as a tender ornamental with several variegated forms available. *Distr.* Japan.

Neohenricia (From Greek *neos*, new, and the genus name *Henricia*.) A genus of 1 species of small, mat-forming herb with creeping stems, groups of 4 succulent leaves and fragrant white flowers that open at night. *Distr.* South Africa. *Fam.* Aizoaceae. *Class* D.

N. sibbetii Grown as a tender ornamental.

Neolitsea (From Greek *neos*, new, and the name of a related genus *Litsea*.) A genus of 80 species of evergreen shrubs and trees with simple, alternate, leathery leaves and umbels of unisexual flowers that are followed by red or black berries. *Distr.* E and SE Asia. *Fam.* Lauraceae. *Class* D.

N. glauca See *N. sericea*

N. sericea Evergreen shrub. Grown as a half-hardy ornamental as well as being used medicinally and as a source of oil. *Distr.* Japan, Korea, China.

Neolloydia (After Francis Ernest Lloyd (1868–1947), American botanist; the Greek prefix *neo-*, new, was added because *Lloydia* had been used for a genus in the family Liliaceae.) A genus of 10–14 species of small cacti with round or cylindrical stems, straight spines, and small, funnel-shaped flowers. Several species are grown as ornamentals. *Distr.* E Mexico, SW Texas. *Fam.* Cactaceae. *Class* D.

N. conoidea Clump-forming. Stems blue-green. Spines white. Flowers purple-violet. *Distr.* E Mexico, SW Texas.

N. grandiflora See *N. conoidea*

N. macdowellii See *Thelocactus macdowellii*

N. schmiedickeana Stems solitary or clustered, blue to grey-green. Flowers white, yellow, pink, or purple. A variable species with a number of varieties in cultivation. *Distr.* NE Mexico.

Neomarica (From Greek *neos*, new, and the genus name *Marica*.) A genus of 15 species of rhizomatous herbs with tough, sword-shaped leaves and clusters of *Iris*-like flowers. Some species may be cultivated as tender ornamentals. *Distr.* Tropical America, W Africa. *Fam.* Iridaceae. *Class* M.

N. caerulea Flowers blue-violet with yellow and orange centres. *Distr.* Brazil.

N. northiana Leaves ribbed. Flowers yellow-white, mottled red or maroon. *Distr.* Brazil.

neomexicanus, -a, -um: of New Mexico.

Neopanax See *Pseudopanax*

N. australasica See *Claytonia australasica*

Neoporteria (After Carlos Porter, a Chilean entomologist; the Greek prefix *neo-*, new, was added because *Porteria* was being used for another genus that is now no longer valid.) A genus of 66 species of small cacti with spherical to column-like, ridged stems and tubular to bell-shaped flowers. Several species are grown as ornamentals. *Distr.* Chile, S Peru, W Argentina. *Fam.* Cactaceae. *Class* D.

N. chilensis Stems column-like, light green. Spines golden. Flowers orange to white. *Distr.* Chile.

N. napina Stems spherical, dark green, tinged red. Spines short, black. Flowers yellow. *Distr.* Chile.

N. nidus Stems spherical to column-like, dark green. Spines soft, grey. Flowers pink. *Distr.* Chile.

N. subgibbosa Stems light green to grey-green. Flowers red-pink. *Distr.* Chile.

N. villosa Stems column-like with soft, curved spines. Flowers pink or white, tubular. *Distr.* Chile (Huasco).

Neoraimondia (After Antonio Raimondi (1825–90), Peruvian geographer and naturalist; the Greek prefix *neo-*, new, was added because *Raimondia* had already been used for a genus in the family Annonaceae.) A genus of 2 species of shrubby or tree-like cacti with branched, ridged stems, long spines, and small, tubular flowers. *Distr.* W South America. *Fam.* Cactaceae. *Class* D.

N. arequipensis Robust shrub. Flowers pink-purple or white. Grown as an ornamental. *Distr.* Peru, N Chile.

Neoregelia (After Eduard Albert von Regel (1815–92), German botanist and director of the St Petersburg botanic garden.) A genus of 71 species of terrestrial herbs with large, funnel-shaped rosettes of rigid, spiny leaves and small flowers borne in dense clusters at the centre of the rosettes. Some of the species are grown as ornamentals. *Distr.* Brazil, E Colombia and Peru. *Fam.* Bromeliaceae. *Class* M.

N. carolinae BLUSHING BROMELIAD. Flowers purple, surrounded by red leaves. *Distr.* Brazil.

N. concentrica Leaves spreading, strap-shaped, tipped red. Inner leaves purple. *Distr.* Brazil.

N. marmorata MARBLE PLANT. Leaves blotched purple. *Distr.* Brazil.

N. spectabilis FINGER-NAIL PLANT, PAINTED FINGER PLANT. Leaves green with a bright red apical spot. *Distr.* Brazil.

Neowerdermannia (After Erich Werdermann (1892–1959), Director of the Botanical Museum and Herbarium, Berlin; the Greek prefix *neo-*, new, was added because *Werdermannia* had been used for a genus in the family Cruciferae.) A genus of 2 species of cacti with spherical stems that bear spirally arranged tubicles and funnel-shaped, white to purple-pink flowers. *Distr.* S South America. *Fam.* Cactaceae. *Class* D.

N. vorwerkii Spines black, turning white. Flowers white with lilac-pink stripes. Grown as an ornamental. *Distr.* N Argentina, S Bolivia, Peru, N Chile.

nepalensis, -e: of Nepal.

Nepenthaceae A family of 1 genus and 70 species. *Class* D.
See: Nepenthes.

Nepenthes (From a Greek plant name used by Homer, meaning grief-assuaging, alluding to supposed medicinal properties.) PITCHER PLANT. A genus of 70 species of carnivorous climbers and epiphytes. The midrib of the leaves is extended into a tendril the apex of which is developed into a pitcher-shaped insect trap. The flowers are small and borne in spike-like inflorescences, the males and the females on separate plants. The stems are used locally in basketry. Several species and numerous cultivars are grown as ornamental curiosities. *Distr.* Madagascar, Seychelles, Sri Lanka, India, Australia. *Fam.* Nepenthaceae. *Class* D.

N. rafflesiana (After Sir Thomas Stamford Raffles (1781–1826), scientific patron and founder of Singapore.) Epiphyte. Pitchers yellow-green marked purple-brown, to 25cm long. *Distr.* SE Asia.

N. ventricosa Terrestrial or epiphytic. Pitchers numerous, contracted around the middle, pale green. *Distr.* Philippines.

nepenthoides: resembling the genus *Nepenthes*.

Nepeta (Possibly after Nepi in Italy.) CATMINT. A genus of 250 species of typically aromatic, perennial herbs with simple, opposite leaves and heads, spikes, or whorls of tubular, 2-lipped flowers. Several species and numerous cultivars are grown as ornamentals and the leaves of some species are used medicinally. *Distr.* Temperate Eurasia, N Africa, mountainous regions of tropical Africa. *Fam.* Labiatae. *Class* D.

N. cataria CATMINT, CATNIP. Perennial herb to 1m high. Leaves grey-hairy beneath. Flowers white, marked blue. Leaves used medicinally and in tea; attractive to cats. *Distr.* Eurasia.

N.×faassenii (After J. H. Faasen, Dutch nurseryman.) A hybrid of *N. mussinii* × *N. nepetella*. The source of most of the garden cultivars. *Distr.* Garden origin.

N. glechoma See *Glechoma hederacea*

N. govaniana Erect, much-branched perennial. Flowers white to pale yellow, with a long narrow tube. *Distr.* W Himalaya.

N. grandiflora Erect perennial. Flowers small, hooded, blue. *Distr.* Caucasus.

N. hederacea See *Glechoma hederacea*

N. nervosa Clump-forming perennial. Flowers blue or yellow, borne in a cylindrical raceme. *Distr.* Kashmir.

nepetoides: resembling the genus *Nepeta*.

Nephrolepis (From Greek *nephros*, kidney, and *lepis*, a scale, alluding to the shape of the indusia.) SWORD FERN, LADDER FERN, BOSTON FERN. A genus of 30 species of epiphytic or terrestrial ferns with pinnately divided leaves. Several species are grown as ornamentals. *Distr.* Tropical and sub-tropical regions. *Fam.* Oleandraceae. *Class* F.

N. exaltata BOSTON FERN. *Distr.* Tropical America, Africa and Polynesia.

nephrolepis with kidney-shaped scales.

Nephthytis (After Nephthys in Greek mythology, mother of Anubis and wife of Typhonis.) A genus of 7 species of rhizomatous herbs that bear arrow-shaped, long-stalked leaves and small flowers borne on a spadix surrounded by a green spathe. Several species are cultivated as ornamentals. *Distr.* W tropical Africa. *Fam.* Araceae. *Class* M.

N. afzelii Cultivated as an ornamental for its tufts of dark green leaves.

neriiflorus, -a, -um: from the genus *Nerium*, and *florum*, flowered.

neriifolius, -a, -um: from the genus *Nerium*, and *folius*, leaved.

Nerine (After Nereis, a sea-nymph in Greek mythology.) A genus of 30 species of bulbiferous perennial herbs with strap-shaped leaves and few to many, funnel-shaped, typically irregular, scented flowers. *Distr.* South Africa. *Fam.* Amaryllidaceae. *Class* M.

N. bowdenii (After Athleston Bowden, who brought it into cultivation in 1902.) Flowers pink, rarely white, appear after the leaves have died off. The most hardy of the species. *Distr.* South Africa (Cape Province, Natal, Orange Free State).

N. crispa See *N. undulata*

N. filifolia Leaves narrow. Flowers white to bright crimson. *Distr.* South Africa (Orange Free State).

N. flexuosa Leaves wide. Flowers white to pale pink. *Distr.* South Africa (Cape Province, Natal, Orange Free State).

N. humilis Flowers 10–20, pink. *Distr.* South Africa (Cape Province).

N. masoniorum Dwarf. Leaves thread-like. Flowers pink with recurved lobes. *Distr.* South Africa (Orange Free State).

N. sarniensis GUERNSEY LILY. Flowers pink, almost regular, appearing before leaves. This species is said to have arrived in Guernsey in the ballast of ships from Japan but was probably presented to islanders by shipwrecked sailors from South Africa in the mid-17th century. *Distr.* South Africa (Cape Province).

N. undulata Flowers pale pink, with crisped lobes, appearing with the leaves. *Distr.* South Africa.

neriniflorus, -a, -um: with flowers resembling those of the genus *Nerine*.

Nerium (Classical Greek name for these plants.) A genus of 1 species of shrub with simple, evergreen, sharply pointed leaves and funnel-shaped, white, pink, or yellow flowers. The leaves are very poisonous. *Distr.* Mediterranean regions to E Asia. *Fam.* Apocynaceae. *Class* D.

N. oleander (From *oleandra*, the Italian vernacular name for this plant.) OLEANDER, ROSE-BAY. There are a very large number of ornamental cultivars of this species available. The leaves are a source of rat poison. *Distr.* Mediterranean regions to E Asia.

Nertera (From Greek *nerteros*, lowly, alluding to the dwarf habit.) A genus of 6 species of perennial creeping herbs with minute, opposite leaves and tiny funnel-shaped flowers that are followed by bead-like berries. *Distr.* Central and South America to Australasia and S China. *Fam.* Rubiaceae. *Class* D.

N. depressa See *N. granadensis*

N. granadensis BEAD PLANT. Grown as tender ground cover or as a pot plant for its orange fruits. *Distr.* South America, Australia, New Zealand, SE Asia.

nerve plant *Fittonia verschaffeltii.*

nervosus, -a, -um: conspicuously veined.

Nesogenaceae family of 1 genus and 7 species of herbs and subshrubs. This family is sometimes included within the family Verbenaceae. Distr. E Africa, Madagascar, Indian Ocean, S Pacific. *Class* D.

nettle *Urtica* **dead** ~ *Lamium* **false** ~ *Boehmeria* **hedge** ~ *Stachys* **painted** ~ *Solenostemon scutellarioides* **Pyrenean read** ~ *Horminum pyrenaicum* **rock** ~ *Loasaceae* **Roman** ~ *Urtica pilulifera* **spotted dead** ~ *Lamium maculatum* **stinging** ~ Urticaceae, *Urtica dioica* **white dead** ~ *Lamium album.*

nettle tree *Celtis* **European** ~ *C. australis.*

Neuradaceae A family of 3 genera and 9 species of prostrate annual herbs with spirally arranged, simple or pinnately divided leaves, and solitary, regular flowers that bear 5 sepals, 5 petals, and 10 stamens. *Distr.* E Mediterranean to India, S Africa. *Class* D.

nevadensis, -e: of the Sierra Nevada, California.

never never plant *Ctenanthe oppenheimiana.*

Neviusia (After the Reverend D. R. Nevius (1827–1913), of Alabama, who collected it.) A genus of 1 species of deciduous shrub with simple, alternate leaves and pale green flowers that bear numerous stamens but lack petals. *Distr.* SE USA. *Fam.* Rosaceae. *Class* D.

N. alabamensis SNOW WREATH. Grown as an ornamental.

Nicandra (After Nikander of Colophon, Greek physician and poet who flourished around 137 BC) A genus of 1 species of erect annual herb with simple, alternate leaves and solitary, 5-lobed, white and purple, bell-shaped flowers. *Distr.* Peru, naturalized in Eurasia and North America. *Fam.* Solanaceae. *Class* D.

N. physalodes APPLE OF PERU. Grown as an ornamental; it allegedly repels flies.

Nicolaia See *Etlingera*

N. elatior See *Etlingera elatior*

Nicotiana (After Jean Nicot (1530–1600), who introduced the tobacco plant into France.) TOBACCO. A genus of 67 species of annual and perennial herbs and shrubs with simple, alternate leaves and panicles of regular or irregular, 5-lobed, funnel-shaped flowers. The foliage is often covered in sticky hairs. A number of species are grown as ornamentals and some are a source of the narcotic and insecticide nicotine. *Distr.* America, S Pacific, Australia, SW Africa. *Fam.* Solanaceae. *Class* D.

N. alata JASMINE TOBACCO, FLOWERING TOBACCO. Perennial herb to 1.5m high. Flowers white, tinged green outside, fragrant at night. *Distr.* South America.

N. glauca TREE TOBACCO, MUSTARD TREE. Shrub to 7m high. Flowers cream-green, hairy outside. *Distr.* Bolivia, Argentina.

N. rustica WILD TOBACCO, AZTEC TOBACCO. Annual herb. The original tobacco introduced into Europe by Raleigh, now grown as a source of insecticide. *Distr.* Mexico, E North America.

N. × sanderae A hybrid of *N. alata* × *N. forgetiana*. This cross has given rise to numerous ornamental cultivars. *Distr.* Brazil.

N. tabacum (From the Caribbean name for a pipe, or from the Haitian *taina*, a roll of tobacco in a maize leaf.) TOBACCO. Annual or biennial herb to 1.5m high. Flowers pendent, green-white to pink. This species is the main source of tobacco for smoking and chewing. *Distr.* Tropical America.

Nidularium (From Latin *nidus*, a nest; the flowers are borne in a nest-like depression in the centre of a cluster of bracts.) A genus of 23 species of stemless, epiphytic or terrestrial herbs with flat rosettes of purple-spotted or banded leaves. The small flowers are clustered in the centre of the leaf rosettes and are surrounded by brightly coloured bracts. Several species are grown as ornamentals. *Distr.* E Brazil. *Fam.* Bromeliaceae. *Class* M.

N. flandria See *Neoregelia carolinae*

N. fulgens BLUSHING BROMELIAD. Leaves pale green, mottled dark green. Bracts bright red.

N. innocentii BIRD'S NEST BROMELIAD. Leaves tinged red-purple below. Bracts red.

nidus: a nest.

Nierembergia (After John Eusebius Nieremberg (1595–1658), Spanish Jesuit and naturalist.) CUPFLOWER. A genus of 23 species of annual and perennial herbs and subshrubs with simple, alternate leaves and 5-lobed, tubular or bell-shaped flowers. Several species are grown as ornamentals. *Distr.* Mexico to Chile. *Fam.* Solanaceae. *Class* D.

N. frutescens Small, much-branched shrub to 80cm high. Flowers pale blue with a yellow centre. *Distr.* Chile.

N. hippomanica Erect perennial. Flowers numerous, blue. *Distr.* Argentina.

N. repens WHITECUP. Mat-forming perennial. Flowers cup-shaped, white with a yellow centre. *Distr.* Warm-temperate South America.

N. rivularis See *N. repens*

Nigella (Diminutive of Latin *niger*, black, alluding to the black seeds.) A genus of 22 species of annual herbs with alternate, typically finely divided leaves, and solitary flowers that bear 5 petal-like sepals, 5–10 tubular, nectariferous petals, and numerous stamens. Several species are grown as ornamentals. *Distr.* Europe, W Asia. *Fam.* Ranunculaceae. *Class* D.

N. damascena LOVE IN A MIST. To 50cm high. Flowers white, tinged pink or purple. A number of cultivars have been raised from this species. *Distr.* S Europe, N Africa.

N. sativa BLACK CUMIN, NUTMEG FLOWER, ROMAN CORIANDER. To 30cm high. Flowers white, tinged blue. The seeds are used in seasoning cakes and bread. *Distr.* SE Europe, SW Asia.

niger: black.

nigerrimus, -a, -um: very black.

nigger hand *Opuntia clavarioides.*

nightshade *Solanum* **climbing** ~ *S. dulcamara* **deadly** ~ *Atropa belladonna, Solanum dulcamara* **enchanter's** ~ *Circaea lutetiana* **poisonous** ~ *Solanum dulcamara* **sticky** ~ *S. sisymbrifolium* **stinking** ~ *Hyoscyamus niger* **woody** ~ *Solanum dulcamara.*

nigrellus, -a, -um: small and black.

nigrescens: becoming black.

nigricans: blackish.

Nigritella (Diminutive of Latin *niger*, black, from the dark-coloured flowers.) A genus of 1 species of terrestrial orchid with finger-like tubers, linear leaves, and a spike of crimson-black or yellow, vanilla-scented flowers. *Distr.* Scandinavia to N Spain and Greece. *Fam.* Orchidaceae. *Class* M.

N. nigra VANILLA ORCHID. Grown as a garden ornamental.

nigrus, -a, -um: black.

nikoensis, -e: of Nikko, Japan.

niloticus, -a, -um: of the Nile river basin.

ninebark *Physocarpus.*

nin sin *Panax ginseng.*

Niphaea (From Greek *niphos*, snow, alluding to the white flowers.) A genus of 5 species of erect herbs with simple crowded leaves and white tubular flowers that bear 5 spreading lobes. *Distr.* Tropical America. *Fam.* Gesneriaceae. *Class* D.

N. oblonga To 15cm high. Flowers to 2.5cm across. Grown as an ornamental. *Distr.* Guatemala.

niphophilus, -a, -um: snow-loving.

nipplewort *Lapsana communis*.

Nipponanthemum (From *Nippon*, the Japanese name for Japan, and Greek, *anthemon*, flower.) A genus of 1 species of perennial herb or subshrub to 1m high with simple leaves and yellow flowers in daisy-like heads that bear distinct white rays. *Distr*. Japan. *Fam*. Compositae. *Class* D.

N. nipponicum Grown as an autumn-flowering hardy ornamental.

nipponicus, -a, -um: of Japan.

nirre *Nothofagus antarctica*.

nirrhe *Eucryphia glutinosa*.

nispero *Eriobotrya japonica*.

nitens: shining.

nitidifolius, -a, -um: with glossy leaves.

nitidus, -a, -um: glossy.

nivalis, -e: of snow.

Nivenia (After James Niven (1774–1826), Scottish gardener who collected in South Africa 1798–1812.) A genus of 9 species of evergreen shrubby perennials with sword-shaped leaves and solitary, tubular, blue flowers. Some species are occasionally grown as ornamentals. *Distr*. South African. *Fam*. Iridaceae. *Class* M.

N. corymbosa Leaves rigid, arranged in a fan. *Distr*. South Africa (Cape Province).

niveus, -a, -um: snow-white.

nivosus, -a, -um: full of snow.

no azami *Cirsium japonicum*.

nobilis, -e: notable.

noctiflorus, -a, -um: flowering at night.

nocturnus, -a, -um: nocturnal, of the night.

nodiflorus, -a, -um: with flowers borne from the nodes.

nodosus, -a, -um: with conspicuous nodes.

Nolana (From Latin *nola*, a small bell, alluding to the shape of the flowers.) CHILEAN BELLFLOWER. A genus of 18 species of annual and perennial herbs and subshrubs with simple, alternate or whorled leaves, and bell-shaped, somewhat irregular, 5-lobed flowers. *Distr*. Chile to Peru. *Fam*. Solanaceae. *Class* D.

N. acuminata See *N. paradoxa*

N. paradoxa Annual or perennial herb. Flowers bright blue with a yellow throat. Grown as a tender ornamental. *Distr*. Chile.

Nolina (After P. C. Nolin, 18th-century French agriculturalist.) A genus of 30 species of trees and subshrubs of dry areas with robust linear leaves and panicles of small, unisexual, cream flowers. Several species are grown as ornamentals. *Distr*. S USA, Mexico. *Fam*. Dracaenaceae. *Class* M.

N. bigelovii Stems thick at base. Leaves in terminal rosettes. *Distr*. Arizona to Mexico.

N. longifolia Leaves to 1m long with rough margins. Leaves used for brooms, thatching, and basketry. *Distr*. Mexico.

N. microcarpa BEARGRASS, SACAHUISTA. Stemless. Leaves tufted, grass-like. *Distr*. SW USA, Mexico.

Nomocharis (From Greek *nomos*, meadow, and *charis*, grace.) A genus of 7 species of bulbiferous herbs with linear leaves and loose racemes of saucer-shaped, white to pink or pale yellow flowers. Several species are grown as ornamentals. *Distr*. W China, Burma, N India. *Fam*. Liliaceae. *Class* M.

N. mairei See *N. pardanthina*

N. pardanthina Flowers flattened, pale pink, blotched purple. *Distr*. W China.

N. saluenensis Flowers 2–5, horizontal or drooping, white with maroon patches. *Distr*. Burma, China, Tibet.

none so pretty *Silene armeria*.

non-scripta: unmarked.

Nopalxochia: See *Discocactus*

norvegicus, -a, -um: of Norway.

nosegay *Plumeria*.

notabilis, -e: notable.

nothofagifolia: from the genus *Nothofagus*, and *folius*, leaved.

Nothofagus (From Greek *nothos*, false, and the genus name *Fagus*, BEECH; it is possible that the author had wanted the name to mean Southern Beech, in which case it should have been *Notofagus*, from *notos*, southern.) SOUTHERN BEECH. A genus of 20 species of deciduous and evergreen shrubs and trees with smooth bark, simple alternate leaves, and small, unisexual, wind pollinated flowers. The fruit consists of 1–7 nuts surrounded by a non-spiny casing (the cupule). A number of species are an important source of timber in the S hemisphere, second only to *Eucalyptus* species. Some species are grown as ornamentals in the N hemisphere. *Distr*. S Temperate and tropical areas. *Fam*. Fagaceae. *Class* D.

N. antarctica NIRRE, ANTARCTIC BEECH. Deciduous conical tree. Leaves small, oval, yellow in autumn. *Distr*. Chile, Argentina.

N. betuloides Evergreen conical tree. Shoots red-brown. Leaves dark green. *Distr*. S Chile, S Argentina.

N. cunninghamii MYRTLE. Evergreen tree. Leaves triangular to nearly round. A commercial source of timber. *Distr*. Australia (Tasmania).

N. dombeyi COIGUE. Deciduous, laxly conical tree. Shoots drooping at tips. *Distr*. Chile, Argentina.

N. fusca RED BEECH. Evergreen tree. Shoots red-brown, often forming a zigzag pattern at the buds. A source of timber for railway sleepers. *Distr*. New Zealand.

N. menziesii (After Archibald Menzies (1754–1842), British naval surgeon and botanist.) SILVER BEECH. Evergreen, conical tree. Leaves very small, almost round. *Distr*. New Zealand.

N. moorei AUSTRALIAN BEECH. Evergreen tree. Leaves pointed and finely serrated. *Distr*. Australia.

N. obliqua ROBLE BEECH. Deciduous tree. Branches arching. Leaves orange-red in autumn. *Distr*. Chile, Argentina.

N. procera RAULI BEECH. Deciduous conical tree. Leaves with impressed veins. *Distr*. Chile, Argentina.

N. solandri (After Daniel Carl Solander (1736–82), botanist on Cook's first voyage.) BLACK BEECH. Evergreen tree. A source of timber for general construction. *Distr*. New Zealand.

Notholaena See *Cheilanthes*

Notholirion (From Greek *nothos*, false, and *leirion*, lily.) A genus of 6 species of bulbiferous herbs with linear to lance-shaped leaves and racemes of trumpet- to star-shaped flowers. Several of the species are grown as ornamentals. *Distr*. Afghanistan to W China. *Fam*. Liliaceae. *Class* M.

N. bulbuliferum To 1.5m high. Flowers held horizontally, trumpet-shaped, pale lilac. *Distr*. Nepal to W China.

N. campanulatum To 80cm high. Flowers nodding, red with green tips. *Distr*. N Burma, W China.

Nothopanax See *Polyscias*

Nothoscordum (From Greek *nothos*, false, and *scordon*, Garlic.) FALSE GARLIC, GRACE GARLIC. A genus of 20 species of bulbiferous herbs with linear leaves and umbels of numerous small flowers. Several species are cultivated as ornamentals. *Distr*. North and South America. *Fam*. Alliaceae. *Class* M.

N. bivalve Flowers star-shaped, white-yellow. *Distr*. S Central and SE USA.

N. gracile Flowers funnel-shaped, white-yellow, fragrant. *Distr*. Subtropical South America, Mexico.

N. neriniflorum See *Caloscordum neriniflorum*

Notocactus See *Parodia*

Notospartium (From Greek *notos*, southern, and the genus name *Spartium*) A genus of 3 species of typically leafless shrubs and trees with flat, green, slender branches and pendulous racemes of pea-like flowers. All the species are grown as ornamentals. *Distr*. New Zealand (South Island). *Fam*. Leguminosae. *Class* D.

N. carmichaeliae Shrub to 5m high. Flowers purple-pink.

N. glabrescens Shrub or round-headed tree to 9m high. Flowers deep purple.

N. torulosum Slender shrub. Flowers purple.

novae-angliae: of New England.

novae-zelandiae: of New Zealand.

novi-belgii: of New York.

nubicola: growing in clouds, above cloud level.

nucifera: nut-bearing.

nudicaulis, -e: bare-stemmed.

nudiflorus, -a, -um: flowering naked, i.e. when leafless.

nudus, -a, -um: naked.

numidicus, -a, -um: of Numidia (now Algeria).

nummulariifolius, -a, -um: with coin-shaped leaves.

nummularioides: coin-like.

nummularius, -a, -um: coin-like.

Nuphar (From *naufar*, the Arabic name for these plants.) A genus of 25 species of aquatic, rhizomatous herbs with large, round, leathery, floating leaves, smaller membranous submerged leaves and solitary, yellow or green, emergent flowers that bear numerous petals and stamens. Several species are grown as ornamentals. *Distr.* N temperate regions. *Fam.* Nymphaeaceae. *Class* D.

N. advena SPATTERDOCK. Some leaves borne erect above the water. Flowers small, yellow, tinged with purple. *Distr.* Mexico, West Indies, naturalized in S England.

N. lutea BRANDY BOTTLE, YELLOW WATER-LILY. All leaves floating. Flowers yellow, with an alcoholic smell. *Distr.* Widespread in N hemisphere.

nut: Brazil ~ *Bertholletia excelsa* **butter** ~ Caryocaraceae, *Caryocar*, *C. nuciferum* **cashew** ~ *Anacardium occidentale* **earth** ~ *Conopodium majus* **goora** ~ *Cola acuminata* **kaya** ~ *Torreya nucifera* **luck** ~ *Thevetia peruviana* **macadamia** ~ *Macadamia integrifolia* **maroochie** ~ *M. ternifolia* **monkey** ~ *Arachis hypogaea* **Para** ~ *Bertholletia excelsa* **pig** ~ *Conopodium majus* **Queensland** ~ *Macadamia tetrphylla* **souari** ~ *Caryocar nuciferum* **swari** ~ *C. amygdaliferum* **tiger** ~ *Cyperus esculentus* **Zulu** ~ *C. esculentus*.

nutans: nodding.

nutmeg Myristicaceae, *Myristica fragrans* **calabash** ~ *Monodora myristica* **California** ~ *Torreya californica* **Jamaica** ~ *Monodora myristica* **Peruvian** ~ *Laurelia sempervirens*.

nutmeg flower *Nigella sativa*.

nuts, soap *Sapindus*.

nyassae: of Nyasaland (now Malawi).

Nyctaginaceae The Bougainvillea family. 38 genera and about 400 species of herbs, shrubs, and trees with opposite or alternate, simple leaves. The flowers are often subtended by petal-like bracts and have a perianth that is not differentiated into sepals and petals but fused into a narrow petal-like tube. *Distr.* Tropical, subtropical, and rarely temperate regions. *Class* D.
See: Bougainvillea, Mirabilis, Pisonia.

Nyctocereus See *Peniocereus*
N. serpentinus See *Peniocereus serpentinus*

Nymphaea (From Greek *nymphe*, water nymph.) WATER-LILY. A genus of 50 species of aquatic, rhizomatous herbs with rounded, leathery, floating leaves and solitary, floating or emergent flowers that bear numerous petals and stamens. Several species and numerous cultivars are grown as ornamentals. *Distr.* Widespread in tropical and temperate regions. *Fam.* Nymphaeaceae. *Class* D.

N. alba WHITE WATER-LILY. Flowers pure white, floating, slightly fragrant. *Distr.* Europe, NW Africa, Caucasus, W Asia.

N. caerulea BLUE EGYPTIAN LOTUS, BLUE LOTUS. Flowers pale blue, emergent. Used as a narcotic in ancient Egypt and found in the wreaths of mummies. *Distr.* N and Central Africa.

N. capensis CAPE BLUE WATER-LILY. Flowers bright blue, sweetly fragrant, emergent. Distr. S and E Africa, Madagascar.

N. gigantea AUSTRALIAN WATER-LILY. Roots tuberous. Flowers blue, scentless, emergent. *Distr.* Tropical Australia, New Guinea.

N. lotus (From a Greek name that was used for a number of different plants.) WHITE EGYPTIAN LOTUS, LOTUS WHITE LILY. Flowers white, emergent. Found in wreaths on the mummies of Rameses II. *Distr.* N and tropical Africa, SE Asia, SE Europe.

N. odorata Flowers white, floating, sweetly fragrant. *Distr.* SE USA, Mexico, Cuba and Guyana.

N. pygmaea See *N. tetragona*

N. rubra INDIAN RED WATER-LILY. Leaves tinged red. Flowers deep purple-red, emergent. *Distr.* India.

N. tetragona PYGMY WATER-LILY. Flowers white, sometimes marked with blue, slightly fragrant. The smallest species in cultivation, much used in hybridization. *Distr*. Eurasia, North America.

Nymphaeaceae The Water-Lily family. 6 genera and 60 species of perennial, aquatic herbs with large rhizomes and round or heart-shaped, floating leaves. The flowers are solitary with 3–6 sepals and many petals. Many species are grown as ornamental aquatics; some have edible seeds and rhizomes. *Distr*. Widespread. *Class* D.
See: Nuphar, Nymphaea, Victoria.

Nymphoides (From the genus name *Nymphaea*, and Greek *-oides*, indicating resemblance.) A genus of 20 species of aquatic, perennial, rhizomatous herbs with long-stalked, round, floating leaves and aerial, yellow or white flowers that bear 5 fused petals and 5 stamens. Several species are grown as ornamentals and some have edible tubers or medicinal seeds. *Distr*. Europe, Asia. *Fam*. Menyanthaceae. *Class* D.

N. peltata FRINGED WATER-LILY, WATER FRINGE. Leaves to 10cm across. Flowers bright, golden yellow. Petals fringed. *Distr*. Europe, Asia.

Nyssa (After Nyssa, a water nymph in Greek mythology.) A genus of 5 species of deciduous trees with simple, alternate leaves and heads of small green flowers. *Distr*. North America, China, Indomalaysia. *Fam*. Cornaceae. *Class* D.

N. aquatica COTTON GUM. A tree of swamps. *Distr*. SE USA.

N. sinensis CHINESE TUPELO. Leaves narrow, purple in spring, deep green in summer, and bright red in autumn. *Distr*. China.

N. sylvatica BLACK GUM, TUPELO, COTTON GUM. Leaves turn yellow, orange, and then red in autumn. An important source of timber. *Distr*. E North America.

O

oak *Quercus* **American red** ~ *Q. falcata* **American white** ~ *Q. prinus* **bear** ~ *Q. ilicifolia* **black** ~ *Q. velutina* **black Jack** ~ *Q. marilandica* **bull** ~ *Casuarina equisetifolia* **bur** ~ *Quercus macrocarpa* **Californian black** ~ *Q. kelloggii* **Californian white** ~ *Q. lobata* **chestnut** ~ *Q. prinus* **chestnut-leaved** ~ *Q. castaneifolia* **common** ~ *Q. robur* **cork** ~ *Q. suber* **daimino** ~ *Q. dentata* **durmast** ~ *Q. petraea* **English** ~ *Q. robur* **evergreen** ~ *Q. ilex* **Gambel** ~ *Q. gambellii* **holly** ~ *Q. ilex* **holm** ~ *Q. ilex* **Hungarian** ~ *Q. frainetto* **Japanese** ~ *Q. mongolica* **Japanese emperor** ~ *Q. dentata* **kermes** ~ *Q. coccifera* **Lebanon** ~ *Q. libani* **live** ~ *Q. chrysolepis* **manna** ~ *Q. cerris* **maul** ~ *Q. chrysolepis* **mossy cup** ~ *Q. macrocarpa* **northern red** ~ *Q. rubra* **Oregon white** ~ *Q. garryana* **page** ~ *Rhus radicans* **pedunculate** ~ *Quercus robur* **pin** ~ *Q. palustris* **poison** ~ *Rhus radicans* **possum** ~ *Quercus nigra* **Quebec** ~ *Q. alba* **red** ~ *Q. rubra* **scarlet** ~ *Q. coccinea* **scrub** ~ *Q. ilicifolia* **sessile** ~ *Q. petraea* **she** ~ Casuarinaceae, *Casuarina* **shin** ~ *Quercus gambellii* **shingle** ~ *Q. imbricaria* **silky** ~ *Grevillea robusta* **southern live** ~ *Quercus virginiana* **southern red** ~ *Q. falcata* **swamp** ~ *Casuarina* **swamp red** ~ *Quercus falcata* **swamp white** ~ *Q. bicolor* **tanbark** ~ *Lithocarpus, L. densiflorus* **Turkey** ~ *Quercus cerris* **Turner's** ~ *Q. x turneri* **valley** ~ *Q. lobata* **water** ~ *Q. nigra* **white** ~ *Q. alba* **willow** ~ *Q. phellos*.

oak of Cyprus, golden *Quercus alnifolia*.

oat, false *Arrhenatherum elatius*.

oatgrass *Helictotrichon*.

oats *Avena* **animated** ~ *A. sterilis* **common** ~ *A. sativa* **North American sea** ~ *Uniola paniculata* **North American wild** ~ *Chasmanthium latifolium* **sea** ~ *C. latifolium* **seaside** ~ *Uniola paniculata* **slender wild** ~ *Avena barabata* **water** ~ *Zizania* **wild** ~ *Uvularia*.

obconicus, -a, -um: shaped like an inverted cone.

obcordatus, -a, -um: cordate at the farthest end of the leaf.

obedience plant *Maranta arundinacea*.

obedient plant *Physostegia, P. virginiana*.

obesus: obese, fat.

obliquinervius, -a, -um: with oblique nerves.

obliquus, -a, -um: oblique, lopsided.

oblongatus, -a, -um: oblong.

oblongifolius, -a, -um: with oblong leaves.

oblongus, -a, -um: oblong.

obovatus, -a, -um: obovate, widest above the middle (usually referring to leaf shape).

Obregonia (After Don Alvaro Obregon (1880–1928), President of Mexico.) A genus of 1 species of cactus with spherical stem, robust taproot, pointed tubicles and bell-shaped flowers. *Distr.* NE Mexico. *Fam.* Cactaceae. *Class* D.

O. denegrii Grown as an ornamental. Now endangered in the wild.

obscurus, -a, -um: indistinct, obscure.

obtusatus, -a, -um: obtuse, blunt.

obtusifolius, -a, -um: with blunt leaves.

obtusilobus, -a, -um: with blunt lobes.

obtusus, -a, -um: blunt.

obvallaris, -e: with a wall.

oca *Oxalis tuberosa.*

occidentalis, -e: western.

ocean spray *Holodiscus discolor.*

ocellatus, -a, -um: marked with concentric spots.

Ochagavia (After Silvestri Ochagavia, 19th-century Chilean statesman.) A genus of 3 species of shrubby, erect herbs with numerous linear, spiny leaves and many-flowered globose heads of pink or yellow flowers sunk into the centre of the leaf rosettes. *Distr.* Chile, including Juan Fernández Islands. *Fam.* Bromeliaceae. *Class* M.

O. carnea Leaves bent backwards. *Distr.* Central Chile.

Ochna (From *ochne*, the ancient Greek name used by Homer for the wild pear, which has similar foliage.) A genus of 86 species of deciduous or semi-evergreen shrubs and trees with simple, alternate, leathery leaves and regular, bisexual flowers that bear 5 sepals, 5 petals, and numerous stamens. The fruit is a 1-seeded drupe attached to a swollen receptacle. *Distr.* Tropical regions of the Old World. *Fam.* Ochnaceae. *Class* D.

O. serrulata Shrub or small tree to 3m high. Grown as an ornamental for its bright red fruits. *Distr.* Tropical regions of the Old World.

Ochnaceae A family of 26 genera and about 400 species of trees, shrubs, and some herbs with alternate, simple leaves and regular bisexual flowers that bear 5–12 petals and 5 to many stamens. Some species are a source of timber. Distr. Widespread in tropical and subtropical regions but centred in South America. Class D.
See: Ochna.

ochraceus, -a, -um ochre-coloured, yellow-brown.

ochroleucus, -a, -um: pale ochre-coloured, yellow-brown.

Ochroma (From Greek *ochros*, pale yellow, alluding to the flower colour.) A genus of 1 species of rapidly growing tree with very light wood. The leaves are simple or palmately lobed. The flowers are large, to 30cm across, with 5 sepals and 5 petals, and are pollinated by bats. *Distr.* Tropical America. *Fam.* Bombacaceae. *Class* D.

O. lagopus BALSA, DOWN TREE. This species is the source of the world's lightest commercial wood, which has numerous applications from insulation to model aeroplanes. The fibres surrounding the seeds are used as an inferior form of KAPOK.

O. pyramidale See *O. lagopus*

Ocimum (Classical Greek name.) A genus of 150 species of aromatic herbs and shrubs with opposite, simple leaves and whorls of 2-lipped flowers that are sometimes arranged into a spike-like inflorescence. *Distr.* Warm-temperate and tropical regions, especially Africa. *Fam.* Labiatae. *Class* D.

O. americanum HOARY BASIL. Annual or short-lived perennial herb. Flowers very small, white or pale-purple. Used as a pot-herb in India and medicinally in Saudi Arabia. *Distr.* Tropical and S Africa, China, India.

O. basilicum BASIL, SWEET BASIL. Annual. Flowers small, white or cream, borne in distinct whorls. One of the most important culinary herbs. *Distr.* Tropical regions of the Old World.

O. canum See *O. americanum*

O. sanctum See *O. tenuiflorum*

O. tenuiflorum HOLY BASIL, SACRED BASIL. Aromatic subshrub to 1m high. Flowers pink or white, very small. A sacred Hindu herb used in funeral rites and as a symbol of good luck. *Distr.* Tropical regions of the Old World.

octandrus, -a, -um: eight-flowered.

Octomeria (From Greek *octo*, eight and *meris*, part, alluding to the 8 pollinia.) A genus of 125 species of epiphytic and terrestrial orchids with erect, fleshy to leathery leaves, and racemes of often small flowers. Some species are cultivated as tender ornamentals. *Distr.* Brazil, W Indies, Central and South America. *Fam.* Orchidaceae. *Class* M.

O. graminifolia Flowers pale yellow-green, fleshy. *Distr.* W Indies, Brazil.

O. grandiflora Flowers to 2cm across, white to straw-yellow, lip yellow marked purple. *Distr.* Venezuela, Trinidad, Surinam, Brazil, Bolivia, Paraguay.

octopetalus, -a, -um: with eight petals.

octopus plant *Aloe arborescens*.

octopus tree *Schefflera actinophylla*.

ocymifolius, -a, -um: from the genus *Ocimum*, and *folius*, leaved.

× ***Odontia*** (From the names of the parent genera.) A genus of hybrids between members of the genera *Miltonia* and *Odontoglossum*. They are grown as tender ornamentals. *Distr.* Garden origin. *Fam.* Orchidaceae. *Class* M.

× ***Odontioda*** (From the names of the parent genera.) A genus of hybrids between members of the genera *Oncidium* and *Odontoglossum*. They are grown as tender ornamentals. *Fam.* Orchidaceae. *Class* M.

Odontoglossum (From Greek *odontos*, tooth, and *glossa*, tongue, alluding to the toothed lip.) A genus of 200 species of epiphytic orchids with 1–3-leaved pseudobulbs and typically large, showy flowers. *Distr.* Central and South America. *Fam.* Orchidaceae. *Class* M.

O. bictoniense See *Lemboglossum bictoniense*

O. brevifolium See *Otoglossum brevifolium*

O. cariniferum Flowers to 5cm across, fleshy, chestnut-brown, lip white. *Distr.* Tropical America, from Costa Rica to Venezuela.

O. cervantesii See *Lemboglossum cervantesii*

O. cirrhosum Flowers white, marked red-brown. *Distr.* Ecuador, Peru.

O. citrosmum See *Cuitlauzina pendula*

O. cordatum See *Lemboglossum cordatum*

O. crispum LACE ORCHID. Flowers to 8.5cm across, white or pink, spotted red or purple, margins undulate or finely toothed. *Distr.* Colombia.

O. cristatum Flowers fleshy, cream-yellow with brown markings. *Distr.* Colombia, Ecuador.

O. grande See *Rossioglossum grande*

O. hallii Flowers yellow with brown-purple markings. *Distr.* Colombia, Ecuador, Peru.

O. harryanum (After Sir Harry Veitch, of the family of British nurserymen.) Flowers 4–22 in erect racemes, variable, brown streaked yellow. *Distr.* Colombia, Venezuela, Peru.

O. nobile Flowers 10–100 in arching racemes, slightly fragrant. *Distr.* Colombia.

O. pendulum See *Cuitlauzina pendula*

O. pescatorei See *O. nobile*

O. pulchellum See *Osmoglossum pulchellum*

O. rossii See *Lemboglossum rossii*

O. schlieperianum (After Adolph Schlieper, an orchid collector.) Flowers fragrant, yellow with brown markings. *Distr.* Costa Rica, Panama.

O. spectatissimum See *O. triumphans*

O. stellatum See *Lemboglossum stellatum*

O. triumphans Flowers to 10cm across, golden-yellow, spotted brown. *Distr.* Colombia.

O. uro-skinneri See *Lemboglossum uro-skinneri*

Odontophorus (From Greek *odontos*, tooth, and *phoros*, bearing, alluding to the toothed leaf margins.) A genus of 3 species of mat-forming, succulent subshrubs with thick, soft, paired leaves on rooting stems and white or yellow, many-petalled flowers. *Distr.* South Africa. *Fam.* Aizoaceae. *Class* D.

O. marlothii (After H. W. R. Marloth (1855–1931), South African botanist.) Leaves grey-green. Flowers yellow. *Distr.* South Africa (W Cape Province).

O. nanus Leaves white-hairy. Flowers white. *Distr.* South Africa (W Cape Province).

odoratissimus, -a, -um: highly scented.

odoratus, -a, -um: scented.

Oeceoclades (From Greek oikeios, private, and *klados*, branch.) A genus of 31 species of terrestrial orchids with woody rhizomes, leaves forming spiny thickets, and racemes or panicles of small flowers. Several species are grown as tender ornamentals for their foliage and flower spikes. *Distr.* Tropical regions, principally Africa and Madagascar. *Fam.* Orchidaceae. *Class* M.

O. maculata Leaves grey-green, mottled dark green. *Distr.* Tropical America, tropical Africa.

O. saundersiana Flowers yellow-green with purple-brown veins. *Distr.* Tropical Africa, from Cameroon to Uganda.

oelandicum: of Öland, Southern Sweden.

Oemleria (After Herr Oemler of Dresden, who supplied numerous, American plants to the German botanist H. G. L. Reichenbach in the mid 19th-century.) A genus of 1 species of deciduous shrub with slender branches, simple leaves, and racemes of fragrant white flowers that are followed by black, plum-like fruits. *Distr.* W North America. *Fam.* Rosaceae. *Class* D.

O. cerasiformis OSO BERRY, OREGON PLUM. Grown as an ornamental and for its edible fruits.

Oenanthe (From Greek *oinanthe*, a grape-like inflorescence.) A genus of 30 species of perennial herbs with pinnate leaves and compound umbels of white flowers. *Distr.* N temperate regions, S Africa, Australia. *Fam.* Umbelliferae. *Class* D.

O. crocata HEMLOCK WATER DROPWORT. This plant is highly toxic in all its parts. *Distr.* Europe, NW Africa.

O. japonica See *O. javanica*

O. javanica Grown as an ornamental and as a vegetable. *Distr.* E Asia to N Australia.

Oenothera (From *oinotheras*, the classical Greek name for another plant.) EVENING PRIMROSE. A genus of 80 species of erect, annual, biennial, and perennial herbs with simple, alternate, or basal leaves and spikes of large, bell-shaped flowers that are white, yellow, or purple. Several species are grown as ornamentals and the seeds are a source of a medicinal oil. *Distr.* Temperate America. *Fam.* Onagraceae. *Class* D.

O. biennis EVENING PRIMROSE. Annual or biennial. Flowers yellow, opening in the evening. The roots and the leaves are edible and the seeds are the major source of Evening Primrose Oil. *Distr.* E North America.

O. fruticosa SUNDROPS. Biennial or perennial herb. Flowers deep yellow, opening during the day. A number of cultivars have been raised from this species. *Distr.* E USA.

O. glaber See *O. biennis*

officinalis, -e: sold as a herb, used in medicine.

officinarus, -a, -um: of herbalists' shops, sold by apothecaries.

oil of dittany plant *Cunila origanoides.*

okra *Abelmoschus esculentus.*

Olacaceae A family of 27 genera and about 200 species of shrubs, trees, and climbers with simple, rough leaves and regular, green or white flowers that bear 4–6 petals. Several species are a source of timber; others have edible fruits. *Distr.* Tropical and subtropical regions. *Class* D.
See: Olax, Ongokea.

old maid *Catharanthus roseus.*

old man *Artemisia abrotanum.*

old man's beard *Clematis vitalba, Tillandsia usneoides.*

old man's whiskers *Geum triflorum.*

old woman *Artemisia stelleriana.*

Olea (From the Greek *elaia*, olive.) OLIVE. A genus of 20 species of evergreen shrubs and trees with simple, opposite, leathery leaves and panicles of inconspicuous, white, 4-lobed flowers. The fruit is a rounded 1-seeded drupe. *Distr.* Tropical and warm-temperate regions of the Old World. *Fam.* Oleaceae. *Class* D.

O. europaea OLIVE. Evergreen tree. This plant is intimately linked with the Mediterranean region and its people, being a source of edible fruit, oil, and timber as well as a symbol of peace. Numerous cultivars have been raised from this species which itself may have been derived from another species, *O. africana*, through cultivation. *Distr.* E Mediterranean regions.

Oleaceae The Olive family. 24 genera and 900 species of deciduous or evergreen trees, shrubs, and climbers with simple, trifoliate or pinnate leaves, and raceme-like inflorescences of small, regular flowers. The most economically important species of this family is *Olea europaea*, OLIVE. Other species are useful as ornamentals and as a source of timber. *Distr.* Widespread. *Class* D.
See: Abeliophyllum, Chionanthus, Forsythia, Fraxinus, Jasminum, Ligustrum, Notelaea, Olea, Osmanthus, X Osmarea, Phillyrea, Syringa.

oleander *Nerium oleander* **yellow** ~ *Thevetia peruviana.*

Oleandra (From the species name *Nerium oleander*, alluding to a slight resemblance the leaves.) A genus of 40 species of epiphytic or terrestrial ferns with mat-forming or straggling rhizomes and simple leaves. Several species are grown as ornamentals. *Distr.* Tropical regions. *Fam.* Oleandraceae. *Class* F.

O. articulata Rhizomes straggling. Fronds leathery. *Distr.* Tropical regions of SE Asia.

Oleandraceae A family of 4 genera and about 90 species of terrestrial or epiphytic ferns with a scaly stem, simple or pinnate leaves and sori that are borne terminally on the veins. *Distr.* Widespread in tropical to warm-temperate regions. *Class* F.
See: Nephrolepis, Oleandra.

Olearia (After Adam Olschlager (Olearius) (1603–71), German botanist.) DAISY BUSH. A genus of 130 species of evergreen herbs, shrubs, and small trees with simple, leathery leaves and flowers in daisy-like heads which usually bear distinct rays. Many species are cultivated as ornamentals and a few for their wood. *Distr.* New Guinea, Australia, New Zealand. *Fam.* Compositae. *Class* D.

O. albida Shrub or small tree. Flowers white. *Distr.* New Guinea, Australia, New Zealand.

O. arborescens Shrub to 4m high. Flowers yellow, rays white. *Distr.* New Guinea, Australia, New Zealand.

O. argophylla MUSKWOOD. Large shrub or tree. Flower-heads in large corymbs. *Distr.* New Guinea, Australia, New Zealand.

O. avicenniifolia Small tree to 7m high. Flower-heads small, flowers white, rays few or absent. *Distr.* New Zealand.

O. capillaris Densely branched shrub. Flowers yellow, rays white. *Distr.* New Zealand.

O. chathamica Densely branched shrub. Flower-heads solitary, flowers purple, rays white. *Distr.* New Zealand (Chatham Islands).

O. cheesemanii (After T. F. Cheeseman (1846–1923), New Zealand botanist.) Shrub or small tree. *Distr.* New Zealand.

O. floribunda Shrub to 2m high. *Distr.* SE Australia.

O. frostii (After Charles Frost.) Straggling shrub. Rays lilac. *Distr.* Australia (Victoria).

O. furfuracea Small tree to 5m high. *Distr.* New Zealand.

O. glandulosa Small aromatic shrub to 2m high. *Distr.* SE Australia.

O. gunniana See *O. phlogopappa*

O. × haastii (After Johann Franz Julius von Haast (1824–87) New Zealand geologist.) A hybrid between *O. avicenniifolia* and *O. moschata*. *Distr.* New Zealand.

O. ilicifolia Musk-scented large shrub or tree. *Distr.* New Zealand.

O. insignis Spreading shrub. *Distr.* New Zealand.

O. lepidophylla Small clump-forming shrub. *Distr.* SE Australia.

O. lirata Shrub or small tree to 3m high. *Distr.* Australia (Tasmania).

O. macrodonta Large shrub or small tree to 6m high. *Distr.* New Zealand.

O. × mollis A hybrid of *O. ilicifolia* and *O. lacunosa*. *Distr.* Garden origin.

O. moschata Somewhat sticky shrub, to 4m high. *Distr.* New Zealand.

O. myrsinoides Straggling small shrub. *Distr.* Australia (Tasmania, Victoria).

O. nummulariifolia Large shrub. Flowers fragrant. *Distr.* New Zealand.

O. odorata Little-branched, wiry shrub. Flowers scented. *Distr.* New Zealand.

O. paniculata Tree to 6m. Flowers fragrant. *Distr.* New Zealand.

O. phlogopappa Much-branched, aromatic shrub, to 3m high. *Distr.* SE Australia (Tasmania).

O. ramulosa Small bristly shrub. *Distr.* Australia (Tasmania).

O. rani See *O. cheesemanii*

O. × scilloniensis A hybrid of *O. lirata* and *O. phlogopappa*. *Distr.* Garden origin.

O. semidentata Rounded shrub to 3m high. *Distr.* New Zealand (Chatham Islands).

O. solandri (After D. C. Solander (1733–82), botanist on Cook's first voyage.) Somewhat sticky shrub or small tree. *Distr.* New Zealand.

O. stellulata *Distr.* Australia (New South Wales).

O. traversii (After W. T. L. Travers (1819–1903).) Tree to 10m high. Rays absent. *Distr.* New Zealand (Chatham Islands).

O. virgata Dense bushy shrub to 5m high. *Distr.* New Zealand.

oleaster Elaeagnaceae, *Elaeagnus*, *E. angustifolia*.

oleifolius, -a, -um: from the genus *Olea*, and *folius*, leaved.

oleoides: resembling the genus *Olea*.

oleraceus, -a, -um: vegetable, of kitchen gardens.

oliganthus, -a, -um: with a few flowers.

oligodon: few-toothed.

oligophyllus, -a, -um: with a few leaves.

oligospermus, -a, -um: with a few seeds.

oligostachyus with a few spikes.

Oliniaceae A family of 1 genus and 8 species of shrubs and trees with simple, opposite leaves on 4-angled branches. The flowers are regular and bisexual with 4–5 sepals and petals. *Distr.* E and S Africa. *Class* D.

olive Oleaceae, *Olea*, *O. europaea* **bastard** ~ *Buddleja saligna* **California** ~ *Umbellularia californica* **Ceylon** ~ *Elaeocarpus serratus* **fragrant** ~ *Osmanthus fragrans* **holly** ~ *O. heterophyllus* **Russian** ~ *Elaeagnus angustifolia* **spurge** ~ Cneoraceae, *Cneorum tricoccon* **wild** ~ *Elaeagnus angustifolia*.

Olsynium (From Greek *olsynion*, hardly united, alluding to the stamens.) A genus of about 12 species of perennial herbs with sword-shaped leaves and bell-shaped flowers that are surrounded by a spathe. Several species are occasionally grown as ornamentals. *Distr.* North and South America. *Fam.* Iridaceae. *Class* M.

O. douglasii (After David Douglas (1798–1834), Scottish plant collector.) GRASS WIDOW, PURPLE-EYED GRASS. Stems flat. Flowers pendent, purple-red, 2–3 per spathe. *Distr.* W North America.

olympicus, -a, -um: from Mount Olympus of ancient Greece.

omeiensis, -e: of the Omei Shan, China.

Omphalodes (From Greek *omphalos*, a navel, and the ending *-oides*, indicating resemblance, alluding to the navel-like depression on the seeds.) NAVELWORT, NAVEL SEED. A genus of 28 species of annual and perennial herbs with simple alternate leaves and cymes of blue or white, tubed flowers. The mouth of the petal tube is almost closed by scale-like structures as in *Myosotis*, FORGET-ME-NOT. Several species are grown as ornamentals. *Distr.* Temperate Eurasia, Mexico. *Fam.* Boraginaceae. *Class* D.

O. cappadocica Creeping perennial. Flowers bright blue. *Distr.* W Asia.

O. linifolia VENUS' NAVELWORT. Erect annual. Flowers white. *Distr.* SW Europe.

O. luciliae (After Lucille Boissier.) Mound-forming perennial. Leaves grey-green. Flowers pink in bud, then blue. *Distr.* Greece, W Asia.

O. verna BLUE-EYED MARY, CREEPING FORGET-ME-NOT. Clump-forming perennial. Leaves relatively large. Flowers blue with a white centre. *Distr.* S Europe.

Omphalogramma (From the Greek *omphalos*, navel, and *gramma*, writing or line.) A genus of 15 species of rhizomatous herbs with simple, glandular leaves and nodding, funnel-shaped flowers. Several species are grown as ornamentals. *Distr.* Himalaya, W China. *Fam.* Primulaceae. *Class* D.

O. vinciflorum Rosette-forming perennial. Flowers violet. *Distr.* China.

Onagraceae The Evening Primrose family. 17 genera and about 500 species of aquatic and terrestrial herbs and a few shrubs. The leaves are simple and opposite or alternate. The flowers may be regular or irregular, bisexual or uni-sexual and are often showy. Many species are grown as ornamentals. *Distr.* Widespread, especially America. *Class* D. *See: Chamaenerion, Circaea, Clarkia, Epilobium, Fuchsia, Gaura, Oenothera.*

Oncidium (From Greek *onkos*, tumour, alluding to a swelling on the lip.) A genus of 412 species of epiphytic, lithophytic, and terrestrial orchids with fleshy leaves and typically fine panicles or racemes of small, yellow or brown flowers. *Distr.* Tropical and subtropical America. *Fam.* Orchidaceae. *Class* M.

O. altissimum Panicles to 3m long. Flowers yellow-green marked maroon. *Distr.* W Indies.

O. ampliatum Inflorescence to 60cm long. Flowers yellow, spotted brown. *Distr.* Tropical America, from Guatemala to Peru.

O. bicallosum Racemes long. Flowers yellow tinged brown. *Distr.* S Mexico, Guatemala, El Salvador.

O. cavendishianum (After William George Spencer Cavendish, 6th Duke of Devonshire.) Panicle to 2m long. Flowers scented, yellow marked red. *Distr.* S Mexico, Guatemala, Honduras.

O. cebolleta Inflorescence to 1.5m long. Flowers green-yellow marked red. *Distr.* Tropical America, from Mexico and the W Indies to N Argentina.

O. cheirophorum COLOMBIAN BUTTERCUP. Dwarf epiphyte. Flowers scented, bright yellow. *Distr.* Costa Rica, Panama, Colombia.

O. concolor Raceme to 30cm long, pendent. Flowers bright golden-yellow. *Distr.* SE Brazil, N Argentina.

O. crispum Panicle to 1m long, erect or pendent. Flowers numerous, brown, spotted yellow. *Distr.* E Brazil.

O. cucullatum Inflorescence to 50cm long. Flowers brown, often edged with yellow *Distr.* Colombia, Ecuador.

O. divaricatum Panicle to 2m long. Flowers yellow, blotched brown. *Distr.* E Brazil.

O. flexuosum DANCING DOLL ORCHID. Panicle to 1m long. Flowers yellow, marked red. *Distr.* SE Brazil, Paraguay, Argentina.

O. forbesii Raceme many-flowered. Flowers brown, marbled yellow, wavy on margins. *Distr.* E Brazil.

O. gardneri Flowers brown with wavy, yellow margins and a small lip. *Distr.* E Brazil.

O. harrisonianum Panicle to 30cm long. Flowers golden-yellow, marked red. *Distr.* Brazil.

O. hastatum Panicles to 1.5m long. Flowers green-yellow, marked maroon. *Distr.* S Mexico.

O. incurvum Panicle to 2m long. Flowers fragrant, white, marked violet-pink. *Distr.* Mexico.

O. leucochilum Panicle to 3m long. Flowers bright green, marked red and white. *Distr.* Mexico, Guatemala, Honduras.

O. longifolium See *O. cebolleta*

O. longipes Raceme to 15cm long. Flowers yellow-brown. *Distr.* SE Brazil.

O. luridum Inflorescence to 1.5m long. Flowers yellow-brown. *Distr.* Tropical America.

O. macranthum Panicle to 3m long, twining. Flowers yellow. *Distr.* Colombia, Ecuador.

O. maculatum Inflorescence erect. Flowers fragrant. *Distr.* Mexico, Guatemala, Honduras.

O. marshallianum (After W. Marshall of Enfield, who grew the type specimen.) Panicles to 2m long. Flowers dull yellow, marked red-brown. *Distr.* E Brazil.

O. microchilum Panicle to 1.5m long, erect. Flowers brown, marked yellow. *Distr.* Mexico, Guatemala.

O. nanum Panicle to 25cm long. Flowers small with a bright yellow lip. *Distr.* Tropical South America, from Colombia to Peru and Central Brazil.

O. onustum Panicle to 14cm long. Flowers golden yellow. *Distr.* Panama, Colombia, Ecuador, Peru.

O. ornithorhyncum Panicle to 50cm long, a ched. Flowers fragrant, white-lilac. *Distr.* Central America, from S Mexico to Costa Rica.

O. papilio BUTTERFLY ORCHID. Panicle erect or sub-erect. Flowers to 15cm across, purple, mottled yellow. *Distr.* Tropical South America, from Trinidad to Peru.

O. phalaenopsis Raceme slender, to 25cm long. Flowers to 3cm across, white marked red. *Distr.* Colombia, Ecuador.

O. pubes Panicle 2-ranked. Flowers red-brown, marked yellow. *Distr.* Brazil.

O. pulchellum Racemes to 50cm long, erect. Flowers usually white. *Distr.* Jamaica, Guyana.

O. pumilum Panicle to 15cm from a creeping rhizome. Flowers yellow, spotted red. *Distr.* Brazil.

O. pusillum Small creeping epiphyte. Inflorescence to 6cm long. *Distr.* Central and South America.

O. sarcodes Leaves with 2 leaflets. Panicle to 2m long. Flowers brown, edged with yellow. *Distr.* Brazil.

O. sphacelatum Panicle to 2m long. Flowers yellow, marked brown. *Distr.* Central America, Venezuela.

O. splendidum Panicle erect, to 1m long. Flowers showy, bright yellow, marked red. *Distr.* Guatemala.

O. superbiens Panicle to 4m long, twining. Flowers to 8cm across, yellow and red. *Distr.* Peru, Colombia, Ecuador, Venezuela.

O. tigrinum Panicle to 90cm long. Flowers to 8cm across. *Distr.* Mexico.

O. triquetrum Panicle to 20cm long. Flowers white-green, spotted purple. *Distr.* Jamaica.

O. varicosum Panicle to 1.5cm long. Flowers to 3cm across, green-yellow. *Distr.* E and Central Brazil.

O. wentworthianum (After Lord Fitzwilliam (1748–1833).) Panicle to 1.5cm long, pendent. Flowers deep-yellow, blotched red. *Distr.* Guatemala, Mexico.

Oncothecaceae A family of 1 genus and 2 species of shrubs and trees with simple, leathery, spirally arranged leaves and clusters of small, regular flowers that bear their parts in fives. *Distr.* New Caledonia. *Class* D.

onion Alliaceae, *Allium cepa* **climbing** ~ *Bowiea volubilis* **Japanese bunching** ~ *Allium fistulosum* **lily** ~ *A. moly* **nodding** ~ *A. cernuum* **sea** ~ *Urginea maritima* **small yellow** ~ *Allium flavum* **tree** ~ *A. cepa* **Welsh** ~ *A. fistulosum* **wild** ~ *A. cernuum* **yellow** ~ *A. moly*.

onites: Greek name for a kind of marjoram.

Onobrychis (From Greek *onos*, ass, and *brycho*, to bray; it is said to be a favourite food of asses.) A genus of 130 species of annual and perennial herbs and shrubs with alternate pinnate leaves and racemes or spikes of pea-like flowers. Several species are grown as ornamentals. *Distr.* Eurasia, Ethiopia. *Fam.* Leguminosae. *Class* D.

O. viciifolia SAINFOIN, HOLY CLOVER. Perennial herb. Flowers pink with purple veins. Grown as fodder and as a bee plant. *Distr.* Asia, naturalized in Europe and North America.

Onoclea (From Greek *onos*, vessel, and *kleio*, to close; the edges of the fertile leaves curl around the sori, enclosing them.) A genus of 1 species of deciduous, terrestrial fern with pinnately divided leaves. *Distr.* N America and E Asia. *Fam.* Woodsiaceae. *Class* F.

O. sensibilis SENSITIVE FERN. Grown as ornamental ground cover. The leaves are sensitive to the first frost and turn brown very quickly.

Ononis (Greek name for these plants.) REST-HARROW. A genus of 75 species of annual and perennial herbs and subshrubs with trifoliate leaves and spikes, racemes, or panicles of pea-like flowers. Several species are grown as ornamentals. *Distr.* Europe, Mediterranean regions, Canaries, Ethiopia, and Iran. *Fam.* Leguminosae. *Class* D.

O. fruticosa SHRUBBY REST-HARROW. Low deciduous shrub. Flowers purple-pink with dark streaks. *Distr.* SE France, Spain, Algeria.

O. natrix GOAT ROOT, LARGE YELLOW REST-HARROW. Dwarf shrub. Flowers yellow with red streaks. *Distr.* S and Central Europe, N Africa.

O. rotundifolia Deciduous upright subshrub. Flowers large, pink, borne in clusters. *Distr.* Central and SW Europe.

Onopordum (From *onopordon*, the Greek name for these plants.) A genus of 40 species of coarse, prickly, biennial herbs with spiny leaves and purple flowers in daisy-like heads that lack distinct rays. *Distr.* Europe, W Asia. *Fam.* Compositae. *Class* D.

O. acanthium SCOTCH THISTLE, COTTON THISTLE. To 3m high. Stem yellow-hairy. *Distr.* Europe, W Asia.

O. arabicum See *O. nervosum*

O. bracteatum To 2m high. Stems with numerous wings. *Distr.* E Mediterranean.

O. nervosum To 3m high. Stem winged, yellow-hairy. *Distr.* Spain, Portugal.

O. salteri To 2m, somewhat sticky. Flowers purple-pink. *Distr.* SE Europe.

Onosma (From Greek *onos*, ass, and *osme*, smell, alluding to the roots.) A genus of 150 species of annual, biennial, and perennial herbs and subshrubs with simple, alternate leaves and cymes of pendent, tubular or bell-shaped flowers. The leaves and stems are often covered in rough hairs that may cause skin irritations. Several species are grown as ornamentals and some are used as a source of dye. *Distr.* Mediterranean to Himalaya and China. *Fam.* Boraginaceae. *Class* D.

O. albo-roseum Perennial herb. Flowers to 3cm long, white turning pink from the tip. *Distr.* W Asia.

O. echioides Tuft-forming perennial herb. Flowers pale yellow. A source of red dye. *Distr.* Italy, SE Europe.

O. stellulata Perennial herb. Leaves covered in star-shaped hairs. Flowers pale yellow. *Distr.* E Central Europe.

O. tauricum GOLDEN DROP. Perennial herb. Leaves often marked white. Flowers white to golden. *Distr.* S Europe, W Asia.

Onychium (From Greek *onychos*, a claw, alluding to the shape of the lobes of the fronds.) A genus of 7 species of small to medium-sized ferns. *Distr.* E Asia and America. *Fam.* Adiantaceae. *Class* F.

O. japonicum Leaves finely divided. Grown as an ornamental. *Distr.* Japan, China.

ookow *Dichelostemma congestum.*

Oophytum (From Greek *oon*, egg, and *phyton*, plant, alluding to the shape made by the fused leaves.) A genus of 2 species of very succulent, small herbs with pairs of fused leaves and solitary flowers which have numerous petals. Both species are grown as ornamentals. *Distr.* South Africa. *Fam.* Aizoaceae. *Class* D.

O. nanum Leaves bright green. Petals white with red tips.

O. oviforme Leaves olive green, often suffused with red. Flowers tinged pink above.

opacus, -a, -um: opaque, dull, not shining.

ophiocarpus, -a, -um: with a snake-like fruit, slender and twisted.

Ophioglossaceae A family of 3 genera and 64 species of terrestrial and epiphytic ferns. The leaves are typically solitary and divided into two parts, a sterile, usually simple, blade and a reduced, often spike-like, fertile portion which bears the sporangia. *Distr.* Widespread. *Class* F.
See: Ophioglossum.

ophioglossifolius, -a, -um: with leaves like ferns of the genus *Ophioglossum.*

Ophioglossum (From Greek *ophis*, snake, and *glossa*, tongue, alluding to the shape of the leaf.) ADDER'S TONGUE, SNAKE'S TONGUE. A genus of 30 species of terrestrial and a few epiphytic ferns with 2-lobed leaves, 1 lobe of which is narrow and bears the sporangia. Some species are edible and some have local medicinal uses. *Distr.* Widespread. *Fam.* Ophioglossaceae. *Class* F.

O. vulgatum COMMON ADDER'S TONGUE. Terrestrial fern to 40cm high. *Distr.* Europe, W Asia, North America.

Ophiopogon (From Greek *ophis*, snake, and *pogon*, beard.) JAPANESE HYACINTH. A genus of 50 species of perennial herbs with evergreen, tufted, grass-like leaves and racemes of numerous, small, white to lilac flowers which are followed by blue berries. *Distr.* Asia to Malaysia and the Philippines. *Fam.* Convallariaceae. *Class* M.

O. graminifolius See *Liriope muscari*

O. japonicus Often grown as hardy ground cover. Tuberous roots are edible. *Distr.* Japan, Korea, China.

Ophrys (Greek name for a 2-leaved plant possibly an orchid.) A genus of 40–50 species of terrestrial, tuberous orchids with a rosette of basal leaves and a spike of yellow-green flowers that bear lips, resembling a large insect. Male insects attracted by the scent, which is similar to that produced by a female insect, attempt to copulate with the lip; the pollen is attached to the insect and so transported to a second flower. A number of species are grown as hardy ornamentals and curiosities. *Distr.* NW Europe, Mediterranean, Middle East. *Fam.* Orchidaceae. *Class* M.

O. apifera BEE ORCHID, WASP ORCHID. Lip 3-lobed, convex, margin red-brown. Mutant form with deformed lip known as a WASP ORCHID. *Distr.* W, S and Central Europe.

O. fusca Lip variable, black to maroon with pale patches, 3-lobed, lateral lobes sometimes very small. *Distr.* Mediterranean, Portugal, SW Romania.

O. holoserica LATE SPIDER ORCHID. A variable species. Lip typically convex, brown, marbled green in the centre. *Distr.* W, SW and Central Europe.

O. insectifera FLY ORCHID. Lip 3-lobed, brown, mid-lobe notched, marked blue in the centre. Tubers sometimes used to make salep. *Distr.* Europe, except SE, uncommon in extreme N and S.

O. lutea YELLOW BEE ORCHID. Lip spreading, 3-lobed, margin yellow or green-yellow. *Distr.* Mediterranean, Portugal.

O. speculum MIRROR ORCHID. Lip with a shiny blue centre and a furry brown margin. *Distr.* Mediterranean regions.

O. sphegodes EARLY SPIDER ORCHID. Lip broad, usually entire, sometimes notched, brown. *Distr.* Europe.

Opiliaceae A family of 10 genera and 28 species of semi-parasitic, evergreen shrubs and trees with simple alternate leaves and small flowers that bear their parts in fours or fives. *Distr.* Tropical and subtropical regions. *Class* D.

Oplismenus (From Greek *hoplismos*, weapon, alluding to the sharp awns.) A genus of 9 species of trailing, annual and perennial grasses with flat, ovate leaves and 1-sided, spike-like inflorescences. Several species are grown as tender or half-hardy ornamentals. *Distr.* Tropical and warm regions. *Fam.* Gramineae. *Class* M.

O. hirtellus BASKET GRASS. The most frequently cultivated species, used in hanging baskets and on the edge of staging in the greenhouse. *Distr.* Tropical America and Africa.

opossum-wood *Halesia tetraptera.*

oppositifolius, -a, -um: with opposite leaves.

opulifolius: with leaves resembling those of *Viburnum opulus.*

Opuntia (Greek name for a spiny plant that grew near Opus (Opuntis).) PRICKLY PEAR. A genus of 200 species of small to large, tree-like cacti with segmented, sometimes flattened stems that occasionally bear cylindrical leaves. The numerous saucer-shaped flowers are followed by a dry or fleshy, pear-shaped fruit. Many species are grown as ornamentals; some are used as hedging while spineless species form a forage crop in hard times. The fruits are sometimes edible and are a source of cochineal. *Distr.* North and South America. *Fam.* Cactaceae. *Class* D.

O. articulata Small shrub. Branches brittle. Spines absent or papery. Flowers white or pale pink, to 4cm across. *Distr.* Argentina.

O. aurantiaca TIGER PEAR. Low jointed shrub. Flowers yellow-orange. This species was introduced into Australia where it became a noxious weed of grazing land and was only brought under control when a moth whose larvae feed on it was also introduced from South America: an early example of biological control. *Distr.* Uruguay and adjacent Argentina.

O. brasiliensis Tree-like. Ultimate branches flattened, bright green, leaf-like. Flowers pale yellow. *Distr.* E South America.

O. clavarioides NIGGER HAND. Low shrub. Roots tuberous. Stems shaped like an inverted cone. Flowers brown. *Distr.* Argentina.

O. compressa Prostrate. Stems flat, round to oval. Flowers large, yellow. *Distr.* USA.

O. cylindrica CANE CACTUS. Small bushy tree with cylindrical stems and short-lived leaves. Flowers red. *Distr.* S Ecuador, Peru.

O. diademata See *O. articulata*

O. ficus-indica INDIAN FIG, BARBARY FIG, PRICKLY PEAR. Bush or small tree. Trunk thick. Stems blue-green. Flowers yellow. Fruit fleshy, yellow to red or purple. Widely cultivated for its edible fruits. *Distr.* Mexico, widely naturalized.

O. humifusa See *O. compressa*

O. imbricata CHAIN-LINK CACTUS. Bushy tree. Stem segments large, covered in tubicles and large spines. Flowers purple. *Distr.* SW USA, Mexico.

O. microdasys BUNNY EARS. Bushy shrub. Stems flattened, oval. Flowers numerous, funnel-shaped, yellow. Several ornamental cultivars are available. *Distr.* Central and N Mexico.

O. robusta Shrub. Stem segments flat, round, blue-green. Flowers saucer-shaped, yellow. *Distr.* Central Mexico.

O. rufida See *O. microdasys*

O. subulata EVE'S PIN CACTUS. Small tree. Stems unsegmented. Leaves cylindrical. Flowers red. *Distr.* S Peru.

O. tunicata Mound-forming. Densely covered in sheathed golden spines. *Distr.* USA (Texas) to Chile.

O. vestita COTTON POLE CACTUS. Low shrub covered in soft white hairs. Flowers violet-red. *Distr.* Bolivia.

orache *Atriplex A. hortensis.*

orange *Citrus sinensis* **bitter** ~ *Poncirus trifoliata* **Mexican** ~ *Choisya ternata* **mock** ~ *Philadelphus P. coronarius*, **Osage** ~ *Maclura pomifera* **Panama** ~ × *Citrofortunella microcarpa* **Seville** ~ *Citrus aurantium* **sour** ~ *C.*

aurantium **sweet** ~ *C. sinensis* **trifoliate** ~ *Poncirus trifoliata.*

orange ball tree *Buddleja globosa.*

orange eye *Buddleja davidii.*

orbicularis, -e: orbicular.

orbiculatus, -a, -um: orbicular.

orchid Orchidaceae **Anatolian** ~ *Orchis anatolica* **bee** ~ *Ophrys apifera* **broad-leaved marsh** ~ *Dactylorhiza majalis* **bug** ~ *Orchis coriophora* **butterfly** ~ *Platanthera, Oncidium papilio, Orchis papilionacea* **Chien Lan** ~ *Cymbidium ensifolium* **clamshell** ~ *Encyclia cochleata* **clown** ~ *Rossioglossum grande* **cockle** ~ *Encyclia cochleata* **common spotted** ~ *Dactylorhiza fuchsii* **dancing doll** ~ *Oncidium flexuosum* **early marsh** ~ *Dactylorhiza incarnata* **early purple** ~ *Orchis mascula* **early spider** ~ *Ophrys sphegodes* **fly** ~ *O. insectifera* **fragrant** ~ *Gymnadenia conopsea* **Fukien** ~ *Cymbidium ensifolium* **greater butterfly** ~ *Platanthera chlorantha* **green-winged** ~ *Orchis morio* **lace** ~ *Odontoglossum crispum* **lady** ~ *Orchis purpurea* **late spider** ~ *Ophrys holoserica* **lesser butterfly** ~ *Platanthera bifolia* **lily of the valley** ~ *Osmoglossum pulchellum* **Madeiran** ~ *Dactylorhiza foliosa* **military** ~ *Orchis militaris* **mirror** ~ *Ophrys speculum* **moth** ~ *Phalaenopsis* **northern marsh** ~ *Dactylorhiza purpurella* **pansy** ~ *Miltonia* **rattlesnake** ~ *Pholidota* **robust marsh** ~ *Dactylorhiza elata* **scented** ~ *Gymnadenia conopsea* **short-spurred fragrant** ~ *G. odoratissima* **slipper** ~ *Paphiopedilum, Cypripedium* **soldier** ~ *Orchis militaris* **southern marsh** ~ *Dactylorhiza praetermissa* **tongue** ~ *Serapias* **vanilla** ~ *Nigritella nigra* **wasp** ~ *Ophrys apifera* **water** ~ *Eichhornia, E. crassipes* **yellow bee** ~ *Ophrys lutea.*

Orchidaceae The Orchid family. 835 genera and about 18000 species of terrestrial and epiphytic herbs. There are a number of features that are characteristic of this family, mostly relating to the flowers; they bear 6 tepals, 5 of which may be more or less similar and are usually petal-like; the sixth is formed into a lip which may act as a landing stage or trap for the insect pollinator or may be very much reduced. The sexual organs are fused into a structure called the column which lies opposite the lip. In most plants pollen is powdery but in the majority of orchids it is fused into sticky masses termed pollinia. The seeds are very small (there may be several hundred thousand in a single pod) and need the help of a fungus, in a special symbiotic relationship, to germinate. Many species are cultivated as ornamentals. The essence vanilla is obtained from pods of *Vanilla planifolia. Distr.* Cosmopolitan. *Class* M.
See: Bletilla, Calanthe, Cuitlauzina, Cymbidium, Cypripedium, Dactylorhiza, Dracula, Dryadella, Encyclia, Epidendrum, Epigeneium, Epipactis, Eria, Eriopsis, Esmeralda, Eulophia, Eulophiella, Eurychone, Galeandra, Gastrochilus, Gomesa, Gongora, Goodyera, Grammangis, Grammatophyllum, Gymnadenia, Habenaria, Lemboglossum, Macodes, Malaxis, Masdevallia, Maxillaria, Meiracyllium, Mendoncella, Miltonia, Miltoniopsis, Mormodes, Mormolyca, Nageliella, Nanodes, Neofinetia, Nigritella, Octomeria, Odontoglossum, × *Odontia,* × *Odontioda, Oeceoclades, Oncidium, Ophrys, Orchis, Ornithocephalus, Osmoglossum, Otoglossum, Pabstia, Paphinia, Paphiopedilum, Paraphalaenopsis, Pecteilis, Peristeria, Pescatorea, Phaius, Phalaenopsis, Pholidota, Phragmipedium, Physosiphon, Platanthera, Pleione, Pleurothallis, Polycycnis, Polystachya, Porroglossum,* × *Potinara Promenaea, Pteroceras, Pterostylis, Renanthera, Rossioglossum, Satyrium, Scaphyglottis, Scuticaria, Serapias.*

orchidiflorus, -a, -um: with orchid-like flowers.

orchidiformis, -e: orchid-shaped.

orchioides: orchid-like.

Orchis (Greek name for testicles, alluding to the shape of the twin tubers of some species.) A genus of 35 species of deciduous terrestrial orchids with 2–3 tubers, linear leaves, and dense racemes of spurred, typically purple-red to white flowers. The lateral petals are often pushed forward to form a hood. The tubers are often used in the production of salep. *Distr.* Europe to Asia. *Fam.* Orchidaceae. *Class* M.

O. anatolica ANATOLIAN ORCHID. Spur very long and conspicuous. Lip pink with white spots. *Distr.* E Mediterranean, W Iran.

O. coriophora BUG ORCHID. Lip 3-lobed, convex, dark. *Distr.* S, Central and E Europe.

O. elata See *Dactylorhiza elata*

O. foliosa See *Dactylorhiza foliosa*

O. fuchsii See *Dactylorhiza fuchsii*

O. maculata See *Dactylorhiza maculata*

O. maderensis See *Dactylorhiza foliosa*

O. majalis See *Dactylorhiza majalis*

O. mascula EARLY PURPLE ORCHID. Flowers purple, lip flared or 3-lobed. *Distr.* Europe to W Iran.

O. militaris SOLDIER ORCHID, MILITARY ORCHID. Lip pink to purple with a pale centre, 3-lobed, mid-lobe notched so that the whole resembles a man in shape. A protected species in Great Britain. *Distr.* Central Europe to the Caucasus.

O. morio GREEN-WINGED ORCHID. Flowers white-green or red-violet. Lateral petals often green-veined. *Distr.* Europe to W Iran.

O. papilionacea BUTTERFLY ORCHID. Flowers white-pink, marked violet or red, lip fan-shaped. *Distr.* S Europe to SW Asia.

O. praetermissa See *Dactylorhiza praetermissa*

O. purpurea LADY ORCHID. Flowers white, marked red or maroon, lip 3-lobed, mid-lobe flared and notched. *Distr.* Europe to N Africa.

oregano *Origanum vulgare*.

oreganus, -a, -um: of Oregon, USA.

Oreocereus (From Greek *oreos*, mountain, and the genus name *Cereus*, alluding to the habitat of these plants.) A genus of 5–7 species of shrubby cacti with cylindrical stems that are often obscured by long white hairs. The flowers are somewhat irregular, tubular to funnel-shaped, red, orange, or purple and open during the day. Several species are grown as ornamentals. *Distr.* W South America. *Fam.* Cactaceae. *Class* D.

O. celsianus Shrub to 3m high. Flowers to 9cm long, dull pink. *Distr.* NW Argentina and Bolivia.

O. doelzianus Small much-branched shrub to 1m high. Flowers purple-pink. *Distr.* Central Peru.

O. hempelianus Small little-branched shrub. Stems grey-green. Flowers red. *Distr.* S Peru, N Chile.

Oreopanax (From Greek *oreos*, mountain, and the genus name *Panax*.) A genus of 80 species of evergreen trees and shrubs with clustered simple or palmately lobed leaves and panicles of small, unisexual flowers. *Distr.* Tropical America. *Fam.* Araliaceae. *Class* D.

O. epremesnilianus Grown as an ornamental.

oreophilus, -a, -um: mountain-loving.

organensis, -e: of the Organ Moun-tains, Brazil.

orientalis, -e: eastern.

origanifolius, -a, -um: from the genus *Origanum*, and *folius*, leaved.

Origanum (From Greek *oreos*, mountain, and *ganos*, joy.) A genus of 36 species of aromatic, rhizomatous herbs and shrubs with simple, opposite leaves and whorls of 2-lipped flowers. Some species are grown for their essential oils, as culinary herbs, or as ornamentals. *Distr.* Eurasia. *Fam.* Labiatae. *Class* D.

O. amanum Deciduous rounded subshrub. Leaves small, heart-shaped. Flowers pink or white. *Distr.* S Turkey.

O. dictamnus CRETAN DITTANY, HOP MARJORAM. Dwarf shrub. Leaves grey-hairy. Flowers purple-pink, surrounded by pink-tinged bracts, borne in pendent heads. Cultivated as a pot-herb and medicinal herb. *Distr.* Crete.

O. × hybridum A hybrid of *O. dictamnus* × *O. sipyleum*. Tufted perennial herb. Flowers pink. *Distr.* Garden origin.

O. kopatdaghense See *O. vulgare*

O. laevigatum Mat-forming subshrub. Flowers in loose inflorescences, tiny, bright pink. *Distr.* Cyprus, Turkey.

O. majorana SWEET MARJORAM, KNOTTED MARJORAM. Annual to perennial herb. Stems tinged red. Flowers small, white-purple. Grown as a culinary herb, particularly for flavouring meat dishes. It was formerly used medicinally. *Distr.* N Africa, SW Asia, naturalized in Europe.

O. onites (From the Greek vernacular name for a kind of marjoram.) POT MARJORAM.

Mound-forming shrub. Used as an inferior substitute for *O. majorana*. *Distr*. Mediterranean regions, W Asia.

O. vulgare WILD MARJORAM, OREGANO, POT MARJORAM. Rhizomatous woody perennial. Leaves strongly aromatic. Flowers purple. An important culinary herb with numerous culinary and ornamental cultivars having been produced. *Distr*. Europe to Central Asia.

Orixa (From the Japanese name for this plant.) A genus of 1 species of aromatic, deciduous shrub with simple, alternate leaves and racemes of small green flowers that bear their parts in fours. *Distr*. E Asia. *Fam*. Rutaceae. *Class* D.

O. japonica Grown as an ornamental, particularly in Japan where it is used for hedging.

ornatus, -a, -um: ornate, adorned, embellished.

Ornithocephalus (From Greek *ornis*, bird, and *kephale*, head, alluding to the shape of the column and anther.) A genus of 50 species of epiphytic orchids with leaves in 2 ranks and racemes of small flowers. Some species are cultivated as tender ornamentals. *Distr*. Tropical regions of the Americas. *Fam*. Orchidaceae. *Class* M.

O. gladiatus Flowers cream with green markings. *Distr*. Central America, NW South America, W Indies.

O. grandiflorus Flowers to 18mm across, white with bright green spots. *Distr*. Brazil.

O. iridifolius Inflorescences to 8cm, flowers small, white. *Distr*. Mexico, Guatemala.

Ornithogalum (From Greek *ornis*, bird, and *gala*, milk.) A genus of 80 species of bulbiferous herbs with a rosette of leaves and short to tall racemes of typically small, white, star-shaped flowers. A number of species are grown as ornamentals. *Distr*. S Africa, Mediterranean regions. *Fam*. Hyacinthaceae. *Class* M.

O. arabicum Flowers cream-white, scented. *Distr*. Mediterranean regions.

O. montanum Clump-forming. Flowers white, borne in stalkless clusters close to the leaves. *Distr*. SE Europe, Turkey, Syria, Lebanon.

O. narbonense To 1m high. Flowers numerous, milk-white. *Distr*. Mediterranean regions, Caucasus, N Iran.

O. nutans DROOPING STAR OF BETHLEHEM. Flowers bell-shaped, pendent. *Distr*. SE Europe, Turkey.

O. pyrenaicum BATH ASPARAGUS. Young inflorescences eaten like asparagus. *Distr*. Europe, Turkey, Caucasus.

O. thyrsoides CHINCHERINCHEE. Flowers white, in dense conical spike on long stem. Grown as a winter cut flower. The common name is said to represent the sound of the flowering stems knocking together in the wind. *Distr*. South Africa (Cape Province).

O. umbellatum STAR OF BETHLEHEM. To 30cm high. Flowers white, borne in flat-topped umbels. *Distr*. Europe, N Africa, SW Asia.

ornithopoda: bird-footed.

ornithorhyncum: like a bird's beak.

ornus: Latin name for the mountain ash.

Orontium (Possibly after the Syrian River Orontes.) A genus of 1 species of aquatic or marginal, rhizomatous herb with floating and aerial leaves that are often tinged purple beneath. The small flowers are borne on a bright yellow spadix that is subtended by a very small green spathe. *Distr*. E USA. *Fam*. Araceae. *Class* M.

O. aquaticum GOLDEN CLUB. Grown as an ornamental. The tubers are edible once they have been cooked.

Orostachys (From Greek *oros*, mountain, and *stachys*, spike.) A genus of 10 species of biennial to perennial, succulent herbs with spherical or hemispherical rosettes of fleshy leaves and tall, spike-like racemes of star-shaped flowers. Each rosette dies after flowering, perennials survive by producing a number of rosettes. Several species are grown as ornamental. *Distr*. Europe, temperate Asia. *Fam*. Crassulaceae. *Class* D.

O. chanetii Biennial. Leaves bear a short, soft spine. Inflorescences pyramidal. Flowers white, tinged pink. *Distr*. China.

O. furusei Perennial. Flowers pale green. *Distr*. Japan.

Oroya See *Oreocereus*

Orphium (After Orpheus, poet and musician in Greek mythology.) A genus of 1 species of erect shrub with opposite, sessile, somewhat succulent leaves and 5-lobed, funnel-shaped, red flowers. *Distr*. South Africa. *Fam*. Gentianaceae. *Class* D.

O. frutescens Grown as a half-hardy ornamental.

orpine *Hylotelephium telephium*.

orthanthus, -a, -um: with straight flowers.

orthocladus, -a, -um: with straight branches.

orthopetalus, -a, -um: with straight petals.

orthophyllus, -a, -um: with straight leaves.

Orthrosanthus (From Greek *orthros*, morning, and *anthos*, flower; the flowers often open in the morning.) A genus of 7 species of rhizomatous herbs with narrow or linear leaves and panicles of regular blue flowers. Several species are occasionally grown as ornamentals. *Distr*. Tropical America, Australia. *Fam*. Iridaceae. *Class* M.

O. chimboracensis Flowers lavender-blue, in loose panicles. *Distr*. Mexico to Peru.

O. multiflorus Flowers light blue with darker veins, in narrow panicles. *Distr*. SW Australia.

Oryza (Greek word for rice, both plant and grain.) RICE. A genus of 19 species of annual or perennial, rhizomatous grasses with flat leaves and loose or dense, paniculate inflorescences. Members of this genus make up the staple food of half the world's population. *Distr*. Pan-tropical. *Fam*. Gramineae. *Class* M.

O. sativa The RICE of commerce. Many cultivars have been produced to provide different qualities of grain and to grow in a number of climates and at various altitudes. Typically grown in seasonally flooded paddy fields in lowland tropical and warm areas or in drier fields at higher altitudes and in cooler regions. Ornamental cultivars are available. *Distr*. SE Asia.

oryzifolius, -a, -um: from the genus *Oryza*, RICE, and *folius*, leaved.

Oryzopsis (From the genus name *Oryza*, RICE, and Greek *opsis*, appearance.) A genus of 50 species of perennial, clump-forming grasses with flat or in-rolled leaves and open paniculate inflorescences. The grains of some species were formerly eaten by North American Indians. Several species are now grown as ornamentals. *Distr*. N temperate and sub-tropical regions. *Fam*. Gramineae. *Class* M.

O. hymenoides SILK GRASS, INDIAN MILLET. *Distr*. SW USA, N America.

O. miliacea SMILO GRASS. *Distr*. Mediterranean.

Osbeckia (After Peter Osbeck (1723–1805), Swedish naturalist.) A genus of 60 species of roughly hairy herbs and shrubs with simple, leathery leaves and showy, red, pink, or violet flowers that bear 5 (or occasionally 4) petals and are borne singly or in clusters. *Distr*. Tropical regions of the Old World. *Fam*. Melastomataceae. *Class* D.

O. stellata Evergreen rounded shrub. Flowers rose-purple, borne in terminal clusters. Petals 4. Grown as an ornamental. *Distr*. India to China.

Oscularia (From Latin *osculuner*, kiss, alluding to the fused leaves.) A genus of 3 species of erect or spreading subshrubs with pairs of succulent fused leaves and clusters of many-petalled flowers. *Distr*. South Africa. *Fam*. Aizoaceae. *Class* D.

O. deltoides Leaves triangular, blue-green. Flowers pink, fragrant. Grown as a tender ornamental. *Distr*. South Africa (Cape Province).

osier *Salix* **purple** ~ *S. purpurea*.

Osmanthus (From Greek *osme*, fragrance, and *anthos*, flower.) A genus of 15 species of evergreen shrubs and trees with simple, opposite, leathery, glandular leaves and clusters or panicles of yellow or white, highly fragrant, tubular flowers. The fruit is a 1-seeded, hard-shelled drupe. Several species and numerous cultivars are grown as ornamentals. *Distr*. Asia to Hawaii. *Fam*. Oleaceae. *Class* D.

O. armatus Evergreen bushy shrub. *Distr*. W China.

O. × burkwoodii (After Burkewood and Skipwith, the raisers.) A hybrid of *O. delavayi* x *O. decorus*. *Distr*. Garden origin.

O. delavayi (After Abbé Jean Marie Delavay (1838–95), French missionary in China who introduced it to France in 1890.) Evergreen shrubs. *Distr*. China.

O. fragrans FRAGRANT OLIVE, SWEET TEA. Evergreen shrub. The male inflorescences are used to flavour tea and confectionery. *Distr.* China.

O. heterophyllus HOLLY OLIVE, CHINESE HOLLY, FALSE HOLLY. Dense evergreen shrub. Leaves variable, often dark green above and yellow-green below. Flowers white, fragrant. *Distr.* Japan.

O. ilicifolius See *O. heterophyllus*

Osmaronia See *Oemleria*

O. cerasiformis See *Oemleria cerasiformis*

Osmoglossum (From Greek *osme*, odour, and *glossa*, lip, alluding to the fragrant flowers.) A genus of 7 species of epiphytic orchids with clustered pseudobulbs and racemes of small, fleshy, white flowers. Members of this genus were formerly included in the genus *Odontoglossum*. *Distr.* Central America. *Fam.* Orchidaceae. *Class* M.

O. pulchellum LILY OF THE VALLEY ORCHID. Flowers fragrant, white, long-lived. *Distr.* Mexico to El Salvador.

Osmorhiza (From the Greek *osme*, smell, and *rhiza*, root.) A genus of 10 species of perennial herbs with thick, fleshy roots, much-divided leaves, and compound umbels of white, green, or purple flowers. Several species are grown as ornamentals and some have edible or medicinal roots. *Distr.* E Asia, North America. *Fam.* Umbelliferae. *Class* D.

O. claytonii WOOLLY SWEET CICELY, SWEET JARVIL. Erect herb to 1m high. Leaves densely hairy. Flowers white. *Distr.* E North America.

Osmunda (Either from Latin *os*, mouth, and *mindare*, to clean, or after Osmunder, the Scandinavian god also known as Thor.) ROYAL FERN. A genus of 10–15 species of terrestrial ferns with large bipinnate leaves. Several species are grown as ornamentals. *Distr.* Widely spread. *Fam.* Osmundaceae. *Class* F.

O. cinnamomea CINNAMON FERN, FIDDLE-HEADS, BUCKHORN. Leaves densely brown-hairy. A source of fibre used in orchid-growing. *Distr.* E North America.

O. claytoniana (After John Clayton (1686–1773), Virginian botanist.) INTERRUPTED FERN. *Distr.* E North America.

O. regalis ROYAL FERN, FLOWERING FERN. Stock forms a short trunk. Flowers to 2m long. Eaten locally. Several ornamental cultivars of this species have been produced. *Distr.* Widespread.

Osmundaceae A family of 3 genera and 18 species of terrestrial ferns with erect, sometimes trunk-like stems covered in old leaf bases. The leaves are pinnate or bipinnate and the sporangia are borne naked on specialized leaf segments. *Distr.* Widespread. *Class* F. *See: Leptopteris, Osmunda, Todea.*

Osteomeles (From Greek *osteon*, bone, and *meles*, apple.) A genus of 2 species of evergreen shrubs and trees with pinnately divided leaves and small white flowers that bear numerous stamens and are followed by hard, apple-like fruits. Both species are grown as ornamentals. *Distr.* China to Hawaii and New Zealand. *Fam.* Rosaceae. *Class* D.

O. schweriniae Arching shrub to 3m high. Fruit blue-black. *Distr.* W China.

Osteospermum (From Greek *osteon*, bone, and Latin *sperma*, seed.) A genus of 70 species of annual and perennial herbs, subshrubs, and shrubs with simple to pinnately divided leaves and flowers in daisy-like heads which have distinct rays. Some species and a number of cultivars are grown as half-hardy ornamentals. *Distr.* S Africa to Middle East. *Fam.* Compositae. *Class* D.

O. barberiae Rhizomatous perennial. Flowers yellow or purple, rays magenta. *Distr.* South Africa.

O. caulescens See *O. ecklonis*

O. ecklonis (After Christian Friedrich Ecklon (1795–1868), German apothecary.) Robust subshrub to 1m high. Flowers blue, rays white above, dark below. *Distr.* South Africa.

O. jucundum Perennial herb. Flowers black-tipped, rays red. *Distr.* South Africa.

O. prostratum See *O. ecklonis*

Ostrowskia (After Michael Nicholazewitsch von Ostrowsky, late 19th-century Russian patron of botany.) A genus of 1 species of perennial, rhizomatous herb with erect stems to 2m high, whorled, simple leaves, and racemes of purple, bell-shaped flowers. *Distr.* Turkestan, central Asia. *Fam.* Campanulaceae. *Class* D.

O. magnifica GIANT BELLFLOWER. Grown as an ornamental.

Ostrya (From Greek *ostrys*, shell, alluding to the inflated floral bracts.) HOP HORNBEAM. A genus of 10 species of deciduous trees with alternate, simple, toothed leaves that are borne in 2 ranks. The tiny unisexual flowers are borne in separate catkins and appear in winter. The fruit is a small nut surrounded by a swollen husk that gives the fruiting catkin a hop-like appearance. Several species are grown as ornamentals and some are exploited for their timber. *Distr.* N Temperate regions. *Fam.* Corylaceae. *Class* D.

O. carpinifolia HOP HORNBEAM. Rounded tree to 20m high. Grown for timber. *Distr.* SE Europe, W Asia, Caucasus.

O. virginiana IRONWOOD, LEVER WOOD, EASTERN HOP HORNBEAM. Rounded tree to 20m high. Grown for timber. *Distr.* E North America.

Ostryopsis (From genus name *Ostrya*, and the Greek suffix *-opsis*, like.) A genus of 2 species of deciduous shrubs to 3m high with alternate, simple, double-toothed leaves. The tiny unisexual flowers are borne in separate catkins and the fruit is a small nut surrounded by a 3-lobed swollen husk. *Distr.* E Mongolia, SW China. *Fam.* Corylaceae. *Class* D.

O. davidiana Grown as a hardy ornamental. *Distr.* E Mongolia, SW China.

Osyris (Classical Greek plant name.) A genus of 10 species of parasitic shrubs with simple alternate leaves and small unisexual flowers, the males being borne in racemes and the females singly. The fruit is a small berry. *Distr.* Mediterranean to China. *Fam.* Santalaceae. *Class* D.

O. alba Shrub to 1.5m high. Flowers white. Fruit red. Grown as a half-hardy ornamental. *Distr.* Mediterranean regions.

Otanthus (From Greek *ous*, ear, and *anthos*, flower, alluding to the shape of the lobes on the flowers.) A genus of 1 species of creeping, perennial, shoreline herb with small, overlapping, hairy leaves and yellow flowers in daisy-like heads that lack distinct rays. *Distr.* Coasts of W Europe to E Asia. *Fam.* Compositae. *Class* D.

O. maritimus COTTON WEED. Occasionally grown as a rock garden plant.

Otatea See *Sinarundinaria*

Othonna (Classical Greek name from *othone*, linen, alluding to the leaves which are soft to the touch.) A genus of 150 species of tuberous herbs and shrubs of dry areas with simple or divided leaves and flowers in daisy-like heads that usually bear distinct rays. *Distr.* Tropical and South African. *Fam.* Compositae. *Class* D.

O. capensis Shrubby herb with long succulent leaves and yellow flowers. Much grown in hanging baskets and in California as ground cover. *Distr.* South Africa.

O. cheirifolia Small spreading shrub. *Distr.* N Africa.

Othonnopsis See *Othonna*

Otoglossum (From Greek *ous*, ear, and *glossa*, lip, alluding to the small lateral lobes of the lip.) A genus of about 7 species of epiphytic orchids with thick rhizomes, fleshy leaves and erect racemes of showy flowers. Several species are cultivated as ornamentals. *Distr.* Central and South America. *Fam.* Orchidaceae. *Class* M.

O. brevifolium Flowers brown with yellow margins. *Distr.* Columbia to Peru.

Ottelia (From the local name for these plants in Malabar, *ottel ambel.*) A genus of 40 species of annual and perennial, submerged aquatic herbs with elliptic or round adult leaves and small, unisexual or bisexual flowers which bear 3 petals and 3–15 stamens. *Distr.* Africa, E Asia, Australia. *Fam.* Hydrocharitaceae. *Class* M.

O. alismoides Grown as an ornamental and as a green vegetable in Malaysia. *Distr.* E Asia, NE Africa, Australia.

Ourisia (After General Ouris (died 1773), Governor of the Falkland Islands, where the first species was found.) A genus of 25 species of rhizomatous herbs and subshrubs with simple, stalked, basal leaves and slightly irregular, 5-lobed, cup-shaped to tubular flowers. Several species are grown as ornamental rock-garden plants. *Distr.* Andes, New Zealand, Australia (Tasmania). *Fam.* Scrophulariaceae. *Class* D.

O. caespitosa Evergreen prostrate perennial. Flowers cup-shaped, white. *Distr.* New Zealand.

O. microphylla Semi-evergreen, mat-forming perennial. Flowers pink with a white centre. *Distr.* New Zealand.

Our Lord's candle *Yucca whipplei.*

ovalifolius, -a, -um: with oval leaves.

ovatus, -a, -um: egg-shaped, with the broad end as the base.

oviferum: egg-bearing.

ovinus, -a, -um: of sheep.

Oxalidaceae The Wood Sorrel family. 6 genera and about 500 species of annual and perennial herbs with simple or compound leaves. The flowers are showy and regular with their parts in fives. *Distr.* Widespread but centred in tropical regions~~. *Class* D~~.
See: Oxalis.

Oxalis (Greek name for sorrel, from *oxys*, acid, alluding to the bitter taste of the leaves.) A genus of 500 species of annual and perennial herbs and shrubs with palmately divided, typically trifoliate leaves and showy flowers that bear 5 petals and 2 whorls of 5 stamens. Numerous species and cultivars are grown as ornamentals. *Distr.* Widespread but especially in South America and South Africa. *Fam.* Oxalidaceae. *Class* D.

O. acetosella WOOD SORREL, CUCKOO BREAD. Rhizomatous perennial herb. Flowers white, tinged purple. The leaves were formerly eaten as a substitute for SORREL *Rumex acetosa*. *Distr.* Europe, N Asia.

O. corniculata CREEPING YELLOW SORREL. Short-lived perennial. Leaves tinged purple. Flowers yellow. A persistent weed of greenhouses. *Distr.* Widespread.

O. deppei See *O. tetraphylla*

O. lactea See *O. magellanica*

O. latifolia Bulbiferous perennial. Flowers tinged pink. A persistent weed. *Distr.* Widespread.

O. magellanica Carpet-forming, rhizomatous perennial. Flowers white. *Distr.* Patagonia, Australia, New Zealand.

O. tetraphylla LUCKY CLOVER, GOOD LUCK LEAF, GOOD LUCK PLANT. Bulbiferous perennial. Leaves bear 3 leaflets, marked purple. Flowers red. Tubers reported to be edible. *Distr.* Mexico.

O. tuberosa OCA. Tuberous, somewhat succulent perennial. Leaves tinged purple. Flowers yellow. Grown as a root vegetable in Peru. *Distr.* Andes.

ox-eye *Heliopsis.*

ox-tongue *Picris, P. echioides.* **bristly** ~ *P. echioides.*

oxlip *Primula elatior.*

oxyacanthus, -a, -um: with pointed spines.

oxycarpus, -a, -um: with pointed fruits.

oxycedrus: pointed cedar.

Oxycoccus See *Vaccinium*

O. macrocarpus See *Vaccinium macrocarpon*

Oxydendrum (From Greek *oxys*, acid and *dendron*, tree, referring to the acid-tasting leaves.) A genus of 1 species of deciduous shrub or tree to 25m high with panicles of white, urn-shaped flowers. *Distr.* E USA. *Fam.* Ericaceae. *Class* D.

O. arboreum SOURWEED, TREE SORREL.

O. caeruleum See *Tweedia caerulea*

oxypetalus, -a, -um: with pointed petals.

oxyphyllus: with sharp-pointed leaves.

Oxyria (From Greek *oxys*, sharp or sour, alluding to the taste of the sap.) A genus of 2 species of perennial herbs with simple basal leaves and panicles of minute flowers. *Distr.* Arctic and mountainous regions of the N hemisphere. *Fam.* Polygonaceae. *Class* D.

O. digyna MOUNTAIN SORREL. Grown as an ornamental rock-garden plant. The leaves are eaten. *Distr.* Eurasia, N America.

oxysepalus, -a, -um: with pointed sepals.

Oxytropis (From Greek *oxys*, sharp, and *tropis*, keel, alluding to the keel of the flowers.) CRAZYWEED, LOCO-WEED. A genus of 300 species of perennial tufted herbs with pinnate leaves and racemes or spikes of pea-like flowers. A few species are grown as ornamentals. Some species are dangerous to grazing animals. *Distr.* N temperate regions, especially Central Asia. *Fam.* Leguminosae. *Class* D.

O. campestris MEADOW MILKVETCH, YELLOW OXYTROPIS. Leaves borne in a basal rosette. Flowers yellow with a violet keel. *Distr.* Europe.

O. halleri Tufted. Flowers purple with a dark purple keel. *Distr.* Europe.

oxytropis, yellow *Oxytropis campestris.*

oyster, vegetable ~ *Tragopogon porrifolius.*

oyster, leaf *Mertensia maritima.*

oyster plant *Tradescantia spathacea, Tragopogon porrifolius.*

Ozothamnus (From Greek *ozo*, smell, and *thamnos*, shrub.) A genus of about 50 species of evergreen shrubs and subshrubs with small, simple leaves and flowers in daisy-like heads that lack distinct rays but have showy, papery bracts. *Distr.* Australasia. *Fam.* Compositae. *Class* D.

O. coralloides Small shrub to 50cm. Leaves tightly appressed to the stem, overlapping. *Distr.* New Zealand.

O. depressum Prostrate or erect shrub. Leaves very small. *Distr.* New Zealand.

O. ericifolius Large shrub to 3m high. *Distr.* Australia (Tasmania).

O. gunnii Large, densely woolly shrub to 3m high. *Distr.* Australia (Tasmania).

O. hookeri Erect shrub to 2m high. *Distr.* SE Australia including Tasmania.

O. ledifolius KEROSENE WEED. Sticky shrub, producing sweet-smelling, yellow exudate. *Distr.* Australia (Tasmania).

O. purpurascens Large shrub with a curry-like odour. *Distr.* Australia (Tasmania).

O. rosmarinifolius Large shrub. Leaves curled under at margins. *Distr.* SE Australia including Tasmania.

O. scutellifolius Medium-sized shrub with scale-like leaves. *Distr.* Australia (Tasmania).

O. selago Small shrub with scale-like leaves. *Distr.* New Zealand.

O. thyrsoideus SNOW IN SUMMER. Large shrub. Leaves dark green, resinous. *Distr.* SE Australia including Tasmania.

P

Pabstia (After Dr Guido Pabst (1914–80), author of numerous articles on Brazilian orchids.) A genus of 5 species of epiphytic orchids with 2-leaved pseudobulbs and large showy flowers. Some species are grown as tender ornamentals. *Distr.* Brazil. *Fam.* Orchidaceae. *Class* M.

P. jugosa Flowers to 7cm across, cream-white, fragrant. *Distr.* Brazil.

Pachycereus (From Greek *pachys*, thick, and the genus name *Cereus*, alluding to the thick stems.) A genus of 9 species of shrub- or tree-like cacti with large erect stems and small, funnel- or bell-shaped flowers. Several species are grown as ornamentals. *Distr.* W Mexico. *Fam.* Cactaceae. *Class* D.

P. pecten-aboriginum NATIVE'S COMB. Stems dark-green, column-like. Spines dark brown. Flowers white, slightly woolly. *Distr.* W Mexico.

P. pringlei (After G. G. Pringle, who collected in Mexico *c.*1887.) Stems blue-green at first, column-like. Spines dark brown. Flowers white, densely woolly. Fruit edible, seeds ground into flour. *Distr.* NW Mexico.

P. schottii Large shrub. Flowers night-scented. *Distr.* NW Mexico, Arizona.

pachyclados: with thick branches.

Pachyphragma (From Greek *pachys*, thick and *phragma*, partition, referring to the stout-ribbed septum of the pod.) A genus of 1 species of rhizomatous perennial herb with bad-smelling, white flowers. *Distr.* Armenia, Caucasus. *Fam.* Cruciferae. *Class* D.

P. macrophyllum Grown as an ornamental for glossy, dark green foliage. *Distr.* Caucasus, Armenia.

pachyphyllus, -a, -um: with thick leaves.

pachyphytoides: from the genus *Pachyphytum*, with the ending *-oides*, indicating resemblance.

Pachyphytum (From Greek *pachys*, thick, and *phyton*, plant, alluding to the succulent habit of these plants.) A genus of 12 species of succulent perennial herbs and subshrubs typically with rosettes of fleshy leaves and pendent racemes of bell- or cup-shaped flowers. This genus is closely related to the genus *Echeveria* with which it forms hybrids. Several species are grown as ornamentals. *Distr.* Mexico. *Fam.* Crassulaceae. *Class* D.

P. amethystinum See *Graptopetalum amethystinium*

P. compactum Leaves round in cross-section, pointed. Sepals green-pink. Petals orange. *Distr.* Mexico.

P. oviferum MOONSTONES. Leaves oval, crowded, purple with a dense white bloom. *Distr.* Mexico.

Pachypodium (From Greek *pachys*, thick, and *podos*, foot.) A genus of 13 species of succulent herbs, shrubs, and trees with fleshy stems, deciduous, leathery leaves, and tubular, white flowers. Several species are grown as ornamentals. *Distr.* Madagascar, S and SW Africa. *Fam.* Apocynaceae. *Class* D.

P. lamerei Tree to 6m high. Trunk cigar-shaped. Leaves leathery, dark green. Flowers white. *Distr.* S Madagascar.

P. succulentum Tree-like succulent. Flowers pink-red. *Distr.* South Africa.

pachypodus, -a, -um: with a thick support.

Pachysandra (From Greek *pachys*, thick, and *andros*, male, alluding to the thick stamens.) A genus of 12 species of evergreen subshrubs with somewhat fleshy stems, alternate, simple leaves and erect spikes of small white flowers that lack petals. Several species are grown as ornamental ground cover. *Distr.* E Asia, E North America. *Fam.* Buxaceae. *Class* D.

P. procumbens ALLEGHENY SPURGE. Stems forming clumps. Leaves grey green. Flowers fragrant, anthers pink. *Distr.* SE North America.

P. terminalis Leaves darkgreen. Several cultivars are available which differ in leaf colour. *Distr.* China, Japan.

pachysanthus, -a, -um: with thick, somewhat fleshy flowers.

pachyscapus, -a, -um: with thick scapes.

Pachystachys (From Greek *pachys*, thick, and *stachys*, spike, alluding to the dense inflorescences.) A genus of 12 species of perennial herbs and shrubs with simple, opposite leaves and dense spikes of 2-lipped flowers. Several species are grown as ornamentals. *Distr.* Tropical America. *Fam.* Acanthaceae. *Class* D.

P. coccinea CARDINAL'S GUARD. Bracts large, green. Flowers red. *Distr.* West Indies, South America.

P. lutea LOLLIPOP PLANT. Flowers tubular, white, in dense spikes with bright yellow bracts. *Distr.* Peru.

Pachystegia See *Olearia*

Pachystima See *Paxistima*

pachytrichus, -a, -um: with thick hair.

× *Pachyveria* (From the names of the parent genera.) A genus of hybrids between members of the genera *Pachyphytum* and *Echeveria*. *Distr.* Mexico. *Fam.* Crassulaceae. *Class* D.

×*P. pachyphytoides* A hybrid of *E. gibbiflora* and *P. bracteosum*. A succulent subshrub to 30cm high. Leaves tinged red-purple. Flowers pink-red. *Distr.* Garden origin.

Pacific grindelia *Grindelia stricta*.

pacificus, -a, -um: of the Pacific.

padus: the Greek name of a wild cherry.

Paederota (From Greek *paideros*, a name applied to a number of different plants.) A genus of 2 species of alpine, perennial herbs with simple, opposite leaves and spike-like inflorescences of 5-lobed, 2-lipped flowers that bear 2 stamens. Both species are grown as ornamentals. *Distr.* S Europe. *Fam.* Scrophulariaceae. *Class* D.

P. bonarota Flowers violet-blue. Stamens exerted. *Distr.* S Europe.

P. lutea YELLOW VERONICA. Flowers yellow. Stamens not exerted. *Distr.* E Alps.

Paeonia (After Paeon, physician of the gods in Greek mythology.) PEONY, TREE PEONY. A genus of 33 species of perennial, rhizomatous herbs and shrubs with compound or lobed leaves and showy globe-shaped flowers that bear 5–10 petals and numerous stamens. Many species and cultivars are grown as ornamentals. *Distr.* Eurasia, W North America. *Fam.* Paeoniaceae. *Class* D.

P. arietina See *P. mascula*

P. cambessedesii (After Jacques Cambessedes (1799–1863), French botanist.) Rhizomatous herb. Leaves dark green with purple veins. Flowers solitary, deep pink. *Distr.* Balearic Islands.

P. caucasica See *P. mascula*

P. corallina See *P. mascula*

P. daurica See *P. mascula*

P. delavayi (After Abbé Jean Marie Delavay (1838–95), French missionary botanist in China.) Small shrub. Flowers maroon. *Distr.* W China.

P. kevachensis See *P. mascula*

P. lutea Small shrub. Flowers yellow. *Distr.* SW China.

P. mascula Rhizomatous herb. Flowers pink or white. *Distr.* Europe to Turkey.

P. potaninii See *P. delavayi*

P. russii See *P. mascula*

Paeoniaceae A family of 1 genus and 33 species. *Class* D.
See: Paeonia.

paeoniifolius, -a, -um: from the genus *Paeonia*, and *folius*, leaved.

Paesia (After Fernando Dias Paes Leme, a 17th-century Portuguese civil servant.) A genus of 12 species of terrestrial or occasionally epiphytic ferns with pinnately divided leaves. *Distr.* SE Asia to New Zealand, Tahiti and Tropical America. *Fam.* Dennstaedtiaceae. *Class* F.

P. scaberula LACE FERN, SCENTED FERN. Leaves to 45cm long, 4-pinnate. Grown as a hardy ornamental. *Distr.* New Zealand.

pagoda flower *Clerodendrum paniculatum*.

pagoda tree *Plumeria* **Japanese** ~ *Sophora japonica.*

paigle *Ranunculus.*

painted drop-tongue *Aglaonema crispum.*

painted leaf *Euphorbia pulcherrima.*

painter's palette *Anthurium andreanum.*

pak choi *Brassica chinensis.*

Palaquium (Latin version of the Philippine name.) A genus of about 120 species of large evergreen trees of lowland tropical rainforests with simple leaves and clusters of white flowers. *Distr.* Taiwan and Malaysia to the Solomon Islands. *Fam.* Sapotaceae. *Class* D.

P. gutta GUTTA PERCHA TREE. Source of latex known as gutta percha which is chemically similar to rubber and is used in dentistry and formerly as an electrical insulator. Originally obtained by ring-barking wild trees *Distr.* Malaysia.

palas *Licuala.*

Paliurus (Classical Greek name for these plants.) A genus of 8 species of spiny deciduous shrubs and small trees with simple alternate leaves and clusters of small yellow flowers that bear their parts in fives. Several species are grown as ornamentals. *Distr.* S Europe to Japan. *Fam.* Rhamnaceae. *Class* D.

P. spina-christi CHRIST'S THORNS. Tree to 7m high. It is believed that twigs of this tree were used to make Christ's crown of thorns. *Distr.* S Europe, W Asia.

pallens: pale.

pallescens: rather pale.

pallidiflorus, -a, -um: with pale flowers.

pallidus, -a, -um: pale.

palm Palmae **African oil** ~ *Elaeis guineensis* **Alexander** ~ *Ptychosperma elegans* **Alexandra** ~ *Archontophoenix alexandrae* **areca** ~ *Chrysalidocarpus lutescens* **Atherton** ~ *Laccospadix australasica* **Australian** ~ *Livistona australis* **Australian fan** ~ *L. australis* **Austalian ivy** ~ *Schefflera actinophylla* **bamboo** ~ *Rhapis excelsa, Chrysalidocarpus lutescens* **bangalow** ~ *Archontophoenix* **Belmore sentry** ~ *Howea belmoreana* **betel** ~ *Areca catechu* **blue fan** ~ *Brahea armata* **blue hesper** ~ *B. armata* **bottle** ~ *Beaucarnea recurvata, Hyophorbe* **Burmese fishtail** ~ *Caryota mitis* **butterfly** ~ *Chrysalidocarpus lutescens* **cabbage** ~ *Livistona australis* **cabbage palmetto** ~ *Sabal palmetto* **Canary Island date** ~ *Phoenix canariensis* **cane** ~ *Chrysalidocarpus lutescens* **cardboard** ~ *Zamia furfuracea* **Chinese fan** ~ *Livistona chinensis* **Chinese windmill** ~ *Trachycarpus* **Christmas** ~ *Veitchia, V. merrillii* **Chusan** ~ *Trachycarpus, T. fortunei* **clustered fishtail** ~ *Caryota mitis* **cotton** ~ *Washingtonia filifera* **curly** ~ *Howea belmoreana* **curly sentry** ~ *H. belmoreana* **date** ~ *Phoenix dactylifera* **desert fan** ~ *Washingtonia filifera* **dwarf date** ~ *Phoenix reclinata* **dwarf fan** ~ *Chamaerops humilis* **fan** ~ *Trachycarpus, T. fortunei* **feather duster** ~ *Rhopalostylis sapida* **Fiji fan** ~ *Pritchardia pacifica* **fishtail** ~ *Caryota* **Forster sentry** ~ *Howea forsteriana* **Gippsland** ~ *Livistona australis* **hemp** ~ *Trachycarpus fortunei* **hesper** ~ *Brahea* **Illawara** ~ *Archontophoenix cunninghamiana* **Japanese peace** ~ *Rhapis excelsa* **jelly** ~ *Butia* **kentia** ~ *Howea forsteriana* **king** ~ *Archontophoenix* **lady** ~ *Rhapis, R. excelsa* **loulu** ~ *Pritchardia* **mabila** ~ *Veitchia, V. merrillii* **Mexican blue fan** ~ *Brahea armata* **miniature** ~ *Rhapis excelsa* **miniature date** ~ *Phoenix roebelenii* **needle** ~ *Rhapidophyllum hystrix, Yucca filamentosa* **nikau** ~ *Rhopalostylis, R. sapida* **Norfolk** ~ *R. baueri* **northern bangalow** ~ *Archontophoenix alexandrae* **northern kentia** ~ *Gronophyllum ramsayi* **oil** ~ *Elaeis guineensis* **panama hat** ~ *Carludovica palmata* **paradise** ~ *Howea forsteriana* **piccabben** ~ *Archontophoenix cunninghamiana* **pignut** ~ *Hyophorbe* **porcupine** ~ *Rhapidophyllum hystrix* **princess** ~ *Dictyosperma album* **pygmy date** ~ *Phoenix roebelenii* **raffia** ~ *Raphia* *farinifera* **rock** ~ *Brahea* **Roebelin** ~ *Phoenix roebelenii* **sago** ~ *Cycas circinalis* **San Jose hesper** ~ *Brahea brandegeei* **sentry** ~ *Howea, H. forsteriana* **slender lady** ~ *Rhapis humilis* **snake** ~ *Amorphophallus, A. rivierei* **solitaire** ~ *Ptychosperma elegans* **soya** ~ *Brahea* **sugar** ~ *Arenga pinnata* **syrup** ~ *Jubaea chilensis* **thatch leaf** ~ *Howea forsteriana* **thread** ~ *Washingtonia robusta* **traveller's** ~ *Ravenala madagascariensis* **tufted fishtail** ~ *Caryota mitis* **umbrella** ~ *Hedyscepe canterburyana, Amorphophallus rivierei* **Washington** ~ *Washingtonia* **wax** ~ *Ceroxylon* **wine** ~ *Pseudophoenix ekmannii* **yatay** ~ *Butia* **yellow** ~ *Chrysalidocarpus lutescens.*

palma christi *Ricinus communis.*

palma *pita* *Yucca treculeana.*

Palmae The Palm family. 202 genera and about 2500 species of plants with unbranched trunks and a crown of feather- or fan-shaped leaves. No secondary thickening occurs in the trunk and so no true wood is formed; instead a mass of fibrous vascular bundles provides the rigidity. The flowers have their parts in threes and may occur in a simple spike or in a large branching inflorescence with as many as a quarter of a million flowers. A very important family whose products include coconuts (*Cocos nucifera*), dates (*Phoenix dactilifera*), palm oil (*Elaeis guineensis*), and raffia fibre (*Raphia ruffia*). *Distr.* Widespread in tropical and subtropical regions, rarely temperate. *Class* M.

See: Archontophoenix, Areca, Arenga, Brahea, Butia, Caryota, Ceroxylon, Chamaedorea, Chamaerops, Chrysalidocarpus, Cocos, Dictyosperma, Didymosperma, Elaeis, Gronophyllum, Hedyscepe, Howea, Hyophorbe, Jubaea, Laccospadix, Licuala, Livistona, Lytocaryum, Microcoelum, Nannorrhops, Neodypsis, Parajubaea, Phoenix, Pritchardia, Pseudophoenix, Ptychosperma, Raphia, Ravenea, Rhapidophyllum, Rhapis, Rhopalostylis, Sabal, Serenoa, Syagrus, Trachycarpus, Veitchia, Washingtonia.

palmatus, -a, -um: palmate, like a hand.

palmella *Yucca elata.*

palmetto *Sabal, S. palmetto* **blue** ~ *Rhapidophyllum hystrix, Sabal palmetto* **bush** ~ *S. minor* **cabbage** ~ *S. palmetto* **common** ~ *S. palmetto* **dwarf** ~ *S. minor* **saw** ~ *Serenoa repens* **scrub** ~ *Sabal minor, Serenoa repens.*

palo colorado *Luma apiculata.*

palo madrono *Amomyrtus luma.*

palustris, -e: growing in marshes.

Pamianthe (After Major Albert Pam (1875–1955), English horticulturalist who was sent *P. peruviana* in 1926.) A genus of 3 species of bulbiferous herbs with linear leaves arising from a long-necked bulb and an umbel of a few large, tubular, white flowers. *Distr.* Tropical South America. *Fam.* Amaryllidaceae. *Class* M.

P. peruviana Flowers white-cream, to 20cm long. *Distr.* Peru.

Panax (From Greek *panakes*, cure-all, alluding to the reputed medicinal qualities of these plants.) A genus of 6 species of perennial herbs with thick, tuberous, branching roots, and 3–7 lobed leaves. The small flowers have 5 sepals, 5 petals, and 5 stamens and are borne in a terminal umbel. The roots have been used medicinally in Europe and Asia since ancient times. *Distr.* N America, E Asia. *Fam.* Araliaceae. *Class* D.

P. ginseng GINSENG, NIN SIN. Roots carrot-shaped. Leaves with 5 leaflets. Fruits red. The ginseng of commerce. *Distr.* Korea, NE China.

P. pseudopanax See *P. ginseng*

P. quinquefolius AMERICAN GINSENG, SANG. Roots cigar-shaped. Leaves with up to 16 leaflets. Fruits bright red. *Distr.* E North America.

pancratioides: from the genus *Pancratium*, with the ending *-oides*, indicating resemblance.

Pancratium (Greek name for a bulbous plant) A genus of 16 species of bulbiferous perennial herbs with 2 ranks of leaves and large, white, scented, funnel-shaped flowers. *Distr.* Canary Islands, Mediterranean regions to tropical Asia and S through W Africa to Namibia. *Fam.* Amaryllidaceae. *Class* M.

P. canariense Bulb large. Leaves broad. Flowering early autumn. *Distr.* Canary Islands.

P. illyricum Bulb large with purple-black scales. Flowering late spring-early summer. *Distr.* W Mediterranean Islands.

P. maritimum SEA DAFFODIL, SEA LILY. Bulb with a long neck. Flowers highly fragrant. Flowering in summer. *Distr.* Mediterranean and SW Europe.

P. zeylanicum Leaves lance-shaped. Flowers solitary. *Distr.* Tropical Asia.

Pandaceae A family of 4 genera and 18 species of trees with alternate, simple leaves arranged in 2 rows on twigs that lack functional axillary or terminal buds. The flowers are regular, unisexual, and bear their parts in fives. *Distr.* Tropical W Africa, Asia. *Class* D.

Pandanaceae The Screw Pine family. 3 genera and 675 species of shrubs, trees, and climbers. The stems are frequently supported by aerial roots. The long narrow leaves are borne in 3 ranks which are twisted about the stem into spirals. The flowers are small and borne in dense spadices which are subtended by a frequently colourful spathe. Some species have edible fruits; others are used locally as building materials and as a source of fibre or essential oils. *Distr.* Old World tropical regions to New Zealand. *Class* M. *See: Pandanus.*

pandanifolius, -a, -um: from the genus *Pandanus*, and *folius*, leaved.

Pandanus (From *pandan*, the Malayan name.) SCREW PINE. A genus of 600 species of evergreen shrubs and trees with simple leaves arranged in 3 twisted ranks. The small unisexual flowers are massed into terminal spikes. Several species have edible fruits and the leaves of some are used as a source of fibre. *Distr.* Tropical regions of the Old World. *Fam.* Pandanaceae. *Class* M.

P. pygmaeus Spreading shrub to 60cm high. Grown as a tender conservatory plant. *Distr.* Madagascar.

P. sanderi To 60cm high. Leaves striped yellow or gold. Grown as a tender conservatory plant. *Distr.* Malaysia.

P. veitchii To 1m high. Leaves to 1m long, dark green with white-silver margins. Often grown as a house-plant. *Distr.* Polynesia.

panda plant *Kalanchoe tomentosa*.

Pandorea (After Pandora, an equivalent of Eve, in Greek, mythology.) A genus of 6 species of woody climbers with pinnate leaves and clusters of funnel-shaped, somewhat 2-lipped flowers. Several species are grown as ornamentals. *Distr.* E Malaysia, Australia, New Caledonia. *Fam.* Bignoniaceae. *Class* D.

P. jasminoides BOWER PLANT. Flowers white, marked pink within. Several cultivars are available, principally differing in flower colour. *Distr.* Australia.

P. lindleyana See *Clytostoma callistegioides*

P. pandorana WONGA WONGA VINE. Flowers small, cream, marked with red. *Distr.* SE Asia to Australia.

paniculatus, -a, -um: with flowers in panicles.

Panicum (Classical Latin name for millet.) PANIC GRASS, CRAB GRASS, MILLET. A genus of 400 species of annual and perennial grasses with flat to thread-like leaves and variable inflorescences. Some species are important fodder and grain crops, others are grown as ornamentals. *Distr.* Tropical and warm regions. *Fam.* Gramineae. *Class* M.

P. capillare WITCH GRASS, OLD WITCH GRASS. Annual. Inflorescence an open panicle. *Distr.* N America.

P. miliaceum PROSO MILLET, COMMON MILLET, RUSSIAN MILLET, HOG MILLET. Grown as a grain crop for milling to flour, fermenting, and for pig food. Ornamental cultivars are available. *Distr.* Origin obscure, possibly NE Asia.

P. virgatum SWITCH GRASS. Grown as an ornamental. *Distr.* North and Central America.

pansy *Viola* **bedding** ~ *V. cornuta*.

Papaver (From the Latin *pappa*, milk, alluding to the milky latex.) POPPY. A genus of 50 species of annual, biennial, and perennial herbs with erect stems, pinnately divided leaves and regular flowers that bear 2 sepals, 4 delicate petals, and numerous stamens. The fruit is a many-seeded capsule that releases its seeds through a ring of pores in a pepper-pot action. Many species and cultivars are grown as ornamentals. *Distr.* Eurasia, S Africa, Australia, W North America. *Fam.* Papaveraceae. *Class* D.

P. alpinum See *P. nudicaule*

P. bracteatum Perennial to 1m high. Flowers red. A commercial source of codeine. *Distr.* W Asia.

P. miyabeanum See *P. nudicaule*

P. nudicaule ARCTIC POPPY, ICELAND POPPY. Perennial to 50cm high. Cultivars of numerous different flower colours have been raised from this species and small amounts of opium have been extracted from the pods. *Distr.* North America, Europe, Asia.

P. orientale Perennial to 1m high. Flowers red or orange with a purple blotch at the base of each petal. The petals have been used as a source of dye. This is possibly 'the flowers of the field' mentioned in the Bible. *Distr.* W Asia.

P. rhoeas CORN POPPY, FLANDERS POPPY, FIELD POPPY. Annual herb to 90cm high. Flowers solitary, blood-red with a black spot at the base of each petal. A weed of disturbed ground that grew in profusion on the battlefields of World War I and has since been used in commemorative services. *Distr.* Temperate regions of the Old World.

P. somniferum OPIUM POPPY. Annual to 1.2m high. Flowers mauve, often double. Fruit-pod rounded. This species is grown from China to Iran as a source of opium which is derived from the latex produced when the immature fruit-pods are lanced. The opium obtained contains a number of narcotic and medicinally important chemicals including morphine. The seeds do not contain opium and are a source of oil as well as being used whole in cooking. *Distr.* W Asia.

Papaveraceae The Poppy family. 41 genera and about 700 species of annual and perennial herbs and a few shrubs that often produce latex. The leaves are usually divided and alternate. The flowers are either large with 4 conspicuous petals that are crumpled in bud or small, irregular, and borne in racemes. This family has previously been considered as two separate families, Papaveraceae (the Poppies) and Fumariaceae (the Fumitories), the latter having much smaller and more complex flowers than the former. Careful examination of Fumariaceae species has shown that they are far more closely related to the Papaveraceae than was previously thought and so they are now included in this family. Many species are grown as ornamentals and *Papaver somniferum* is the source of opium. *Distr.* Widespread, but chiefly N temperate regions. *Class* D.

See: Adlumia, Argemone, Bocconia, Chelidonium, Corydalis, Dendromecon, Dicentra, Dicranostigma, Eomecon, Fumaria, Glaucium, Hylomecon, Macleaya, Meconopsis, Papaver, Platystemon, Pseudofumaria, Pteridophyllum, Romneya, Rupicapnos, Sanguinaria, Sarcocapnos, Stylomecon, Stylophorum.

papaya *Carica papaya.*

paper bush *Edgeworthia, E. papyrifera.*

paper flower *Bougainvillea glabra.*

paper plant, glossy-leaved *Fatsia japonica.*

Paphinia (After Paphios, a city in Cyprus sacred to Aphrodite (or Paphia).) A genus of 7 species of epiphytic orchids with 1–3 leaved pseudobulbs and large showy flowers with 3-lobed lips. Some species are grown as ornamentals. *Distr.* Central and South America. *Fam.* Orchidaceae. *Class* M.

P. cristata Flowers yellow-white, striped red, lip brown-purple. *Distr.* Tropical South America from Colombia and Trinidad to N Brazil.

Paphiopedilum (From Paphios, site of a temple on Cyprus, sacred to the goddess Aphrodite (or Paphia), and *pedilon*, a slipper, alluding to the shape of the lip.) SLIPPER ORCHID, VENUS' SLIPPER. A genus of 60 species of terrestrial, occasionally epiphytic or lithophytic orchids with leathery leaves and no pseudobulbs. The flowers are large and waxy with a pouch-shaped lip. Pollination occurs though a capture-and-release mechanism whereby an insect is attracted to the fleshy column and on alighting slips off into the pouched lip. The only escape route available is through an aperture at the base of the column where pollinia become attached to the prisoner's back. Some species and numerous hybrids are cultivated as tender ornamentals. *Distr.* Himalaya, S India, SE Asia, Philippines, New Guinea and the Solomon Islands. *Fam.* Orchidaceae. *Class* M.

P. acmodontum Lateral petals spreading, to 5cm long. Lip deeply pouched, bronze-green. *Distr.* Philippines.

P. appletonianum Lateral petals to 6cm, half-twisted, lip pale-purple to ochre. *Distr.* Laos, Thailand, Cambodia.

P. argus Lateral petals recurved, white-green with maroon spots and hairs, lip green. *Distr.* Philippines.

P. barbatum Lateral petals with hairy maroon warts. *Distr.* Malaya, S Thailand.

P. bellatulum Lithophyte on calcareous rocks. Flowers round, cream-white. *Distr.* Burma, Thailand.

P. bullenianum Lip ochre-green, notched at tip. *Distr.* Tioman Isles (off the Malay peninsula), Borneo, Sulawesi, Ambon.

P. callosum Lateral petals white, maroon at tips. Lip incurved. *Distr.* Thailand, Cambodia, S Vietnam.

P. charlesworthii Upper sepal showy, pink, lip wide-mouthed. *Distr.* Burma, NE India.

P. concolor Terrestrial or lithophytic on limestone. Flowers yellow, spotted purple. *Distr.* Burma, Thailand, Cambodia, S Laos, S China.

P. dayanum Flowers to 8cm across, petals purple-pink, lip maroon. *Distr.* Indonesia.

P. delenatii Flowers to 8cm across, pale pink with red and yellow markings. *Distr.* Central Vietnam.

P. exul Terrestrial or lithophytic. Flowers white and yellow-green with darker markings. *Distr.* S Thailand.

P. fairrieanum (After Mr Fairrie of Liverpool, who bought it in a sale of Assam plants.) Flowers white, marked purple, lip yellow-green, deep. Assam, Sikkim, Bhutan, NE India.

P. godefroyae Terrestrial or lithophytic. Flowers ivory-white, spotted purple. *Distr.* Burma, Thailand, S Vietnam.

P. hirsutissimum Terrestrial or epiphytic. Flowers yellow-green, marked red-brown, petals ciliate. *Distr.* NE India, Indo-Burmese border, Thailand, S China.

P. hookerae Flowers pale yellow-green, marked purple. *Distr.* N Borneo.

P. insigne A very variable species. Flowers pale green with maroon spots. *Distr.* Assam, Himalaya.

P. javanicum Lip short-hairy on outer surface. *Distr.* Java, Borneo.

P. lawrenceanum Petals with maroon warts and purple ciliate margin. *Distr.* N Borneo.

P. lowii Epiphyte, rarely lithophytic. Petals to 9cm long, often twisted. *Distr.* Malaysia, Indonesia.

P. mastersianum Flowers to 10cm across, lip pale red-brown. *Distr.* Indonesia (Borneo, Moluccas).

P. niveum Flowers 6cm across, white. *Distr.* Thailand, Malaya.

P. parishii Petals tapering and spirally twisted to tip. *Distr.* Burma, NW Thailand, S China.

P. philippinense Terrestrial or lithophytic. Lip small. *Distr.* Philippines.

P. purpuratum Flowers purple-maroon. *Distr.* Hong Kong, S China.

P. rothschildianum Terrestrial or lithophytic. Flowers large to 30cm across. *Distr.* N Borneo.

P. spicerianum (After Mr Spicer, a tea planter who introduced many orchids into cultivation.) Terrestrial or lithophytic. Flowers pale yellow-green. *Distr.* Assam, N Burma, N India.

P. stonei Lip yellow, flushed pink. *Distr.* NW Borneo.

P. sukhakulii (After P. Sukhakuli, a Thai nurseryman.) Flowers green, spotted maroon. *Distr.* NE Thailand.

P. superbiens Flowers white, lip maroon. *Distr.* Philippines.

P. tonsum Flowers white and yellow-green, lip flushed pink. *Distr.* Sumatra.

P. venustum Leaves dark slate-green. Flowers white and green with purple markings. *Distr.* Nepal, Bhutan, Bangladesh, NE India.

P. victoria-reginae Flowers numerous, opening in succession. *Distr.* Sumatra.

P. villosum Scape hairy. Lip tapering to apex. *Distr.* NE India, Burma, N Thailand, Laos.

papilio a butterfly.

papilionaceus, -a, -um: butterfly-like.

papoose root *Caulophyllum thalictroides*.

papyraceus, -a, -um: with a papery texture.

papyriferus, -a, -um: paper-bearing.

papyrus *Cyperus papyrus*.

Parabenzoin See *Lindera*

Paracryphiaceae A family of 1 genus and 1 species of tree with simple leaves and spikes of small flowers. *Distr.* New Caledonia. *Class* D.

paradise flower *Caesalpinia pulcherrima, Solanum wendlandii.*

paradise plant *Justicia carnea.*

Paradisea (After Count Giovanni Paradisi (1760–1826).) A genus of 2 species of rhizomatous perennial herbs with tufted basal leaves and tall stems of bell- or funnel-shaped flowers. Several species are grown as ornamentals. *Distr.* S Europe. *Fam.* Asphodelaceae. *Class* M.

P. liliastrum ST BRUNO'S LILY, PARADISE LILY. Flowers white with green markings. *Distr.* S Europe.

P. lusitanica Flowers white, borne in 2 distinct ranks. *Distr.* Portugal.

paradoxus, -a, -um: unusual.

paraguayensis, -e: of Paraguay.

Parahebe (From Greek *para*, close to, and the genus name *Hebe*.) A genus of 30 species of dwarf evergreen shrubs and herbs with simple opposite leaves and short racemes of small, tubular, typically 5-lobed flowers that bear 2 stamens. This genus is closely related to the genus *Hebe* but differs in having a laterally compressed capsule. Several species are grown as ornamentals. *Distr.* New Zealand. *Fam.* Scrophulariaceae. *Class* D.

P. catarractae Dwarf sprawling shrub. Flowers white, marked with pink and purple. *Distr.* New Zealand.

P. lyallii (After David Lyall (1817–95), naval surgeon and naturalist who collected the type specimen.) Dwarf prostrate shrub. Flowers white with pink veins. *Distr.* New Zealand.

P. perfoliata See *Veronica perfoliata*

Parajubaea (From Greek *para*, close to, and the genus name *Jubaea*.) A genus of 2 species of medium-sized palms with feather-shaped leaves and little-branched, erect or pendent inflorescences. *Distr.* Andes. *Fam.* Palmae. *Class* M.

P. coccoides Grown as an ornamental. *Distr.* Ecuador, Colombia.

Paramongaia (After Paramonga, Peru, where *P. weberbaueri* was collected.) A genus of 1 species of deciduous, bulbiferous herb with narrow leaves and solitary, bright yellow, fragrant flowers. *Distr.* Peru. *Fam.* Amaryllidaceae. *Class* M.

P. weberbaueri Tender cultivated ornamental.

para para *Pisonia umbellifera.*

Paraphalaenopsis (From Greek *para*, close to, and the genus name *Phalaenopsis*.) A genus of a 3 species of epiphytic orchids with cylindrical leaves and a 3-lobed lip to the flowers. *Distr.* W Borneo. *Fam.* Orchidaceae. *Class* M.

P. denevei Lip white spotted red.

P. serpentilingua Flowers fragrant, lip lemon-yellow.

Paraquilegia (From Greek *para*, close to, and the genus name *Aquilegia*.) A genus of 6 species of tuft-forming perennial herbs with compound alternate leaves and large, white to lilac flowers that bear 5 petal-like sepals and numerous stamens, the true petals being represented by a whorl of nectaries. *Distr.* Mountainous areas of Central Asia to W China. *Fam.* Ranunculaceae. *Class* D.

P. anemonoides Densely tufted perennial to 18cm high. Grown as an ornamental. *Distr.* Himalaya, W China.

P. grandiflora See *P. anemonoides*

P. microphylla See *P. anemonoides*

parasol tree, Chinese *Firmiana simplex.*

Parasyringa See *Ligustrum*

pardalinus, -a, um: spotted like a leopard.

pardanthinus, -a, -um: with flowers resembling those of the genus *Pardanthus* (now *Belamcanda*).

Pardanthopsis (From Greek *pardos*, panther, *anthos*, flower, and *-opsis*, indicating resemblance, alluding to the similarity of these plants to the now suppressed genus *Pardanthus*.) A genus of 2 species of rhizomatous herb with linear leaves in a basal fan and short-lived, white, *Iris*-like flowers. *Distr.* Siberia, N China, Mongolia. *Fam.* Iridaceae. *Class* M.

P. dichotoma Occasionally grown as a hardy ornamental.

pareira brava *Chondrodendron tomentosum.*

pareira root *Chondrodendron tomentosum.*

Parietaria (From Latin *paritinae*, old walls, alluding to the habitat of some species.) PELLITORY. A genus of 20 species of annual and perennial herbs with simple, alternate leaves and clusters of small, unisexual, green flowers. *Distr.* Widespread. *Fam.* Urticaceae. *Class* D.

P. judaica PELLITORY OF THE WALL. Hairy perennial to 40cm high. A weed of rocks and walls. *Distr.* W and S Europe.

P. officinalis See *P. judaica*

parilla, yellow *Menispermum canadense.*

Paris (From Latin *par*, equal, alluding to the regularity of the parts of the plant.) A genus of 5 species of rhizomatous herbs with erect stems and whorls of 4–12 leaves borne near the tops of the stems. The solitary terminal flowers bear 4–6 sepals and 4–6 yellow petals. All the species are grown as ornamentals. *Distr.* Temperate Eurasia. *Fam.* Trilliaceae. *Class* M.

P. polyphylla See *Daiswa polyphylla*

P. quadrifolia HERB PARIS. Leaves 4, sepals 4, petals 4. Occasionally bearing 3 or 5 leaves in which case the number of sepals and petals also changes. Has been used medicinally although the berries are poisonous. *Distr.* Europe, Caucasus, Siberia.

Parkeriaceae A family of 1 genus and 4 species of small, floating, short-lived ferns with lobed to tripinnate sterile leaves and once to many times pinnate fertile leaves. The sporangia are spread along the veins and more or less covered by a marginal indusium. *Distr.* Widespread in tropical regions. *Class* F.

Parkinsonia (After John Parkinson (1567–1650), London apothecary and botanical author.) A genus of 19 species of shrubs and trees with pinnate or bipinnate leaves and many short racemes of pea-like flowers. *Distr.* Tropical America. *Fam.* Leguminosae. *Class* D.

P. aculeata JERUSALEM THORN. Spiny shrub or small tree. Flowers yellow with orange dots. Grown as an ornamental. *Distr.* Tropical America.

Parnassia (After Mount Parnassus in Greece.) GRASS OF PARNASSUS. A genus of 15 species of perennial herbs with rosettes of long-stalked leaves and large, white to yellow flowers that bear their parts in fives. Several species are grown as bog-garden ornamentals. *Distr.* N temperate and Arctic regions. *Fam.* Parnassiaceae. *Class* D.

P. palustris GRASS OF PARNASSUS. To 15cm high. Flowers white with green veins. *Distr.* N temperate regions.

Parnassiaceae A family of 2 genera and 16 species of herbs. This family is sometimes included within the family Saxifragaceae. *Distr.* N temperate regions to Mexico, Chile, and Uruguay. *Class* D.
See: Parnassia.

parnassifolius, -a, -um: from the genus *Parnassia*, and *folius*, leaved.

Parochetus (From Greek *para*, close to, and *ochetus*, brook.) A genus of 1 species of prostrate perennial herb with trifoliate leaves and deep blue, pea-like flowers that are borne singly or in pairs. *Distr.* Mountains of tropical Africa and Asia. *Fam.* Leguminosae. *Class* D.

P. communis SHAMROCK PEA. Grown as a rock-garden ornamental.

Parodia (After Lorenzo Raimondo Parodi (1895–1966), Argentine botanist, or Dr Domingo Parodi (1823–90), a pharmacist and student of the flora of Paraguay.) A genus of 35–50 species of small, spherical or cylindrical cacti with rows of tubercles and brightly coloured, funnel-shaped flowers. Many species are grown as ornamentals. *Distr.* Central South America. *Fam.* Cactaceae. *Class* D.

P. chrysacanthion Stems solitary, spherical, densely covered in golden spines. Flowers funnel-shaped, yellow. *Distr.* N Argentina.

P. nivosa Stems ovoid, covered by white spines. Flowers bright red. *Distr.* N Argentina.

P. sanguiniflora Stems clustered, densely covered in red-brown spines. Flowers numerous, blood red. *Distr.* N Argentina.

Paronychia (From Greek *para*, close to, and *onyx*, nail, alluding to the original use of the plant to treat whitlows.) A genus of 50 species of tufted, annual, and perennial herbs with pairs of linear leaves and inconspicuous heads of small, axillary flowers that lack petals. Several species are grown as ornamentals. *Distr.* Cosmopolitan, introduced into Australia. *Fam.* Illecebraceae. *Class* D.

P. argentea Mat-forming perennial. Leaves small. Grown as a rock-garden ornamental. Formerly used medicinally. *Distr.* N Africa, SW Asia.

parrot feather *Myriophyllum aquaticum.*

Parrotia (After F. W. Parrot (1792–1841), German naturalist.) A genus of 1 species of deciduous shrub or tree with simple, alternate, hairy leaves and dense clusters of small flowers that bear 5–7 sepals, no petals and 15 conspicuous, white and red stamens. The flowers are borne before the leaves in spring and the leaves turn deep red, yellow, and orange in autumn. *Distr.* N Iran. *Fam.* Hamamelidaceae. *Class* D.

P. persica PERSIAN IRONWOOD. Grown as an ornamental for its spectacular autumn colour and flaking bark.

Parrotiopsis (From the genus name *Parrotia*, and Greek *-opsis* indicating resemblance.) A genus of 1 species of deciduous tree with simple, sharply toothed, hairy leaves and heads of small flowers that bear numerous, conspicuous, yellow stamens. *Distr.* Himalaya. *Fam.* Hamamelidaceae. *Class* D.

P. jacquemontiana (After Victor Jacquemont (1801–32), French naturalist.) Grown as an ornamental. The twigs are used in basketry. *Distr.* W Himalaya

parrot's beak *Clianthus puniceus*, *Lotus berthelotii*.

Parrya (After Sir William Edward Parry (1790–1855), Arctic explorer.) A genus of 25 species of perennial herbs with large white or purple flowers. *Distr.* N temperate regions. *Fam.* Cruciferae. *Class* D.

P. menziesii See *Phoenicaulis cheiranthoides*

P. nudicaulis Tufted perennial with violet, dark-centred flowers. *Distr.* Eurasia to Alaska.

parsley *Petroselinum*, *P. crispum* **black** ~ *Melanoselinum decipiens* **Chinese** ~ *Coriandrum sativum* **cow** ~ *Anthriscus sylvestris* **horse** ~ *Smyrnium olusatrum.*

parsnip *Pastinaca sativa* **water** ~ *Sium*, *S. suave* **wild** ~ *Angelica archangelica.*

Parsonsia (After Dr John Parsons (1705–70), Scottish physician and writer on natural history.) A genus of 80 species of climbing shrubs with simple opposite leaves and panicles of tubular flowers. Some are used locally as medicines. *Distr.* E Asia, W Pacific. *Fam.* Apocynaceae. *Class* D.

P. capsularis Leaves evergreen. Flowers bell-shaped. Grown as a tender ornamental. *Distr.* New Zealand

Parthenium (Classical Greek name for a plant related to *Matricaria*.) A genus of 15 species of perennial herbs and shrubs with simple to pinnately divided leaves and flowers in daisy-like heads which have distinct rays. *Distr.* North America. *Fam.* Compositae. *Class* D.

P. integrifolium WILD QUININE, AMERICAN FEVERFEW, PRAIRIE DOCK. Perennial aromatic herb to 1m high. *Distr.* E USA.

Parthenocissus (From Greek *parthenos*, virgin, and *kissos*, ivy.) VIRGINIA CREEPER. A genus of 10 species of deciduous vines with tendrils that are often tipped with adhesive disks, palmately lobed or compound leaves, and clusters of small green flowers. The fruit is a dry black berry. Several species are grown as ornamentals. *Distr.* Temperate Asia, North America. *Fam.* Vitaceae. *Class* D.

P. quinquefolia VIRGINIA CREEPER. Leaves bear 5 leaflets. *Distr.* North America.

P. tricuspidata BOSTON IVY, JAPANESE CREEPER, VIRGINIA CREEPER. Leaves simple to 3-lobed. Numerous cultivars have been raised from this species. *Distr.* Japan, China.

parviflorus, -a, -um: small-flowered.

parvifolius, -a, -um: small-leaved.

parvula, -a, -um: small.

Pasithea (After Pasithea, one of the Graces, who was also called Aglaia.) A genus of 1 species of rhizomatous perennial herb with grass-like leaves and panicles of blue flowers. *Distr.* SW South America. *Fam.* Anthericaceae. *Class* M.

P. caerulea A cultivated ornamental. *Distr.* Chile, Peru.

Paspalum (From *paspalos*, the Greek name for millet.) A genus of 250 species of annual or perennial grasses with inflorescences of branched racemes. Several species are important as fodder and grain crops and some are grown as ornamentals for their branched inflorescences. *Distr.* Tropical and warm America. *Fam.* Gramineae. *Class* M.

P. ceresia Inflorescence covered in silver hairs. A half-hardy ornamental. *Distr.* Tropical South America.

pasque flower *Pulsatilla*, *P. vulgaris*.

Passiflora (From the Latin *passio*, passion, and *flos*, a flower; Spanish missionaries in South America compared the parts of this plant to aspects of Christ's Crucifixion.) PASSION FLOWER. A genus of 350 species of evergreen lianas with axillary tendrils and entire or palmately lobed leaves (representing the hands of Christ's persecutors). The flowers bear 5 sepals and 5 petals (the apostles, less Judas and Peter), a whorl or crown of sterile stamens (the crown of thorns), 5 fused, fertile stamens (the 5 wounds), and 3 stigmas (the 3 nails). The fruit is a juicy berry. Many species are grown as ornamentals and some have edible fruits. *Distr.* Tropical and warm regions of the New World. *Fam.* Passifloraceae. *Class* D.

P. caerulea BLUE PASSION FLOWER. Flowers blue-green, around 20cm across. Grown as a hardy climber in the British Isles. *Distr.* S Brazil.

P. chinensis See *P. caerulea*

P. coccinea RED GRANADILLA, RED PASSION FLOWER. Flowers red. *Distr.* N South America.

P. edulis GRANADILLA, PURPLE GRANADILLA, PASSION FRUIT. Flowers purple with white bands. Cultivated for fruit in Mexico, Australia, and North America. *Distr.* Brazil.

P. incarnata MAY APPLE, MAY POPS, APRICOT VINE, WILD PASSION FLOWER. Flowers white, lavender, and purple. Grown for edible fruit locally. Hardy in the British Isles. *Distr.* S USA.

P. laurifolia YELLOW GRANADILLA, JAMAICA HONEYSUCKLE, WATER LEMON, BELLE APPLE, VINEGAR PEAR. Flowers red, green, and purple. Fruit edible. *Distr.* W Indies, N South America.

P. ligularis SWEET GRANADILLA. Flowers somewhat bell-shaped, pink, white, and purple. Fruit edible, sweet to the taste. *Distr.* Tropical America.

P. maliformis SWEET CALABASH, SWEET CUP. Flowers green and white, fragrant. Grown for fruit juice. *Distr.* Tropical America.

P. mayana See *P. caerulea*

P. mollissima BANANA PASSION FRUIT. Flowers pendent, pink. Fruit edible. *Distr.* N South America.

P. quadrangularis GIANT GRANADILLA. Flowers blue-green and pink. The large yellow fruit is eaten as a vegetable when immature. *Distr.* South America.

Passifloraceae The Passion Flower family. 17 genera and 530 species of herbs, shrubs, vines, and trees with alternate, entire or lobed leaves. The vines climb by means of sterile pedicels. The flowers are regular and typically bisexual with 5 sepals and 5 petals; they may also bear a whorl of filaments that form a corona or crown. The stamens are frequently fused into a column. Many species have edible fruits or are grown as ornamentals. *Distr.* Widespread in tropical and warm-temperate regions. *Class* D.
See: Passiflora.

passion flower Passifloraceae, *Passiflora* **blue** ~ *P. caerulea* **red** ~ *P. coccinea* **wild** ~ *P. incarnata*.

passion fruit *Passiflora edulis* **banana** ~ *P. mollissima*.

Pastinaca (Classical Latin name for PARSNIP and CARROT, from *pastus*, food.) A genus of 14 species of biennial and perennial herbs with thick rootstocks, simple or pinnately divided leaves, and compound umbels of yellow flowers. *Distr.* Temperate Eurasia. *Fam.* Umbelliferae. *Class* D.

P. sativa PARSNIP. Aromatic biennial herb. The substantial white taproot is widely eaten as a vegetable, with numerous cultivars having been raised. *Distr.* Eurasia.

patagonicus, -a, -um: of Patagonia.

pate *Schefflera digitata*.

patens: spreading.

patience plant *Impatiens*, *I. walleriana*.

Patrinia (After E. L. M. Patrin (1742–1814), French traveller in Siberia.) A genus of 15 species of perennial herbs with opposite or basal, pinnately lobed or divided leaves, and heads of small 5-lobed flowers. Several species are grown as rock-garden ornamentals. *Distr.* Europe to Himalaya and E Asia. *Fam.* Valerianaceae. *Class* D.

P. gibbosa Perennial herb to 25cm high. Flowers yellow. *Distr.* Japan.

P. triloba Perennial herb to 60cm high. Leaves 3- or 5-lobed. Flowers golden, fragrant. *Distr.* Japan.

patulus, -a, -um: spreading.

pauciflorus, -a, -um: with few flowers.

paucifolius, -a, -um: with few leaves.

paucinervis, -e: with few nerves.

pauciramosus, -a, -uma: with few branches.

Paulownia (After Anna Paulowna (1795–1865), daughter of Tsar Paul I of Russia.) A genus of 6 species of deciduous trees with large, palmately lobed, opposite leaves and panicles of large, white, bell-shaped, 5-lobed flowers. Several species are grown as ornamentals. *Distr.* E Asia. *Fam.* Scrophulariaceae. *Class* D.

P. fortunei Spreading tree to 20m high. Flowers white with purple spots, fragrant. *Distr.* China, Japan.

P. tomentosa FOXGLOVE TREE, PRINCESS TREE. Spreading tree to 15m high. Flowers pink-white, fragrant. *Distr.* China.

paulus, -a, -um: little.

Pavonia (After Jose Antonio Pavon (1754–1840), Spanish botanist.) A genus of 150 species of herbs and shrubs with simple, entire or palmately lobed leaves and regular bisexual flowers that bear 5 petals and numerous fused stamens. Several species are grown as tender ornamentals and some are used locally as a source of fibre. *Distr.* Tropical and warm regions. *Fam.* Malvaceae. *Class* D.

P. multiflora Tender evergreen shrub. Flowers solitary, purple-red. *Distr.* Brazil.

pavoninus, -a, -um: peacock-like.

pawpaw Caricaceae, *Carica papaya* **mountain** ~ *C. pubescens*.

Paxistima (From Greek *pachys*, thick, and *stigma*, stigma.) A genus of 2 species of small evergreen shrubs with simple opposite leaves and minute flowers that bear 4 sepals, 4 petals, and 4 stamens. The fruit is a compressed capsule. Both species are grown as ornamentals. *Distr.* North America. *Fam.* Celastraceae. *Class* D.

P. canbyi (After William Marriot Canby (1831–1904), who collected it.) CLIFF GREEN. Shrub to 40cm high. Flowers green. *Distr.* E USA.

P. myrtifolia OREGON BOXWOOD. Much-branched shrub to 1m high. *Distr.* NW North America.

pea Leguminosae, *Pisum* **asparagus** ~ *Lotus tetragonolobus* **balloon** ~ *Sutherlandia frutescens* **beach** ~ *Lathyrus japonicus* **black coral** ~ *Kennedia nigricans* **blue** ~ *Psoralea pinnata* **broad-leaved everlasting** ~ *Lathyrus latifolius* **Canada** ~ *Vicia cracca* **chick** ~ *Cicer arietinum* **circumpolar** ~ *Lathyrus japonicus* **coral** ~ *Kennedia* **desert** ~ *Clianthus formosus* **dogtooth** ~ *Lathyrus sativus* **earth nut** ~ *L. tuberosus* **everlasting** ~ *L. grandiflorus* **flat** ~ *L. sylvestris* **Fyfield** ~ *L. tuberosus* **garden** ~ *Pisum sativum* **glory** ~ *Clianthus formosus*, *C. puniceus* **golden** ~ *Thermopsis macrophylla* **holly flame** ~ *Chorizema ilicifolium* **Indian** ~ *Lathyrus sativus* **Lord Anson's blue** ~ *L. nervosus* **narrow-leaved everlasting** ~ *L. sylvestris* **perennial** ~ *L. latifolius* **Persian everlasting** ~ *L. rotundifolius* **purple coral** ~ *Hardenbergia violacea* **Riga** ~ *Lathyrus sativus* **sea** ~ *L. japonicus* **shamrock** ~ *Parochetus communis* **Sturt's desert** ~ *Clianthus formosus* **sweet** ~ *Lathyrus odoratus* **tuberous** ~ *L. tuberosus* **two-flowered** ~ *L. grandiflorus* **Western Australia coral** ~ *Hardenbergia comptoniana* **wild** ~ *Lathyrus* **winged** ~ *Lotus tetragonolobus* **yellow** ~ *Crotalaria capensis*.

peach *Prunus persica* **David's** ~ *P. davidiana*.

peacock flower *Caesalpinia pulcherrima*, *Tigridia pavonia*, *Delonix regia*.

peacock plant *Calathea makoyana.*

peanut *Arachis hypogaea.*

pear *Pyrus* **alligator** ~ *Persea americana* **avocado** ~ *P. americana* **Chinese** ~ *Pyrus pyrifolia* **common** ~ *P. communis* **evergreen** ~ *P. kawakmii* **Japanese** ~ *P. pyrifolia* **melon** ~ *Solanum muricatum* **prickly** ~ *Opuntia O. ficus-indica*, **sand** ~ *Pyrus pyrifolia* **tiger** ~ *Opuntia aurantiaca* **vegetable** ~ *Sechium edule* **vinegar** ~ *Passiflora laurifolia.*

pearlbush *Exochorda.*

pearl fruit *Margyricarpus pinnatus.*

pearlwort *Sagina.*

pea tree *Caragana* **Russian** ~ *C. frutex* **Siberian** ~ *C. arborescens.*

pecan *Carya, C. illinoinensis* **bitter** ~ *C. aquatica.*

Pecteilis (From Latin *pecten*, comb, alluding to the lateral lobes of the lip.) A genus of 4 species of terrestrial orchids with linear leaves and erect racemes of yellow-white flowers. *Distr.* Tropical E Asia. *Fam.* Orchidaceae. *Class* M.

P. radiata Flowers 1–3, tinged green. *Distr.* Japan, Korea.

pectinatus, -a, -um: comb-like.

pectinifera: comb-bearing.

Pedaliaceae The Sesame family. 17 genera and 95 species of annual and perennial herbs and a few shrubs with opposite simple leaves and irregular tubular flowers. *Distr.* Widespread, mainly tropical. *Class* D.
See: Ceratotheca, Proboscidea, Sesamum.

pedalis, -e a foot long (approximately 30 centimetres).

pedatifidus, -a, -um: resembling the foot of a bird.

pedatisectus, -a, -um: divided like a foot.

pedatus, -a, -um: foot-like.

pedemontanus, -a, -um: of the foot of mountains.

pedicellatus, -a, -um: with a pedicel (flower-stalk).

Pedicularis (From Greek *pediculus*, a louse; the presence of these plants in pasture was thought to cause sheep to become infested with lice.) LOUSEWORT. A genus of about 350 species of semi-parasitic, annual, biennial, and perennial herbs with toothed or bipinnately divided leaves and spike-like racemes of 2-lipped flowers. The upper lip of the flowers is often laterally compressed and/or beaked. *Distr.* N hemisphere, especially in mountainous regions of Europe and Asia. *Fam.* Scrophulariaceae. *Class* D.

P. canadensis Perennial herb. Flowers cream or purple or bicoloured. Grown as an ornamental. *Distr.* N America.

Pedilanthus (From Greek *pedilon*, slipper, and *anthos*, flower, alluding to the shape of the involucre.) A genus of 14 species of succulent shrubs with simple, somewhat fleshy leaves and small flowers that are borne in irregular inflorescences and surrounded by a slipper-shaped whorl of bracts. This genus is closely related to the genus *Euphorbia*. *Distr.* N America to tropical S America. *Fam.* Euphorbiaceae. *Class* D.

P. tithymaloides SLIPPER FLOWER, RIBBON CACTUS, BIRD CACTUS. Evergreen or deciduous shrub. Flowers surrounded by red and yellow bracts. *Distr.* Central and N South America.

Pediocactus (From Greek *pedion*, plain, and the old genus name *Cactus*, alluding to the Great Plains of North America where the type specimen grew.) A genus of 6 species of small cacti with round, tubicle-covered stems and bell-shaped flowers. Several species are grown as ornamentals. *Distr.* W and SW USA. *Fam.* Cactaceae. *Class* D.

P. knowltonii Stems solitary or clustered. Tubercles very small. Flowers pale yellow. *Distr.* USA (Colorado, New Mexico).

P. simpsonii Stems solitary or clustered. Tubercles spiralled. Flowers white, pink, or yellow. *Distr.* W USA.

peduncularis, -e: with a peduncle, an inflorescence stalk.

pedunculatus, -a, -um: with a peduncle, an inflorescence stalk.

Peganum (Classical Greek name for a species of *Ruta*.) A genus of 5–6 species of perennial herbs and subshrubs with pinnate leaves and 5-lobed white flowers. *Distr.* Mediterranean to Mongolia, S North America. *Fam.* Zygophyllaceae. *Class* D.

P. harmala HARMAL. Perennial herb to 1m high. Flowers white. Grown as an ornamental and used medicinally to treat eye diseases, rheumatism, and Parkinson's disease. *Distr.* Mediterranean to Asia.

pegwood *Cornus sanguinea*.

pekinensis, -e: of Peking.

pelargoniiflorus, -a, -um: with flowers like Pelargonium.

Pelargonium (From Greek *pelargos*, stork, alluding to the bill-like fruits.) GERANIUM. A genus of 280 species of perennial herbs, subshrubs, and succulents with simple or divided leaves that are often hairy or aromatic. The typically irregular flowers bear 5 sepals and 5 petals, 2 of which are larger than the others. There are 10 stamens, 7 of which are fertile. This genus is closely related to the genus *Geranium* but differs in having irregular flowers. A number of species and hundreds of cultivars are grown as ornamentals. *Distr.* S and tropical Africa, Australia. *Fam.* Geraniaceae. *Class* D.

P. asperum See *P. graveolens*

P. crispum LEMON-SCENTED PELARGONIUM. Erect evergreen subshrub. Leaves lemon-scented, numerous, small with a wavy margin. Flowers small, pink. Fragrant oil is extracted from the leaves. *Distr.* South Africa.

P. × domesticum REGAL PELARGONIUM. The name given to a complex group of hybrids with rounded leaves and large flowers of varying colours. *Distr.* South Africa.

P. × fragrans NUTMEG PELARGONIUM. A group of shrubby hybrids with spicy-scented foliage; sometimes considered a distinct species. *Distr.* South Africa.

P. gibbosum GOUTY GERANIUM, KNOTTED GERANIUM. Scrambling perennial. Stems swollen at nodes. Leaves slightly fleshy. Flowers dull yellow. *Distr.* W Coast of S Africa.

P. graveolens ROSE PELARGONIUM, ROSE GERANIUM, SWEET-SCENTED GERANIUM. Soft-hairy subshrub. Leaves rose-scented. Flowers white, tinged pink. *Distr.* South Africa.

P. × hortorum ZONAL PELARGONIUM, BEDDING GERANIUM. A group of complex hybrids with distinctly marked leaves. *Distr.* South Africa.

P. odoratissimum APPLE PELAR-GONIUM. Low-growing perennial. Leaves apple-scented, round. Flowers small, white with red veins. Commercial source of Wawah oil. *Distr.* South Africa.

P. peltatum IVY LEAVED PELARGONIUM, IVY GERANIUM, HANGING GERANIUM. Stems trailing and hanging. Leaves bright green with dark markings. Flowers large, pink. Commonly grown in hanging baskets. *Distr.* South Africa.

P. quercifolium OAK-LEAVED PELARGONIUM, OAK-LEAVED GERANIUM, ALMOND GERANIUM, VILLAGE OAK-GERANIUM. Erect sticky subshrub, smelling of balsam. Leaves rough-hairy. Flowers small, purple-pink. *Distr.* South Africa.

P. tomentosum PEPPERMINT-SCENTED PELARGONIUM, PEPPERMINT GERANIUM. Spreading perennial. Leaves velvety, smelling of peppermint. Flowers small, bright red. *Distr.* South Africa (NE Natal).

pelargonium: apple ~ *Pelargonium odoratissimum* **ivy-leaved** ~ *P. peltatum* **lemon-scented** ~ *P. crispum* **nutmeg** ~ *P. × fragrans* **oak-leaved** ~ *P. quercifolium* **peppermint-scented** ~ *P. tomentosum* **regal** ~ *P. × domesticum* **rose** ~ *P. graveolens* **zonal** ~ *P. × hortorum*.

Pelecyphora (From Greek *pelekus*, hatchet, and *phoros*, bearing, alluding to the shape of the tubercles.) A genus of 2 species of small cacti with rounded to top-shaped stems that are covered in keeled, compressed tubercles. The bell- or funnel-shaped flowers are purple to pink. Both species are grown as ornamentals. *Distr.* NE Mexico. *Fam.* Cactaceae. *Class* D.

P. aselliformis Stems solitary or clustered, grey-green. *Distr.* NE Mexico.

P. strobiliformis Stems typically solitary, spherical. Tubercles scale-like. Flowers funnel-shaped, pale pink-purple. *Distr.* NE Mexico.

pelican flower *Aristolochia grandiflora*.

pelican's beak *Lotus berthelotii.*

Peliosanthes (From Greek *pelios*, livid, and *anthos*, flower.) A genus of 1 species of rhizomatous herb with simple, tufted, basal leaves and racemes of green, blue, or purple flowers. *Distr.* E Himalaya to SE Asia, Taiwan and China. *Fam.* Convallariaceae. *Class* M.

P. teta A greenhouse ornamental.

Pellaea (From Greek *pellaios*, dark, alluding to the colour of the stalks.) CLIFF BRAKE. A genus of 80 species of medium-sized ferns of rocky places. Some species are cultivated as ornamentals. *Distr.* Tropical and warm-temperate regions. *Fam.* Adiantaceae. *Class* F.

P. atropurpurea PURPLE CLIFF BRAKE. Leaves tufted, with a hard, dark purple rachis. *Distr.* North America.

P. rotundifolia BUTTON FERN. Pinnae more or less round, dark green. *Distr.* New Zealand.

P. viridis GREEN CLIFF BRAKE. Leaves bright green, occasionally blue-green. *Distr.* Africa, Mascarene Islands.

Pellicieraceae A family of 1 genus and 1 species of tree with simple, leathery, spirally arranged leaves and regular bisexual flowers that bear their parts in fives. *Distr.* Costa Rica to Ecuador. *Class* D.

Pellionia (After Alphonse Odet Pellion (1796–1868), French admiral.) A genus of 80 species of evergreen perennial herbs and subshrubs with 2 ranks of simple alternate leaves and dense clusters of small green flowers. Several species are grown for their colourful foliage. *Distr.* Tropical or subtropical Asia. *Fam.* Urticaceae. *Class* D.

P. pulchra RAINBOW VINE. Creeping herb. Stem and underside of leaves tinged green. *Distr.* Vietnam.

P. repens Creeping, somewhat succulent herb. Leaves tinged bronze. *Distr.* SE Asia.

pellitory *Parietaria.*

pellitory of the wall *Parietaria judaica.*

Peltandra (From Greek *pelta*, target or shield, and *aner*, stamen, alluding to the shape of the stamens.) ARROW ARUM. A genus of 4 species of aquatic rhizomatous herbs that bear large leaves on long stalks. The small unisexual flowers are borne on a spadix that is partially surrounded by a green or white spathe. The starchy rhizomes were formerly an important food of American Indians. *Distr.* E North America. *Fam.* Araceae. *Class* M.

P. sagittifolia WHITE ARROW ARUM. Limb of spathe white. Berries red. *Distr.* SE USA.

P. undulata See *P. virginica*

P. virginica GREEN ARROW ARUM. Spathe limb green with yellow-white margins. Berries green. *Distr.* E USA.

peltatus, -a, -um: peltate, i.e. with the leaf stalk attached to the centre of the leaf blade rather than to the margin.

Peltiphyllum See *Darmera*

P. peltatum See *Darmera peltata*

Peltoboykinia (From Greek *pelte*, shield, and the genus name *Boykinia*.) A genus of 1 species of deciduous rhizomatous herb with peltate basal leaves and bell-shaped, cream-yellow flowers. *Distr.* Japan. *Fam.* Saxifragaceae. *Class* D.

P. tellimoides Ornamental herb of damp, shaded areas. *Distr.* Japan.

pelviformis, -e: concave, like a shallow dish.

pemakoensis, -e: of Pemako, Tibet.

Penaeaceae A family of 7 genera and 21 species of heath-like shrubs with small sessile leaves. The flowers are regular with 2–4 coloured bracts and 4 sepals fused into a tube but no petals. *Distr.* South Africa. *Class* D.

Penang lawyers *Licuala.*

pendens: pendulous, hanging down.

penduliflorus, -a, -um: with pendulous flowers.

pendulus, -a, -um: pendulous.

peninsularis, -e: of a peninsula.

Peniocereus (From Greek *penios*, thread, and the genus name *Cereus*, alluding to the slender stems.) A genus of 20 species of spreading, shrubby cacti with thick roots, narrow, ribbed stems, and tubular flowers that open at night. Several species are grown as ornamentals. *Distr.* Central America to SW USA. *Fam.* Cactaceae. *Class* D.

P. greggii Root turnip-like. Stems hairy. Flowers white. *Distr.* N Mexico and S USA.

P. serpentinus Stems green. Flowers white, tinged red. Fruit edible. *Distr*. Mexico.

P. viperinus Root tuberous. Stems dark grey, woolly. Flowers pink-red. *Distr*. Central S Mexico.

Pennantia (After Thomas Pennant (1726–98), British naturalist.) A genus of 3 species of trees with spirally arranged, simple leaves and racemes or panicles of regular bisexual flowers that bear their parts in fours or fives. *Distr*. Australia, New Zealand, Norfolk Islands. *Fam*. Icacinaceae. *Class* D.

P. corymbosa A source of hard timber for tool handles and, formerly, for fire-making by friction. *Distr*. Australia, New Zealand, Norfolk Islands.

pennatus, -a, -um: pinnate, feathery.

pennigera: bearing feathers.

Pennisetum (From Latin *penna*, feather, and *seta*, bristle, alluding to the feathery bristles of the inflorescences of some species.) A genus of 120 species of annual and perennial, rhizomatous or stoloniferous grasses to 3.5m high with flat leaves and spike-like, cylindrical or round inflorescences. Some species are important as fodder or grain crops and for erosion control. A few species are grown as ornamentals. *Distr*. Tropical and warm-temperate regions. *Fam*. Gramineae. *Class* M.

P. alopecuroides CHINESE FOUNTAIN GRASS, CHINESE PENNISETUM, SWAMP FOXTAIL GRASS. Tufted perennial. Leaves narrow. Inflorescence purple or yellow. *Distr*. E Asia to E Australia.

P. compressum See *P. alopecuroides*

P. glaucum BULRUSH MILLET, PEARL MILLET, SPIKED MILLET. Important food source in drier areas of India and the African Sahel; grain is cooked like rice or made into a gruel or unleavened bread. Grain is also malted to produce beer in Africa and to feed poultry in Europe. *Distr*. Derived from a number of species in W Africa.

P. longistylum See *P. villosum*

P. ruppelii See *P. setaceum*

P. setaceum FOUNTAIN GRASS. Perennial. Inflorescences cylindrical, copper-red or purple. *Distr*. Tropical Africa, Arabia, SW Asia.

P. villosum ABYSSINIAN FEATHERTOP, AFRICAN FOUNTAIN GRASS. Perennial to 60cm high. Inflorescence rounded or cylindrical, tinged purple. *Distr*. Mountains of NE tropical Africa.

pennisetum, Chinese *Pennisetum alopecuroides*.

pennispinosus, -a, -um: with feathery spines.

pennivenius, -a, -um: with pinnate veins

pennsylvanicus, -a, -um: of Pennsylvania, USA.

pennycress *Thlaspi arvense*. **mountain** ~ *T. montanum*.

penny flower *Lunaria annua*.

penny pies *Umbilicus rupestris*.

pennyroyal *Mentha pulegium*.

pennywort *Umbilicus rupestris*, *Hydrocotyle*, *Cymbalaria muralis* **wall** ~ *Umbilicus rupestris*.

pensilis, -e: pendulous.

Penstemon (From Greek *pente*, five, and *stemon*, stamen, alluding to the 4 fertile and 1 sterile stamens.) A genus of 250 species of perennial herbs and subshrubs with simple, typically alternate leaves, and tubular, 2-lipped, 5-lobed flowers. Numerous species and cultivars are grown as ornamentals. *Distr*. North America, 1 species NE Asia. *Fam*. Scrophulariaceae. *Class* D.

P. barbatus Rosette-forming perennial. Flowers pink-red. *Distr*. SW USA, Mexico.

P. davidsonii Evergreen prostrate shrub. Flowers violet to red. *Distr*. W North America.

P. diffusus See *P. serrulatus*

P. fruticosus Evergreen upright shrub. Flowers lavender-blue. *Distr*. W North America.

P. menziesii See *P. davidsonii*

P. scouleri See *P. fruticosus*

P. serrulatus CASCADE PENSTEMON. Subshrub. Flowers purple-blue. *Distr*. NW North America.

penstemon, Cascade *Penstemon serrulatus*.

Pentaglottis (From Greek *pente*, five, and *glotta*, tongue, probably alluding to the 5 petal lobes.) A genus of 1 species of large, rough-hairy, perennial herb with simple alternate leaves and cymes of blue funnel-shaped flowers which bear 5 spreading lobes. *Distr.* SW Europe. *Fam.* Boraginaceae. *Class* D.

P. sempervirens Grown as a hardy ornamental.

pentandrus, -a, -um: with five stamens.

pentapetalus, -a, -um: with five petals.

Pentaphragmataceae A family of 1 genus and 25 species of succulent herbs with simple, spirally arranged leaves and tubular flowers that bear their parts in fives. This family is sometimes included within the family Campanulaceae. *Distr.* S Asia. *Class* D.

Pentaphylacaceae A family of 1 genus and 1 species of dwarf tree with simple, spirally arranged leaves and racemes of small regular flowers that bear their parts in fives. *Distr.* China to Sumatra. *Class* D.

pentaphyllus, -a, -um: with five leaves.

Pentapterygium See *Agapetes*

Pentas (From Greek *pentas*, a series of five, alluding to the floral parts.) A genus of 34 species of perennial herbs and shrubs with simple, opposite or whorled leaves, and cymes or corymbs of funnel-shaped, 5-lobed flowers. Several species are grown as ornamentals. *Distr.* Africa, Arabia, Madagascar. *Fam.* Rubiaceae. *Class* D.

P. lanceolata EGYPTIAN STAR CLUSTER, STAR CLUSTER. Perennial herb or subshrub to 2m high. Flowers pink, magenta, or blue, borne in dense clusters. *Distr.* E Africa, Arabia.

Penthoraceae The Ditch Stonecrop family. 1 genus and 2 species of rhizomatous perennial herbs with simple leaves and regular flowers that typically bear 5 sepals, 5 petals, and 10 stamens. This family is sometimes included within the Crassulaceae or the Saxifragaceae. *Distr.* Tropical Asia, E USA. *Class* D.

pen wiper *Kalanchoe marmorata.*

peony *Paeonia* **tree** ~ *Paeonia.*

Peperomia (From Greek *peper*, pepper, and *homios*, resembling.) A genus of 1000 species of typically epiphytic herbs and subshrubs with simple leaves and dense spikes of minute flowers. Many species are cultivated as ornamental house-plants. *Distr.* Tropical and warm regions. *Fam.* Piperaceae. *Class* D.

P. argyreia WATERMELON PEPEROMIA, RUGBY FOOTBALL PLANT. Compact bushy perennial. Leaves fleshy, dark green, striped with silver. *Distr.* N South America.

P. caperata EMERALD RIPPLE PEPPER. Perennial herb to 40cm high. Leaves dark green with deeply sunken veins. *Distr.* Brazil.

P. fraseri MIGNONETTE PEPEROMIA, FLOWERING PEPPER. Perennial herb. Leaves somewhat succulent, with purple-red veins. *Distr.* Ecuador.

P. glabella WAX PRIVET. Perennial herb. Leaves somewhat fleshy, waxy. *Distr.* West Indies, Central and South America.

P. griseoargentea IVY-LEAF PEPEROMIA. Perennial herb. Leaves marked silver-grey. *Distr.* Brazil.

P. magnoliifolia See *Peperomia obtusifolia*

P. obtusifolia BABY RUBBER PLANT. Stoloniferous herb. Leaves notched. *Distr.* N South America, Central America, Mexico, West Indies.

P. redisiflora See *Peperomia fraseri*

P. sandersii See *Peperomia argyreia*

peperomia: ivy-leaf ~ *Peperomia griseoargentea* **mignonette** ~ *P. fraseri* **watermelon** ~ *P. argyreia.*

peperomioides: from the genus *Peperomia*, with the ending *-oides*, indicating resemblance.

pepino *Solanum muricatum.*

pepper Piperaceae, *Capsicum* **bell** ~ *C. annum* **betel** ~ *Piper betle* **black** ~ *P. nigrum* **bush** ~ *Clethra alnifolia* **Cayenne** ~ *Capsicum annum* **Celebes** ~ *Piper ornatum* **cherry** ~ *Capsicum annum* **chilli** ~ *C. annum* **common** ~ *Piper nigrum* **emerald ripple** ~ *Peperomia caperata* **flowering** ~ *P. fraseri* **hot** ~ *Capsicum frutescens* **Jamaica** ~ *Pimenta dioica* **Japan** ~ *Zanthoxylum piperitum* **Madagascar** ~ *Piper nigrum* **spur** ~ *Capsicum frutescens* **sweet** ~ *C. annum* **Tabasco** ~ *C. frutescens* **wall** ~ *Sedum acre* **white** ~ *Piper nigrum.*

pepper bush, sweet *Clethra alnifolia.*

peppermint *Mentha* × *piperita*
Mount Wellington ~ *Eucalyptus coccifera.*

pepper tree *Macropiper excelsum*
Brazilian ~ *Schinus terebinthifolius*
Californian ~ *S. molle.*

pepperwood *Umbellularia californica.*

pepperwort *Marsilea, Lepidium sativum.*

peregrinus, -a, -um: foreign.

perennis, -e: perennial.

Pereskia (After Nicolas Claude Fabri de Peiresc (1580–1637), French naturalist.) A genus of 16 species of shrubby, climbing, and tree-like cacti with woody stems and more or less succulent, deciduous leaves. The rose-like flowers are solitary or borne in panicles or cymes. Several species are grown as ornamentals. *Distr.* Mexico, W Indies Central and South America. *Fam.* Cactaceae. *Class* D.

P. aculeata BARBADOS GOOSEBERRY, LEMON VINE. Liana to 10m with recurved spines. Flowers numerous, cream, fragrant. Cultivated for edible fruit and leaves. *Distr.* Tropical America.

P. grandifolia ROSE CACTUS. Bushy or climbing shrub with black spines. Flowers solitary, pink. Leaves edible. *Distr.* Brazil, widely cultivated throughout the tropics.

Perezia (After Lazarus Perez, an apothecary in Toledo during the 16th century.) A genus of 30 species of annual and perennial herbs and shrubs with simple to pinnate leaves and flowers in daisy-like heads which lack distinct rays. *Distr.* Andes. *Fam.* Compositae. *Class* D.

P. recurvata Mat-forming perennial. Grown as a rock-garden ornamental. *Distr.* S South America.

perfoliatus, -a, -um: with leaves joined around the stem.

perforatus, -a, -um: perforated with small holes or clear spots.

Pericallis (From Greek *peri*, around, and *kallos*, beauty.) FLORIST'S CINERARIA. A genus of 14 species of perennial herbs and shrubs with simple, palmately veined leaves and flowers in daisy-like heads that bear distinct rays. *Distr.* Macaronesia. *Fam.* Compositae. *Class* D.

P.* × *hybrida FLORIST'S CINERARIA. A group of hybrids (probably with *P. lanata* and *P. cruenta* in the parentage) that are perennial but grown as annuals and flower profusely. Important cut flowers of which numerous cultivars are available. *Distr.* Garden origin.

P. lanata Subshrub to 1m high. Flowers violet-scented. *Distr.* Macaronesia, naturalized in California.

periclymenum from *periklymenon*, the Greek name for honeysuckle.

Peridiscaceae A family of 2 genera and 2 species of trees with simple, spirally arranged leaves and clusters of small flowers that bear 4–7 sepals and numerous stamens but no petals. *Distr.* Tropical South America. *Class* D.

Perilla (Derivation obscure, possibly from the Hindu name.) A genus of 6 species of annual herbs with simple, opposite, often variegated leaves and pairs of 2-lipped flowers borne in spikes. *Distr.* India to Japan. *Fam.* Labiatae. *Class* D.

P. frutescens YEGOMA OIL PLANT. Annual herb to 1m high. Leaves marked purple. Flowers white. Much cultivated in E Asia and SE Europe as an ornamental and for yegoma oil which is used in much the same way as linseed oil. *Distr.* Himalaya to E Asia.

Periploca (From Greek *peri*, around, and *ploke*, twining, alluding to the climbing habit of some species.) A genus of 11 species of shrubs and climbers with simple opposite leaves and corymbs or cymes of more or less funnel-shaped, bad-smelling flowers. *Distr.* Mediterranean, E Asia, tropical Africa. *Fam.* Asclepiadaceae. *Class* D.

P. graeca SILK VINE. Twining climber to 10m high. Leaves deciduous. Grown as a hardy ornamental. *Distr.* SE Europe, W Asia.

Peristeria (From Greek *peristera*, dove, alluding to the shape of the column.) A genus of 10 species of epiphytic and terrestrial orchids with fleshy pseudobulbs, stalked leaves, and racemes of numerous, fleshy, fragrant flowers. *Distr.* Central and South America. *Fam.* Orchidaceae. *Class* M.

P. elata DOVE FLOWER, HOLY GHOST FLOWER. Flowers white, waxy. The national flower of Panama. *Distr.* Tropical America from Costa Rica to Venezuela and Colombia.

P. pendula Epiphyte. Inflorescence pendulous. Flowers yellow-green, spotted red. *Distr.* Tropical South America from Trinidad to N Brazil and N Peru.

Peristrophe (From Greek *peri*, around, and *strophe*, turning, alluding to the twisted petal tube.) A genus of 15 species of perennial herbs and subshrubs with simple leaves and solitary or clustered long-tubed, 2-lipped flowers. *Distr.* Tropical regions of the Old World. *Fam.* Acanthaceae. *Class* D.

P. angustifolia See *P. hyssopifolia*

P. hyssopifolia Leaves tapering to a long point. Flowers pink, in small clusters. Grown as an ornamental with variegated cultivars available. *Distr.* Java.

P. speciosa Leaves dark green. Flowers violet to crimson in clusters of 2–3. *Distr.* India.

periwinkle Apocynaceae, *Vinca*, *Catharanthus* **greater** ~ *Vinca major* **lesser** ~ *V. minor* **Madagascar** ~ *Catharanthus roseus*.

Pernettya See *Gaultheria*

Perovskia (After V. A. Perovsky (1794–*c.*1857), Turkestani statesman.) A genus of 7 species of aromatic subshrubs and shrubs with simple, opposite, deeply toothed leaves and panicles of 2-lipped flowers. Several species are grown as ornamentals. *Distr.* NE Iran to NW India. *Fam.* Labiatae. *Class* D.

P. abrotanoides Subshrub to 1m high. Flowers numerous, violet. *Distr.* W Asia.

P. atriplicifolia Subshrub to 1.5m high. Flowers soft blue. *Distr.* W Himalaya, Afghanistan.

perpusillus, -a, -um: very small.

Persea (From Greek persea, originally applied to *Cordia uryseo*.) A genus of 150 species of evergreen shrubs and trees with simple alternate leaves and panicles of small, yellow-green, uni- and bisexual flowers that are followed by berry-like fruits. Some species have edible fruits, others are grown for their timber. *Distr.* Tropical and subtropical regions of E Asia and America. *Fam.* Lauraceae. *Class* D.

P. americana AVOCADO PEAR, ALLIGATOR PEAR, AGUACATE. Tree or large shrub. Grown for its highly nutritious pear-shaped fruits. *Distr.* Central America.

Persian shield *Strobilanthes dyerianus*.

Persicaria affinis See *Polygonum affine*

Persicaria amphibia See *Polygonum amphibium*

persicus, -a, -um: of Persia (now Iran).

persimmon *Diospyros virginiana* **American** ~ *D. virginiana* **Japanese** ~ *D. kaki*.

perulatus, -a, -um: with conspicuous bud scales.

peruvianus, -a, -um: of Peru.

pes-caprae: like a goat's foot.

Pescatorea (After V. Pescatore, 19th-century French orchid collector.) A genus of 17 species of epiphytic orchids with leaves arranged in a fan and showy solitary flowers, without pseudobulbs. *Distr.* S Central America and W tropical South America. *Fam.* Orchidaceae. *Class* M.

P. cerina Flowers white with yellow-green markings, highly fragrant. *Distr.* Costa Rica, Nicaragua, Panama.

P. dayana Flowers highly fragrant, white, lip flushed purple. *Distr.* Colombia, Panama.

P. lehmannii Flowers waxy, white, lip dark purple. *Distr.* Ecuador, Colombia.

Petamenes See *Gladiolus*

Petasites (From Greek *petasos*, a large brimmed hat, alluding to the large leaves.) BUTTERBUR, SWEET COLTSFOOT. A genus of 15 species of perennial rhizomatous herbs with large leaves and flowers, which are produced before the leaves, in spikes of daisy-like heads. The male and female flowers are borne on separate plants. *Distr.* N temperate regions. *Fam.* Compositae. *Class* D.

P. albus Leaves near-round to 40cm across. Flowers white. *Distr.* Europe to W Asia.

P. fragrans WINTER HELIOTROPE. Flowers vanilla-scented. *Distr.* Mediterranean.

P. hybridus BOG RHUBARB, BUTTERBUR. Leaves to 60cm across. Flowers lilac-pink or yellow. Used since the Middle Ages as an anticonvulsive and for other medicinal purposes. *Distr.* Eurasia.

P. japonicus Leaves to 80cm across. Leaf stalks long. The leaf stalks are eaten like RHUBARB by the Japanese and the flower buds are used as a condiment. *Distr.* Korea, China, Japan.

P. paradoxus Leaves delta-shaped. Flowers red-pink. *Distr.* Mountains of Europe.

petasitis: from the classical Greek for a large-leaved plant.

Petermanniaceae A family of 1 genus and 1 species of liana with a starchy rhizome, simple leaves, and small flowers that bear 2 whorls of 3 tepals. This family is sometimes included within the families Smilacaceae or Dioscoreaceae. *Distr.* E Australia. *Class* M.

petiolaris, -e: with a petiole, a leaf stalk.

petiolatus, -a, -um: with a petiole, a leaf stalk.

petiolulus, -a, -um: with a petiolule, a small leaf stalk.

petraea: of rocky places.

Petrea (After Robert James, 8th Baron Petre (1713–43), English botanist and horticultural patron.) A genus of 30 species of evergreen or deciduous shrubs, trees, and lianas with simple, opposite or whorled leaves and many long racemes of more or less bell-shaped, blue, purple, or white flowers. Several species are grown as tender ornamentals. *Distr.* Tropical America. *Fam.* Verbenaceae. *Class* D.

P. volubilis PURPLE WREATH, QUEEN'S WREATH, SANDPAPER VINE. Liana to 10m high. Leaves rough-hairy. Flowers pale lilac to purple. *Distr.* Mexico, Central America.

Petrocallis (From Greek *petros*, rock and *callis*, beauty.) A genus of 2 species of perennial, cushion-forming herbs with small lilac-pink flowers. *Distr.* Mountains of S Europe, N Iran. *Fam.* Cruciferae. *Class* D.

P. pyrenaica Cultivated as an ornamental rock-garden plant. *Distr.* S Europe.

Petrocoptis (From Greek *petros*, rock, and *kopto*, to break, alluding to their habit of growing in rock crevices. The equivalent derivation in Latin is found in the genus name *Saxifraga*.) A genus of 7 species of perennial, tussock-forming herbs with simple leaves and flowers with 5 sepals, 5 petals, and 5 styles. This genus is closely related to the genus *Silene*. Several species are grown as rock-garden ornamentals. *Distr.* Pyrenees, mountains of N Spain. *Fam.* Caryophyllaceae. *Class* D.

P. glaucifolia Tuft-forming. Flowers pink-purple. *Distr.* Pyrenees, mountains of N Spain.

P. pyrenaica Rosette-forming. Flowers white or pale pink. *Distr.* W Pyrenees.

Petrocosmea (From Greek *petros*, rock, and *kosmos*, ornament; the plants grow on rocks.) A genus of 29 species of rhizomatous herbs with rosettes of simple, stalked leaves and 5-lobed, 2-lipped, tubular flowers on erect stems. Several species are grown as ornamentals. *Distr.* E Asian mountains. *Fam.* Gesneriaceae. *Class* D.

P. kerrii Leaves to 10cm long. Flowers small, white, blotched yellow. *Distr.* Thailand.

Petrophytum (From Greek *petros*, rock, and *phyton*, plant, alluding to the normal habitat of these plants.) ROCK SPIRAEA. A genus of 3 species of tuft-forming evergreen shrubs with crowded leathery leaves and dense racemes of white flowers. Several species are grown as ornamentals. *Distr.* W North America. *Fam.* Rosaceae. *Class* D.

P. cinerascens Small dense shrub. Foliage grey-green. *Distr.* USA (Washington State).

Petrorhagia (From Greek *petros*, rock, and *rhagas*, chink, alluding to the principal habitat of these species.) A genus of 25 species of annual and perennial herbs with simple opposite leaves and panicles of numerous small flowers that bear 5 sepals and 5 petals. This genus is closely related to the genus *Gypsophila* but shows some resemblance to *Dianthus* in having several sepal-like bracts below each flower. *Distr.* Europe, Asia, N Africa. *Fam.* Caryophyllaceae. *Class* D.

P. saxifraga Perennial mat-forming herb. Leaves grass-like. Flowers numerous, white or pale pink. Grown as a rock-garden ornamental. *Distr.* Central and S Europe to Central Asia.

Petroselinum (From Greek *petros*, rock, and the genus name *Selinum*.) PARSLEY. A genus of 3 species of biennial herbs with pinnately divided leaves and compound umbels of white, yellow, or pink flowers. *Distr.* Europe, Mediterranean regions. *Fam.* Umbelliferae. *Class* D.

P. crispum PARSLEY. Grown as a culinary herb and for its fruits which are used medicinally. The leaves of cultivated forms are wavy (or crisped) while those of the wild form are smooth. Numerous cultivars of this species have been raised. *Distr.* Europe, W Asia.

pe tsai *Brassica pekinensis.*

Petteria (After Franz Petter (1798–1853), professor at Spalato, Croatia.) A genus of 1 species of upright shrub with trifoliate leaves and clusters of yellow pea-like flowers. *Distr.* Balkans. *Fam.* Leguminosae. *Class* D.

P. ramentacea Grown as a hardy ornamental.

pettigrue *Ruscus aculeatus.*

petty whin *Genista anglica.*

Petunia (From *petun*, the Brazilian vernacular name for tobacco.) A genus of 35 species of annual and perennial herbs with simple alternate leaves and solitary, salver- or funnel-shaped, 5-lobed flowers. Several species and numerous cultivars are grown as ornamentals, typically as annuals for their colourful flowers. *Distr.* Tropical and warm regions of South America. *Fam.* Solanaceae. *Class* D.

P. axillaris LARGE WHITE PETUNIA. Annual herb to 60cm high. Flowers white, fragrant at night. *Distr.* S South America.

P. × hybrida PETUNIA. This is a complex group of hybrids thought to be derived from *P. axillaris* × *P. integrifolia* that has given rise to over 200 ornamental cultivars. *Distr.* Garden origin.

P. integrifolia VIOLET-FLOWERED PETUNIA. Annual herb or perennial subshrub. Flowers violet, tinged with red outside. *Distr.* Argentina.

petunia *Petunia x hybrida* **large white** ~ *P. axillaris* **Mexican** ~ *Strobilanthes atropurpureus* **violet-flowered** ~ *Petunia integrifolia.*

Pfeiffera (After Louis Carl George Pfeiffer (1805–77), German botanist.) A genus of 1 species of terrestrial or epiphytic, shrubby cactus with spreading or pendent stems and small, bell-shaped, purple-tinged flowers. *Distr.* N Argentina. *Fam.* Cactaceae. *Class* D.

P. ianthothele Grown as an ornamental.

Phacelia (From Greek *phakelos*, bundle, alluding to the clustered flowers.) A genus of 150 species of annual to perennial herbs with pinnately lobed or divided leaves and cymes or racemes of 5-lobed, tubular flowers. *Distr.* W North America. *Fam.* Hydrophyllaceae. *Class* D.

P. campanularia CALIFORNIAN BLUEBELL. Annual. Flowers bell-shaped, dark blue, spotted white. *Distr.* W USA.

P. tanacetifolia FIDDLENECK, Annual. Flowers bell-shaped, mauve to lilac. Grown as bee-fodder in Europe. *Distr.* SW North America, Mexico.

P. viscida Densely hairy annual. Flowers broadly bell-shaped, blue, flushed red at base. *Distr.* USA (California).

phaeacanthus, -a, -um: with dark-coloured spines.

phaeantherus, -a, -um: with dark-coloured anthers.

Phaedranassa (From Greek *phaedros*, bright, and *anassa*, queen, alluding to the beauty of the flowers.) QUEEN LILY. A genus of 6 species of bulbiferous perennial herbs with stalked oblong leaves and drooping, funnel-shaped flowers. Several species are grown as half-hardy ornamentals. *Distr.* South America. *Fam.* Amaryllidaceae. *Class* M.

P. carmioli Flowers crimson with a yellow fringe. *Distr.* South America.

Phaedranthus See *Distictis*

phaeochrysus, -a, -um: dark golden-coloured.

phaeus, -a, -um: dark-coloured.

Phaiophleps See *Olsynium*

Phaius (From Greek *phaios*, grey, alluding to the flowers which darken with age.) A genus of 30–50 species of terrestrial orchids with a few large leaves and erect racemes of showy flowers. Some species are grown as tender ornamentals. *Distr.* Tropical Africa to Australia. *Fam.* Orchidaceae. *Class* M.

P. humblotii Flowers pink, blotched white and red. *Distr.* Madagascar.

P. mishmensis Flowers purple-brown, lip pink-white. *Distr.* NE India, Burma, Thailand, Taiwan, Philippines.

P. tankervilliae (After Lady Emma Tankerville (*c.*1750–1836).) Inflorescence of brown and yellow flowers to 1m high. *Distr.* Central China, N India, Sri Lanka, SE Asia, N Australia.

Phalaenopsis (From Greek *phalaina*, moth, and the ending *-opsis*, indicating resemblance, alluding to the flowers.) MOTH ORCHID. A genus of 40–55 species of epiphytic orchids with purple-tinged leaves and moth-like flowers with a 3-lobed lip. A number of species and hybrids are grown as tender ornamentals. *Distr.* Himalaya, SE Asia, N Australia. *Fam.* Orchidaceae. *Class* M.

P. amabilis Inflorescence to 1m long, arching. Flowers fragrant. *Distr.* E Indies to N Australia.

P. amboinensis Flowers few, orange or cream to yellow. *Distr.* Indonesia (Amboina).

P. aphrodite See *P. amabilis*

P. cornu-cervi Flowers waxy, yellow-green, marked red, lip fleshy, white. *Distr.* SE Asia.

P. equestris Flowers white-pink, lip deeper pink or purple. *Distr.* Philippines, Taiwan.

P. fasciata Flowers fleshy, yellow, marked red. *Distr.* Philippines.

P. fimbriata Flowers numerous, white-cream, lip fringed. *Distr.* Sumatra, Java.

P. heiroglyphica Flowers white, lined red-purple. *Distr.* Philippines.

P. lueddemanniana (After Lueddemann, a Paris orchid-grower.) Raceme to 30cm. Petals white, lip red. *Distr.* Philippines.

P. maculata Flowers few, white. *Distr.* Malay peninsula, Sarawak.

P. mannii Flowers numerous, fragrant. *Distr.* Himalaya, Vietnam.

P. mariae Flowers cream-white, marked red, lip purple-mauve. *Distr.* Philippines.

P. parishii (After Charles Samuel Parish, an army surgeon who collected orchids in Burma.) Flowers small, white, lip purple. *Distr.* E Himalaya, India, Burma.

P. sanderiana (After M. C. F. Sander (1847–1920), who had orchid nurseries in England and Belgium.) Flowers variable, white and pink. *Distr.* Philippines.

P. schilleriana (After Herr Schiller of Hamburg, an orchid-grower.) Inflorescence pendent. Flowers very numerous, fragrant, pink-white. *Distr.* Philippines.

P. stuartiana Flowers white, lip spotted orange. *Distr.* Philippines.

P. sumatrana Flowers white-yellow, lip striped red. *Distr.* Thailand, Malay Peninsula, Java, Sumatra, Borneo.

P. violacea Small. Flowers green-white, lip tipped violet. *Distr.* Malay peninsula, Borneo, Sumatra.

phalaenopsis: moth-like.

Phalaris (Classical Greek name for a grass.) A genus of 15 species of annual and perennial grasses with flat leaves and compact inflorescences. Several species are important as fodder and grain crops. A few are grown as ornamentals. *Distr.* Temperate regions, chiefly around the Mediterranean. *Fam.* Gramineae. *Class* M.

P. arundinacea REED CANARY GRASS. Used as fodder grass and for erosion control. *Distr.* Europe, Asia, North America, S Africa.

P. canariensis CANARY GRASS. The canary seed of commerce. *Distr.* Canary Islands, N Africa, W Mediterranean.

Phanerophlebia (From Greek *phaneros*, clear, and *phlebion*, vein.) A genus of 20 species of terrestrial ferns with papery to leathery leaves. Several species are grown as half-hardy ornamentals. *Distr.* Montane tropical regions. *Fam.* Dryopteridaceae. *Class* F.

P. juglandifolia Leaves to 90cm long, pinnate. *Distr.* Mexico to Venezuela.

Pharbitis See *Ipomoea*

Phaseolus (Latin diminutive of the Greek name *phaselos*) that was applied to a type of bean.) BEAN. A genus of 50 species of annual and perennial, typically climbing herbs with trifoliate leaves and irregular pea-like flowers. Several species are grown for their edible seeds, beans. *Distr.* Tropical America. *Fam.* Leguminosae. *Class* D.

P. coccineus RUNNER BEAN, DUTCH CANE KNIFE BEAN. Perennial climbing herb. Flowers red, borne in many-flowered racemes. Pods are a popular vegetable. *Distr*. Tropical America.

P. lunatus LIMA BEAN. Erect or twining annual. Flowers yellow-green. Pods to 10cm long with 2–4 seeds. Seeds eaten but must be cooked. Numerous cultivars available. *Distr*. Tropical South America.

P. vulgaris FRENCH BEAN, KIDNEY BEAN, HARICOT BEAN, CANELLINI BEAN. Annual, climbing or erect herb. A very important food plant, the source of numerous cultivars which provide us with many different edible beans including the ubiquitous 'baked bean'. *Distr*. Tropical America.

pheasant's eye *Narcissus poeticus*, *Adonis*.

Phegopteris (From Greek *phegos*, acorn, and *pteris*, fern.) A genus of 3 species of medium-sized terrestrial ferns with pinnately divided leaves. *Distr*. Temperate Eurasia. *Fam*. Thelypteridaceae. *Class* F.

P. connectilis BEECH FERN, LONG BEECH FERN. Leaves emerald green, to 20cm long. Grown as an ornamental. *Distr*. Eurasia.

Phellinaceae A family of 1 genus and 12 species of trees with simple leaves and clusters of small flowers. This family is sometimes included within the family Aquifoliaceae. *Distr*. New Caledonia. *Class* D.

Phellodendron (From Greek *phelos*, cork, and *dendron*, tree, alluding to the corky bark.) A genus of 10 species of tall, aromatic, deciduous trees with pinnate, opposite leaves and panicles of small yellow-green flowers. Several species are grown as ornamentals and some are used locally as a source of timber and medicines. *Distr*. E Asia. *Fam*. Rutaceae. *Class* D.

P. amurense Tree to 15m high. Bark thick, corky. *Distr*. NE Asia.

P. chinense Tree to 10m high. Bark thin. *Distr*. China.

phellomanus, -a, -um: with corky shoots.

phellos: corky.

philadelphicus, -a, -um: of Philadelphia, USA.

Philadelphus (Greek name meaning brotherly love.) MOCK ORANGE. A genus of 65 species of deciduous shrubs with peeling bark, simple opposite leaves, and strongly scented flowers that bear 4 sepals, 4 petals, and numerous stamens. Numerous species, hybrids, and cultivars are grown as ornamentals for their fragrant flowers. *Distr*. N temperate regions. *Fam*. Hydrangeaceae. *Class* D.

P. coronarius MOCK ORANGE. Shrub to 3m high. Flowers cream-white, very fragrant. The flowers are used in garlands and numerous cultivars have been raised. *Distr*. SE Europe, W Asia.

P. delavayi (After Abbè Jean Marie Delavay (1838–95), a French missionary in China who introduced it into cultivation in 1887.) Shrub to 4m high. Flowers pure white, in short racemes. *Distr*. W China, SE Tibet, N Burma.

P.* × *lemoinei A hybrid of *P. coronarius* x *P. microphyllus*. Low compact shrub. Petals notched. The source of numerous hybrids. *Distr*. Garden origin.

×*Philageria* (From the names of the parent genera.) A genus of hybrids of members of the genera *Lapageria* and *Philesia*. *Distr*. Garden origin. *Fam*. Philesiaceae. *Class* M.

×*P. veitchii* A hybrid of *Lapageria rosea* and *Philesia magellanica*. Scrambling shrub with leathery leaves and pendent bell-shaped flowers that bear 3 fleshy, dull red, outer tepals and 3 longer, bright red, inner tepals. *Distr*. Garden origin.

Philesia (From Greek *phileo*, to love, alluding to the attractive flowers.) A genus of 1 species of evergreen shrub with small, crowded, leathery leaves and narrow, bell-shaped, pendulous flowers that bear 3 pink outer tepals and 3 purple-red inner tepals. *Distr*. Chile. *Fam*. Philesiaceae. *Class* M.

P. buxifolia See *P. magellanica*

P. magellanica A very attractive half-hardy ornamental.

Philesiaceae A family of 6 genera and 9 species of shrubs and climbers with simple leathery leaves and flowers that bear 6 free tepals, 6 stamens, and a superior ovary of 3 carpels. This family is sometimes included within the Liliaceae. *Distr*. Temperate S hemisphere. *Class* M.

See: Eustrephus, Geitonoplesium, Lapageria, Luzuriaga, X Philageria, Philesia.

philippinensis, -e: of the Philippines.

Phillyrea (Classical Greek name for these plants.) A genus of 4 species of evergreen shrubs and trees with simple opposite leaves and clusters of small, 4-lobed, fragrant flowers. Several species are grown as ornamentals and some are used locally for timber. *Distr.* Mediterranean, W Asia. *Fam.* Oleaceae. *Class* D.

P. angustifolia Bushy shrub. *Distr.* S Europe, N Africa.

P. latifolia Tree or large shrub. *Distr.* Mediterranean regions.

P. media See *P. latifolia*

Philodendron (From Greek *phileo*, to love, and *dendron*, tree, alluding to their tree-climbing habit.) A genus of about 500 species of epiphytic or terrestrial lianas and shrubs with large, entire to deeply lobed leaves. The small unisexual flowers are borne on a spadix that is surrounded at the base by an often fleshy spathe. Several species are cultivated as tender ornamentals, others have local medicinal or food value, but as house-plants they have been the cause of poisoning in children and pets. *Distr.* Tropical America. *Fam.* Araceae. *Class* M.

P. bipinnatifidum Large tree-like shrub with bipinnate leaves. *Distr.* SE Brazil.

P. domesticum SPADE-LEAF PHILODENDRON. Slow-growing climber with bright green, arrow-shaped leaves. Variegated cultivars available. *Distr.* Known only in cultivation.

P. epipremnum See *Epipremnum pinnatum*

P. erubescens BLUSHING PHILODENDRON, RED-LEAF PHILODENDRON. Underside of leaves and young growth tinged purple-red. *Distr.* Colombia.

P. laciniatum See *Philodendron pedatum*

P. melanochrysum BLACK-GOLD PHILODENDRON. Leaves copper at first, large, velvety, dark green when mature. *Distr.* Colombia.

P. pedatum Leaves oval, deep green, pinnately cut into 5–7 lobes. *Distr.* S Venezuela to Surinam and SE Brazil.

P. pinnatifidum Stem erect. Leaves to 60cm long, pinnately cut. Spathe limb white, boat-shaped. *Distr.* Venezuela, Trinidad.

P. sagittifolium Slow-growing climber. Leaves bright green. *Distr.* S Mexico.

P. scandens HEART-LEAF PHILODENDRON. A slender climber with luscious green, heart-shaped leaves. One of the most commonly cultivated Philodendrons; several subspecies and cultivars are available. *Distr.* Mexico and W Indies to N South America.

P. selloum See *Philodendron bipinnatifidum*

philodendron: black-gold ~ *Philodendron melanochrysum* **blushing** ~ *P. erubescens* **heart-leaf** ~ *P. scandens* **red-leaf** ~ *P. erubescens* **spade-leaf** ~ *P. domesticum*.

Philydraceae A family of 3 genera and 4 species of perennial rhizomatous herbs with linear leaves and solitary irregular flowers. *Distr.* Tropical E Asia to Australia. *Class* M.

Phlebodium (Diminutive of the Greek *phlebion*, vein.) A genus of 10 species of epiphytic ferns with pinnately lobed, leathery leaves. *Distr.* Tropical regions of the New World. *Fam.* Polypodiaceae. *Class* F.

P. aureum GOLDEN POLYPODY, RABBIT'S FOOT FERN. Fronds to 1m long, grey-green. *Distr.* Tropical regions of the New World.

Phleum (From Greek *phleos*, a type of reed.) CAT-TAIL GRASS. A genus of 15 species of annual and perennial grasses with flat leaves and cylindrical or rounded, spike-like inflorescences. *Distr.* Temperate regions, chiefly Europe and the Mediterranean. *Fam.* Gramineae. *Class* M.

P. pratense TIMOTHY, MEADOW CAT'S TAIL. Perennial to 1.5m high. A fodder grass that first became important in North America; it was introduced into the British Isles in the late 18th century by Timothy Hansen, hence its common name. It is now grown as an ornamental. *Distr.* Europe.

phlogopappa: with phlox-like leaves.

Phlomis (Classical Greek name for another plant, possibly mullein.) A genus of 100 species of perennial herbs and evergreen shrubs with narrow, often woolly, opposite leaves and whorls of 2-lipped flowers. *Distr.* Mediterranean to Central Asia and China. *Fam.* Labiatae. *Class* D.

P. cashmeriana Erect evergreen shrub. Flowers numerous, pale-lilac. *Distr.* Kashmir.

P. chrysophylla Rounded evergreen shrub. Flowers golden-yellow. *Distr.* W Asia.

P. fruticosa JERUSALEM SAGE. Spreading subshrub. Flowers deep golden-yellow. *Distr.* Mediterranean regions.

P. italica Erect subshrub. Flowers pink to lilac. *Distr.* Balearic Islands.

P. longifolia Evergreen bushy shrub. Flowers deep yellow. *Distr.* SW Asia.

P. russeliana Evergreen sub-shrub. Leaves large, rough. Flowers in whorls on long stems, deep yellow. *Distr.* W Asia.

phlomoides: from the genus *Phlomis*, with the ending *-oides*, indicating resemblance.

Phlox (From Greek *phlox*, flame.) A genus of 67 species of annual and perennial herbs and shrubs with simple, typically opposite leaves and clustered or solitary, salver-shaped flowers that bear their parts in fives. Many species and numerous cultivars are grown as ornamentals. *Distr.* North America, NE Asia. *Fam.* Polemoniaceae. *Class* D.

P. drummondii (After Thomas Drummond (*c.*1790–1835), who collected in North America.) PRIDE OF TEXAS, ANNUAL PHLOX, DRUMMOND PHLOX. Annual herb to 50cm high. Numerous cultivars with different coloured flowers have been raised from this species. *Distr.* USA (Texas).

P. paniculata PERENNIAL PHLOX, SUMMER PHLOX, AUTUMN PHLOX, FALL PHLOX. Erect perennial herb. Flowers fragrant. Over 100 cultivars of this species are currently available. *Distr.* E USA.

P. subulata ALEXANDER'S SURPRISE, MOSS PHLOX, MOUNTAIN PHLOX, MOSS PINK. Mat-forming perennial. Flowers pink, lavender or white. Grown as a rock garden ornamental. *Distr.* North America.

phlox Polemoniaceae **annual** ~ *Phlox drummondii* **autumn** ~ *P. paniculata* **Drummond** ~ *P. drummondii* **fall** ~ *P. paniculata* **moss** ~ *P. subulata* **mountain** ~ *P. subulata* **night** ~ *Zaluzianskya capensis* **perennial** ~ *Phlox paniculata* **summer** ~ *P. paniculata*.

Phoebe (From a Greek female name.) A genus of 150 species of evergreen shrubs and trees with simple alternate leaves and panicles of small fragrant flowers. *Distr.* E Asia, tropical and subtropical America. *Fam.* Lauraceae. *Class* D.

P. formosana Small tree. Flowers white. Flowers purple-black. Grown as a tender ornamental. *Distr.* Taiwan, China.

P. shearii See *P. formosana*

Phoenicaulis (From Greek *phoinix*, purple-red, and *kaulos*, stem.) A genus of 1 species of tufted perennial herb with small, pink-purple, occasionally white flowers. *Distr.* Mountains of W North America. *Fam.* Cruciferae. *Class* D.

P. cheiranthoides Cultivated as an ornamental rock-garden plant. *Distr.* Mountainous areas of W North America.

phoeniceus, -a, -um: scarlet red-coloured.

phoenicolasius, -a, -um: with purple hairs.

Phoenix (Greek name for the date plam.) DATE. A genus of 17 species of small to medium-sized palms with feather-shaped leaves and erect or arching inflorescences of yellow or orange flowers. *Distr.* Drier tropical and subtropical areas of the Old World. *Fam.* Palmae. *Class* M.

P. canariensis CANARY ISLAND DATE PALM. Trunk to 15m high. Leaves to 6m long. Fruit yellow, tinged red. *Distr.* Canary Islands, widely cultivated elsewhere.

P. dactylifera DATE PALM. The fruit is the staple of many nomadic peoples in Arabia and N Africa and the tree has been cultivated for at least 5000 years. Today it is a commercial crop in Iraq, N Africa, and California. The leaves are used locally for matting and thatch. *Distr.* W Asia, N Africa,.

P. reclinata DWARF DATE PALM. Stems short. Grown as a hedge plant because of its vigorous suckering habit. *Distr.* Africa.

P. roebelenii MINIATURE DATE PALM, PYGMY DATE PALM, ROEBELIN PALM. Grown as a pot plant. *Distr.* Laos.

P. sylvestris WILD DATE, INDIAN DATE. Cultivated as a source of palm sugar and molasses. *Distr.* India.

P. threophrasti A species closely related to *P. dactylifera* that is endangered in the wild. *Distr.* Turkey, Crete.

phoenix tree *Firmiana simplex*.

Pholidota (From Greek *pholidos*, scale, and *karpos*, fruit, alluding to the scaly covering of the fruit.) RATTLESNAKE ORCHID. A genus of 40 species of epiphytic or terrestrial orchids with solitary or paired leaves and racemes that bear 2 ranks of overlapping bracts and small white flowers. *Distr.* India, China, and SE Asia. *Fam.* Orchidaceae. *Class* M.

P. articulata Flowers numerous, fragrant. *Distr.* Himalaya, SE Asia.

Phormiaceae A family of 7 genera and about 40 species of rhizomatous herbs and subshrubs with grass-like or sword-shaped leaves. The regular flowers have 6 free tepals, 6 free stamens, a superior ovary, and are borne in open panicles. This family is sometimes included in the family Liliaceae. *Distr.* Asia, Australasia and South America. *Class* M. *See: Dianella, Phormium, Stypandra, Xeronema.*

Phormium (From Greek *phormion*, mat, alluding to the use of the fibre.) FLAX LILY. A genus of 2 species of rhizomatous herbs with 2-ranked, sword-shaped leaves and erect panicles of brown to yellow-green flowers that bear 6 regular free tepals. *Distr.* New Zealand. *Fam.* Phormiaceae. *Class* M.

P. colensoi MOUNTAIN FLAX. Leaves to 2.5m long. A popular ornamental with numerous cultivars bearing strikingly coloured leaves available. *Distr.* New Zealand.

P. cookianum See *P. colensoi*

P. tenax NEW ZEALAND FLAX. Leaves to 4.5m long. The leaves are a source of bush flax used by Maoris for textiles, cordage, and nets. There are also a number of variegated ornamental cultivars available. *Distr.* New Zealand.

Photinia (From Greek *photos*, light, alluding to the shiny leaves of some species.) CHRISTMAS BERRY. A genus of 40 species of deciduous and evergreen shrubs and trees with simple alternate leaves and clusters of white flowers that are followered by round red 'fruits'. *Distr.* Himalaya to Japan and Sumatra. *Fam.* Rosaceae. *Class* D.

P. arbutifolia See *Heteromeles arbutifolia*

P. beauverdiana (After Gustav Beauverd (1867–1942), botanist and artist.) Deciduous shrub or narrow tree. *Distr.* W China.

P. davidiana (After Armand David (1826–1900), French missionary who introduced it to cultivation from China in 1869.) Large shrub or small tree, evergreen. *Distr.* China.

P. × fraseri (After the Fraser nurseries, Alabama, where it was raised.) A group of evergreen hybrids from *P. glabra* x *P. serrulata* that have given rise to a number of ornamental cultivars. *Distr.* Garden origin.

P. serratifolia Evergreen shrub or tree to 12m high. *Distr.* China.

P. serrulata See *P. serratifolia*

P. villosa Erect deciduous shrub. Noted for its brilliant autumn colour. *Distr.* Japan, China, Korea.

Phragmipedium (From Greek *phragma*, partition, and *pedilon*, slipper, alluding to the slipper-shaped lip and the 3-locular ovary.) LADY'S SLIPPER. A genus of 15–20 species of terrestrial, epiphytic or lithophytic orchids with leathery leaves. The flowers are large and showy with a pouch-shaped lip. Pollination occurs though a capture-and-release mechanism whereby an insect is attracted to the fleshy column and on alighting slips into the pouched lip. The only escape route available is through an aperture at the base of the column, where pollinia become attached to the prisoner's back. This genus has previously been included within the genera *Cypripedium* and *Paphiopedilum*. Some species are cultivated as tender ornamentals. *Distr.* Central and South America. *Fam.* Orchidaceae. *Class* M.

P. caricinum Flowers bronze, edged purple. *Distr.* Peru, Bolivia and Brazil.

P. caudatum Flowers largest in genus, to 1.25m across, lip undulate. *Distr.* Mexico to Peru, Ecuador, Colombia, and Venezuela.

P. lindenii Flowers very large, yellow-green. *Distr.* Colombia, Ecuador, and Peru.

P. lindleyanum Flowers green-yellow. *Distr.* Venezuela, Guyana.

P. longifolium Flowers green-yellow, with rose margins. *Distr.* Costa Rica to Colombia.

P. schlimii Flowers white, flushed pink, lip rose-pink. *Distr.* Colombia.

Phragmites (From Greek *phragma*, partition or screen; it is sometimes planted along ditches to form a fence.) REED. A genus of 3–4 species of fast-growing, rhizomatous, perennial grasses with robust, hollow stems to 3m high, flat leaves, and large, plumed, terminal inflorescences. They are characteristic of fens and riversides. *Distr.* Temperate and tropical regions. *Fam.* Gramineae. *Class* M.

P. australis REED. A very fast-growing plant that forms floating fens. The stems are harvested for cellulose, paper pulp, and numerous other purposes.

P. communis See *P. australis*

Phrymaceae A family of 1 genus and 2 species of erect perennial herbs with simple opposite leaves and irregular flowers in spike-like racemes. *Distr.* India to Japan and North America. *Class* D.

Phuopsis (From *phou*, the classical Greek name for a species of *Valeriana*, and the ending *-opsis*, indicating resemblance.) A genus of 1 species of mat-forming perennial herb with whorls of simple leaves and a profusion of tiny fragrant flowers. *Distr.* Caucasus. *Fam.* Rubiaceae. *Class* D.

P. stylosa Grown as ornamental ground cover. *Distr.* Caucasus.

Phygelius (Probably from Greek *phyge*, flight, and *helios*, the sun.) A genus of 2 species of evergreen or semi-evergreen shrubs and subshrubs with simple toothed leaves and 1-sided inflorescences of tubular 5-lobed flowers. Both species and numerous cultivars are grown as ornamentals. *Distr.* South Africa. *Fam.* Scrophulariaceae. *Class* D.

P. aequalis Flowers pink with a yellow throat and red lobes. *Distr.* South Africa.

P. capensis Flowers orange to deep red, lobes reflexed. *Distr.* South Africa.

P. × rectus A hybrid of *P. aequalis* × *P. capensis*. Numerous cultivars have been raised from this cross. *Distr.* Garden origin.

Phyla (From Greek *phyla*, tribe, alluding to the compound flower heads.) A genus of 15 species of low perennial herbs and subshrubs with simple opposite leaves and dense spikes of small tubular flowers. *Distr.* Tropical and warm-temperate regions. *Fam.* Verbenaceae. *Class* D.

P. nodiflora Creeping herb. Flowers white with a yellow centre. Grown as tender ornamental ground-cover. *Distr.* Tropical and subtropical regions.

phyllanthoides: from the genus *Phyllanthus*, with the ending *-oides*, indicating resemblance.

× Phylliopsis (From the names of the parent genera.) A genus of hybrids between members of the genera *Phyllodoce* and *Kalmiopsis*. *Fam.* Ericaceae. *Class* D.

×P. hillieri (After Messrs Hillier and Sons, in whose nursery it was produced.) An evergreen shrub from *Kalmiopsis leachiana* × *Phyllodoce breweri*.

Phyllitis See *Asplenium*

phyllochlamys with a leafy covering.

Phyllocladaceae A family of 1 genus and 4 species. *Class* G.
See: Phyllocladus.

Phyllocladus (From Greek *phyllon*, leaf, and *klados*, a branch, alluding to the flattened leaf-like branches.) CELERY VINE. A genus of 4 species of evergreen, coniferous shrubs and trees with tiny scale-like leaves and flattened, leaf-like shoots. *Distr.* Australasia, Malaysia. *Fam.* Phyllocladaceae. *Class* G.

P. trichomanoides TANEKAHA. Tree to 20m high Occasionally grown as a half-hardy ornamental. Used as a source of red dye and timber by Maoris. *Distr.* New Zealand.

Phyllodoce (After Phyllodoce, a sea nymph in Greek mythology.) A genus of 6–7 species of dwarf evergreen shrubs with leathery leaves and bell- or urn-shaped flowers. *Distr.* Arctic and Alpine regions of the N Hemisphere. *Fam.* Ericaceae. *Class* D.

P. aleutica Scrambling or horizontal shrub. *Distr.* Japan to USA (Alaska).

P. breweri (After William Henry Brewer (1828–1910), professor of agriculture at Yale University, who worked on the botany of California.) PURPLE HEATHER, BREWER'S MOUNTAIN HEATHER. Semi-prostrate shrub. *Distr.* USA (California).

P. caerulea Erect or scrambling shrub. A protected species in Great Britain. *Distr.* Arctic and Alpine N Hemisphere.

P. empetriformis PINK MOUNTAIN HEATHER. Low, diffuse, mat-forming shrub. *Distr.* W North America.

P. glanduliflora YELLOW MOUNTAIN HEATHER. Erect shrub with fragrant flowers. *Distr.* W North America.

P. × intermedia A hybrid of *P. empetriformis* × *P. glanduliflora*. *Distr.* W North America.

P. nipponica Sub-erect shrub. *Distr.* Japan.

Phyllostachys (From Greek *phyllon*, leaf, and *stachys*, spike, alluding to the leafy inflorescences.) BAMBOO. A genus of 60 species of medium to large, rhizomatous bamboos with grooved stems, zigzagging branches, and narrow-linear leaves. The stems are used as timber and for making paper. The young shoots are edible. *Distr.* India, China, Burma. *Fam.* Gramineae. *Class* M.

P. aurea FISH-POLE BAMBOO, GOLDEN BAMBOO. Stems growing to 10m high in the wild. The stems are use for fishing-rods and plant stakes. One of the more commonly cultivated hardy ornamentals bamboos. *Distr.* SE China.

P. aureosulcata YELLOW GROOVE BAMBOO. Grown as a half-hardy ornamental. *Distr.* NE China.

P. bambusoides GIANT TIMBER BAMBOO. The most important timber bamboo in Asia, capable of far higher production rates than PINE. The edible shoots are often used in Chinese cooking. It is also frequently grown as a hardy ornamental with numerous cultivars available. *Distr.* China.

P. nigra BLACK BAMBOO, WANGEE CANE. Stems relatively thin, turning black in second season. Grown as a hardy ornamental. *Distr.* E China.

phyllostachyus, -a, -um: with leafy spikes.

× *Phyllothamnus* A genus of hybrids of members of the genera *Phyllodoce* and *Rhodothamnus*. *Fam.* Ericaceae. *Class* D.

×*P. erectus* A hybrid of *Phyllodoce empetriformis* × *Rhodothamnus chamaecistus*.

Phymosia A genus of 8 species of shrub and small trees with simple, palmately lobed leaves and regular, bisexual, pink or red flowers that bear 5 petals and numerous fused stamens. *Distr.* Central America. *Fam.* Malvaceae. *Class* D.

P. abutiloides BAHAMAS PHYMOSIA. Shrub to 3m high. Flowers pink with white veins and a dark centre. Petals notched. Grown as a tender ornamental. *Distr.* Bahamas.

phymosia, Bahamas *Phymosia abutiloides*.

Physalis (From Greek *physa*, bladder, alluding to the inflated whorls of sepals that surround the fruit.) A genus of 80 species of annual and perennial herbs with simple to pinnately lobed leaves and solitary bell-shaped flowers. The fruit is a small yellow to green or purple berry surrounded by a papery, often coloured whorl of sepals. Several species are grown as ornamentals. *Distr.* Widespread especially in the Americas. *Fam.* Solanaceae. *Class* D.

P. alkekengi CHINESE LANTERN, WINTER CHERRY, BLADDER CHERRY. Rhizomatous herbs. Sepals bright red in fruit. The berry is edible and was formerly used medicinally. *Distr.* SE Europe to Japan.

P. peruviana CAPE GOOSEBERRY, PURPLE GROUND CHERRY. Perennial herb. Berry purple, edible. *Distr.* South America.

P. pruinosa STRAWBERRY TOMATO, GROUND CHERRY, DWARF CAPE GOOSEBERRY. Fruit used to make jams, jellies, and tarts. Gathered from the wild in N America, planted in Europe. *Distr.* E and Central N America.

physalodes: resembling the genus *Physalis*.

Physaria (From Greek *physa*, bladder, alluding to the inflated fruit.) BLADDERPOD. A genus of 14 species of perennial herbs with small, typically yellow flowers. *Distr.* W North America. *Fam.* Cruciferae. *Class* D.

P. alpestris Cultivated as an ornamental rock-garden plant.

Physenaceae A family of 1 genus and 2 species of shrubs and trees. This family is sometimes included within the family Capparidaceae. *Distr.* Madagascar. *Class* D.

Physocarpus (From Greek *physa*, bladder, and *karpon*, fruit, alluding to the inflated fruits.) NINEBARK. A genus of 10 species of deciduous shrubs with peeling bark, large, palmately lobed leaves, and clusters of white or pink flowers that are followed by whorls of inflated fruits. Several species are grown as ornamentals. *Distr.* North America, NE Asia. *Fam.* Rosaceae. *Class* D.

P. opulifolius Dense arching shrub. Flowers tiny, white. *Distr.* E North America.

P. ribesifolius See *P. opulifolius*

physocarpus, -a, -um: with bladdery fruit.

Physoplexis (From Greek *physa*, bladder, and *plexis*, plaiting.) A genus of 1 species of tuft-forming perennial herb with deeply toothed leaves and rounded heads of purple-blue, bottle-shaped flowers. *Distr.* Alps. *Fam.* Campanulaceae. *Class* D.

P. comosa DEVIL'S CLAW. Grown as a rock garden ornamental.

Physosiphon See *Pleurothallis*

Physostegia (From Greek *physa*, bladder, and *stege*, covering.) OBEDIENT PLANT, FALSE DRAGON-HEAD. A genus of 12 species of erect perennial herbs with opposite simple leaves and branched racemes of 2-lipped flowers. Several species are grown as annuals. *Distr.* North America. *Fam.* Labiatae. *Class* D.

P. virginiana OBEDIENT PLANT. If the flowers are moved sideways they do not spring back to their original position. *Distr.* E North America.

Phyteuma (Classical Greek name for an aphrodisiac plant, possibly *Reseda phyteuma.*) HORNED RAMPION. A genus of 40 species of perennial herbs with simple alternate leaves and heads or spikes of blue bottle-shaped flowers. The tips of the petals are often fused into a tube that surrounds the anthers. Pollen is released when the style elongates and pushes it through and out of the tip of the tube. Several species are grown as hardy ornamentals. *Distr.* Mediterranean, Europe. *Fam.* Campanulaceae. *Class* D.

P. comosum See *Physoplexis comosa*

P. orbiculare RAMPION, ROUND-HEADED RAMPION. Stems erect to 50cm high. Flowers in dense spherical heads, dark blue. Roots sometimes eaten in salads. *Distr.* Europe.

P. scheuchzeri Leaves entire. Stems to 20cm high. Flowers in small heads, light to dark blue. *Distr.* Alps.

P. spicatum SPIKED RAMPION. Stems erect to 90cm high. Flowers in dense spikes, white to blue. *Distr.* Europe.

Phytolacca (From Greek *phyton*, plant, and Latin *lac*, red pigment.) POKEWEED. A genus of 25 species of herbs, shrubs, and trees with simple leaves and spike-like racemes of small flowers that are followed by fleshy berries. A red dye is obtained from the berries. *Distr.* Tropical and subtropical regions. *Fam.* Phytolaccaceae. *Class* D.

P. acinosa INDIAN POKEWEED. Perennial herb to 1.5m high. Flowers green-white. *Distr.* Himalaya to China and Japan.

P. americana POKEWEED, PIGEON BERRY, INKBERRY. Bad-smelling, perennial herb to 4m high. Flowers white-purple. The juice from the black berries is used as a dye and to help control bilharzia. *Distr.* E North America.

P. decandra See *P. americana*

Phytolaccaceae The Pokeweed family. 15 genera and about 60 species of herbs, woody climbers, shrubs, and trees with simple alternate leaves and small flowers that bear 4–5 tepals. Some species are grown as ornamentals, others have medicinal qualities or are a source of dye. *Distr.* Tropical and warm-temperate regions. *Class* D.
See: Ercilla, Phytolacca.

Picea (Classical Latin name for a pine that produced pitch, from *pix*, pitch.) SPRUCE. A genus of 40 species of evergreen coniferous trees with 4-sided, needle-like leaves that are joined to a distinct projection, known as the pulvinus. When they fall from the tree they leave this projection behind, making the leafless twigs rough. The woody fruiting cones mature in their first year and open to release their seeds. These trees are an important source of timber for chip-board, pulp, and cellulose. *Distr.* Temperate regions of the N hemisphere. *Fam.* Pinaceae. *Class* G.

P. abies NORWAY SPRUCE, CHRISTMAS TREE. Tree to 55m high. An important source of timber, resins, and tannins. Commonly brought indoors as a Christmas tree, although the needles soon fall. *Distr.* Europe.

P. asperata DRAGON SPRUCE. Tree to 40m high. *Distr.* NW China.

P. breweriana (After William Henry Brewer (1828–1910), American botanist, who collected it.) BREWER SPRUCE. Tree to 35m high. Branches of mature tree drooping. *Distr.* USA (California, Oregon).

P. glauca WHITE SPRUCE. Tree to 20m high. A source of timber and pulp. Several ornamental cultivars have been raised from this species. *Distr.* North America.

P. mariana BLACK SPRUCE, BOG SPRUCE. Tree to 20m high. A source of chewing gum, known as spruce gum, also grown as an ornamental. *Distr*. North America.

P. sitchensis SITKA SPRUCE. Fast-growing tree to 90m high. Much planted for timber and pulp. *Distr*. W North America.

pichi *Fabiana imbricata*.

pickerel weed Pontederiaceae, *Pontederia cordata*, *Potamogeton natans*.

Picrasma (From Greek *pikra*, bitter taste, alluding to the bitter leaves and wood.) A genus of 6 species of deciduous trees with alternate, pinnate leaves and cymes of small flowers that bear their parts in fours or fives. *Distr*. Tropical America. *Fam*. Simaroubaceae. *Class* D.

P. ailanthoides Tree to 14m high. Grown as a tender ornamental. *Distr*. Himalaya, E Asia.

P. quassioides See *P. ailanthoides*

Picris (Classical Greek name for a type of bitter herb.) OX-TONGUE, BITTERWEED. A genus of 45 species of annual to perennial herbs with milky sap, simple to pinnately lobed leaves, and yellow flowers in dandelion-like heads. *Distr*. Europe, Mediterranean, Asia, African mountains. *Fam*. Compositae. *Class* D.

P. echioides OX-TONGUE, BRISTLY OX-TONGUE. Spiny annual or biennial to 1m high. *Distr*. Mediterranean, naturalized elsewhere.

pictus, -a, -um: painted.

pie plant *Rheum* × *cultorum*.

Pieris (From Pierides, another name for the Muses (goddesses of the arts).) A genus of 7 species of evergreen shrubs, trees, and lianas with simple leathery leaves and white pitcher-shaped flowers. *Distr*. E Asia, E North America. *Fam*. Ericaceae. *Class* D.

P. floribunda FETTER BUSH. Evergreen bushy shrub. *Distr*. SE North America.

P. formosa Shrub or small tree. *Distr*. Himalaya and China.

P. japonica LILY OF THE VALLEY BUSH. Shrub or small tree to 4m high. *Distr*. Japan.

P. taiwanensis See *P. japonica*

piggy-back plant *Tolmiea menziesii*.

pignut *Carya glabra*, *Simmondsia chinensis*.

Pilea (From Latin *pileus*, cap, alluding to the enlarged sepal borne on the fruits of some species.) A genus of 600 species of annual and perennial herbs with simple opposite leaves and clusters of very small green flowers. Several species are grown for their ornamental foliage. *Distr*. Tropical regions. *Fam*. Urticaceae. *Class* D.

P. cardierei ALUMINIUM PLANT. Leaves with silvery stripes. *Distr*. Vietnam.

P. involucrata FRIENDSHIP PLANT. Trailing or erect herb. Leaves marked bronze. *Distr*. Central and South America.

P. microphylla ARTILLERY PLANT, GUNPOWDER PLANT, PISTOL PLANT. Annual or short-lived perennial herb. Leaves somewhat succulent. Anthers release their pollen explosively. *Distr*. Tropical America, from Mexico to Brazil.

P. nummulariifolia CREEPING CHARLIE. Trailing herb. Leaves pale green. *Distr*. Central and tropical South America.

pileatus, -a -um: cap-shaped.

Pileostegia See *Schizophragma*

pilewort *Ranunculus ficaria*.

Pilgerodendron uviferum See *Libocedrus uvifera*

pillwort *Pilularia*.

pilosulus, -a, -um: with small soft hairs.

pilosus, -a, -um: with long soft hairs.

Pilularia (From Latin *pilula*, diminutive of *pila*, ball.) PILLWORT. A genus of 6 species of aquatic to terrestrial ferns with pill-like sporocarps. *Distr*. Temperate regions. *Fam*. Marsileaceae. *Class* F.

P. globulifera Semi-aquatic creeping fern. Occasionally grown as an ornamental. *Distr*. W Europe.

pilulifera, -um: bearing small balls.

Pimelea (From Greek *pimele*, fat, alluding to the very oily seeds.) RICE FLOWER. A genus of 80 species of evergreen shrubs with small simple leaves and tubular flowers that are borne in heads surrounded by a whorl of leafy coloured bracts. Several species are grown as ornamentals. *Distr*. New Zealand, Australia, Timor, and Lord Howe Islands. *Fam*. Thymelaeaceae. *Class* D.

P. longiflora Upright evergreen shrub to 2m high. Flowers red. *Distr*. W Australia.

pimeleoides: from the genus *Pimelea*, with the ending *-oides*, indicating resemblance.

Pimenta (From Spanish *pimiento*, the local name for the most common species.) A genus of 2–5 species of aromatic trees with simple glandular leaves and small regular flowers that bear 5 petals and numerous stamens. *Distr.* Tropical America. *Fam.* Myrtaceae. *Class* D.

P. dioica ALLSPICE, PIMENTO, JAMAICA PEPPER. Whole tree smells of cloves, nutmeg and cinnamon. Spice derived from unripe fruit used as flavouring and medicinally. *Distr.* Central America.

pimento *Pimenta dioica.*

pimpernel *Anagallis* **blue** ~ *A. monelli* **bog** ~ *A. tenella* **common** ~ *A. arvensis* **scarlet** ~ *A. arvensis* **water** ~ *Samolus* **yellow** ~ *Lysimachia nemorum.*

Pimpinella (Derivation obscure, possibly from the mediaeval Latin *bipinella*, 2-winged.) A genus of 150 species of annual, biennial, and perennial herbs with simple or much-divided leaves and compound umbels of white or yellow to purple flowers. *Distr.* Eurasia, Africa. *Fam.* Umbelliferae. *Class* D.

P. anisum ANISE. Aromatic annual herb. Grown since ancient times as the source of the flavouring anise. *Distr.* Greece to Egypt.

P. major GREATER BURNET SAXIFRAGE. Perennial to 1m high. Leaves pinnate. Flowers white to pink. *Distr.* Europe.

P. saxifraga BURNET SAXIFRAGE. Perennial to 1m high. Leaves pinnate. Flowers white occasionally pink or purple. *Distr.* Europe.

pimpinellifolius, -a, -um: from the genus *Pimpinella*, and *folius*, leaved.

pimpinelloides: from the genus *Pimpinella*, and with the ending *-oides*, indicating resemblance.

Pinaceae The Pine family. 12 genera and about 200 species of typically evergreen, resinous, coniferous trees and shrubs with spirally arranged needle-like leaves that are often densely grouped into short lateral shoots. The fruiting cones are usually large and woody and take several seasons to mature. This family is one of the most important sources of timber and pulp for paper. It also provides turpentine, resin, and edible seeds. *Distr.* N hemisphere regions S to Malaysia and Central America. *Class* G.
See: Abies, Cedrus, Larix, Picea, Pinus, Pseudolarix, Pseudotsuga, Tsuga.

pinang *Areca catechu.*

pincushion, rose *Mammillaria zielmanniana.*

pincushion flower *Scabiosa, S. atropurpurea.*

pincushion tree *Hakea.*

pine Pinaceae *Pinus* **beach** ~ *Pinus contorta* **Bhutan** ~ *P. wallichiana* **Blue Mountain** ~ *Microstrobos fitzgeraldii* **Brazilian** ~ *Araucaria augustifolia* **Chile** ~ *A. araucana* **cypress** ~ *Callitris* **dammar** ~ *Agathis* **ground** ~ *Lycopodium clavatum* **Himalayan** ~ *Pinus wallichiana* **house** ~ *Araucaria heterophylla* **Huon** ~ *Lagarostrobos franklinii* **Illawara** ~ *Callitris rhomboidea* **kauri** ~ *Agathis, A. australis* **lodgepole** ~ *Pinus contorta* **Monterey** ~ *P. radiata* **mountain** ~ *Halocarpus bidwillii* **Norfolk Island** ~ *Araucaria heterophylla* **Oregon** ~ *Pseudotsuga menziesii* **Parana** ~ *Araucaria augustifolia* **parasol** ~ *Sciadopitys verticillata* **ponderosa** ~ *Pinus ponderosa* **Port Jackson** ~ *Callitris rhomboidea* **red** ~ *Dacrydium cupressinum* **running** ~ *Lycopodium clavatum* **shore** ~ *Pinus contorta* **stone** ~ *P. pinea* **umbrella** ~ *P. pinea* **western yellow** ~ *P. ponderosa* **whistling** ~ *Casuarina equisetifolia.*

pinea: the Latin name for pine nuts.

pineapple Bromeliaceae, *Ananas comosus.*

Pinellia (After Giovanni Vincenzo Pinelli (1535–1601), director of the Naples botanic garden.) A genus of 6 species of tuberous herbs with simple or compound basal leaves. The small flowers are borne on a spadix that is surrounded by a green spathe. Several species are cultivated as ornamentals. *Distr.* E Asia. *Fam.* Araceae. *Class* M.

P. ternata Leaves divided into 3 leaflets. Leaf stalks producing small bulblets. *Distr.* China, Korea, Japan.

Pinguicula (From Latin *pinguis*, fat, alluding to the greasy appearance of the leaves.) BUTTERWORT. A genus of 46 species of small, carnivorous, perennial herbs with rosettes of sticky leaves and tubular, 2-lipped flowers that are borne singly or in twos on an erect stem. Small insects are caught and digested on the sticky leaves. *Distr.* N temperate regions. *Fam.* Lentibulariaceae. *Class* D.

P. grandiflora Leaves yellow-green. Flowers with a long straight spur, purple, occasionally with a white throat. *Distr.* W Europe.

P. vulgaris BUTTERWORT, BOG VIOLET. Leaves yellow-green, somewhat curled along margins. Flowers violet with a short spur. *Distr.* North America, Europe, Asia.

pinguifolius, -a, -um: with fat leaves.

pinifolius, -a, -um: from the genus name *Pinus*, and *folius*, leaved.

pink Caryophyllaceae, *Dianthus* **Cheddar** ~ *gratianopolitanus* **Chinese** ~ *D. chinensis* **clove** ~ *D. caryophyllus*, *D. plumarius* **Deptford** ~ *D. armeria* **Indian** ~ *Ipomoea quamoclit*, *Lobelia cardinalis*, *Spigelia marilandica* **maiden** ~ *Dianthus deltoides* **meadow** ~ *D. deltoides* **moss** ~ *Phlox subulata* **mullein** ~ *Lychnis coronaria* **sea** ~ *Armeria*, *A. maritima* **swamp** ~ *Helonias bullata*.

pink pot *Hypoestes phyllostachya*

pink spire *Clethra alnifolia*.

pinkroot spigelia maryland ~ *S. marilandica*.

pinkwood *Eucryphia lucida*, *E. moorei*.

pinnatifidus, -a, -um: pinnately divided.

pinnatifolius, -a, -um: with pinnate leaves.

pinnatisectus, -a, -um: with pinnate divisions.

pinnatus, -a, -um: pinnate.

pinnifolius, -al, -um: with pinnate leaves.

Pinus (Classical Latin name for these trees.) PINE. A genus of 100 species of evergreen trees and occasionally shrubs with long, needle-like leaves in clusters of 2–5 and woody fruiting cones that open to release their seeds. These trees are one of the most important sources of timber, pulp, and resins. *Distr.* N hemisphere. *Fam.* Pinaceae. *Class* G.

P. contorta BEACH PINE, SHORE PINE, LODGEPOLE PINE. Erect tree to 30m high or twisted shrub of poor coastal areas. Needles in clusters of 2. *Distr.* W North America.

P. griffithii See *P. wallichiana*

P. pinea STONE PINE, UMBRELLA PINE. Dome-shaped tree to 25m high. Needles in clusters of 2. Cultivated for edible seeds (pine nuts). *Distr.* S Europe, W Asia.

P. ponderosa WESTERN YELLOW PINE, PONDEROSA PINE. Tree to 50m high. Needles in clusters of 3. A valuable timber. *Distr.* W North America.

P. radiata MONTEREY PINE. Fast-growing tree to 40m high. Needles in clusters of 3. Planted in Mediterranean regions for timber. *Distr.* USA (California).

P. wallichiana (After Nathaniel Wallich (1786–1854), Danish surgeon and botanist with the East India Company.) BHUTAN PINE, HIMALAYAN PINE. Tree to 50m high. Needles in clusters of 5. *Distr.* W Himalaya.

pinwheel *Aeonium haworthiae*.

Piper (Classical Latin name for PEPPER.) A genus of 2000 species of fragrant shrubs, lianas, and small trees with simple, alternate, often asymmetrical leaves and dense spikes of minute flowers that are followed by 1-seeded fruits. Several species are grown as ornamentals. *Distr.* Tropical regions. *Fam.* Piperaceae. *Class* D.

P. betle BETEL PEPPER. Climbing shrub to 5m high. The leaves are used to wrap the betel nut (*Areca catechu*), a masticatory drug. *Distr.* SE Asia.

P. nigrum BLACK PEPPER, WHITE PEPPER, COMMON PEPPER, MADAGASCAR PEPPER. Climbing shrub to 4m high. The source of the spice pepper; the fruits are ground whole or with their husks removed to produce black or white pepper, respectively. *Distr.* S India and Sri Lanka, naturalized elsewhere in Asia, widely cultivated throughout the tropics.

P. ornatum CELEBES PEPPER. Spreading or climbing shrub to 5m high. *Distr.* Indonesia (Sulawesi).

Piperaceae The Pepper family. 10 genera and about 2000 species of shrubs, small trees, and woody climbers with alternate gland-dotted leaves and tiny flowers. The source of pepper (*Piper nigrum*) is in this family. *Distr.* Tropical regions, particularly in rainforests. *Class* D.
See: Macropiper, Peperomia, Piper.

piperianus, -a, -um: peppery.

piperitus, -a ,-um: pepper-like, tasting hot and peppery.

pipe tree *Syringa vulgaris*.

pipewort Eriocaulaceae.

Piptanthus (From Greek *pipto*, to fall, and *anthos*, flower, alluding to the flowers which fall intact.) A genus of 2 species of shrubs and small trees with trifoliate leaves and racemes of yellow pea-like flowers. *Distr.* Himalaya. *Fam.* Leguminosae. *Class* D.

P. forrestii See *P. nepalensis*

P. laburnifolius See *P. nepalensis*

P. nepalensis Evergreen shrub. Flowers bright yellow with a purple standard. Grown as an ornamental. *Distr.* Himalaya.

Piptatherum See *Oryzopsis*

Pisonia (After Wilhelm Piso (died 1678), Dutch naturalist.) A genus of 50 species of shrubs, trees, and climbers with simple, opposite leaves and panicles or cymes of funnel-shaped flowers. *Distr.* Tropical and warm regions, especially the Americas. *Fam.* Nyctaginaceae. *Class* D.

P. brunoniana See *P. umbellifera*

P. umbellifera BIRD CATCHER TREE, PARA PARA. Tree to 20m high. Flowers pink or yellow. *Distr.* Islands of the Indian Ocean, Malaysia, Australia, New Zealand.

pistachio *Pistacia vera*.

Pistacia (From Greek *pistake*, pistachio nut.) A genus of 9 species of deciduous or evergreen shrubs and trees with pinnate or occasionally simple, leathery leaves. The small unisexual flowers are borne in racemes or panicles and are followed by a fleshy fruit. Some species are grown as ornamentals and some are a source of resins, oils, and edible seeds. *Distr.* Mediterranean, Asia, and Malaysia. *Fam.* Anacardiaceae. *Class* D.

P. chinensis Deciduous shrub or tree. Young shoots eaten as a vegetable in China. *Distr.* China.

P. lentiscus MASTIC, LENTISCO. Shrub or tree. Leaves evergreen, with 4–6 leaflets. Resin used as a chewing gum and a source of varnish. *Distr.* Mediterranean regions.

P. terebinthus CYPRUS TURPENTINE. Tree or shrub. Leaves with 6–12 leaflets. A source of a type of turpentine. *Distr.* SW Europe, N Africa, Turkey.

P. vera PISTACHIO, GREEN ALMOND, FUSTUQ. Tree to 10m high. Leaves with 3–5 leaflets, deciduous. Edible seeds used in desserts, ice-cream, and confectionery. *Distr.* W Asia, much cultivated in Mediterranean and N America.

Pistia (From Greek *pistos*, water, alluding to its habitat.) A genus of 1 species of free-floating herb that forms numerous rosettes of leaves joined by fine stolons over the surface of lakes and rivers. The leaves are coated with water-repellent hairs and move together at night. The small unisexual flowers are borne on a minute spadix that is hidden among the leaves. *Distr.* Tropical and subtropical regions. *Fam.* Araceae. *Class* M.

P. stratiotes WATER LETTUCE, TROPICAL DUCKWEED. This plant is a serious water-weed, blocking reservoirs, canals, and drainage ditches (see also *Eichhornia crassipes*). It is sometimes used as animal fodder.

pistol plant *Pilea microphylla*.

Pisum (Classical Latin name for these plants.) PEA. A genus of 5 species of climbing annual herbs with pinnate leaves and solitary irregular flowers that bear 5 fused sepals and 5 petals. The upper petal or standard is held erect, the 2 lateral petals or wings are usually spread, and the lower 2 petals are held together to form the keel. *Distr.* Mediterranean, W Asia. *Fam.* Leguminosae. *Class* D.

P. sativum GARDEN PEA. Annual herb. Flowers white-lilac, sometimes marked purple. The seeds are eaten as a vegetable, fresh, tinned, frozen, dried, or as a flour. *Distr.* Europe, Asia.

Pitcairnia (After William Pitcairn (1711–91), London doctor.) A genus of 260 species of terrestrial, rarely epiphytic herbs with rosettes of often spiny leaves and showy, somewhat irregular flowers. Some species are grown as ornamentals. *Distr.* Central and South America, W Africa. *Fam.* Bromeliaceae. *Class* M.

P. andreana Leaves narrow, arching. Flowers tubular, orange-red. *Distr.* Colombia.

P. feliciana The only member of the family Bromeliaceae found outside the Americas. *Distr.* W Africa

P. heterophylla Leaves forming a bulb at their base. *Distr.* Central and South America.

pitcher, sun *Heliamphora.*

pitcher plant *Nepenthes*, *Sarracenia* **California** ~ *Darlingtonia californica* **common** ~ *Sarracenia purpurea* **New World** ~ Sarraceniaceae **northern** ~ *Sarracenia purpurea* **yellow** ~ *S. flava.*

pitchfork *Bidens*

Pittosporaceae A family of 9 genera and 240 species of evergreen shrubs, trees, and occasionally climbers with simple leathery leaves and more or less regular flowers which have their parts in fives. Some species are used locally for timber. *Distr.* Tropical and warm-temperate regions of the Old World, especially Australasia. *Class* D.
See: Billardiera, Bursaria, Hymenosporum, Pittosporum, Sollya.

Pittosporum (From Greek *pitta*, pitch, and *sporum*, seed, alluding to the sticky seeds.) A genus of 200 species of evergreen shrubs and trees with simple alternate leaves and fragrant, cup-shaped flowers that bear their parts in fives. The fruit is a woody capsule. Several species are grown as street trees and ornamentals in frost-free areas. *Distr.* Tropical and S Africa to New Zealand and Pacific. *Fam.* Pittosporaceae. *Class* D.

P. tenuifolium Tree to 10m high. Numerous cultivars have been raised from this species. *Distr.* New Zealand.

P. undulatum VICTORIAN BOX, CHEESE WOOD. Tree to 14m high. A source of timber and fragrant oils. *Distr.* E Australia.

Pityrogramma (From Greek *pituron*, husk or chaff, and *gramma*, letter, alluding to the appearance of the fronds.) A genus of 40 species of terrestrial ferns with white or yellow waxy powder on the underside of the leaves. *Distr.* Tropical America, Africa, Madagascar. *Fam.* Adiantaceae. *Class* F.

P. calomelanos SILVER FERN. Leaves silver-white, scaly beneath. Cultivated as an ornamental. *Distr.* Tropical America, naturalized throughout the Old World tropics, Australia, and the Pacific.

P. chrysophylla GOLDEN FERN. Leaves usually bright golden-yellow, scaly beneath. Cultivated as an ornamental. *Distr.* South America.

P. triangularis The black leaf stalks are used in Indian basketry. *Distr.* USA (California) to Mexico.

Plagianthus (From Greek *plagios*, oblique, and *anthos*, flower.) A genus of 2 species of deciduous shrubs and trees with simple leaves and inconspicuous, fragrant, yellow or white flowers that bear 5 petals and numerous fused stamens. *Distr.* New Zealand. *Fam.* Malvaceae. *Class* D.

P. betulinus See *P. regius*

P. divaricatus Deciduous bushy shrub. Flowers tiny, yellow, highly fragrant. Grown as an ornamental. *Distr.* New Zealand.

P. lyallii See *Hoheria lyallii*

P. regius RIBBON WOOD. Deciduous tree to 15m high. Flowers white, usually unisexual. The bark is a source of fibre used as a substitute for RAFFIA. *Distr.* New Zealand.

Plagiogyriaceae A family of 1 genus and 37 species of terrestrial medium-sized ferns with creeping to erect stems and pinnate leaves. The sporangia are not enclosed in sori but borne on specialized leaves with narrow segments. *Distr.* Tropical and subtropical regions, excluding Africa. *Class* F.

Plagiopteraceae A family of 1 genus and 2 species of climbing shrubs with simple leaves and small flowers. This family is sometimes included within the family Flacourtiaceae. *Distr.* Burma, Thailand, China. *Class* D.

plane *Platanus* **American** ~ *P. occidentalis* **London** ~ *P.* × *acerifolia* **Oriental** ~ *P. orientalis.*

Planera (After J. J. Planer (1743–89), German botanist.) A genus of 1 species of deciduous shrub or tree to 15m high with 2-ranked, simple leaves and clusters of inconspicuous, 4- or 5-lobed, green flowers that lack petals. *Distr.* North America. *Fam.* Ulmaceae. *Class* D.

P. aquatica WATER ELM, PLANER TREE. Grown as an ornamental *Distr.*

planifolius, -a, -um: with flat leaves.

planipes: with a flat stalk.

planiscapus, -a, -um: with a flat flower stalk.

Plantaginaceae The Plantain family. 3 genera and 255 species of annual to perennial herbs with alternate, simple or divided leaves and typically dense spikes of small, wind-pollinated flowers. *Distr.* Widespread. *Class* D.
See: Plantago.

Plantago (From Latin *planta*, sole of the foot, alluding to the leaves of *P. major* that are broad and flat and pressed to the ground.) PLANTAIN, RIBWORT. A genus of 250 species of perennial herbs with rosettes of simple leaves and dense erect spikes of small wind-pollinated flowers. Several species may be grown as ornamentals. *Distr.* Cosmopolitan. *Fam.* Plantaginaceae. *Class* D.

P. coronopus BUCK'S HORN PLANTAGO. Leaves deeply pinnately lobed. *Distr.* Sea-cliffs in Europe.

P. lanceolata RIBWORT, RIBWORT PLANTAIN. Leaves narrow. Flowering stem deeply furrowed. *Distr.* Eurasia.

P. maritima SEA PLANTAGO. Leaves very narrow. *Distr.* Salt marshes of Europe.

plantago: buck's horn ~ *Plantago coronopus* **sea** ~ *P. maritima*.

plantain Plantaginaceae, *Plantago*, *Musa* × *paradisiaca* **French** ~ *M.* × *paradisiaca* **mud** ~ *Heteranthera* **ribwort** ~ *Plantago lanceolata* **water** ~ Alismataceae, *Alisma*, *A. plantago-aquatica* **wild** ~ *Heliconia*.

planus, -a, -um: flat.

Platanaceae A family of 1 genus and 6 species. *Class* D.
See: Platanus.

platanifolius, -a, -um: from the genus *Platanus*, and *folius*, leaved.

Platanthera (From Greek *platys*, wide, and the Latin *anthera*, anther.) BUTTERFLY ORCHID. A genus of 80–100 species of terrestrial tuberous orchids with a few basal leaves and cylindrical spikes of white-green flowers. Some species are grown as ornamentals. *Distr.* Temperate Europe, Asia, North and South America. *Fam.* Orchidaceae. *Class* M.

P. bifolia LESSER BUTTERFLY ORCHID. Flowers white, fragrant. *Distr.* Europe, N Africa, SW Asia.

P. chlorantha GREATER BUTTERFLY ORCHID. Flowers green-white, slightly fragrant. *Distr.* Europe, SW Asia.

Platanus (From *platanos*, the classical Greek name for *P. orientalis*) PLANE, BUTTERWOOD TREE. A genus of 6–7 species of deciduous trees with palmately lobed leaves and round heads of unisexual flowers. Some species are grown as ornamentals, some are a source of fine-grained timber. *Distr.* N Hemisphere. *Fam.* Platanaceae. *Class.*

P. × acerifolia LONDON PLANE. A hybrid of *P. occidentalis* × *P. orientalis*. Large deciduous tree. Bark peeling. Frequently planted as a street tree *Distr.* Raised near Oxford, before 1700.

P. × hispanica See *P.* × *acerifolia*

P. occidentalis BUTTONWOOD, AMERICAN PLANE, AMERICAN SYCAMORE. Large tree to 50m high. *Distr.* SW North America.

P. orientalis ORIENTAL PLANE. Tree to 30m high. *Distr.* SE Europe to N Iran.

platyanthus, -a, -um: with broad flowers.

platycalyx with a broad whorl of sepals.

Platycarya (From Greek *platys*, broad, and *karyon*, nut.) A genus of 1 species of deciduous tree with alternate, pinnate leaves and small unisexual flowers, the male flowers being borne in clustered erect catkins and the females in solitary cone-like structures. The fruit is a small winged nut. *Distr.* Japan, Korea, China. *Fam.* Juglandaceae. *Class* D.

P. strobilacea Grown as an ornamental. The fruit is used as a source of dye in E Asia.

Platycerium (From Greek *platys*, broad, and *keras*, horn, alluding to the flattened, horn-like leaves) ELK'S HORN FERN, STAG'S HORN FERN. A genus of 12 species of dramatic epiphytic ferns with flat, lobed leaves appressed to the support. *Distr.* Tropical regions. *Fam.* Polypodiaceae. *Class* F.

P. alcicorne See *P. bifurcatum*

P. bifurcatum ELK'S HORN FERN, COMMON STAG'S HORN FERN. Leaves rounded, heart- or kidney-shaped, to about 70cm across. Grown as an ornamental. *Distr.* Australia, Polynesia.

platyclados: with a broad stem.

Platycladus See *Thuja*

Platycodon (From Greek *platys*, broad, and *kodon*, bell, alluding to the flowers.) A genus of 1 species of perennial herb with erect branched stems, simple blue-tinged leaves, and large spherical buds that open into blue, bell-shaped flowers. *Distr.* NE Asia. *Fam.* Campanulaceae. *Class* D.

P. grandiflorus BALLOON FLOWER, CHINESE BELLFLOWER. Grown as an ornamental with numerous cultivars available.

platylobus, -a, -um: with broad lobes.

platypetalus, -a, -um: with broad petals.

platyphyllus, -a, -um: with broad leaves.

Platystemon (From Greek *platys*, broad, and *stemon*, stamen.) A genus of 1 species of annual herb with simple, densely hairy leaves and solitary regular flowers that bear 3 sepals, 6 yellow petals, and numerous unequal stamens. *Distr.* W North America. *Fam.* Papaveraceae. *Class* D.

P. californicus CREAM CUPS, CALIFORNIAN POPPY. Grown as an ornamental *Distr.*

platystigmus, -a, -um: with a broad stigma.

Platyzomataceae A family of 1 genus and 1 species of small terrestrial fern of dry areas that grow in dense colonies. *Distr.* N and NE Australia. *Class* F.

Plectostachys (From Greek *plecos*, to plait, and *stachys*, spike, alluding to the inflorescence.) A genus of 2 species of annual herbs or subshrubs with simple leaves and flowers in daisy-like heads that bear distinct rays. *Distr.* South Africa. *Fam.* Compositae. *Class* D.

P. serpyllifolia Straggling subshrub to 1.5m high. *Distr.* South Africa.

Plectranthus (From Greek *plectron*, spur, and *anthos*, flower, alluding to the spurred flowers of the type species.) A genus of 300 species of annual and perennial herbs, shrubs, and succulents with opposite leaves and 2-lipped flowers. Many species are grown as ornamentals and several have edible tubers. *Distr.* Tropical and warm regions of the Old World. *Fam.* Labiatae. *Class* D.

P. australis SWEDISH IVY. Evergreen perennial. Leaves rounded, waxy, glossy. Flowers white or pale mauve, borne in racemes. A very popular house-plant. *Distr.* SE Australia.

P. coleoides Bushy perennial. Flowers white. *Distr.* SW India.

P. oertendahlii CANDLE PLANT. Somewhat succulent, perennial herb. Cultivated in hanging baskets. *Distr.* Natal.

Pleioblastus (From Greek *pleios*, many, and *blastos* buds.) A genus of 20 species of small to medium-sized, rhizomatous bamboos with hollow stems and rough leaves. Several species are grown as ornamentals. *Distr* China, Japan. *Fam.* Gramineae. *Class* M.

P. auricoma Stems purple. Leaves pale green-yellow. *Distr.* Japan.

P. pygmaeus DWARF FERN-LEAF BAMBOO. Stems thread-like, leaves in 2 ranks. *Distr.* Not known in wild. Originally cultivated in Japan.

P. variegatus DWARF WHITE-STRIPED BAMBOO. Stems to 1m high. Leaves dark green with a white stripe, downy below. *Distr.* Japan.

P. viridistriatus See *P. auricoma*

Pleione (After Pleione, wife of Atlas and mother of the Pleiades in Greek mythology.) INDIAN CROCUS. A genus of 12 species of small, epiphytic or terrestrial orchids with leaves rolled when young and flowers with large, rolled tubular lips. Some species and many hybrids are cultivated as ornamentals. *Distr.* Himalaya. *Fam.* Orchidaceae. *Class* M.

P. aurita Flowers pink to purple, lip 3-lobed at apex. *Distr.* China (Yunnan).

P. bulbocodioides Flowers pink to magenta, lip spotted dark purple. *Distr.* W China, Taiwan, Burma.

P. formosana Flowers white to magenta, lip white, marked yellow. *Distr.* E China.

P. forrestii Flowers white to yellow, lip marked red. *Distr.* China.

P. hookeriana Flowers pink-lilac, lip marked yellow-purple. *Distr.* NE India, Bhutan, Nepal, Thailand, Laos.

P. humilis Flowers white, lip marked red and with a central yellow patch. *Distr.* NE India, Burma, Nepal.

P. maculata Flowers cream, fragrant. *Distr.* NE India, Bhutan, Burma, Thailand.

P. pogonioides See *P. bulbocodioides*

P. praecox Flowers white to purple, lip yellow in centre. *Distr.* Nepal, NE India, China, Burma.

P. pricei See *P. formosana*

P. speciosa Flowers magenta, lip marked orange. *Distr.* China.

P. yunnanensis Flowers lavender-pink, occasionally white. *Distr.* China (Yunnan).

pleiospermus, -a, -um: with more seeds than usual.

Pleiospilos (From Greek *pleios*, many, and *spilos*, spot, alluding to the spotted leaves.) LIVING ROCK, CLEFTSTONE PLANT, LIVING GRANITE, SPLITROCK. A genus of 7 species of very succulent, clump-forming herbs with pairs of thick fused leaves that resemble small rocks. The flowers are relatively large, yellow to orange and are often coconut-scented. All 4 species are grown as ornamentals. *Distr.* South Africa (Cape Province). *Fam.* Aizoaceae. *Class* D.

P. bolusii (After Harry Bolus (1834–1911), South African botanist.) Leaves tinged red or brown. Flowers golden yellow. *Distr.* South Africa (Cape Province).

pleniflorus, -a, -um: with full flowers, double flowers.

plenus, -a, -um: full.

Pleomele See *Dracaena*

Pleopeltis (From Greek *pleos*, full, and *pelte*, shield.) A genus of 40 species of epiphytic ferns with simple or pinnately divided leaves and sori that are covered by shield-like scales. *Distr.* Tropical regions. *Fam.* Polypodiaceae. *Class* F.

P. macrocarpa Leaves to 30cm long, simple, borne on a black stalk. *Distr.* Tropical regions.

Pleurospermum (From Greek *pleuron*, rib, and *sperma*, seed, alluding to the ridged seeds.) A genus of 3 species of biennial and perennial herbs with pinnately divided leaves and compound umbels of white flowers. The umbels are usually subtended by a whorl of toothed bracts. *Distr.* Temperate Eurasia. *Fam.* Umbelliferae. *Class* D.

Pleurothallis (From Greek *pleuron*, side vein or rib, and *thallo*, to stem.) A genus of 1000 species of epiphytic and lithophytic orchids with fleshy or leathery leaves and racemes of small-lipped flowers. *Distr.* Mountainous regions of the New World tropics and subtropics. *Fam.* Orchidaceae. *Class* M.

P. gelida Flowers yellow, downy. *Distr.* USA (Florida), S through Central America and W Indies to Venezuela, Colombia, Ecuador and Peru.

P. grobyi Racemes with few, white, green, or orange flowers. *Distr.* W Indies to Peru and Brazil.

P. immersa Racemes with numerous yellow-green to purple flowers. *Distr.* Central America, W Indies.

P. quadrifida Flowers numerous, pendent, yellow, fragrant. *Distr.* Central America to Columbia and W Indies.

plicatilis: with small folds or pleats.

plicatus, -a, -um3: pleated.

plum *Prunus domestica* **cherry** ~ *P. cerasifera* **coco** ~ Chrysobalanaceae **Davidson's** ~ *Davidsonia pruriens* **Natal** ~ *Carissa grandiflora* **Oregon** ~ *Oemleria cerasiformis* **sugar** ~ *Amelanchier*.

Plumbaginaceae The Leadwort family. 25 genera and about 500 species of annual to perennial herbs and climbers with simple glandular leaves in basal rosettes or alternately along the stem. The flowers are bisexual and regular with their parts in fives. Some species are grown as ornamentals, others have medicinal properties. *Distr.* Widespread. *Class* D.

See: Acantholimon, Armeria, Ceratostigma, Dictyolimon, Goniolimon, Limonium, Plumbago, Psylliostachys.

plumbaginoides: resembling plants of the family Plumbaginaceae.

Plumbago (From Latin *plumbum*, lead; it was thought that it was a cure for lead poisoning.) LEADWORT. A genus of 10 species of annual and perennial herbs and shrubs with spike-like racemes of tubular flowers that bear their parts in fives. Several species are grown as ornamentals and some are used medicinally. *Distr*. Tropical and warm regions. *Fam*. Plumbaginaceae. *Class* D.

P. auriculata CAPE LEADWORT. Evergreen arching shrub. Flowers pale blue. Commonly grown as a tender ornamental. *Distr*. S Africa.

P. capensis See *P. auriculata*

P. indica SCARLET LEADWORT. Climbing or scrambling subshrub. Flowers red-purple. *Distr*. SE Asia.

plumbago, Japanese *Ceratostigma willmottianum*.

plume: Brazilian ~ *Justicia carnea* **Prine of Wales's** ~ *Leptopteris superba* **scarlet** ~ *Euphorbia fulgens*.

plume bush *Calomeria amaranthoides*.

plume flower *Justicia carnea*.

plume plant *J. carnea*.

Plumeria (After Charles Plumier (1646–1706), French monk and botanist.) TEMPLE TREE, NOSEGAY, WEST INDIAN JASMINE, PAGODA TREE. A genus of 7 species of shrubs and small trees with succulent branches, leathery, deciduous leaves, and showy, fragrant, tubular flowers. Several species are grown as ornamentals. *Distr*. Tropical America. *Fam*. Apocynaceae. *Class* D.

P. rubra FRANGIPANI. Flowers yellow-orange. The flowers are offered in Buddhist temples and the bark is used as a purgative. *Distr*. Mexico, Central America.

plumosus, -a, -um: feathery.

plum tree *Eucryphia moorei*.

pluricaulis, -e: many-stemmed.

pluriflorus, -a, -um: many-flowered.

plush plant *Echeveria pulvinata*.

pluvialis, -e: of rain.

Pneumatopteris (From Greek *pneuma*, wind, and *pteris*, fern.) A genus of 80 species of terrestrial ferns with pinnately divided leaves. *Distr*. Tropical regions of the Old World, and Hawaii. *Fam*. Thelypteridaceae. *Class* F.

P. pennigera Rhizome erect, trunk-like. Leaves to 1m long. Grown as a tender ornamental. *Distr*. Australia, New Zealand.

Poa (Ancient Greek name for grass or fodder.) MEADOW GRASS, BLUE GRASS, SPEAR GRASS. A genus of 250 species of annual and perennial grasses with narrow, flat or folded leaves and paniculate inflorescences. Many species are important as pasture, forage, and lawn grasses. *Distr*. Temperate and cold regions. *Fam*. Gramineae. *Class* M.

P. annua ANNUAL MEADOW GRASS. A weed. *Distr*. Europe, North America.

P. arachnifera TEXAS BLUE GRASS. A winter fodder and lawn grass. *Distr*. W USA.

P. pratensis KENTUCKY BLUE GRASS, COMMON MEADOW GRASS. A pasture and lawn grass. *Distr*. Europe, naturalized in North America.

Poaceae See Gramineae

poached egg flower *Limnanthes douglasii*.

Podoaceae A family of 2 genera and 3 species of herbs and shrubs with small flowers that bear their parts in fives. This family is sometimes included within the Anacardiaceae. *Distr*. Nepal to S China and Thailand. *Class* D.

Podocarpaceae A family of 17 genera and 200 species of evergreen coniferous shrubs and trees with simple, linear to scale-like leaves. The female cones usually have a fleshy growth around or supporting the seeds, which facilitates bird dispersal. This family is of some importance as a source of timber and for its ornamental species. *Distr*. S hemisphere, N to Japan and Central America. *Class* G. *See: Dacrycarpus, Dacrydium, Halocarpus, Lagarostrobos, Lepidothamnus, Microcachrys, Microstrobos, Podocarpus, Prumnopitys, Saxegothaea*.

Podocarpus (From Greek *podos*, foot, and *karpos*, fruit, alluding to the fleshy receptacles on which the seeds are borne.) A genus of 100 species of evergreen coniferous shrubs and trees with flattened linear leaves that are borne in spirals but twisted so as to be arranged in 2 ranks. The fruiting cones bear few scales but have fleshy red receptacles. Several species are grown as ornamentals. *Distr*. S Hemisphere N to E Asia. *Fam*. Podocarpaceae. *Class* G.

P. chilinus See *P. salignus*

P. macrophyllus Tree to 15m high. Used as a hedging plant in Japan. *Distr*. S Japan, S China.

P. nivalis Small shrub to 1m high. Grown as a rock-garden ornamental. *Distr.* New Zealand.

P. salignus Tree to 20m high. Occasionally grown as an ornamental. *Distr.* Chile.

Podophyllum (From Greek *podos*, foot, and *phyllon*, leaf.) A genus of 9 species of rhizomatous herbs with large, peltate, palmately lobed leaves and solitary or clustered red-yellow or white flowers that bear 6 sepals, 6–9 petals, and 6–18 stamens. The fruit is a large fleshy berry. Several species are grown as ornamentals. The rhizomes are used medicinally. *Distr.* E North America, Himalaya, China. *Fam.* Berberidaceae. *Class* D.

P. emodi See *P. hexandrum*

P. hexandrum Shrub to 30cm high. Leaves to 25cm across. Flowers solitary, white-pink. *Distr.* Himalaya.

P. peltatum AMERICAN MANDRAKE, MAY APPLE. Leaves to 30cm across. Flowers nodding, large, white. The fruit is edible when ripe and has long been used medicinally by North American Indians. It is now being used against certain forms of cancer. *Distr.* E North America.

podophyllus, -a, -um: with stoutly stalked leaves.

Podostemaceae A family of 50 genera and 275 species of small aquatic herbs of rapidly flowing water. The vegetative organs are reduced to a thallus with a root-like holdfast. The flowers are regular and bisexual with 2–3 sepals and no petals. *Distr.* Tropical regions. *Class* D.

Podranea (Anagram of the genus name *Pandorea*.) A genus of 1–2 species of woody climbers with pinnate leaves and pyramidal clusters of fragrant, pink, funnel-shaped flowers. Several species are grown as ornamentals. *Distr.* Tropical and South Africa. *Fam.* Bignoniaceae. *Class* D.

P. brycei QUEEN OF SHEBA, ZIMBABWE CLIMBER. Leaflets 9–11 with finely toothed margins. Flowers purple-pink. *Distr.* Zimbabwe.

P. ricasoliana PINK TRUMPET VINE. Leaflets 5–11 with entire margins. Flowers pale pink with darker stripes. *Distr.* Tropical and South Africa.

Poellnitzia (After Karl von Poellnitz (1896–1945), a specialist in succulent plants.) A genus of 1 species of succulent, perennial herb with 4 ranks of leaves and a raceme of pale red flowers. *Distr.* South Africa. *Fam.* Aloaceae. *Class* M.

P. rubriflora Cultivated as a succulent ornamental. *Distr.* South Africa (Cape Province).

poeticus: of poets.

pogonanthus, -a, -um: with bearded flowers.

pogonioides: beard-like.

pogonostylus, -a, -um: with bearded styles.

Poinsettia See *Euphorbia*

poinsettia *Euphorbia pulcherrima.*

pokeweed Phytolaccaceae, *Phytolacca, P. americana* **Indian** ~ *P. acinosa.*

polaris: polar.

polecat bush *Rhus aromatica.*

polecat weed *Symplocarpus foetidus.*

Polemoniaceae The Phlox family. 19 genera and 275 species of annual to perennial herbs, rarely lianas, shrubs, or small trees, with simple or compound leaves and showy flowers that bear 5 fused petals. Many species are grown as ornamentals. *Distr.* Temperate Eurasia, temperate and tropical America. *Class* D.

See: Cantua, Collomia, Gilia, Ipomopsis, Linanthus, Phlox, Polemonium.

Polemonium (From *polemonion*, the Greek name for an unknown plant.) JACOB'S LADDER. A genus of 25 species of annual and perennial herbs with alternate pinnate leaves and funnel-shaped flowers that bear their parts in fives. Several species are grown as ornamentals. *Distr.* N temperate regions, Mexico, Chile. *Fam.* Polemoniaceae. *Class* D.

P. caeruleum JACOB'S LADDER, GREEK VALERIAN, CHARITY. Perennial herb to 1m high. Flowers blue, rarely white. *Distr.* Europe, Asia.

Polianthes (From Greek *polios*, bright, and *anthos*, flower.) A genus of 13 species of herbs with thick, bulb-like bases, succulent leaves and tubular flowers. Several species are grown as ornamentals. *Distr.* Mexico. *Fam.* Agavaceae. *Class* M.

P. tuberosa TUBEROSE. Often grown as an ornamental for its very fragrant, white flowers. This species is known only in cultivation and is believed to have arisen in pre-Columbian Peru, where the flowers were used to flavour chocolate.

policeman's helmet *Impatiens glandulifera*

polifolius, -a, -um: grey-leaved.

Poliothyrsis (From Greek *polios*, grey-white, and *thyrsos*, panicle.) A genus of 1 species of deciduous tree with simple alternate leaves and loose, white-hairy inflorescences of unisexual, yellow-white, fragrant flowers. The fruit is a 3–4-valved capsule that opens to release numerous winged seeds. *Distr.* China. *Fam.* Flacourtiaceae. *Class* D.

P. sinensis Grown as a hardy ornamental.

politus, -a, -um: polished, elegant.

polka-dot plant *Hypoestes phyllostachya.*

polyacanthus, -a, -um: with many spines.

polyandrus, -a, -um: with many stamens.

polyanthus, -a, -um: with many flowers.

polycarpus, -a, -um: with many fruits.

polychromus, -a, -um: multi-coloured.

polycladus, -a, -um: with many branches.

Polycycnis (From Greek *polys*, many, and *kyknos*, swan, the lip and column together bearing some resemblance to a swan.) A genus of 12 species of epiphytic orchids with large leathery leaves and racemes of showy flowers. *Distr.* Costa Rica, Colombia. *Fam.* Orchidaceae. *Class* M.

P. barbata Racemes pendent. Flowers translucent yellow, marked red. *Distr.* Costa Rica, Colombia.

Polygala (From Greek *polys*, much, and *gala*, milk; they were thought to enhance milk production.) MILKWORT, SENECA, SNAKEROOT. A genus of 500 species of annual and perennial herbs, shrubs, and rarely trees with simple leaves and racemes of irregular flowers that bear 5 sepals, 2 of which are large and petal-like, and 3–5 petals which are usually fused. Several species are grown as ornamentals. *Distr.* Widespread. *Fam.* Polygalaceae. *Class* D.

P. calcarea CHALK MILKWORT. Perennial herb to 20cm high. Flowers blue or white. *Distr.* Europe.

P. chamaebuxus BASTARD BOX. Evergreen shrub. Leaves leathery. Flowers cream with a yellow keel. Grown as a rock garden plant. *Distr.* Europe.

P. vulgaris GAND FLOWER, COMMON MILKWORT. Perennial herb. Flowers blue-pink. Formerly used medicinally. *Distr.* Europe and Mediterranean.

Polygalaceae The Milkwort family. 18 genera and about 1000 species of herbs, shrubs, small trees, and climbers with simple leaves and small irregular flowers that often bear 2 large petal-like sepals. *Distr.* Widespread, except New Zealand and Polynesia. *Class* D.
See: Polygala.

polygalifolius, -a, -um: from the genus *Polygala*, and *folius*, leaved.

Polygonaceae The Rhubarb family. 49 genera and about 1000 species of herbs, shrubs, and a few trees with simple leaves and small flowers that bear 3–6 sepals and 6–9 stamens but no petals. A number of species are cultivated as ornamentals and several are crop plants including Rhubarb (*Rheum*) and Buckwheat (*Fagopyrum esculentum*). *Distr.* Widespread. *Class* D.
See: Aconogonon, Antigonon, Atraphaxis, Bistorta, Eriogonum, Fagopyrum, Fallopia, Muehlenbeckia, Oxyria, Persicaria, Polygonum, Rheum, Rumex.

Polygonatum (From Greek *polys*, many, and *gony*, knee, alluding to the jointed rhizomes.) SOLOMON'S SEAL. A genus of 50 species of rhizomatous herbs with erect or arching stems, simple leaves, and pendent, green, white or pink, bell-shaped flowers. Several species are grown as ornamentals and some have edible rhizomes which are reported to have medicinal qualities. *Distr.* N temperate regions. *Fam.* Convallariaceae. *Class* M.

P. biflorum GREAT SOLOMON'S SEAL. Stems arching to 2m long. Flowers pendent, white, borne in clusters. *Distr.* E North America.

P. canaliculatum See *P. biflorum*

P. commutatum See *P. biflorum*

P. giganteum See *P. biflorum*

P. hirtum Stems erect, then arching. Leaves hairy below. Flowers white, green tipped. *Distr.* Siberia, China, Japan, Korea.

P. hookeri (After Sir Joseph Hooker (1817–1911), English botanist.) Stems short. Leaves linear. Flowers solitary, pink-violet. *Distr.* Himalaya, W China.

P. × hybridum A hybrid of *P. multiflorum* × *P. odoratum*. Perhaps the most commonly grown ornamental *Polygonatum*. *Distr.* Garden origin.

P. latifolium See *P. hirtum*

P. multiflorum COMMON SOLOMON'S SEAL, DAVID'S HARP, LADY'S SEAL. Stems arching. Flowers white, borne in clusters in the leaf axils. *Distr.* Europe, temperate Asia.

P. odoratum ANGLED SOLOMON'S SEAL. Stems arching, angled. Flowers white, in pairs, fragrant. A number of cultivars of this species are available. *Distr.* Europe, temperate Asia.

P. officinale See *P. odoratum*

P. verticillatum WHORLED SOLOMON'S SEAL. Stem erect. Leaves whorled. Flowers green, drooping. *Distr.* Europe, temperate Asia.

Polygonum (From Greek *polys*, many, and *gony*, knee, alluding to the jointed stems.) KNOTWEED, SMARTWEED, SILVER LACE VINE. A genus of 300 species of annual and perennial, occasionally aquatic herbs and scramblers with simple, alternate leaves and small, funnel- or bell-shaped flowers. Several species are grown as ornamentals, others are pernicious weeds. *Distr.* N temperate regions. *Fam.* Polygonaceae. *Class* D.

P. affine Mat-forming perennial. Flowers in dense spikes, pink or red. *Distr.* Afghanistan, Himalaya, Tibet.

P. amphibium WILLOW GRASS. Semi-aquatic herb. *Distr.* N temperate regions.

P. bistorta (From Latin *bis*, twice, and *tortus*, twisted, alluding to the twisted roots.) BISTORT, SNAKEWEED, EASTER LEDGER. Perennial to 60cm high. Flowers in a dense cylindrical spike, pink to white. The rhizome has been used medicinally and is still used in a pudding traditionally made in the Lake District of Great Britain, possibly as a fertility rite. *Distr.* Mountains of Eurasia.

P. japonicum JAPANESE KNOTWEED, MEXICAN BAMBOO. Rhizomatous perennial to 2m high. Flowers cream-white, borne in panicles. A very persistent weed of cultivated ground that was originally introduced into Great Britain as an ornamental. *Distr.* E Asia.

polylepis: with many scales.

polymorphus, -a, -um: with many forms.

polyneura: with many veins.

polypetalus, -a, -um: with many petals.

polyphyllus, -a, -um: with many leaves.

Polypodiaceae A family of 47 genera and about 500 species of terrestrial or epiphytic ferns with scaly rhizomes and simple, pinnate or rarely bipinnate leaves. The sori are rounded and borne on the veins or on specialized leaves. *Distr.* Widespread. *Class* F.

See: Aglaomorpha, Goniophlebium, Lemmaphyllum, Merinthosorus, Microgramma, Microsorum, Notholaena, Phlebodium, Platycerium, Pleopeltis, Polypodium, Pyrrosia, Selliguea.

Polypodium (From Greek *polys*, many, and *podos*, foot, alluding to the branched rhizomes.) POLYPODY. A genus of 75 species of epiphytic or terrestrial ferns with scaly rhizomes and pinnately divided leaves. Several species are grown as ornamentals. *Distr.* Widespread but chiefly in N temperate regions. *Fam.* Polypodiaceae. *Class* F.

P. cambricum WELSH POLYPODY. Epiphyte. Leaves broad, soft. A number of ornamental cultivars have been raised from this species. *Distr.* Europe.

P. glycyrrhiza LIQUORICE FERN. Epiphytic or terrestrial. The liquorice-flavoured rhizomes are eaten by North American Indians. *Distr.* North America.

P. vulgare COMMON POLYPODY, ADDER'S FERN, GOLDEN MAIDENHAIR FERN. Epiphytic or terrestrial. Leaves to 30cm long, deeply pinnately lobed. A number of cultivars have been raised from this species. *Distr.* Europe, Asia.

polypody *Polypodium* **common** ~ *P. vulgare* **golden** ~ *Phlebodium aureum* **limestone** ~ *Gymnocarpium robertianum* **Welsh** ~ *Polypodium cambricum.*

Polypogon (From Greek *polys*, many, and *pogon*, beard, alluding to the bristly inflorescences.) A genus of 10 species of annual and perennial grasses with thin stems, flat leaves, and dense, spike-like inflorescences. Several species are grown for their silky inflorescences. *Distr.* Warm-temperate regions. *Fam.* Gramineae. *Class* M.

P. monspeliensis ANNUAL BEARD GRASS, RABBIT'S FOOT GRASS. Annual. Inflorescences cylindrical, yellow-green, silky-hairy. *Distr.* Mediterranean regions.

polyrhizus: with many roots.

Polyscias (From Greek *polys*, many, and *skias*, umbel, alluding to the numerous umbels in the inflorescence) UMBRELLA TREE. A genus of 100 species of evergreen shrubs and trees with simple or pinnate, anise-scented leaves and large compound inflorescences of small flowers. Several species are grown as ornamentals. *Distr.* Tropical regions of the Old World. *Fam.* Araliaceae. *Class* D.

P. filicifolia FERN-LEAF ARALIA. Large shrub. Leaves with numerous narrow leaflets, sometimes variegated. *Distr.* W Pacific Islands.

P. fruticosa Shrub or small tree. Used as a hedging plant in frost-free regions. *Distr.* Indomalaysia.

P. guilfoylei (After W. R. Guilfoyle (1840–1912), who collected in the South Pacific Islands.) WILD COFFEE, COFFEE TREE. Slow-growing, rounded tree. Leaves deep green. *Distr.* Polynesia.

Polystachya (From the Greek *polys*, many, and *stachys*, spike, alluding to the inflorescence of some species) A genus of 200 species of epiphytic, occasionally terrestrial, orchids with woody rhizomes and many-flowered racemes or panicles. Some species produce false pollen, rich in protein or starch, to attract pollinators. A few species are cultivated as ornamentals. *Distr.* Tropical regions, centred in Africa. *Fam.* Orchidaceae. *Class* M.

P. affinis Flowers numerous, yellow-white, fragrant. *Distr.* W Africa.

polystachyus: many-spiked.

Polystichum (From Greek *polys*, many, and *stichos*, a row, alluding to the arrangement of the sori.) A genus of 200 species of terrestrial, woodland ferns with erect, divided leaves. Several species are grown as ornamentals. *Distr.* Cosmopolitan. *Fam.* Dryopteridaceae. *Class* F.

P. acrostichoides CHRISTMAS FERN, HOLLY FERN. *Distr.* E North America.

P. aculeatum HARD SHIELD FERN. *Distr.* N temperate Eurasia.

P. lonchitis NORTHERN HOLLY FERN. *Distr.* N temperate regions.

P. munitum AMERICAN SWORD FERN. Leaves leathery, sharp-edged. The pith of the rhizome is reported to taste of banana. *Distr.* W North America.

P. proliferum See *P. setiferum*

P. setiferum SOFT SHIELD FERN. A number of cultivars have been raised from this species. *Distr.* Atlantic Islands, W Europe.

polytrichoides: from the genus *Polytrichum*, with the ending *-oides*, indicating resemblance.

polytrichus, -a, -um: with many hairs.

Polyxena (After the daughter of Priam, beloved of Achilles.) A genus of 5–6 species of small bulbiferous herbs with 2 grooved, linear leaves and corymbs of tubular fragrant flowers. Several species are grown as ornamentals. *Distr.* S African. *Fam.* Hyacinthaceae. *Class* M.

P. odorata Flowers more or less sessile, white. *Distr.* South Africa.

pomegranate *Punica granatum.*

pomelo *Citrus maxima, C. grandis.*

pomifera: apple-bearing.

pompelmous *Citrus maxima.*

Poncirus (From French poncire, for a kind of CITRON.) A genus of 1 species of deciduous, spiny shrub or small tree with trifoliate leaves and solitary, scented flowers that bear 5 white petals and numerous stamens. The fruit is fleshy and similar to that found in the genus *Citrus*, but is inedible unless made into marmalade. *Distr.* Central and N China. *Fam.* Rutaceae. *Class* D.

P. trifoliata TRIFOLIATE ORANGE, BITTER ORANGE. Grown as a hedging plant and as a rootstock for species of *Citrus*.

ponderosus, -a, -um: heavy.

ponticus, -a, -um: of N Turkey and the S Black sea.

pondweed Potamogetonaceae, *Elodea*, *Potamogeton* **broad-leaved** ~ *P. natans* **Canadian** ~ *Elodea canadensis* **Cape** ~ *Aponogeton*, *A. distachyos* **horned** ~ *Zannichelliaceae*, *Zannichellia palustris*.

Pontederia (After Guilo Pontedera (1688–1757), professor of botany at Padua.) A genus of 5 species of rhizomatous aquatic herbs with lance- and kidney-shaped leaves and spike-like inflorescences of 2-lipped, purple or white flowers. *Distr.* the Americas. *Fam.* Pontederiaceae. *Class* M.

P. cordata PICKEREL WEED, WAMPEE. Grown as a hardy, marginal, aquatic ornamental. *Distr.* North and South America.

P. lanceolata See *P. cordata*

Pontederiaceae The Pickerel Weed family. 9 genera and about 35 species of annual and perennial, rhizomatous, aquatic herbs which may be submerged, free-floating, or emergent. In the free-floating species buoyancy is provided by inflated leaf stalks. The flowers are usually regular, often showy, and bear 2 whorls of 3 tepals, 1–6 fertile stamens, and a superior ovary. Some species are important water-weeds. *Distr.* Widespread in tropical regions, especially New World, rarely temperate. *Class* M.
See: Eichhornia, Heteranthera, Monochoria, Pontederia.

pony tail *Beaucarnea recurvata*.

poor man's weather glass *Anagallis arvensis*.

poplar *Populus* **balsam** ~ *P. balsamifera* **black** ~ *P. nigra* **necklace** ~ *P. deltoides* **tulip** ~ *Liriodendron tulipifera* **western balsam** ~ *Populus trichocarpa* **white** ~ *P. alba* **yellow** ~ *Liriodendron tulipifera*.

poppy Papaveraceae, *Papaver* **Arctic** ~ *P. nudicaule* **Californian** ~ *Platystemon californicus* **Californian tree** ~ *Romneya coulteri* **celandine** ~ *Stylophorum diphyllum* **corn** ~ *Papaver rhoeas* **field** ~ *P. rhoeas* **flaming** ~ *Stylomecon heterophylla* **Flanders** ~ *Papaver rhoeas* **Himalayan blue** ~ *Meconopsis betonicifolia* **horned** ~ *Glaucium*, *G. flavum* **Iceland** ~ *Papaver nudicaule* **Matilija** ~ *Romneya coulteri* **Mexican** ~ *Argemone mexicana* **opium** ~ *Papaver somniferum* **plume** ~ *Macleaya cordata* **prickly** ~ *Argemone*, *A. mexicana* **red horned** ~ *Glaucium corniculatum* **satin** ~ *Meconopsis napaulensis* **sea** ~ *Glaucium flavum* **snow** ~ *Eomecon chionanthum* **water** ~ *Hydrocleys nymphoides*, Limnocharitaceae **Welsh** ~ *Meconopsis cambrica* **wind** ~ *Stylomecon heterophylla* **yellow horned** ~ *Glaucium flavum*.

populeus, -a, -um: of *Populus*, poplar.

populifolius, -a, -um: from the genus *Populus*, and *folius*, leaved.

Populus (Classical Latin name for these trees.) POPLAR, COTTONWOOD, ASPEN. A genus of 35–40 species of deciduous trees with simple alternate leaves and small unisexual flowers that are borne in pendent catkins. The male and female flowers are borne on separate trees and are wind-pollinated. The seeds are covered in white hairs and wind-dispersed. Several species are planted as ornamentals; they are also used for fast-growing shelter belts. *Distr.* Temperate regions of the N hemisphere. *Fam.* Salicaceae. *Class* D.

P. alba ABELE, WHITE POPLAR. Deciduous tree. Underside of leaves covered in white hairs. A source of flexible, split-resistant wood. *Distr.* Central and SE Europe to Central Asia.

P. balsamifera BALSAM POPLAR. A source of plywood and pulp. *Distr.* North America.

P. deltoides COTTONWOOD, NECKLACE POPLAR. *Distr.* E North America.

P. nigra BLACK POPLAR. A source of timber and medicinal resin. *Distr.* Europe, N Africa, Central Asia.

P. tacamahaca See *P. balsamifera*

P. tremula ASPEN. Much-branched tree. A source of timber for gunpowder charcoal. *Distr.* Temperate Europe and Asia.

P. trichocarpa COTTONWOOD, WESTERN BALSAM POPLAR. *Distr.* W North America.

porcelain flower *Hoya.*

porrifolius, -a, -um: with leaves like *Allium porrum.*

Porroglossum (From Greek *porro*, forward, and glossa, tongue, alluding to the lip.) A genus of 8–9 species of epiphytic or terrestrial orchids with erect, fleshy, or leathery leaves. The flowers are borne in erect racemes and have their sepals fused into a bell-shaped tube that exceeds the other floral parts; their apices are extended into short or long tails. *Distr.* Venezuela to Peru. *Fam.* Orchidaceae. *Class* M.

P. echidnum Flowers small, green-brown. The lip of the flower is sensitive and when triggered by a pollinator lifts, carrying the insect up into the funnel formed by the united sepal-bases; the insect can only escape through the tube between the lip and column where the pollinia and stigma are located. The trap remains closed for around 30 minutes. *Distr.* Colombia and Ecuador.

P. muscosum Sepals brown, petals yellow-white. *Distr.* Columbia, Ecuador, Venezuela.

Portuguese broom *Chamaecytisus albus.*

Portulaca (Classical Latin name for *P. oleracea.*) PURSLANE, MOSS ROSE. A genus of 100 species of succulent, trailing, annual and perennial herbs with simple, alternate or opposite leaves and purple or yellow flowers that typically bear 5 petals and are often surrounded by a whorl of coloured leaves. *Distr.* Tropical and warm regions. *Fam.* Portulacaceae. *Class* D.

P. grandiflora SUN PLANT, ELEVEN O'CLOCK PLANT, MOSS ROSE. Annual. Foliage moss-like. Flowers short-lived, pink, red, yellow, or white. Grown as a tender ornamental. *Distr.* Brazil, Uruguay, naturalized in parts of Central Europe.

P. oleracea COMMON PURSLANE. Succulent, mat-forming annual. Flowers bright yellow. A weed of cultivation. Sometimes eaten as a vegetable. *Distr.* India.

Portulacaceae The Purslane family. 25 genera and 400 species of annual or perennial herbs and subshrubs with more or less fleshy, entire leaves. The flowers are typically small with 2 sepals and 5 petals. Several larger-flowered species are grown as ornamentals. *Distr.* Widespread. *Class* D.
See: Calandrinia, Calyptridium, Claytonia, Lewisia, Montia, Neopaxia, Portulaca, Spraguea, Talinum.

Posidonia (Name of Greek settlement in Italy, better known as Poertuem.) A genus of 3 species of marine perennial herbs with grass-like, fibrous leaves. *Distr.* Mediterranean and Australia. *Fam.* Posidoniaceae. *Class* M.

P. australis Source of posidonia fibre, cellonia, or lanmar used in making coarse fabric and as a stuffing, often mixed with wool. *Distr.* S coast of Australia.

Posidoniaceae A family of 1 genus and 3 species. *Class* M.
See: Posidonia.

possum haw *Ilex decidua.*

possumwood *Diospyros virginiana.*

pot *Cannabis sativa.*

Potamogeton (From Greek *potamos*, river, and *geiton*, neighbour.) PONDWEED. A genus of 100 species of rhizomatous, aquatic herbs with submerged or floating leaves and spikes of small green flowers. Several species are grown as ornamentals. *Distr.* Cosmopolitan. *Fam.* Potamogetonaceae. *Class* M.

P. natans PICKEREL WEED, BROAD-LEAVED PONDWEED. Floating leaves formerly believed, in East Anglia, to give rise to young pikes, hence the vernacular name. *Distr.* Eurasia.

Potamogetonaceae The Pond-weed family. 3 genera and about 100 species of perennial, occasionally annual, aquatic herbs with spikes of inconspicuous flowers that bear 4 free tepals, 4 free stamens, and 4 free carpels. *Distr.* Widespread. *Class* M.
See: Potamogeton.

potato Solanaceae, *Solanum tuberosum*, **air** ~ *Dioscorea bulbifera* **duck** ~ *Sagittaria latifolia* **Indian** ~ *Apios americana* **sweet** ~ *Ipomoea batatas* **telingo** ~ *Amorphophallus paeoniifolius* **yam** ~ *Dioscorea esculenta.*

potato bush, blue *Lycianthes rantonnetii.*

potato vine *Solanum wendlandii* **giant** ~ *S. wendlandii.*

Potentilla (From Latin *potens*, powerful, alluding to the medicinal properties of some of the species.) CINQUEFOIL. A genus of 500 species of annual to perennial herbs and shrubs with palmate, pinnate, or trifoliate leaves and saucer-shaped flowers. A large number of species and cultivars are grown as ornamentals and a few have medicinal uses locally. *Distr.* N temperate and cold regions, a few S temperate regions. *Fam.* Rosaceae. *Class* D.

P. alba Spreading perennial. Flowers clustered, white. *Distr.* Europe, Caucasus.

P. anserina SILVERWEED, GOOSE GRASS, GOOSE TANSY. Stoloniferous perennial. Flowers solitary, yellow. Roots edible. *Distr.* N temperate regions.

P. arbuscula See *P. fruticosa*

P. erecta TORMENTIL. Erect perennial herb. Leaves palmate. Flowers small. The dried rhizomes of this plant have been used medicinally. *Distr.* Eurasia.

P. fruticosa SHRUBBY CINQUEFOIL, GOLDEN HARDHACK, WIDDY. Deciduous bushy shrub. Flowers yellow or white. Numerous garden cultivars have been raised from this species. *Distr.* N temperate regions.

P. reptans CINQUEFOIL. Trailing perennial. Flowers yellow. Used medicinally. *Distr.* Eurasia, naturalized in North America.

P. sterilis BARREN STRAWBERRY. Perennial herb. Flowers white, borne on very slender stems. *Distr.* Europe, Mediterranean.

P. tormentilla See *P. erecta*

potentilloides from the genus *Potentilla*, with the ending *-oides*, indicating resemblance.

poteriifolius, -a, -um: from the genus *Poterium*, and Latin *folius*, leaved.

Poterium See *Sanguisorba*

× *Potinara* (After M. Potin, president of the French orchid society in 1922.) A genus of hybrids from crosses between the genera *Brassavola*, *Cattleya*, *Laelia*, and *Sophronitis*. They are grown as tender ornamentals. *Distr.* Garden origin. *Fam.* Orchidaceae. *Class* M.

powder puff *Mammillaria bocasana*.

powder puff tree *Calliandra*.

praecox early.

praemorsus, -a, -um: appearing to have been bitten off.

praestans distinguished.

praeteritus, -a, -um: past, overlooked.

praetermissus, -a, -um: overlooked, left out.

praevernus, -a, -um: before the spring, very early.

prairie smoke *Geum triflorum*.

pratensis, -e: of meadows.

Pratia (After C. L. Prat-Bernon (died 1817), French naval officer.) A genus of about 20 species of small, mat-forming, perennial herbs with wiry stems, alternate leaves, and numerous, small, star-shaped flowers. Several species are grown as ornamentals, chiefly in rock gardens. *Distr.* Tropical Asia, Australia, New Zealand and South America. *Fam.* Campanulaceae. *Class* D.

P. angulata Leaves dark green, oval. Flowers white. *Distr.* New Zealand.

P. pendunculata Leaves dark green. Flowers pale to dark purple-blue. *Distr.* Australia.

prayer plant *Maranta leuconeura*.

prickly Moses *Acacia verticillata*.

pride of Bolivia *Tipuana tipu*.

pride of India *Koelreuteria paniculata*.

pride of Texas *Phlox drummondii*.

primrose Primulaceae, *Primula vulgaris* **Cape** ~ *Streptocarpus S. rexii*. **evening** ~ Onagraceae, *Oenothera O. biennis* **peerless** ~ *Narcissus* × *medioluteus*.

Primula (From Latin *primus*, first, alluding to the early flowers.) A genus of 400 species of rhizomatous, perennial herbs with rosettes of simple, often waxy leaves and umbels or whorls of tubular or bell-shaped flowers that bear their parts in fives. Very many species and cultivars are grown as ornamentals. *Distr.* Temperate regions of the N hemisphere, tropical African mountains, and S South America. *Fam.* Primulaceae. *Class* D.

P. acaulis See *P. vulgaris*

P. allionii (After Carlo Allioni (1705–1804), Italian botanist.) Perennial herb. Leaves small, fleshy. Flowers pale pink to red-purple with a white centre. Numerous cultivars have been raised from this species and it has been used as a parent in many hybrid crosses. *Distr.* Alps.

P. altaica See *P. elatior*

P. amoena See *P. elatior*

P. auricula (From Latin *auricula*, ear; the leaves have been likened to bears' ears.) AURICULA, BEAR'S EARS. Rhizomatous perennial. Leaves somewhat fleshy, waxy. Flowers deep yellow, fragrant. This species, probably by hybridization with another, gave rise to a large number of cultivars that were very popular during the 18th and 19th centuries, particularly as show plants in British mining communities. Many of these cultivars have now been lost. *Distr.* Europe.

P. elatior OXLIP. Perennial herb. Leaves erect, spreading later. Flowers yellow with a green-orange throat, borne in an umbel atop an erect stem, odourless. *Distr.* Europe, W Asia.

P. farinosa Leaves small, covered in scaly wax. Flowers pink. *Distr.* Eurasia.

P. sibthorpii See *P. vulgaris*

P. uralensis See *P. veris*

P. veris COWSLIP. Perennial herb. Leaves large. Flowers fragrant, yellow, borne in an umbel atop an erect stalk. The flowers have been used to make wine and the leaves were formerly used to make a medicinal tea. *Distr.* Europe, W Asia.

P. vernalis See *P. vulgaris*

P. vulgaris PRIMROSE. Perennial herb. Leaves large. Flowers pale yellow, sometimes fragrant, borne in an umbel at ground level. *Distr.* Europe.

Primulaceae The Primrose family. 23 genera and 800 species of annual or perennial herbs usually with rhizomes or tubers and simple leaves. The flowers are typically borne on a leafless stalk and have their parts in fives. There are numerous important ornamental plants in this family, notably in the genera *Primula* and *Cyclamen*. *Distr.* Widespread but most occur in N temperate regions. *Class* D.
See: Anagallis, Androsace, Coris, Cortusa, Cyclamen, Dionysia, Dodecatheon, Douglasia, Glaux, Hottonia, Lysimachia, Omphalogramma, Primula, Samolus, Soldanella, Trientalis, Vitaliana.

primuliflorus, -a, -um: with flowers resembling those of the genus *Primula*.

primulifolius, -a, -um: from the genus *Primula*, and Latin *folius*, leaved.

primulinus, -a, -um: primrose yellow-coloured.

primuloides: from the genus *Primula*, with the ending *-oides*, indicating resemblance.

princeps: the most distinguished.

princess tree *Paulownia tomentosa*.

prismaticus, -a, -um: shaped like a prism.

prismatocarpus, -a, -um: with fruit shaped like a prism.

Pritchardia (After W. T. Pritchard, 19th-century British official in Polynesia.) LOULU PALM. A genus of 36 species of medium-sized palms with fan-shaped leaves and branched inflorescences of yellow or orange flowers. *Distr.* Pacific Islands. *Fam.* Palmae. *Class* M.

P. pacifica FIJI FAN PALM. To 10m high. Leaves to 1m across, Flowers yellow. Fruit round, black. Grown as a tender ornamental. *Distr.* Tonga, introduced into Fiji.

privet *Ligustrum* **Californian** ~ *L. ovalifolium* **Chinese** ~ *L. sinense*, *L. lucidum* **common** ~ *L. vulgare* **European** ~ *L. vulgare* **glossy** ~ *L. lucidum* **Japanese** ~ *L. japonicum* **wax** ~ *Peperomia glabella*.

Proboscidea (From Greek *proboskis*, snout or nose, alluding to the long, curved beak of the fruit.) A genus of 9 species of fragrant, annual and perennial herbs with simple opposite leaves and loose inflorescences of 5-lobed, tubular flowers. Several species are grown as ornamentals and some have edible fruits. *Distr.* Warm regions of the New World. *Fam.* Pedaliaceae. *Class* D.

P. fragrans SWEET UNICORN PLANT. Annual. Flowers purple to red, fragrant. *Distr.* Mexico.

P. lousianica Annual. Flowers cream to purple. Fruit to 6cm long with horns to 15cm long. The fruits become entangled in sheep's wool and so make shearing difficult; they are also edible and sometimes grown for pickling. *Distr.* S USA.

proboscideus, -a, -um: with a proboscis, like an elephant's trunk.

procerus, -a, -um: tall.

procumbens: prostrate, low-growing.

procurrens: extending or spreading out.

proliferus, -a, -um: spreading readily from offshoots.

prolificus, -a, -um: prolific, producing a large amount of fruit.

Promenaea (After Promeneia, a priestess of the temple of Dodona, mentioned by Herodotus.) A genus of 15 species of small, epiphytic orchids with clustered, fleshy pseudobulbs and showy, yellow or cream flowers. Several species are grown as tender ornamentals. *Distr*. Brazil. *Fam*. Orchidaceae. *Class* M.

P. rollissonii Flowers pale yellow with a spotted lip.

P. stapelioides Flowers to 5cm across, cream with maroon banding.

P. xanthina Flowers pure yellow, strongly fragrant.

pronus, -a, -um: prostrate, bending forwards.

prophet flower *Arnebia pulchra*.

propinquus, -a, -um: related to.

Prosopis (From the Greek name for BURDOCK.) A genus of 44 species of spiny shrubs and trees with bipinnate leaves and spike-like racemes of minute flowers that each bear 10 long stamens. An important genus of desert and semi-desert plants that are used to stabilize dunes, as a source of fodder, and as bee plants. *Distr*. Warm America, SW Asia, Africa. *Fam*. Leguminosae. *Class* D.

P. chilensis ALGAROBA. Locally grown as a cattle fodder and for its edible, sweet-tasting pods. *Distr*. Argentina, Chile.

Prostanthera (From Greek *prosthema*, appendage, and *anthera*, anther, alluding to the spurred anthers.) A genus of 50 species of aromatic, evergreen shrubs and small trees with simple opposite leaves and 2-lipped flowers borne in leafy racemes or panicles. Several species are grown as half-hardy ornamentals. *Distr*. Australia. *Fam*. Labiatae. *Class* D.

P. cuneata Densely branched shrub. Leaves shiny green, very fragrant. Flowers white with violet markings. *Distr*. Australia (Victoria, Tasmania).

P. lasianthos VICTORIA DOGWOOD, VICTORIA CHRISTMAS BUSH. Shrub to 8m high. Flowers large, violet or lilac, borne in leafless racemes. *Distr*. E Australia, including Tasmania.

P. melissifolia BALM MINT BUSH. Slender shrub to 2m high. Flowers pink-purple. *Distr*. SE Australia.

P. ovalifolia Rounded shrub. Leaves small. Flowers purple. *Distr*. SE Australia.

P. rotundifolia ROUND LEAVED MINT BUSH. Bushy rounded shrub. Leaves tiny, deep green. Flowers bell-shaped, purple to lavender. *Distr*. SE Australia, including Tasmania.

prostratus, -a, -um: prostrate, low-growing.

Protea (After Proteus, a sea-god with the ability to change shape, in Greek mythology; the species look very diffent.) A genus of 115 species of evergreen shrubs with tough, leathery leaves and small, tubular flowers in large heads that are surrounded by a whorl of colourful bracts. Several species are grown as ornamentals. *Distr*. South Africa. *Fam*. Proteaceae. *Class* D.

P. cynaroides KING PROTEA. Rounded shrub. Flower-heads to 20cm across, surrounded by pink bracts. Grown for long-lasting cut flowers. *Distr*. South Africa (Cape Province).

P. neriifolia Erect shrub to 3m high. Leaves narrow. Inflorescences to 14cm across, cup-shaped. *Distr*. South Africa (Cape Province).

protea Proteaceae **king** ~ *Protea cynaroides*.

Proteaceae The Protea family. 69 genera and about 1000 species of shrubs and trees with alternate, simple or divided leaves, and racemes, spikes, or heads of irregular flowers that bear 4 tepals and 4 stamens. *Distr*. S hemisphere to tropical regions, rare in the N hemisphere. *Class* D.
See: Banksia, Dryandra, Embothrium, Gevuina, Grevillea, Hakea, Isopogon, Knightia, Leucadendron, Leucospermum, Lomatia, Macadamia, Protea, Telopea.

proteiflorum: with flowers resembling those of the genus *Protea*.

proteoides: from the genus *Protea*, with the ending *-oides*, indicating resemblance.

pruinatus, -a, -um: with a white, glistening bloom.

pruinosus, -a, -um: with a white, glistening bloom.

Prumnopitys (From Greek *prumme*, stern, and *pitys* pine.) A genus of 10 species of evergreen coniferous trees with flat linear leaves that are borne in spirals but twisted so as to be arranged in 2 ranks. The fruit is fleshy and resembles a drupe. Several species are grown as ornamentals. *Distr.* Tropical Asia, Pacific, South America. *Fam.* Podocarpaceae. *Class* G.

P. andina PLUM-FRUITED YEW. Tree to 25m high. The seed is edible. *Distr.* Chile.

P. elegans See *P. andina*

Prunella (From the pre-Linnean name *Brunella*, which may have been derived from the German *Breaume*, quinsy, a throat infection these plants were supposed to cure.) SELFHEAL, HEAL ALL. A genus of 7 species of prostrate herbs with simple, opposite, stalked leaves and 2-lipped flowers in which the upper lip forms a hood. *Distr.* N temperate regions, NW Africa. *Fam.* Labiatae. *Class* D.

P. grandiflora Leaves sparsely hairy. Flowers large, deep violet. *Distr.* Europe.

P. incisa See *P. vulgaris*

P. vulgaris SELFHEAL. Leaves glabrous or sparsely hairy. Flowers dark blue or purple, occasionally white. Formerly used medicinally, especially for sore throats. *Distr.* Eurasia.

pruniflorus, -a, -um: with flowers resembling those of the genus *Prunus*.

prunifolius, -a, -um: from the genus *Prunus*, and Latin *folius*, leaved.

Prunus (Classical Latin name for a plum tree.) A genus of about 400 species of deciduous and occasionally evergreen shrubs and trees with simple alternate leaves and white or pink flowers that bear 5 petals and numerous stamens. The fruit is a fleshy drupe. This genus is important for its fruit trees and ornamentals. *Distr.* Temperate regions. *Fam.* Rosaceae. *Class* D.

P. amygdalus See *P. dulcis*

P. armeniaca APRICOT. Tree to 10m high. Flowers white, occasionally pink. The fruit is eaten fresh, tinned, or dried and the stone is a source of oil that is used in the cosmetics industry. *Distr.* N China.

P. avium MAZZARD, WILD CHERRY, SWEET CHERRY, BIRD CHERRY, CHERRY. Tree to 25m high. Bark purple. Flowers white. This species has given rise to a number of cultivars which provide the cherries of commerce that are used in preserves, the manufacture of liqueurs, and as a dessert fruit. The trees are also a source of timber for veneers. *Distr.* Europe.

P. cerasifera CHERRY PLUM, MIRABELLE. Deciduous shrub or small tree. Flowers white. Fruit small, round, red, borne on a short stalk. Numerous ornamental cultivars have been raised from this species. The fruits are edible and often used to make preserves and liqueurs, but are not commercially important. The tree is important, however, as a rootstock for other species of plum. *Distr.* Central Asia to the Balkans.

P. cerasus SOUR CHERRY, MORELLO CHERRY, AMARELLE CHERRY. Large shrub or small tree. Flowers white. The long-stalked, dark red fruit is used in cooking, particularly in the preparation of preserves. *Distr.* SW Asia, widely naturalized.

P. davidiana (After Armand David (1826–1900), French missionary who introduced it to cultivation in 1865.) DAVID'S PEACH. Deciduous tree to 9m high. Grown as an ornamental. *Distr.* China.

P. domestica PLUM, GAGE, DAMSON, ST JULIEN. A very variable, deciduous tree to 9m high. The fruit is also variable but includes the familiar plum of commerce. *Distr.* This species arose in cultivation from a number of other species.

P. dulcis ALMOND. Small, bushy, deciduous tree. Bark very dark. Flowers pink. Fruit fleshy with a large stone. The almond is the most widely grown of all the nuts, the seed being used in confectionery and as a source of oil. *Distr.* SE Europe, N Africa, W Asia.

P. fruticosa STEPPE CHERRY. Small shrub. Grown as an ornamental. *Distr.* Central Europe to Siberia.

P. glandulosa DWARF FLOWERING ALMOND. Small shrub. Flowers clustered, white, turning pink. *Distr.* NE Asia.

P. laurocerasus CHERRY LAUREL, VERSAILLES LAUREL. Evergreen shrub or small tree. Leaves leathery. Flowers small, borne in dense racemes. A very commonly grown hedging plant with numerous cultivars available. *Distr.* SE Europe, W Asia.

P. lusitanica PORTUGAL LAUREL, PORTUGUESE CHERRY LAUREL. Evergreen shrub. Flowers white, borne in erect racemes. *Distr.* Spain, Portugal.

P. myrobalana See *P. cerasifera*

P. padus BIRD CHERRY, HAGBERRY. Deciduous tree to 15m high. A source of timber. *Distr.* N Europe, N Asia.

P. persica PEACH, NECTARINE. Small deciduous tree. Flowers pink to white. Fruit fleshy, delicious. After APPLE the world's most widely grown fruit crop. *Distr.* China.

P. pissardii See *P. cerasifera*

P. sargentii (After Charles Sprague Sargent (1841–1927), the first director of the Arnold Arboretum, Boston.) Deciduous tree to 19m high. Flowers purple-pink. *Distr.* N Japan.

P. sato-zakura JAPANESE FLOWERING CHERRY. This name is applied to a very large group of complex hybrids and cultivars that are grown for their spectacular displays of flowers, particularly in Japan. *Distr.* Garden origin.

P. serotina AMERICAN CHERRY, BLACK CHERRY, PLUM CHERRY, RUM CHERRY. Large tree. Flowers small, borne in racemes. A source of timber, medicinal bark, and a flavouring for rum. *Distr.* E North America.

P. spinosa SLOE, BLACKTHORN. Deciduous, spiny shrub or small tree. Flowers white, appearing before the leaves. Fruit small, black. These trees are a source of hard-wearing wood, their leaves were formerly used as a tea substitute, and the fruits are used to flavour alcoholic drinks, notably gin. *Distr.* Europe, N Asia.

pseudacorus: false *Acorus*.

pseudarmeria: false *Armeria*.

Pseuderanthemum: (From Greek *pseudo*, false, and the genus name *Eranthemum*.) A genus of 60 species of herbs, subshrubs, and shrubs with simple leaves and racemes or cymes of long-tubed flowers. *Distr.* Polynesia. *Fam.* Acanthaceae. *Class* D.

P. atropurpureum Leaves tinged purple. Flowers 2-lipped, white, marked purple, borne in short spikes. Grown as a tender ornamental. *Distr.* Polynesia.

pseudocamellia: false *Camellia*.

pseudocapsicum: false *Capsicum*.

pseudochrysanthum: false *Rhododendron chrysanthum*.

Pseudocydonia (From Greek *pseudo*, false, and the genus name *Cydonia*.) A genus of 1 species of semi-evergreen shrub or small tree with peeling bark, simple leaves, and solitary, pale pink flowers. *Distr.* China. *Fam.* Rosaceae. *Class* D.

P. sinensis Grown as an ornamental. The large fruits are used as air fresheners in China.

pseudocyperus: false *Cyperus*.

pseudodictamnus: false *Dictamnus*.

Pseudolarix (From Greek *pseudo*, false, and the genus name *Larix*.) A genus of 1 species of deciduous, coniferous tree with needle-like leaves and clusters of woody fruiting cones that disintegrate at maturity. *Distr.* China. *Fam.* Pinaceae. *Class* G.

P. amabilis GOLDEN LARCH. Grown as an ornamental. *Distr.* S and E China.

P. kaempferi See *P. amabilis*

pseudomuscari: false *Muscari*.

pseudonarcissus: false *Narcissus*.

Pseudopanax (From Greek *pseudo*, false, and the genus name *Panax*.) LANCEWOOD. A genus of 6 species of shrubs and trees with different shaped leaves depending on the age of the plant and small flowers in clusters of umbels. Several species are grown as ornamentals. *Distr.* New Zealand, Chile. *Fam.* Araliaceae. *Class* D.

P. arboreus FIVE FINGERS. Robust tree to 8m high. *Distr.* New Zealand.

P. crassifolius LANCEWOOD. Evergreen tree to 10m high, little branched at first but becoming much branched. Adult leaves narrow. *Distr.* New Zealand.

P. ferox TOOTHED LANCEWOOD. Leaves long, thin, toothed, dark bronze-green. *Distr.* New Zealand.

pseudopetiolatus, -a, -um: with false petioles.

Pseudophegopteris (From Greek *pseudo*, false, and the genus name *Phegopteris*.) A genus of 20 species of medium-sized terrestrial ferns with pinnately divided leaves and sori that lack indusia. *Distr.* Tropical regions of the Old World. *Fam.* Thelypteridaceae. *Class* F.

P. levingei Grown as a tender ornamental. *Distr.* Tropical regions of the Old World.

Pseudophoenix (From Greek *pseudo*, false, and the genus name *Phoenix*.) A genus of 4 species of small to medium-sized palms with feather-shaped leaves and erect or arching inflorescences. *Distr.* Caribbean. *Fam.* Palmae. *Class* M.

P. ekmannii WINE PALM. Formerly tapped for fermentable sap. Now thought to be extinct.

pseudoplatanus: false *Platanus*.

Pseudosasa (From Greek *pseudo*, false, and the genus name *Sasa*.) A genus of 6 species of rhizomatous bamboos with erect stems and glabrous leaves. Several species are grown as ornamentals. *Distr.* E Asia. *Fam.* Gramineae. *Class* M.

P. japonica METAKE. A much cultivated bamboo used for screening and hedges. *Distr.* Japan, Korea.

Pseudotsuga (From Greek *pseudo*, false, and the genus name *Tsuga*.) DOUGLAS FIR. A genus of 6 species of evergreen, coniferous trees with spreading branches and somewhat flattened, needle-like leaves. The woody fruiting cones are pendulous and open to release the seeds. *Distr.* E Asia, W North America. *Fam.* Pinaceae. *Class* G.

P. menziesii (After Archibald Menzies (1754–1842), naval surgeon and botanist.) DOUGLAS FIR, OREGON PINE. A major source of timber. *Distr.* W North America.

P. taxifolia See *P. menziesii*

pseudoviola: false *Viola*.

Pseudowintera (From Greek *pseudo*, false, and the old genus name *Wintera*.) A genus of 3 species of evergreen shrubs and small trees with alternate, leathery, often coloured leaves and clusters of regular flowers that bear 5–6 petals and 5–15 stamens. *Distr.* New Zealand. *Fam.* Winteraceae. *Class* D.

P. colorata Grown as an ornamental. *Distr.* New Zealand.

Psidium (From *psidion*, the Greek name for a pomegranate.) GUAVA. A genus of 100 species of evergreen shrubs and trees with simple opposite leaves and large white flowers that bear 5 petals and many stamens. The fruit is a large, rounded, typically edible berry. *Distr.* Tropical America. *Fam.* Myrtaceae. *Class* D.

P. guajava (From the local vernacular name.) GUAVA. The pear-shaped, edible fruit is tinned or made into jelly, jam or chutney. *Distr.* Tropical America.

psilostemon: with glabrous stamens.

Psilotaceae A family of 2 genera and 6 species of small to medium-sized, often epiphytic, fern-like plants with spirally arranged leaves or scales and no true roots. *Distr.* Widespread in tropical and warm-temperate regions. *Class* F.
See: Psilotum.

Psilotum (From Greek *psilos*, naked, alluding to the sporangia which are not covered by an indusium.) FORK FERN, WHISK FERN. A genus of 3 species of terrestrial or epiphytic, clump-forming, fern-like plants that consist of pendent or arching, branching stems covered in scale-like leaves. *Distr.* Tropical and sub-tropical regions. *Fam.* Psilotaceae. *Class* F.

P. nudum SKELETON FORK FERN. Grown as an ornamental, particularly in Japan. *Distr.* Tropical regions of the Old World.

psittacinus, -a, -um: with parrot-like colours.

Psoralea (From Greek *psoraleos*, warty or scruffy, alluding to the glandular, dotted leaves of some species.) A genus of 20 species of perennial herbs and shrubs with alternate pinnate leaves that are occasionally reduced to a single leaflet and spikes, racemes, or heads of irregular, pea-like flowers. *Distr.* S Africa. *Fam.* Leguminosae. *Class* D.

P. pinnata BLUE PEA. Shrub to 2m high. Flowers numerous, blue with white wings. Grown as an ornamental. *Distr.* South Africa.

Psychotria (From Greek *psyche*, life, alluding to the medicinal properties of many of the species.) A genus of 1400 species of perennial herbs, shrubs, and trees with simple, opposite or whorled leaves and clusters of funnel- or bell-shaped flowers that are followed by red, blue, or black berries. Several species are grown as ornamentals. *Distr.* Tropical regions. *Fam.* Rubiaceae. *Class* D.

P. capensis Shrub or small tree. Flowers cream to yellow. Fruit black. *Distr.* South Africa.

Psylliostachys (From the specific name of *Plantago psyllion*, and Greek *stachys*, spike, alluding to the similar inflorescences.) A genus of 6 species of annual herbs with pinnately lobed leaves and spikes of white or pink flowers. *Distr.* W and Central Asia. *Fam.* Plumbaginaceae. *Class* D.

P. spicata Annual herb. Flowers pale pink. Grown as an ornamental. *Distr.* W and Central Asia.

Ptaeroxylaceae A family of 2 genera and 3 species of shrubs and trees with pinnate leaves and small, regular flowers. *Distr.* Tropical and S Africa, Madagascar. *Class* D.

Ptelea (Greek name for an elm alluding to the winged fruits.) A genus of 11 species of aromatic shrubs and small trees typically with trifoliate leaves and corymbs or panicles of green-white flowers that bear their parts in fours or fives and are followed by winged fruits. Several species are grown as ornamentals. *Distr.* North America. *Fam.* Rutaceae. *Class* D.

P. trifoliata HOP TREE, STINKING ASH, WATER ASH. Shrub or tree to 8m high. The fruits have been used as a hop substitute in brewing beer. *Distr.* E North America.

Pteridaceae A family of 6 genera and about 290 species of terrestrial, lithophytic, or aquatic ferns with erect or creeping stems and pinnate, entire, or palmate leaves. The sporangia are not borne in sori. *Distr.* Widespread. *Class* F.
See: Pteris.

pteridifolius, -a, -um: with fern-like leaves.

Pteridium (Classical Greek name for a fern, from *pteron*, feather or wing, alluding to the shape of the leaves.) A genus of 1 species of terrestrial fern with long, creeping rhizomes and stalked, divided leaves. *Distr.* N hemisphere. *Fam.* Dennstaedtiaceae. *Class* F.

P. aquilinum BRACKEN, EAGLE FERN, HOG BRAKE. A very widespread, invasive weed. The young fronds and rhizomes are eaten in Japan and were formerly widely eaten by North American Indians but are now known to promote stomach cancer. The leaves have been used as thatch and the rhizomes in the tanning industry.

Pteridophyllum (From Greek *pteris*, fern, and *phyllon*, leaf.) A genus of 1 species of rhizomatous perennial herb with pinnately divided, fern-likc, basal leaves and racemes of white flowers that bear 2 sepals, 4 petals, and 4 stamens. *Distr.* Japan. *Fam.* Papaveraceae. *Class* D.

P. racemosum Grown as a hardy ornamental Distr.

Pteris (Classical Greek name for a fern, from *pteron*, feather or wing, alluding to the shape of the leaves.) BRAKE, DISH FERN, TABLE FERN. A genus of 280 species of small to large terrestrial ferns with leaves that are pinnately to 4-pinnately divided. Several species are grown as tender ornamentals. *Distr.* Tropical and subtropical regions extending as far S as New Zealand. *Fam.* Pteridaceae. *Class* F.

P. cretica CRETAN BRAKE. Medium-sized fern. A commonly grown pot plant with many cultivars available. *Distr.* Tropical and subtropical regions.

P. multifida SPIDER FERN, SPIDER BRAKE. Used medicinally locally. *Distr.* Japan, China, Taiwan.

Pterocactus (From Greek *pteron*, a wing, and the old genus name *Cactus*, alluding to the winged seeds.) A genus of 9 species of shrubby cacti with tuberous roots, cylindrical or club-shaped stems, and small, short-lived leaves. The small, spreading, yellow or pink flowers are borne terminally. Several species are grown as ornamentals. *Distr.* Argentina. *Fam.* Cactaceae. *Class* D.

P. kuntzei Stems segmented, marked with violet lines. Flowers pale yellow, tinged orange-brown. *Distr.* Argentina.

P. tuberosus See *P. kuntzei*

Pterocarya (From Greek *pteron*, wing, and *karyon*, nut, alluding to the winged fruit.) WING-NUT. A genus of 10 species of large deciduous trees with alternate pinnate leaves and small unisexual flowers, the male and female flowers occurring in separate pendent catkins. The fruit is a winged nut. Several species are grown as ornamentals. *Distr*. Asia. *Fam*. Juglandaceae. *Class* D.

P. fraxinifolia CAUCASIAN WING-NUT. Tree to 38m, often with numerous stems. *Distr*. Caucasus, N Iran.

P. stenoptera Tree to 25m high. *Distr*. China.

pterocaulis, -e: with a winged stem

Pterocephalus (From Greek *pteron*, wing, and *kephale*, head, alluding to the feathery fruiting heads.) A genus of 25 species of annual and perennial herbs, subshrubs, and shrubs with opposite, simple or pinnately lobed leaves and flattened heads of purple-pink tubular flowers. The sepals have long terminal awns that are elongated so as to aid wind dispersal of the seeds. *Distr*. Mediterranean to Central Asia. *Fam*. Dipsacaceae. *Class* D.

P. parnassi See *P. perennis*

P. perennis Mat-forming perennial. Flowers purple-pink. *Distr*. Greece.

pterocephalus, -a, -um: with a winged head.

Pterostemonaceae A family of 1 genus and 2 species of shrubs with simple leaves. This family is sometimes included within the family Grossulariaceae. *Distr*. Mexico. *Class* D.

Pterostylis (From Greek *pteron*, wing, and *stylis*, column; the column is broadly winged.) GREENHOOD. A genus of 95 species of terrestrial orchids with rosettes of basal leaves and green flowers that bear a hood-like upper sepal and mobile lip. *Distr*. Australia to New Zealand, New Guinea, and New Caledonia. *Fam*. Orchidaceae. *Class* M.

P. banksii Flowers pale green-lined dark green, lip with an arching appendage. *Distr*. New Zealand.

P. curta Flowers white, striped green, lip twisted at apex. *Distr*. Australia, New Caledonia.

Pterostyrax (From Greek *pteron*, wing, and the genus name *Styrax*.) EPAULETTE TREE. A genus of 4 species of deciduous shrubs and trees with simple alternate leaves and loose panicles of bisexual flowers that bear their parts in 5s. Several species are grown as ornamentals. *Distr*. Burma to Japan. *Fam*. Styracaceae. *Class* D.

P. hispida FRAGRANT EPAULETTE TREE. Deciduous shrub or small tree to 15m high. Bark grey, peeling. Flowers white, fragrant. *Distr*. Japan, China.

Ptilostemon (From Greek *ptilon*, down, and *stema*, stamen.) A genus of 14 species of annual to perennial herbs and shrubs with simple or pinnately lobed, often spiny leaves and purple flowers in daisy-like heads that lack distinct rays. *Distr*. Mediterranean. *Fam*. Compositae. *Class* D.

P. afer Spiny perennial to 1m high. *Distr*. Balkans.

Ptilotrichum See *Alyssum*

Ptychosperma (From Greek *ptyche*, fold, and *sperma*, seed; the flesh of the seeds appears folded.) A genus of 30 species of medium-sized palms with solitary or clump-forming stems, feather-shaped leaves, and branched inflorescences. *Distr*. Australia, New Guinea, Solomon Islands, Micronesia. *Fam*. Palmae. *Class* M.

P. elegans ALEXANDER PALM, SOLITAIRE PALM. Stems solitary, to 12m high. Grown as an ornamental. *Distr*. Australia.

P. macarthurii Stems clump-forming, to 7m high. Grown as an ornamental. *Distr*. New Guinea.

pubescens: hairy.

pubicalyx: with hairy sepals.

pubiflorus, -a, -um: with hairy flowers.

pubigera, -um: softly hairy.

puccoon, red *Sanguinaria canadensis*.

pudding pipe tree *Cassia fistula*.

pudicus, -a, -um: shy, modest.

puka *Meryta sinclairii*.

pulchellus, -a, -um: pretty.

pulcher: pretty.

Pulicaria (From Latin *pulex*, flea.) FLEABANE. A genus of 40 species of annual to perennial herbs with simple leaves and yellow flowers in daisy-like heads that typically bear distinct yellow rays. *Distr*. Temperate and warm Eurasia, tropical and South Africa. *Fam*. Compositae. *Class* D.

P. dysenterica Stoloniferous perennial. Sometimes grown as an ornamental. *Distr.* Europe and N Africa.

Pulmonaria (From Latin *pulmo*, lung; the leaves of *Pulmonaria officinalis* were said resemble a diseased lung and so used to treat bronchitis under the doctrine of signatures.) LUNGWORT. A genus of 14 species of rhizomatous, hairy herbs with simple, occasionally white-spotted leaves, and cymes of funnel-shaped flowers. Several species and numerous cultivars are grown as ornamentals. *Distr.* Europe. *Fam.* Boraginaceae. *Class* D.

P. angustifolia Leaves narrow, plain green. Flowers bright blue. Numerous cultivars of this species are available. *Distr.* Central and E Europe.

P. officinalis JERUSALEM SAGE, LUNGWORT. Leaves white-spotted. Flowers changing from red to blue as they mature. Used medicinally for lung complaints. *Distr.* Europe.

P. saccharata JERUSALEM SAGE. Leaves narrow, spotted cream. Flowers pink, turning red with age. *Distr.* SE France, N Italy.

pulmonarioides: from the genus *Pulmonaria*, with the ending *-oides*, indicating resemblance.

Pulsatilla (From Latin *pellere*, *pulsare* to beat; the flowers more easily.) PASQUE FLOWER. A genus of 30 species of clump-forming, often densely hairy, perennial herbs with pinnately or palmately divided leaves and large regular flowers that typically bear 6 petal-like tepals (that are hairy on the outside), a whorl of nectaries, and numerous stamens. Several species are grown as ornamentals. *Distr.* N temperate regions. *Fam.* Ranunculaceae. *Class* D.

P. vernalis Densely hairy, perennial herb. Flowers pearly-white. *Distr.* Europe.

P. vulgaris PASQUE FLOWER. Perennial herb. Leaves feathery. Flowers purple, occasionally white. This species has given rise to a number of ornamental cultivars and was formerly used medicinally. *Distr.* Europe.

pulverulentus, -a, -um: dusty, covered with a glaucous bloom.

pulvinaris: cushion-like.

pumilus, -a, -um: dwarf.

Pummelo See *Citrus*

pumpkin *Cucurbita pepo*, *C. maxima* **foetid wild** ~ *C. foetidissima*.

punctatus, -a, -um: spotted.

pungens: sharp-pointed.

Punica (From Latin *malum punicum*, Pliny's name for the fruit.) A genus of 2 species of deciduous shrubs or trees with simple, opposite or clustered leaves and groups of funnel-shaped, red or yellow flowers that bear 5–7 crumpled petals. The fruit is a large berry with numerous fleshy seeds and a leathery shell. *Distr.* SE Europe to Himalaya. *Fam.* Lythraceae. *Class* D.

P. granatum POMEGRANATE. Shrub or tree. Leaves bear solitary apical nectaries. Flowers bright red. The fruit is eaten fresh or fermented into the liqueur grenadine. The bark and flowers have been used medicinally. *Distr.* Iran, Afghanistan.

puniceus, -a, -um: purple-red-coloured.

punjabensis, -e: of the Punjab, India.

purple top *Verbena bonariensis*.

purple wreath *Petrea volubilis*.

purpurascens: becoming purplish.

purpuratus, -a, -um: purple.

purpureus, -a, -um: purple.

Purshia (After Frederick T. Pursh (1774–1820), German botanist who travelled in North America.) ANTELOPE BUSH. A genus of 2 species of deciduous shrubs and small trees with 3-lobed leaves and small solitary flowers. *Distr.* W North America. *Fam.* Rosaceae. *Class* D.

P. tridentata Grown as a hardy ornamental.

purslane *Claytonia*, *Portulaca*, Portulacaceae **common** ~ *Portulaca oleracea* **rock** ~ *Calandrinia* **tree** ~ *Atriplex halimus* **white** ~ *Claytonia australasica* **winter** ~ *Montia perfoliata*.

Puschkinia (After Count Apollos Mussin-Puschkin (died 1815), Russian chemist who collected in the Caucasus.) A genus of 1 species of bulbiferous herb with linear leaves and loose racemes of small, bell-shaped, blue-striped flowers. *Distr.* Caucasus, Turkey, N

Iran, N Iraq, Lebanon. *Fam.* Hyacinthaceae. *Class* M.

P. libanotica See *P. scilloides*

P. scilloides STRIPED SQUILL, LEBANON SQUILL. Grown as an ornamental. *Distr.* Caucasus, Turkey, N Iran, N Iraq and Lebanon.

pusillus, -a, -um: dwarf.

pussy ears *Kalanchoe tomentosa.*

pussy paws *Spraguea umbellata.*

pussy toes *Antennaria.*

pustulatus, -a, -um: covered in pustules or blisters.

Putoria (From Latin *putor*, strong smell, alluding to the malodorous leaves.) A genus of 3 species of small foetid shrubs with simple, opposite, leathery leaves and funnel-shaped, 4- or 5-lobed flowers. *Distr.* Europe. *Fam.* Rubiaceae. *Class* D.

P. calabrica STINKING MADDER. Mat-forming shrub. Flowers pink, borne in dense clusters. Occasionally grown as an ornamental. *Distr.* Europe.

Puya (Native name of the plants in Chile.) A genus of 170 species of terrestrial herbs with rosettes of large, leathery, spiny leaves borne at ground level or on stout stems. The showy flowers are borne in tall branched or simple inflorescences. *Distr.* South America. *Fam.* Bromeliaceae. *Class* M.

P. alpestris Leaves with hooked marginal teeth. Flowers metallic blue. *Distr.* S Chile.

P. chilensis Leaf rosettes borne on short woody stems. The leaves yield a fibre used in the manufacture of fishing nets. Their hooked marginal spines are deemed a hazard to sheep which become entangled in the leaf rosettes. *Distr.* Central Chile.

Pycnanthemum (From Greek *pyknos*, dense, and *anthos*, flower, alluding to the inflorescences.) A genus of 17 species of perennial herbs with simple opposite leaves and dense heads of tubular 2-lipped flowers. Some species are used locally as medicines and flavourings. *Distr.* North America. *Fam.* Labiatae. *Class* D.

P. pilosum Stems erect, to 1.5m high. Flowers in dense clusters, pink. Grown as an ornamental. *Distr.* Central and E North America.

pycnanthus, -a, -um: with dense flowers.

Pycnostachys (From Greek *pyknos*, dense, and *stachys*, spike, alluding to the dense inflorescences.) A genus of 37 species of erect perennial herbs and shrubs with simple, opposite or whorled leaves and dense terminal spikes of tubular to bell-shaped, 2-lipped flowers. Several species are grown as ornamentals. *Distr.* Tropical and S Africa, Madagascar. *Fam.* Labiatae. *Class* D.

P. dawei Bushy perennial herb. Leaves red-glandular beneath. Flowers in dense heads, blue. *Distr.* Tropical Africa.

P. urticifolia Erect herb or subshrub. Leaves sharply toothed. Flowers small, bright blue. *Distr.* Tropical and S Africa.

pycnostachyus, -a, -um: with dense spikes.

pygmaeus: dwarf.

Pyracantha (From Greek *pyr*, fire, and *akantha*, thorn, alluding to the spiny shoots and red berries of some species.) FIRE THORN. A genus of 6 species of evergreen shrubs with simple alternate leaves and small white flowers that are followed by showy red to orange berries. Several of the species and numerous cultivars are grown as ornamentals. *Distr.* Eurasia. *Fam.* Rosaceae. *Class* D.

P. angustifolia Dense bushy shrub. Fruits yellow-orange. *Distr.* W China.

P. atalantioides Arching shrub. Fruits bright red. *Distr.* Central China.

P. coccinea FIRE THORN. Dense bushy shrub. Berries bright red. Numerous cultivars have been raised from this species. *Distr.* S Europe, W Asia.

P. gibbsii See *P. atalantioides*

pyramidalis, -e: pyramid-shaped.

pyramidatus, -a, -um: pyramid-shaped.

pyramid tree, Queensland *Lagunaria patersonii.*

pyrenaeus, -a, -um: of the Pyrenees.

pyrenaicus, -a, -um: of the Pyrenees.

pyrethrifolius, -a, -um: from the genus *Pyrethrum*, and *folius*, leaved.

Pyrethropsis (From the genus name *Pyrethrum*, and Greek *opsis*, appearance.) A genus of about 10 species of tufted perennial herbs and subshrubs with small leaves and flowers in solitary daisy-like heads which bear a few distinct rays. *Distr.* NW Africa. *Fam.* Compositae. *Class* D.

P. hosmariensis Bushy perennial herb. Flowers yellow, rays white. *Distr.* Morocco.

P. mariesii Perennial herb, to 30cm high. Foliage silver-grey. Rays yellow, turning purple. *Distr.* Algeria.

Pyrethrum See *Tanacetum*

pyrethrum *Tanacetum coccineum*, *T. cinerariifolium* **Dalmatia** ~ *T. cinerariifolium* **garden** ~ *T. coccineum*.

pyrifolius, -a, -um: from the genus *Pyrus*, PEAR, and *folius*, leaved.

Pyrola (Diminutive of *Pyrus*, PEAR, alluding to the similar leaves.) WINTERGREEN. A genus of 35 species of perennial, rhizomatous, dwarf herbs with simple basal leaves and nodding flowers. Some species have medicinal qualities, others are cultivated as ornamentals. *Distr.* N Hemisphere, Sumatra, and temperate South America. *Fam.* Ericaceae. *Class* D.

P. asarifolia Flowers pink-crimson. *Distr.* North America.

P. elliptica Flowers white. *Distr.* North America and Japan.

P. media Flowers white, tinged with red. *Distr.* Europe, W Asia.

P. rotundifolia Flowers white. *Distr.* Europe, W Asia.

pyroliflorus, -a, -um: with flowers resembling those of the genus *Pyrola*.

pyroloides: from the genus *Pyrola*, with the ending *-oides*, indicating resemblance.

Pyrostegia (From Greek *pyr*, fire, and *stega*, roof, alluding to the flower colour.) A genus of 3–4 species of evergreen woody climbers with flowers that consist of 2 leaflets and a tendril. The flowers are tubular and appear club-shaped before they are fully open. *Distr.* Tropical South America. *Fam.* Bignoniaceae. *Class* D.

P. venusta FLAME VINE, GOLDEN SHOWER, FLAME FLOWER. Climbs to 10m. Flowers orange. Grown as a tender ornamental. *Distr.* Brazil, Paraguay.

Pyrrosia (From Greek *pyrros*, red, alluding to the colour of the hairs.) FELT FERN. A genus of 100 species of epiphytic or terrestrial ferns with simple or lobed, fleshy or leathery leaves that are covered in a dense red-tinged felt. Several species are grown as ornamentals. *Distr.* Tropical regions of the Old World. *Fam.* Polypodiaceae. *Class* F.

P. lingua JAPANESE FELT FERN. Fronds to 30cm long, simple, leathery. *Distr.* S Japan.

Pyrus (Classical Latin name for a pear.) PEAR. A genus of about 20 species of deciduous and occasionally evergreen shrubs and trees with simple alternate leaves and regular white flowers that bear 5 petals and around 20 stamens. The fruit is technically termed a pome: see APPLE, *Malus*. The flesh differs from that of an apple in that it contains numerous hard cells which give it a gritty texture. Several species are grown for their fruit and numerous ornamental cultivars have been raised. *Distr.* Eurasia. *Fam.* Rosaceae. *Class* D.

P. calleryana (After J. Callery, a French missionary who collected the type specimen.) Deciduous tree. Grown as an ornamental with a number of cultivars having been raised. It is frequently planted as a street tree in parts of North America. *Distr.* China.

P. communis COMMON PEAR. Deciduous tree to 15m high. This species is widely grown for its fruits which are eaten fresh or tinned or fermented into perry, a drink not unlike cider; it is also grown as an ornamental. *Distr.* Not known in the wild, probably the result of complex hybridization in cultivation.

P. kawakamii EVERGREEN PEAR. Evergreen shrub or small tree. *Distr.* Chin, Taiwan.

P. pyrifolia CHINESE PEAR, JAPANESE PEAR, SAND PEAR. Deciduous tree to 12m high. *Distr.* China, naturalized in Japan, and grown in tropical Asia.

pyxidatus, -a, -um: with a lid, a pyxidium.

Q

Qiongzhuea See *Chimonobambusa*.

quadrangularis, -e: four-angled.

quadrangulus, -a, -um: four-angled.

quadriauritus, -a, -um: four-eared.

quadribracteolus, -a, -um: with four bracts.

quadrifidus, -a, -um: divided into four.

quadrifolius, -a, -um: four-leaved.

quake *Briza*.

quaker ladies *Houstonia caerulea*.

quamash *Camassia quamash*.

quartervine *Bignonia capreolata*.

quassioides: from the genus *Quassia*, with the ending *-oides*, indicating resemblance.

Queen of Sheba *Podranea brycei*.

queen of the meadows *Filipendula ulmaria*.

queen of the prairie *F. rubra*.

queen's tears *Billbergia nutans*.

queen wreath *Petrea volubilis*.

Quercifilix See *Tectaria*

quercifolius, -a, -um: from the genus *Quercus*, OAK, and *folius*, leaved.

Quercus (Classical Latin name for these trees.) OAK. A genus of about 600 species of deciduous and evergreen trees and shrubs with simple, alternate, lobed or toothed leaves and small, unisexual, wind-pollinated flowers. The fruits are characteristic, consisting of a single-seeded nut, the acorn, sitting in a scale-covered cup, the cupule. This is an important economic genus with the trees being a source of high-quality timber, the acorns eaten by humans and fed to pigs, the bark and galls a source of tannin, and the leaves used as fodder for both goats and silkworms. *Distr.* Temperate and warm regions of the N Hemisphere and mountainous regions of the tropics. *Fam.* Fagaceae. *Class* D.

Q. alba WHITE OAK, QUEBEC OAK. Deciduous tree to 50m high. Important timber and fuel source. *Distr.* SE North America.

Q. alnifolia GOLDEN OAK OF CYPRUS. Evergreen shrub or small tree, to 10m. *Distr.* Cyprus.

Q. bicolor SWAMP WHITE OAK. Deciduous tree to 25m high. *Distr.* NE North America.

Q. borealis See *Q. rubra*

Q. castaneifolia CHESTNUT-LEAVED OAK. Deciduous, large-crowned tree, to 35m high. *Distr.* N Iran, Caucasus.

Q. cerris TURKEY OAK, MANNA OAK. Deciduous, somewhat conical tree to 45m high. Source of timber. *Distr.* Central and S Europe, E Mediterranean region.

Q. chrysolepis MAUL OAK, CANYON OAK, LIVE OAK. Evergreen tree to 25m high. *Distr.* W North America.

Q. coccifera KERMES OAK. Evergreen bushy shrub. Used for feeding cochineal insects. *Distr.* W Mediterranean region, N Africa.

Q. coccinea SCARLET OAK. Deciduous tree to 30m high. Leaves bright red in autumn. Source of timber. *Distr.* E North America.

Q. dentata DAIMIO OAK, JAPANESE EMPEROR OAK. Deciduous tree to 25m high. *Distr.* Japan, Korea, China.

Q. falcata AMERICAN RED OAK, SOUTHERN RED OAK, SWAMP RED OAK. Deciduous tree to 30m high. Bark red-brown, sometimes falling in large scales. *Distr.* S and E North America.

Q. frainetto HUNGARIAN OAK. Deciduous tree to 30m. *Distr.* S and SE Europe, Turkey.

Q. gambellii SHIN OAK, GAMBEL OAK. Deciduous shrub or small tree. Acorns edible. *Distr.* S North America.

Q. garryana OREGON WHITE OAK. Deciduous tree to 30m high. An important source of timber and fuel. *Distr.* W North America.

Q. glabra See *Lithocarpus glaber*

Q. ilex HOLM OAK, HOLLY OAK, EVERGREEN OAK. Evergreen tree. Leaves variable but sometimes sharply toothed. Acorns edible, galls used in tanning. *Distr.* Mediterranean region, SW Europe.

Q. ilicifolia BEAR OAK, SCRUB OAK. Deciduous shrub or small tree. *Distr.* E North America.

Q. imbricaria (From Latin *imbrex*, tile; the wood was used for roof tiles.) SHINGLE OAK. Deciduous tree to 20m high. Timber used for clapboards and tiles. *Distr.* Central and E North America.

Q. kelloggii CALIFORNIAN BLACK OAK. Deciduous shrub or tree to 35m. *Distr.* W North America.

Q. libani LEBANON OAK. Deciduous slender shrub or tree. *Distr.* Turkey, Syria, N Iran, N Iraq.

Q. lobata CALIFORNIAN WHITE OAK, VALLEY OAK. Deciduous tree to 35m. Acorns eaten by Indians. *Distr.* California.

Q. macrocarpa BUR OAK, MOSSY CUP OAK. Deciduous tree to 35m high. Timber used for construction and furniture. *Distr.* Central and E North America.

Q. marilandica BLACK JACK OAK. Small deciduous tree. Wood used to make charcoal. *Distr.* SE North America.

Q. mongolica JAPANESE OAK. Large deciduous tree. Locally important for timber. *Distr.* E Asia.

Q. nigra POSSUM OAK, WATER OAK. Deciduous tree. Acorns edible. *Distr.* SE North America.

Q. palustris PIN OAK. Fast-growing deciduous tree. *Distr.* E North America.

Q. pedunculata See *Q. robur*

Q. petraea DURMAST OAK, SESSILE OAK. Deciduous tree. Acorns sessile. *Distr.* Europe, W Asia.

Q. phellos WILLOW OAK. Deciduous tree. *Distr.* E North America.

Q. prinus CHESTNUT OAK, AMERICAN WHITE OAK. Deciduous tree to 30m high. *Distr.* E North America.

Q. pumila See *Q. prinus*

Q. robur (Classical Latin name for hard wood, especially oak.) COMMON OAK, PEDUNCULATE OAK, ENGLISH OAK. Deciduous tree to 40m. Acorns on stalks. The most important British hardwood timber. Acorns used as pig fodder and as a coffee substitute. *Distr.* Europe, Caucasus, SW Asia.

Q. rubra RED OAK, NORTHERN RED OAK. Deciduous tree to 30m high. Leaves turn red in winter. Timber used for general construction. *Distr.* E North America.

Q. sessiliflora See *Q. petraea*

Q. suber CORK OAK. Evergreen tree to 25m high. The thick bark is the cork of commerce, sheets of it being removed, in rotation, from the living trees. *Distr.* S Europe, N Africa.

Q. × turneri (After Spencer Turner (*c.*1728–76), in whose Essex nursery it was raised.) TURNER'S OAK. A hybrid of *Q. ilex* × *Q. robur*. Semi-evergreen tree to 20m high. *Distr.* Garden origin.

Q. velutina BLACK OAK. Deciduous tree to 25m high. A source of yellow dye used in printing calico. *Distr.* E North America.

Q. virginiana SOUTHERN LIVE OAK. Evergreen tree to 45m. Formerly an important source of ship-building timber. *Distr.* S USA.

Quesnelia (After M. Quesnel, a 19th-century French consul and botanist in Guiana, who introduced the genus to Europe.) A genus of 14 species of rhizomatous herbs with rosettes of spiny leaves. The flowers are borne in tall inflorescences with brightly coloured bracts. Several species are cultivated as ornamentals. *Distr.* SE Brazil. *Fam.* Bromeliaceae. *Class* M.

Q. marmorata GRECIAN URN PLANT. Leaves blue-green, forming a nearly circular rosette.

Q. seideliana Leaf rosettes stemless. Flowers blue.

quickbeam *Sorbus aucuparia.*

Quiinaceae A family of 4 genera and 44 species of trees and large shrubs with opposite or whorled, simple to pinnate leaves and small, typically bisexual flowers that bear 4–5 sepals and petals. *Distr.* Tropical forests of Central and South America and the West Indies. *Class* D.

quill, pink *Tillandsia cyanea.*

quillai *Quillaja saponaria.*

Quillaja (From *quillay*, the Chilean name for these plants.) SOAPBARK TREE. A genus of 4 species of evergreen shrubs and trees with simple opposite leaves and regular white flowers that bear 5 petals and 10 stamens. *Distr.* Temperate South America. *Fam.* Rosaceae. *Class* D.

Q. saponaria QUILLAI, SOAPBARK TREE, SOAP BUSH. Shrub or small tree to 10m high. The bark is used as a source of soap and in the manufacture of fire extinguishers. *Distr.* Chile.

quinatus, -a, -um: with parts in fives.

quince *Cydonia oblonga* **common** ~ *C. oblonga* **flowering** ~ *Chaenomeles* **Japanese** ~ *C. japonica* **Maule's** ~ *C. japonica* **Portuguese** ~ *Cydonia oblonga.*

quinine, wild *Parthenium integrifolium.*

quinine bush *Garry fremontii.*

quinoa *Chenopodium quinoa.*

quinquefolius, -a, -um: five-leaved.

quinquelobatus, -a, -um: with five lobes, usually referring to the leaves.

quinquelocularis, -e: with five locules.

quinquenervis, -e: with five nerves.

Quintinia (After Jean de La Quintinie (1626–88), French gardener.) A genus of 2 species of shrubs and trees with simple leathery leaves and racemes or panicles of numerous small flowers that bear their parts in fours or fives. *Distr.* New Zealand. *Fam.* Escalloniaceae. *Class* D.

Q. acutifolia Tree to 12m high. Flowers lilac, borne in short racemes. Grown as a tender ornamental. *Distr.* New Zealand.

quintuplinervius, -a, -um: five-veined.

Quisqualis (From Latin *quis*, who, and *qualis*, what kind, either alluding to the difficulty that was encountered in placing the genus in a family or from the Malay name *udani* which was punned with the Dutch *hoedanig*, how or what.) A genus of 16 species of scrambling shrubs with simple opposite leaves and leaf stalks that persist after the blade has fallen and act like thorns. The showy trumpet-shaped flowers are borne in racemes or panicles and have their parts in fives. *Distr.* Tropical Asia and Africa, S Africa. *Fam.* Combretaceae. *Class* D.

Q. indica RANGOON CREEPER. Rampant climber. Flowers fragrant, white, turning pink with age. Grown as a tender ornamental. *Distr.* Tropical Asia and Africa.

quitensis, -e: of Quito, Ecuador.

Quivisia See *Turraea*

R

rabbit bush *Chrysothamnus*.

racemiflorus, -a, -um: with flowers in racemes.

racemosus, -a, -um: racemose, referring to the type of inflorescence.

Racosperma See *Acacia*

radiatus, -a, -um: radiating.

radicans: with rooting stems.

radicatus, a, -um: with roots.

radiiflorus, -a, -um: with radial flowers.

radish *Raphanus sativus*.

radula: with a hard rough surface; the same word is used for a snail's rasping tongue.

Rafflesiaceae A family of 9 genera and 50 species of root or stem parasites with a thin network of underground parts that invade the host plant's tissues. Flower buds develop within the host and then push through to the surface. The flowers are unisexual, lack petals, but have a fleshy calyx of 4–6 lobes. The flowers of *Rafflesia* are the largest known. *Distr*. Tropical and occasionally warm-temperate regions. *Class* D.

ragged robin *Lychnis flos-cuculi*.

ragwort *Senecio jacobaea* **marsh** ~ *S. aquaticus* **Oxford** ~ *S. squalidus*.

rainbow star *Cryptanthus bromelioides*.

rain poll *Ranunculus repens*.

rain tree *Brunfelsia undulata* **Chinese** ~ *Koelreuteria* **golden** ~ *K. paniculata* **white** ~ *Brunfelsia undulata*.

raisin tree *Hovenia dulcis* **Chinese** ~ *H. dulcis* **Japanese** ~ *H. dulcis*.

rakkyo *Allium chinense*.

ramentaceus, -a, -um: covered with thin scales.

ramiflorus, -a, -um: flowering on older branches.

ramine *Boehmeria nivea*.

Ramonda (After Louis Francis Ramond (1753–1827), French botanist.) A genus of 3 species of low-growing, perennial, stemless herbs with rosettes of simple leaves and flowers that have a short basal tube and 4–6 spreading lobes. All 3 species and several cultivars are grown as ornamental rock-garden plants. *Distr*. Mountainous areas of S Europe. *Fam*. Gesneriaceae. *Class* D.

R. myconi (After Francisco Mico (born 1528), a Spanish physician and botanist.) Flowers violet to pink or white with a yellow centre. *Distr*. Pyrenees, N Spain.

R. nathaliae Flowers white to lilac or purple with an orange-yellow centre. *Distr*. Balkans.

R. pyrenaica See *R. myconi*

R. serbica Flowers somewhat bell-shaped, white to lilac-purple. *Distr*. SE Europe.

ramosissimus, -a, -um: much branched.

ramosus, -a, -um: branched.

rampion *Campanula rapunculus*, *Phyteuma orbiculare* **horned** ~ *Phyteuma* **round-headed** ~ *P. orbiculare* **spiked** ~ *P. spicatum*.

ramsons *Allium ursinum*.

ramulosus, -a, -um: with slender branches.

Ranunculaceae The Buttercup family. 52 genera and about 1500 species of terrestrial and aquatic herbs, woody climbers, and small shrubs with simple or compound leaves and bisexual, typically regular flowers that bear spirally arranged tepals and numerous stamens. Numerous species are grown as ornamentals, some have medicinal properties, and some are poisonous. *Distr*. Widespread but centred in temperate and cool regions. *Class* D.

See: Aconitum, Actaea, Adonis, Anemone, Anemonella, Anemonopsis, Aquilegia, Callianthemum, Caltha, Cimicifuga, Clematis, Consolida, Coptis, Delphinium, Eranthis, Helleborus, Hepatica, Hydrastis, Isopyrum, Myosorus, Nigella, Paraquilegia, Pulsatilla, Ranunculus, Semiaquilegia, Thalictrum, Trollius, Xanthorhiza.

ranunculoides from the genus *Ranunculus*, BUTTERCUP, with the ending *-oides*, indicating resemblance.

Ranunculus (Classical Latin name, from *rana*, frog, as many species grow in wet places.) BUTTERCUP, CROWFOOT, PAIGLE. A genus of about 400 species of annual to perennial herbs with simple or compound leaves and yellow, white, or red flowers that bear 3–8 petal-like sepals and numerous stamens. True petals are either absent or very much reduced. Many species are grown as ornamentals. All are poisonous. *Distr.* Widespread but chiefly in temperate and cool regions. *Fam.* Ranunculaceae. *Class* D.

R. acris YELLOW BACHELOR'S BUTTONS, MEADOW BUTTERCUP. Perennial herb. Flowers glossy yellow. *Distr.* Europe, SW Asia, naturalized elsewhere.

R. amplexicaulis Perennial. Roots fleshy. Flowers white, occasionally pink. *Distr.* Pyrenees, N Spain.

R. aquatilis WATER CROWFOOT. Submerged aquatic herb. Leaves finely divided below the water or lobed and floating on the surface. Flowers emergent, white. *Distr.* Europe.

R. ficaria LESSER CELANDINE, PILEWORT. Tuberous perennial. Flowers golden yellow, fading to white. Blanched leaves sometimes eaten. Tubers said, through the doctrine of signatures, to cure piles. *Distr.* Europe, NW Africa, SW Asia, naturalized North America.

R. flammula LESSER SPEARWORT. Variable semi-aquatic perennial. Flowers yellow. *Distr.* Europe to temperate Asia.

R. lingua GREATER SPEARWORT. Semi-aquatic stoloniferous perennial; flowers bright yellow. *Distr.* Europe, Caucasus, Siberia and Central Asia.

R. repens CREEPING BUTTERCUP, RAIN POLL. Stoloniferous perennial. Flowers golden-yellow. A persistent weed although ornamental cultivars are available. *Distr.* Europe, Asia, naturalized North America, New Zealand.

Ranzania (After Ono Ranzon, Japanese naturalist.) A genus of 1 species of rhizomatous herb with trifoliate leaves and nodding purple flowers that appear before the leaves. The fruit is a many-seeded berry. *Distr.* Japan (Honshu). *Fam.* Berberidaceae. *Class* D.

R. japonica Grown as an ornamental.

Raoulia (After Edward Raoul (1815–52), French surgeon who studied New Zealand plants.) A genus of about 20 species of tufted or creeping perennial herbs and sub-shrubs with small crowded leaves and flowers in daisy-like heads that lack distinct rays. Some species are cultivated ornamentally. *Distr.* New Zealand. *Fam.* Compositae. *Class* D.

R. australis Mat-forming hairy perennial. Flowers yellow. *Distr.* New Zealand (North and South Island).

R. glabra Prostrate perennial herb. Flowers yellow. *Distr.* New Zealand (North and South Islands).

R. haastii Cushion-forming perennial herb. Flowers yellow. *Distr.* New Zealand (South Island).

R. hectoris Mat-forming. Flowers pale yellow. *Distr.* New Zealand (South Island).

R. hookeri (After Sir Joseph Hooker (1817–1911), English botanist.) Perennial herb forming flat mats. *Distr.* New Zealand (North and South Island).

R. lutescens See *R. australis*

R. monroi Perennial mat-forming herb. *Distr.* New Zealand (South Island).

R. parkii Perennial mat-forming herb. *Distr.* New Zealand (South Island).

R. × petrimea Perennial herb forming hemispherical cushions. Flowers red. *Distr.* New Zealand (North and South Island).

R. subsericea Perennial mat-forming herb. *Distr.* New Zealand (South Island).

R. tenuicaulis Perennial, white-hairy, mat-forming herb. *Distr.* New Zealand (North and South Islands).

Rapateaceae A family of 17 genera and 85 species of perennial, often large, rhizomatous herbs frequently found in swampy habitats. *Distr.* Mainly tropical South America but with one genus occurring in W tropical Africa. *Class* M.

rape: forage ~ *Brassica napus* **oilseed** ~ *B. napus* **turnip** ~ *B. campestris*.

raphanifolius, -a, -um: from the genus *Raphanus*, and *-folius*, leaved.

Raphanus (Classical Greek name for RADISH, *R. sativus*.) A genus of 8 species of annual, biennial, or perennial herbs typically with succulent roots and pinnate leaves. *Distr.* W and Central Europe to Central Asia. *Fam.* Cruciferae. *Class* D.

R. sativus RADISH. Root eaten raw in salads, some varieties cooked and eaten as a vegetable, others grown as winter fodder for animals. *Distr.* Cultivated for more than 4500 years; probably derived from *R. landra* of Italy.

Raphia (From Greek *raphis*, needle, alluding to the beaked fruits.) A genus of 28 species of medium-sized palms with large feather-shaped leaves and clustered, erect or pendent inflorescences. *Distr.* Tropical America, Africa and Madagascar. *Fam.* Palmae. *Class* M.

R. farinifera RAFFIA PALM. The young leaves are the source of the fibre raffia, used in horticulture and basketry. *Distr.* Tropical Africa, Madagascar.

R. ruffia See *R. farinifera*

rapunculoides from the genus *Rapunculus* (now *Campanula*), with the ending *-oides*, indicating resemblance.

raspberry *Rubus idaeus* **European** ~ *R. idaeus* **flowering** ~ *R. odoratus* **Mauritius** ~ *R. rosifolius* **purple flowering** ~ *R. odoratus* **red** ~ *R. idaeus* **Rocky Mountain** ~ *R. deliciosus* **Rocky Mountian flowering** ~ *R. deliciosus* **wild** ~ *R. idaeus*.

rata *Metrosideros excelsa*.

Ratibida PRAIRIE CONEFLOWER, MEXICAN HAT. A genus of about 6 species of coarse-hairy biennial and perennial herbs with pinnate leaves and flowers in solitary daisy-like heads. Several species are cultivated as ornamentals. *Distr.* North America, Mexico. *Fam.* Compositae. *Class* D.

R. columnifera LONG-HEAD CONEFLOWER, PRAIRIE CONEFLOWER. Glandular perennial. Rays yellow to red or brown-purple. *Distr.* W North America, naturalized in E North America.

rattan, ground *Rhapis excelsa*.

rattle weed *Baptisia tinctoria*.

Ravenala (From the Madagascan vernacular name.) A genus of 1 species of giant tree-like herb, to 16m high. The sheaths of the large banana-like leaves form a pseudostem and the flowers appear from within large boat-shaped bracts. Both the leaf sheaths and the floral bracts are said to collect water that can be drunk, in an emergency, by a traveller. *Distr.* Madagascar. *Fam.* Strelitziaceae. *Class* M.

R. madagascariensis TRAVELLER'S PALM, TRAVELLER'S TREE. Widely planted in tropical regions as an ornamental.

Rechsteineria See *Sinningia*

recognitus, -a, -um: recognized.

rectifolius, -a, -um: with erect or ascending leaves.

rectus, -a, -um: erect, straight, upright.

recurvatus, -a, -um: curved back.

recurvifolius, -a, -um: with recurved leaves.

recurvus, -a, -um: curved back.

recutitus , -a, -um: circumcised.

redbud *Cercis canadensis* **Californian** ~ *C. occidentalis* **Chinese** ~ *C. chinensis* **western** ~ *C. occidentalis*.

red-hot poker *Kniphofia*.

redivivus, -a, -um: coming back to life.

redolens: scented.

red ribbons *Clarkia concinna*.

red star *Rhodohypoxis baurii*.

reductus, -a, -um: reduced, dwarf.

red, white, and blue flower *Cuphea ignea*.

red-veined pie plant *Rheum australe*.

redwood, Californian *Sequoia sempervirens* **coast** ~ *S. sempervirens* **dawn** ~ *Metasequoia glyptostroboides* **giant** ~ *Sequoiadendron giganteum.*

reed *Phragmites, P. australis, P. australis* **bur** ~ *Sparganium* **cane** ~ *Arundinaria gigantea* **feather** ~ *Calamagrostis x acutiflora* **giant** ~ *Arundinaria gigantea, Arundo, A. donax* **coast bur** ~ *Sparganium minimum* **Mauritania vine** ~ *Ampelodesmus mauritanicus* **unbranched bur** ~ *Sparganium emersum.*

reedmace Typhaceae **narrow-leaved** ~ *Typha angustifolia.*

read rhapis *Rhapis humilis.*

reflexipetalus, -a, -um: with reflexed petals.

reflexus, -a, -um: reflexed, bent backwards.

refractus, -a, -um: bent or curved back.

regalis, -e: royal.

reginae: of the Queen.

regius, -a, -um: royal.

Regnellidium (After Anders Fedrik Regnell (died 1884), Swedish botanist.) A genus of 1 species of aquatic or semi-aquatic fern with creeping rhizomes and 2-lobed leaves. *Distr.* S Brazil. *Fam.* Marsileaceae. *Class* F.

R. diphyllum Occasionally grown as an ornamental. *Distr.* S Brazil.

Rehderodendron (After Alfred Rehder (1863–1949), botanist at the Arnold Arboretum, Boston and the Greek *dendron*, tree.) A genus of 9 species of deciduous shrubs or trees with simple alternate leaves and racemes or panicles of white flowers that bear their parts in fives. *Distr.* S and W China. *Fam.* Styracaceae. *Class* D.

R. macrocarpum Tree to 9 meters. Branches red. Fruit red. Grown as a hardy ornamental. *Distr.* W China.

Rehmannia (After Joseph Rehmann (1753–1831), German physician.) A genus of 9–10 species of perennial herbs with simple, toothed, alternate leaves and racemes of spreading, 2-lipped, 5-lobed flowers. Several species are grown as ornamentals. *Distr.* E Asia. *Fam.* Scrophulariaceae. *Class* D.

R. angulata Stem to 30cm high. Leaves pinnately lobed. Flowers yellow to red. *Distr.* Central China.

R. elata Stem to 2m high. Flowers bright purple with a yellow throat. *Distr.* China.

R. glutinosa Stem to 30cm high, softly purple-hairy. Flowers red-brown to yellow. *Distr.* China.

Reineckea (After Joseph Heinrich Julius Reinecke (1799–1871), German gardener.) A genus of 1 species of clump-forming herb with prostrate stems, 2-ranked, linear to lance-shaped leaves, and dense spikes of fragrant pink flowers. *Distr.* China, Japan. *Fam.* Convallariaceae. *Class* M.

R. carnea Grown as ground cover for shaded areas. *Distr.* China, Japan.

Reinwardtia (After Caspar Rein-wardt (1773–1854), founder of Bogor Botanic Garden, Java.) A genus of 1 species of subshrub or shrub to 90cm high with simple alternate leaves and large yellow flowers that bear their parts in fives. *Distr.* Himalaya, E Asia. *Fam.* Linaceae. *Class* D.

R. indica YELLOW FLAX. Grown as an ornamental.

R. trigyna See *R. indica*

Renanthera (From Latin *renes*, kidney, and *anthera*, anther, alluding to the kidney-shaped pollinia.) A genus of 10 species of robust epiphytic orchids with 2-ranked leathery leaves and dense racemes or panicles of small flowers. Cultivated ornamentally. *Distr.* SE Asia. *Fam.* Orchidaceae. *Class* M.

R. coccinea Flowers red, lip yellow at base. *Distr.* Thailand, Laos, Vietnam, S China, Java, Burma.

R. imschootiana (After A. van Imschoot of Ghent.) Stem scrambling. Flowers red and orange. *Distr.* NE India, Burma, Laos.

R. storiei Flowers orange, mottled red, margin yellow. *Distr.* Philippines.

reniformis, -e: kidney-shaped.

repandus, -a, -um: with wavy margins.

repens: creeping.

replicatus, -a, -um: turned or folded back on itself.

reptans: creeping.

Reseda (From Latin *resedo*, to heal or calm, alluding to their medicinal properties.) MIGNONETTE. A genus of 55 species of annual and perennial herbs with pinnately cut leaves and racemes of small irregular flowers that bear 4–8 fused petals. Several species are grown as ornamentals. *Distr.* Europe to Central Asia. *Fam.* Resedaceae. *Class* D.

R. alba WILD MIGNONETTE. Annual or occasionally perennial herb to 80cm high. Flowers cream or white. *Distr.* Mediterranean, widespread as a weed.

R. luteola DYER'S ROCKET, DYER'S WELD. Annual or perennial. Flowers yellow or yellow-green. A source of yellow dye. *Distr.* Europe, Central Asia.

Resedaceae The Mignonette family. 7 genera and 75 species of herbs and a few shrubs, mostly of dry areas. The leaves are alternate, entire or divided, and the flowers are irregular, small, and arranged in spikes or racemes with 2–8 sepals and petals. *Distr.* Mostly N temperate regions, centred in the Mediterranean, but also to India and in S Africa and California. *Class* D.
See: Reseda.

resinifera: resin-bearing.

resinosus, -a, -um: resinous.

restharrow *Ononis* **large yellow** ~ *O. natrix* **shrubby** ~ *O. fruticosa.*

Restio (From Latin *restio*, rope-maker.) A genus of 88 species of rhizomatous herbs with leaves that are reduced to sheaths around the flowering stems. The small unisexual flowers are borne in panicles or spikes. Male and female flowers are borne on separate plants. *Distr.* Tropical and S Africa, Madagascar, Australia. *Fam.* Restionaceae. *Class* M.

R. tetraphyllus Occasionally grown as an ornamental. *Distr.* E Australia, Tasmania.

Restionaceae A family of 40 genera and about 500 species of rush-like herbs with tough unbranched to many-branched green stems and leaves reduced to small scales. Flowers are small and arranged in spikelets in loose inflorescences. Various species are used in matting, thatching, or as brooms. *Distr.* S hemisphere, rarely tropical regions. *Class* M.
See: Ischyrolepis, Restio.

resurrection plant *Selaginella lepidophylla.*

Retama (From Arabic *retam*, a desert shrub, first mentioned in Biblical times.) A genus of 4 species of upright, much-branched shrubs with simple leaves and large, yellow or white, pea-like flowers. *Distr.* Europe, Canaries, and Mediterranean to W Asia. *Fam.* Leguminosae. *Class* D.

R. raetam WHITE BROOM, JUNIPER BUSH. Large shrub. Flowers white, borne in axillary clusters. *Distr.* Middle East.

R. sphaerocarpa Deciduous shrub. Flowers yellow, crowded into a raceme. *Distr.* Spain, N Africa.

reticulatus, -a, -um: netted, reticulate, marked with a network.

retortus, -a, -um: twisted, bent back.

retusus, -a, -um: retuse, with a rounded slightly notched end.

Retziaceae A family of 1 genus and 1 species of shrub with simple opposite leaves and small flowers. This family is sometimes included within the Loganiaceae. *Distr.* South Africa. *Class* D.

revolutus, -a, -um: revolute, rolled back.

rewa rewa *Knightia excelsa.*

rex: king.

Rhabdodendraceae A family of 1 genus and 3 species of thick-stemmed shrubs with simple leathery leaves and racemes of regular flowers that bear 10 tepals and numerous stamens. *Distr.* N South America. *Class* D.

Rhamnaceae The Buckthorn family. 51 genera and about 850 species of trees, shrubs, and some climbers with simple leaves and often bearing thorns. The flowers are small and inconspicuous and bear their parts in fours or fives. Some species are a source of dyes others have edible fruits or are grown as ornamentals. *Distr.* Widespread. *Class* D.
See: Berchemia, Ceanothus, Colletia, Discaria, Hovenia, Paliurus, Rhamnus, Sageretia.

rhamnoides: from the genus *Rhamnus*, with the ending *-oides*, indicating resemblance.

Rhamnus (From *rhamnos*, the classical Greek name for these plants.) BUCKTHORN. A genus of 125–150 species of deciduous or evergreen, typically thorny shrubs and trees with spirally arranged or opposite simple leaves and clusters of small flowers that typically bear their parts in fours. The fruit is a globose drupe. These plants are a source of dyes and medicines as well as being grown as ornamentals. *Distr.* N hemisphere, Brazil, E and S Africa. *Fam.* Rhamnaceae. *Class* D.

R. cathartica COMMON BUCKTHORN, PURGING BUCKTHORN, EUROPEAN BUCKTHORN. Shrub or small tree to 6m high, twigs grey-green or brown, spiny. Flowers yellow-green. The fruits are used medicinally and as a source of the artist's pigment sap green. *Distr.* Temperate Europe and Asia, naturalized in NW Africa, introduced in E USA.

R. frangula ALDER BUCKTHORN. Shrub or small tree, to 2–5m. Fruit red, black when mature. *Distr.* Europe, Turkey, N Africa.

Rhaphiolepis (From Greek *raphis*, needle, and *lepis*, scale, alluding to the needle-like bracts on the inflorescence.) A genus of 14–15 species of evergreen shrubs and small trees with simple leathery leaves and racemes or panicles of white or pink flowers that bear 5 petals and 15–20 stamens. The apple-like fruit is purple-black and 1-seeded. Several species are cultivated ornamentally. *Distr.* Warm temperate regions, subtropical E Asia. *Fam.* Rosaceae. *Class* D.

R. indica INDIAN HAWTHORN. Shrub to 1m high. Flowers white, tinted pale pink towards centre. *Distr.* S China.

R. ovata See *R. umbellata*

R. umbellata Shrub to 4m high. Flowers fragrant. Fruit pear-shaped. *Distr.* Japan, Korea.

Rhaphithamnus (From Greek *raphis*, needle, and *thamnos*, shrub, alluding to the spiny habit.) A genus of 2 species of spiny shrubs and small trees with simple, opposite or alternate, leathery leaves, and short racemes of funnel-shaped, 2-lipped flowers that are followed by blue berries. *Distr.* S South America. *Fam.* Verbenaceae. *Class* D.

R. cyanocarpus See *R. spinosus*

R. spinosus PRICKLY MYRTLE. Grown as an ornamental. *Distr.* S South America.

Rhapidophyllum (From Greek *raphis*, needle, and *phyllon*, leaf, alluding to the needle-like appendages on the leaf sheaths.) A genus of 1 species of medium-sized palms with fan-shaped leaves and branched, spike-like inflorescences of yellow or orange to purple flowers. *Distr.* E coastal United States. *Fam.* Palmae. *Class* M.

R. hystrix NEEDLE PALM, PORCUPINE PALM, BLUE PALMETTO. Grown as an ornamental.

Rhapis (From Greek *raphis*, needle, alluding to the slender leaf segments.) LADY PALM. A genus of 9–12 species of small to medium-sized palms with fan-shaped leaves and short inflorescences of cream flowers. *Distr.* S China, Thailand, Japan. *Fam.* Palmae. *Class* M.

R. excelsa JAPANESE PEACE PALM, GROUND RATTAN, BAMBOO PALM, LADY PALM, MINIATURE PALM, FERN RHAPIS. Much cultivated in Japan where numerous cultivars are available. *Distr.* S China, introduced Japan.

R. humilis REED RHAPIS, SLENDER LADY PALM. Stems slender, reed-like. Grown as a tender ornamental. *Distr.* S China.

Rheum (From *rheon*, the Greek name for rhubarb.) RHUBARB. A genus of 50 species of coarse perennial herbs with tough or woody rhizomes, large long-stalked leaves, and clusters of small wind-pollinated flowers. Several species are grown as ornamentals, others are edible or used medicinally. *Distr.* Temperate and subtropical Asia, W to Europe. *Fam.* Polygonaceae. *Class* D.

R. acuminatum SIKKIM RHUBARB. To 1m high. Flowers bright red or maroon. *Distr.* Nepal to SE Tibet.

R. australe HIMALAYAN RHUBARB, INDIAN RHUBARB, RED-VEINED PIE PLANT. To 3m high. Leaves to 1m across. Flowers white to wine-red. The roots are used medicinally. *Distr.* Central Asia, Himalaya.

R. × cultorum RHUBARB, GARDEN RHUBARB, PIE PLANT. Leaves large, round. Flowers red. The red-tinged leaf stalks are stewed as a pudding or used in preserves and wines. Other parts of the plant are poisonous. *Distr.* An ancient garden hybrid of unknown origin.

R. emodi See *R. australe*

R. palmatum CHINESE ROOT, TURKISH ROOT. Leaves nearly round, palmately lobed. Flowers deep red. Used medicinally. *Distr.* NW China.

rheumatism root *Jeffersonia diphylla.*

Rhinephyllum (From Greek *rhine*, file, and *phyllon*, leaf, alluding to the rough-textured leaves.) A genus of 12 species of succulent perennial herbs and subshrubs with pairs of rough semi-cylindrical leaves and solitary yellow or white flowers that open at night. *Distr.* South Africa (Cape Province). *Fam.* Aizoaceae. *Class* D.

R. broomii Short compact shrublet. Flowers yellow. Grown as a tender ornamental.

Rhipogonaceae A family of 1 genus and 8 species of erect or climbing shrubs. This family is sometimes included within the family Liliaceae. *Distr.* New Guinea, Australia, New Zealand. *Class* M.

Rhizophoraceae A family of 14 genera and 130 species of shrubs, climbers, and trees with simple entire leaves and typically bisexual, regular flowers. The genera *Rhizophora*, *Bruguiera*, and *Ceriops* are ecologically important mangroves. *Distr.* Tropical regions, especially the Old World. *Class* D.

rhodanthus, -a, -um: with rosy red flowers.

Rhodiola (From Greek *rhodon*, rose, alluding to the rose-scented root of *R. rosea*.) A genus of 30–50 species of rhizomatous succulent herbs with simple fleshy leaves and cymes of star-shaped flowers. The leaves of the sterile stems and fertile stems may vary considerably on the same plant. The genus as a whole shows wide morphological variation and is sometimes included within the genus *Sedum*. A number of species are grown as ornamentals. *Distr.* N temperate regions, especially N Africa. *Fam.* Crassulaceae. *Class* D.

R. fastigiata To 15cm high, deciduous. Flowers white. *Distr.* Himalaya, Tibet, SW China.

R. rosea ROSEROOT. Rhizome thick, branching, rose-scented. A very variable plant. The leaves are sometimes eaten. *Distr.* N Arctic and Alpine regions.

Rhodochiton (From Greek *rhodon*, rose, and *chiton*, cloak, alluding to the inflated sepal tube.) A genus of 1–3 species of perennial climbing herbs with twining stems, simple alternate leaves, and pendulous flowers that bear a cup-shaped calyx surrounding a narrow 5-lobed petal tube. *Distr.* Mexico. *Fam.* Scrophulariaceae. *Class* D.

R. atrosanguineus PURPLE BELL VINE. Stem climbing to 6m. Calyx membranous, green, tinged maroon. Petals red to dark maroon. *Distr.* Mexico.

R. volubilis See *R. atrosanguineus*

Rhododendron (Originally a Greek name for *Nerium oleander*, from *rhodos*, rose, and *dendron*, tree.) A genus of 700–850 species of evergreen and deciduous shrubs, trees, and epiphytes with simple leaves and tubular, funnel- or bell-shaped flowers. The name 'azalea' is used for the members of several subgenera which are typically small deciduous shrubs. This genus contains numerous ornamental species which have given rise to hundreds of hybrids and cultivars of great importance in the garden, especially on acid soils. *Distr.* N hemisphere, especially Himalaya and SE Asia to Australia. *Fam.* Ericaceae. *Class* D.

R. arboreum Tree to 50m high (usually much less in gardens), with pink to deep crimson, rarely white, fleshy flowers. *Distr.* Himalaya, China to Thailand, S India and Sri Lanka.

R. barbatum Shrub or small tree with smooth bark and fleshy, crimson, rarely white flowers. *Distr.* Himalaya.

R. beesianum Shrub or small tree with white, occasionally pink-flushed flowers. *Distr.* China (SE Xizang, NW Yunnan, SW Sichuan).

R. calendulaceum Much-branched deciduous shrub with orange, red, or yellow flowers. *Distr.* E North America.

R. delavayi See *R. arboreum*

R. eximium See *R. falconeri*

R. falconeri (After Hugh Falconer (1808–65), Scottish doctor and botanist with the East India Company.) Tree to 12m high. Flowers obliquely campanulate. *Distr.* Himalaya.

R. impeditum Compact, much-branched, scaly shrub with violet-purple flowers. One of the most frequently grown of the small rhododendrons and an important parent species of a number of hybrids. *Distr.* W China (N Yunnan, SW Sichuan).

R. indicum Low, sometimes prostrate shrub with red flowers. *Distr.* S Japan.

R. macranthum See *R. indicum*

R. ponticum Large shrub with purple to lilac-pink flowers. Often used as a grafting stock for other species and cultivars. *Distr.* W Asia, Balkans, widely naturalized in many parts of Britain and Ireland and often considered a noxious weed.

R. viscosum SWAMP HONEYSUCKLE. Shrub with white to pink, sticky flowers. *Distr.* E North America.

Rhodohypoxis (From Greek *rhodon*, rose, and the genus name *Hypoxis*.) A genus of 6 species of perennial herbs with fleshy roots, broad to thread-like leaves and red to white flowers which bear 6 tepals, the inner 3 of which meet in the centre so that the flowers do not have a clear 'eye'. *Distr.* SE Africa. *Fam.* Hypoxidaceae. *Class* M.

R. baurii (After the Reverend R. Baur, who collected *Rhodohypoxis* in South Africa.) RED STAR. Leaves to 1cm wide. Flowers white, pink, or red, to 4cm across. Sometimes grown as a rock-garden plant. *Distr.* South Africa.

Rhodophiala (From Greek *rhodon*, red, and *phiala*, a drinking vessel.) A genus of 31 species of bulbiferous perennial herbs with narrow leaves and an umbel of 1–6 funnel-shaped flowers. *Distr.* Andes. *Fam.* Amaryllidaceae. *Class* M.

R. advena Flowers horizontal, red-pink or yellow. *Distr.* Chile.

R. bagnoldii Flowers erect or ascending, open, yellow. *Distr.* Chile.

R. bifida Flowers erect or ascending, bright red. *Distr.* Argentina and Uruguay.

R. pratensis Flowers ascending, bright red or purple-violet. *Distr.* Chile.

R. roseum Flowers horizontal, 1–2, bright red. *Distr.* Chile.

Rhodothamnus (From Greek *rhodon*, rose, and *thamnos*, shrub.) A genus of 1 species of evergreen dwarf shrub to 40cm with simple leaves and rose-coloured, saucer-shaped flowers. *Distr.* E Alps. *Fam.* Ericaceae. *Class* D.

R. chamaecistus Grown as an ornamental. *Distr.* E Alps.

Rhodotypos (From Greek *rhodon*, rose, and *typos*, type, alluding to the rose-like flowers.) A genus of 1 species of deciduous shrub to 5m high with simple opposite leaves and solitary white flowers that bear 4 round petals and numerous stamens, followed by black berries. *Distr.* China, Japan. *Fam.* Rosaceae. *Class* D.

R. kerrioides See *R. scandens*

R. scandens Grown as a hardy ornamental.

Rhoeo See *Tradescantia*

Rhoicissus (From Latin *rhoia*, pomegranate, and the genus name *Cissus*.) A genus of 10–12 species of scrambling shrubs and vines with simple to palmately divided leaves and clusters of small inconspicuous flowers. *Distr.* Tropical and S Africa. *Fam.* Vitaceae. *Class* D.

R. capensis Evergreen vine. Leaves simple. Grown as an ornamental. *Distr.* S Africa.

Rhoipteleaceae A family of 1 genus and 1 species of aromatic deciduous tree with pinnate, spirally arranged leaves and small wind-pollinated flowers. *Distr.* SW China, Vietnam. *Class* D.

rhombeus, -a, -um: rhombic, of a four-sided shape with opposite angles and sides equal.

rhombifolius, -a, -um: with rhombic leaves.

Rhopalostylis (From Greek *rhopalon*, club, and *stylis*, small pillar, alluding to the spadix.) NIKAU PALM. A genus of 3 species of medium-sized palms with feather-shaped leaves and pendent to somewhat rigid inflorescences. *Distr.* New Zealand, Norfolk Island, and Raoul Island. *Fam.* Palmae. *Class* M.

R. baueri NORFOLK PALM. To 6m high. Flowers white. Fruit red. Grown as an ornamental. *Distr.* Norfolk Island.

R. sapida NIKAU PALM, FEATHER DUSTER PALM. To 10m high. Flowers lilac to cream. Leaves once used for hut-building by Maoris. *Distr.* New Zealand.

rhubarb Polygonaceae, *Rheum*, *R.* × *cultorum* **bog** ~ *Petasites hybridus* **garden** ~ *Rheum* × *cultorum* **giant** ~ *Gunnera manicata* **Himalayan** ~ *Rheum* *australe* **Indian** ~ *Darmera peltata, Rheum australe* **monk's** ~ *Rumex alpinus* **mountain** ~ *R. alpinus* **Sikkim** ~ *Rheum acuminatum.*

Rhus (Classical Latin name for *Rhus coriaria.*) SUMACH, SUMAC. A genus of about 200 species of trees, shrubs, and lianas with simple or pinnate, evergreen or deciduous leaves, and panicles or spikes of very small flowers. This genus is a source of dyes, lacquers, and tanning materials as well as some cultivated ornamentals. *Distr.* Temperate and subtropical regions. *Fam.* Anacardiaceae. *Class* D.

R. aromatica FRAGRANT SUMAC, LEMON SUMAC, POLECAT BUSH. Low, bushy, deciduous shrub to 1m high. Flowers yellow. *Distr.* E North America.

R. copallina DWARF SUMAC, MOUNTAIN SUMAC, SHINING SUMAC. To 6m high. Leaves deciduous. Flowers green. Leaves used for tanning and dyeing. *Distr.* E North America.

R. coriaria SICILIAN SUMAC, TANNER'S SUMAC, ELM-LEAVED SUMAC. Shrub to 3m. Leaves deciduous. Flowers yellow-green. *Distr.* S Europe.

R. cotinus See *Cotinus coggygria*

R. glabra SMOOTH SUMAC, SCARLET SUMAC, VINEGAR TREE. Shrub or tree, to 3m. Leaves deciduous. Flowers green. Fruits a source of dye. Fruits made into a drink. *Distr.* E North America.

R. integrifolia LEMONADE BERRY, LEMONADE SUMAC, SOURBERRY. Shrub or tree, to 10m. Leaves evergreen. Flowers white to pink. *Distr.* S Carolina.

R. potaninii Rounded tree, to 8m. Leaves deciduous, turning bright red in autumn. Flowers off-white. *Distr.* Central and W China.

R. radicans POISON IVY, PAGE OAK, POISON OAK. Liana, shrub, or tree. The leaves produce an irritant that promotes dermatitis; over 350,000 cases are reported each year in North America. *Distr.* E North America.

R. succedanea JAPANESE TALLOW, WAX TREE. Wax from fruits used as substitute for beeswax in polishes. *Distr.* E Asia.

R. toxicodendron See *R. radicans*

R. trilobata SKUNK BUSH. Shrub, to 2m high. Leaves deciduous. Flowers yellow. Stems used in basketry. *Distr.* California.

R. typhina STAG'S HORN SUMAC, VELVET SUMAC, VIRGINIAN SUMAC. Shrub or tree, to 10m. Leaves deciduous, brightly coloured in autumn. Flowers green. An important source of tanning agents. The young shoots are used as pipes for tapping SUGAR MAPLE. *Distr.* E North America.

R. verniciflua CHINESE LACQUER TREE, JAPANESE LACQUER TREE. A source of lacquers and candle wax. *Distr.* Temperate E Asia.

Rhynchocalycaceae A family of 1 genus and 1 species of tree with simple opposite leaves and small flowers. This family is sometimes included within the family Crypteroniaceae. *Distr.* South Africa. *Class* D.

rhytidophyllus, -a, um: with wrinkled leaves.

ribbon plant *Dracaena sanderiana, Chlorophytum comosum, Hypoestes aristata.*

ribbon wood *Plagianthus regius, Hoheria sexstylosa.*

Ribes (From Arabic or Persian *ribas*, acid-tasting, alluding to the fruit.) CURRANT, GOOSEBERRY. A genus of 150 species of deciduous or sometimes evergreen, prickly or smooth shrubs with simple, palmately lobed leaves, and small regular flowers that bear their parts in fours or fives, the petals being borne on the throat of the sepal tube and sometimes very much reduced. The fruit is a juicy, typically red or black, many-seeded berry. Many species have edible fruits and a few are grown as ornamentals. Numerous ornamental and edible cultivars have been produced. *Distr.* N temperate regions, Andes. *Fam.* Grossulariaceae. *Class* D.

R. alpinum MOUNTAIN CURRANT, ALPINE CURRANT. Dense bushy shrub to 2m high. Flowers small, green-yellow, in upright racemes. *Distr.* N Europe.

R. americanum AMERICAN BLACK CURRANT. Shrub to 1.5m. Flowers numerous, yellow-white, borne in pendulous racemes. *Distr.* E North America.

R. atrosanguineum See *R. sanguineum*

R. aureum GOLDEN CURRANT, BUFFALO CURRANT. Erect loose shrub to 2m high. Flowers spicily aromatic, yellow, in pendent racemes. Fruit formerly eaten with buffalo meat. *Distr.* W USA to Mexico.

R. divaricatum WORCESTERBERRY. Bristly shrub to 3 meters. Flowers green-purple, in small clusters. The parent of many mildew-resistant hybrids. Fruit formerly dried. *Distr.* W North America.

R. gayanum (After Clande Gay (1800–73), French botanist.) Evergreen shrub to 1.5m high. Flowers in upright racemes, downy, yellow, honey-scented. *Distr.* Chile.

R. glandulosum SKUNK CURRANT, FETID CURRANT. Low shrub to 40cm high. Flowers red-white, downy, borne in erect racemes. *Distr.* North America.

R. grossularioides GOOSEBERRY, CATBERRY. Spiny shrub to 2m high. Flowers bell-shaped, red-green. *Distr.* Japan.

R. laurifolium Evergreen spreading shrub to 1.5m high. Leaves leathery. Flowers green-yellow. *Distr.* W China.

R. nigrum BLACK CURRANT, EUROPEAN BLACK CURRANT. Aromatic shrub to 2m high. Flowers in downy pendent racemes, red-green. The fruit has a high vitamin C content and is made into preserves, soft alcoholic drinks. *Distr.* Europe to Central Asia.

R. odoratum BUFFALO CURRANT, CLOVE CURRANT, MISSOURI CURRANT. Erect shrub to 2m high. Flowers large, scented, yellow. *Distr.* Central North America.

R. sanguineum FLOWERING CURRANT, AMERICAN CURRANT, WINTER CURRANT. Deciduous shrub to 2m high. Flowers red or rosy-red. *Distr.* W North America.

R. speciosum CALIFORNIAN FUCHSIA, FUCHSIA-FLOWERED GOOSEBERRY. Evergreen thorny shrub to 4m high. Flowers red, pendent. *Distr.* USA (California).

R. uva-crispa GOOSEBERRY, FEABERRY. Spiny bushy shrub to 1m high. Flowers clustered, pink-green. Fruit pale-green. Fruit much used in pies and jam. *Distr.* N and Central Europe, N Africa.

ribesifolius, -a, -um: from the genus *Ribes*, CURRANTS, and *folius*, leaved.

ribifolius, -a, -um: from the genus *Ribes*, and *folius*, leaved.

ribwort *Plantago*, *P. lanceolata*.

rice *Oryza* **annual wild** ~ *Zizania aquatica* **Canadian wild** ~ *Z. aquatica* **Manchurian wild** ~ *Z. latifolia* **water** ~ *Z. aquatica*, *Z. latifolia* **wild** ~ *Zizania*

rice flower *Pimelea*

rice paper plant *Tetrapanax papyriferus*.

Richea (After Colonel A. G. Riche (died 1791), French naturalist.) A genus of 11 species of evergreen shrubs and small trees with parallel-veined, strap-shaped leaves and small tubular flowers that bear their parts in fives. The petal tube is closed at the apex but for a very small hole. Several species are grown as ornamentals. *Distr.* SE Australia, Tasmania. *Fam.* Epacridaceae. *Class* D.

R. scoparia Small tree. Flowers in dense racemes, pink, orange or white. *Distr.* Tasmania.

Ricinus (From Latin *ricinus*, tick, alluding to the appearance of the seeds.) A genus of 1 species of large shrub with palmately lobed, peltate leaves and panicles of small unisexual flowers. The fruit is a spiny or smooth capsule containing 3 seeds. *Distr.* E and NE Africa to Middle East, naturalized throughout the tropics. *Fam.* Euphorbiaceae. *Class* D.

R. communis CASTOR OIL PLANT, PALMA CHRISTI, CASTOR BEAN PLANT. Fast-growing shrub often grown as an annual. This plant has been grown for its seed-oil for over 6000 years, the ancient Egyptians having used it to fuel their lamps. The oil is now used in a variety of products from the familiar purgative castor oil to dyes and waterproofing for textiles; it has also formed the basis of a man-made fibre. The residue from the oil extraction is used as a fertilizer and the plant stems are a source of fibre. The seeds contain a number of toxic principles including ricin, one of the most toxic plant compounds known. 2–6 seeds contain enough toxins to kill. *Distr.* Tropical Africa.

rigescens: rather stiff, becoming stiff.

Rigidella (Diminutive of the Latin *rigidus*, rigid, alluding to the stiff, erect, fruit stalks.) A genus of 4 species of bulbiferous herbs with broad sword-shaped leaves and nodding, bright red, cup-shaped flowers that are subtended by a spathe. Several of the species are grown as ornamentals. *Distr.* Mexico to Guatemala. *Fam.* Iridaceae. *Class* M.

R. flammea To 1.5m high. Flowers pendent, to 12cm across, scarlet, marked purple-black. *Distr.* Mexico.

R. orthantha Flowers scarlet, unmarked, erect. *Distr.* Mexico, Guatemala.

rigidipilus, -a, -um: with stiff hairs.

rigidulus, -a, -um: somewhat rigid.

rigidus, -a, -um: rigid.

rimu *Dacrydium cupressinum* **mountain** ~ *Lepidothamnus laxifolium.*

ringens: gaping.

riparius, -a, -um: of river banks.

rivalis, -e: growing by streams.

rivularis, -e: of small streams.

rivulosus, -a, -um: with fine wavy grooves.

roast beef plant *Iris foetidissima.*

Robinia (After Jean Robin (1550– 1629), herbalist and botanist to Henry IV of France.) LOCUST. A genus of about 15 species of spiny deciduous shrubs and trees with pinnate leaves and pendent racemes of fragrant pea-like flowers. Several species and hybrids are grown ornamentally. *Distr.* North America. *Fam.* Leguminosae. *Class* D.

R. hispida ROSE ACACIA, MOSS LOCUST, BRISTLY LOCUST. Shrub to 2.5m high. Flowers rose or purple. *Distr.* SE USA.

R. kelseyi (After Harlan P. Kelsey, who introduced it to cultivation.) ALLEGHENY MOSS. Shrub to 3m high. Flowers lilac to rose. *Distr.* SE USA.

R. pseudoacacia FALSE ACACIA, BASTARD ACACIA, BLACK LOCUST, YELLOW LOCUST. Spreading tree. Flowers white, blotched yellow. Numerous ornamental cultivars have been raised from this species. *Distr.* Central and E USA, naturalized Central Europe.

roble de Chile *Eucryphia cordifolia.*

robur: the Latin name for hard wood, especially oak.

robustus, -a, -um: robust.

rocambole *Allium scorodoprasum.*

Rochea See *Crassula*

rock brake *Cryptogramma.*

rockcress *Arabis.*

rocket *Hesperis matronalis* **dame's** ~ *H. matronalis* **dyer's** ~ *Reseda luteola* **garden** ~ *Eruca vesicaria* **salad** ~ *Eruca, E. vesicaria* **sweet** ~ *Hesperis matronalis* **tansy-leaved** ~ *Hugueninia tanacetifolia* **yellow** ~ *Barbarea vulgaris.*

rockfoil *Saxifraga* **great alpine** ~ *S. cotyledon.*

rock rose *Cistus, Helianthemum* **Montpelier** ~ *Cistus monspeliensis* **white-leaved** ~ *C. albidus.*

Rodgersia (After Admiral John Rodgers (1812–82), American commander of the expedition on which *R. podophylla* was first collected.) A genus of 6 species of perennial herbs with stout rhizomes, peltate, palmate, or pinnate leaves and large terminal inflorescences of small white flowers. *Distr.* Nepal and E Asia, China, Japan. *Fam.* Saxifragaceae. *Class* D.

R. aesculifolia Leaves palmate. Flowers white. *Distr.* China.

R. henrici Leaves palmate. Flowers red-purple. *Distr.* China, N Burma, SE Tibet.

R. pinnata Leaves palmate but appearing pinnate. Flowers red. *Distr.* China (Yunnan).

R. podophylla Leaves bronze in winter. Flowers cream. *Distr.* China, Japan, Korea.

R. sambucifolia Leaves pinnate. Flowers white-pink. *Distr.* China.

R. tabularis See *Astilboides tabularis*

Rohdea (After Michael Rohde (1782– 1812), German physician and botanist.) A genus of 1 species of rhizomatous herb with basal rosettes of leathery leaves and dense spikes of bell-shaped, foul-smelling, white or yellow-green flowers. The flowers are pollinated by slugs and snails which feed off their fleshy tepals. *Distr.* China, Japan. *Fam.* Convallariaceae. *Class* M.

R. japonica LILY OF CHINA. Cultivated ornamentally especially in Japan. *Distr.* China, Japan.

roman candle *Yucca gloriosa.*

romanicus, -a, -um: of Romania.

Roman's root *Gillenia trifoliata*.

romanus, -a, -um: Roman.

Romanzoffia (After Nicholas Romanzoff, who initiated a Russian transglobal expedition, 1816–17.) A genus of 4 species of small perennial herbs with long-stalked, broad, palmately lobed leaves and bell-shaped flowers which bear their parts in fives. Several species are cultivated ornamentally. *Distr.* W North America to Aleutian Islands. *Fam.* Hydrophyllaceae. *Class* D.

R. sitchensis Leaves 5cm across. Flowers white. *Distr.* W North America.

R. tracyi Leaves succulent. Flowers white. *Distr.* W North America.

Romneya (After Dr Thomas Romney Robinson (1792–1882), Irish astronomer.) A genus of 1 species of perennial subshrub with pinnately divided leaves and large, fragrant, white flowers that have 3 sepals, 2 whorls of 3 silky petals, and numerous stamens. *Distr.* S California, Mexico. *Fam.* Papaveraceae. *Class* D.

R. coulteri (After Dr Thomas Coulter (1793–1843), who collected it in 1833.) MATILIJA POPPY, CALIFORNIAN TREE POPPY. Grown as an ornamental.

R. × hybrida See *R. coulteri*

Romulea (After Romulus, legendary founder and first king of Rome.) A genus of about 80 species of small *Crocus*-like herbs with narrow leaves and funnel-shaped flowers that are sometimes borne on aerial stems. The flowers open and close with changes in temperature. Some of the species are grown as ornamentals. *Distr.* South Africa (Cape Province), mountainous areas of E Africa, Mediterranean basin. *Fam.* Iridaceae. *Class* M.

R. bulbocodium Flowers on short stems, upward-facing, violet with white to yellow centres. *Distr.* Portugal, NW Spain, Bulgaria.

R. flava Leaves narrow, hollow. Flowers yellow with a deeper yellow centre. *Distr.* South Africa (Cape Province).

R. longituba See *R. macowanii*

R. macowanii Leaves very thin, thread-like. Flowers upward facing, yellow. *Distr.* South Africa (Drakensberg), Lesotho.

R. sabulosa Leaves erect, thread-like. Flowers bright red with a black centre. *Distr.* South Africa (Cape Province).

roquette *Eruca*.

Roridulaceae A family of 1 genus and 2 species of insectivorous subshrubs with spirally arranged sticky leaves. This family is sometimes included within the family Byblidaceae. *Distr.* South Africa. *Class* D.

Rorippa (From Anglo-Saxon, *rorippen*.) A genus of 70 species of annual and rhizomatous perennial herbs with small flowers. *Distr.* Cosmopolitan except Antarctica. *Fam.* Cruciferae. *Class* D.

R. nasturtium-aquaticum WATERCRESS. Aquatic perennial herb. Salad plant grown in inundated beds. *Distr.* Europe, naturalized in North America.

Rosa (Classical Latin name for these plants.) ROSE. A genus of 100–150 species of deciduous and occasionally perennial, thorny shrubs with erect, scrambling or climbing stems, simple to pinnately divided, opposite leaves, and regular flowers that typically bear 5 petals and numerous stamens. These plants are very important ornamentals and thousands of hybrids and cultivars have been produced, most with double flowers, which are not only cherished for their decorative value but also for their scent and as a source of essential oil. *Distr.* N temperate regions and the mountains of tropical and subtropical areas. *Fam.* Rosaceae. *Class* D.

R. × alba WHITE ROSE, WHITE ROSE OF YORK. A hybrid of complex parentage. *Distr.* Garden origin.

R. altaica See *R. pimpinellifolia*

R. arkansana PRAIRIE ROSE. Erect shrub. Flowers deep pink. *Distr.* Central USA.

R. arvensis FIELD ROSE. Trailing or climbing shrub. Flowers white to pink. *Distr.* SW and Central Europe, S Turkey.

R. banksiae (After Dorothea, Lady Banks, wife of Sir Joseph Banks (1743–1820), English botanist.) BANKSIAN ROSE. Climbing shrub. Flowers yellow or white. The bark of the roots is used in tanning in China. *Distr.* W and Central China.

R. bracteata MACARTNEY ROSE. Scrambling or climbing shrub. Flowers white with a fruity odour. *Distr.* SE China, Taiwan, naturalized in USA.

R. brunonii (After Robert Brown (1773–1858), Scottish botanist.) HIMALAYAN

MUSK ROSE. Arching or climbing shrub. Flowers densely clustered, fragrant, white. *Distr.* Himalaya, W China.

R. canina DOG ROSE, COMMON BRIER, DOG BRIER. Arching or climbing shrub. Flowers solitary, fragrant, pale pink. The hips have been used as a source of vitamin C and as a vermifuge. *Distr.* Europe, SW Asia, NW Africa, naturalized in North America.

R. carolina PASTURE ROSE, CAROLINA ROSE. Erect, suckering shrub. Flowers solitary, pale pink. *Distr.* E and Central North America.

R. × centifolia PROVENCE ROSE, CABBAGE ROSE, HOLLAND ROSE. A complex hybrid. Lax-branched, suckering shrub. Flowers very fragrant, typically pink. *Distr.* Garden origin.

R. chinensis FAIRY ROSE, CHINA ROSE, BENGAL ROSE. Dwarf or climbing shrub. Spines soft. Flowers pale pink to scarlet. *Distr.* China, long cultivated.

R. cinnamomea See *R. majalis*

R. ecae See *R. gallica*

R. eglanteria See *R. rubiginosa*

R. foetida AUSTRIAN BRIER, AUST-RIAN YELLOW ROSE. Erect shrub. Flowers deep yellow, malodorous. *Distr.* SW and W Central Asia.

R. foliolosa LEAFY ROSE, WHITE PRAIRIE ROSE. Suckering shrub. Stems red. Flowers fragrant, white to pale pink. *Distr.* W China.

R. gallica RED ROSE, FRENCH ROSE. Erect shrub. Flowers fragrant, pink to crimson. *Distr.* S and Central Europe, W Asia, naturalized E North America.

R. glutinosa See *R. eglanteria*

R. majalis CINNAMON ROSE, MAY ROSE. Suckering erect shrub. Flowers purple-pink. *Distr.* N and Central Europe to Siberia.

R. moschata MUSK ROSE. Robust arching or scrambling shrub. Flowers musk-scented, white to cream. *Distr.* S Europe, SW Asia.

R. multiflora BABY ROSE, JAPANESE ROSE. Arching or trailing shrub. Flowers cream with a fruity odour. *Distr.* Japan, Korea, naturalized in USA.

R. officinalis See *R. gallica*

R. pimpinellifolia BURNET ROSE, SCOTCH ROSE. Densely branched, suckering shrub. Flowers cream-white. *Distr.* W and S Europe E Asia.

R. roxburghii CHESTNUT ROSE, CHINQUAPIN ROSE. Spreading shrub. Flowers fragrant, solitary, pink. *Distr.* W China.

R. rubiginosa SWEET BRIER, EGLANTINE ROSE. Densely branched shrub. Foliage fragrant in damp weather. Flowers pink, fragrant. *Distr.* Europe, N Africa, W Asia, naturalized North America.

R. rubra See *R. gallica*

R. rugosa JAPANESE ROSE, TURKESTAN ROSE. Stout erect shrub. Flowers fragrant, purple-pink. The hips are eaten locally. *Distr.* E USSR, NE Asia, Japan, Korea, naturalized in the British Isles and NE USA.

R. sempervirens EVERGREEN ROSE. Prostrate shrub. Leaves evergreen or semi-evergreen. Flowers white. *Distr.* S Europe, NW Africa, Turkey.

R. setigera PRAIRIE ROSE, SUNSHINE ROSE, CLIMBING ROSE. Slender scrambling shrub. Flowers deep pink. *Distr.* North, Central and South America.

R. spinosissima See *R. pimpinellifolia*

R. suffulta See *R. arkansana*

Rosaceae The Rose family. 107 genera and about 3000 species of herbs, shrubs, trees, and a few climbers with alternate or compound leaves and frequently large showy flowers. A large number of the species are actual or potential ornamentals while others provide the major tree fruits of the N hemisphere, notably the APPLE, PEAR, CHERRY, and PLUM. *Distr.* Widespread, especially in the N hemisphere. *Class* D.
See: Acaena, Agrimonia, Alchemilla, Amelanchier, Aphanes, Aronia, Aruncus, Cercocarpus, Chaenomeles, Chamaebatiaria, Cotoneaster, Cowania, Crataegus, Cydonia, Dendriopoterium, Dryas, Duchesnea, Eriobotrya, Exochorda, Filipendula, Fragaria, Geum, Gillenia, Holodiscus, Ivesia, Kelseya, Kerria, Luetkea, Lyonothamnus, Malus, Margyricarpus, Mespilus, Neillia, Neviusia, Oemleria, Osteomeles, Petrophytum, Photinia, Physocarpus, Potentilla, Prunus, Pseudocydonia, Purshia, Pyracantha, Pyrus, Quillaja, Rhaphiolepis, Rhodotypos, Rosa, Rubus, Sanguisorba, Sibbaldia, Sibbaldiopsis, Sibiraea, Sorbaria, Sorbus, Spiraea, Stephanandra, × Stravinia, Waldsteinia.

rosaceus, -a, -um: rosaceous, resembling a rose flower.

rosa de Montana: *Antigonon leptopus.*

rosa-sinensis: rose of China.

Roscoea (After William Roscoe (1753–1831), founder of Liverpool Botanic Garden.) A genus of about 17 species of rhizomatous herbs with lance-shaped leaves and dense spikes of irregular purple flowers. Several species are cultivated as ornamentals. *Distr.* China and the Himalaya. *Fam.* Zingiberaceae. *Class* M.

R. cautleoides Flowers yellow. *Distr.* SW China.

R. humeana (After David Hume of Edinburgh Botanic Garden.) Flowers purple. *Distr.* SW China.

R. purpurea Flowers long-tubed, purple. *Distr.* Himalaya, Sikkim.

rose Rosaceae, *Rosa* **Austrian yellow** ~ *R. foetida* **baby** ~ *R. multiflora* **Banksian** ~ *R. banksiae* **Bengal** ~ *R. chinensis* **burnet** ~ *R. pimpinellifolia* **cabbage** ~ *R.* × *centifolia* **Carolina** ~ *R. carolina* **chestnut** ~ *R. roxburghii* **China** ~ *Hibiscus rosa-sinensis, Rosa chinensis* **chinquapin** ~ *R. roxburghii* **Christmas** ~ *Helleborus niger* **cinnamon** ~ *Rosa majalis* **cliff** ~ *Cowania, C. mexicana* **climbing** ~ *Rosa setigera* **dog** ~ *R. canina* **eglantine** ~ *R. rubiginosa* **Egyptian** ~ *Scabiosa atropurpurea* **evergreen** ~ *Rosa sempervirens* **fairy** ~ *R. chinensis* **field** ~ *R. arvensis* **French** ~ *R. gallica* **giant velvet** ~ *Aeonium canariense* **fuelder** ~ *Viburnum opulus* **Himalayan musk** ~ *Rosa brunonii* **Holland** ~ *R.* × *centifolia* **Japanese** ~ *Kerria japonica, Rosa multiflora, R. rugosa* **leafy** ~ *R. foliolosa* **lenten** ~ *Helleborus orientalis* **Macartney** ~ *Rosa bracteata* **May** ~ *R. majalis* **moss** ~ *Portulaca, P. grandiflora* **musk** ~ *Rosa moschata* **pasture** ~ *R. carolina* **prairie** ~ *R. arkansana, R. setigera* **Provence** ~ *R.* × *centifolia* **red** ~ *R. gallica* **rock** ~ Cistaceae **Scotch** ~ *Rosa pimpinellifolia* **stock** ~ *Sparmannia africana* **sunshine** ~ *Rosa setigera* **Turkestan** ~ *R. rugosa* **velvet** ~ *Aeonium canariense* **white** ~ *Rosa* × *alba* **white Mexican** ~ *Echeveria elegans* **white prairie** ~ *Rosa foliolosa* **wood** ~ *Merremia tuberosa.*

rose imperial Cochlospermaceae

rose mandarin *Streptopus roseus*

rosemary *Rosmarinus, R. officinalis* **Australian** ~ *Westringia fruticosa* **bog** ~ *Andromeda, A. glaucophylla* **common bog** ~ *A. polifolia* **marsh** ~ *Limonium.*

rose of China *Hibiscus rosa-sinensis.*

rose of Jericho *Selaginella lepidophylla*

rose of Sharon *Hypericum calycinum.*

rose of York, white *Rosa* × *alba.*

roseroot *Rhodiola rosea.*

roseus, -a, -um: rose-coloured, rosy.

rose wood *Tipuana tipu.*

rosifolius, -a, -um: from the genus *Rosa*, ROSE, and *folius*, leaved.

rosin weed *Silphium, Grindelia.*

rosmarinifolius, -a, -um: from the genus *Rosmarinus*, and *folius*, leaved.

rosmariniformis, -e: resembling the genus *Rosmarinus*, ROSEMARY.

Rosmarinus (Latin name, from *ros*, dew, and *marinus*, of the sea.) ROSEMARY. A genus of 2 species of evergreen shrubs with linear aromatic leaves and whorls of 2-lipped flowers. *Distr.* S Europe, N Africa. *Fam.* Labiatae. *Class* D.

R. angustifolius See *R. officinalis*

R. corsicus See *R. officinalis*

R. officinalis ROSEMARY. Spreading or occasionally prostrate shrub. Leaves leathery, white-hairy below. Flowers blue. An important culinary herb. Numerous ornamental cultivars have been produced from this species. *Distr.* Mediterranean region.

Rossioglossum (After J. Ross, who collected orchids in Mexico in the 1830s.) A genus of about 6 species of epiphytic orchids with clustered pseudobulbs and racemes of large showy flowers. Members of this genus were formerly included in the genus *Odontoglossum*. *Distr.* Mexico to Panama *Fam.* Orchidaceae. *Class* M.

R. grande CLOWN ORCHID. Flowers long-lived, waxy, yellow with red-brown markings. *Distr.* Guatemala, Mexico.

Rostraria (From Latin *rostratus*, curved.) A genus of about 10 species of annual grasses with flat leaves and branched inflorescences. *Distr*. Temperate Eurasia and N Africa. *Fam*. Gramineae. *Class* M.

R. cristata CAT TAIL GRASS. Grown as an ornamental. *Distr*. Mediterranean region and NE Africa.

rostratus, -a, -um: beaked or curved.

rostriflorus, -a, -um: with beaked flowers.

Rosularia (From Latin *rosula*, little rose, alluding to the leaf rosettes.) A genus of 25–36 species of succulent perennials with rosettes of fleshy leaves and panicles of tubular or funnel-shaped flowers. Some species are cultivated as ornamentals. *Distr*. N Africa, Europe, Central Asia. *Fam*. Crassulaceae. *Class* D.

R. aizoon Rosettes to 3.5cm across. Flowers yellow. *Distr*. Turkey.

R. pallida See *R. aizoon*

rosularis, -e: with rosettes.

rosulatus, -a, -um: forming a rosette.

rotatus, -a, -um: wheel-shaped, spreading almost flat and circular.

rotundatus, -a, -um: rounded.

rotundifolius, -a, -um: with round leaves.

rotundus, -a, -um: rounded.

roundwood *Sorbus americana*.

rowan *Sorbus aucuparia*.

royal paint brush *Scadoxus puniceus*.

rubber, Assam *Ficus elastica*.

rubber plant *Ficus elastica* **baby** ~ *Peperomia obtusifolia* **creeping** ~ *Ficus pumila* **mistletoe** ~ *F. deltoidea* **small-leaved** ~ *F. benjamina*.

rubber tree, india *F. elastica* **para** ~ *Hevea brasiliensis*.

rubellus, -a, -um: reddish.

ruber: red.

rubescens: turning red.

Rubia (From Latin *ruber*, red, alluding to the red dye extracted from *R. tinctoria*.) A genus of 40–60 species of erect and climbing herbs and shrubs with simple whorled leaves and panicles or cymes of funnel-shaped, 4- or 5-lobed flowers. Several species are grown as ornamentals; some are a source of dye and some are used medicinally. *Distr*. Eurasia, Africa. *Fam*. Rubiaceae. *Class* D.

R. peregrina LEVANT MADDER, WILD MADDER. Climbing or scrambling, evergreen herb. Flowers yellow-green. Fruit black. *Distr*. W and S Europe, N Africa, Middle East.

R. tinctorum MADDER. Climbing or scrambling evergreen herb. Flowers pale yellow-green. Fruit dark red-brown. Formerly grown as a source of red and brown dye that was extracted from the roots. *Distr*. S Europe, W Asia.

Rubiaceae The Coffee family. 606 genera and about 10000 species of trees, shrubs, lianas and herbs, sometimes ant-inhabited, epiphytic or aquatic. The leaves are simple and usually entire and the flowers are typically regular and bisexual with 4–5 fused petals. The most economically important genus of the family is *Coffea*, the source of coffee; other members are important medicinally or as ornamentals. *Distr*. Widespread. *Class* D. *See: Asperula, Bouvardia, Cephalanthus, Coffea, Coprosma, Crucianella, Cruciata, Emmenopterys, Galium, Gardenia, Guettarda, Hedyotis, Houstonia, Luculia, Manettia, Mitchella, Nertera, Paederia, Pentas, Phuopsis, Psychotria, Putoria, Rubia, Serissa, Sherardia.*

rubicaulis, -e: red-stemmed.

rubicundus, -a, -um: red, reddish.

rubidus, -a, -um: red.

rubifolius, -a, -um: red-leaved.

rubiginosus, -a, -m: rusty red-coloured.

rubra-compacta: red and compact.

rubriflorus, -a, -um: with red flowers.

rubrifolius, -a, -um: with red leaves.

rubrocyaneus, -a, -um: red-blue-coloured.

rubromarginatus, -a, -um: with red edges.

rubromucronatum: with a red mucro or short pointed tip.

rubrus, -a, -um: red.

Rubus (Classical Latin name for a blackberry.) BLACKBERRY, BOYSENBERRY. A genus of 250 species of deciduous and evergreen, typically thorny shrubs with simple or compound leaves and regular flowers that usually bear 4–5 petals and numerous stamens. The fruit is made up of a number of fused drupes. Many cultivated forms have been raised for their fruit and as ornamentals. *Distr.* Cosmopolitan, but especially found in N temperate regions. *Fam.* Rosaceae. *Class* D.

R. arcticus CRIMSON BRAMBLE, ARCTIC BRAMBLE. Low, thornless, soft shrub. Flowers pink or red. *Distr.* Circumpolar.

R. caesius DEWBERRY, EUROPEAN DEWBERRY. Deciduous creeping shrub. Flowers white. Fruit large. Fruit eaten much as that of *R. fruticosus*. *Distr.* N Europe.

R. deliciosus ROCKY MOUNTAIN FLOWERING RASPBERRY, ROCKY MOUNTAIN RASPBERRY. Deciduous thornless shrub. Flowers pure white. Fruit dark purple. *Distr.* W North America.

R. fruticosus BRAMBLE, BLACKBERRY. Scrambling shrub. Flowers white. Fruit blue-black. This name is applied to a large number of very similar species known as microspecies that have occurred because of the breeding systems in this plant. Fruit eaten raw as well as in jams and baking. *Distr.* Europe, Mediterranean.

R. idaeus RASPBERRY, WILD RASPBERRY, RED RASPBERRY, EUROPEAN RASPBERRY. Erect shrub. Leaves pinnate. Flowers white. Fruit red or orange. The edible fruit is used in preserves and baking as well as being eaten fresh. *Distr.* Europe, N Asia, Japan.

R. longanobaccus (After Judge J. H. Logan (1841–1928) a Californian who discovered it.) LOGANBERRY. Erect or procumbent, evergreen shrub. The fruit is used in desserts, jams, and wines but is a little too acid to eat raw. *Distr.* W North America.

R. odoratus FLOWERING RASPBERRY, PURPLE FLOWERING RASPBERRY, THIMBLE BERRY. Deciduous shrub to 3m high. Leaves simple. Flowers fragrant. *Distr.* E North America.

R. parviflorus THIMBLE BERRY, SALMON BERRY. Small deciduous shrub. Flowers bright pink. Fruit red. *Distr.* North America, N Mexico.

R. phoenicolasius WINEBERRY. Deciduous shrub to 3m high. Fruits small, edible. *Distr.* E Asia.

R. rosifolius MAURITIUS RASPBERRY. Trailing shrub. Grown in the tropics for its edible fruits. The roots were formerly used medicinally. *Distr.* E Asia.

R. ulmifolius BRAMBLE. Arching, very thorny shrub. Flowers white with crumpled petals. Fruits edible but not good. *Distr.* W and Central Europe, Mediterranean.

Rudbeckia (After Olof Rudbeck the elder (1630–1702) and the younger (1660–1740), Swedish physicians and botanists.) CONEFLOWER, BLACK-EYED SUSAN, MEXICAN HAT. A genus of 15 species of annual, biennial, and perennial herbs with simple to bipinnately divided leaves and flowers in daisy-like heads that bear distinct rays. Several species are cultivated ornamentally. *Distr.* North America, now widespread as garden escape. *Fam.* Compositae. *Class* D.

R. californica Perennial herb, to 2m high. *Distr.* USA (California).

R. echinacea See *Echinacea purpurea*

R. fulgida Perennial herb. Flowers purple-brown, rays orange-yellow. *Distr.* SE USA.

R. gloriosa See *R. fulgida*

R. hirta BLACK-EYED SUSAN. Biennial or short-lived perennial, to 2m high. *Distr.* Central North America.

R. laciniata Tall perennial, to 3m high. *Distr.* North America.

R. maxima Perennial, to 1.5m high. *Distr.* Central and S USA.

R. newmannii See *R. fulgida*

R. nitida Perennial herb, to 2m high. Flowers green, rays yellow. *Distr.* North America.

R. occidentalis Perennial herb, to 2m high. Flower-heads solitary. *Distr.* W USA.

R. purpurea See *Echinacea purpurea*

R. subtomentosa SWEET CONEFLOWER. Perennial herb. Flowers purple-brown, rays yellow. *Distr.* Central North America.

ruddles *Calendula officinalis.*

rudis, -e: rough, coarse.

rue *Ruta*, *R. graveolens* **meadow** ~ *Thalictrum* **wall** ~ *Asplenium rutamuraria* **yellow meadow** ~ *Thalictrum flavum*.

rue-anemone *Anemonella thalictroides*.

Ruellia (After Jean de la Ruel of Soissons (1474–1537), French herbalist, botanist, and physician to Francis I.) A genus of about 150 species of perennial herbs, subshrubs, and shrubs with simple leaves and funnel-shaped flowers borne singly or in terminal panicles. *Distr.* Tropical America, temperate North America, Africa, Asia. *Fam.* Acanthaceae. *Class* D.

R. amoena See *R. graecizas*

R. devosiana (After Cornelius de Vos (1806–95), Belgian plant-collector in Brazil.) Leaves hairy, with pale veins above, purple beneath. Flowers solitary, white, tinged purple-blue. Grown as a tender ornamental. *Distr.* Brazil.

R. graecizas Leaves glabrous, pale green. Flowers in clusters, long-stalked, narrow, red. *Distr.* South America.

R. makoyana (After Jacob Makoy, Belgian nurseryman.) TRAILING VELVET PLANT, MONKEY PLANT. Stems spreading. Leaves marked violet and purple. Flowers solitary, rosy. *Distr.* Brazil.

rufescens: turning red.

rufus, -a, -um: red.

rugby football plant *Peperomia argyreia*.

rugospermus, -a, -um: with wrinkled seeds.

rugosus, -a, -um: wrinkled.

rugulosus, -a, -um: somewhat wrinkled.

Rumex (Classical Latin name for *R. acetosa*.) DOCK, SORREL. A genus of 150–200 species of annual, biennial, and perennial herbs with simple to deeply lobed, typically basal leaves, and racemes or panicles of small wind-pollinated flowers that are followed by papery fruits. Several species are grown as ornamentals and some are eaten. The leaves of some species are traditionally rubbed on nettle stings. *Distr.* N temperate regions. *Fam.* Polygonaceae. *Class* D.

R. acetosa COMMON SORREL, GARDEN SORREL, SOUR DOCK. Tufted perennial herb. The leaves are eaten raw or cooked. The juice will remove rust stains from linen. *Distr.* N temperate and arctic regions.

R. alpinus MONK'S RHUBARB, MOUNTAIN RHUBARB. Rhizomatous perennial to 1.5m high. The leaves are edible. *Distr.* Mountains of Europe and SW Asia.

R. scutatus FRENCH SORREL, GARDEN SORREL. Perennial herb to 2m high. The leaves are eaten like those of *R. acetosa* but the plant will grow in much drier conditions. *Distr.* Europe, W Asia, N Africa.

Rumohra (After Karl von Rumohr Holstein, 19th-century German patron of botany.) A genus of about 6 species of terrestrial and epiphytic ferns with pinnately divided leaves. *Distr.* S hemisphere, Madagascar. *Fam.* Davalliaceae. *Class* F.

R. adiantiformis LEATHER FERN, LEATHERLEAF FERN, IRON FERN. Grown as an ornamental. *Distr.* Widespread in the S hemisphere.

run away robin *Glechoma hederacea*.

runcinatus, -a, -um: toothed, the teeth pointing towards the base.

running postman *Kennedia prostrata*.

rupestris, -e: growing in rocky places.

Rupicapnos (From Latin *ruper*, rock, and Greek *kapnos*, smoke, FUMITORY.) A genus of 32 species of evergreen herbs with pinnately divided leaves and racemes of small, spurred flowers that bear 2 sepals, 4 petals. *Distr.* N Africa, Spain. *Fam.* Papaveraceae. *Class* D.

R. africanus Foliage blue-green. Flowers rose with maroon tips. Grown as a rock garden ornamental. *Distr.* N Africa.

rupicola: growing on rocks.

rupifragus, -a, -um: rock-breaking, growing in rock crevices.

rupturewort *Herniaria*, *H. glabra*.

Ruscaceae The Butcher's Broom family. 3 genera and 8 species of rhizomatous, clump-forming or climbing subshrubs with much reduced leaves that are replaced by stiff leaf-like branches (cladophylls). The flowers are very small, unisexual or bisexual, and borne on the cladophylls. The fruit is a

relatively large, red to orange-red berry. *Distr.* Macaronesia, Europe, Mediterranean, W Asia. *Class* M.
See: Danae, Ruscus, Semele.

Ruschia (After Ernest Julius Rusch (1867–1957), farmer in Namaqualand, South Africa.) A genus of 350–360 species of succulent shrubs, with 3-angled, clasping or fused, blue-green leaves and small clustered or solitary flowers. Several species are cultivated as ornamentals. *Distr.* S and SW Africa. *Fam.* Aizoaceae. *Class* D.

R. schollii Stems erect or creeping. Flowers solitary, red. *Distr.* Cape Province.

ruscifolius, -a, -um: from the genus *Ruscus*, and *folius*, leaved.

Ruscus (Classical Latin name used for these plants.) BUTCHER'S BROOM. A genus of 6 species of rhizomatous clump-forming subshrubs with much reduced leaves that are replaced by stiff leaf like branches (cladophylls) with sharply pointed tips. The flowers are very small, unisexual and borne on the midrib of the cladophylls. The fruit is a relatively large red berry. *Distr.* N Africa, W Europe to Caspian Sea. *Fam.* Ruscaceae. *Class* M.

R. aculeatus BUTCHER'S BROOM, PETTIGRUE, BOX HOLLY, JEW'S MYRTLE. Cultivated as ornamentals and formerly used to decorate meat. The young shoots can be eaten like ASPARAGUS. *Distr.* W Europe to Black Sea.

R. hypoglossum Cultivated ornamentally. *Distr.* S Europe, N Turkey.

rush Juncaceae, *Juncus* **black bog** ~ *Schoenus nigricans* **club** ~ *Scirpus* **common** ~ *Juncus effusus* **creeping spike** ~ *Eleocharis palustris* **flowering** ~ *Butomus umbellatus* **grassy** ~ *B. umbellatus* **hard** ~ *Juncus inflexus* **needle spike** ~ *Eleocharis acicularis* **scouring** ~ *Equisetum, E. hyemale* **slender spike** ~ *Eleocharis acicularis* **snow** ~ *Luzula nivea* **soft** ~ *Juncus effusus* **spike** ~ *Eleocharis* **wood club** ~ *Scirpus sylvaticus*.

russatus, -a, -um: russet.

Russelia (After Dr Alexander Russell (1715–68), Scottish physician and naturalist.) A genus of 52 species of evergreen subshrubs and shrubs with erect to pendent stems, simple, often scale-like leaves and tubular, 2-lipped, 5-lobed flowers. *Distr.* Cuba, Mexico, Colombia. *Fam.* Scrophulariaceae. *Class* D.

R. equisetiformis CORAL PLANT, FIRECRACKER PLANT, FOUNTAIN PLANT. Much-branched weeping shrub. Flowers red. Grown as an ornamental. *Distr.* Mexico, widely naturalized.

R. juncea See *R. equisetiformis*

rusticanus, -a, -um: of the country.

rusty leaf *Menziesia ferruginea*.

Ruta (Classical Latin name for these plants.) RUE. A genus of 7–8 species of aromatic evergreen subshrubs with alternate, pinnately divided leaves and cymes of yellow flowers with 4–5 petals and 8–10 stamens. Several species are grown as ornamentals. *Distr.* Macaronesia, Mediterranean, SW Asia. *Fam.* Rutaceae. *Class* D.

R. chalepensis Probably the rue of the Bible. *Distr.* Mediterranean, Arabia, Somalia, S Europe.

R. graveolens RUE, HERB OF GRACE. Subshrub to 50cm high. Foliage blue-green. Several ornamental cultivars have been raised from this species and it was formerly much used as a flavouring and in medicinal teas; it is sometimes believed to have magical properties. *Distr.* SE Europe.

R. prostrata See *R. chalepensis*

Rutaceae The Citrus family. 158 genera and about 1500 species of aromatic trees, shrubs, lianas, and herbs with alternate, typically trifoliate leaves and 4–5 lobed, green-yellow flowers. This family is the source of citrus fruits, notably the LEMON, ORANGE, GRAPEFRUIT, and LIME. *Distr.* Widespread, but mostly in tropical and subtropical regions. *Class* D.
See: Acradenia, Aegle, Boenninghausenia, Boronia, Choisya, Citrus, Coleonema, Correa, Crowea, Dictamnus, Diosma, Euodia, Fortunella, Melicope, Murraya, Orixa, Phellodendron, Poncirus, Ptelea, Ruta, Skimmia, Tetradium, X Citrofortunella, X Citroncirus, Zanthoxylum, Zieria.

ruthenicus, -a, -um: of Ruthenia, SW Russia.

rutifolius, -a, -um: with leaves resembling those of the family Rutaceae.

rutilans: reddish.

Ruttya (After Dr John Rutty (1697–1775), Irish naturalist.) A genus of 3 species of shrubs with simple leaves and short spikes of tubular 2-lipped flowers. *Distr.* Tropical and S Africa. *Fam.* Acanthaceae. *Class* D.

R. fruticosa Flowers yellow to scarlet with dark markings. Occasionally grown as a tender annual. *Distr.* E Africa.

rye: Canadian wild ~ *Elymus canadensis* **French** ~ *Arrhenatherum elatius* **wild** ~ *Elymus*.

S

Sabal (Possibly derived from an American vernacular name.) PALMETTO. A genus of about 14 species of small to large palms with fan-shaped leaves and much-branched inflorescences of solitary cream flowers. Some species are cultivated as ornamentals; some are commercially important for their fibres and as thatch. *Distr.* Tropical Central and North America. *Fam.* Palmae. *Class* M.

S. minor DWARF PALMETTO, SCRUB PALMETTO, BUSH PALMETTO. Trunk subterranean, occasionally to 3m high. Leaves blue-green. Grown as a half-hardy pot plant. *Distr.* SE USA.

S. palmetto PALMETTO, CABBAGE PALMETTO PALM, PALM CABBAGE, CABBAGE PALMETTO, CABBAGE TREE, COMMON PALMETTO, BLUE PALMETTO. Trunk to 30m high. Timber used in construction and furniture-making, leaves for thatch and basketry. Can be grown as a half-hardy tub plant. *Distr.* SE USA to Bahamas.

sabal *Gaultheria shallon.*

Sabiaceae A family of 1 genus and 19 species of shrubs, trees, and lianas with spirally arranged, simple or pinnate leaves and clusters of small flowers. *Distr.* Tropical and E Asia to the Solomon Isles. *Class* D.

sabulosus, -a, -um: growing in sandy places.

sacahuista *Nolina microcarpa.*

saccaticupulus, -a, -um: with a bag-shaped cupule.

saccharatus, -a, -um: appearing to be sprinkled with sugar.

saccharifera, -um: sugar-bearing.

sacchariflorus, -a, -um: with flowers resembling those of the genus *Saccharum*, SUGAR CANE.

saccharinus, -a, -um: sugary.

saccharoideus, -a, -um: resembling the genus *Saccharum*, SUGAR CANE.

Saccharum (From Greek *sakchar*, sugar.) SUGAR CANE. A genus of about 35 species of clump-forming rhizomatous grasses with cane-like stems, flat, 2-ranked leaves, and a plumed paniculate inflorescence. *Distr.* Tropical and warm temperate regions of the Old World. *Fam.* Gramineae. *Class* M.

S. officinarum SUGAR CANE. The source of over half the world supply of sugar. By-products are used as cattle feed or distilled to give alcoholic drinks and a petrol substitute; the fibres are used to make fibre-board for the construction industry. Until the 20th century this was the only species used for sugar production but breeding has resulted in the introduction of other species, hybrids, and cultigens. *Distr.* Tropical SE Asia.

Saccifoliaceae A family of 1 genus and 1 species of subshrub with small, spirally arranged leaves and small tubular flowers. *Distr.* Guyana Highlands of South America. *Class* D.

sachalinensis, -e: of Sakhalin, an island north of Japan.

Sadleria (After Joseph Sadler (1791–1841), professor of botany at Budapest.) A genus of 6–7 species of small tree ferns. *Distr.* Hawaii. *Fam.* Blechnaceae. *Class* F.

S. cyatheoides Occasionally grown as an ornamental as well as being a source of red dye and packing material. *Distr.* Hawaii.

safflower *Carthamus tinctorius.*

saffron *Crocus sativus* **false** ~ *Carthamus tinctorius* **meadow** ~ *Colchicum autumnale* **wild** ~ *Crocus cartwrightianus.*

saffron spike *Aphelandra squarrosa.*

sage *Salvia, S. officinalis* **common** ~ *S. officinalis* **Jerusalem** ~ *Phlomis fruticosa, Pulmonaria officinalis, P. saccharata* **mealy** ~ *Salvia farinacea* **mealy-cup** ~ *S. farinacea* **pineapple**

~ *S. rutilans* **pineapple-scented** ~ *S. elegans*, *S. rutilans* **pitcher** ~ *Lepechinia calycina* **South African wood** ~ *Buddleja salviifolia* **white** ~ *Artemisia ludoviciana* **wood** ~ *Teucrium*, *T. scorodonia*, *T. canadense* **yellow** ~ *Lantana camara.*

sage brush *Artemisia.*

Sageretia A genus of 35 species of spiny or spineless shrubs with simple opposite leaves and inconspicuous flowers that typically bear their parts in fives. *Distr.* SW Asia, E Africa, warm regions of the New World. *Fam.* Rhamnaceae. *Class* D.

S. thea Leaves used to make tea in Vietnam. *Distr.* Central and E China.

S. theezans See *S. thea*

Sagina (From Latin *sagina*, fodder, the name for a related plant that was fed to sheep.) PEARLWORT. A genus of 20–25 species of annual and perennial, usually tufted herbs with pairs of basally fused leaves and small flowers that bear 4–5 sepals, 4–5 or no petals, and 4–10 stamens. Several species are grown as rock-garden ornamentals. *Distr.* N temperate regions, tropical mountains. *Fam.* Caryophyllaceae. *Class* D.

S. boydii Dense cushion-forming perennial. Flowers rarely produced. *Distr.* Origin unknown, believed to be Scotland.

S. glabra See *S. subulata*

S. subulata Mat-forming. Flowers very small, white. *Distr.* W and Central Europe.

Sagittaria (From Latin *sagittarius*, armed with arrows, alluding to the shape of the leaves.) ARROWHEAD. A genus of about 20 species of aquatic perennial herbs with fine submerged leaves, broad floating leaves, and white flowers. Some species are cultivated as ornamentals, others have edible tubers. *Distr.* Temperate and tropical regions. *Fam.* Alismataceae. *Class* M.

S. japonica See *S. sagittifolia*

S. latifolia WAPATO, DUCK POTATO. Tubers formerly eaten by North American Indians, now eaten by Chinese North Americans as substitute for *S. sagittifolia. Distr.* North America.

S. sagittifolia ARROWHEAD, OLD WORLD ARROWHEAD. Cultivated in China and Japan for its edible tubers which must not be eaten raw. *Distr.* Eurasia.

sagittatus, -a, -um: sagittate, shaped like an arrow-head.

sagittifolius, -a, -um: with arrow-shaped leaves.

sago, false *Cycas*, *C. circinalis.*

sainfoin *Onobrychis viciifolia.*

St Barbara's herb *Barbarea.*

St John's wort Guttiferae, *Hypericum* **perforated** ~ *H. perforatum.*

St Julien *Prunus domestica.*

Saintpaulia (After Baron Walter von Saint Paul-Illaire (1860–1910), who collected *S. ionantha.*) AFRICAN VIOLET. A genus of 20 species of perennial herbs with rosettes of succulent hairy leaves and clustered or solitary, 5-lobed, 2-lipped flowers. Several species are grown as very popular house-plants with over 2000 different cultivars having been produced. *Distr.* Tropical E Africa. *Fam.* Gesneriaceae. *Class* D.

S. ionantha AFRICAN VIOLET, USAMBARA VIOLET. Leaves 5cm across. Flowers to 2.5cm across, purple-blue or white with a blue centre. Numerous cultivars of this species have been produced. It is now rare in the wild. *Distr.* Tanzania.

sakaki *Cleyera japonica.*

salal *Gaultheria shallon.*

salep *Dactylorhiza* **Indian** ~ *Eulophia.*

Salicaceae The Willow family. 2 genera and 335 species of shrubs and trees with simple, alternate, typically deciduous leaves. The flowers are small and unisexual, the sexes being borne on separate plants in catkin-like inflorescences. This family is an important source of timber for cricket bats and young wood for basket-making. *Distr.* Widespread, except Australasia. *Class* D.
See: Populus, Salix.

salicaria: resembling the genus *Salix*, WILLOW.

salicifolius, -a, -um: from the genus *Salix*, and *folius*, leaved.

salicornioides: from the genus *Salicornia*, with the ending *-oides*, indicating resemblance.

salignus, -a, -um: willow-like.

saligot *Trapa natans.*

Salix (Classical Latin name for these plants.) WILLOW, OSIER, SALLOW. A genus of about 300 species of deciduous shrubs and trees with simple alternate leaves and small, unisexual, insect-pollinated flowers that are borne in erect catkins. The male and female flowers are borne on separate plants. Several species are grown as ornamentals and some are used as a source of flexible timber, most notably for cricket bats. *Distr.* Temperate regions of the N hemisphere. *Fam.* Salicaceae. *Class* D.

S. aegyptiaca MUSK WILLOW. Tall shrub or small tree. *Distr.* W and Central Asia.

S. alba WHITE WILLOW. Tree to 25m high. Leaves white-hairy below. A source of timber and young shoots for basketry. The leaves were formerly used to make a kind of tea. *Distr.* Europe, N Africa to Central and W Asia.

S. arctica ARCTIC WILLOW. Creeping shrub to 10cm high. *Distr.* Arctic Europe, Asia and America.

S. aurita EARED WILLOW. Shrub to 3m high. Leaves with small auricles at the base. *Distr.* Central and N Europe.

S. babylonica BABYLON WEEPING WILLOW. Tree to 12m high. Branches pendulous. The willow depicted on willow-pattern china. *Distr.* Asia, China, Manchuria.

S. candida SAGE WILLOW, HOARY WILLOW. Shrub to 3m high. *Distr.* North America.

S. caprea GOAT WILLOW, PUSSY WILLOW, FLORIST'S WILLOW, SALLOW. Shrub or tree of damp places. Wood used for charcoal. Young twigs used as decoration. *Distr.* Europe to NE and Central Asia.

S. daphnoides VIOLET WILLOW. *Distr.* N Europe to Central Asia.

S. fragilis CRACK WILLOW, BRITTLE WILLOW. Tree to 25m high. Leaves make a distinct snap when they break off. *Distr.* Europe to N Asia, naturalized in North America.

S. herbacea DWARF WILLOW. Creeping shrub to 12cm high. *Distr.* Europe and Asia above the Arctic circle.

S. humilis PRAIRIE WILLOW. *Distr.* North America.

S. japonica See *S. babylonica*

S. lanata WOOLLY WILLOW. *Distr.* Arctic and subarctic Asia and Europe.

S. lapponum DOWNY WILLOW, LAPLAND WILLOW. *Distr.* N Europe to N Asia.

S. matsudana See *S. babylonica*

S. myrsinifolia WHORTLE WILLOW. *Distr.* N Europe, N Turkey, W Siberia.

S. myrtilloides SWAMP WILLOW. *Distr.* N Europe, N Asia, NW America.

S. nigricans See *S. myrsinifolia*

S. occidentalis See *S. humilis*

S. pentandra BAY WILLOW, LAUREL WILLOW. *Distr.* N and Central Europe, naturalized E North America.

S. polaris POLAR WILLOW. *Distr.* Arctic Europe and Asia.

S. purpurea PURPLE OSIER, BASKET WILLOW, PURPLE WILLOW. Shrub to 5m high. Young shoots used for weaving. Roots are a source of dye. *Distr.* Europe, N Africa, Central Asia.

S. repens CREEPING WILLOW. Low creeping shrub. *Distr.* Europe, Turkey, Siberia.

S. × sepulcralis WEEPING WILLOW. A hybrid of *S. alba* × *S. babylonica*. A large tree with pendulous branches. Frequently planted in parks and gardens. *Distr.* Garden origin.

S. tristis See *S. humilis*

S. vitellina See *S. alba*

sallow *Salix, S. caprea.*

salmoneus, -a, -um: salmon-pink-coloured.

salsify *Tragopogon porrifolius.*

salsuginosus: growing in brackish places.

salt bush *Atriplex.*

salt bush tree *Halimodendron halodendron.*

saltwort, black *Glaux maritima.*

saluenensis, -e: of the Nu Jiang (Salween) river, Western Yunnan Province, China.

Salvadoraceae A family of 3 genera and 11 species of shrubs and trees of arid and saline areas with simple leaves and dense axillary clusters of flowers. *Distr.* Tropical and subtropical regions of the Old World. *Class* D.

Salvia (From Latin *salvare*, to save or heal, alluding to the medicinal properties of many of the species.) SAGE. A genus of 900 species of annual and perennial herbs and shrubs with opposite, simple to pinnately divided leaves, and whorls of tubular 2-lipped flowers. The 2 stamens work on a rocker mechanism, a visiting bee having pollen pressed on to its head as it pushes against a sterile projection from the anther. Many species are grown as ornamentals and as medicinal and culinary herbs. *Distr.* Tropical and temperate regions. *Fam.* Labiatae. *Class* D.

S. argentea Rosette-forming perennial. Leaves silver-hairy. Flowers white. *Distr.* Mediterranean.

S. elegans PINEAPPLE-SCENTED SAGE. Perennial herb or subshrub. Flowers red. *Distr.* Mexico, Guatemala.

S. farinacea MEALY-CUP SAGE, MEALY SAGE. Perennial herb to 50cm high. Flowers purple. *Distr.* SW North America.

S. fruticosa Small shrub. Flowers pink or mauve. Formerly used medicinally and in local beverages. *Distr.* Canary Islands, N Africa, Middle East.

S. grahamii See *S. microphylla*

S. haematodes See *S. pratensis*

S. hispanica Annual herb to 1m high. Flowers blue. *Distr.* Central Mexico, widely naturalized.

S. horminum See *S. viridis*

S. leucantha Perennial herb or subshrub. Flowers white. *Distr.* Mexico and Tropical America.

S. microphylla Shrub. Flowers red. *Distr.* Mexico.

S. officinalis COMMON SAGE, SAGE. Perennial shrub. Flowers white, pink, or purple. An important culinary herb. Packets of dried leaves are sometimes adulterated with those of *S. fruticosa*. *Distr.* S Europe, N Africa.

S. pratensis MEADOW CLARY. Perennial herb to 1m high. Flowers in branched spikes, violet. *Distr.* Europe.

S. rutilans PINEAPPLE SAGE, PINEAPPLE-SCENTED SAGE. Clump-forming subshrub. Flowers pink or red. *Distr.* Origin obscure, probably Mexico.

S. scabra See *S. elegans*

S. sclarea CLARY. Biennial or perennial. Flowers white to mauve or lilac. A source of aromatic oil used in soaps, scent, vermouth, and liqueurs. *Distr.* S Europe to Central Asia.

S. triloba See *S. fruticosa*

S. verbenaca VERVAIN, WILD CLARY. Perennial herb. Flowers lavender to purple. *Distr.* Europe to W Asia, widely naturalized.

S. viridis BLUEBEARD. Annual or biennial herb. Flowers white to purple. *Distr.* S Europe.

salviifolius, -a, -um: from the genus *Salvia*, SAGE, and *folius*, leaved.

Salvinia (After Professor Antonio Maria Salvini (1633–1729), professor of Greek at Florence.) A genus of 10–12 species of free-floating aquatic ferns that bear whorls of 2 floating simple leaves and 1 submerged, finely divided leaf. Several species are grown as tender ornamentals. *Distr.* Tropical and warm temperate regions. *Fam.* Salviniaceae. *Class* F.

S. auriculata FLOATING FERN. *Distr.* Tropical America, widely naturalized.

Salviniaceae A family of 1 genus and 12 species. *Class* F.
See: Salvinia.

sambucifolius, -a, -um: from the genus *Sambucus*, ELDER, and *folius*, leaved.

Sambucus (Latin name for these plants, possibly derived from *sambuca*, a kind of harp made of elder wood.) ELDER, ELDERBERRY. A genus of 20–25 species of deciduous robust herbs, shrubs, and small trees with pithy stems, opposite pinnate leaves, and flat-topped inflorescences of small, regular, white flowers. The fruits are berry-like and fleshy. Several species are grown as ornamentals and some have edible fruits. *Distr.* Temperate and subtropical N Hemisphere, South America, E Australia, Tasmania. *Fam.* Caprifoliaceae. *Class* D.

S. caerulea BLUE ELDER, BLUE ELDERBERRY. Shrub or small tree. Flowers tinged yellow. Fruit black with a blue-grey bloom. *Distr.* W North America.

S. canadensis AMERICAN ELDER, SWEET ELDER. Grown for purple-black fruit used in jellies, pies, sauces, and wines. *Distr.* E North America.

S. ebulus DANEWORT, DWARF ELDER, DANE'S ELDER, WALLWORT. Perennial rhizomatous herb. Foliage smells of cooking meat.

Black fruit a source of dye. *Distr.* Eurasia, Mediterranean.

S. nigra COMMON ELDER, ELDERBERRY, ELDER, BLACK ELDER, EUROPEAN ELDER. Shrub or small tree. Fruit shiny purple-black. Flowers were formerly used medicinally and the fruit and flowers are used in wine-making. The pith is used for holding specimens when sectioning botanical materials. *Distr.* Europe, N Africa, SW Asia.

S. pubens AMERICAN RED ELDER, RED-BERRIED ELDER, STINKING ELDER. Shrub. Flowers tinged yellow. Fruit red, inedible. *Distr.* North America.

S. racemosa RED-BERRIED ELDER, EUROPEAN RED ELDER. Shrub. Flowers tinged green. Fruit bright red. Numerous cultivars are available. *Distr.* Europe, Turkey, W Asia, Siberia.

samia: of the island of Samos in the Aegean Sea.

Samolus (Classical Latin name.) WATER PIMPERNEL. A genus of 15 species of perennial herbs of salt marshes with simple leaves and racemes or corymbs of smal bell-shaped flowers that bear their parts in fives. *Distr.* Cosmopolitan. *Fam.* Primulaceae. *Class* D.

S. valerandii Leaves borne in a basal rosette. Flowers white, borne in a raceme. Grown as an ornamental. *Distr.* Europe to China.

samphire *Crithmum maritimum* **sea** ~ *C. maritimum.*

Sanchezia (After Josef Sanchez, 18th-century Spanish botanist.) A genus of 20 species of perennial herbs and subshrubs with large simple leaves and spikes or panicles of long-tubed flowers. Some species are cultivated as ornamentals. *Distr.* Tropical Central and South America. *Fam.* Acanthaceae. *Class* D.

S. nobilis Leaves dark green with sunken, white to yellow veins. Bracts red. Flowers yellow. *Distr.* Ecuador, Peru.

sanctae-rosae: holy rose.

sanctus, -a, -um: holy

sandalwood Santalaceae.

Sandersonia (After John Sanderson (1820–91), honorary secretary of the Horticultural Society of Natal.) A genus of 1 species of tuberous herb with an erect stem and simple leaves, the tips of which are often developed into tendrils. The urn-shaped orange flowers are borne in the axils of the upper leaves and have 6 fused tepals and 6 stamens. *Distr.* South Africa (Natal). *Fam.* Colchicaceae. *Class* M.

S. aurantiaca CHRISTMAS BELLS, CHINESE LANTERN LILY. Cultivated as a tender ornamental.

sandwort *Minuartia, Arenaria* **Teesdale** ~ *Minuartia stricta.*

sang *Panax quinquefolius.*

Sanguinaria (From Latin *sanguis*, blood, alluding to the red latex.) A genus of 1 species of rhizomatous perennial herb with large, palmately lobed, blue-green leaves and solitary regular flowers that bear 2 sepals, 8–12 white petals, and many stamens. *Distr.* E North America. *Fam.* Papaveraceae. *Class* D.

S. canadensis BLOODROOT, RED PUCCOON. Grown as an ornamental and used medicinally.

sanguineus, -a, -um: blood-red.

sanguiniflorus, -a, -um: with blood-red flowers.

Sanguisorba (From Latin *sanguis*, blood, and *sorbeo*, to absorb; it was said to stop bleeding.) BURNET. A genus of about 18 species of rhizomatous perennial herbs and small shrubs with pinnate leaves and spikes of small 4-lobed flowers that lack petals. Several species are grown as ornamentals and some are eaten as vegetables. *Distr.* N temperate regions. *Fam.* Rosaceae. *Class* D.

S. canadensis CANADIAN BURNET. Clump-forming perennial. Flowers white. *Distr.* North America.

S. minor BURNET, GARDEN BURNET. Erect perennial to 90cm high. Flowers green, borne in somewhat rounded heads. *Distr.* Europe, Canary Islands, SW and Central Asia.

S. officinalis BURNET, GREAT BURNET, BURNET BLOODWORT. Creeping perennial. Stems erect, to 1m high. Flowers dark purple. *Distr.* Eurasia, naturalized in North America.

sanicle *Sanicula europaea* **wood** ~ *S. europaea.*

Sanicula (From Latin *sonus*, healthy, alluding to the former medical use of some species.)

A genus of 37 species of biennial and perennial herbs with palmately divided leaves and clusters of umbels of white to red or purple flowers. The fruits are often covered in prickles to aid dispersal. Several species are grown as ornamentals and the fruits of some have been used medicinally. *Distr.* Widespread, except Australasia. *Fam.* Umbelliferae. *Class* D.

S. europaea WOOD SANICLE, SANICLE. Erect perennial. Rootstock thick. Flowers white to pink. *Distr.* Eurasia.

Saniella A genus of 1 species of perennial herb with fleshy roots, linear, fleshy, basal leaves and solitary regular flowers that are around 5cm across. *Distr.* South Africa, Lesotho. *Fam.* Hypoxidaceae. *Class* M.

S. verna Occasionally grown as a rock-garden plant.

Sansevieria (After Raimond de Sangro, Prince of Sanseviero (1710–71), Italian patron of horticulture.) SNAKE PLANT, MOTHER IN LAW'S TONGUE, BOW STRING HEMP. A genus of 50–70 species of perennial rhizomatous herbs with tough, succulent, spine-tipped leaves and tall racemes of fragrant, tubular, cream flowers. Several species are grown as tender ornamentals, others are a source of fibre for ropes, sails, and paper. *Distr.* Tropical and subtropical Africa, Madagascar, Arabia, India. *Fam.* Dracaenaceae. *Class* M.

S. cylindrica Leaves in a large basal rosette, often striped. *Distr.* S Africa.

S. hyacinthoides AFRICAN BOWSTRING HEMP. An important fibre crop. *Distr.* S Africa.

S. trifasciata MOTHER IN LAW'S TONGUE, SNAKE PLANT. Leaves few, large, in a basal rosette. A commonly cultivated pot plant with several variegated cultivars available. *Distr.* W tropical Africa, Nigeria.

Santalaceae The Sandalwood family. 37 genera and 500 species of herbs, shrubs, and trees most of which are semi-parasitic. The leaves are simple and usually spirally arranged. The flowers are small, inconspicuous and borne in spikes or heads. *Distr.* Widespread. *Class* D.
See: Osyris.

Santolina (From Latin *sanctum linum*, holy flax, a name used for *S. rosmarinifolia* and *S. virens*.) HOLY FLAX, LAVENDER COTTON. A genus of 18 species of small, evergreen, aromatic shrubs with simple or pinnately divided leaves and flowers in small daisy-like heads that lack distinct rays. *Distr.* Mediterranean. *Fam.* Compositae. *Class* D.

S. chamaecyparissus LAVENDER COTTON. Silvery grey shrub. Cultivated as an ornamental, used as a vermifuge and in scent. *Distr.* Mediterranean.

S. elegans Creeping grey-hairy shrub. *Distr.* S Spain.

S. incana See *S. chamaecyparissus*

S. pectinata See *S. rosmarinifolia*

S. pinnata Densely leafy shrub, to 80cm high. *Distr.* Italy.

S. rosmarinifolia HOLY FLAX. Small erect shrub. Leaves used as flavouring. *Distr.* Mediterranean region, Iberian Peninsula to S France.

S. tomentosa See *S. pinnata*

S. virens See *S. rosmarinifolia*

S. viridis See *S. rosmarinifolia*

Sapindaceae The Lychee family. 134 genera and about 1300 species of shrubs, trees, and lianas with simple or compound leaves and small, regular or irregular, flowers that bear 5 sepals, 5 petals, and 2 whorls of 5 stamens. Various species have fruits which are edible or used to produce beverages. *Distr.* Tropical, subtropical, and occasionally temperate regions. *Class* D.
See: Alectryon, Cardiospermum, Dodonaea, Koelreuteria, Litchi, Sapindus, Xanthoceras.

Sapindus (From Latin *sapo*, soap, and *indicus*, Indian, alluding to the use of the fruit of *S. saponaria* as soap by North American Indians.) SOAP NUTS, SOAP BERRY. A genus of about 13 species of evergreen shrubs and trees with spirally arranged, simple or pinnately divided leaves, and racemes or panicles of small white flowers that are followed by fleshy fruits. Several species are grown as ornamentals. The fruits are used as a soap substitute. *Distr.* Tropical and warm regions. *Fam.* Sapindaceae. *Class* D.

S. drummondii WILD CHINA TREE. Deciduous tree to 16m high. *Distr.* SW USA.

Sapium (Name used by Pliny for a resinous pine, alluding to the greasy latex of these plants.) TALLOW TREE. A genus of 100 species of shrubs and trees with simple alternate leaves and poisonous latex. The small unisexual flowers are borne in terminal or axillary spikes and the fruit is a fleshy capsule. Some species are exploited for their latex and a few are grown as ornamentals or shade trees. *Distr.* Tropical and warm temperate regions. *Fam.* Euphorbiaceae. *Class* D.

S. japonicum Small tree. Leaves blue-green below. Inflorescences catkin-like, yellow-green. *Distr.* China, Korea, Japan.

saponaceus, -a, -um: soapy.

Saponaria (From Latin *sapo*, soap; *S. officinalis* is a source of soap.) SOAPWORT. A genus of 20–30 species of annual and perennial herbs with simple opposite leaves and clusters of small flowers that bear 5 sepals and 5 petals. The petals are often notched at the apex, sometimes deeply so. This genus is closely related to the genus *Silene* but differs in usually having 2 styles and a fruit capsule that opens into 4 teeth. Several species are grown as ornamentals. *Distr.* Mountains of S Europe and SW Asia. *Fam.* Caryophyllaceae. *Class* D.

S. caespitosa Mat-forming perennial. Petals entire, pink. *Distr.* E Mediterranean region.

S. lutea Tufted perennial. Flowers yellow. *Distr.* SW and Central Alps, restricted to limestone.

S. officinalis BOUNCING BET, SOAPWORT, FULLER'S HERB. Robust, upright herb. Flowers few, large, pink, red or white. Formerly grown as a medicinal herb and a source of soap. *Distr.* Europe, widely naturalized.

Sapotaceae A family of 53 genera and about 1000 species of trees with simple, spirally arranged leaves and clusters of scented white-cream flowers which are bat-pollinated. This family is a source of timber and latex. *Distr.* Widespread in tropical regions. *Class* D.
See: Palaquium, Sideroxylon.

sapree wood *Widdringtonia nodiflora.*

sapu *Michelia champaca.*

Sarcocapnos (From Greek *sarx*, flesh, and *kapnos*, smoke, FUMITORY, alluding to the fleshy leaves.) A genus of 4 species of small, tufted, perennial herbs with simple or compound leaves and racemes of small spurred flowers. *Distr.* W Mediterranean region. *Fam.* Papaveraceae. *Class* D.

S. enneaphylla Leaves compound with round leaflets. Flowers yellow. Grown as a half-hardy ornamental. *Distr.* N Africa, SW Europe.

sarcocaulis, -e: with a fleshy stem.

Sarcococca (From Greek *sarx*, flesh, and *kokkos*, berry, alluding to the fleshy fruits.) CHRISTMAS BOX, SWEET BOX. A genus of 11–14 species of evergreen shrubs and small trees with spirally arranged, leathery leaves and clusters or spikes of small unisexual flowers. The fruit is red, purple, or black and fleshy. Several species are cultivated as ornamentals, particularly as ground cover. *Distr.* SE Asia, Himalaya, W China, W Malaysia. *Fam.* Buxaceae. *Class* D.

S. confusa Dense bushy shrub. Flowers very fragrant. Fruit becoming black when ripe. *Distr.* Of obscure origin, possibly from China, not known in the wild.

S. hookeriana Clump-forming. Leaves dark green, pointed. Flowers fragrant. Fruit black. *Distr.* Himalaya.

S. ruscifolia Spreading shrub. Leaves pointed, dark green. Flowers cream. Fruit red. *Distr.* W China, Tibet.

sarcodes: flesh-like.

Sarcolaenaceae A family of 9 genera and 40 species of trees and shrubs with simple alternate leaves and regular, often showy flowers. This family once formed the dominant vegetation in areas of Madagascar but has now been largely replaced by grassland. *Distr.* Madagascan rainforest. *Class* D.

Sargentodoxaceae A family of 1 genus and 1 species of twining climber with trifoliate leaves and pendent racemes of small flowers. *Distr.* India to China. *Class* D.

sarmentosus, -a, -um: creeping.

sarniensis: of Guernsey (Sarnia).

Sarracenia (After Michel Sarrasin de l'Etang (1659–1734), French botanist who sent plants from Quebec to Tournefort in Paris.) PITCHER PLANT. A genus of 8 species of carnivorous perennial herbs with rosettes of pitcher-shaped leaves that trap and digest insects. The flowers are borne on a long stalk above the pitchers and bear their parts in fives. Several species and numerous hybrids are grown as ornamentals. *Distr.* E North America. *Fam.* Sarraceniaceae. *Class* D.

S. flava YELLOW PITCHER PLANT, YELLOW TRUMPET, TRUMPETS. Pitchers to 1m long, yellow-green. Flowers yellow. *Distr.* SE US.

S. purpurea NORTHERN PITCHER PLANT, HUNTSMAN'S CUP, HUNTSMAN'S HORN, COMMON PITCHER PLANT. Pitchers tinged red. Flowers red-purple. *Distr.* E North America, naturalized in Ireland.

Sarraceniaceae The New World Pitcher Plant family. 3 genera and 15 species of carnivorous perennial herbs. The leaves are modified into pitcher-shaped insect traps and the large nodding flowers have 5 white or coloured petals. Some species are grown as ornamentals and curiosities. *Distr.* North America and the Guyana Highlands of South America, particularly in marshy habitats. *Class* D.
See: Darlingtonia, Heliamphora, Sarracenia.

sarsparilla, false ~ *Hardenbergia violacea* **Jamaica** ~ *Smilax regelii* **wild** ~ *S. glauca.*

Sasa (From *zasa*, the Japanese name for the small bamboos.) A genus of 20–40 species of small and medium-sized, rhizomatous bamboos with ascending stems and small leaves. Several species are cultivated as ornamentals. *Distr.* E Asia. *Fam.* Gramineae. *Class* M.

S. disticha See *Pleioblastus pygmaeus*

S. nana See *S. veitchii*

S. palmata Cultivated as a fast-spreading ornamental. *Distr.* Japan.

S. tessellata See *Indocalamus tessellatus*

S. veitchii KUMA ZASA. Leaves with a broad, papery, white margin. *Distr.* Japan.

Sasaella See *Sasa*

Sasamorpha See *Sasa*

Sassafras (From Spanish *salsafras*, which is probably derived from an American Indian name.) A genus of 2–3 species of aromatic deciduous trees with simple alternate leaves and racemes of numerous, small, yellow-green, uni- and bisexual flowers. *Distr.* 1 species E North America, 1 or 2 species E Asia. *Fam.* Lauraceae. *Class* D.

S. albidum SASSAFRAS. Tree to 20m high. A source of timber and an essential oil that is used medicinally, particularly on insect bites, and in perfumery. *Distr.* E North America.

S. tzumu CHINESE SASSAFRAS. An important local source of timber. *Distr.* Central China.

sassafras *Sassafras albidum* **Chinese** ~ *S. tzumu.*

satin wood *Murraya paniculata.*

sativus, -a, -um: cultivated.

satsuma *Citrus reticulata.*

Satureja (From a Latin name used by Pliny.) SAVORY. A genus of 30 species of herbs and subshrubs with linear opposite leaves and whorls of tubular 2-lipped flowers. Several species are grown as ornamentals and some are used as culinary herbs. *Distr.* N temperate regions. *Fam.* Labiatae. *Class* D.

S. hortensis SUMMER SAVORY. Aromatic shrubby annual. Flowers white or pink. The dried leaves are used as a flavouring in a similar way to SAGE. but especially with BROAD BEANS. *Distr.* Mediterranean region.

S. montana WINTER SAVORY. Dwarf shrub. Flowers white to pale violet, borne in dense whorls. The dried leaves are used in a similar way to those of *S. hortensis* but are considered to have a poorer flavour. *Distr.* S Europe.

saturejoides: from the genus *Satureja*, with the ending *-oides*, indicating resemblance.

Satyrium (From *satyrion*, the ancient Greek name for an orchid.) A genus of 100–135 species of terrestrial tuberous orchids with spikes of flowers in which the lip forms a hood over the column. Some are cultivated. *Distr.* Africa, particularly South Africa, a few species extending to Asia. *Fam.* Orchidaceae. *Class* M.

S. nepalense Leaves restricted to a basal rosette. Flowers bright pink. *Distr.* N India to Sri Lanka, Burma, and SW China.

Sauromatum (From Greek *sauros*, lizard, alluding to the spotted spadix and its long tail-like appendage.) A genus of about 2 species of tuberous herbs with solitary leaves that appear after the inflorescence. The small unisexual flowers are borne on a spadix with a long foul-smelling appendage. *Distr.* Tropical Old World. *Fam.* Araceae. *Class* M.

S. guttatum See *S. venosum*

S. nubicum Tubers are roasted and eaten. *Distr.* Tropical Africa.

S. venosum VOODOO LILY, MONARCH OF THE EAST, RED CALLA. Grown as an ornamental curiosity which flowers from an apparently dry tuber without having to be watered. *Distr.* Tropical regions of the Old World.

Saururaceae A family of 5 genera and 7 species of perennial herbs with alternate simple leaves and small flowers borne in terminal spikes. Some species are grown as ornamentals. *Distr.* E Asia and E North America. *Class* D.
See: Houttuynia, Saururus.

Saururus (From Greek *sauros*, lizard, and *oura*, tail, alluding to the tail-like inflorescence.) BOG PLANT, LIZARD'S TAIL. A genus of 1–4 species of perennial rhizomatous herbs with simple, stalked, alternate leaves and spikes of small flowers that lack sepals and petals but are each subtended by a small scale-like bract. *Distr.* North America, E Asia. *Fam.* Saururaceae. *Class* D.

S. cernuus SWAMP LILY. Flowers white, fragrant. Grown as an ornamental. *Distr.* E North America.

Saussurea (After Horace Benedict de Saussure (1740–99), Swiss naturalist and philosopher, and his son Nicholas Theodor (1767–1845), botanist and chemist.) A genus of about 300 species of perennial herbs with simple to pinnately lobed leaves and flowers in daisy-like heads which lack distinct rays. Some species are cultivated as ornamentals. *Distr.* Eurasia, North America. *Fam.* Compositae. *Class* D.

S. alpina Flowers purple, sweetly scented. *Distr.* Europe to NW Asia.

savory *Satureja* **summer** ~ *S. hortensis* **winter** ~ *S. montana.*

saxatilis, -e: growing on rocks.

Saxegothaea (After Prince Albert of Saxe-Coburg-Gotha (1819–61), Prince Consort of Queen Victoria.) A genus of 1 species of evergreen coniferous shrub or tree to 20m high with flattened linear leaves that are borne spirally but are twisted so as to be arranged in 2 ranks. The fruiting cones bear fleshy fertile scales. *Distr.* Chilean Andes, Patagonia. *Fam.* Podocarpaceae. *Class* G.

S. conspicua PRINCE ALBERT'S YEW. Grown as a ornamental. *Distr.* Chile.

saxicola: growing among rocks, in rocky places.

Saxifraga (From Latin *saxum*, rock, and *frango*, to break; by growing in crevices they appear to break rocks, and through the doctrine of signatures they were thought to cure kidney stones. The equivalent name in Greek is used for the genus *Petrocoptis*.) SAXIFRAGE, ROCKFOIL. A genus of 300–400 species of perennial, occasionally annual herbs, forming cushions or mats or growing taller. The flowers have 4–5 sepals, an equal number of petals, and 8–10 stamens. Many species are cultivated as ornamentals, some are eaten locally in salads. *Distr.* N temperate to subarctic regions, Asia, plus South American and N African mountains. *Fam.* Saxifragaceae. *Class* D.

S. aizoon See *S. paniculata*

S. andersonii Cushion-forming perennial with white or pink flowers. *Distr.* Sikkim, Nepal, Bhutan, Tibet.

S. × andrewsii A hybrid of *S. paniculata* × *S. spathularis*. *Distr.* Garden origin.

S. × apiculata A hybrid of *S. marginata* × *S. sancta*. *Distr.* Garden origin.

S. aquatica See *S. irrigua*

S. × arco-valleyi A hybrid of *S. lilacina* × *S. marginata*. *Distr.* Garden origin.

S. × arendsii A hybrid of *S. rosacea* × *S. exarata*. *Distr.* Garden origin.

S. aretioides Evergreen cushion-forming perennial with yellow flowers. *Distr.* Balkans.

S. × borisii A hybrid of *S. marginata* × *S. ferdinandicoburgi*. *Distr.* Garden origin.

S. brunoniana See *S. brunonis*

S. brunonis (After Robert Brown (1773–1858), Scottish botanist.) Deciduous mat-forming perennial with yellow flowers. *Distr.* Himalaya.

S. bryoides MOSSY SAXIFRAGE. Ever-green mat-forming perennial with white flowers spotted with orange. *Distr.* European mountains.

S. × burnatii A hybrid of *S. cochlearis* × *S. paniculata*. *Distr.* Garden origin.

S. burseriana (After Joachim Burser (1583–1649), German physician and botanist.) BURSER'S SAXIFRAGE. Evergreen cushion-forming perennial with large white flowers. *Distr.* E Alps.

S. caesia BLUE-GREEN SAXIFRAGE. Evergreen, low, cushion-forming perennial with white flowers. *Distr.* S Central Europe.

S. callosa LIMESTONE SAXIFRAGE. Evergreen perennial with lime-secreting leaves. *Distr.* W Mediterranean.

S. canaliculata Evergreen perennial with white flowers. *Distr.* N Spain.

S. × canis-dalmatica See *S.* × *gaudinii*

S. catalaunica See *S. callosa*

S. caucasica Mat-forming perennial with yellow flowers. *Distr.* Caucasus.

S. cebennensis Evergreen perennial forming a large cushion. *Distr.* S France.

S. chrysosplenifolia See *S. rotundifolia*

S. cochlearis Evergreen tuft-forming perennial with fleshy lime-secreting leaves. *Distr.* Alps.

S. corsica Rosette-forming perennial with white flowers. *Distr.* Corsica and Sardinia.

S. cortusifolia Deciduous, rosette- or tuft-forming perennial. *Distr.* E Asia, Japan.

S. corymbosa Dense, cushion-forming, evergreen perennial. *Distr.* Mountains of SW Europe.

S. cotyledon GREAT ALPINE ROCKFOIL, GREATER EVERGREEN SAXIFRAGE. Evergreen perennial with leaf rosettes to 12cm across. *Distr.* Alps, Pyrenees, NW Europe, Scandinavia, Iceland.

S. crustata Evergreen perennial forming deep cushions. *Distr.* E Alps to the Balkans.

S. cuneata Evergreen perennial forming loose tufts of old leaves.

S. cuneifolia SHIELD-LEAVED SAXIFRAGE. Evergreen perennial with fleshy leaves that are often red-purple beneath. *Distr.* Mountains of Central and S Europe.

S. cuscutiformis DODDER-LIKE SAXIFRAGE. Perennial herb with long stolons and small round leaves. *Distr.* China.

S. cymbalaria Self-sowing annual with yellow flowers. *Distr.* SE Europe, SW Asia.

S. × elisbethae A hybrid of *S. burseriana* × *S. sancta*. *Distr.* Garden origin.

S. exarata FURROWED SAXIFRAGE. Compact, cushion-forming, evergreen perennial with cream flowers. *Distr.* Mountains of Central and S Europe.

S. federici-augusti Mat- or cushion-forming evergreen perennial with a dark red inflorescence. *Distr.* Balkan Peninsula.

S. ferdinandi-coburgii Small-leaved, cushion-forming, evergreen perennial with yellow flowers. *Distr.* Balkans.

S. ferruginea Short-lived perennial covered in red hairs. *Distr.* N America (Alaska to Montana).

S. flagellaris Deciduous perennial with long runners and yellow flowers. *Distr.* N circumpolar.

S. fortunei (After Robert Fortune (1812–80), who collected in China and introduced the tea plant from China to India.) Deciduous clump-forming perennial with white flowers. *Distr.* Japan.

S. × gaudinii A hybrid of *S. cotyledon* × *S. umbrosa*.

S. × geum A hybrid of *S. hirsuta* × *S. umbrosa*.

S. granulata MEADOW SAXIFRAGE, BULBOUS SAXIFRAGE, FAIR MAIDS OF FRANCE. Rosette-forming perennial with kidney-shaped leaves and showy white flowers. *Distr.* Europe.

S. hirsuta KIDNEY SAXIFRAGE. Sprawling evergreen perennial with long-stalked leaves. *Distr.* SW Europe and Ireland.

S. × hornibrookii A hybrid of *S. stribrnyi* × *S. lilacina*.

S. hostii Evergreen rosette-forming perennial. *Distr.* E Alps.

S. hypnoides MOSSY SAXIFRAGE. Prostrate mat-forming perennial with small sparse leaves and white flowers. *Distr.* Europe.

S. iranica Cushion-forming perennial with purple to white flowers. *Distr.* Caucasus.

S. irrigua Compact tuft-forming perennial with white flowers. *Distr.* N Iran, Caucasus.

S. × irvingii A hybrid of *S. burseriana* × *S. lilacina*. *Distr.* Garden origin.

S. juniperifolia Evergreen perennial forming dense cushions. *Distr.* Bulgaria.

S. × kellereri A hybrid of *S. burseriana* × *S. stribrnyi*. *Distr.* Garden origin.

S. × landaueri A hybrid of *S. burseriana* × *S. stribrnyi* × *S. marginata*. *Distr.* Garden origin.

S. lilacina Mat-forming perennial with solitary lavender flowers. *Distr.* Himalaya.

S. lingulata See *S. callosa*

S. longifolia PYRENEAN SAXIFRAGE. Evergreen perennial with large rosettes and showy inflorescences of white flowers. *Distr.* Pyrenees.

S. luteoviridis See *S. corymbosa*

S. lyallii Rosette-forming perennial with white flowers. *Distr.* N America (Rocky Mountains).

S. × macnabiana A hybrid of *S. callosa* × *S. cotyledon*. *Distr.* Garden origin.

S. mandschuriensis Perennial herbs with cymes of pink flowers. *Distr.* N China, Korea, Manchuria.

S. marginata Cushion-forming evergreen perennial. *Distr.* S Italy, Balkan peninsula.

S. media Evergreen perennial with bright pink-purple flowers. *Distr.* Pyrenees.

S. × megaseiflora A complex hybrid involving *S. burseriana*. *Distr.* Garden origin.

S. moschata See *S. exarata*

S. mutata Evergreen perennial with yellow or orange flowers. *Distr.* Alps.

S. oppositifolia PURPLE SAXIFRAGE. Perennial forming dense mats or loose cushions with solitary pink-purple flowers. *Distr.* Europe.

S. paniculata LIFELONG SAXIFRAGE. Mat-forming perennial herbs with cream-white flowers. Numerous garden cultivars are available. *Distr.* N temperate regions.

S. pedemontana Evergreen perennial forming loose cushions. *Distr.* Mountains of S Europe and N Africa.

S. petraea Biennial with sticky glandular hairs. *Distr.* Alps.

S. × petraschii A hybrid of *S. burseriana* × *S. tombeanensis*. *Distr.* Garden origin.

S. poluniniana Perennial with solitary flowers. *Distr.* Nepal.

S. porophylla Dense, cushion-forming, evergreen perennial with pink-green flowers. *Distr.* Italy (Apennines).

S. pubescens Evergreen perennial forming flat rosettes. *Distr.* Pyrenees.

S. retusa Dense cushion-forming perennial with pink-purple flowers. *Distr.* Central and S Europe.

S. rotundifolia ROUND-LEAVED SAXIFRAGE. Loose clump-forming perennial. *Distr.* SW France to Caucasus.

S. sancta Cushion-forming perennial. *Distr.* Balkans.

S. sarmentosa See *S. stolonifera*

S. scardica Evergreen perennial forming firm cushions. *Distr.* SW Balkans.

S. sempervivum Evergreen, mat- or cushion-forming perennial. *Distr.* Balkans.

S. spruneri Evergreen perennial forming firm cushions. *Distr.* Balkans.

S. squarrosa Evergreen perennial forming tall cushions. *Distr.* E Alps.

S. stolitzkae Cushion-forming perennial. *Distr.* W Himalaya.

S. stolonifera MOTHER OF THOUSANDS, WANDERING JEW, STRAWBERRY GERANIUM. Perennial with long, thin, red stolons. Often grown in hanging baskets. *Distr.* E Asia.

S. stribrnyi Evergreen perennial with violet-purple flowers. *Distr.* Balkans.

S. taygetea Rhizomatous perennial forming loose clumps of leaf rosettes. *Distr.* Balkans.

S. tenella Mat-forming perennial. *Distr.* E Alps.

S. tombeanensis Cushion-forming perennial. *Distr.* SE Alps.

S. trifurcata Evergreen perennial forming large loose cushions. *Distr.* N Spain.

S. umbrosa Evergreen perennial with flat leaf rosettes. *Distr.* Pyrenees.

S. × urbium LONDON PRIDE. A hybrid of *S. umbrosa* × *S. hirsuta*. *Distr.* W Europe.

S. vayredana Cushion-forming perennial. *Distr.* NE Spain.

Saxifragaceae The Saxifroge family. 31 genera and 450 species of perennial herbs and shrubs, a few annuals, and some trees. The leaves are typically simple but may be compound. The flowers have 5 sepals and usually 5 petals although these may sometimes be absent. A number of genera include important ornamentals; others provide soft fruits including the GOOSEBERRY and CURRANT. *Distr.* N temperate regions, rarer in mountainous areas in the tropics and in the S hemisphere. *Class* D.

See: Aceriphyllum, Astilbe, Astilboides, Bensoniella, Bergenia, Boykinia, Chrysosplenium, Darmera, Elmera, Francoa, Heuchera, × Heucherella, Lithophragma, Mitella, Mukdenia, Peltiphyllum, Peltoboykinia, Rodgersia, Saxifraga, Tanakaea, Tellima, Tiarella, Tolmiea.

saxifrage *Saxifraga* **blue-green** ~ *S. caesia* **bulbous** ~ *S. granulata* **Burser's** ~ *S. burseriana* **dodder-like** ~ *S. cuscutiformis* **furrowed** ~ *S. exarata* **golden** ~ *Chrysosplenium, C. oppositifolium* **greater evergreen** ~ *Saxifraga cotyledon* **kidney** ~ *S. hirsuta* **lifelong** ~ *S. paniculata* **limestone** ~ *S. callosa* **meadow** ~ *S. granulata* **mossy** ~ *S. bryoides, S. hypnoides* **purple** ~ *S. oppositifolia* **Pyrenean** ~ *S. longifolia* **round-leaved** ~ *S. rotundifolia* **shield-leaved** ~ *S. cuneifolia* **Siberian** ~ *Bergenia.*

saxorum: growing on rocks.

saxosus, -a, -um: growing in rocky places.

scaberrimus, -a, -um: very rough or gritty.

scaberulus, -a, -um: somewhat rough or gritty, minutely rough.

scaber: scabrous, rough or gritty to touch.

Scabiosa (From Latin *scabies*, itch; the rough leaves were said to cure skin complaints.) SCABIOUS, PINCUSHION FLOWER. A genus of 60–80 species of annual and perennial herbs and subshrubs with rosettes of simple or pinnate leaves and hemispherical heads of unequal tubular flowers. The fruit is surrounded by a persistent whorl of bracts which aids the wind dispersal of the seed. Several species and cultivars are grown as ornamentals. *Distr.* Eurasia, Mediterranean, E African mountains, S Africa. *Fam.* Dipsacaceae. *Class* D.

S. atropurpurea SWEET SCABIOUS, MOURNFUL WIDOW, PINCUSHION FLOWER, EGYPTIAN ROSE. Annual. Flowers dark purple to white or red, fragrant. *Distr.* S Europe.

S. caucasica Perennial. Flowers light blue. *Distr.* Caucasus.

S. columbaria SMALL SCABIOUS. Perennial. Flowers blue-lilac, rarely white. *Distr.* Europe, N Africa, Asia.

S. gigantea See *Cephalaria gigantea*

S. graminifolia Evergreen tuft-forming perennial. Leaves grass-like. Flowers blue-violet. *Distr.* S Europe.

S. lucida Clump-forming perennial. Flower-heads borne on long stems. Flowers mauve. *Distr.* Central Europe.

S. ochroleuca See *S. columbaria*

S. parnassi See *Pterocephalus perennis*

S. pterocephala See *Pterocephalus perennis*

S. rumelica See *Knautia macedonica*

S. succisa See *Succisa pratensis*

S. tatarica See *Cephalaria gigantea*

scabiosifolius, -a, -um: from the genus *Scabiosa*, and *folius*, leaved.

scabious *Scabiosa* **devil's bit** ~ *Succisa pratensis* **field** ~ *Knautia arvensis* **sheep's bit** ~ *Jasione laevis* **shepherd's** ~ *J. laevis* **small** ~ *Scabiosa columbaria* **sweet** ~ *S. atropurpurea.*

scabridus, -a, -um: somewhat rough.

scabrifolius, -a, -um: with rough leaves.

scabrosus, -a, -um: rough.

scabrus, -a, -um: rough, gritty.

Scadoxus A genus of about 9 species of bulbiferous or rhizomatous herbs with spirally arranged, elliptic-ovate leaves and numerous tubular flowers in a conical or spherical head. This genus is closely related to *Haemanthus*. Several species are cultivated ornamentally. *Distr.* Tropical Africa. *Fam.* Amaryllidaceae. *Class* M.

S. multiflorus BLOOD FLOWER. Flowers scarlet, fading to pink. *Distr.* Tropical and S Africa.

S. pole-evansii Flowers 30–70, pink or scarlet. *Distr.* E Zimbabwe.

S. puniceus ROYAL PAINT BRUSH. Flowers number over 100 and are borne in a dense inflorescence that resembles a single flower. *Distr.* E and S Africa.

Scaevola (From Latin *scaevus*, left-handed, alluding to the one-sided, fan-shaped petal-tube.) A genus of 90–130 species of herbs and shrubs with alternate leaves and 5-lobed tubular flowers, the petal tube being split to the base on the upper side. Several species may be grown as tender ornamentals. *Distr.* Australia, 2 species widespread on beaches in warm regions. *Fam.* Goodeniaceae. *Class* D.

S. aemula Perennial herb. Flowers solitary, white to pale blue. *Distr.* S and E Australia.

S. calendulacea Procumbent perennial herb. Flowers solitary, blue with a yellow throat, fragrant. *Distr.* SE and E Australia.

S. hookeri Creeping subshrub. Flowers solitary, blue-purple, hairy. *Distr.* SE Australia.

S. suaveolens See *S. calendulacea*

scalaris, -e: ladder-like.

scandens: climbing, twining.

Scaphyglottis (From Greek *skaphe*, boat, and *glotta*, tongue, alluding to the shape of the lip.) A genus of 30–40 species of epiphytic or lithophytic orchids with narrow leaves and racemes or clusters of small flowers. Several species are cultivated ornamentally. *Distr.* Tropical America. *Fam.* Orchidaceae. *Class* M.

S. amethystina Flowers violet to white. *Distr.* Guatemala to Panama.

scapigera: bearing a leafless flowering stem or scape.

scaposus, -a, -um: with a leafless flowering stem or scape.

scariosus, -a, -um: shrivelled.

scarlet runner *Kennedia prostrata*.

sceleratus, -a, -um: acrid.

sceptrum: sceptre.

Schefflera (After J. C. Scheffler (1722–1811), a physician in Danzig.) IVY TREE, UMBRELLA TREE. A genus of about 700 species of shrubs, trees, and vines with palmate leaves and small flowers borne in complex, paniculate, or umbellate inflorescences. Many species are grown as ornamental trees and houseplants. *Distr.* Tropical and warm regions. *Fam.* Araliaceae. *Class* D.

S. actinophylla QUEENSLAND UMBRELLA TREE, AUSTRALIAN IVY PALM, OCTOPUS TREE. Shrub or small tree, to 12m high. *Distr.* Queensland.

S. arboricola Epiphytic shrub or liana. A popular house-plant. *Distr.* Hainan, Taiwan.

S. digitata SEVEN FINGERS, PATE. Shrub to 8m. Tender. *Distr.* New Zealand.

Scheuchzeriaceae A family of 1 genus and 1 species of slender perennial herb found in marshes and bogs, with rush-like leaves and regular bisexual flowers that bear 6 tepals, 6 stamens, and a superior ovary. *Distr.* N temperate to Arctic zones. *Class* M.

Schima (From the Arabic vernacular name.) A genus of 1 species of large evergreen tree with simple, spirally arranged leaves, solitary or clustered, fragrant, white or purple-cream flowers. *Distr.* Tropical and warm regions. *Fam.* Theaceae. *Class* D.

S. argentea See *S. wallichii*

S. khasiana See *S. wallichii*

S. wallichii Occasionally grown as an ornamental.

Schinus (From Greek *schinos*, the name used for *Pistacia lentiscus*.) A genus of 27–30 species of shrubs and trees with simple or pinnate, evergreen leaves and racemes or panicles of small regular flowers. *Distr.* Tropical America. *Fam.* Anacardiaceae. *Class* D.

S. molle CALIFORNIAN PEPPER TREE, PERUVIAN MASTIC TREE. Tree. Branches slender, pendent. Leaves pinnate, with fine narrow leaflets. Flowers yellow. Grown as a street tree in frost-free regions. *Distr.* South America.

S. terebinthifolius BRAZILIAN PEPPER TREE, CHRISTMAS BERRY TREE. Spreading trees. Flowers white. Berries red. *Distr.* South America.

Schisandra (From Greek *schizo*, to divide, and *aner*, man, alluding to the divided anthers.) A genus of 25 species of evergreen and deciduous climbers with simple opposite leaves and small solitary or clustered flowers that bear 5–12 tepals and 5–15 stamens. Several species are grown as ornamentals. *Distr.* E Asia, 1 species SE USA. *Fam.* Schisandraceae. *Class* D.

S. grandiflora Deciduous climber to 7m high. Flowers white, sometimes tinged pink, fragrant. *Distr.* N India, Bhutan, Nepal.

Schisandraceae A family of 2 genera and 47 species of trailing or twining woody shrubs with simple alternate leaves and small unisexual flowers that may occur on the same or different plants. Some species are cultivated as ornamentals. *Distr.* India to Moluccas, Japan and SE USA. *Class* D.
See: Kadsura, Schisandra.

Schizaeaceae A family of 4 genera and 150 species of terrestrial ferns with dichotomous or pinnate, rarely simple leaves. The

sporangia are typically embedded in specialized leaf segments or leaves and are rarely grouped into sori. *Distr.* Tropical and S warm-temperate regions. *Class* F.
See: Anemia, Lygodium.

Schizocentron See *Heterocentron*
S. elegans See *Heterocentron elegans*

schizopetalus, -a, -um: with split petals.

Schizophragma (From Greek *schizo*, to divide, and *phragma*, wall, alluding to the dehiscing fruit.) A genus of 4 species of deciduous climbers with aerial clinging rootlets, simple, opposite, long-stalked leaves and flat-topped heads of small white flowers. Several species are grown as ornamentals. *Distr.* Himalaya to Japan and Taiwan. *Fam.* Hydrangeaceae. *Class* D.
S. hydrangeoides Climber to 10m high. Flowers irregular with one large sepal. *Distr.* Korea, Japan.

Schizostylis (From Greek *schizo*, to divide, and *stylis*, style; the style is deeply divided into 3.) A genus of 1 species of rhizomatous herb with sword-shaped leaves and spikes of cup-shaped red flowers. *Distr.* South Africa. *Fam.* Iridaceae. *Class* M.
S. coccinea KAFFIR LILY. Cultivated as an ornamental with several cultivars available. *Distr.* E South Africa.

Schoenus (From Greek *schoinos*, rush, alluding to the resemblance to the true rush (*Juncus*).) A genus of 80 species of perennial and occasionally annual, grass-like herbs with narrow leaves and heads of very small, wind-pollinated flowers. *Distr.* Cosmopolitan, but chiefly SE Asia and Australia. *Fam.* Cyperaceae. *Class* M.
S. nigricans BLACK BOG RUSH. Used as a thatching material in Ireland. *Distr.* Europe, S Africa, North America.
S. pauciflorus Grown as a hardy ornamental. *Distr.* New Zealand (North Island).

scholar tree, Chinese *Sophora japonica.*

Schotia (After Richard van der Schot (died 1819), head gardener at Schön-brunn, Austria.) BOERBOON. A genus of 4–5 species of deciduous shrubs and trees with alternate pinnate leaves and panicles of irregular flowers that bear 5 petals and 10 stamens of 2 different lengths. Several species are grown as ornamentals. *Distr.* Africa. *Fam.* Leguminosae. *Class* D.
S. brachypetala WEEPING BOERBOON, TREE FUCHSIA, AFRICAN WALNUT. Large shrub or spreading tree. Flowers fragrant, red. *Distr.* S Africa.

Sciadopitys (From Greek *skiados*, umbrella, and *pitys*, fir tree; the leaves appear in whorls like the ribs of an umbrella.) A genus of 1 species of evergreen coniferous tree to 40m high with needle-like green leaves and small, brown, scale-like leaves. The fruiting cones are leathery and open to release their seeds. *Distr.* Japan (Central Honshu). *Fam.* Taxodiaceae. *Class* G.
S. verticillata PARASOL PINE. Planted as an ornamental, particularly around temples in Japan. A source of timber.

Scilla (Greek name for the sea squill, *Urginea maritima.*) A genus of about 90 species of bulbiferous herbs with a few linear leaves and spikes of small, typically blue flowers. Several species are grown as ornamentals. *Distr.* Europe, Asia and S Africa. *Fam.* Hyacinthaceae. *Class* M.
S. adlamii See *Ledebouria cooperi*
S. amethystina See *S. litardieri*
S. autumnalis AUTUMN SQUILL. Flowers lilac-pink. Several cultivars available. *Distr.* Europe, NW Africa, SW Asia.
S. bifolia Leaves 2, strap-shaped. *Distr.* Central and S Europe, Caucasus, Turkey.
S. campanulata See *Hyacinthoides hispanica*
S. japonica See *S. scilloides*
S. litardieri Flowers flat, star-shaped, violet. *Distr.* Central Europe.
S. natalensis Flowers numerous, flat, blue, on a long spike. *Distr.* E South Africa.
S. non-scripta See *Hyacinthoides non-scripta*
S. nutans See *Hyacinthoides non-scripta*
S. peruviana CUBAN LILY. Flowers violet, in large heads of up to 50. *Distr.* SW Europe, NW Africa.
S. pratensis See *S. litardieri*
S. scilloides Flowers pink, in slender, dense spikes. *Distr.* China, Korea, Taiwan, Japan.

S. verna SPRING SQUILL. Seeds reportedly ant-dispersed. *Distr.* W Europe.

S. violacea See *Ledebouria socialis*

scillifolius, -a, -um: from the genus *Scilla*, and *folius*, leaved.

scilloides: from the genus *Scilla*, with the ending *-oides*, indicating resemblance.

Scindapsus (From a Greek name used for an ivy-like climbing plant.) A genus of 24–40 species of evergreen climbers with simple leaves. The small hermaphrodite flowers are borne on a spadix that is subtended by a white or green boat-shaped spathe. Several species are cultivated as ornamentals while others have medicinal properties. *Distr.* Tropical Asia and the Pacific with 1 species in Brazil. *Fam.* Araceae. *Class* M.

S. aureus See *Epipremnum aureum*

S. pictus SILVER VINE. Often cultivated in its juvenile phase whilst the leaves are silvery-green. *Distr.* Java to Borneo.

scintillans: gleaming.

scirpoides: from the genus *Scirpus*, with the ending *-oides*, indicating resemblance.

Scirpus (Classical Latin name.) BULRUSH, CLUB RUSH. A genus of 100–200 species of rhizomatous perennial herbs with jointed stems and grass-like leaves. The very small wind-pollinated flowers are borne in spikelets that are arranged in much-branched panicles. Some species are grown as ornamentals, the stems of others are used for thatch, matting, and basketry. The rhizomes of several species are edible. *Distr.* Cosmopolitan. *Fam.* Cyperaceae. *Class* M.

S. cyperinus Spikelets appear woolly in fruit. *Distr.* E North America.

S. sylvaticus WOOD CLUB RUSH. Leaves to 2cm wide. *Distr.* Europe to Siberia.

Scleranthus (From Greek *scleros*, hard, and *anthos*, flower.) KNAWEL. A genus of about 10 species of annual and perennial herbs with pairs of pointed linear leaves and solitary or clustered, minute, inconspicuous flowers that lack petals. *Distr.* Temperate regions. *Fam.* Illecebraceae. *Class* D.

S. biflorus Mat-forming yellow-green perennial. Occasionally grown as an ornamental. *Distr.* New Zealand, Australia (Tasmania), South America.

S. brockiei See *S. biflorus*

S. uniflorus See *S. biflorus*

Scoliopus (From Greek *skolios*, crooked, and *pous*, foot, alluding to the twisted flower-stalks.) A genus of 2 species of rhizomatous herbs with 2 broad basal leaves and an umbel of bad-smelling flowers which bear 2 whorls of 3 tepals and 3 stamens. Both species are grown as ornamentals. *Distr.* W North America. *Fam.* Trilliaceae. *Class* M.

S. bigelovii FOETID ADDER'S TONGUE, STINK POD, BROWNIES. Flowers green with red veins. *Distr.* W USA (California, SW Oregon).

S. hallii OREGON FOETID ADDER'S TONGUE. Flowers yellow-green to purple. *Distr.* W USA (Oregon).

Scolopendrium See *Asplenium*

Scolymus (Classical Greek name.) A genus of 3 species of spiny, annual to perennial herbs with pinnately lobed leaves and flowers in dandelion-like heads. *Distr.* Mediterranean including Europe. *Fam.* Compositae. *Class* D.

S. hispanicus GOLDEN THISTLE, SPANISH OYSTER THISTLE. Biennial to perennial. Leaves, stalks, and roots eaten as vegetables. *Distr.* S Europe.

scoparius, -a, -um: broom-like.

Scopolia (After Giovanni Antonio Scopoli (1723–88), professor of natural history at Pavia.) A genus of 5 species of rhizomatous herbs with simple membranous leaves and solitary, pendent, 5-lobed, bell-shaped flowers. *Distr.* Mediterranean to Himalaya. *Fam.* Solanaceae. *Class* D.

S. carniolica Clump-forming perennial. Flowers red to brown, yellow-green within. Grown as a hardy ornamental. *Distr.* Central and SE Europe.

scopulorus, -a, -um: growing on cliffs.

scordiifolius, -a, -um: from the genus *Scordium* (now *Teucrium*), and *folius*, leaved.

Scordium: the name of a genus now included in *Teucrium*.

scorodoprasum: from the Greek *skorodon*, garlic, and *prason*, leek.

scorpioides: resembling a scorpion.

Scorzonera (From French *scorzon*, viper (the root was said to cure snake bites), or possibly from the Italian *scorza*, bark, and *nera*, black, alluding to the roots.) A genus of 150 species of annual, biennial, and perennial herbs and subshrubs with flowers in dandelion-like heads. *Distr.* Mediterranean to Central Asia. *Fam.* Compositae. *Class* D.

S. hispanica COMMON VIPER'S GRASS. Perennial, to 1.3m high. Roots black. The roots are edible. *Distr.* Europe to Siberia.

S. purpurea PURPLE VIPER'S GRASS. Leaves grass-like. Flowers purple. *Distr.* Central Europe to Balkans.

Scotch attorney *Clusia major.*

scoticus, -a, -um: of Scotland.

screw pine *Pandanus*, Pandanaceae.

scriptus, -a, -um: written.

Scrophularia (From Latin *scrofulae*, swelling of the neck glands, alluding to supposed medicinal properties of these plants.) FIGWORT. A genus of 200 species of perennial herbs and shrubs with 4-angled stems, simple or compound leaves and cymes of somewhat inflated, 2-lipped, yellow-green or purple flowers. A few species are grown as ornamentals and some are used medicinally. *Distr.* N temperate regions, tropical America. *Fam.* Scrophulariaceae. *Class* D.

S. aquatica See *S. auriculata*

S. auriculata WATER FIGWORT, WATER BETONY. Perennial herb to 1m high. Flowers small, purple-brown. *Distr.* W Europe, N Africa.

S. nodosa COMMON FIGWORT. Perennial to 1m high. Flowers small, green to purple-brown. *Distr.* Europe.

Scrophulariaceae The Foxglove family. 292 genera and about 5000 species of herbs, shrubs, and lianas, with simple or pinnately lobed leaves and bisexual flowers that are almost regular or distinctly 2-lipped and typically bear their parts in fives. The family contains numerous cultivated ornamentals. *Distr.* Widespread. *Class* D.
See: Alonsoa, Anarrhinum, Antirrhinum, Asarina, Bartsia, Bowkeria, Calceolaria, Centranthera, Chaenorrhinum, Chelone, Chionohebe, Cymbalaria, Dermatobotrys, Diascia, Digitalis, Erinus, Freylinia, Gratiola, Hebe, Hebenstretia, Isoplexis, Jovellana, Keckiella, Kickxia, Linaria, Maurandya, Mazus, Mimulus, Misopates, Nemesia, Ourisia, Paederota, Parahebe, Paulownia, Pedicularis, Penstemon, Phygelius, Rehmannia, Rhodochiton, Russelia, Scrophularia, Selago, Sutera, Synthyris, Verbascum, Veronica, Veronicastrum, Wulfenia, Zaluzianskya.

scutatus, -a, -um: shield-bearing.

Scutellaria (From Latin *scutella*, small dish, alluding to the appearance of the sepals in fruit.) SKULL CAP, HELMET FLOWER. A genus of 300 species of rhizomatous herbs and subshrubs with opposite, simple or pinnately lobed leaves, and curved, tubular, 2-lipped flowers. Some species are grown as ornamentals and some are used medicinally. *Distr.* Cosmopolitan except S Africa. *Fam.* Labiatae. *Class* D.

S. hastata See *S. hastifolia*

S. hastifolia Leaves narrow. Flowers paired, violet-blue. *Distr.* Central Europe and Turkey.

S. indica Slender perennial herb. Leaves white-hairy, toothed. Flowers blue. *Distr.* China, Japan.

S. orientalis Perennial herb. Flowers yellow. *Distr.* SE Europe, mountains of SE Spain.

S. scordiifolia Mat-forming perennial herb. Flowers purple. *Distr.* Asia.

scutellarioides: from the genus *Scutellaria*, with the ending *-oides*, indicating resemblance.

scutellifolius, -a, -um: from the genus *Scutellaria*, and *folius*, leaved.

Scuticaria (From Latin, *scutica*, lash, alluding to the drooping whip-like leaves.) A genus of 4–5 species of epiphytic or lithophytic orchids with small pseudobulbs, fleshy leaves, and typically large, showy flowers. Several species are grown as tender ornamentals. *Distr.* Tropical South America. *Fam.* Orchidaceae. *Class* M.

S. steelei Flowers fragrant, yellow-green, marked red, lip white. *Distr.* Brazil, Guyana, Colombia and Venezuela.

Scyphostegiaceae A family of 1 genus and 1 species of tree with simple, spirally arranged leaves and spikes of very small flowers that bear their parts in threes. *Distr.* Borneo. *Class* D.

Scyphularia (From Greek *skyphos*, cup.) A genus of about 8 species of small epiphytic ferns with pinnate leaves. *Distr.* SE Asia, Polynesia and Malaysia. *Fam.* Davalliaceae. *Class* F.

S. pentaphylla BLACK CATERPILLAR FERN. Grown as an ornamental. *Distr.* Indonesia to New Guinea.

Scytopetalaceae A family of 5 genera and 22 species of shrubs and trees with alternate, often 2-ranked, simple, leaves and regular bisexual flowers. *Distr.* Tropical W Africa. *Class* D.

Sechium (From the West Indian name for these plants, *chacha*.) A genus of 6–8 species of climbing herbs with branched tendrils and palmately lobed leaves. The white bell-shaped flowers are deeply 5-lobed and the fruit is a large berry which may bear soft fleshy spines. *Distr.* Central America. *Fam.* Cucurbitaceae. *Class* D.

S. edule CHAYOTE, CHOCHO, CHOW CHOW, CHRISTOPHINE, VEGETABLE PEAR. A somewhat woody perennial. Fruit pear-shaped, furrowed, green or white, with white flesh and a single seed. Fruits and roots are eaten as vegetables and the young shoots are eaten like spinach. The mature stems yield a silver-white fibre used in hat-making. *Distr.* Central America, Mexico.

sectus, -a, -um: deeply divided.

secundatus, -a, -um: with parts arranged along one side.

secundiflorus, -a, -um: with flowers on one side of the stalk.

secundus, -a, -um: with parts arranged along one side.

Securigera See *Coronilla*

Sedastrum See *Sedum*

sedge *Carex*, Cyperaceae **common** ~ *Carex nigra* **greater pond** ~ *C. riparia* **leatherleaf** ~ *C. buchananii* **mace** ~ *C. grayi* **pendulous** ~ *C. pendula* **tufted** ~ *C. elata*.

sedifolius, -a, -um: from the genus *Sedum*, and *folius*, leaved.

sedoides: from the genus *Sedum*, with the ending *-oides*, indicating resemblance.

Sedum (Classical Latin name used for several succulent plants, from *sedo*, to sit.) STONECROP. A genus of about 300 species of succulent herbs and small shrubs with fleshy simple leaves and cymes of star-shaped flowers that typically have their parts in fives. Many species, hybrids, and cultivars are grown as ornamentals, some are edible, and some have medicinal qualities. *Distr.* N temperate regions, tropical mountains. *Fam.* Crassulaceae. *Class* D.

S. acre BITING STONECROP, WALL PEPPER, STONECROP. Loosely tufted or mat-forming perennial herb. Leaves small, triangular. Flowers yellow. *Distr.* Europe, N Africa, W and N Asia, naturalized in North America.

S. adolphi (After Adolf Engler (1844–1930), German botanist.) GOLDEN STONECROP. Loose, bushy, evergreen perennial. Leaves yellow-green with a red margin. Flowers white. *Distr.* Mexico.

S. dendroidium Shrub to 2m high with a thick trunk. Flowers yellow. *Distr.* Mexico.

S. integrifolium See *Rhodiola rosea*

S. maweanum See *S. acre*

S. maximum See *S. telephium*

S. morganianum (After Dr Meredith Morgan, who grew it soon after its discovery.) DONKEY'S TAIL. Subshrub. Stems trailing or pendent, tail-like. Leaves small, over-lapping. Flowers pendent, deep pink. *Distr.* Mexico.

S. pachyphyllum JELLY BEANS. Subshrub. Leaves more or less sausage-shaped, tips red. Flowers bright yellow, borne in flat-topped cymes. *Distr.* S Mexico.

S. reflexum Mat-forming perennial herb. Leaves cylindrical, eaten in salads. *Distr.* Europe.

S. rhodiola See *Rhodiola rosea*

S. rosea See *Rhodiola rosea*

S. sarmentosum See *Crassula sarcocaulis*

S. spectabile See *Hylotelephium spectabile*

S. telephium See *Hylotelephium telephium*

S. weinbergii See *Graptopetalum paraguayense*

Seemannia See *Gloxinia*

seersucker plant *Geogenanthus poeppigii*.

segetus, -a, -um: of cornfields.

Selaginella (Diminutive of *selago*, from the species name *Lycopodium selago*.) LITTLE

CLUB MOSS, SPIKE MOSS. A genus of 600–700 species of terrestrial or occasionally epiphytic, moss-like plants with creeping to erect stems and spirally arranged or 4-ranked scale-like leaves. Sporangia are borne in terminal strobili. Several species are grown as ornamentals. *Distr.* Cosmopolitan but mostly from wet tropical regions. *Fam.* Selaginellaceae. *Class* F.

S. kraussiana (After C. Ferdinand F. Krauss (1812–90), who introduced it to cultivation.) SPREADING CLUBMOSS, TRAILING SPIKE MOSS, MAT SPIKE MOSS. Trailing. Leaves bright green, 3–4mm long. *Distr.* Tropical and S Africa, Azores, naturalized in SW Britain.

S. lepidophylla RESURRECTION PLANT, ROSE OF JERICHO. Tuft-forming. Branches curl up into a ball when dry but re-open when wet. Used locally as a medicine. *Distr.* Tropical and subtropical America.

Selaginellaceae A family of 1 genus and 700 species. *Class* F.
See: Selaginella.

Selago (Latin name used by Pliny for an unknown plant gathered by Druids.) A genus of 150 species of herbs, subshrubs, and shrubs with simple, often heath-like leaves and clusters of small, tubular, somewhat 2-lipped flowers. Several species are cultivated as ornamentals. *Distr.* Tropical and S Africa. *Fam.* Scrophulariaceae. *Class* D.

S. thunbergii Much-branched perennial herb. Flowers blue. *Distr.* South Africa.

selago: resembling the club moss species *Lycopodium selago.*

selfheal *Prunella*, *P. vulgaris.*

Selinum (From Greek *selinon*, CELERY.) A genus of 6 species of perennial herbs with pinnately divided leaves and simple umbels of white or purple flowers. Several species are grown as ornamentals. *Distr.* Europe to Central Asia. *Fam.* Umbelliferae. *Class* D.

S. tenuifolium Erect perennial herb. Leaves very finely divided. Umbels flat-topped. *Distr.* Himalaya, Assam, Tibet.

Selliguea A genus of about 5 species of epiphytic ferns with simple leathery leaves. Several species are grown as ornamentals. *Distr.* SE Asia. *Fam.* Polypodiaceae. *Class* F.

S. feei Frond pointed, to 15cm long. *Distr.* Malaysia, Philippines.

Semele (After Semele, daughter of Cadmus, mother of Bacchus.) A genus of 1 species of rhizomatous climbing subshrub with reduced leaves that are replaced by leaf-like branches (cladophylls). The small, cream, unisexual flowers are borne on the margins of the cladophylls and are followed by orange-red berries. *Distr.* Canary Islands. *Fam.* Ruscaceae. *Class* M.

S. androgyna CLIMBING BUTCHER'S BROOM. Cultivated ornamentally.

semi-inferus, -a, -um: with a semi-inferior ovary, with floral parts inserted halfway up the ovary.

Semiaquilegia (From Latin *semi-*, half, and the genus name *Aquilegia.*) A genus of 6–7 species of perennial herbs with divided leaves and regular flowers that bear 5 petal-like sepals, 5 pouched petals, and numerous stamens. The flowers are similar to those of the genus *Aquilegia* but the petals have a short pouch rather than a distinct spur. Several species are grown as ornamentals. *Distr.* E Asia. *Fam.* Ranunculaceae. *Class* D.

S. adoxoides Flowers pink. *Distr.* E China, Korea, Japan.

S. ecalcarata Flowers red-purple. *Distr.* China.

S. simulatrix See *S. ecalcarata*

Semiarundinaria (From its similarity to the genus *Arundinaria.*) A genus of 10–20 species of tall rhizomatous bamboos with grooved stems, short branches, and small leaves. Several species are cultivated as ornamentals. *Distr.* E Asia. *Fam.* Gramineae. *Class* M.

S. fastuosa NARIHIRA BAMBOO. Stems to 12m high, erect, marked with purple-brown lines. Commonly grown as an ornamental. *Distr.* Japan.

semicordatus, -a, um: with one cordate lobe.

semideciduus, -a, -um: semideciduous.

semidentatus, -a, -um: half toothed.

semipinnatus, -a, -um: half pinnate.

semperflorens: ever-flowering.

sempervirens: evergreen.

Sempervivella See *Rosularia*

sempervivoides: from the genus name *Sempervivum*, with the ending *-oides*, indicating resemblance.

Sempervivum (From Latin *semper*, always, and *vivus*, alive.) HOUSELEEK. A genus of 42 species of succulent perennial herbs with fleshy leaves crowded into basal rosettes and cymes of white star-shaped flowers that bear 6–20 petals. Many species and hybrids are grown as rock-garden ornamentals. *Distr.* Europe, Morocco, W Asia. *Fam.* Crassulaceae. *Class* D.

S. arachnoideum COBWEB HOUSELEEK. Rosettes forming mats. Leaves red-tipped, joined by numerous, long, cobweb-like hairs. Flowers pink. *Distr.* Alps, Pyrenees, S Europe.

S. arvernense See *S. tectorum*

S. ciliosum Mat- or cushion-forming. Leaves short white-hairy. Flowers yellow. *Distr.* Bulgaria.

S. densum See *S. tectorum*

S. giuseppii Mat-forming. Rosettes somewhat spherical. Leaves hairy. Flowers deep pink or red. *Distr.* Spain.

S. helveticum See *S. montanum*

S. hirtum See *Jovibarba hirta*

S. montanum Rosettes large, mat-forming. Leaves dark green. Flowers wine-red. *Distr.* Europe.

S. patens See *Jovibarba heuffelii*

S. soboliferum See *Jovibarba sobolifera*

S. stansfieldii See *S. arachnoideum*

S. tectorum COMMON HOUSELEEK, ROOF HOUSELEEK. Fast-growing. Leaves red-tipped or sometimes completely red. Flowers red-purple. Formerly much planted on roofs to keep slates in place and believed to ward off thunder. *Distr.* Pyrenees to SE Alps.

seneca *Polygala*.

Senecio (From Latin *senex*, old man, alluding to the fluffy white seed-heads in many species.) A genus of 1000–1500 species of herbs, lianas, shrubs, and trees with simple or lobed leaves and yellow flowers in daisy-like heads that typically bear distinct rays. Some are cultivated as ornamentals, others are weeds and toxic to livestock. *Distr.* Cosmopolitan except Antarctic. *Fam.* Compositae. *Class* D.

S. abrotanifolius Rhizomatous perennial herb. Rays yellow to red. *Distr.* Mountains of Central and E Europe.

S. aquaticus MARSH RAGWORT. Biennial herb of wet meadows and ditches. *Distr.* Europe.

S. bicolor See *S. cineraria*

S. bidwillii See *Brachyglottis bidwillii*

S. chrysanthemoides See *Euryops chrysanthemoides*

S. cineraria DUSTY MILLER. Small shrub, to 50cm high. *Distr.* Mediterranean.

S. compactus See *Brachyglottis compacta*

S. doria Perennial herb, to 1m high. *Distr.* Europe to W Asia.

S. elaeagnifolius See *Brachyglottis elaeagnifolia*

S. glastifolius Subshrub, to 1m high. Rays purple. *Distr.* South Africa.

S. greyi See *Brachyglottis greyi*

S. heritieri See *Pericallis lanata*

S. herreianus (After H. Herre of Stellenbosch.) Succulent perennial. Flowers white. *Distr.* South Africa.

S. jacobaea RAGWORT. Biennial or perennial herb, to 150cm. A weed of pasture, poisonous to cattle. *Distr.* Europe to W Asia, introduced to Australasia and the Americas.

S. kirkii See *Brachyglottis kirkii*

S. laxifolius See *Brachyglottis laxifolia*

S. leucostachys See *S. vira-vira*

S. macroglossus CAPE IVY, NATAL IVY, WAX VINE. Slender, twining perennial. Rays cream. *Distr.* South Africa.

S. maritimus See *S. cineraria*

S. monroi See *Brachyglottis monroi*

S. nudicaulis Perennial, to 1m high. *Distr.* Temperate Himalaya, 1500–3000 meters.

S. przewalskii See *Ligularia przewalskii*

S. pulcher Perennial herb, to 60cm high. Flower-heads large. *Distr.* S Brazil, Uruguay, Argentina.

S. reinoldii See *Brachyglottis rotundifolia*

S. rowleyanus (After Gordon D. Rowley (born 1921), British succulent plant enthusiast.) STRING OF BEADS. Mat-forming. Stems slender. Leaves bead-like. *Distr.* SW Africa.

S. scandens Large subshrub, to 5m high. *Distr.* E Asia.

S. serpens BLUE CHALKSTICKS. Succulent shrub. Rays missing. *Distr.* South Africa.

S. smithii Perennial herb to 1.2m high. *Distr.* Island Chile, Falkland Islands.

S. speciosus Perennial herb, to 70cm high. Rays pink-purple. *Distr.* SE Africa.

S. squalidus OXFORD RAGWORT. Annual, occasionally perennial herb. *Distr.* S and Central Europe, thought to have been introduced to Britain after escaping from Oxford Botanic Garden in 1794 and spreading throughout the country chiefly by migrating along railway lines.

S. vira-vira Small subshrub, to 40cm high. Rays absent. *Distr.* Central Argentina.

S. vulgaris GROUNDSEL. Annual, occasionally overwintering, weed of waste ground. *Distr.* Europe, Asia and N Africa.

senecioides: from the genus *Senecio*, with the ending *-oides*, indicating resemblance.

senescens: becoming old.

senilis, -e: appearing old.

Senna (From the Arabic name for these plants, *sana*.) A genus of 260 species of shrubs and trees with pinnate leaves and racemes or panicles of pea-like flowers. Several species are grown as ornamentals. *Distr.* Tropical and subtropical regions. *Fam.* Leguminosae. *Class* D.

S. artemisioides WORMWOOD SENNA, SOLVER CASSIA, FEATHERY CASSIA. Shrub to 3m high. Flowers yellow, fragrant. *Distr.* Australia.

S. didymobotrya GOLDEN WONDER. Evergreen erect shrub. Flowers yellow, borne in large spikes. *Distr.* Tropical Africa.

S. marilandica WILD SENNA. Perennial subshrub to 2m high. Flowers borne in racemes, yellow. *Distr.* W North America.

senna, bastard *Coronilla valentina* **bladder** ~ *Colutea arborescens* **scorpion** ~ *Coronilla emerus* **wild** ~ *Senna marilandica* **wormwood** ~ *S. artemsioides*.

sensibilis, -e: sensitive.

sensitive plant *Mimosa pudica*.

senticosus, -a, -um: thorny.

sepium: of hedges.

septemfida: divided into seven.

septemlobus: with seven lobes.

septentrionalis, -e: northern.

Sequoia (After Sequoiah (1770–1843), son of a British merchant by a Cherokee Indian woman and inventor of a system for transcribing the Cherokee language.) A genus of 1 species of evergreen coniferous tree to 110m high with scale-like and linear leaves and small oblong fruiting cones. *Distr.* W North America. *Fam.* Taxodiaceae. *Class* G.

S. sempervirens COAST REDWOOD, CALIFORNIAN REDWOOD. Formerly an important timber tree, now restricted to a few small areas, owing to over-exploitation.

Sequoiadendron (From the genus name *Sequoia* and the Greek *dendron*, tree.) A genus of 1 species of evergreen coniferous tree to 95m high with small scale-like leaves and small fruiting cones that open to release their seed. *Distr.* Restricted to a few areas in the mountains of California. *Fam.* Taxodiaceae. *Class* G.

S. giganteum BIG TREE, GIANT REDWOOD, WELLINGTONIA. One of the largest living organisms, some specimens weighing over 2000 tonnes.

Serapias (After the Egyptian god Serapis.) TONGUE ORCHID. A genus of 6–10 species of terrestrial tuberous orchids with erect leaves and spikes of flowers in which the sepals and petals form a hood and the central lobe of the lip forms a pendent or reflexed tongue. Several species are cultivated as ornamentals. *Distr.* Azores to the Mediterranean and Caucasus. *Fam.* Orchidaceae. *Class* M.

S. cordigera Flowers grey-purple, tongue large, very dark purple. *Distr.* Spain, Portugal, Mediterranean.

S. lingua Flowers pale purple, tongue yellow to magenta or white with dark markings. *Distr.* Portugal, N Africa, Mediterranean.

S. vomeracea Flowers grey-red, tongue ochre with a dark purple margin. In Israel bees sleep inside the flowers and are 3 degrees warmer than the ambient temperature in the morning. *Distr.* S Europe, Mediterranean.

serbicus, -a, -um: of Serbia, southeast Europe.

Serenoa (After Sereno Watson (1826–92), North American botanist.) A genus of 1 species of small clump-forming palm to 4m high, with fan-shaped leaves and an erect, much-branched inflorescence of cream flowers. *Distr.* SE USA. *Fam.* Palmae. *Class* M.

S. repens SAW PALMETTO, SCRUB PALMETTO. One of the more hardy species of ornamental palms. The fruit is edible and is reported to have medicinal properties.

serl *Cymbopogon citratus.*

sericeus, -a, -um: silky.

sericifera, -um: silk-bearing.

Seriphidium (Diminutive of Greek *seriphon*, a type of wormwood.) A genus of 60 species of annual to perennial herbs and shrubs with deeply pinnately lobed leaves and yellow flowers in daisy-like heads that lack distinct rays. Several species are grown for their attractive, often aromatic foliage. *Distr.* N temperate regions. *Fam.* Compositae. *Class* D.

S. maritimum SEA WORMWOOD. Strongly aromatic, hairy, perennial herb. *Distr.* Coastal NW Europe.

S. nutans Perennial herb, to 2m high. *Distr.* SE Russia.

S. palmeri Subshrub to 3m high. *Distr.* USA (California).

S. tridentata Shrub with toothed, silver-hairy leaves. *Distr.* Mexico (Baja California).

S. vallesiacum Strongly aromatic perennial, to 2m high. *Distr.* Central Europe.

Serissa (From the Indian vernacular name.) A genus of 1 species of evergreen shrub with simple foetid leaves and profuse clusters of funnel-shaped white flowers. *Distr.* SE Asia. *Fam.* Rubiaceae. *Class* D.

S. foetida Grown as an ornamental with a number of cultivars available, often bearing double flowers and variegated leaves.

S. japonica See *S. foetida*

serotinus, -a, -um: late-flowering.

serpens: creeping.

serpentilingua: with a snake's tongue.

serpentinus, -a, -um: snake-like.

serpyllifolius, -a, -um: with leaves resembling those of the species *Thymus serpyllum*.

serpyllum: the Latin name for THYME.

serratifolius, -a, -um: with serrated leaves.

serratipetalus, -a, -um: with serrated petals.

Serratula (From Latin *serratula*, little saw, alluding to the leaf margins.) A genus of about 70 species of erect perennial herbs with pinnate toothed leaves and flowers in daisy-like heads that lack distinct rays. Some species are cultivated ornamentally. *Distr.* Europe to N Africa and Japan. *Fam.* Compositae. *Class* D.

S. seoanei To 70cm high. Flowers pink-purple. *Distr.* SW Europe.

S. shawii See *S. seoanei*

S. tinctoria To 1m high. Leaves finely toothed to deeply pinnately lobed. Flowers pink-purple. *Distr.* Eurasia, N Africa.

serratus, -a, -um: serrated, saw-toothed.

serrulatus, -a, -um: with small serrations.

service tree *Sorbus domestica* **wild** ~ *S. torminalis.*

sesame *Sesamum indicum*, Pedaliaceae.

Sesamum (From Greek *sesame*.) A genus of 15 species of annual and perennial herbs with long-stalked, palmately lobed leaves and 5-lobed tubular flowers that are followed by a deeply grooved capsule. *Distr.* Tropical regions of the Old World. *Fam.* Pedaliaceae. *Class* D.

S. indicum SESAME, SIMSIM, GINGELLY. Annual to 2m high. Grown as a seed crop. The seeds are pressed for oil which is used like olive oil, or they are used whole, sprinkled on bread and cakes like poppy seeds, or made into sticky sweetmeats. *Distr.* Tropical regions of the Old World, widely naturalized.

Sesbania (From the Arabic name, *sesban*, for *S. sesban*.) A genus of 50 species of herbs, shrubs, and small trees with pinnate leaves and racemes of irregular pea-like flowers. Some species are cultivated as ornamentals. *Distr.* Tropical and subtropical regions. *Fam.* Leguminosae. *Class* D.

S. grandiflora SCARLET WISTERIA TREE, VEGETABLE HUMMINGBIRD. Tree to 12m high. Flowers red, pink, or white. *Distr.* Tropical Asia, N Australia, naturalized West Indies.

S. punicea Shrub to 2m high. Flowers bright red, 2cm long. *Distr.* South America, naturalized SE USA.

Seseli (Classical Greek name of an unbelliferous plant.) MOON CARROT. A genus of 65 species of biennial or perennial herbs with firmly divided leaves and dense umbels of small pink or white flowers. *Distr.* Europe to Central Asia. *Fam.* Umbelliferae. *Class* D.

S. gummiferum Flowers white or red. Stem yields gum. *Distr.* Crimea, S Aegean.

S. libanotis Flowers white or pink. *Distr.* Europe, Iran, Siberia.

S. pallarii (After P. S. Pallas (1741–1811), Russian naturalist.) Flowers white. *Distr.* E. Europe, Ukraine.

Sesleria A genus of 33 species of perennial wiry grasses with blue-grey paniculate inflorescences. *Distr.* Europe, W Asia. *Fam.* Gramineae. *Class* M.

S. albicans PURPLE MOOR GRASS. A grass of chalk hills and pastures. *Distr.* Central and W Europe to Iceland.

sessiliflorus, -a, -um: with sessile flowers.

sessilifolius, -a, -um: with sessile leaves.

sessilis, -e: sessile, without a stalk.

setaceus, -a, -um: bristled.

Setaria (From Latin *seta*, bristle, alluding to the bristly inflorescence.) A genus of 100–125 species of annual and perennial grasses with flat leaves and spiky inflorescences. *Distr.* Tropical and warm temperate regions, 3 species in Europe. *Fam.* Gramineae. *Class* M.

S. italica FOXTAIL, GERMAN MILLET, HUNGARIAN MILLET, ITALIAN MILLET, JAPANESE MILLET. Annual with large flower-heads. Grown as a pasture grass and as a cereal as well as an ornamental. *Distr.* Derived in cultivation probably from *S. viridis*.

S. viridis GREEN BOTTLE GRASS. Annual. Inflorescence compact, very bristly, tinged purple-green. *Distr.* Eurasia.

Setcreasea See *Tradescantia*

setiferum: bearing bristles.

setigerum: bearing bristles.

setipodus, -a, -um: with a bristly stem.

setosus, -a, -um: with many bristles.

setterwort *Helleborus foetidus.*

sexstylosus, -a, -um: with six styles.

shad *Amelanchier.*

shad bush *Amelanchier.*

shaddock *Citrus maxima.*

shallon *Gaultheria shallon.*

shallot *Allium cepa.*

shamrock *Trifolium repens.*

shawl, Spanish *Heterocentron elegans.*

Shawnee salad *Hydrophyllum virginianum.*

sheepberry *Viburnum lentago.*

shellflower *Chelone*, *Moluccella lavis.*

Shepherdia (After John Shepherd (1764–1836), curator of Liverpool Botanic Garden.) BUFFALO BERRY. A genus of 3 species of deciduous or evergreen shrubs with scaly twigs, simple leaves, and racemes of small unisexual flowers that lack petals. The fruit is a fleshy drupe. *Distr.* North America. *Fam.* Elaeagnaceae. *Class* D.

S. argentea BUFFALO BERRY, BEEF SUET TREE, SILVERBERRY. Grown as an ornamental and as a hedging plant. The edible fruits are used in preserves or traditionally dried by Indians and eaten in winter with buffalo meat. *Distr.* E North America.

shepherd's clock *Anagallis arvensis, Tragopogon pratensis.*

shepherd's rod *Dipsacus pilosus.*

shepherd's weather glass *Anagallis arvensis.*

Sherardia (After William Sherard (1659–1728), English botanist.) A genus of 1 species of much-branched annual herb with small simple leaves and clusters of tiny pale blue or lilac flowers. *Distr.* Europe, N Africa, W Asia. *Fam.* Rubiaceae. *Class* D.

S. arvensis FIELD MADDER, SPURWORT. Occasionally grown as an ornamental.

Shibataea (After Keita Shibata (1877–1949), Japanese botanist.) A genus of 5–8 species of rhizomatous short bamboos with a flat side to the stems and very short branches. *Distr.* China, Japan. *Fam.* Gramineae. *Class* M.

S. kumasasa (From the Japanese vernacular name for a species of BAMBOO.) Grown as an ornamental ground cover especially in Japan. *Distr.* Japan.

shingle plant *Monstera acuminata.*

shoe black *Hibiscus rosa-sinensis.*

shoe black flower *H. rosa-sinensis.*

shooting star *Dodecatheon meadia.*

Shortia (After Charles W. Short (1794–1863), Kentucky botanist.) A genus of 6 species of evergreen rhizomatous herbs with round or heart-shaped, long-stalked leaves and bell-shaped flowers that bear their parts in fives. Several species are grown as ornamentals. *Distr.* E Asia, 1 species in North America. *Fam.* Diapensiaceae. *Class* D.

S. galacifolia Clump-forming. Flowers open bell-shaped, white, flushed pink. Petals toothed. *Distr.* North America.

S. soldanelloides FRINGED GALAX, FRINGEBELL. Flowers deep pink, fringed. *Distr.* Japan.

S. uniflora NIPPON BELLS. Mat-forming. Flowers pink, veined white, broadly bell-shaped. *Distr.* Japan.

shrimp bush *Justicia brandegeana.*

shrimp plant *Justicia brandegeana*
Mexican ~ *J. brandegeana.*

shrimp tree *Koelreuteria.*

Sibbaldia (After Sir Robert Sibbald (1641–1722), first professor of medicine at the University of Edinburgh.) A genus of 8 species of low woody perennials with compound leaves and small flowers that bear their parts in fives. *Distr.* N temperate regions. *Fam.* Rosaceae. *Class* D.

S. parviflora Occasionally grown as ornamental ground cover. *Distr.* Balkans to Iran.

Sibiraea (After Siberia, where the type species occurs.) A genus of 2 species of deciduous shrubs with simple alternate leaves and racemes of small, 5-lobed, cream flowers. *Distr.* SE Europe to Siberia. *Fam.* Rosaceae. *Class* D.

S. altaiensis Robust bushy shrub. Grown as a hardy ornamental. *Distr.* Siberia to W China.

S. laevigata See *S. altaiensis*

siculus, -a, -um: of Sicily.

Sida (From a Greek name for a water plant.) A genus of 150 species of annual and perennial herbs and subshrubs with simple, toothed or lobed leaves and regular bisexual flowers that bear 5 petals and numerous fused stamens. *Distr.* Tropical and warm regions, especially America. *Fam.* Malvaceae. *Class* D.

S. hermaphrodita VIRGINIA MALLOW. Perennial herb to 3m high. Flowers white, borne in loose cymes. Grown as an ornamental. *Distr.* E USA.

Sidalcea (From the names of two closely related genera, *Sida* and *Alcea*.) PRAIRIE MALLOW, CHECKER MALLOW. A genus of 20 species of annual and perennial herbs with palmately lobed or dissected leaves and spikes or racemes of regular showy flowers that bear 5 petals and numerous fused stamens. Several species and numerous cultivars are grown as ornamentals. *Distr.* North America. *Fam.* Malvaceae. *Class* D.

S. candida WHITE PRAIRIE MALLOW. Perennial herb to 80cm high. Flowers white or cream, borne in a dense terminal spike. *Distr.* USA (Wyoming to New Mexico).

S. malviflora CHECKERBLOOM. Grey-hairy perennial herb to 80cm high. Flowers lilac or pink with white veins, borne in an elongated raceme. Numerous cultivars have been raised from this species. *Distr.* USA (Oregon, California), Mexico (Baja California).

Sideritis (From Greek *sideros*, iron, alluding to supposed medicinal properties in the treatment of iron-inflicted injuries.) A genus of 100 species of annual and perennial herbs and shrubs with simple opposite leaves and whorls of 2-lipped flowers. Some species are cultivated as ornamentals, others are used as tea in Turkey. *Distr.* N temperate regions of the Old World, Macaronesia. *Fam.* Labiatae. *Class* D.

S. hyssopifolia Small shrub to 80cm high. Flowers yellow, marked purple. *Distr.* S Europe.

S. syriaca Perennial white-hairy herb. Flowers yellow, marked brown. *Distr.* E Mediterranean.

sidra *Cucurbita ficifolia.*

sikkimensis: of Sikkim, eastern Himalaya.

sikokianus, -a, -um: of Shikoku, Japan.

Silene (Classical Latin name.) CAMPION, CATCHFLY. A genus of about 500 species of annual, biennial, and perennial herbs, rarely shrubby at the base, with simple opposite leaves. The flowers bear 5 sepals that are fused into a tube and 5 petals which have spreading, entire, notched, or deeply divided limbs. There are 10 stamens and typically 3–4 styles. The fruit is a many-seeded capsule that typically opens into 6 teeth. The sepal tube is often characteristically inflated. Numerous species and cultivars are grown as ornamentals. *Distr.* N temperate regions, especially the Mediterranean area and European mountains. *Fam.* Caryophyllaceae. *Class* D.

S. acaulis MOSS CAMPION. Tuft- or dense mat-forming perennial. Flowers small, pink. *Distr.* N arctic and alpine regions.

S. alba See *S. latifolia*

S. alpestris Perennial forming loose tufts. Flowers white. Petals fringed. *Distr.* E Alps.

S. armeria NONE SO PRETTY. Annual or biennial. Flowers in flat-topped cymes, bright pink. Petals shallowly notched. *Distr.* Europe.

S. dioica RED CAMPION. Spreading perennial of woods and hedges. Flowers red. Petals deeply notched. *Distr.* NW Europe, naturalized North America.

S. latifolia WHITE CAMPION, WHITE COCKLE. Annual or short-lived perennial. Flowers large, white, opening in the evening and lightly scented at night. A weed of open ground. *Distr.* Eurasia, N Africa, introduced North America.

S. maritima See *S. uniflora*

S. pendula NODDING CATCHFLY. Annual. Flowers in large clusters, pendent in fruit, pink. *Distr.* N Africa, naturalized around Mediterranean.

S. schafta Spreading perennial. Flowers magenta. *Distr.* Caucasus.

S. uniflora SEA CAMPION. Stems prostrate from a woody base, forming loose mats. Flowers large, white. Petals deeply divided. *Distr.* Coasts of Atlantic and Arctic Europe.

S. vulgaris BLADDER CAMPION. Perennial. Flowers with large swollen sepal tubes. Petals small, white. *Distr.* Eurasia, Mediterranean.

S. wallichiana See *S. vulgaris*

sileniflorus, -a, -um: with flowers resembling those of the genus *Silene*.

silk cotton tree *Bombax* **red** ~ *B. ceiba.*

silk plant, Chinese *Boehmeria nivea.*

silk tassel *Garrya.*

silk tree *Albizia juilibrissin.*

silkweed *Asclepias* **purple** ~ *A. purpurascens.*

Silphium (From *silphion*, a classical Greek name for a plant that produced resin.) ROSIN WEED, PRAIRIE DOCK. A genus of 23 species of robust perennial herbs with flowers borne in daisy-like heads which bear distinct yellow or white rays. *Distr.* E North America. *Fam.* Compositae. *Class* D.

S. laciniatum COMPASS PLANT. Young plants bear leaves tipped vertically, avoiding full mid-day sun, the surfaces facing N and S in a similar way to *Lactuca serriola*. Has been used medicinally. *Distr.* Prairies of North America, naturalized in Great Britain.

S. perfoliatum CUP PLANT. Leaves fused in opposite pairs around the stem so as to form a small cup. *Distr.* E North America.

S. trifoliatum To 2m high. Leaves in whorls of 3–4. *Distr.* S and E USA.

silver dollar *Lunaria annua.*

silverballs *Styrax.*

silverbell tree *Halesia.*

silverberry *Shepherdia argentea.*

silverbush *Convolvulus cneorum.*

silver jade plant *Crassula arborescens.*

silver lace *Tanacetum ptarmiciflorum.*

silver nerve *Fittonia verschaffeltii.*

silver net plant *Fittonia verschaffeltii.*

silver rod *Solidago bicolor.*

silver threads *Fittonia verschaffeltii.*

silver tree *Leucadendron argenteum.*

silver vase *Aechmea fasciata*.

silverweed *Potentilla anserina*.

Silybum (From *silybon*, the Greek name for a similar plant.) A genus of 2 species of erect annual to biennial herbs with white-veined or variegated leaves and flowers in solitary daisy-like heads that lack distinct rays. *Distr.* Mediterranean, naturalized in North and South America. *Fam.* Compositae. *Class* D.

S. marianum (Of the Virgin Mary, who is said to have caused the white mottling of the leaves by dropping milk on them.) OUR LADY'S MILK THISTLE, HOLY THISTLE, MILK THISTLE, BLESSED THISTLE, KENGUEL SEED. Annual or biennial herb, to 1.5m high. Leaves blotched white. Cultivated ornamentally; the fruit (kenguel seed) was formerly used as a coffee substitute and medicinally.

Simaroubaceae The Tree of Heaven family. 24 genera and about 200 species of shrubs and trees with alternate, usually pinnate leaves, and numerous small regular flowers. Some species are grown as ornamentals, others are a source of timber. *Distr.* Tropical, subtropical, and occasionally temperate regions. *Class* D.
See: Ailanthus, Picrasma.

Simethis (After the nymph Simethis, mistress of Acis in Greek mythology.) A genus of 1 species of rhizomatous herb with basal grass-like leaves and a panicle of white flowers. *Distr.* W Europe, N Africa. *Fam.* Asphodelaceae. *Class* M.

S. planifolia Occasionally grown as an ornamental.

similis, -e: similar.

Simmondsia (After Thomas William Simmonds (died 1804), English botanist, who died in Trinidad.) A genus of 1 species of evergreen shrub to 2.5m high with simple, opposite, leathery leaves and small unisexual flowers that bear 5–6 sepals but lack petals. *Distr.* SW North America. *Fam.* Simmondsiaceae. *Class* D.

S. chinensis JOJOBA, GOAT NUT, PIG NUT. The seed is the source of jojoba oil and was formerly ground as a coffee substitute. This shrub is also grown as an ornamental and in soil conservation schemes. The leaves are used as fodder.

Simmondsiaceae A family of 1 genus and 1 species. *Class* D.
See: Simmondsia.

simplex: simple.

simplicicaulis, -e: with simple stems (unbranched).

simplicifolius, -a, -um: with simple leaves (entire).

simplicissimus, -a, -um: very simple, unbranched.

simsim *Sesamum indicum*.

simul *Bombax ceiba*.

simulatus, -a, -um: imitating.

Sinarundinaria (From Latin *sinae* Chinese, and the genus name *Arundinaria*.) A genus of 12 species of clump-forming, medium-sized bamboos with numerous branches and delicate, often weeping leaves. Several species are grown as ornamentals. *Distr.* E Asia. *Fam.* Gramineae. *Class* M.

S. nitida FOUNTAIN BAMBOO. Produces spectacular thickets with cascades of branches. Stems slender, to 4m high, becoming tinged purple-brown. Often grown as an ornamental. *Distr.* N China.

sinensis, -e: of China.

singuliflorus, -a, -um: with solitary flowers.

sinicus, -a, -um: of China.

Sinningia (After William Sinning (1794–1874), Prussian botanist.) A genus of 40 species of perennial, often tuberous herbs and shrubs with simple opposite or whorled leaves and tubular to bell-shaped, 2-lipped flowers. Many species and cultivars are grown as ornamentals. *Distr.* Tropical America, Brazil. *Fam.* Gesneriaceae. *Class* D.

S. cardinalis CARDINAL FLOWER. Flowers trumpet-shaped, bright red, bird-pollinated. *Distr.* Brazil.

S. speciosa FLORIST'S GLOXINIA. Stems to 30cm long. Leaves flushed red. Flowers nodding, funnel-shaped, fleshy, red, violet or white. A great variety of cultivars of this species have been produced. *Distr.* Brazil.

Sinobambusa (From Latin *sinae*, Chinese, and the genus name *Bambusa*.) A genus of 8–20 species of rhizomatous bamboos with erect stems to 12m high and rough leaves. *Distr.* E Asia. *Fam.* Gramineae. *Class* M.

S. tootsik Grown as an ornamental. *Distr.* China.

Sinofranchetia (After Adrien René Franchet (1834–1900), French botanist, and Latin *sinae*, Chinese.) A genus of 1 species of deciduous twining shrub with trifoliate leaves and pendulous racemes of white unisexual flowers in which the petals have been replaced by nectaries and the sepals have become petal-like. The fruit is a blue berry. *Distr.* Central and W China. *Fam.* Lardizabalaceae. *Class* D.

S. chinensis Grown as a hardy ornamental.

Sinojackia (After J. G. Jack (1861–1949), who studied Chinese plants at the Arnold Arboretum, Boston and the Latin *sinae* Chinese.) A genus of 2 species of deciduous shrubs or small trees with simple leaves and racemes of white flowers that bear their parts in fives, sixes or sevens. Both species are grown as ornamentals. *Distr.* S China. *Fam.* Styracaceae. *Class* D.

S. xylocarpa To 6m high. Leaves to 7cm long. Flowers 1.2cm across. *Distr.* E China.

sinoornatus, -a, -um: showy and Chinese.

sinopurpurascens: purple and Chinese.

sinopurpurea: purple and Chinese.

Sinowilsonia (After Dr E. H. Wilson (1876–1930), English plant collector, who introduced many Chinese plants to the West.) A genus of 1 species of deciduous shrub or small tree with simple, stalked, toothed leaves and racemes of small, inconspicuous, unisexual flowers that lack petals. *Distr.* Central and W China. *Fam.* Hamamelidaceae. *Class* D.

S. henryi Grown as an ornamental.

sinuatus, -a, -um: with a wavy margin.

siphiliticus, -a, -um: a reference to supposed medicinal properties, being a cure for syphilis.

sipho: tube.

sisal Agavaceae, *Agave sisalana*.

sisymbrifolius, -a, -um: from the genus *Sisymbrium*, and *folius*, leaved.

Sisymbrium (From Greek *sisymbrion*, watercress, also a name given to aromatic herbs sacred to Venus.) A genus of 80 species of annual, biennial, or perennial herbs with white to yellow flowers. *Distr.* Eurasia, South Africa, North America, Andes. *Fam.* Cruciferae. *Class* D.

S. luteum Perennial with yellow flowers. *Distr.* Japan, Korea, Manchuria.

Sisyrinchium (Classical Greek name of a plant related to Iris.) BLUE-EYED GRASS. A genus of 70–200 species of annual and perennial, rhizomatous herbs with linear or sword-shaped leaves and clusters of regular, star- or cup-shaped flowers that are enclosed by a pair of spathes. Some of the species are cultivated as ornamentals. *Distr.* New World (except North and Central America) (some authors consider that 1 or 2 species may also be native to W Europe) with 1 species extending to Ireland, 1 in New Zealand, Australia, and New Guinea. *Fam.* Iridaceae. *Class* M.

S. × anceps See *S. angustifolium*

S. angustifolium BLUE-EYED GRASS. Stems flattened and narrowly winged. Flowers small, blue with a yellow centre. *Distr.* E North America, naturalized in W Europe.

S. bellum CALIFORNIAN BLUE-EYED GRASS. Stems branched, tuft-forming. Flowers violet with purple veins. *Distr.* USA (California, New Mexico).

S. bermudiana See *S. angustifolium*

S. boreale See *S. californicum*

S. brachypus See *S. californicum*

S. californicum GOLDEN-EYED GRASS. Tuft-forming. Flowers star-shaped, bright yellow. *Distr.* W North America.

S. douglasii See *Olsynium douglasii*

S. graminoides See *S. angustifolium*

S. grandiflorum See *S. douglasii*

S. striatum Leaves tuft-forming, yellow-striped. Flowers borne in narrow spikes, yellow, striped purple. *Distr.* Chile and Argentina, naturalized Isles of Scilly.

sitchensis, -e: of Sitka, Alaska.

Sium (Classical Greek name.) WATER PARSNIP. A genus of 10 species of perennial herbs of wet places with pinnate leaves and compound umbels of white flowers. Some species are cultivated as ornamentals, others have leaves that are eaten locally. *Distr.* Widespread in the N hemisphere. *Fam.* Umbelliferae. *Class* D.

S. sisarum SKIRRET. Grown for its edible roots, eaten, as a vegetable or as a coffee substitute. *Distr.* E Asia.

S. suave WATER PARSNIP. Stems to 1.2m high. Leaves occasionally simple. *Distr.* N North America.

Skimmia (From the Japanese name for these plants, *Miyami-Shikimi*.) A genus of 5–6 species of evergreen shrubs and small trees with whorls of simple leaves and panicles of small fragrant flowers. Several species are grown as ornamentals. *Distr.* China, Japan. *Fam.* Rutaceae. *Class* D.

S. japonica Upright or creeping shrub. Flowers sweetly scented. Fruit a red drupe. Numerous cultivars have been raised from this species. *Distr.* Japan, Philippines.

S. laureola Creeping shrub or small tree. Fruit used as an abortifacient in India. *Distr.* Himalaya, W China.

skirret *Sium sisarum.*

skull cap *Scutellaria.*

skunk bush *Garrya fremontii*, *Rhus trilobata.*

skyflower *Thunbergia grandiflora.*

sky plant *Tillandsia ionantha.*

skyrocket *Ipomopsis aggregata.*

slipper flower *Pedilanthus tithymaloides*, *Calceolaria.*

sloe *Prunus spinosa.*

smartweed *Polygonum.*

Smilacaceae A family of 2 genera and about 220 species of climbing shrubs and a few herbs. The leaves often have 2 tendrils at the base of the leaf-stalk. The flowers have 2 whorls of 3 tepals and 6 stamens. This family is sometimes include within the family Liliaceae. *Distr.* Widespread, mostly tropical. *Class* M.
See: Smilax.

Smilacina See *Maianthemum*

Smilax (Classical Greek name for this plant.) GREEN BRIER, CAT BRIER, BAMBOO VINE. A genus of about 200 species of rhizomatous vines with spiny stems and simple alternate leaves. The flowers are unisexual, regular, and bear 6 free tepals. Dried rhizomes of some tropical American species are the source of the sarsaparilla of commerce. A number of species are grown as ornamentals. *Distr.* Tropical regions, temperate Asia, North America. *Fam.* Smilacaceae. *Class* M.

S. asparagoides See *Asparagus asparagoides*

S. aspera ROUGH BINDWEED. *Distr.* Mediterranean region to W Asia.

S. china Climber. Leaves broad, deciduous. Flowers yellow, borne in umbels. *Distr.* China, Korea, Japan.

S. glauca WILD SARSAPARILLA. *Distr.* E North America.

S. regelii JAMAICA SARSAPARILLA. Formerly used medicinally as a tonic. *Distr.* N Central America, tropical South America.

S. rotundifolia HORSE BRIER, BULL BRIER. *Distr.* E North America.

Smithiantha (After Matilda Smith (1854–1926), botanical artist.) TEMPLE BELLS. A genus of 4 species of perennial herbs with creeping scaly rhizomes, simple leaves, and bell-shaped flowers. *Distr.* Mexico. *Fam.* Gesneriaceae. *Class* D.

S. cinnabarina TEMPLE BELLS. Robust erect perennial. Leaves dark green with red hairs. Flowers orange-red with yellow lines. Grown as a tender ornamental. *Distr.* Mexico.

smoke bush *Cotinus coggygria.*

smoke tree *Continus coggygria.*

smokewood *Cotinus* **American** ~ *C. obovatus.*

Smyrnium (From Greek *smyrna*, MYRRH, alluding to the smell of the plants.) A genus of 7 species of biennial herbs with divided leaves and compound inflorescences of yellow flowers. *Distr.* Europe, Mediterranean. *Fam.* Umbelliferae. *Class* D.

S. olusatrum ALEXANDERS, BLACK LOVAGE, HORSE PARSLEY. Formerly used as a vegetable in a similar way to CELERY. *Distr.* W Europe, Mediterranean.

S. perfoliatum Grown as an ornamental. *Distr.* S Europe.

snail seed *Cocculus, C. carolinus.*

snakeberry *Actaea rubra.*

snakehead *Chelone glabra.*

snake plant *Sansevieria, S. trifasciata.*

snake root *Liatris, Aristolochia Polygala* **black** ~ *Cimicifuga racemosa* **button** ~ *Liatris, L. pycnostachya* **white** ~ *Ageratina altissima.*

snake's head *Fritillaria meleagris.*

snakeskin plant *Fittonia verschaffeltii.*

snake's tongue *Ophioglossum.*

snake tree *Ficus elastica.*

snakeweed *Polygonum bistorta, Stachytarpheta.*

snakewort, button *Liatris spicata.*

snapdragon *Antirrhinum majus* **spurred** ~ *Linaria* **twining** ~ *Asarina* **violet twining** ~ *A. antirrhiniflora* **wild** ~ *Linaria vulgaris.*

sneezeweed *Achillea ptarmica, Helenium, H. hoopesii, H. autumnale.* **bitter** ~ *H. amarum.*

sneezewort *Achillea ptarmica.*

snowball: Mexican *Echeveria elegans* **wild** ~ *Ceanothus americanus.*

snowbell *Soldanella* **Alpine** ~ *S. alpina.*

snowbells *Styrax.*

snowberry *Symphoricarpos, S. albus* **common** ~ *S. albus.*

snow bush *Breynia nivosa.*

snowdrop *Galanthus* **Barbados** ~ *Zephyranthes tubispatha* **common** ~ *Galanthus nivalis.*

snowdrop tree *Halesia, H. tetraptera* **mountain** ~ *H. monticola.*

snowflake *Leucojum* **autumn** ~ *L. autumnale* **spring** ~ *L. vernum* **summer** ~ *L. aestivum.*

snowflower *Spathiphyllum floribundum.*

snow in summer *Cerastium tomentosum, Ozothamnus thyrsoideus.*

snow queen *Synthyris reniformis.*

snow wreath *Neviusia alabamensis.*

soapbark tree *Quillaja, Q. saponaria.*

soap bush *Q. saponaria.*

soap tree *Yucca elata.*

soap weed *Y. elata.*

soapwood *Caryocar glabrum.*

soapwort *Saponaria, S. officinalis.*

sobolifera, -um: bearing a creeping underground stem or stolon.

socialis, -e: growing in colonies.

socotrana: of Socotra.

Solanaceae The Potato family. 96 genera and about 2500 species of annual to perennial herbs, rarely shrubs and trees, with simple or dissected, typically alternate leaves and regular bisexual flowers that normally bear their parts in 5s. A very important family containing not only numerous ornamentals but many food plants such as the potato, tomato, aubergine and chilli pepper. *Distr.* Widespread. *Class* D.

See: Acnistus, Atropa, Browallia, Brugmansia, Brunfelsia, Capsicum, Cestrum, Cyphomandra, Datura, Dunalia, Fabiana, Hyoscyamus, Iochroma, Jaborosa, Juanulloa, Lycianthes, Lycium, Lycopersicon, Mandragora, Nicandra, Nicotiana, Nierembergia, Nolana, Petunia, Physalis, Scopolia, Solandra, Solanum, Streptosolen, Vestia, Withania.

Solandra (After Daniel Carlsson Solander (1733–82), Swedish botanist who worked in London from 1760.) CHALICE VINE. A genus of 8 species of shrubs and lianas with simple alternate leaves and showy funnel-shaped flowers that are fragrant at night. Several species are grown as ornamentals and some are used as hallucinogenic drugs. *Distr.* Tropical America. *Fam.* Solanaceae. *Class* D.

S. hartwegii See *S. maxima*

S. maxima GOLDEN CHALICE VINE. Scrambling climber. Flowers golden-yellow, marked purple within. *Distr.* Mexico.

Solanum (Classical Latin name for these plants.) NIGHTSHADE. A genus of 1400 species of annual and perennial herbs, shrubs, climbers, and trees with simple or lobed leaves and bell- or funnel-shaped, 5-lobed flowers. A number of species are grown as ornamentals, some are weeds, and others are important food plants. *Distr.* Widespread, particularly in tropical America. *Fam.* Solanaceae. *Class* D.

S. aviculare KANGAROO APPLE. Shrub to 3m high. Grown in Europe as a source of solasodine, an alkaloid used in the manufacture of contraceptives. *Distr.* New Zealand, Australia.

S. dulcamara BITTERSWEET, WOODY NIGHTSHADE, CLIMBING NIGHTSHADE, DEADLY NIGHTSHADE, POISONOUS NIGHTSHADE. Perennial climber. Flowers blue or white. Fruit bright red. Formly used medicinally. Highly poisonous. *Distr.* Eurasia, naturalized in North America.

S. jasminoides POTATO VINE. Scrambling climber. Flowers small, pale grey-blue. *Distr.* South America.

S. melongena EGG PLANT, AUBERGINE, BRINJAL, JEW'S APPLE, MAD APPLE. Perennial, erect, branching herb. Fruit variable, from the size and shape of a hen's egg to 20cm long, black, yellow, or white; eaten as a vegetable. *Distr.* SE Asia.

S. muricatum PEPINO, MELON PEAR, MELON SHRUB. Perennial herb or shrub. Fleshy yellow fruit eaten fresh. *Distr.* Andes.

S. pseudocapsicum JERUSALEM CHERRY, CHRISTMAS CHERRY, WINTER CHERRY, MADEIRA WINTER CHERRY. Bushy shrub. Grown as an ornamental pot plant for its bright red fruits in winter. The fruits are poisonous. *Distr.* Tropical regions of the Old World, widely naturalized.

S. sisymbrifolium STICKY NIGHTSHADE. A sticky annual herb. *Distr.* South America, naturalized coastal North America.

S. tuberosum POTATO. The most important vegetable in the world today, the main source of carbohydrate in Europe after cereals. *Distr.* Andes, at altitudes above 2000m, cultivated there since pre-Columbian times, introduced to Europe 1570, became established as a food crop in 17th-century Ireland, but not in the rest of Europe until the 19th century.

S. wendlandii POTATO VINE, GIANT POTATO VINE, PARADISE FLOWER. Scrambling climber. Flowers lavender. *Distr.* Costa Rica.

Soldanella (Diminutive of the Italian *soldo*, coin, alluding to the rounded leaves.) SNOWBELL. A genus of 10 species of small perennial herbs with rosettes of rounded or kidney-shaped leaves and nodding, funnel-shaped, fringed flowers that bear their parts in fives. These alpine plants are often found emerging from melting snow. Several species are grown as ornamentals. *Distr.* Central and S European mountains. *Fam.* Primulaceae *Class* D.

S. alpina ALPINE SNOWBELL. Clump-forming perennial. Flowers lavender to purple. *Distr.* Europe.

S. villosa Softly hairy, clump-forming perennial. Flowers purple. *Distr.* W Pyrenees.

soldanelloides: from the genus *Soldanella*, with the ending *-oides*, indicating resemblance.

soldier's cap *Aconitum napellus*.

Soleirolia (After Joseph François Soleirol (1796–1863), French soldier who collected plants in Corsica.) A genus of 1 species of mat-forming perennial herb with alternate round leaves and solitary, small, white flowers. *Distr.* Islands of the W Mediterranean, Italy. *Fam.* Urticaceae. *Class* D.

S. soleirolii BABY'S TEARS, MIND YOUR OWN BUSINESS, HELXINE, MOTHER OF THOUSANDS, POLLYANNA VINE, ANGEL'S TEARS, IRISH MOSS, CORSICAN CURSE. A popular tender ground-cover plant prized for its neat foliage.

Solenomelus (From Greek *solen*, tube, and *melos*, limb, alluding to the tubular perianth.) A genus of 2 species of rhizomatous herbs with 2 ranks of linear leaves, rounded stems, and clusters of yellow or blue tubular flowers that are enclosed by 2 spathes. Both species are grown as ornamentals. *Distr.* Temperate South America. *Fam.* Iridaceae. *Class* M.

S. penduncululatus To 30cm high. Flowers yellow.

S. segethii To 30cm high. Flowers blue.

S. sisyrinchium See *S. segethii*

Solenostemon (From Greek *solen*, tube, and *stemon*, stamen, alluding to the fused stamens.) A genus of 60 species of perennial, often succulent, herbs and subshrubs with simple opposite leaves and 2-lipped flowers. This genus is closely related to *Plectranthus* but separated from it on the basis of the stamens being fused to the base of the petal tube. Some species are cultivated as ornamentals and some have edible tubers. *Distr.* Tropical regions of the Old World. *Fam.* Labiatae. *Class* D.

S. scutellarioides COLEUS, PAINTED NETTLE. Aromatic perennial herb. Stems succulent, 4-angled. Leaves very variable, usually brightly coloured. Numerous ornamental cultivars of this species have been produced and are grown as popular house and bedding plants. *Distr.* SE Asia.

Solidago (From Latin *solido*, to make whole or strengthen, alluding to medicinal properties.) GOLDEN ROD. A genus of 100 species of perennial herbs with leaves typically in rosettes and yellow flowers in small daisy-like heads which bear distinct yellow rays. Several species and some hybrids are cultivated as ornamentals, although the species have a tendency to become weedy; other species are used medicinally. *Distr.* North and South America, Eurasia. *Fam.* Compositae. *Class* D.

S. altissima See *S. canadensis*

S. bicolor SILVER ROD. To 1m high. Rays pale yellow or white. *Distr.* E North America.

S. brachystachys See *S. cutleri*

S. caesia Rhizomatous. To 1m high. *Distr.* Central and E North America.

S. canadensis To 1.5m. Flower-heads in a pyramidal panicle. Most frequently cultivated species in Europe. *Distr.* North America.

S. cutleri To 50cm high. Flower-heads few, borne in corymbs . *Distr.* NE USA.

S. flexicaulis To 1.3m high. Flower-heads in small clusters. *Distr.* E North America.

S. glomerata To 1.2m high. Flower-heads borne in clusters on ascending branches. *Distr.* E USA.

S. graminifolia Rhizomatous. Stems much-branched. Possible source of latex for rubber production. *Distr.* North America.

S. hybrida See × *Solidaster luteus*

S. latifolia See *S. flexicaulis*

S. microcephalus Rhizomatous. Leaves thin. *Distr.* E USA.

S. multiradiata Rhizome woody. Stems to 40cm high. *Distr.* North America to E Siberia.

S. odora To 2m high. Leaves scent of aniseed when crushed. *Distr.* E and SE North America.

S. rigida Stems robust, to 1.5m high. Possible source of latex for rubber production. *Distr.* USA.

S. sempervirens To 2m high. Flower-heads in a panicle. *Distr.* NE America, Azores.

S. spathulata Rhizome woody. Stems to 60cm high. *Distr.* W USA (California, Oregon).

S. virgaurea To 1m high. Flower-heads borne on ascending branches. Formerly of medicinal importance as a diuretic. *Distr.* Europe, N Africa, W Asia.

S. vulgaris See *S. virgaurea*

×*Solidaster* (From the names of the parent genera.) A genus of hybrids between members of the genera *Solidago* and *Aster*. *Distr.* Garden origin. *Fam.* Compositae. *Class* D.

×*S. hybridus* See *X. luteus*

×*S. luteus* Perennial herb to 1m high. Leaves simple. Flowers small, yellow, borne in daisy-like heads that bear distinct rays. Grown as a hardy ornamental. *Distr.* Garden origin.

solidus, -a, -um: solid.

Sollya (After Richard Horseman Solly (1778–1858), English botanist.) A genus of 3 species of evergreen climbers with simple alternate leaves and small, bright blue, bell-shaped flowers. *Distr.* SW Australia. *Fam.* Pittosporaceae. *Class* D.

S. fusiformis See *S. heterophylla*

S. heterophylla AUSTRALIAN BLUEBELL CREEPER, BLUEBELL CREEPER, AUSTRALIAN BLUEBELL. Slender climbing or scrambling shrub. Grown as a half-hardy ornamental. *Distr.* W Australia.

Solomon's seal *Polygonatum*
angled ~ *P. odoratum* **common** ~ *P. multiflorum* **great** ~ *P. biflorum*
whorled ~ *P. verticillatum*.

somaliensis, -e: of Somalia.

somniferum, -a: sleep-inducing.

sonchifolius, -a, -um: from the genus *Sonchus*, and *folius*, leaved.

Sonchus (Classical Greek plant name.) MILK THISTLE. A genus of 60–62 species of annual to perennial herbs and shrubs with simple to pinnately lobed, occasionally spiny leaves and yellow flowers in dandelion-like heads. *Distr.* Eurasia, tropical Africa. *Fam.* Compositae. *Class* D.

S. arvensis FIELD SOW THISTLE. Stoloniferous perennial, to 150cm high. A plant of stream sides, salt marshes, and arable land. *Distr.* Europe, SW Asia, introduced Americas, Australia, and Africa.

S. palustris MARSH SOW THISTLE. Tall perennial, to 3m high. A plant of marshes, fens and stream sides. Becoming rare in Britain. *Distr.* Europe to Central Russia and the Caucasus.

Sophora (From Arabic *sophera*, used for a tree with pea-like flowers.) A genus of 52 species of deciduous and evergreen subshrubs, shrubs, and trees with racemes or panicles of pea-like flowers. Several species and numerous cultivars are grown as ornamentals. *Distr.* Tropical and temperate regions. *Fam.* Leguminosae. *Class* D.

S. japonica JAPANESE PAGODA TREE, CHINESE SCHOLAR TREE. Deciduous tree with panicles of creamy-white flowers. The fruit is a source of yellow dye and the dried flowers are used medicinally. Numerous ornamental cultivars have been raised from this species. *Distr.* China, Korea.

S. microphylla Evergreen tree. Flowers golden-yellow, borne in clustered pendulous racemes. *Distr.* New Zealand.

S. tetraptera KOWHAI. Evergreen or semi-deciduous, large shrub or tree. Flowers yellow, borne in short pendulous racemes. Used as a source of timber. *Distr.* New Zealand, Chile.

Sorbaria (From *Sorbus*, the name of a related genus.) FALSE SPIRAEA. A genus of 4–7 species of deciduous shrubs with pinnately divided, alternate leaves and panicles of small white flowers that bear their parts in fives. Several species are grown ornamentally. *Distr.* E Asia. *Fam.* Rosaceae. *Class* D.

S. aitchisonii See *S. tomentosa*

S. arborea See *S. kirilowii*

S. kirilowii Shrub to 7m high. Leaves bear 7–32 leaflets. *Distr.* Central and W China.

S. lindleyana See *S. tomentosa*

S. sorbifolia Stout shrub to 2m high. Flowers borne in erect panicles. *Distr.* N and E Asia.

S. tomentosa Spreading shrub to 6m high. Flowers tinged yellow. *Distr.* Himalaya.

sorbet *Cornus mas.*

sorbifolius, -a, -um: from the genus *Sorbus*, and *folius*, leaved.

Sorbus (Classical Latin name for *S. domestica.*) A genus of about 100 species of deciduous shrubs and trees with simple or pinnately divided, alternate leaves and corymbs of white or occasionally pink flowers that bear their parts in fives. The fruits are often showy, red, pink, or white, and are strictly termed pomes: see APPLE, *Malus*. Many species, hybrids, and cultivars are grown as ornamentals and some have edible fruits. *Distr.* Temperate and warm regions of the N Hemisphere. *Fam.* Rosaceae. *Class* D.

S. americana ROUNDWOOD, AMERICAN MOUNTAIN ASH. Bushy tree to 10m high. Leaves pinnate. Fruit scarlet-red. The fruit is used medicinally. *Distr.* E North America.

S. aria WHITEBEAM. Tree to 12m high, often having numerous trunks. Leaves simple, white-hairy below. Fruit orange-red. The fruit is sometimes used in brandy. *Distr.* Europe.

S. aucuparia (From Latin *avis*, bird, and *capere*, to catch, alluding to the fruits which attract birds.) MOUNTAIN ASH, ROWAN, QUICKBEAM, COMMON MOUNTAIN ASH. Tree to 15m high. Leaves pinnate. Fruit bright red. This tree is a source of timber and the fruits are used in preserves. *Distr.* Europe, SW Asia.

S. chamaemespilus DWARF WHITEBEAM. Shrub to 2m high. Leaves simple. Flowers red. Fruit red. *Distr.* Central Europe.

S. domestica SERVICE TREE. Tree to 20m high. Leaves pinnate. Fruit pear-shaped, yellow, ripening red. This tree is a useful source of timber and tannin. The fruits are eaten raw after they have been slightly frosted or are made into a cider-like drink. *Distr.* S and E Europe, Caucasus, N Africa, SW Asia.

S. latifolia Tree to 15m. Leaves simple. Fruit brown. *Distr.* Europe.

S. moravica See *S. aucuparia*

S. scopulina See *S. aucuparia*

S. torminalis WILD SERVICE TREE. Tree to 15m high. Leaves simple. Fruit brown. The fruit is used like that of *S. domestica*. *Distr.* Europe, N Africa, SW Asia, Mediterranean.

sordidus, -a, -um: dirty.

Sorghastrum (From the genus name *Sorghum*, and the Latin suffix *-astrum*, a poor imitation.) A genus of 13–20 species of annual and perennial clump-forming grasses with narrow leaves and narrow, simple or branched inflorescences. *Distr.* Warm temperate and tropical North and South America and tropical Africa. *Fam.* Gramineae. *Class* M.

S. avenaceum See *S. nutans*

S. nutans INDIAN GRASS, WOOD GRASS. An important forage grass with ornamental varieties available. *Distr.* USA.

Sorghum (From the Italian name for this plant, *sorgo*.) GUINEA CORN, MILLET. A genus of 20–24 species of annual and perennial rhizomatous grasses with robust stems, flat leaves and a dense paniculate inflorescence. A number of species and hybrids are grown as a source of grain and fodder with *S. bicolor* being the most important. *Distr.* Warm temperate and tropical regions of the Old World, 1 species in Mexico. *Fam.* Gramineae. *Class* M.

S. bicolor SORGHUM, GREAT MILLET. The world's fourth most important cereal after rice, wheat, and maize. Forms a staple food in parts of Africa, India, and China. Some cultivars are a source of grain, others are grown for their stems which have a very high sugar content. Some are grown as fodder or as a source of paper pulp. *Distr.* Ethiopia and Sudan.

S. halepense JOHNSON GRASS, ALEPPO GRASS, MEANS GRASS. Sometimes grown as a perennial ornamental. *Distr.* Mediterranean.

sorghum *Sorghum bicolor*.

sororius, -a, -um: sisterly, being closely related.

sorrel *Rumex* **common** ~ *R. acetosa* **creeping yellow** ~ *Oxalis corniculata* **French** ~ *Rumex scutatus* **garden** ~ *R. scutatus*, *R. acetosa* **mountain** ~ *Oxyria digyna* **tree** ~ *Oxydendrum arboreum* **wood** ~ Oxalidaceae, *Oxalis acetosella*.

sotol *Dasylirion* **Texas** ~ *D. texanum*.

sourberry *Rhus integrifolia*.

sourweed *Oxydendrum arboreum*.

southernwood *Artemisia abrotanum*.

sowbread *Cyclamen hederifolium*.

sow thistle: Alpine ~ *Cicerbita alpina* **mountain** ~ *C. alpina*.

soybean *Glycine max*.

Spaniard, golden *Aciphylla aurea*.

Spanish bayonet *Yucca aloifolia*.

Spanish dagger *Yucca treculeana*, *Y. gloriosa*.

Sparaxis (From Greek *sparasso*, to tear, alluding to the lacerated spathes.) WANDFLOWER. A genus of 6 species of perennial herbs with small corms, ribbed leaves, and spikes of large funnel-shaped flowers. Several species are cultivated as ornamentals. *Distr.* South Africa (Cape Province). *Fam.* Iridaceae. *Class* M.

S. elegans Flowers orange or white with a yellow centre surrounded by a black band.

S. tricolor HARLEQUIN FLOWER. Flowers borne in a loose spike, with a yellow tube and red lobes that are marked black.

Sparganium (Classical Greek name for a waterside plant, probably *Butomus*.) BUR REED. A genus of 12–20 species of aquatic herbs with 2 ranks of linear leaves and racemes or heads of unisexual flowers. The fruit is an important part the diet of wildfowl in autumn and spring. *Distr.* N temperate regions, Malaysia to Australia and New Zealand. *Fam.* Typhaceae. *Class* M.

S. emersum UNBRANCHED BUR REED. Erect or floating perennial. Inflorescence simple. *Distr.* Europe, W and Central Asia, N America.

S. minimum LEAST BUR REED. Stems floating. Inflorescence simple. *Distr.* N temperate regions.

Sparmannia (After Dr Anders Sparmann (1748–1820), Swedish traveller and member of Captain Cook's second expedition.) A genus of 3 species of shrubs and small trees with simple or lobed alternate leaves and umbels of white flowers that bear 4 petals and numerous stamens. *Distr.* Africa, Madagascar. *Fam.* Tiliaceae. *Class* D.

S. africana AFRICAN HEMP, HOUSE LIME, STOCK ROSE. Shrub to 6m high. Flowers profuse. Often grown as a house plant. The bark is a source of fibre. *Distr.* South Africa.

Spartina (From the Greek name for ESPARTO GRASS (*Stipa tenacissima*), *spartion.*) MARSH GRASS, CORD GRASS. A genus of 14–17 species of rhizomatous perennial grasses of salt and freshwater marshes as well as prairies, with tough, flat or rolled leaves and one-sided inflorescences. *Distr.* W and S Europe, NW and S Africa, N America. *Fam.* Gramineae. *Class* M.

S. alterniflora SMOOTH CORD GRASS. *Distr.* N America, naturalized Europe.

S. anglica COMMON CORD GRASS. Thought to have arisen by hybridization between *S. maritima* and *S. alterniflora* at the end of the 19th century when the latter was introduced to Europe. An aggressive weed which is rapidly replacing *S. maritima.*

S. maritima SMALL CORD GRASS. *Distr.* W Europe and N Africa.

S. pectinata PRAIRIE CORD GRASS, FRESHWATER CORD GRASS, SLOUGH GRASS. Grown as an ornamental for its slender inflorescence. *Distr.* North America.

Spartium (From Greek *spartion*, a type of grass; they were both used for cordage.) A genus of 1 species of almost leafless, deciduous shrub with rush-like stems, sparse simple leaves, and fragrant, yellow, pea-like flowers. *Distr.* SW Europe, naturalized USA (California). *Fam.* Leguminosae. *Class* D.

S. junceum SPANISH BROOM, WEAVER'S BROOM. Grown as an ornamental. The stems are used in weaving.

Spartocytisus See *Cytisus*

spathaceus, -a, -um: spathe-like.

Spathicarpa (From Greek *spatha*, spathe, and *karpos*, fruit, alluding to the fusion of the spadix and spathe.) A genus of about 6 species of evergreen rhizomatous herbs with simple or 3-lobed leaves. The small unisexual flowers are borne in 4–5 rows on a spadix which is fused to a horizontal green spathe around 5cm long. *Distr.* Tropical South America. *Fam.* Araceae. *Class* M.

S. sagittifolia Grown as a tender ornamental foliage plant. *Distr.* Brazil, Paraguay, Argentina.

spathiflorus, -a, -um: with spathe-like flowers.

Spathiphyllum (From Greek *spatha*, spathe, and *phyllon*, leaf, alluding to the leaf-like spathe.) PEACE LILY. A genus of about 36 species of evergreen rhizomatous herbs with simple, dark green leaves. The small hermaphrodite flowers are borne on a short spadix that is subtended by, and partially fused to, a green or white spathe. Several species and numerous cultivars are grown as ornamentals. *Distr.* Tropical America, Philippines, Indonesia. *Fam.* Araceae. *Class* M.

S. floribundum SNOWFLOWER. Leaves somewhat velvety above. Spathe white. Frequently cultivated with several cultivars available. *Distr.* Colombia.

S. wallisii (After Gustave Wallis, who introduced it from Colombia in 1824.) PEACE LILY, WHITE SAILS. Spathe white. Flowers fragrant. The original species from which the very popular cultivar 'Clevelandii' was probably derived. *Distr.* Central America.

spathularis, -e: spathulate, broad and flat, rounded in the upper part.

spathulatus, -a, -um: spathulate, broad and flat, rounded in the upper part.

spathulifolius, -a, -um: with spatula-shaped leaves.

spatterdock *Nuphar advena.*

speargrass *Aciphylla squarrosa.*

spearmint *Mentha spicata.*

spearwort: greater ~ *Ranunculus lingua* **lesser** ~ *R. flammula.*

speciosissimus, -a, -um: very showy.

speciosus, -a, -um: showy.

spectabilis, -e: spectacular.

spectatissimus, -a, -um: very spectacular.

speculus, -a, -um: like a mirror.

speedwell *Veronica* **common** ~ *V. officinalis* **digger's** ~ *V. perfoliata*

germander ~ *V. chamaedrys* **rock** ~ *V. fruticans* **silver** ~ *V. incana.*

speedy Jenny *Tradescantia fluminensis.*

Speirantha (From Greek *speira*, wreath, and *anthos*, flower, alluding to the inflorescence.) A genus of 1 species of rhizomatous herb with basal rosettes of lance-shaped leaves and loose racemes of star-shaped white flowers. *Distr.* China. *Fam.* Convallariaceae. *Class* M.

S. convallarioides Grown as a half-hardy ornamental. *Distr.* China.

S. gardenii See *S. convallarioides*

spelta *Triticum spelta.*

Spergula (Probably from the German name *Spörgel.*) A genus of 5–6 species of small annual and perennial herbs with linear opposite leaves and small flowers that bear 5 white petals. *Distr.* Mediterranean, 1 species cosmopolitan. *Fam.* Caryophyllaceae. *Class* D.

S. arvensis CORN SPURREY. Annual. Flowers foetid. A weed of acidic soils, formerly grown as a fodder crop. *Distr.* Cosmopolitan.

Spergularia (From the name of a closely related genus, *Spergula.*) SAND SPURREY, SEA SPURREY. A genus of 40 species of annual and perennial herbs of sandy soils with opposite linear leaves and cymes of small flowers that bear 5 sepals and 5 or sometimes no petals. *Distr.* Cosmopolitan, especially in salty areas. *Fam.* Caryophyllaceae. *Class* D.

S. rupicola Perennial. Flowers star-shaped, deep pink. Sometimes grown as an ornamental. *Distr.* Atlantic coast of Europe, Spain to Scotland.

Sphacele See *Lepechinia*

Sphaeralcea (From Greek *sphaira*, globe, and *Alcea*, the name of a related genus, alluding to the spherical fruits.) GLOBE MALLOW, FALSE MALLOW. A genus of about 60 species of hairy annual and perennial herbs and subshrubs with simple or occasionally palmately lobed leaves and cup-shaped bisexual flowers that bear 5 petals and numerous fused stamens. Several species are grown as ornamentals. *Distr.* North, Central, and S America. *Fam.* Malvaceae. *Class* D.

S. ambigua DESERT MALLOW. Shrubby perennial herb. Flowers coral-orange. *Distr.* S USA.

S. munroana Shrubby perennial. Flowers brilliant coral-pink. *Distr.* W USA.

sphaerandrus, -a, -um: with spherical anthers.

sphaeranthus, -a, -um: with spherical flowers.

sphaerocarpus, -a, -um: with spherical fruit.

sphaerocephalus, -a, -um: with a spherical head.

Sphaerosepalaceae A family of 2 genera and 17 species of deciduous shrubs and trees with simple leaves and regular bisexual flowers that typically bear 4 sepals, 4 petals, and numerous stamens. *Distr.* Madagascar. *Class* D.

sphaerostachyus, -a, -um: with a round spike.

sphegodes wasp-like.

sphenantherus, -a, -um: with wedge-shaped flowers.

Sphenomeris (From Greek *sphen*, wedge, and *meris*, part or division, alluding to the shape of the final leaf divisions.) A genus of 11–20 species of terrestrial and epiphytic ferns with pinnately divided leaves. *Distr.* Tropical regions and N Japan. *Fam.* Dennstaedtiaceae. *Class* F.

S. chusana Grown as an ornamental. *Distr.* E Asia to Madagascar and Polynesia.

Sphenostemonaceae A family of 1 genus and 7 species of evergreen shrubs and trees with simple, spirally arranged leaves and small regular flowers. This family is sometimes included within the family Aquifoliaceae. *Distr.* Indonesia to Australia and New Caledonia. *Class* D.

spicant: tufted.

spicatus, -a, -um: spiky.

spiceberry *Ardisia crenata.*

spice bush *Lindera benzoin.*

spiculifolius, -a, -um: with spiky leaves.

spider flower *Grevillea*, *Cleome hassleriana.*

spider plant *Chlorophytum comosum.*

spiderwort *Commelinaceae*, *Tradescantia*, *T. virginiana.*

Spigelia (After Adrian van der Spiegel (1578–1625), Dutch professor of anatomy at Padua.) PINK ROOT. A genus of 50 species of annual and perennial herbs and subshrubs with simple opposite leaves and clusters of tubular or salver-shaped flowers which bear their parts in fives. Several species are grown as ornamentals. *Distr.* Tropical and warm regions of America. *Fam.* Loganiaceae. *Class* D.

S. marilandica MARYLAND PINKROOT, INDIAN PINK, WORM GRASS. Perennial herb. Flowers red outside, yellow inside, borne in a spike. Used medicinally as a vermifuge. *Distr.* SE United States.

spignel *Meum athamanticum.*

spikenard Valerianaceae, *Aralia californica* **American** ~ *A. racemosa* **Japanese** ~ *A. cordata* **Manchurian** ~ *A. continentalis* **ploughman's** ~ *Inula conyzae.*

spinach *Spinacia oleracea* **Chinese** ~ *Amaranthus tricolor* **Cuban** ~ *Montia perfoliata* **New Zealand** ~ *Tetragonia tetragonioides* **water** ~ *Ipomoea aquatica* **wild** ~ *Chenopodium bonus-henricus.*

spina-christi Christ's thorn.

Spinacia (From Latin *spina*, spine; the fruits of some species are prickly.) A genus of 3 species of annual or biennial glabrous herbs with large flat leaves and dense spikes of small flowers. *Distr.* SW Asia. *Fam.* Chenopodiaceae. *Class* D.

S. oleracea SPINACH. An annual herb with broad dark green leaves. The most nutritious of all leaf vegetables. *Distr.* Origin unclear, probably Iran.

spinalbus, -a, -um: white-spined.

spindle Celastraceae.

spindle tree *Euonymus europaeus* **common** ~ *E. europaeus* **winged** ~ *E. alatus.*

spinescens: spiny.

spinks *Cardamine pratensis.*

spinosissimus, -a, -um: very spiny.

spinosus, -a, -um: spiny.

spinulifera, -um: bearing small spines.

spinulosus, -a, -um: with small spines.

Spiraea (From Greek *speiraia*, a plant used in garlands.) SPIRAEA, BRIDAL WREATH. A genus of 70–80 species of deciduous shrubs with simple alternate leaves and racemes or panicles of small, white to red flowers. Many species and cultivars are grown as ornamentals. *Distr.* N temperate regions to Mexico and Himalaya. *Fam.* Rosaceae. *Class* D.

S. albiflora See *Spiraea japonica*

S. arborea See *Sorbaria kirilowii*

S. × bumalda See *Spiraea japonica*

S. callosa See *Spiraea japonica*

S. crispifolia See *Spiraea japonica*

S. japonica JAPANESE SPIRAEA. Shrub to 1.5m high. Flowers clustered, rose-pink. Numerous cultivars have been raised from this species. *Distr.* Japan, China.

S. ulmaria See *Filipendula ulmaria*

S. venusta See *Filipendula rubra*

spiraea *Spiraea*, *Astilbe*, *A. japonica.* **false** ~ *Sorbaria* **Japanese** ~ *Spiraea japonica* **rock** ~ *Petrophytum*, *Holodiscus dumosus.*

spiralis, -e: spiralled.

Spirodela (From Greek *speir*, spiral, and *delos*, obvious.) DUCKWEED. A genus of 3–4 species of small, floating, aquatic herbs that consist of undifferentiated leaf-like thalli. The rarely produced flowers are formed in pouches on the thallus. Grown on waste water from dairy farms and used as a substitute for alfalfa in cattle and pig farms in the USA. *Distr.* Cosmopolitan. *Fam.* Lemnaceae. *Class* M.

S. polyrrhiza GREAT DUCKWEED, WATER FLAXSEED. Thallus to 10mm across, flowers rare. *Distr.* More or less cosmopolitan.

spleenwort *Asplenium*, Aspleniaceae **black** ~ *Asplenium adiantumnigrum* **maidenhair** ~ *A. trichomanes* **mother** ~ *A. bulbiferum.*

splendidus, -a, -um: splendid.

splitrock *Pleiospilos.*

Spodiopogon (From Greek *spodios*, ashen, and *pogon*, beard, alluding to the grey-hairy inflorescences.) A genus of 8–9 species of annual and perennial grasses with flat leaves and open or compact, paniculate inflorescences. *Distr.* Temperate and tropical Asia. *Fam.* Gramineae. *Class* M.

S. sibiricus Occasionally grown for its pretty inflorescences. *Distr.* E Asia.

Spraguea (After Isaac Sprague (1811–1895), American botanical and zoological artist.) A genus of about 9 species of annual and perennial herbs with rosettes of simple basal leaves and compact heads of small flowers that bear 4 petals and 3 stamens. *Distr.* W North America. *Fam.* Portulacaceae. *Class* D.

S. umbellata PUSSY PAWS. Annual to perennial herb. Flowers white or pink. Grown as a hardy ornamental. *Distr.* W North America.

Sprekelia (After J. H. von Sprekelsen (died 1764), German botanist.) A genus of 1 species of bulbiferous perennial herb with deciduous linear leaves and solitary, showy, irregular, red or white flowers. *Distr.* Mexico, Guatemala. *Fam.* Amaryllidaceae. *Class* M.

S. formosissima JACOBEAN LILY, ST JAMES LILY, AZTEC LILY. Cultivated as a tender ornamental.

spring beauty *Claytonia*, *C. virginica.*

sprouts, Brussel *Brassica oleracea.*

spruce *Picea* **black** ~ *P. mariana* **bog** ~ *P. mariana* **Brewer** ~ *P. breweriana* **dragon** ~ *P. asperata* **hemlock** ~ *Tsuga* **Norway** ~ *Picea abies* **Sitka** ~ *P. sitchensis* **white** ~ *P. glauca.*

spurge *Euphorbia*, Euphorbiaceae **Allegheny** ~ *Pachysandra procumbens* **caper** ~ *Euphorbia lathyris* **cypress** ~ *E. cyparissias* **Irish** ~ *E. hyberna* **myrtle** ~ *E. lathyris* **Portland** ~ *E. portlandica* **wood** ~ *E. amygdaloides.*

spurius, -a, -um: false.

spurrey: sand ~ *Spergularia* **sea** ~ *Spergularia.*

spurwort *Sherardia arvensis.*

squalidus, -a, -um: squalid, dirty.

squamatus, -a, -um: scaly.

squameus, -a, -um: scaly.

squamigera, -um: bearing scales.

squamosus, -a, -um: full of scales.

squarrosus, -a, -um: with parts spreading horizontally.

squash: summer ~ *Cucurbita pepo* **winter** ~ *C. maxima.*

squaw root *Caulophyllum thalictroides.*

squill *Urginea maritima* **autumn** ~ *Scilla autumnalis* **indigo** ~ *Camassia scilloides* **Lebanon** ~ *Puschkinia scilloides* **sea** ~ *Urginea maritima* **silver** ~ *Ledebouria socialis* **spring** ~ *Scilla verna* **striped** ~ *Puschkinia scilloides.*

stachyoides from the genus *Stachys*, with the ending *-oides*, indicating resemblance.

Stachys (From Greek *stachys*, spike, alluding to the inflorescence.) HEDGE NETTLE, GROUNDWORT, BETONY, WOUNDWORT. A genus of 300 species of perennial herbs and shrubs with simple, opposite, wrinkled leaves and small 2-lipped flowers that are usually gathered into whorls or long spikes. Numerous species are grown as ornamentals and some are used medicinally. *Distr.* Widespread, particularly in temperate regions. *Fam.* Labiatae. *Class* D.

S. affinis CHINESE ARTICHOKE, JAPANESE ARTICHOKE, KNOTROOT. Erect perennial herb. Flowers white or pink. The slender white tubers are eaten salted, pickled, or boiled. *Distr.* China.

S. byzantina LAMB'S TONGUE, LAMB'S EARS, LAMB'S TAILS, WOOLLY BETONY. Perennial. Leaves densely silver-hairy. *Distr.* SW Asia.

S. coccinea Clump-forming perennial. Flowers bright red, borne in spikes. *Distr.* Texas to Mexico.

S. germanica DOWNY WOUNDWORT. Densely white-woolly perennial. Flowers pink-purple, borne in distinct whorls. *Distr.* Europe, N Africa, Central Asia.

S. grandiflora See *S. macrantha*

S. lanata See *S. byzantina*

S. macrantha BISHOP'S WORT. Clump-forming perennial. Flowers large, pink-purple, borne in distinct whorls. *Distr.* Caucasus.

S. officinalis BETONY, BISHOP'S WORT, WOOD BETONY. Mat-forming to erect perennial. Flowers red-purple, borne in dense spikes. Formerly used medicinally. *Distr.* Eurasia.

S. olympica See *S. byzantina*

S. palustris MARSH BETONY. Perennial. Flowers small, dull purple, bore in whorls. The tubers were formerly eaten. *Distr.* Eurasia, North America.

S. spicata See *S. grandiflora*

S. tuberifera See *S. affinis*

Stachytarpheta (From Greek *stachys*, spike, and *tarphys*, thick, alluding to the inflorescence.) FALSE VERVAIN, SNAKE WEED, BASTARD VERVAIN. A genus of 65 species of herbs and small shrubs with simple, opposite or alternate leaves and thick-stemmed spikes of salver-shaped flowers. Several species are grown as ornamentals. *Distr.* Tropical and warm regions. *Fam.* Verbenaceae. *Class* D.

S. cayennensis BRAZILIAN TEA. Shrubby herb to 2m high. Flowers blue, pink or purple. *Distr.* Central and South America.

Stachyuraceae A family of 1 genus and 5 species. *Class* D.
See: Stachyurus.

Stachyurus (From Greek *stachys*, spike, and *oura*, tail, alluding to the slender inflorescences.) A genus of 5–10 species of deciduous shrubs and small trees with simple alternate leaves and spike-like racemes of small flowers that bear their parts in fours. The flowers appear before the leaves expand in spring. Several species are grown as ornamentals. *Distr.* Himalaya to Japan. *Fam.* Stachyuraceae. *Class* D.

S. praecox Shrub to 2m high. Flowers bell-shaped, yellow. *Distr.* Japan.

Stackhousiaceae A family of 3 genera and 28 species of rhizomatous herbs of dry areas with simple, leathery or succulent leaves and regular flowers that bear their parts in fives. *Distr.* Indonesia to Australia, New Zealand and the Pacific. *Class* D.

Staehelina (After Stahelinus, a friend of Sebastien Vaillant (1669–1722), French botanist.) A genus of 8 species of dwarf tufted shrubs and subshrubs with small leaves crowded towards the ends of the branches and flowers in daisy-like heads that lack distinct rays. *Distr.* Mediterranean. *Fam.* Compositae. *Class* D.

S. uniflosculosa Small rounded shrub. Flowers clear pink. *Distr.* S and W Balkans.

staff tree *Celastrus scandens.*

stamineus, -a, -um: with prominent stamens.

Stangeria (After Dr Stanger, Surveyor-General of Natal, who collected this plant in 1851.) A genus of 1 species of fern-like cycad with a turnip-like stock and a rosette of large pinnate leaves. *Distr.* South Africa (Cape Province to Natal). *Fam.* Stangeriaceae. *Class* G.

S. eriopus Occasionally grown as an ornamental.

Stangeriaceae A family of 1 genus and 1 species. *Class* G.
See: Stangeria.

stans: erect.

stapeliformis, -e: resembling the genus *Stapelia*.

stapelioides: from the genus *Stapelia*, with the ending *-oides*, indicating resemblance.

Staphylea (From Greek *staphyle*, cluster, alluding to the inflorescences.) BLADDERNUT. A genus of 11 species of deciduous shrubs and trees with opposite pinnate leaves and panicles of white flowers. The fruit is a membranous inflated capsule. Several species are grown as ornamentals. *Distr.* N temperate regions. *Fam.* Staphyleaceae. *Class* D.

S. pinnata BLADDERNUT. Shrub to 5m high. Flowers white. Fruit to 3cm long. *Distr.* SE Europe, W Asia.

S. trifolia *Distr.* E USA.

Staphyleaceae The Bladdernut family. 5 genera and 27 species of trees and shrubs with trifoliate or pinnate leaves and regular, bisexual or occasionally unisexual flowers that bear their parts in fives. *Distr.* N temperate regions, tropical Asia and America. *Class* D.
See: Staphylea.

star, shooting *Dodecatheon.*

star cluster *Pentas lanceolata*
Egyptian ~ *P. lanceolata.*

starfish plant *Cryptanthus acaulis.*

starflower, spring *Ipheion uniflorum.*

star-glory *Ipomoea quamoclit.*

stargrass *Aletris.*

star of Bethlehem *Eucharis x grandiflora*, *Campanula isophylla*, *Ornithogalum umbellatum* **drooping** ~ *O. nutans* **yellow** ~ *Gagea lutea*.

star of Persia *Allium christophii*.

starwort Callitrichaceae, *Callitriche*, *Stellaria* **water** ~ *Callitriche stagnalis*.

star zygadene *Zigadenus fremontii*.

Statice See *Limonium*

statice *Goniolimon tataricum*, *Limonium* **Tartarian** ~ *Goniolimon tataricum*.

Stauntonia (After Sir George Leonard Staunton (1737–1801), who travelled in China.) A genus of about 10 species of evergreen twining shrubs with alternate palmate leaves and racemes of somewhat fleshy, unisexual flowers in which the petals have been replaced by nectaries and the sepals have become petal-like. The fruit is a berry. *Distr.* E Asia. *Fam.* Lardizabalaceae. *Class* D.

S. hexaphylla Flowers fragrant, white. Berry purple. Grown as an ornamental. *Distr.* Japan, China, Korea, Taiwan.

Stegnospermataceae A family of 1 genus and 3 species of small trees and lianas with simple, spirally arranged leaves and small flowers. This family is sometimes included within the family Phytolaccaceae. *Distr.* Mexico, Central America, West Indies. *Class* D.

Steirodiscus (From Greek *steiros*, barren, and *diskos*, disk.) A genus of 5 species of annual herbs with leathery simple or pinnately lobed leaves and yellow flowers in daisy-like heads that bear distinct yellow-orange rays. Some species are cultivated as ornamentals. *Distr.* South Africa. *Fam.* Compositae. *Class* D.

S. tagetes Stems arching, to 30cm high. *Distr.* South Africa.

Stellaria (From Latin *stella*, star, alluding to the star-shaped flowers of some species.) STITCHWORT, CHICKWEED, STARWORT. A genus of 120 species of annual and perennial herbs with simple opposite leaves and cymes of flowers that bear 5 free sepals, 5 white petals, 1–10 stamens, and 3 styles. The petals may be entire, notched, or deeply divided and are absent in some species. *Distr.* Cosmopolitan. *Fam.* Caryophyllaceae. *Class* D.

S. holostea GREATER STITCHWORT, ADDER'S MEAT, STITCHWORT. Upright perennial. Petals white, deeply divided. *Distr.* Europe, Mediterranean.

stellaris: star-like.

stellatus, -a, -um: starry.

stellulatus, -a, -um: with small stars.

Stemonaceae A family of 4 genera and 32 species of rhizomatous or tuberous climbing herbs. The flowers have 2 whorls of 2 tepals. *Distr.* Tropical Asia, Japan, Australia, E North America. *Class* M.

stenantherus, -a, -um: with narrow flowers.

Stenanthium (From Greek *stenos*, narrow, and *anthos*, flower, alluding to the narrow inflorescences and tepals.) A genus of 4–5 species of bulbiferous herbs with grass-like leaves and racemes or panicles of narrow bell-shaped flowers. Several species are grown as ornamentals. *Distr.* E Asia, North America. *Fam.* Melanthiaceae. *Class* M.

S. gramineum Flowers white to purple, fragrant, borne in panicles. *Distr.* SE USA.

S. occidentale Flowers brown-purple, borne in loose racemes. *Distr.* W North America.

stenanthus, -a, -um: with narrow flowers.

stenaulus, -a, -um: with narrow grooves.

stenocephalus, -a, -um: with a narrow head.

Stenochlaena (From Greek *stenos*, narrow, and *chlaina*, cloak, alluding to the absent indusia.) A genus of 5–6 species of large climbing and epiphytic ferns. Young fronds of some species are edible and some are cultivated as ornamentals. *Distr.* Tropical regions. *Fam.* Blechnaceae. *Class* F.

S. palustris CLIMBING FERN. Rhizome climbing. Leaves pinnate. *Distr.* Malaysia to Polynesia.

Stenomesson (From Greek *stenos*, narrow, and *mesos*, middle, alluding to the shape of the perianth.) A genus of about 20 species of bulbiferous perennial herbs with linear leaves and pendent tubular flowers. Some species are cultivated as tender ornamentals. *Distr.* Mostly native to the high Andes. *Fam.* Amaryllidaceae. *Class* M.

S. miniatum Flowers red, tubular below, pitcher-shaped above. *Distr.* Peru, Bolivia.

S. variegatum Flower colour variable from red to pink, with yellow or green banding. *Distr.* Bolivia, Ecuador, Peru.

stenopetalus, -a, -um: with narrow petals.

stenophyllus, -a, -um: with narrow leaves.

stenopterus, -a, -um: with narrow wings.

stenostachyus, -a, -um: with narrow spikes.

Stenotaphrum (From Greek *stenos*, narrow, and *taphros*, trench, alluding to the grooved raceme axis.) A genus of 7 species of annual and perennial grasses with ascending rooting stems, flat or folded leaves, and racemose inflorescences. Several species are important as fodder grasses. *Distr.* Tropical and warm regions. *Fam.* Gramineae. *Class* M.

S. secundatum ST AUGUSTINE GRASS, BUFFALO GRASS. Used as a lawn grass and for binding sand in tropical regions or as a half-hardy ornamental in temperate zones. Variegated versions are available. *Distr.* Warm America.

Stephanandra (From Greek *stephanos*, crown, and *andros*, man, alluding to the stamens which form a wreath around the capsule.) A genus of 4 species of deciduous shrubs with simple alternate leaves, and racemes or panicles of small, white, 5-lobed flowers. Several species are grown as ornamentals. *Distr.* E Asia. *Fam.* Rosaceae. *Class* D.

S. incisa Densely branched, weeping shrub to 2.5m high. *Distr.* Japan, Korea.

Stephanotis (From Greek *stephanos*, crown, and *otos*, ear, alluding to the auricles that make up the staminal crown of the flowers.) A genus of 5–15 species of twining, evergreen, woody climbers with leathery opposite leaves and cymes of funnel- or cup-shaped flowers. *Distr.* Tropical regions of the Old World. *Fam.* Asclepiadaceae. *Class* D.

S. floribunda WAX FLOWER, MADAGASCAR JASMINE, BRIDAL WREATH, CHAPLET FLOWER, FLORADORA. Flowers large, white, waxy, fragrant. Grown as an ornamental and for cut flowers. *Distr.* Madagascar.

Sterculiaceae A family of 67 genera and about 1500 species of shrubs, trees, and a few climbers with alternate, simple or lobed leaves. The flowers are unisexual or bisexual with 3–5 sepals and 5 free or united petals which may be absent in some species. The two important products derived from this family are cocoa from *Theobroma cacao* and cola chiefly from *Cola acuminata*. *Distr.* Tropical, subtropical, and occasionally temperate regions. *Class* D.
See: Cola, Firmiana, Fremontodendron, Guichenotia, Hermannia, Theobroma, Trochetiopsis.

sterculiaceus, -a, -um: resembling the genus *Sterculia*.

sterilis, -e: sterile, barren.

Sternbergia (After Count Kaspar Moritz von Sternberg (1761–1838), Austrian botanist.) AUTUMN DAFFODIL. A genus of 7–8 species of bulbiferous perennial herbs with simple or lobed leaves and solitary white or yellow flowers. Some species are cultivated as ornamentals. *Distr.* Turkey, W to Spain and E to Kashmir. *Fam.* Amaryllidaceae. *Class* M.

S. candida Flowers white. *Distr.* SW Turkey.

S. clusiana (After Charles de l'Ecluse (Carolus Clusius) (1526–1609), Flemish botanist.) Flowers green-yellow to bright yellow. *Distr.* W Asia.

S. colchiciflora Flowers pale yellow. *Distr.* SE Spain to W Asia.

S. fischeriana (After Fischer, who collected the type specimen.) Flowers bright yellow. *Distr.* Azerbaijan and Turkey to Kashmir.

S. lutea Flowers deep yellow. *Distr.* Spain to Iran and Central Asia.

S. macrantha See *S. clusiana*

S. sicula Flowers yellow, borne on a stem to 7cm high. *Distr.* Mediterranean.

Stewartia (After John Stuart (1713–92), Earl of Bute, British botanist and Prime Minister.) A genus of 6–9 species of deciduous trees and shrubs with simple alternate leaves and solitary cream flowers that bear

5–8 petals and numerous stamens. Several species are grown as ornamentals. *Distr.* E North America, E Asia. *Fam.* Theaceae. *Class* D.

S. malacodendron SILKY CAMELLIA. Shrub or small tree to 5m high. Shoots woolly-hairy. *Distr.* SE USA.

S. ovata MOUNTAIN CAMELLIA. Shrub to 5m. Flowers cup-shaped. *Distr.* SE USA.

S. pseudocamellia JAPANESE STEWARTIA. Tree to 20m high. *Distr.* Japan.

stewartia, Japanese *Stewartia pseudocamellia.*

stick tight *Bidens.*

Stictocardia (From Greek *stiktos*, spotted, and *kardia*, heart; the leaves are glandular below.) A genus of 12 species of climbing herbs and shrubs with simple leaves and funnel-shaped red or purple flowers. *Distr.* Tropical regions. *Fam.* Convolvulaceae. *Class* D.

S. beraviensis Liana. Flowers crimson. Cultivated ornamentally. *Distr.* Africa.

Stilbaceae A family of 5 genera and 13 species of shrubs with simple to pinnate leaves and flowers that typically bear their parts in fives. This family is sometimes included within the family Verbenaceae. *Distr.* South Africa. *Class* D.

stink pod *Scoliopus bigelovii.*

stinkweed *Thlaspi arvense.*

stinkwood *Eucryphia moorei.*

stinkwort *Helleborus foetidus.*

Stipa (From Latin *stipa*, oakum or loose bunch of fibres, alluding both to the feathery inflorescences and the use of *S. tenacissima* as a source of cordage.) FEATHER GRASS, NEEDLE GRASS, SPEAR GRASS. A genus of 150–300 species of annual and perennial, clump-forming grasses with narrow, often in-rolled, rough leaves and narrow paniculate inflorescences. Some species are cultivated as ornamentals, others are grown for their fibre, as fodder, or as erosion protection. *Distr.* Tropical and temperate, often dry regions (40 species in Europe). *Fam.* Gramineae. *Class* M.

S. arundinacea NEW ZEALAND WIND GRASS, PHEASANT'S TAIL GRASS. Leaves turn orange in autumn. Inflorescences pendent, purple-green. Grown as an ornamental. *Distr.* E Australia, New Zealand.

S. gigantea Perennial to 2.5m high. Inflorescences open silvery panicles frequently cultivated as an ornamental. *Distr.* SW Europe.

S. pennata EUROPEAN FEATHER GRASS. Perennial. Stems to 60cm high. Inflorescences compact, yellow. *Distr.* S and Central Europe to Himalaya.

S. tenacissima ESPARTO, ALGERIAN GRASS, ALFA, ESPARTO GRASS. Stems used as a source of fibre for ropes, sails, mats and even paper making. *Distr.* W Mediterranean.

stipulaceus, -a, -um: with stipules.

stipularis, -e: with stipules.

stipulaceus, -a, -um: with stipules.

stitchwort *Stellaria, S. holostea*
greater ~ *S. holostea.*

stock, Brompton *Matthiola incana*
night-scented ~ *M. longipetala*
Virginian ~ *Malcolmia maritima.*

stoechas: of the Stoechades (now the Iles d'Hyères) off the S coast of France.

Stokesia (After Dr Jonathan Stokes (1755–1831), Edinburgh physician and friend of Linnaeus the younger.) A genus of 1 species of perennial herb to 1m high with spiny leaves and blue flowers in large daisy-like heads which bear distinct, 5-lobed, blue rays. *Distr.* SE USA. *Fam.* Compositae. *Class* D.

S. laevis STOKE'S ASTER. Cultivated ornamentally and for cut flowers, a possible commercial oilseed crop.

stolonifera, um: bearing stolons, creeping underground stems.

stonecrop *Sedum, S. acre*, Crassulaceae
biting ~ *Sedum acre* **ditch** ~
Penthoraceae **golden** ~ *Sedum adolphi*
mossy ~ *Crassula tillaea.*

stone root *Collinsonia canadensis.*

stopper *Eugenia.*

storax *Styrax.*

stork's bill *Erodium.*

Stranvaesia See *Photinia*

S. davidiana See *Photinia davidiana*

Strasburgeriaceae A family of 1 genus and 1 species of evergreen tree with simple leaves and regular bisexual flowers. This family is sometimes included within the family Ochnaceae. *Distr.* New Caledonia. *Class* D.

Stratiotes (Classical Greek name for *Pistia stratiotes*, from *stratiotes*, soldier, perhaps alluding to the sword-shaped leaves.) A genus of 1 species of perennial, submerged, aquatic herb with rosettes of linear to sword-shaped leaves. The unisexual white flowers are borne above the water surface in the summer. The plants then sink, owing to an increase in the level of calcium carbonate in their tissues during the growing season. The production of new leaves in the spring brings them back to the surface again. *Distr.* Eurasia. *Fam.* Hydrocharitaceae. *Class* M.

S. aloides WATER SOLDIER. Cultivated ornamentally and where abundant used as manure.

× ***Stravinia*** See *Photinia*

strawbell *Uvularia sessilifolia*.

strawberry *Fragaria* **Alpine** ~ *F. vesca* **barren** ~ *Potentilla sterilis*, *Waldsteinia fragarioides* **beach** ~ *Fragaria chiloensis* **cultivated** ~ *F. x ananassa* **garden** ~ *F. x ananassa* **hautbois** ~ *F. moschata* **Indian** ~ *Duchesnea indica* **mock** ~ *D. indica* **Plymouth** ~ *Fragaria moschata* **scarlet** ~ *F. virginiana* **wild** ~ *F. vesca*.

strawberry bush *Euonymus americanus*.

strawberry shrub *Calycanthus floridus*.

strawberry tree *Arbutus*, *A. unedo*.

strawflower *Helichrysum bracteatum*.

Strelitzia (After Charlotte of Mecklenberg-Strelitz (1744–1818), wife of George III.) BIRD OF PARADISE. A genus of 4–5 species of evergreen herbs with 2 ranks of large leaves on long, grooved, sheathing stalks. The inflorescence is enclosed in a boat-shaped spathe from which flowers emerge one at a time. The flowers are irregular, often brightly coloured, and bear 3 sepals, 2 petals, and 5 stamens. All 4 species are grown as ornamental greenhouse plants. *Distr.* S Africa. *Fam.* Strelitziaceae. *Class* M.

S. reginae BIRD OF PARADISE FLOWER, CRANE LILY, CRANE FLOWER. Flowers bright orange and purple, pollinated by sunbirds. The most popular species. *Distr.* South Africa.

Strelitziaceae The Bird of Paradise family. 3 genera and 7 species of banana-like herbs and trees with 2-ranked leaves and irregular flowers that emerge from characteristic boat-shaped bracts. *Distr.* S Africa, Madagascar, South America. *Class* M.
See: Ravenala, Strelitzia.

Streptocarpus (From Greek *streptos*, twisted, and *karpos*, fruit, alluding to the spirally twisted fruits.) CAPE PRIMROSE. A genus of 120–130 species of annual and perennial herbs and subshrubs with simple hairy leaves and tubular, 5-lobed, 2-lipped flowers. Several species and numerous cultivars are grown ornamentally. *Distr.* Tropical and S Africa, Madagascar. *Fam.* Gesneriaceae. *Class* D.

S. caulescens Erect herb. Leaves small. Flowers white with violet stripes. *Distr.* Tanzania, Kenya.

S.* × *hybridus A complex group of rosette-forming hybrids with large purple or red flowers. *Distr.* Garden origin.

S. rexii CAPE PRIMROSE. Leaves in basal rosettes. Flowers 5cm long, funnel-shaped, pale-blue. *Distr.* South Africa.

S. saxorum FALSE AFRICAN VIOLET. Stems pendent. Leaves fleshy, rounded, woolly. Flowers lilac. *Distr.* E Africa.

streptocarpus, -a, -um: with twisted fruits.

streptopetalus, -a, -um: with twisted petals.

streptophyllus, -a, -um: with twisted leaves.

Streptopus (From Greek *streptos*, twisted, and *pous*, foot, alluding to the twisted or bent flower-stalks.) TWISTED STALK. A genus of 4–10 species of rhizomatous herbs with erect stems, clasping simple leaves, and nodding, bell- or star-shaped flowers. Several species are grown as ornamentals. *Distr.*

Temperate N hemisphere. *Fam.* Convallariaceae. *Class* M.

S. amplexifolius LIVER BERRY, SCOOT BERRY, WHITE MANDARIN. Leaves clasping. Flowers paired, green-white. *Distr.* N temperate regions.

S. roseus ROSE MANDARIN. Flowers solitary, rose-purple. *Distr.* NE North America.

Streptosolen (From Greek *streptos*, twisted, and *solen*, tube, alluding to the shape of the flowers.) A genus of 1 species of evergreen shrub with simple alternate leaves and corymbs of orange-red funnel-shaped flowers which have a long twisted petal-tube. *Distr.* Peru. *Fam.* Solanaceae. *Class* D.

S. jamesonii (After Dr William Jameson (1796–1873), professor of botany at Quito.) MARMALADE BUSH, ORANGE BROWALLIA, FIREBUSH. Grown as an ornamental.

striatus, -a, -um: striped.

strictus, -a, -um: upright.

strigillosus, -a, -um: with small bristles.

strigosus, -a, -um: with bristles.

stringbark, yellow *Eucalyptus muelleriana.*

string of beads *Senecio rowleyanus.*

string of hearts *Ceropegia linearis.*

strobiformis: cone-shaped.

strobilaceus, -a, -um: cone-like.

Strobilanthes (From Greek *strobilos*, cone, and *anthos*, flower, alluding to the dense inflorescence.) A genus of 250 species of perennial herbs and subshrubs with simple leaves and cone-like clusters of 2-lipped flowers. Several species are grown as ornamentals. *Distr.* Asia. *Fam.* Acanthaceae. *Class* D.

S. atropurpureus MEXICAN PETUNIA. Flowers indigo to deep purple. *Distr.* Siberia.

S. dyerianus (After Sir William Thiselton-Dyer (1843–1928), who introduced it to Kew.) PERSIAN SHIELD. Leaves flushed purple. Flowers pale blue. *Distr.* Burma.

strobiliformis, -e: cone-shaped.

Stromanthe (From Greek *stroma*, bed, and *anthos* flower, alluding to the inflorescence.) A genus of 13 species of perennial herbs with 2-ranked leaves and racemes or panicles of small irregular flowers, each cluster of flowers being subtended by a large folded or boat-shaped bract. Several species are cultivated as ornamentals. *Distr.* Tropical South America. *Fam.* Marantaceae. *Class* M.

S. sanguinea Leaves dark green above, purple beneath. Bracts red, flowers red. *Distr.* SE Brazil.

Stromatopteridaceae A family of 1 genus and 1 species of terrestrial ferns with erect, pinnate, leathery leaves. The sori usually occur singly on each leaf segment. *Distr.* New Caledonia. *Class* F.

strumosus, -a, -um: with cushion-like swellings.

Stuartia See *Stewartia*

Stylidiaceae A family of 6 genera and about 180 species of annual to perennial herbs and a few shrubs with simple alternate or whorled leaves and irregular flowers. *Distr.* Mainly Australasia, also tropical Asia and temperate South America. *Class* D.
See: Stylidium.

Stylidium (From Greek *stylos*, column, alluding to the fused style and stamens.) TRIGGER PLANT. A genus of 100–150 species of perennial herbs with simple grass-like leaves and irregular 2-lipped flowers. The stamens are spring-loaded so as to release pollen on to a visiting insect explosively. Several species are cultivated as ornamentals. *Distr.* SE Asia to New Zealand. *Fam.* Stylidiaceae. *Class* D.

S. graminifolium TRIGGER PLANT. To 60cm high. Flowers pink, borne in a spike-like raceme. *Distr.* Australia.

Stylobasiaceae A family of 1 genus and 2 species of shrubs and trees with simple, spirally arranged leaves and bisexual flowers that bear their parts in fives. This family is sometimes included within the family Surianaceae or Sapindaceae. *Distr.* W Australia. *Class* D.

Stylomecon (From Greek *stylos*, style, and *mekon*, poppy.) A genus of 1 species of annual herb with yellow latex, pinnately divided leaves, and fragrant orange-red flowers that bear 2 sepals, 4 petals, and numerous stamens. *Distr.* W North America. *Fam.* Papaveraceae. *Class* D.

S. heterophylla WIND POPPY, FLAMING POPPY. Grown as an ornamental.

Stylophorum (From Greek *stylos*, style, and *phoros*, bearing, alluding to the long style, a distinctive character of the genus.) A genus of 2–3 species of perennial herbs with red or yellow latex, basal rosettes of pinnately lobed leaves, and poppy-like flowers that bear 2 sepals, 4 petals, and many stamens. *Distr.* E Asia, 1 species E North America. *Fam.* Papaveraceae. *Class* D.

S. diphyllum CELANDINE POPPY. Downy perennial to 50cm high. Flowers yellow. Grown as an ornamental. *Distr.* E North America.

stylosus, -a, -um: with a prominent style.

Stypandra (From Greek *stypeion*, bundle of fibres, and *aner*, man, stamen, alluding to the downy stamens.) A genus of 6 species of rhizomatous herbs with 2-ranked grass-like leaves and loose cymes of blue, yellow, or white flowers that bear 6 free tepals and 6 downy stamens. *Distr.* Temperate Australia. *Fam.* Phormiaceae. *Class* M.

S. caespitosa Flowers blue. Grown as a tender ornamental. *Distr.* SE Australia.

Styphelia (From Greek *styphelos*, hard or rough, alluding to the sharp pointed leaves.) A genus of 12 species of evergreen shrubs with alternate pointed leaves and typically solitary, tubular flowers which bear their parts in fives. Several species are grown as ornamentals and some have edible fruits. *Distr.* Australia, Java, New Guinea. *Fam.* Epacridaceae. *Class* D.

S. viridis GREEN FIVECORNER. Shrub to 1.5m high. Flowers solitary, 2cm long, green, tinged yellow, tube 5-angled, petal lobes hairy. *Distr.* SW Australia.

Styracaceae A family of 11 genera and 170 species of shrubs and trees with simple alternate leaves and tubular flowers. Several species are grown as ornamentals, others are a source of gums and resins. *Distr.* Tropical and warm-temperate regions. *Class* D.
See: Halesia, Pterostyrax, Rehderodendron, Sinojackia, Styrax.

Styrax (Classical Greek name for *S. officinalis*.) SNOWBELLS, SILVERBALLS, STORAX. A genus of 100–120 species of evergreen and deciduous shrubs and trees with simple leaves and racemes or clusters of 5- or 10-lobed, cup- or bell-shaped flowers. The tropical species are a source of resin that is used medicinally and in incense. Several species are grown as ornamentals. *Distr.* Mediterranean, SE Asia, W Malaysia, tropical America. *Fam.* Styracaceae. *Class* D.

S. hemsleyana (After William Botting Hemsley (1843–1924), English botanist.) Tree to 10m high. *Distr.* China.

S. wilsonii (After Ernest Henry Wilson (1876–1930), who brought it into cultivation in 1908.) Shrub or small tree to 4m high. *Distr.* China (W Sichuan).

suaveolens: with a sweet scent.

subacaulis, -e: almost without a stem.

subalpinus, -a, -um: subalpine, growing just below alpine mountain regions.

subauriculatus, -a, -um: slightly eared.

subcordatus, -a, -um: slightly cordate.

subcrenulatus, -a, -um: slightly crenulate, with a round toothed margin.

suber: cork.

suberectus, -a, -um: almost upright.

suberosus, -a, -um: with corky bark.

subgibbosus, -a, -um: somewhat swollen on one side.

submammillaris: with small nipples.

suboppositus, -a, -um: almost opposite.

subpeltatus, -a, -um: almost peltate.

subrigidus, -a, -um: slightly rigid.

subsessilis, -e: almost sessile.

subsimilis: almost similar.

subsimplex: almost undivided.

subspinosus, -a, -um: somewhat spiny.

subtomentosus, -a, -um: somewhat hairy.

subulatus, -a, -um: awl-shaped.

Succisa (From Latin *succisus*, carved out underneath.) A genus of 1 species of perennial herb with rosettes of simple leaves and hemispherical heads of tubular, deep purple, or occasionally white flowers. *Distr.* Europe to W Siberia, N Africa. *Fam.* Dipsacaceae. *Class* D.

S. pratensis DEVIL'S BIT SCABIOUS, BLUE BUTTONS. Occasionally grown as an ornamental; reported to have medicinal properties.

succisus, -a, -um: cut off below.

succory *Cichorium intybus*.

suffrutescens: slightly shrubby.

suffruticosus, -a, -um: slightly shrubby.

sugarberry *Celtis*.

sugar cane *Saccharum*, *S. officinarum*.

sugar scoop *Tiarella*.

sulcatus, -a, -um: sulcate, with grooves or furrows.

sulfureus, -a, -um: sulphur-yellow-coloured.

sulphur flower *Eriogonum umbellatum*.

sultan's flower *Impatiens walleriana*.

sultana *Impatiens*, *I. walleriana*.

sumac *Rhus* **dwarf** ~ *R. copallina* **fragrant** ~ *R. aromatica* **Hungarian** ~ *Cotinus coggygria* **Indian** ~ *C. coggygria* **lemon** ~ *Rhus aromatica* **lemonade** ~ *R. integrifolia* **mountain** ~ *R. copallina* **scarlet** ~ *R. glabra* **shining** ~ *R. copallina* **smooth** ~ *R. glabra* **stag's horn** ~ *R. typhina* **Turkish** ~ *Cotinus coggygria* **Tyrolean** ~ *C. coggygria* **velvet** ~ *Rhus typhina* **Venetian** ~ *Cotinus coggygria* **Virginian** ~ *Rhus typhina*.

sumach *Rhus*.

sumatranus, -a, -um: of Sumatra.

summer sweet *Clethra*.

sundew Droseraceae, *Drosera* **Cape** ~ *D. capensis* **English** ~ *D. anglica* **great** ~ *D. anglica* **round-leaved** ~ *D. rotundifolia* **spoon-leaf** ~ *D. spathulata*.

sundrops *Oenothera fruticosa*.

sunflower *Helianthus*, *H. annuus* **dark-eyed** ~ *H. atrorubens* **Mexican** ~ *Tithonia*, *T. rotundifolia* ~ **thin-leaf** *Helianthus decapetalus* **woolly** ~ *Eriophyllum*.

sun plant *Portulaca grandiflora*.

sun rose *Helianthemum*.

superbus, -a, -um: superb.

supinus, -a, -um: prostrate, low-growing.

supranubius: above the clouds.

surculosus, -a, -um: with suckers.

Surianaceae A family of 3 genera and 4 species of shrubs and trees with simple, spirally arranged leaves and clusters of bisexual flowers that bear their parts in fives. *Distr.* Widespread in tropical regions. *Class* D.

susianus, -a, -um: of Shush (Susa), Iran.

suspensus, -a, -um: hanging.

sutchuenensis, -e: of Sichuan (Szechwan).

Sutera (After Johann Rudolf Suter (1766–1827), Swiss botanist and professor at Berne.) A genus of 130 species of annual and perennial herbs and shrubs with simple or lobed opposite leaves and tubular 5-lobed Swiss botanist and flowers which are occasionally 2-lipped. *Distr.* Macaronesia, tropical and South Africa. *Fam.* Scrophulariaceae. *Class* D.

S. grandiflora PURPLE GLORY PLANT. Perennial herb. Flowers lavender to deep purple, borne in racemes to 30cm long. Grown as an ornamental. *Distr.* S Africa.

Sutherlandia (After James Sutherland (died 1719), superintendent of the Edinburgh Botanic Garden.) A genus of 5 species of shrubs with alternate pinnate leaves and slender racemes of red or purple pea-like flowers that are followed by bladder-like pods that float in water. *Distr.* S Africa. *Fam.* Leguminosae. *Class* D.

S. frutescens BALLOON PEA, DUCK PLANT. Grown as an ornamental. *Distr.* S Africa.

swallow wort *Asclepias curassavica*, *Chelidonium majus*, *Vincetoxicum*.

swan flower *Aristolochia grandiflora.*

Swan plant *Asclepias physocarpa.*

swede *Brassica napus.*

sweet cup *Passiflora maliformis.*

sweet gum *Liquidambar styraciflua* **American** ~ *L. styraciflua* **Oriental** ~ *L. orientalis.*

sweetleaf, Asiatic *Symplocos paniculata.*

sweetsop *Annona squamosa.*

sweetspire *Itea virginica.*

sweet William *Dianthus barbatus.*

sweetwood *Glycyrrhiza glabra.*

Swietenia (After Gerard van Swieten (1700–22), Dutch botanist.) A genus of 3 species of large trees with pinnate leaves. All three species are exploited for their fine timber. *Distr.* Tropical America. *Fam.* Meliaceae. *Class* D.

S. mahogoni MAHOGANY. A very important timber tree used in cabinet-making and shipbuilding. The natural populations have been depleted through genetic erosion, the felling of the good trees leaving inferior ones to provide seed for the next generation. *Distr.* Tropical S and Central America.

Swiss cheese plant *Monstera deliciosa.*

sword plant *Echinodorus.*

Syagrus (Latin name used by Pliny for a palm.) A genus of 32–34 species of medium-sized palms with feather-shaped leaves that appear to be arranged in 3 ranks and solitary spike-like inflorescences. Several species are grown as tender ornamentals. *Distr.* Tropical America. *Fam.* Palmae. *Class* M.

S. coronata The fruits are a source of palm oil. *Distr.* Brazil.

sycamore *Acer pseudoplatanus* **American** ~ *Platanus occidentalis.*

×*Sycoparrotia* (From the names of the parent genera.) A genus of hybrids between members of the genera *Sycopsis* and *Parrotia*. *Distr.* Garden origin. *Fam.* Hamamelidaceae. *Class* D.

×*S. semidecidua* A hybrid of *P. persica* x *S. sinensis*. Semi-evergreen shrub with leathery leaves and clusters of small woolly flowers. Grown as an ornamental. *Distr.* Garden origin.

Sycopsis (From Greek *sykon*, fig, and *-opsis*, indicating resemblance.) A genus of 7 species of evergreen shrubs and small trees with simple alternate leaves and clusters of inconspicuous male and bisexual flowers that lack petals. *Distr.* China, Himalaya, Central Malaysia, New Guinea. *Fam.* Hamamelidaceae. *Class* D.

S. sinensis Grown as an ornamental. *Distr.* W and Central China.

S. tutcheri See *Distylium racemosum*

sylvaticus, -a, -um: of woods.

Symphoricarpos (From Greek *symphorein*, to bear together, and *karpos*, fruit, alluding to the clustered fruits.) SNOWBERRY, CORALBERRY. A genus of 17 species of deciduous suckering shrubs with simple opposite leaves and small regular flowers that are borne singly or in racemes. The fruits are spherical and fleshy. Several species are grown as ornamentals. *Distr.* North America, 1 species in China. *Fam.* Caprifoliaceae. *Class* D.

S. albus SNOWBERRY, COMMON SNOWBERRY, WAXBERRY. Fruit *c.*1cm across, white. *Distr.* E North America.

S. × chenaultii A hybrid of *S. microphyllus* × *S. orbiculatus*. Fruit red with white markings. *Distr.* Garden origin.

S. occidentalis Fruit white, tinged green. *Distr.* W North America.

S. orbiculatus INDIAN CURRANT, CORALBERRY. Fruit white, turning red. *Distr.* North America.

Symphyandra (From Greek *symphyo*, to grow together, and *aner*, anther, alluding to the fused anthers.) RING BELLFLOWER. A genus of 12 species of perennial herbs with simple leaves and racemes or panicles of bell-shaped flowers. This genus is very closely related to the genus *Campanula* but differs in that the anthers are fused into a tube. Several species are grown as ornamentals, chiefly in rock gardens. *Distr.* E Mediterranean to Central Asia. *Fam.* Campanulaceae. *Class* D.

S. armena Stems erect or spreading. Flowers borne in panicles, white or blue, erect or pendulous. *Distr.* Caucasus.

S. pendula Stems arching. Flowers pendent, cream-white. *Distr.* Caucasus.

S. wanneri Clump-forming. Flowers pendent, purple-blue. *Distr.* Alps.

Symphytum (From Greek *symphytis*, growing together of bones, and *phyton*, plant; it was reputed to heal broken bones.) COMFREY. A genus of 35 species of perennial, often hairy herbs with simple, opposite or basal leaves, and curved cymes of pendent, tubular to club-shaped flowers. Several species and hybrids are grown as ornamentals and as forage crops. *Distr.* Europe, Mediterranean to Caucasus. *Fam.* Boraginaceae. *Class* D.

S. asperum PRICKLY COMFREY. To 1.5m high. Flowers pink, turning blue with age. Cultivated as a forage crop. *Distr.* Europe, Caucasus, Iran.

S. caucasicum To 60cm high. Leaves rough-hairy. Flowers blue. *Distr.* Caucasus, Iran.

S. grandiflorum Creeping. Stems horizontal to ascending. Flowers cream. *Distr.* Caucasus.

S. ibericum See *S. grandiflorum*

S. officinale Stems winged, to 1.2m high. Leaves large. Flowers white to purple-blue. Young shoots eaten like asparagus. Used medicinally to stop bleeding. *Distr.* Eurasia, naturalized in North America.

S. peregrinum See *S.* × *uplandicum*

S. tuberosum Rhizome tuberous. Flowers pale yellow. *Distr.* Europe, W Asia.

S.* × *uplandicum RUSSIAN COMFREY, BLUE COMFREY. A hybrid of *S. officinale* × *S. asperum*. Stems to 2m high. Flowers pink, becoming purple. Grown as a fodder crop. *Distr.* Caucasus, also produced artificially, widely naturalized.

Symplocaceae A family of 1 genus and 250 species. *Class* D.
See: Symplocos.

Symplocarpus (From Greek *symploke*, connection, and *karpos*, fruit; the ovaries combine to make a single fruit.) A genus of 1 species of rhizomatous herb with large clumps of soft, musk-scented leaves. The small bisexual flowers are borne on a short barrel-shaped spadix which is surrounded by a large, shoe-shaped, yellow-green spathe. The inflorescence usually appears before the leaves in spring and uses solar radiation to raise the temperature of the spadix to well above the ambient air temperature. *Distr.* NE Asia, NE America. *Fam.* Araceae. *Class* M.

S. foetidus SKUNK CABBAGE, POLECAT WEED. An interesting, if somewhat malodorous, ornamental for the garden.

Symplocos (From Greek *symploke*, connection; the stamens are fused together.) A genus of 250 species of evergreen and deciduous shrubs and trees with simple, alternate, yellow-green leaves and racemes or clusters of 5-lobed flowers that are followed by blue fruits. Some species are a source of dye. *Distr.* Tropical and warm Asia, America, and New Caledonia. *Fam.* Symplocaceae. *Class* D.

S. paniculata SAPPHIRE BERRY, ASIATIC SWEETLEAF. Deciduous shrub or small tree to 5m high. Flowers fragrant. Grown as a hardy ornamental. *Distr.* Himalaya, China, Japan.

Syngonium (From Greek *syn*, together, and *gone*, womb, alluding to the fused ovaries.) A genus of about 33 species of epiphytic or terrestrial, perennial herbs with simple juvenile and lobed adult leaves. The small unisexual flowers are borne on spadices that are subtended by a spathe and carried in clusters in the axils of the leaves. Several species are grown as pot plants in their juvenile stage but rarely have enough room to reach maturity in the greenhouse or home. *Distr.* Tropical regions of the New World. *Fam.* Araceae. *Class* M.

S. angustatum ARROWHEAD VINE. Adult leaves 5–11-lobed. Spathes green. *Distr.* Mexico to Costa Rica.

S. auritum FIVE FINGERS. Adult leaves 5-lobed, each lobe bearing 2 auricles at the base. *Distr.* West Indies.

S. podophyllum ARROWHEAD VINE. Perhaps the most commonly cultivated species with numerous cultivars available. *Distr.* Mexico to Guyana, Brazil and Bolivia.

Synnotia See *Sparaxis*

Synthyris (From Greek *syn*, with, and *thyris*, window, alluding to the valves of the capsule.) A genus of 14–15 species of tufted rhizomatous herbs with heart- or kidney-shaped leaves and racemes of 4-lobed, more or less bell-shaped flowers. Several species are grown as ornamentals. *Distr.* W North America. *Fam.* Scrophulariaceae. *Class* D.

S. missurica Leaves leathery, bluntly toothed. Flowers bright purple-blue. *Distr.* N North America.

S. reniformis SNOW QUEEN, ROUND-LEAVED SYNTHYRIS. Leaves rounded. Flowers blue. *Distr.* W North America.

synthyris, round-leaved *Synthyris reniformis.*

syriacus: of Syria.

Syringa (From Greek *syrinx*, pipe, alluding to the hollow stems.) LILAC. A genus of 20–25 species of deciduous, occasionally evergreen shrubs and small trees with opposite, simple to pinnate leaves and showy panicles of 4-lobed, tubular, highly fragrant flowers. Several species and numerous hybrids and cultivars are grown as ornamentals. *Distr.* SE Europe to E Asia. *Fam.* Oleaceae. *Class* D.

S. emodi HIMALAYAN LILAC. Vigorous, deciduous, arching shrub. Flowers deep pink. *Distr.* Afghanistan, Himalayas.

S. meyeri (After F. N. Meyer, who introduced it to America in 1908.) DWARF LILAC. Compact shrub. Flowers purple. Often seen in cultivation as a dwarf. *Distr.* N China.

S. palibiniana See *S. meyeri*

S. reticulata JAPANESE LILAC, JAPANESE TREE LILAC. Broad conical shrub or tree. Flowers creamy white. *Distr.* E Asia.

S. vulgaris COMMON LILAC, PIPE TREE. Small tree. Stems formerly used as pipes. Numerous ornamental cultivars have been raised from this species. *Distr.* SE Europe.

S. yunnanensis Deciduous upright shrub. Flowers pink. *Distr.* China (Yunnan).

syringanthus, -a, -um: with flowers resembling those of the genus *Syringa*.

Syringodea (From Greek *syringodes*, like a small pipe, alluding to the slender perianth tube.) A genus of 7–8 species of small perennial herbs with thread-like leaves and 1–2 funnel-shaped flowers surrounded by a pair of spathes. Some of the species are cultivated as ornamentals. *Distr.* South Africa. *Fam.* Iridaceae. *Class* M.

S. pulchella To 15cm high. Flowers pale purple. *Distr.* South Africa.

szechuanicus, -a, -um: of Szechwan.

T

tabularis, -e: table-like, flat.

tabuliformis, -e: table-shaped, flat-topped.

Tacca (From *taka*, the Indonesian name.) A genus of 10 species of rhizomatous herbs with rosettes of entire or lobed leaves and umbels of nodding, typically bell-shaped flowers. The tubers are a source of starch used for bread-making once bitter compounds have been removed. Several species are grown as ornamentals. *Distr*. Tropical regions of the Old World. *Fam*. Taccaceae. *Class* M.

T. chantrieri (After Chantrier Frères, French nurserymen.) CAT'S WHISKERS, DEVIL FLOWER, BAT FLOWER. Similar to *T. integrifolia* but smaller *Distr*. SE Asia.

T. integrifolia BAT PLANT, BAT FLOWER. Inflorescence subtended by 2 large, dark bracts. *Distr*. NE India to S China and Java.

T. leontopetaloides EAST INDIAN ARROWROOT, TAHITI ARROWROOT. Leaves used in hat-making. *Distr*. Tropical regions of the Old World.

Taccaceae A family of 1 genus and 10 species. *Class* M.
See: Tacca.

Tacitus See *Graptopetalum*

Tagetes (From Tages, an Etruscan deity, the grandson of Jupiter, who sprang from the ploughed earth.) MARIGOLD. A genus of about 50 species of annual and perennial foetid herbs with pinnately lobed, glandular leaves and flowers in daisy-like heads with distinct rays. Some species and a very large number of cultivars are grown as ornamentals. *Distr*. Mexico, Central America, Africa. *Fam*. Compositae. *Class* D.

T. erecta AFRICAN MARIGOLD, AZTEC MARIGOLD, BIG MARIGOLD. Robust annual, to 1.5m high. Rays few, yellow to orange. This species has given rise to a very large number of very different cultivars with different coloured flowers and numbers of flower parts. *Distr*. Mexico and Central America.

T. lucida SWEET MACE, SWEET-SCENTED MARIGOLD, SWEET-SCENTED MEXICAN MARIGOLD. Glabrous perennial. Rays few, yellow. A hallucinogen. *Distr*. Mexico, Guatemala.

T. patula FRENCH MARIGOLD. Glabrous annual, to 50cm high. This species has given rise to a large number of cultivars. *Distr*. Mexico to Guatemala.

tail flower *Anthurium, A. andreanum.*

tailwort *Borago officinalis.*

taiwanensis, -e: of Taiwan.

Taiwania (After Taiwan.) A genus of 1 species of evergreen coniferous tree to 55m high with long, stout, juvenile leaves, small scale-like adult leaves and small fruiting cones. *Distr*. Yunnan, Taiwan. *Fam*. Taxodiaceae. *Class* G.

T. cryptomerioides Grown as an ornamental.

taliensis, -e: of the Tali mountains, Yunnan.

Talinum (Derivation obscure, possibly from an African vernacular name.) FAME-FLOWER. A genus of 50 species of succulent annual and perennial herbs with simple leaves and short-lived showy flowers that bear 2 sepals and typically 5 petals. Some species are cultivated as ornamentals. *Distr*. Tropical and subtropical regions. *Fam*. Portulacaceae. *Class* D.

T. spinescens Low cushion-forming perennial. Flowers pink to magenta. *Distr*. W North America.

tallow, Japanese *Rhus succedanea.*

tallow tree *Sapium.*

tamarack *Larix laricina.*

Tamaricaceae The Tamarisk family. 5 genera and 87 species of heath-like shrubs and small trees with reduced scale-like leaves and flowers that bear 4–5 sepals and petals and 5 to many stamens. Some species are

grown as ornamentals, others yield dyes. *Distr.* Eurasia and Africa, mostly in warm-temperate regions. *Class* D.
See: Myricaria, Tamarix.

tamarilla *Cyphomandra crassicaulis.*

tamarisk Tamaricaceae, *Tamarix*
false ~ *Myricaria.*

Tamarix (Classical Latin name for these plants.) TAMARISK, SALT CEDAR. A genus of 54 species of deciduous shrubs and small trees with small scale- or needle-like leaves and racemes of 4- or 5-lobed flowers that typically bear 4–5 free stamens. Grown as ornamentals and for the production of galls that were formerly used in the tanning industry. *Distr.* Eurasia, Africa. *Fam.* Tamaricaceae. *Class* D.

T. gallica FRENCH TREE, MANNA PLANT. Small tree to 10m high. Galls used in leather tanning. *Distr.* W Europe, N Africa.

T. germanica See *Myricaria germanica*

T. parviflora See *Myricaria germanica*

T. pentandra See *T. ramosissima*

T. ramosissima Shrub or small tree to 6m high. *Distr.* Central Asia to China.

tampala *Amaranthus tricolor.*

Tamus (From Latin *tammus* a climbing plant, probably T. *communis.*) A genus of 4–5 species of tuberous herbs with annual twining stems, heart-shaped leaves, and racemes of small yellow-green flowers. *Distr.* Europe, Mediterranean, N Africa and SW Asia. *Fam.* Dioscoreaceae. *Class* M.

T. communis BLACK BRYONY. Reported to have medicinal properties. *Distr.* W and S Europe, Mediterranean, N Africa, SW Asia.

tanacetifolius, -a, -um: from the genus *Tanacetum*, and *folius*, leaved.

Tanacetum (From mediaeval Latin *tanazeta.*) A genus of about 70 species of annual to perennial aromatic herbs with simple to lobed leaves and flowers in daisy-like heads that may or may not bear distinct rays. Several species are cultivated as ornamentals. *Distr.* N temperate regions especially in the Old World. *Fam.* Compositae. *Class* D.

T. argenteum Perennial. Roots woody. Rays white. *Distr.* Turkey.

T. balsamita ALECOST, COSTMARY, MINT GERANIUM. Aromatic rhizomatous perennial. Rays white. *Distr.* Europe to E Asia.

T. cinerariifolium PYRETHRUM, DALMATIA PYRETHRUM. Glaucous perennial. Flowerheads are the source of the insecticide pyrethrum, cultivated especially in the highlands of E Africa. *Distr.* Balkans.

T. coccineum GARDEN PYRETHRUM, PYRETHRUM, PAINTED DAISY. Glabrous perennial. Rays white to red. Numerous cultivars are available. *Distr.* Caucasus, Iran.

T. corymbosum Perennial, to 1m high. Rays white. *Distr.* Europe.

T. densum Erect subshrub. Rays yellow. *Distr.* W Asia.

T. haradjanii Subshrub, to 30cm high. Rays absent. *Distr.* Turkey.

T. macrophyllum Perennial. Flowerheads numerous, rays small, white. *Distr.* Central Europe, S Russia.

T. parthenium (From *parthenion*, the Greek name for an unknown plant.) BACHELOR'S BUTTONS, FEVERFEW. Aromatic perennial. Rays white. Dried capitula are used medicinally and as tea. A number of ornamental cultivars are available. *Distr.* SE Europe to Caucasus.

T. praeteritum Subshrub. Leaves grey-hairy. Rays white. *Distr.* Turkey.

T. ptarmiciflorum DUSTY MILLER, SILVER LACE. Subshrub. Leaves silver-grey-hairy. Rays white. *Distr.* Canary Islands.

T. vulgare TANSY, GOLDEN BUTTONS. Aromatic perennial. Rays absent. This species is used medicinally as a tonic tea and in cooking. *Distr.* Europe, Asia/ Eurasia.

Tanakaea (After Yoshio Tanaka (1836–1916), Japanese botanist.) A genus of 1 species of evergreen perennial herb with long-stalked leaves and dense panicles of tiny white flowers. *Distr.* E Asia. *Fam.* Saxifragaceae. *Class* D.

T. radicans JAPANESE FOAM FLOWER. Cultivated ornamentally.

tanekaha *Phyllocladus trichomanoides.*

tangarine *Citrus reticulata.*

tangelo *C.* × *tangelo.*

tannia *Xanthosoma X. sagittifolium.*

tansy *Tanacetum vulgare* **goose** ~ *Potentilla anserina.*

Tapeinochilos (From Greek *tapeinos*, low, and *cheilos*, lip, alluding to the small lip of the flowers.) A genus of about 15 species of clump-forming rhizomatous herbs with spirally arranged, somewhat fleshy leaves. The inflorescences bear bright red overlapping bracts that more or less conceal the yellow-brown flowers. *Distr.* Malaysia to Australia. *Fam.* Costaceae. *Class* M.

T. ananassae Inflorescence resembles a red pineapple. Sometimes grown as a tender ornamental. *Distr.* Indonesia.

tarajo *Ilex latifolia.*

Taraxacum (From Persian *talkh chakok*, bitter herb, through medieval Latin *tarasacon.*) DANDELION, BLOWBALLS. A genus of 60 species of biennial and perennial herbs with stout taproots, milky latex, and a basal rosette of simple to lobed leaves. The numerous small flowers are borne in a dense head that resembles a single large flower; they are not differentiated into distinct rays and fertile flowers like a daisy (see *Bellis*). Owing to the nature of the breeding system in these plants, many of the species are very variable and sometimes split into small units termed microspecies. Most species are considered weeds although a few are cultivated. *Distr.* N temperate regions and temperate South America. *Fam.* Compositae. *Class* D.

T. officinale COMMON DANDELION. Perennial herb. Leaves lobed. Flowers deep yellow. Fruits forming a rounded, white-haired seed-head known as a dandelion clock. Although this plant is familiar to most people who live in temperate regions of the N hemisphere, it is very variable and there is still some doubt as to its correct scientific name. *T. officinale* is thought to be technically incorrect. A persistant weed of cultivation, especially in lawns, this plant has a number of uses; the leaves are eaten as a salad vegetable, the taproots roasted and ground as a coffee substitute, the immature flower-heads are eaten like capers, and the mature flowers are made into wine. *Distr.* Widespread in the N hemisphere.

tardiflorus, -a, -um: late-flowering.

tardivus, -a, -um: late.

tardus, -a, -um: late.

tare *Vicia*, *V. sativa.*

taro *Colocasia esculenta* **giant ~** *Alocasia macrorrhiza* **Imperial ~** *Colocasia esculenta.*

taronensis, -e: of the Taron gorge, Yunnan.

tarragon *Artemisia dracunculus.*

tarweed *Grindelia.*

tarwood *Halocarpus bidwillii.*

tasmanicus, -a, -um: of Tasmania.

Tasmannia See *Drimys*

tassel flower *Emilia coccinea*, *Amaranthus caudatus.*

tassel tree *Garrya.*

tataricus, -a, -um: of central Asia, formerly Tartary.

tauricus, -a, -um: of the Crimea.

Taxaceae The Yew family. 5 genera and 18 species of evergreen coniferous shrubs and trees with simple, linear, spirally arranged leaves. The male and female cones are borne on separate plants. The female cone consists of a solitary ovule which develops a fleshy aril at maturity. Members of the family are valued for their timber and as ornamentals. *Distr.* N hemisphere to Malaysia and New Caledonia. *Class* G.
See: Taxus, Torreya.

taxifolius, -a, -um: from the genus *Taxus*, and *folius*, leaved.

Taxodiaceae A family of 10 genera and 13 species of resinous, coniferous trees with linear to scale-like, evergreen or deciduous leaves. *Distr.* E Asia, Tasmania, North America. *Class* G.
See: Athrotaxis, Cryptomeria, Cunninghamia, Glyptostrobus, Metasequoia, Sciadopitys, Sequoia, Sequoiadendron, Taiwania, Taxodium.

Taxodium (From the genus name *Taxus*, and Greek *oidos*, resemblance.) SWAMP CYPRESS, BALD CYPRESS. A genus of 2–3 species of deciduous, coniferous trees with alternate linear leaves that are borne in 2 ranks and small round fruiting cones that disintegrate to release their seeds. *Distr.* North America, Mexico. *Fam.* Taxodiaceae. *Class* G.

T. distichum SWAMP CYPRESS, SOUTHERN CYPRESS, BALD CYPRESS. Tree 45m high. A source of rot-resistant timber and an ornamental. *Distr.* S and SE USA.

Taxus (Classical Latin name for these plants.) YEW. A genus of 3–10 species of evergreen coniferous trees with flat linear leaves that are borne spirally but twisted so as to be arranged in 2 ranks. The female cones consists of a single ovule that develops a fleshy red aril. *Distr.* North America, Europe, E Asia. *Fam.* Taxaceae. *Class* G.

T. baccata COMMON YEW, ENGLISH YEW, YEW. Slow-growing tree to 20m high. Grown as an ornamental and a source of timber. *Distr.* Mediterranean, Europe, and N Africa E to Iran.

tazetta: an Italian name meaning a small cup.

tea *Camellia sinensis* **Brazilian** ~ *Stachytarpheta cayennensis* **mountain** ~ *Gaultheria procumbens* **New Jersey** ~ *Ceanothus americanus* **Oswego** ~ *Monarda didyma* **sweet** ~ *Osmanthus fragrans*.

teaberry *Gaultheria procumbens*.

teak Verbenaceae **bastard** ~ *Butia monosperma*.

teasel *Dipsacus*, Dipsacaceae **common** ~ *Dipsacus fullonum* **fuller's** ~ *D. sativus* **small** ~ *D. pilosus*.

tea tree *Leptospermum*, *L. scoparium* **Duke of Argyll's** ~ *Lycium barbarum*.

Tecoma (From the Mexican name for these plants, *tecomaxochitl*.) TRUMPET BUSH, YELLOW BELLS. A genus of 12 species of shrubs and trees with pinnate leaves and racemes or panicles of yellow or orange, bell- or funnel-shaped flowers. Several species are grown as ornamentals in frost-free areas. *Distr.* Tropical America especially the Andes, S. Africa. *Fam.* Bignoniaceae. *Class* D.

T. capensis CAPE HONEYSUCKLE. Scrambling shrub. Flowers red. *Distr.* Tropical America.

T. ricasoliana See *Podranea ricasoliana*

T. stans YELLOW ELDER, YELLOW BELLS. Flowers bright yellow. Used locally as a diuretic. *Distr.* SE USA, Central and South America.

Tecomanthe (From the genus name *Tecoma*, and Greek *anthe*, flower.) A genus of 5 species of evergreen woody climbers with pinnate leaves and pendent racemes of pink funnel-shaped flowers. Several species are grown as ornamentals in frost-free areas. *Distr.* Malaysia, Australia, New Zealand. *Fam.* Bignoniaceae. *Class* D.

T. speciosa Flowers cream-yellow, borne in dense clusters. Very rare in the wild. *Distr.* New Zealand (Three Kings Island).

Tecomaria See *Tecoma*

T. capensis See *Tecoma capensis*

T. nyassae See *Tecoma capensis*

Tecophilaea (After Tecofila Billioti, 19th-century Italian botanical artist.) A genus of 2 species of perennial herbs with spreading leaves and solitary blue flowers. *Distr.* Chile (Andes). *Fam.* Tecophilaeaceae. *Class* M.

T. cyanocrocus CHILEAN CROCUS. Flowers royal blue, tinted white in the throat. *Distr.* Chile (Andes).

Tecophilaeaceae A family of 6 genera and 19 species of perennial herbs with rhizomes or corms. *Distr.* Africa, South America to California. *Class* M.
See: Cyanella, Tecophilaea.

Tectaria (From Latin *tectum*, roof, alluding to the indusia.) A genus of 200 species of terrestrial and epiphytic ferns with simple to tripinnately divided leaves. Some are cultivated as ornamentals. *Distr.* Damp tropical regions. *Fam.* Dryopteridaceae. *Class* F.

T. cicutaria BUTTON FERN. Leaves to 60cm long. Grown as an ornamenttal. *Distr.* West Indies.

tectorum: growing on house roofs.

tectus, -a, -um: covered.

t'ef *Eragrostis tef*.

teff *Eragrostis tef*.

Telanthophora (From Greek *telos*, end, *anthera*, anther, and *phoros*, bearing.) A genus of 14 species of shrubs and trees with pinnately nerved leaves and yellow flowers in daisy-like heads that may or may not have rays. *Distr.* Central America. *Fam.* Compositae. *Class* D.

T. grandifolia Shrub, to 2m high. Grown as a tender ornamental. *Distr.* Mexico.

Telekia (After Samuel Teleki di Szék (fl.1816), Hungarian nobleman and botanical patron.) A genus of 2 species of coarse perennial herbs with simple leaves and yellow flowers borne in daisy-like heads that are arranged in loose racemes. Several species are cultivated as ornamentals. *Distr.* Central Europe to Caucasus. *Fam.* Compositae. *Class* D.

T. speciosa Strongly scented. To 2m high. *Distr.* SE Europe to S Russia, Caucasus, W Asia.

Telesonix See *Boykinia*

T. jamesii See *Boykinia jamesii*

Teline See *Genista*

Tellima (Anagram of the genus name *Mitella*.) A genus of 1 species of perennial glandular-hairy herb with racemes of small white flowers. *Distr.* W North America. *Fam.* Saxifragaceae. *Class* D.

T. grandiflora FRINGE CUPS. Cultivated as an ornamentals ground cover.

Telopea (From Greek *telopos*, seen from afar, alluding to the showy flowers.) A genus of 3–4 species of evergreen shrubs with alternate leathery leaves and dense heads of bright red flowers. Several species are cultivated as tender ornamentals. *Distr.* E Australia. *Fam.* Proteaceae. *Class* D.

T. speciosissima WARATAH. Erect straggling shrub. Flowers red, borne in a head that is surrounded by red bracts. *Distr.* Australia (New South Wales).

T. truncata TASMANIAN WARATAH. Bushy shrub. Flowers red, borne in heads that are surrounded by green bracts. *Distr.* Australia (Tasmania).

temperate *Hedychium densiflorum*.

temple tree *Plumeria*.

temu *Luma apiculata*.

ten commandments *Maranta leuconeura*.

tenax: tough, matted.

tenellus, -a, -um: dainty, delicate.

tenerifus, -a, -um: of Tenerife.

tenerus, -a, -um: tender, delicate.

tenuicaulis, -e: slender-stemmed.

tenuiflorus, -a, -um: slender-flowered.

tenuifolius, -a, -um: slender-leaved.

tenuis, -e: slender.

tenuisectus, -a, -um: with slender divisions.

tenuissimus, -a, -um: very slender.

tepa *Laurelia sempervirens*.

tephropeplum: grey-cloaked.

Tepuianthaceae A family of 1 genus and 5 species of shrubs and trees with simple, opposite or alternate leaves and regular flowers that bear their parts in fives. *Distr.* Colombia, Venezuela, N Brazil. *Class* D.

teretifolius, -a, -um: with round leaves in cross-section.

Terminalia (From Latin *terminus*, end, alluding to the leaves which are often borne in terminal clusters.) A genus of 150–200 species of large trees with simple, often glandular leaves and spikes of bell-shaped flowers. The trunks are often buttressed and the branches are often arranged in a pagoda shape. These trees are grown as ornamentals and as a source of timber, dyes, tannin, and gums. *Distr.* Tropical regions. *Fam.* Combretaceae. *Class* D.

T. catappa WILD ALMOND, INDIAN ALMOND, BARBADOS ALMOND. Planted as a street tree as well as for timber, edible seeds, and leaves which are fed to silkworms. *Distr.* Malay Peninsula.

terminalis, -e: terminal.

terniflorus, -a, -um: with flowers in threes.

ternifolius, -a, -um: with leaves in threes.

Ternstroemia (After Christopher Ternstroem (died 1748), Swedish naturalist and traveller in China.) A genus of 85 species of evergreen shrubs and trees with simple leathery leaves and solitary or clustered flowers that bear 5 petals and numerous stamens. *Distr.* Tropical regions. *Fam.* Theaceae. *Class* D.

T. japonica Shrub or small tree. Flowers white. Occasionally grown as a tender ornamental. *Distr.* S Japan.

terrestris, -e: of the ground.

tessellatus, -a, -um: tessellated, with regular squared venation.

tetracanthus, -a, -um: four-spined.

Tetracentraceae A family of 1 genus and 1 species. *Class* D.
See: Tetracentron.

Tetracentron (From Greek *tetra*, four, and *kentron*, spur, alluding to the four appendages on the fruit.) A genus of 1 species of deciduous shrub or tree to 30m high with simple leaves and pendulous spikes of small yellow flowers. *Distr.* NE India, Nepal, N Burma, SW and Central China. *Fam.* Tetracentraceae. *Class* D.

T. sinense Grown as a tender ornamental.

Tetrachondraceae A family of 1 genus and 2 species of small creeping herbs with simple opposite leaves and regular, solitary, 4-lobed flowers. *Distr.* New Zealand, S Chile, S Argentina. *Class* D.

Tetradium (From Greek *tetradion*, quarter, alluding to the number of floral parts.) A genus of 9 species of evergreen and deciduous shrubs and trees with opposite pinnate leaves and corymbs or panicles of flowers that typically bear their parts in fours. Several species are grown as ornamentals. *Distr.* Himalaya to Japan, Sumatra and Java. *Fam.* Rutaceae. *Class* D.

T. daniellii Shrub or tree to 20m high. Flowers white. *Distr.* N China, Korea.

Tetragonia (From Greek *tetra*, four, and *gonia*, angle, alluding to the shape of the fruits.) A genus of 50–60 species of somewhat fleshy herbs with long leaves and small yellow flowers. *Distr.* E Asia, Africa, Australia, New Zealand and temperate S America. *Fam.* Aizoaceae. *Class* D.

T. tetragonioides NEW ZEALAND SPINACH. Grown as a substitute for spinach, *Spinacea oleracea*, in arid areas. *Distr.* Japan and Pacific to New Zealand and temperate S America.

tetragonioides from the genus *Tetragonia*, with the ending *-oides*, indicating resemblance.

Tetragonolobus See *Lotus*

tetragonus, -a, -um: four-angled.

Tetrameristaceae A family of 2 genera and 2 species of shrubs and trees with simple, spirally arranged leaves and racemes of bisexual flowers that bear their parts in fours or fives. *Distr.* SE Asia, South America (Guyana Highlands). *Class* D.

tetramerus, -a, -um: with floral parts in fours.

tetrandrus, -a, -um: with four stamens.

Tetrapanax (From Greek *tetra*, four, and the genus name *Panax*.) A genus of 1 species of large shrub or small tree with large, palmately lobed leaves and small round flowers that are borne in stalked umbels. All the aerial parts are typically coated in flossy hairs. *Distr.* S China, Taiwan. *Fam.* Araliaceae. *Class* D.

T. papyriferus RICE PAPER PLANT. Grown as an ornamental. Pith is the source of rice-paper. *Distr.* Taiwan.

Tetrapathaea See *Passiflora*

tetraphyllus, -a, -um: four-leaved.

tetrapterus, -a, -um: four-winged.

tetraquetra: four-angled.

tetrasepalus, -a, -um: with four sepals.

Tetrastigma (From Greek *tetra*, four, and *kentron*, spur, alluding to the fruit.) JAVAN GRAPE. A genus of 90 species of deciduous or evergreen trees and lianas with palmately lobed leaves and cymes of small flowers. Several species are grown as house plants and some are the hosts for *Rafflesia arnoldii*. *Distr.* India to tropical Australia. *Fam.* Vitaceae. *Class* D.

T. voinierianum (After M. Voinier, chief veterinary surgeon with the French army in Hanoi.) CHESTNUT VINE, LIZARD PLANT. Fast-growing vine. Grown as a screening plant in California. *Distr.* Laos.

teucrioides from the genus *Teucrium*, with the ending *-oides*, indicating resemblance.

Teucrium (From Greek *teucrion*, possibly after the Trojan king, Teucer, who used the plant medicinally.) GERMANDER, WOOD SAGE. A genus of 100–300 species of perennial herbs and shrubs with typically aromatic, simple or pinnately lobed leaves and whorls of 2-lipped flowers in which the upper lip is often much reduced. Several species and numerous cultivars are grown as ornamentals. *Distr.* Cosmopolitan, especially the Mediterranean region. *Fam.* Labiatae. *Class* D.

T. aroanum Procumbent, much-branched subshrub. Leaves hairy. Flowers purple. *Distr.* Greece.

T. canadense AMERICAN GERMANDER, WOOD SAGE. Perennial to 1m high. Flowers cream to purple. *Distr.* E North America.

T. chamaedrys (From *chamaidrys*, dwarf oak, the classical Greek name.) GERMANDER, WALL GERMANDER. Herb or subshrub. Flowers pink or purple, rarely white. The leaves have been used locally as a medicinal tea. *Distr.* Europe, SW Asia.

T. fruticans SHRUBBY GERMANDER, TREE GERMANDER. Shrub to 2.5m high. Flowers blue or lilac. *Distr.* W Mediterranean region.

T. marum CAT THYME. Small shrub. Flowers in dense cylindrical racemes, purple. *Distr.* W Mediterranean islands.

T. scordium A source of yellow-green dye for cloth. *Distr.* Eurasia.

T. scorodonia WOOD GERMANDER, WOOD SAGE. Subshrub to 1m high. Flowers yellow-green. Formerly medicinal. *Distr.* Europe, naturalized in North America.

texanus, -a, -um: of Texas, USA.

pride of Texas, *Phlox drummondii.*

Texas mud baby *Echinodorus cordifolius*.

texensis, -e: of Texas, USA.

Thalia (After Johann Thal (1542– 1583), German botanist.) A genus of 7–12 species of aquatic or semi-aquatic herbs with rosettes of long-stalked leaves and panicles of irregular flowers that bear 3 sepals, 3 petals, several petal-like sterile stamens, and 1 fertile stamen. Several species are grown as tender ornamentals. *Distr.* Tropical and warm-temperate America, tropical Africa. *Fam.* Marantaceae. *Class* M.

T. dealbata Leaves grey-green, to 50cm long, on stalks 2m long. Flowers violet. *Distr.* S USA.

T. geniculata Leaves lance-shaped, to 65cm long, on stalks 2m long. Flowers violet, in pendulous inflorescences. *Distr.* Tropical America, W tropical Africa.

thalictrifolius, -a, -um: from the genus *Thalictrum*, and *folius*, leaved.

thalictroides: from the genus *Thalictrum*, with the ending *-oides*, indicating resemblance.

Thalictrum (Classical Greek name for an unknown plant.) MEADOW RUE. A genus of 85–130 species of tuberous or rhizomatous herbs with compound leaves and small white, pink, or purple flowers that bear 4–5 petal-like sepals, no true petals, and numerous conspicuous stamens. Some species are insect-pollinated others are wind-pollinated. Several species are grown as ornamentals and some have medicinal uses. *Distr.* Widespread. *Fam.* Ranunculaceae. *Class* D.

T. aquilegiifolium Flowers numerous. Stamens club-shaped, long, exceeding the sepals. *Distr.* Europe to temperate Asia.

T. delavayi (After Abbé Jean Marie Delavay (1838–95), French missionary in China who introduced it into cultivation.) Sepals red or lilac. Stamens yellow. *Distr.* W China.

T. dipterocarpum See *T. delavayi*

T. flavum YELLOW MEADOW RUE. Flowers numerous, fragrant, borne in panicles. *Distr.* Europe, SW Asia.

T. speciosissimum See *T. flavum*

Thamnocalamus (From Greek *thamnos*, shrub, and *kalamos*, reed.) A genus of 6 species of clump-forming bamboos with thick-walled stems and glabrous leaves. *Distr.* E Asia. *Fam.* Gramineae. *Class* M.

T. khasianus See *Raoulia hectoris*

T. spathaceus UMBRELLA BAMBOO, MURIEL BAMBOO. Leaves apple-green, drawn out into a long thin point. Grown as an ornamental. *Distr.* Central China.

T. spathiflorus Stems grey, later flushed pink. Flowers every 10–20 years. Grown as an ornamental. *Distr.* NW Himalaya.

Theaceae The Camellia family. 25 genera and about 500 species of shrubs, trees and occasionally scrambling vines with alternate, leathery, and frequently evergreen leaves. The flowers are regular, bisexual, and bear 4–7 petals and typically numerous stamens. The most economically important species of this family is *Camellia sinensis*, TEA, although other species are cultivated as ornamentals. *Distr.*

Tropical to warm-temperate regions. *Class* D.

See: Camellia, Cleyera, Eurya, Franklinia, Gordonia, Pyrenaria, Schima, Stewartia, Stuartia, Ternstroemia, Tutcheria.

Theligonaceae A family of 1 genus and 3 species of annual and perennial herbs with simple fleshy leaves that bear an apical gland. The flowers are small and unisexual, both sexes occurring on the same plant. Young shoots are eaten as a vegetable. *Distr.* Mediterranean to Japan. *Class* D.

Thelocactus (From Greek *thele*, nipple, and the old genus name *Cactus*.) A genus of about 20 species of small cacti with simple, clustered, spherical to cylindrical stems, and apical funnel-shaped flowers. Several species are grown as ornamentals. *Distr.* Mexico, SW North America. *Fam.* Cactaceae. *Class* D.

T. bicolor Stems spherical or cylindrical, yellow to deep green. Flowers pink to magenta. *Distr.* N Mexico, S Texas.

T. macdowellii Stems spherical, spines yellow-white. Flowers around 5cm across, magenta. *Distr.* NE Mexico.

Thelypteridaceae A family of 30 genera and 900 species of terrestrial, rarely epiphytic ferns with creeping or erect trunk-like stems and simple to bi-pinnate leaves. The sporangia are typically borne in discrete round sori which may lack an indusium. *Distr.* Tropical, subtropical and temperate areas. *Class* F.

See: Macrothelypteris, Phegopteris, Pneumatopteris, Pseudophegopteris, Thelypteris.

Thelypteris (Classical Greek name for a fern, from *thelys*, female, and *pteris*, fern.) A genus of about 2 species of terrestrial ferns with bipinnately divided, grey-hairy leaves. *Distr.* Widespread in temperate regions. *Fam.* Thelypteridaceae. *Class* F.

T. palustris MARSH FERN, MEADOW FERN, SNUFF BOX FERN. Leaves to 60cm long, tinged yellow-green. Occasionally planted as an ornamental. *Distr.* Widespread in temperate regions.

T. phegopteris See *Phegopteris connectilis*

Theobroma (From Greek *theos*, god, and *broma*, food, alluding to *T. cacao*, the source of chocolate.) A genus of 20 species of trees with simple leathery leaves and flowers that are often borne in clusters directly on the trunk, that is, they are cauliflorous. *Distr.* Tropical regions of the New World. *Fam.* Sterculiaceae. *Class* D.

T. cacao (From the local vernacular name.) COCOA, CHOCOLATE PLANT. Evergreen tree to 8m high with simple, thin, leathery leaves and clusters of flowers borne on the thicker branches and trunk. The fruit is a ridged green, yellow, or purple pod containing many seeds embedded in a mucilaginous pulp. The seeds were originally used as a form of currency and medicinally; their most important role today is as the source of chocolate. *Distr.* Central and South America, much cultivated throughout the tropics.

Theophrastaceae A family of 4 genera and 80 species of often thick-stemmed shrubs and trees with spirally arranged, simple leaves and large bisexual flowers. *Distr.* Tropical America. *Class* D.

Thermopsis (From Greek *thermos*, LUPIN, and *-opsis*, indicating resemblance.) A genus of about 20 species of erect rhizomatous herbs with trifoliate leaves and racemes of irregular pea-like flowers. Several species are grown as ornamentals. *Distr.* E Asia, North America. *Fam.* Leguminosae. *Class* D.

T. caroliniana See *T. villosa*

T. macrophylla FALSE LUPIN, GOLDEN PEA. Bushy perennial. Flowers yellow. *Distr.* W USA.

T. villosa CAROLINA LUPIN. Straggling herb. Flowers yellow. *Distr.* SE USA.

Thevetia (After André Thevet (1502–92), French monk and traveller in South America.) A genus of 8 species of evergreen shrubs and trees with spirally arranged, simple leaves and cymes of showy funnel-shaped flowers. *Distr.* Tropical America. *Fam.* Apocynaceae. *Class* D.

T. neriifolia See *T. peruviana*

T. peruviana YELLOW OLEANDER, LUCKY BEAN, LUCK NUT. Shrub or small tree. Flowers yellow or orange, scented. Grown as a tender ornamental. *Distr.* Tropical America.

thibetanus, -a, -um: of Tibet.

thibeticus, -a, -um: of Tibet.

thistle *Carduus* **blessed** ~ *Cnicus benedictus*, *Silybum marianum* **bull** ~ *Cirsium vulgare* **carline** ~ *Carlina*, *C. acaulis* **common carline** ~ *C. vulgaris* **cotton** ~ *Onopordum acanthium* **field sow** ~ *Sonchus arvensis* **globe** ~ *Echinops* **golden** ~ *Scolymus hispanicus* **holy** ~ *Silybum marianum* **marsh sow** ~ *Sonchus palustris* **milk** ~ *Silybum marianumm*, *Sonchus* **musk** ~ *Carduus nutans* **Our Lady's milk** ~ *Silybum marianum* **plume** ~ *Cirsium* **Scotch** ~ *Carduus nutans*, *Onopordum acanthium* **Spanish oyster** ~ *Scolymus hispanicus* **spear** ~ *Cirsium vulgare* **star** ~ *Centaurea* **swamp** ~ *Cirsium palustre*.

Thlaspi (From Greek *thlao*, to flatten, and *aspis*, shield, alluding to the shape of the fruits.) A genus of 60 species of annual to perennial, low herbs with white to purple-red flowers. *Distr*. N temperate regions. *Fam*. Cruciferae. *Class* D.

T. arvense PENNYCRESS, FRENCH WEED, STINKWEED, MITHRIDATE MUSTARD. Bad-smelling annual with white flowers. *Distr*. SW and Central Europe.

T. bellidifolium Tufted perennial with lilac flowers. *Distr*. Balkans.

T. bulbosum Biennial or perennial with deep violet flowers. *Distr*. Greece, Aegean islands.

T. densiflorum Biennial with white to violet flowers. *Distr*. Turkey.

T. montanum MOUNTAIN PENNYCRESS. Loose mat-forming perennial with white flowers. *Distr*. Central Europe.

T. stylosum Dwarf tufted perennial with lilac petals. *Distr*. Central Italy.

thorn: box ~ *Lycium barbarum* **branch** ~ *Erinacea anthyllis* **christ** ~ *Euphorbia milii* **cockspur** ~ *Crataegus crus-galli* **fire** ~ *Pyracantha*, *P. coccinea* **goat's** ~ *Astragalus* **Jerusalem** ~ *Parkinsonia aculeata* **kangaroo** ~ *Acacia paradoxa* **quick set** ~ *Crataegus laevigata* **sallow** ~ *Hippophae rhamnoides* **Washington** ~ *Crataegus phaenopyrum* **white** ~ *C. laevigata*.

thorn apple, *Datura stranonium* **downy** *D. metel*.

thorns, crown of *Euphorbia milii*.

thoroughwort *Eupatorium perfoliatum*

thorow wax *Bupleurum*, *B. rotundifolium*.

three birds flying *Linaria triornithophora*.

thrift *Armeria*, *A. maritima* **Jersey** ~ *A. alliacea* **prickly** ~ *Acantholimon*.

throatwort *Campanula trachelium*, *Trachelium caeruleum*.

Thuja (From Greek *thyia*, scented gum.) ARBOR VITAE, THUJA, RED CEDAR. A genus of 5–6 species of evergreen coniferous trees with minute scale-like leaves. Several species are grown as ornamentals and some are a source of timber. *Distr*. E Asia and North America. *Fam*. Cupressaceae. *Class* G.

T. occidentalis WESTERN WHITE CEDAR, AMERICAN ARBOR VITAE, WHITE CEDAR. Tree to 20m high. A source of medicinal fragrant oil and timber. Many ornamental cultivars have been raised from this species. *Distr*. E North America.

T. plicata WESTERN RED CEDAR, WESTERN ARBOR VITAE. Tree to 70m high. A source of water-resistant wood. Much planted as hedging. *Distr*. W North America.

thuja *Thuja*.

Thujopsis (From the genus name *Thuja*, and Greek *-opsis*, resemblance.) A genus of 1 species of evergreen coniferous tree to 40m high with minute, scale-like, waxy leaves. *Distr*. Japan. *Fam*. Cupressaceae. *Class* G.

T. dolabrata Grown as an ornamental and a source of durable timber.

Thunbergia (After Carl Peter Thunberg (1743–1828), professor of botany at Uppsala University who also travelled in E Asia.) A genus of 100 species of erect or twining herbs and shrubs with simple leaves and trumpet-shaped flowers that are borne singly or in pendent racemes. Several species are grown as tender ornamentals. *Distr*. Tropical regions of the Old World. *Fam*. Acanthaceae. *Class* D.

T. alata BLACK-EYED SUSAN. Climber. Flowers cream, white, or orange with a dark centre. A commonly grown greenhouse ornamental. *Distr*. Tropical Africa, precise origin unclear as long cultivated in Africa.

T. erecta KING'S MANTLE, BUSH CLOCK VINE. Climber. Flowers solitary, yellow with blue tips. *Distr.* Tropical W and S Africa.

T. grandiflora BENGAL CLOCK VINE, BLUE TRUMPET VINE, CLOCK VINE, SKY VINE, SKYFLOWER, BLUE SKYFLOWER. Climber. Flowers blue, 2-lipped. *Distr.* India.

Thurniaceae A family of 1 genus and 2 species of perennial sedge-like herbs with leathery leaves and pendulous wind-pollinated flowers that bear 2 whorls of 3 tepals and 2 whorls of 3 stamens. *Distr.* N South America. *Class* M.

thyme *Thymus* **basil** ~ *Acinos arvensis* **caraway** ~ *Thymus herba-barona* **cat** ~ *Teucrium marum* **common** ~ *Thymus vulgaris* **curly water** ~ *Lagarosiphon* **French** ~ *Thymus vulgaris* **garden** ~ *T. vulgaris* **home of** ~ *Acinos arvensis* **lemon** ~ *Thymus x citriodorus* **wild** ~ *T. serpyllum*.

Thymelaeaceae The Daphne family. 58 genera and about 1000 species of shrubs with entire simple leaves and cup- or tube-shaped flowers that bear their parts in fours or fives. *Distr.* Widespread. *Class* D.
See: Daphne, Drapetes, Edgeworthia, Pimelea.

thymifolius, -a, -um: from the genus *Thymus*, and *folius*, leaved.

thymoides: from the genus *Thymus*, with the ending *-oides*, indicating resemblance.

Thymus (Classical Greek name *thymos*.) THYME. A genus of about 350 species of aromatic perennial herbs and dwarf shrubs with simple opposite leaves and crowded heads of tubular 2-lipped flowers. Numerous species and hybrids are grown as ornamentals as well as a few that are used as culinary herbs. *Distr.* Temperate Eurasia. *Fam.* Labiatae. *Class* D.

T. caespititius Mat-forming aromatic shrub. Flowers small, lilac-pink. *Distr.* Portugal, NW Spain, Azores.

T. × citriodorus LEMON THYME. A hybrid of *T. pulegioides* × *T. vulgaris*. This cross has given rise to numerous ornamental cultivars. *Distr.* Garden origin.

T. drucei See *T. praecox*

T. erectus See *T. vulgaris*

T. herba-barona CARAWAY THYME. Wiry carpet-forming subshrub. Leaves very small, dark green. Flowers pink or mauve. *Distr.* Corsica, Sardinia.

T. praecox Creeping perennial. Leaves small. Flowers purple to white. *Distr.* SW and Central Europe.

T. serpyllum WILD THYME. Dwarf shrub. Flowers crowded, pink or purple. This species has given rise to numerous ornamental cultivars. *Distr.* Europe.

T. vulgaris COMMON THYME, FRENCH THYME, GARDEN THYME. Erect subshrub. Leaves linear. Flowers crowded, white or purple. An important culinary herb and formerly of medicinal importance. *Distr.* S Europe.

thyrsiflorus, -a, -um: with flowers in a thyrse, a type of inflorescence.

thyrsoides: with flowers in a thyrse, a type of inflorescence.

Thyrsopteris (From Greek *thyrsos*, panicle or raceme, and *pteris*, fern, alluding to the arrangement of the sori.) A genus of 1 species of tree-fern with fronds to 2m long. *Distr.* Chile (Juan Fernández). *Fam.* Dicksoniaceae. *Class* F.

T. elegans Grown as an ornamental.

Thysanotus (From Greek *thysanos*, fringed, and *ous*, ear, alluding to the fringed inner perianth segments.) A genus of 47 species of rhizomatous or tuberous perennials with grass-like leaves and panicles or umbels of red-blue, blue, or lavender flowers. Several species are cultivated as ornamentals. *Distr.* Australia, SE Asia. *Fam.* Anthericaceae. *Class* M.

T. tuberosus Flowers purple. *Distr.* E Australia.

Tiarella (Diminutive of Greek *tiara*, small crown, referring to the shape of the fruit.) FALSE MITREWORT, SUGAR SCOOP. A genus of 3–7 species of perennial, rhizomatous herbs with trifoliate or palmate leaves and racemes of small white or red flowers. Several species are cultivated as ornamentals. *Distr.* Mountains and coasts of North America, 1 species in E Asia. *Fam.* Saxifragaceae. *Class* D.

T. collina See *T. wherryi*

T. cordifolia FOAM FLOWER, COOLWORT. Stoloniferous perennial. Grown as an ornamental but tends to be invasive. *Distr.* E North America.

T. polyphylla Leaves 5-lobed. Flowers nodding. *Distr.* China, Japan, Himalaya.

T. trifoliata Leaves trifoliate. Rhizome slender. *Distr.* NW North America, Alaska to Oregon.

T. unifoliata Leaves 3–5-lobed. *Distr.* USA (Montana to Alberta and Alaska).

T. wherryi (After its discoverer, Edgar Theodore Wherry (born 1885), American botanist.) Similar to *T. cordifolia* but lacks invasive stolons. *Distr.* SE USA.

tiarelloides: from the genus *Tiarella*, with the ending *-oides*, indicating resemblance.

tibeticus, -a, -um: of Tibet.

Tibouchina (From a Guyanan vernacular name.) A genus of 350 species of perennial herbs, subshrubs, shrubs, and climbers with large leathery leaves and panicles of showy flowers that bear 5 somewhat irregular petals and 10 stamens. *Distr.* Tropical regions of the New World. *Fam.* Melastomataceae. *Class* D.

T. semidecandra See *T. urvilleana*

T. urvilleana (After Jules Sébastian César Dumont d'Urville (1790–1844), French naval officer and explorer.) GLORY BUSH. Slender evergreen shrub. Flowers large, purple to violet, with purple anthers. Grown as an ornamental. *Distr.* Brazil.

tickseed *Bidens, Coreopsis.*

Ticodendraceae A family of 1 genus and 1 species of recently discovered tree. *Distr.* Costa Rica, Nicaragua, Panama. *Class* D.

tiger flower *Tigridia pavonia.*

tiger's jaws *Faucaria, F. tigrina.*

Tigridia (From Latin *tigris*, tiger, alluding to the markings on the flowers.) A genus of 23–27 species of bulbiferous herbs with a basal fan of folded leaves and clusters of large *Iris*-like flowers. Several species are grown as ornamentals. *Distr.* Mexico, Guatemala, Andes of Peru and Chile. *Fam.* Iridaceae. *Class* M.

T. pavonia TIGER FLOWER, PEACOCK FLOWER. Flowers to 15cm across, red with yellow and purple spots. Cultivars are available with white, yellow, pink, and orange flowers, each lasting only a few hours. The starchy bulbs have been eaten since Aztec times. *Distr.* Mexico, naturalized in South America.

tigrinus, -a, -um: with tiger-like markings.

Tilia (Classical Latin name for these trees.) LIME, LINDEN, BASSWOOD. A genus of 45 species of deciduous trees with simple alternate leaves that are borne in 2 ranks and small fragrant flowers that are borne in clusters subtended by a large bract. A number of species and hybrids are grown as ornamentals and some are important as timber trees. *Distr.* N temperate regions. *Fam.* Tiliaceae. *Class* D.

T. americana AMERICAN LIME, AMERICAN BASSWOOD, WHITEWOOD. Broad-crowned tree to 40m high. Several ornamental cultivars have been raised from this species and it is an important source of timber. *Distr.* E North America.

T. cordata SMALL-LEAVED LIME, LITTLE-LEAF LINDEN. Spreading tree to 35m high. This tree was formerly an important forest tree in S England and is still important as a source of pale timber. *Distr.* Europe.

T. × europaea See *T. × vulgaris*

T. japonica JAPANESE LIME. Tree to 20m high. *Distr.* Japan.

T. platyphyllos BROAD-LEAVED LIME, LARGE-LEAVED LIME. Large conical tree. A source of timber. Numerous cultivars have been produced. *Distr.* Europe, SW Asia.

T. tomentosa EUROPEAN WHITE LIME, SILVER LIME. Tree to 35m high. Branches pendent. Leaves grey-hairy below. *Distr.* SE Europe, Asia.

T. × vulgaris LIME, COMMON LIME, EUROPEAN LIME. A hybrid of *T. cordata* × *T. plathyphyllos*. Much planted as a street tree and noted for the sticky exudate that drops from the large numbers of aphids it harbours. *Distr.* Garden origin.

Tiliaceae The Lime family. 53 genera and about 1000 species of trees and shrubs with typically simple, deciduous, asymmetric leaves borne in 2 ranks. The small bisexual flowers are borne in complex cymes. A number of species are important sources of timber. Jute is obtained from *Corchorus capsularis*. *Distr.* Widespread. *Class* D.
See: Corchorus, Entelea, Grewia, Sparmannia, Tilia.

tiliifolius, -a, -um: from the genus *Tilia*, LIME OR LINDEN, and *folius* leaved.

Tillaea See *Crassula*

Tillandsia (After Elias Tillands (1640–93), Swedish botanist and physician.) A genus of about 400 species of terrestrial or epiphytic herbs with leaves in rosettes or spread along a stem, sometimes coated in water-absorbing scales. The inflorescence usually takes the form of an erect condensed spike with the flowers held in 2 to many distinct rows. Numerous species and cultivars are grown as ornamentals. *Distr.* Tropical America. *Fam.* Bromeliaceae. *Class* M.

T. argentea Leaves very thin, grey, scaly, forming a bulb-like rosette at base. *Distr.* West Indies, Central America.

T. caput-medusae Leaves contorted, finely pointed above, broad and fused into a bulb-like rosette at the base.

T. cyanea PINK QUILL. Leaves arching, channelled. Flowers dark violet, appearing from between pink bracts. *Distr.* Ecuador, Peru.

T. fasciculata Leaves narrow, grooved, forming a bulb-like rosette at the base. Flowers purple-blue, appearing from between yellow-red bracts. *Distr.* N South America to Mexico and West Indies.

T. ionantha SKY PLANT. Leaves narrow, clump-forming. Flowers violet, appearing from between white bracts. *Distr.* Central America.

T. lindenii (After J. J. Linden (1817–98), Belgian nurseryman.) BLUE-FLOWERED TORCH. Leaves linear, channelled, arching. Flowers blue. *Distr.* NW Peru.

T. recurvata BALL MOSS. Leaves narrow, in rosettes on a thin stem, forming ball-shaped clusters. *Distr.* Central America.

T. stricta Leaves narrow above, forming a bulb-like rosette below. Flowers small, pale blue-green. *Distr.* N South America.

T. usneoides SPANISH MOSS, FLORIDA MOSS, OLD MAN'S BEARD. An unusual plant hanging in festoons from trees and overhead wires like a lichen such as *Usnea*, which it resembles. Roots are present only as a holdfast, × nutrients being absorbed over the entire surface of the plant. The solitary fragrant flowers are very rarely seen, reproduction usually occurring through fragmentation. Dried plants are used as packing material and like horsehair in upholstery. *Distr.* Florida to Argentina, over 5000 miles of latitude, an almost unique distribution.

timothy *Phleum pratense.*

tinctorius, -a, -um: used in dyeing.

tinctus, -a, -um: coloured.

tinco *Weinmannia trichosperma.*

tingitanus, -a, -um: of Tangiers, North Africa.

tinker's weed *Triosteum perfoliatum.*

Tipuana (After the Tipuani Valley in Bolivia, where it occurs.) A genus of 1 species of evergreen tree with alternate pinnate leaves and pendulous panicles of yellow pea-like flowers that are followed by a winged pod. *Distr.* Bolivia, Argentina. *Fam.* Leguminosae. *Class* D.

T. tipu YELLOW JACARANDA, TIPU TREE, ROSEWOOD, PRIDE OF BOLIVIA. Grown as an ornamental and widely planted as a street tree in tropical regions.

tipu tree *Tipuana tipu.*

Tithonia (After Tithonus, a beautiful youth in Greek mythology. He was loved by Aurora, goddess of the dawn, who turned him into a grasshopper.) MEXICAN SUNFLOWER. A genus of 10 species of annual and perennial herbs and shrubs with alternate, typically pinnately lobed leaves and flowers borne in daisy-like heads. Several species are grown as ornamentals. *Distr.* Mexico and Central America. *Fam.* Compositae. *Class* D.

T. rotundifolia MEXICAN SUNFLOWER. Annual herb to 4m high. Flower-heads to 7cm across, rays orange-red. Several ornamental cultivars have been raised from this species. *Distr.* Mexico, Central America.

tithymaloides: from the genus *Tithymalus* (now *Euphorbia*), with the ending *-oides*, indicating resemblance.

titoki *Alectryon excelsus.*

ti tree *Leptospermum.*

toadflax *Linaria* **Alpine** ~ *L. alpina* **common** ~ *L. vulgaris* **ivy-leaved** ~ *Cymbalaria muralis* **purple** ~ *Linaria* *purpurea* **striped** ~ *L. repens* **three birds** ~ *L. triornithophora.*

tobacco *Nicotiana, N. tabacum.* **Aztec** ~ *N. rustica* **flowering** ~ *N. alata* **Indian** ~ *Lobelia inflata* **jasmine** ~ *Nicotiana alata* **ladies** ~ *Antennaria* **mountain** ~ *Arnica montana* **tree** ~ *Nicotiana glauca* **wild** ~ *N. rustica.*

Todea (After Henry Julius Tode (1733–1797), German mycologist.) A genus of 1–2 species of tree-like ferns with leathery bipinnate leaves. *Distr.* S temperate regions of the Old World. *Fam.* Osmundaceae. *Class* F.

T. barbara CRAPE FERN, CRÊPE FERN, KING FERN. Rhizome forms an erect trunk. Leaves to 1m long. Occasionally grown as a tender ornamental. *Distr.* S Africa, E Australia, Tasmania, New Zealand.

Tofieldia (After Thomas Tofield (1730–79), British botanist.) FALSE ASPHODEL. A genus of 17–18 species of rhizomatous herbs with narrow, tufted, basal leaves and racemes of small star-shaped flowers. Some species are cultivated ornamentally. *Distr.* N Temperate regions. *Fam.* Melanthiaceae. *Class* M.

T. calyculata To 30cm high. Flowers yellow-green. *Distr.* Europe.

T. pusilla SCOTCH ASPHODEL. To 25cm high. Flowers white or green-white. *Distr.* N Europe.

Tolmiea (After Dr William Fraser Tolmie (1830–86), Scottish physician and botanist.) A genus of 1 species of perennial herb with kidney-shaped leaves and chocolate-coloured flowers. Small plantlets are often formed at the junction of the leaf-stalk and blade. *Distr.* W North America. *Fam.* Saxifragaceae. *Class* D.

T. menziesii (After Archibald Menzies (1754–1842), naval surgeon and botanist.) PIGGY-BACK PLANT. Cultivated as a house plant and garden ornamental.

Tolpis A genus of 20 species of annual and perennial herbs with milky latex, lobed basal leaves, and small yellow flowers in dandelion-like heads. Several species are grown as ornamentals. *Distr.* Macaronesia, Mediterranean to Ethiopia and Somalia. *Fam.* Compositae. *Class* D.

T. barbata Annual herb to 1m high. *Distr.* Mediterranean.

tomato *Lycopersicon esculentum* **strawberry** ~ *Physalis pruinosa* **tree** ~ *Cyphomandra crassicaulis.*

tomentellus, -a, -um: with small woolly hairs.

tomentosus, -a, -um: hairy.

Tonestus (Anagram of the genus name *Stenotus.*) A genus of 5 species of perennial herbs with simple alternate leaves and small flowers in daisy-like heads that bear yellow rays. Several species are grown as ornamentals. *Distr.* North America. *Fam.* Compositae. *Class* D.

T. lyallii Mat-forming perennial herb. *Distr.* NW North America.

tongaensis, -e: of Tonga.

tonga plant *Epipremnum mirabile.*

tonsus, -a, -um: smooth.

Toona A genus of 6 species of evergreen or deciduous trees with large pinnate leaves and panicles of fragrant white flowers. *Distr.* Tropical Asia to N Australia. *Fam.* Meliaceae. *Class* D.

T. sinensis Deciduous spreading tree. Bark shaggy. Flowers cream, fragrant. Grown as a street tree and exploited for its timber. *Distr.* China to mountainous regions of Malaysia.

toquilla *Carludovica palmata.*

torch: blue-flowered ~ *Tillandsia lindenii* **summer** ~ *Billbergia pyramidalis.*

torches *Verbascum thapsus.*

torch plant *Aloe aristata, A. arborescens.*

tormentil *Potentilla erecta.*

torminalis, -e: effective against colic.

Torreya (After John Torrey (1796–1873), American botanist.) NUTMEG YEW. A genus of 5–6 species of evergreen coniferous shrubs and trees with linear, rigid, sharp-tipped leaves that are borne spirally but twisted so as to be arranged in 2 ranks. The female cones consists of a single ovule that develops a fleshy red aril. *Distr.* E Asia, North America. *Fam.* Taxaceae. *Class* G.

T. californica CALIFORNIA NUTMEG, CALIFORNIA NUTMEG YEW. Tree to 20m high. Grown as an ornamental. *Distr.* USA (California).

T. nucifera KAYA NUT, JAPANESE NUTMEG YEW. Tree to 25m high. Seeds a source of cooking oil. *Distr.* Japan.

Torricelliaceae A family of 1 genus and 3 species of small trees with simple leaves and small flowers. This family is sometimes included within the family Cornaceae. *Distr.* Himalaya, W China. *Class* D.

tortifrons: with twisted leaves.

torulosus, -a, um: cylindrical with bulges and constrictions.

totter *Briza media.*

touch me not *Mimosa pudica, Impatiens.*

tovarensis, -e: of Tovar, Venezuela.

Tovariaceae A family of 1 genus and 2 species of pungent-smelling annual herbs and shrubs with trifoliate leaves and flowers that bear their parts in eights. *Distr.* Tropical and temperate regions of the New World. *Class* D.

Townsendia (After David Townsend (1787–1858), American botanist.) A genus of 21 species of annual and perennial herbs with simple alternate leaves and flowers in daisy-like heads that bear distinct rays. Some species are cultivated as ornamentals. *Distr.* W North America. *Fam.* Compositae. *Class* D.

T. exscapa EASTERN DAISY. Biennial herb. Flowers yellow. Rays purple-red. *Distr.* W North America.

T. parryi (After Charles Christopher Parry (1823–90), American botanical explorer.) Biennial or perennial herb. Leaves fleshy. Flowers yellow. Rays violet-purple. *Distr.* W North America.

toxicodendron: poisonous tree, poison ivy (see *Rhus radicans*).

Trachelium (From Greek *trachelos*, neck, alluding to supposed medicinal properties.) A genus of 7 species of perennial herbs with simple alternate leaves and tubular to star-shaped flowers that are borne singly or in dense corymbs. Several species are grown as ornamentals. *Distr.* Mediterranean. *Fam.* Campanulaceae. *Class* D.

T. asperuloides Mat-forming. Leaves minute. Flowers purple-blue, more or less star-shaped. *Distr.* Greece.

T. caeruleum THROATWORT. Stems erect, to 1m high. Flowers white, borne in dense umbel-like clusters. *Distr.* Mediterranean region.

Trachelospermum (From Greek *trachelos*, neck, and *sperma*, seed, alluding to the narrow seeds.) A genus of 20 species of evergreen climbing shrubs with simple leathery leaves, milky sap, and funnel-shaped scented flowers. Several species are cultivated as ornamentals. *Distr.* India to Japan, SE USA. *Fam.* Apocynaceae. *Class* D.

T. asiaticum Flowers yellow-white, scented. *Distr.* Japan, Korea.

T. jasminoides Flowers white. Flowers very highly scented. *Distr.* China, Japan.

T. majus See *T. jasminoides*

Trachycarpus (From Greek *trachys*, rough, and *karpos*, fruit.) FAN PALM, CHINESE WINDMILL PALM, CHUSAN PALM. A genus of 4–6 species of large palms with fan-shaped leaves and erect or arching inflorescences of yellow or cream flowers. *Distr.* Subtropical Asia, Himalaya. *Fam.* Palmae. *Class* M.

T. fortunei (After Robert Fortune (1812–80), who collected in China and introduced the tea plant from China to India.) CHUSAN PALM, FAN PALM, CHUSAN, HEMP PALM. To 20m high. Leaves around 1m across. Flowers yellow. One of the few palms that are hardy in the British Isles. The fibre (Chinese coir) can be used for cordage. *Distr.* Origin unknown, probably N Burma or China; widely cultivated and becoming naturalized in China and Japan.

Trachymene (From Greek *trachys*, rough, and *menix*, membrane, alluding to the fruits.) A genus of about 12 species of annual, biennial, and perennial herbs with divided leaves and simple umbels of white, pink, or blue flowers. Several species are grown as ornamentals. *Distr.* Australia and the W Pacific. *Fam.* Umbelliferae. *Class* D.

T. coerulea BLUE LACE FLOWER. Annual or biennial herb. Leaves pale green. Flowers blue. *Distr.* W Australia.

Trachystemon (From Greek *trachys*, rough, and *stemon*, stamen, alluding to the rough stamens of one species.) A genus of 2 species of robust, hairy, perennial herbs with simple leaves and loose panicles of pendent, funnel-shaped, blue flowers. *Distr.* Mediterranean region. *Fam.* Boraginaceae. *Class* D.

T. orientalis Cultivated as ornamentals. *Distr.* SE Europe, W Asia.

Tradescantia (After John Trades-cant (1608–62), English botanist, gardener and plant collector.) SPIDER LILY, SPIDERWORT, WANDERING JEW. A genus of 65–70 species of herbs typically with fibrous or tuberous roots, creeping or trailing stems, and simple, somewhat clasping, striped leaves. The somewhat irregular flowers have 3 petals and 6 stamens. The inflorescence is subtended by a leaf-like bract. A number of species and very many cultivars and hybrids are grown as ornamentals. *Distr.* North and South America. *Fam.* Commelinaceae. *Class* M.

T. albiflora See *T. fluminensis*

T. × andersoniana A name wrongly applied to hybrids of *T. virginiana* × *T. ohiensis* × *T. subaspera*.

T. cerinthoides FLOWERING INCH PLANT. Stems trailing and ascending, purple. Leaves densely white-hairy below. *Distr.* N Argentina.

T. fluminensis SPEEDY JENNY, WANDERING JEW. Rhizomatous. Leaves broad, ovate, green, often striped white. Flowers white. A species with numerous variants in cultivation and a number of cultivars erroneously attributed to it. *Distr.* SE Brazil to N Argentina.

T. multiflora See *Tripogandra multiflora*

T. pallida Leaves and stems intense violet-purple. Often grown as a bedding plant. *Distr.* NE Mexico.

T. sillamontana WHITE VELVET. Stem erect. Leaves boat-shaped, densely white-hairy. *Distr.* N Mexico.

T. spathacea BOAT LILY, OYSTER PLANT, CRADLE LILY, MOSES IN HIS CRADLE, MOSES ON A RAFT, MOSES IN THE BULRUSHES, MEN IN A BOAT. Stem thick. Leaves sheathing and overlapping. Flowers small, white, enclosed in a boat-shaped bract. *Distr.* Central America, S Mexico, Belize, Guatemala, naturalized in West Indies.

T. virginiana SPIDERWORT. Stems tufted. Leaves somewhat fleshy. Flowers purple. The central species in a group of species and cultivars that are important as hardy garden ornamentals. *Distr.* E North America.

T. zebrina Stems trailing. Leaves with 2 silvery bands separated by green above, purple below. Flowers purple-blue. One of the most popular of all house plants. *Distr.* Mexico, widely naturalized throughout the tropical regions.

tragophyllus, -a, um: from Greek *tragos*, goat, and *phyllon*, leaf.

Tragopogon (From Greek *tragos*, goat, and *pogon*, beard, alluding to the hairy fruits.) GOAT'S BEARD. A genus of 50 species of annual to perennial herbs with stout taproots, narrow lance-shaped leaves, and dandelion-like flower-heads. The fruiting head resembles a very large dandelion clock. *Distr.* Temperate Eurasia. *Fam.* Compositae. *Class* D.

T. porrifolius SALSIFY, VEGETABLE OYSTER, OYSTER PLANT. Biennial herb. Cultivated for edible roots and for young flowering shoots. *Distr.* S Europe.

T. pratensis GOAT'S BEARD, JOHNNY GO TO BED AT NOON, SHEPHERD'S CLOCK. Annual to perennial herb. Flower-heads open only in the morning. *Distr.* Europe, naturalized in North America.

tranquillans: tranquil, calm.

transcaucasicus, -a, -um: of the Caucasus mountains.

transilvanicus, -a, -um: of Transylvania.

transitorius, -a, -um: transitory, intermediate.

transparens: transparent, see-through.

Trapa (From Latin *calcitrappa*, caltrop, a four-pointed weapon that when thrown on the ground always has one spike pointing upwards, alluding to the shape of the fruits.) WATER CHESTNUT. A genus of about 15 species of aquatic perennial herbs with floating leaves that are supported by broad spongy leaf-stalks. The flowers are small, white, and bear their parts in fours. The fruits are edible once cooked and are rich in carbohydrate and fat. Not to be confused with Chinese water chestnut, *Eleocharis dulcis*. *Distr.* Central and SE Europe, Asia, Africa. *Fam.* Trapaceae. *Class* D.

T. natans WATER CHESTNUT, WATER CALTROP, SALIGOT, HORN NUT, JESUIT'S NUT. Previously grown for food, this species is an aquatic weed in some areas. *Distr.* Eurasia, Africa, naturalized in North America.

Trapaceae A family of 1 genus and 15 species. *Class* D.
See: Trapa.

traveller's joy *Clematis vitalba.*

traveller's tree *Ravenala madagascariensis.*

treasure flower *Gazania, G. rigens.*

tree: big ~ *Sequoiadendron giganteum* **Christmas** ~ *Picea abies* **planer** ~ *Planera aquatica.*

tree of heaven *Ailanthus, A. altissima,* Simaroubaceae.

trefoil: bird's foot ~ *Lotus corniculatus* **marsh** ~ *Menyanthes trifoliata* **moon** ~ *Medicago arborea.*

Tremandraceae A family of 3 genera and 43 species of small shrubs with simple, opposite or whorled leaves, and numerous, regular, red or purple flowers. *Distr.* Australia. *Class* D.

tremulus, -a, -um: trembling.

triacanthos: three-spined.

triandrus, -a, -um: with three stamens.

triangularis, -e: triangular.

trichanthus, -a, -um: with hairy flowers.

trichocalyx: with hairy sepals.

trichocarpus, -a, -um: with hairy fruit.

trichocaulon: with a hairy stem.

trichocladus, -a, -um: with hairy branches.

trichogynus, -a, -um: with a hairy ovary.

tricholepis: with a hairy stalk.

trichomanifolius: with leaves resembling those of the fern genus *Trichomanes*.

trichomanoides: from the fern genus *Trichomanes*, with the ending *-oides*, indicating resemblance.

trichophyllus, -a, -um: with hairy leaves.

Trichopodaceae A family of 1 genus and 1 species of rhizomatous herbs. This family is sometimes included within the family Dioscoreaceae. *Distr.* India, Sri Lanka, Malay Peninsula. *Class* M.

trichosanthus, -a, -um: with hairy flowers.

trichospermus, -a, -um: with hairy seeds.

trichostomus, -a, -um: with a hairy mouth.

trichotomus, -a, -um: branching into three.

tricolor, -um: three coloured.

Tricuspidaria See *Crinodendron*

T. dependens See *Crinodendron hookerianum*

T. lanceolata See *Crinodendron hookerianum*

tricuspidatus, -a, -um: three-pointed.

Tricyrtis (From Greek *tri*, three, and *kyrtos*, humped, alluding to the swollen bases of the three outer tepals.) TOAD LILY. A genus of 10–16 species of rhizomatous herbs with erect or arched stems, ovate to lance-shaped leaves, and white or yellow, more or less bell-shaped flowers. Most of the species are grown as ornamentals. *Distr.* Mountainous regions of E Asia. *Fam.* Convallariaceae. *Class* M.

T. formosana Flowers somewhat flared, pink with purple-pink spots and a yellow throat. *Distr.* Taiwan.

T. hirta TOAD LILY, JAPANESE TOAD LILY. Flowers white with purple spots. A number of cultivars of this species are available. *Distr.* Japan.

T. japonica See *T. hirta*

T. macrantha Stems arched or drooping. Flowers pendulous, yellow with chocolate spots. *Distr.* Japan.

T. stolonifera See *T. formosana*

tridentatus, -a, -um: three-toothed.

Trientalis (From Latin *trientalis*, one-third of a foot in length.) CHICKWEED, WINTERGREEN. A genus of 4 species of perennial rhizomatous herbs with terminal tufts of 4–7 simple leaves and flat 5-lobed flowers. Several species are grown as ornamentals. *Distr.* N temperate regions. *Fam.* Primulaceae. *Class* D.

T. europaea Stems to 30cm high. Flowers white, tinged pink. *Distr.* N Europe and mountainous regions of the S.

trifasciatus, -a, -um: in three bundles.

trifidus, -a, -um: three-lobed.

triflorus, -a, -um: three-flowered.

trifoliatus, -a, -um: with three leaflets.

Trifolium (From Latin *tres*, three, and *folius*, leaf.) CLOVER. A genus of about 230 species of annual, biennial, and perennial herbs with creeping stems, trifoliate or palmate leaves, and racemes of pea-like flowers. A number of species are important as fodder. *Distr.* Temperate and subtropical regions except Australia. *Fam.* Leguminosae. *Class* D.

T. alpinum ALPINE CLOVER. Densely tufted perennial. Leaves trifoliate. Flowers pink or cream, fragrant. *Distr.* S Europe.

T. incarnatum ITALIAN CLOVER, CRIMSON CLOVER. Erect or ascending annual. Flowers yellow to deep red. *Distr.* S and W Europe.

T. pannonicum HUNGARIAN CLOVER. Bushy perennial. Flowers yellow. Grown as a drought-resistant fodder crop. *Distr.* E Europe.

T. pratense RED CLOVER, PURPLE CLOVER. Erect or decumbent, short-lived perennial. Flowers red or pink, occasionally cream. Several ornamental cultivars have been raised from this species. *Distr.* Europe, naturalized North America.

T. repens WHITE CLOVER, DUTCH CLOVER, SHAMROCK. Rhizomatous perennial. Flowers fragrant, white or pink. Perhaps the most important fodder and rotational crop plant as well as being a valuable bee plant and the source of a number of ornamental cultivars. *Distr.* Europe, naturalized North America.

trifolius, -a, -um: three-leaved.

trigger plant *Stylidium, S. graminifolium.*

Trigonella (From Greek *treis*, three, and *gonu*, joint.) A genus of 80 species of annual and perennial, often aromatic herbs with trifoliate leaves and heads of pea-like flowers. *Distr.* Mediterranean, Canary Islands, South Africa, Australia. *Fam.* Leguminosae. *Class* D.

T. foenum-graecum FENUGREEK, GREEK CLOVER, GREEK HAY. Annual, to 60cm high. Flowers cream. Grown since the time of the Assyrians for seeds which are used like lentils and medicinally. *Distr.* S Europe, W Asia.

Trigoniaceae A family of 4 genera and about 30 species of shrubs and trees with simple leaves and irregular bisexual flowers. *Distr.* Madagascar, Indonesia and tropical America. *Class* D.

trigynus, -a, -um: with three pistils.

Trilliaceae A family of 5 genera and about 60 species of rhizomatous herbs with whorls of simple leaves often borne on top of an erect stem. The flowers are regular, bisexual, and have 3–6 sepals, 3–6 petals, and 3–10 stamens. *Distr.* N temperate regions. *Class* M.
See: Daiswa, Paris, Scoliopus, Trillium.

Trillium (From Latin *tri*, three, alluding to the number of leaves.) WAKE ROBIN, WOOD LILY, BIRTHROOT, STINKING BENJAMIN. A genus of about 30 species of rhizomatous herbs with erect or short stems and an apical whorl of 3 simple leaves. The solitary showy flowers are borne at the junction of the leaves and stem and have 3 sepals and 3 large petals. Numerous species are cultivated as ornamentals and some have medicinal properties. *Distr.* N America, W Himalaya, NE Asia. *Fam.* Trilliaceae. *Class* M.

T. chloropetalum Used in chromosome studies. *Distr.* W North America.

T. erectum STINKING BENJAMIN. Several ornamental cultivars available. Reported to have medicinal properties. *Distr.* E North America.

T. grandiflorum WAKE ROBIN. Flowers large. Unusual flower mutations often occur in wild plants. *Distr.* E North America.

T. undulatum PAINTED WOOD LILY. Flowers pink or white with a maroon centre. *Distr.* E North America.

trilobatus, -a, -um: three-lobed.

trilobus, -a, -um: three-lobed.

Trimeniaceae A family of 1 genus and 3 species of trees and scrambling shrubs with simple opposite leaves and cymes of small flowers. *Distr.* Indonesia and Malaysia to Australia and Pacific. *Class* D.

Trimezia (From Greek *treis*, three, and *meizon*, greater, alluding to the outer tepals which are much larger than the inner ones.) A genus of 5–13 species of rhizomatous herbs with rush-like leaves and short-lived *Iris*-like flowers. *Distr.* Central and tropical South America and the West Indies. *Fam.* Iridaceae. *Class* M.

T. martinicensis Flowers yellow, spotted brown or purple. Grown as a tender ornamental. *Distr.* South America, West Indies; naturalized elsewhere in the tropics.

trinervis, -e: three-nerved.

Triosteum (From Greek *tri-*, three, and *osteon*, bone, alluding to the 3 very hard seeds.) HORSE GENTIAN, FEVERWORT. A genus of 5–6 species of perennial herbs with woody rootstocks, simple or pinnately lobed leaves, and spikes of small 2-lipped flowers. Several species are grown as ornamentals. *Distr.* Asia, E North America. *Fam.* Caprifoliaceae. *Class* D.

T. perfoliatum WILD COFFEE, FEVER ROOT, TINKER'S WEED. Stems to 1.2m high. Flowers purple to dull red. Used as a coffee substitute. *Distr.* North America.

tripartitus, -a, um: with three parts.

Tripetaleia (From Greek *tri-*, three, and *petalon*, petal.) A genus of 2 species of deciduous shrubs with alternate simple leaves and 4–5-lobed flowers. *Distr.* Japan. *Fam.* Ericaceae. *Class* D.

tripetalus, -a, -um: with three petals.

triphyllus, -a, -um: with three leaves.

Tripleurospermum (From Greek *tri-*, three, *pleuron*, rib, and *sperma*, seed, alluding to the 3-ribbed nutlets.) SCENTLESS FALSE CAMOMILE, TURFING DAISY. A genus of 30 species of annual and perennial herbs with pinnately divided leaves and flowers in daisy-like heads that may or may not bear distinct rays. Several species are grown as ornamentals. *Distr.* N temperate regions. *Fam.* Compositae. *Class* D.

T. inodorum Annual herb to 80cm high. Flowers white, borne in heads with distinct rays. *Distr.* Europe.

triplinervis, -e: three-nerved.

Triplostegiaceae A family of 1 genus and 2 species of foetid herbs. This family is sometimes included within the family Valerianaceae. *Distr.* Himalaya to China and New Guinea. *Class* D.

Tripogandra (From Greek *tri-*, three, *pogon*, beard, and *aner*, man, alluding to the three bearded stamens in the type species.) A genus of 21–22 species of erect or trailing herbs with fibrous roots and simple leaves. The small flowers are irregular and bear 3 pink or white petals and 6 stamens. Several species are grown as ornamentals. *Distr.* Tropical America. *Fam.* Commelinaceae. *Class* M.

T. multiflora Stems trailing and rooting. Flowers very small, numerous. *Distr.* Jamaica, Trinidad, Costa Rica, W tropical South America to Argentina.

tripteris: three-winged.

tripteron: three-winged.

Tripterygirum (From Greek *tri-*, three, and *pteryx*, wing, alluding to the 3-winged fruit.) A genus of 2 species of deciduous scrambling shrubs with large alternate leaves and large panicles of small white flowers. *Distr.* E Asia. *Fam.* Celastraceae. *Class* D.

T. regelii Grown as an ornamental. *Distr.* E Asia.

T. regelii *Distr.* Japan, Korea, Manchuria.

triquetrus, -a, -um: three-angled.

tristachyus, -a, -um: with three spikes.

Tristagma (From Greek *tri*, three, and *stagma*, something which drips, alluding to the three nectar pores on the ovary.) A genus of 5 species of small bulbous herbs with a few linear leaves and umbels of a few tubular flowers. Several species are cultivated as ornamentals. *Distr.* Chile, S Argentina. *Fam.* Alliaceae. *Class* M.

T. nivale Flowers olive-green, erect. *Distr.* S South America.

Tristania (After Jules de Tristan (1776–1861), French biologist.) A genus of 1 species of shrub or small tree with simple, narrow, opposite leaves and clusters of small yellow flowers that bear 5 petals and numerous stamens. *Distr.* Australia (New South Wales). *Fam.* Myrtaceae. *Class* D.

T. neriifolia Grown as an ornamental.

tristis, -e: sad, sombre-coloured.

trisulcatus, -a, -um: with three grooves.

Triteleia (From Greek *tri*, three, and *teleios*, perfect; the floral parts are in threes.) A genus of 16–17 species of perennial herbs arising from corms with few linear leaves and umbels of small flowers. Some species are cultivated as ornamentals. *Distr.* W North America. *Fam.* Alliaceae. *Class* M.

T. bridgesii (After Thomas Bridges (1807–65), who collected plants in South America and California.) Flowers lilac or blue. *Distr.* W USA.

T. grandiflora Flowers bright blue, rarely white, to 3cm long. *Distr.* British Columbia to Utah.

T. hendersonii Tepals yellow, with a dark-purple midrib. *Distr.* SW Oregon.

T. hyacinthina Flowers usually white, sometimes blue. *Distr.* British Columbia to Idaho and California.

T. ixioides Flowers golden yellow. *Distr.* California.

T. laxa GRASSNUT, TRIPLET LILY. Flowers deep violet-blue. *Distr.* California to Oregon.

T. peduncularis Flowers funnel-shaped, white to blue. *Distr.* California.

T. uniflora See *Ipheion uniflorum*

triternatus, -a, -um: with three lots of three leaflets or other parts.

Triticum (Classical Latin name for wheat.) WHEAT. A genus of 20 species of robust annual and perennial grasses to 1.2m tall with flat leaves and spike-like inflorescences. Possibly the most important cereal crop in the world, wheat is grown almost entirely for human consumption. Numerous species, hybrids, and cultivars have arisen over the past 10,000 years, but the precise origin of all the modern wheats is still not fully understood. *Distr.* Europe, Mediterranean, W Asia. *Fam.* Gramineae. *Class* M.

T. aestivum BREAD WHEAT. The most important modern group of wheats with very many cultivars available.

T. dicoccon EMMER WHEAT. A very ancient group of wheats cultivated by the Babylonians.

T. durum DURUM, MACARONI WHEAT, SEMOLINA WHEAT. A wheat with a high level of gluten, used for pasta and semolina.

T. monococcum EINKORN WHEAT. A fodder wheat of the Palaeolithic period, little grown today.

T. spelta SPELTA. An ancient group of wheats tolerant of poor soils.

Tritonia (From Greek *triton*, considered to mean weathercock; the stamens point in different directions.) A genus of 28–30 species of perennial herbs with 2 ranks of linear leaves and spikes of more or less irregular flowers. Some species are cultivated as ornamentals. *Distr.* S tropical Africa. *Fam.* Iridaceae. *Class* M.

T. crocata Flowers wide, cup-shaped, orange-pink. *Distr.* South Africa (Cape Province).

T. rosea See *T. rubrolucens*

T. rubrolucens Flowers more or less funnel-shaped, pink. *Distr.* South Africa (Cape Province, Natal).

triumphans: splendid.

Triuridaceae A family of 6 genera and 42 species of small, red-purple, saprophytic herbs with scale-like leaves and racemes of inconspicuous flowers. *Distr.* Widespread in tropical regions, extending as far N as Japan. *Class* M.

Trochetiopsis (From the name of the genus *Trochetia* and Greek *-opsis*, resembling.) A genus of 2 species of evergreen shrubs and trees with alternate simple leaves and clusters of small bisexual flowers. *Distr.* St Helena. *Fam.* Sterculiaceae. *Class* D.

T. melanoxylon This species was believed to be extinct until it was rediscovered in 1970.

Trochocarpa (From Greek *trochos*, wheel, and *karpos*, fruit, alluding to the spoke-like cells of the fruit.) A genus of 12 species of evergreen shrubs and trees with simple leathery leaves and short racemes of tubular or bell-shaped flowers that bear their parts in fives. The fruit is a fleshy drupe. *Distr.* Malaysia, E Australia. *Fam.* Epacridaceae. *Class* D.

T. laurina Shrub or small tree. Leaves glossy green. Flowers small, white, hairy within. Fruit fleshy, blue-black. Grown as an ornamental. *Distr.* Australia.

Trochodendraceae A family of 1 genus and 1 species. *Class* D.
See: Trochodendron.

Trochodendron (From Greek *trochos*, wheel, and *dendron*, tree, alluding to the arrangement of the stamens.) A genus of 1 species of large forest tree with evergreen leathery leaves and small flowers that lack sepals and petals. *Distr.* Korea and Japan to Taiwan. *Fam.* Trochodendraceae. *Class* D.

T. aralioides Occasionally cultivated as an ornamental. The bark is used locally as bird lime.

trochopteranthus, -a, um: with wheel-shaped, winged flowers.

trojanus, -a, -um: of Troy.

trolliifolius, -a, -um: from the genus *Trollius*, and *folius*, leaved.

Trollius (From the Swiss-German name for these plants, *Trollblume*, globeflower.) GLOBEFLOWER. A genus of 20–30 species of perennial herbs with palmately lobed or divided leaves and regular, white, yellow or orange flowers that bear 5–15 petal-like sepals, a similar number of smaller true petals, and numerous stamens. Several species are grown as ornamentals. *Distr.* Eurasia/N temperate regions (2 species in Europe). *Fam.* Ranunculaceae. *Class* D.

T. europaeus GLOBEFLOWER. Flowers spherical, lemon-yellow. Several garden cultivars have been raised from this species. *Distr.* Europe, Caucasus, N North America.

T. ledebourii To 1m high. Flowers bowl-shaped, orange-yellow. *Distr.* E Siberia, N Korea, N Mongolia.

trompetilla, scarlet *Bouvardia ternifolia.*

Tropaeolaceae The Nasturtium family. 3 genera and 88 species of climbing, somewhat succulent herbs with peltate, sometimes deeply lobed leaves. The flowers are showy, irregular, and spurred. This family contains the cultivated ornamental *Tropaeolum majus*, the garden nasturtium. *Distr.* Mexico to temperate South America. *Class* D.
See: Tropaeolum.

Tropaeolum (From Greek *tropaion*, trophy, originally used for the trunk of a tree on which were fixed the shields and helmets of a defeated enemy. The leaves represent the shields and the flowers the helmets.) NASTURTIUM, INDIAN CRESS, CANARY BIRD VINE, CANARY BIRD FLOWER, FLAME FLOWER. A genus of 86 species of annual and perennial climbing or bushy herbs with alternate peltate leaves and irregular spurred flowers that typically bear 5 sepals, 5 petals, and 8 stamens. Several species are grown as ornamentals. *Distr.* S Mexico to Brazil and Patagonia. *Fam.* Tropaeolaceae. *Class* D.

T. majus NASTURTIUM, INDIAN CRESS. Annual climber. Leaves round. Flowers large, variable in colour. Numerous ornamental cultivars have been raised from this species. *Distr.* S America, cultivated since antiquity.

T. peregrinum CANARY CREEPER. Half-hardy annual or perennial climber. Flowers orange or yellow, fringed. Cultivated as an ornamentals. *Distr.* S America.

trumpet: angel's ~ *Datura inoxia, Brugmansia arborea* **golden** ~ *Allamanda cathartica* **herald's** ~ *Beaumontia grandiflora* **hummingbird's** ~ *Epilobium canum* **yellow** ~ *Sarracenia flava.*

trumpet bush *Tecoma*

trumpet climber *Campsis radicans*

trumpet creeper *Campsis.*

trumpet flower *Bignonia capreolata* **Chinese** ~ *Campsis grandiflora* **evening** ~ *Gelsemium sempervirens* **Nepal** ~ *Beaumontia grandiflora.*

trumpets *Sarracenia flava.*

trumpet weed *Eupatorium purpureum.*

truncatulus, -a, -um: somewhat abruptly cut off.

truncatus, -a, -um: truncated, abruptly cut off.

tsangpoensis, -e: of near the Tsangpo River, Tibet.

Tsuga (From the Japanese name for these trees.) HEMLOCK, HEMLOCK SPRUCE. A genus of 9–10 species of evergreen coniferous trees with narrow flattened leaves that are spirally arranged but twisted so as to lie in 2 distinct ranks. The woody cones are pendulous and ripen in their first year. *Distr.* N temperate regions. *Fam.* Pinaceae. *Class* G.

T. canadensis EASTERN HEMLOCK, CANADA HEMLOCK. Many ornamental cultivars have been raised from this species and it is a source of rough timber and tannin. *Distr.* E North America.

T. heterophylla WESTERN HEMLOCK. Tree to 65m high. An important source of timber and pulp. *Distr.* W North America.

T. menziesii See *Pseudotsuga menziesii*

Tsusiophyllum (From *tsutsuji*, the Japanese verncular name for a section of the genus *Rhododendron*, and Greek *phyllon*, leaf.) A genus of 1 species of prostrate semi-evergreen shrub with dense clusters of white flowers. *Distr.* Japan. *Fam.* Ericaceae. *Class* D.

T. tanakae Cultivated ornamentally as a rock-garden shrub.

tubatus, -a, -um: trumpet-shaped.

Tuberaria A genus of 12 species of annual and perennial herbs with basal rosettes of simple leaves and cymes of saucer-shaped yellow flowers that bear 5 sepals, 5 petals, and numerous stamens. This genus is closely related to the genus *Helianthemum*. Several species are grown as ornamentals. *Distr.* W and Central Europe, Mediterranean. *Fam.* Cistaceae. *Class* D.

T. globulariifolia Perennial. Flowers to 5cm across. Petals with a red-brown basal spot. *Distr.* NW Spain, Portugal.

T. guttata Annual herb. Leaves hairy. Petals spotted maroon at base. *Distr.* Central and S Europe.

T. lignosa Perennial. Flowers to 3cm across. *Distr.* W Mediterranean.

tuberculatus, -a, -um: covered with warts.

tuberifera, -um: bearing tubers.

tuberose *Polianthes tuberosa*.

tuberosus, -a, -um: tuberous.

tuber root *Asclepias tuberosa*.

tubiflorus, -a, -um: with tubular flowers.

tubiformis, -e: tubular.

tubulosus, -a, -um: tubular.

tuft, yellow *Alyssum murale*.

Tulbaghia (After Ryk Tulbagh (died 1771), governor of the Cape of Good Hope.) WILD GARLIC, SOCIETY GARLIC. A genus of 20–26 species of perennial rhizomatous or bulbous herbs with linear leaves and umbels of many star-shaped, often fragrant, flowers. The flowers bear a small corona, similar to that found in *Narcissus*. Some species are cultivated as ornamentals. *Distr.* S Africa. *Fam.* Alliaceae. *Class* M.

T. alliacea Flowers green-white with an orange corona. *Distr.* South Africa, Zimbabwe.

T. capensis Flowers olive-green with a maroon corona. *Distr.* Cape Province.

T. cominsii Flowers white, fragrant at night. *Distr.* Cape Province.

T. fragrans SWEET GARLIC, PINK AGAPANTHUS. Flowers purple, sweet-scented. *Distr.* Transvaal.

T. natalensis Flowers white with a yellow- or green-tinged corona, fragrant. *Distr.* Natal.

T. violacea Flowers sweet-scented, lilac. *Distr.* Cape Province, Transvaal.

tulip *Tulipa* **Cape** ~ *Haemanthus coccineus* **golden globe** ~ *Calochortus amabilis* **horned** ~ *Tulipa acuminata* **lady** ~ *T. clusiana* **red Cape** ~ *Haemanthus* **water-lily** ~ *Tulipa kaufmanniana*.

Tulipa (From Turkish *tulband*, a turban, in allusion to the shape of the flowers.) TULIP. A genus of about 100 species of bulbiferous herbs with linear to broadly ovate leaves and typically solitary, bell- to cup-shaped flowers that bear 2 whorls of 3 free tepals, 6 stamens, and a 3-sided superior ovary. Many of the species and numerous hybrids and cultivars are grown as ornamentals. Tulips have been so popular at times that single bulbs have changed hands for high prices and have been the subject of financial speculation. The two main periods when this occurred were 1634–7 in Holland and 1702–20 in Turkey. Tulips are currently produced on a large scale, for both the cut-flower and the bulb markets, in Holland, E England, and parts of the W USA. Most species in cultivation today are of complex hybrid origin and only a few of the rock-garden plants are true wild species. *Distr.* Temperate regions of the Old World, especially Central Asia. *Fam.* Liliaceae. *Class* M.

T. acuminata HORNED TULIP. Flowers yellow, striped red. Tepals tapering to a long tip. *Distr.* Cultivated.

T. aitchisonii See *T. clusiana*

T. chrysantha See *T. clusiana*

T. clusiana (After Charles de l'Ecluse (Carolus Clusius) (1526–1609), Flemish botanist.) LADY TULIP. Leaves grey-green. Flowers cream but the outer tepals have a broad red stripe on the outside. *Distr.* N India to Iran, naturalized in S Europe and Turkey.

T. fosteriana Flowers scarlet with a central black blotch. This species forms the basis of a complex group of hybrids with *T. greigii* and *T. kaufmanniana*. *Distr.* Soviet Central Asia.

T. greigii (After General Greig, President of the Imperial Russian Horticultural Union.) Leaves streaked purple-brown. Flowers usually red, occasionally yellow-white. *Distr.* Tadjikistan.

T. kaufmanniana (After General von Kaufmann, governor-general of the region in which it was found.) WATER-LILY TULIP. Flowers opening flat, yellow-white to brick red. *Distr.* Central Asia.

T. stellata See *T. clusiana*

tulipifera: tulip-bearing.

tulip tree *Liriodendron tulipifera* **Chinese** ~ *L. chinense*.

tuna Mexican name for the *Opuntia* fruit.

Tunica See *Petrorhagia*

T. saxifraga See *Petrorhagia saxifraga*

tunicatus, -a, -um: coated, sheathed.

tuolumnensis, -e: of Tuolumne County, California, USA.

tupelo *Nyssa sylvatica* **Chinese** ~ *N. sinensis*.

turbinatus, -a, -um: shaped like an upside-down cone, wider near the top.

turgidus, -a, -um: turgid, full.

Turk's cap *Aconitum napellus*.

turkey beard *Xerophyllum asphodeloides*.

Turkish root *Rheum palmatum*.

turmeric *Curcuma longa*.

turmeric root *Hydrastis canadensis*.

Turneraceae A family of 10 genera and 110 species of shrubs, small trees, and a few herbs with alternate leaves that often have glandular teeth on the margin. The flowers are regular with their parts in fives and are typically borne singly in the axils of the leaves. *Distr.* Tropical and warm-temperate regions. *Class* D.

turnip *Brassica rapa* **Indian** ~ *Arisaema, A. triphyllum*.

turnsole *Heliotropium*.

turpentine, Cyprus *Pistacia terebinthus*.

Turraea (After Giorgio della Turra (1607–88), professor of botany at Padua.) A genus of 60 species of shrubs and trees with alternate, simple or lobed leaves and clusters of white flowers that bear 5 narrow petals and 10 fused stamens. *Distr.* Tropical and S Africa, Tropical Asia, Australia. *Fam.* Meliaceae. *Class* D.

T. obtusifolia Evergreen bushy shrub. Flowers fragrant, borne in autumn and spring, followed by orange fruits. Grown as a tender ornamental. *Distr.* South America.

turtlehead *Chelone obliqua, C. glabra*.

Tussilago (From Latin *tussis*, a cough, alluding to medicinal uses of the leaves.) COLTSFOOT. A genus of 15 species of rhizomatous herbs with round to kidney-shaped, stalked leaves and yellow flowers in dandelion-like heads. *Distr.* N temperate regions. *Fam.* Compositae. *Class* D.

T. farfara COLTSFOOT. Flowers borne in spring before leaves appear. The leaves were formerly smoked as a treatment for asthma; they are reported as having a synergistic effect with *Cannabis sativa*. *Distr.* Eurasia, Mediterranean, naturalized in E NorthAmerica.

Tutcheria (After William James Tutcher (1867–1920), of the Botanical and Forestry

Department Hong Kong.) A genus of 2 species of evergreen trees with simple alternate leaves and white flowers that bear 5 petals and numerous fused stamens. *Distr.* SE China. *Fam.* Theaceae. *Class* D.

T. spectabilis Small tree. Grown as a half-hardy ornamental. *Distr.* SE China.

Tweedia (After J. Tweedie (1775–1862), of Glasgow and Argentina.) A genus of 1 species of climbing subshrub with simple white-hairy leaves and somewhat fleshy, blue, star-shaped flowers. *Distr.* Temperate South America. *Fam.* Asclepiadaceae. *Class* D.

T. caerulea Grown as a tender ornamental.

twinberry *Lonicera involucrata.*

twinflower *Linnaea borealis.*

twin leaf *Jeffersonia.*

twisted stalk *Streptopus.*

twistwood *Viburnum lantana.*

Typha (Classical Greek name for these plants.) REEDMACE, CAT TAIL, BULRUSH. A genus of 10–15 species of rhizomatous herbs of shallow water with erect simple stems and 2 ranks of leathery linear leaves. The small unisexual flowers are borne in dense, poker-like, terminal spikes and surrounded by numerous hairs. Several species are grown as ornamentals. The rhizomes have been consumed as an emergency food and the leaves used for matting. Several species are of ecological importance in stabilizing marshes. *Distr.* Widespread. *Fam.* Typhaceae. *Class* M.

T. angustifolia LESSER BULLRUSH, NARROW-LEAVED REEDMACE, SOFT FLAG. *Distr.* Europe, Asia, N America.

T. latifolia CAT'S TAIL, BULLRUSH, NAILROD. *Distr.* N America, Eurasia, N Africa.

Typhaceae The Reedmace family. 2 genera and 24 species of tall aquatic herbs with simple erect stems usually submerged at base. Leaves arise from the subterranean part of the stem and are strap-shaped. Flowers are crowded into a characteristic terminal club-shaped spadix. Rhizomes of some species are eaten, leaves of others are used in weaving. *Distr.* Widespread in freshwater habitats. *Class* M.

See: Sparganium, Typha.

typhinus, -a, -um: resembling the genus *Typha.*

U

udo *Aralia cordata.*

ugandensis, -e: of Uganda, East Africa.

ugli fruit *Citrus* × *tangelo.*

Ugni (From the Chilean vernacular name, *uñi.*) A genus of 5–15 species of shrubs with simple, leathery, opposite leaves and nodding bell-shaped flowers that bear 5 petals and numerous stamens. *Distr.* Tropical and warm America. *Fam.* Myrtaceae. *Class* D.

U. molinae CHILEAN GUAVA. Grown as an ornamental and for its blue-black berries that are used for jam. *Distr.* Chile, W Argentina.

ukranicus, -a, -um: of Ukraine.

Ulex (Latin name used by Pliny for a type of heather.) FURZE, GORSE, WHIN. A genus of 20 species of spiny shrubs with trifoliate leaves that only occur in the seedlings and are replaced by green spiny branches in adult plants. The yellow pea-like flowers are borne singly or in small clusters and are usually highly fragrant. Several species are of ecological significance and some are grown as ornamentals. *Distr.* W Europe, N Africa. *Fam.* Leguminosae. *Class* D.

U. europaeus COMMON GORSE, [illegible], FURZE, GORSE, WHIN. Upright shrub to 2m high. Several ornamental cultivars have been raised from this species. *Distr.* W and Central Europe.

U. minor DWARF GORSE. Densely branched, dwarf shrub to 1m high. *Distr.* W Europe.

U. nanus See *U. minor*

uliginosus, -a, -um: marshy, growing in marshes.

Ulmaceae The Elm family. 15 genera and 140 species of shrubs and trees with simple alternate leaves and clusters of small, green, inconspicuous flowers. Trees of this family are an important source of decay-resistant timber. *Distr.* Widespread. *Class* D.

See: Celtis, Planera, Ulmus, Zelkova.

ulmaria: like *Ulmus*, ELM, referring to the leaflets.

ulmifolius, -a, -um: from the genus *Ulmus*, ELM, and *folius*, leaved.

ulmo *Eucryphia cordifolia.*

ulmoides: from the genus *Ulmus*, ELM, with the ending *-oides*, indicating resemblance.

Ulmus (Classical Latin name for these trees.) ELM. A genus of 18–45 species of deciduous trees with simple alternate leaves and inconspicuous 4- or 5-lobed, green flowers that lack petals and are followed by winged fruits. Many species are vulnerable to Dutch Elm Disease which considerably reduced their numbers in Great Britain during the 1970s and 80s. These trees are an important source of timber. *Distr.* N temperate regions to N Mexico. *Fam.* Ulmaceae. *Class* D.

U. glabra WYCH ELM. Open tree to 40m high. *Distr.* Europe, W Asia.

U. × hollandica DUTCH ELM. A hybrid of *U. minor* × *U. glabra*. Numerous cultivars and clones have been raised from this cross.

U. montana See *U. glabra*

U. parvifolia CHINESE ELM, LACEBARK. Semi-deciduous tree to 12m high. *Distr.* Japan, China.

U. procera ENGLISH ELM. Tree to 30m high. Prior to the onset of Dutch Elm Disease this tree played a major part in the British landscape. *Distr.* S and Central England.

ultramontana: of the farther side of the mountains.

umbellatus, -a, -um: with flowers arranged in an umbel, a type of inflorescence with individual flower-stalks arising from one place and forming a flat head, as in COW PARSLEY, for example.

Umbelliferae The Carrot family. 428 genera and 3000 species of annual and perennial herbs and subshrubs typically with hollow stems and trifoliate or pinnate leaves. The flowers are borne in umbels or compound umbels, are typically small, and may vary in sex and size across the inflorescence. This family contains many important herbs, spices, and flavourings as well as ornamentals and medicinal plants. *Distr.* Widespread. *Class* D. *See: Aciphylla, Actinotus, Aegopodium, Anethum, Angelica, Anisotome, Anthriscus, Apium, Astrantia, Athamanta, Azorella, Berula, Bolax, Bupleurum, Carum, Chaerophyllum, Conium, Conopodium, Coriandrum, Crithmum, Cryptotaenia, Cuminum, Daucus, Eryngium, Ferula, Foeniculum, Hacquetia, Heracleum, Hydrocotyle, Laserpitium, Levisticum, Ligusticum, Lilaeopsis, Lomatium, Melanoselinum, Meum, Myrrhis, Oenanthe, Osmorhiza, Pastinaca, Petroselinum, Pimpinella, Pleurospermum, Sanicula, Selinum, Silaum, Sium, Smyrnium, Trachymene, Trinia, Zizia.*

umbelliferus, -a, -um: bearing umbels.

umbelliformis, -e: with the shape of an umbel.

Umbellularia (From Latin *umbella*, parasol, alluding to the inflorescences.) A genus of 1 species of evergreen aromatic tree with simple, alternate, leathery leaves and dense umbels of small bisexual flowers that are followed by red berries. *Distr.* W North America. *Fam.* Lauraceae. *Class* D.

U. californica CALIFORNIA LAUREL, CALIFORNIA OLIVE, CALIFORNIA BAY, PEPPERWOOD, OREGON MYRTLE. Grown as an ornamental as well as being a source of good timber.

umbilicatus, -a, -um: with a navel.

Umbilicus (From Latin *umbilicus*, navel, alluding to the leaf shape.) A genus of 18 species of succulent perennial herbs with round peltate leaves which have a dimple in the centre of the upper surface opposite the point of attachment of the stalk on the lower surface. The numerous tubular or bell-shaped flowers are borne in racemes or panicles. Some are cultivated as ornamentals. *Distr.* Europe, W Asia, African mountains. *Fam.* Crassulaceae. *Class* D.

U. rupestris NAVELWORT, PENNY PIES, WALL PENNYWORT, PENNYWORT. Base tuberous. Flowers in simple racemes, white-green. Often grows in cracks of rocks and walls. *Distr.* British Isles to Macaronesia and SW Asia.

umbonatus, -a -um: with a raised central point.

umbrella leaf *Diphylleia cymosa.*

umbrella plant *Eriogonum allenii, Darmera peltata, Cyperus involucratus.*

umbrella tree *Magnolia tripetala, Polyscias, Schefflera* **Queensland** ~ *S. actinophylla.*

umbrosus, -a, -um: growing in shade.

uncifolius, -a, -um: with hooked leaves.

uncinatus, -a, -um: hooked.

Uncinia (From Latin *uncinatus*, hooked at the end, alluding to the hooked fruit.) A genus of 35 species of grass-like perennial herbs with very small, wind-pollinated flowers. The female flowers are enclosed in a structure known as a utricle which has a hooked tip. *Distr.* Tropical regions. *Fam.* Cyperaceae. *Class* M.

U. egmontiana Leaves dull-red or green. Grown as a hardy ornamental. *Distr.* New Zealand.

undulatifolius, -a, -um: with wavy leaves.

undulatus, -a, -um: wavy.

unguicularis, -e: clawed.

unicolor: one-coloured.

unicorn plant, sweet *Proboscidea fragrans.*

unicorn root *Aletris farinosa, Chamaelirium luteum.*

uniflorus, -a, -um: single-flowered, with solitary flowers.

uniflosculosus, -a, -um: with one floret.

unifolius, -a, -um: with one leaf.

Uniola (Classical Latin name for a different (unknown) plant.) SPANGLE GRASS, SPIKE GRASS. A genus of 4 species of rhizomatous, stoloniferous grasses with flat leaves that roll up when dry. Several species are grown as ornamentals, others are used to prevent erosion. *Distr.* North and South America. *Fam.* Gramineae. *Class* M.

U. latifolia See *Chasmanthium latifolium*

U. paniculata NORTH AMERICAN SEA OATS, SEASIDE OATS. A grass of sand dunes and salt marshes that is sometimes grown as an ornamental. *Distr.* E USA to the West Indies.

unvarifolius, -a, -um: with uniform leaves.

uralensis, -e: from the Ural Mountains.

urbanus, -a, -um: of towns or built-up areas.

urceolatus, -a, -um: urn-shaped.

Urceolina (From Latin *urceolus*, a small pitcher, alluding to the shape of the flowers.) A genus of 2–6 species of bulbiferous perennial herbs with oblong to elliptic leaves and umbels of pitcher-shaped flowers. Several species are cultivated as ornamentals. *Distr.* Andes. *Fam.* Amaryllidaceae. *Class* M.

U. miniata See *Stenomesson miniatum*

U. peruviana Flowers scarlet, pendulous. *Distr.* Peru and Bolivia.

U. urceolata Flowers red with white margins. *Distr.* Peru.

urens: stinging.

Urginea (From Beni Urgin, the name of an Arab tribe in Algeria.) A genus of about 100 species of bulbiferous herbs with narrow or oblong leaves and many-flowered racemes of star-shaped flowers. Several species are grown as ornamentals. *Distr.* Africa, Mediterranean. *Fam.* Hyacinthaceae. *Class* M.

U. maritima SEA SQUILL, SEA ONION, SQUILL. Flowers white with green or pink stripes. *Distr.* Mediterranean coasts and Portugal, W Asia.

urn plant *Aechmea fasciata*.

Urospermum (From Greek *oura*, tail, and *sperma*, seed, alluding to the beaked nutlets.) A genus of 2 species of annual and perennial herbs with milky latex, simple to pinnately lobed leaves, and small yellow flowers that are borne in large dandelion-like heads. *Distr.* Mediterranean region. *Fam.* Compositae. *Class* D.

U. dalechampii Biennial or perennial clump-forming herb. Grown as a hardy ornamental. *Distr.* Mediterranean.

Ursinia (After Johannes Heinrich Ursinus (1608–67), German botanical author.) A genus of 40 species of annual and perennial herbs and shrubs with pinnately divided or lobed leaves and small flowers in daisy-like heads that bear distinct rays. Several species are grown as ornamentals. *Distr.* South Africa, Ethiopia. *Fam.* Compositae. *Class* D.

U. chrysanthemoides See *Euryops chrysanthemoides*

U. sericea Procumbent shrub. Flowers yellow. *Distr.* South Africa (Cape Province).

Urtica (From Latin *uro*, to burn, alluding to their stinging hairs.) NETTLE. A genus of 45 species of annual or perennial herbs or subshrubs with simple opposite leaves and clusters of small, green, unisexual flowers. These plants are armed with stinging hairs. *Distr.* More or less cosmopolitan especially in the N hemisphere. *Fam.* Urticaceae. *Class* D.

U. dioica STINGING NETTLE. Perennial herb to 1m high. Eaten as a vegetable after blanching to remove stings. A persistent weed of cultivation that thrives on nitrogen-rich soils. *Distr.* N hemisphere.

U. pilulifera ROMAN NETTLE. Annual herb to 60cm high. *Distr.* S Europe.

Urticaceae The Stinging Nettle family. 48 genera and about 1000 species of herbs, shrubs, and a few trees with alternate or opposite leaves and small, green, typically unisexual flowers. Some species are a source of fibre. *Distr.* Widespread. *Class* D.
See: Boehmeria, Debregeasia, Elatostema, Myriocarpa, Parietaria, Pellionia, Pilea, Soleirolia, Urtica.

urticifolius, -a, -um: from the genus *Urtica*, NETTLE, and *folius*, leaved.

uruguayensis, -e: of Uruguay.

usneoides: from the genus *Usnea*, a lichen that hangs from trees, with the ending *-oides*, indicating resemblance.

utahensis, -e: of Utah, USA.

utilis, -e: useful.

Utricularia (From Latin *utriculus*, small bottle, alluding to the bladders.) BLADDERWORT. A genus of about 200 species of aquatic, epiphytic or terrestrial, carnivorous herbs with vegetative parts that are not clearly differentiated into roots, stems, and leaves but with clearly defined racemes of 2-lipped flowers. Small bladder-like traps are used to capture and digest small aquatic animals. The water pressure within the bladder is reduced so that when an external trigger hair is touched a small hinged door swings open and the passing animal is sucked into the bladder, where it is digested. *Distr.* Widespread. *Fam.* Lentibulariaceae. *Class* D.

U. exoleta See *U. gibba*

U. gibba Annual or perennial, floating, aquatic herb. Flowers bright yellow with red veins. *Distr.* Widespread.

U. vulgaris GREATER BLADDERWORT. Floating, perennial, aquatic herb. Leaves numerous, finely divided. Flowers yellow. *Distr.* Europe, N Africa, temperate Asia.

utriculatus, -a, -um: bladder-like.

uva-crispa: a medieval name meaning crisp grape.

uvaria resembling a bunch of grapes.

uva-ursi: bear's grape.

uva-vulpis: fox's grape.

Uvularia (From Latin *uvula*, the soft hanging structure in the human throat, referring to the hanging flowers.) BELLWORT, MERRYBELLS, WILD OATS. A genus of 5 species of rhizomatous herbs with arching stems, 2 ranks of clasping or perforated leaves, and bell-shaped, yellow, pendulous flowers. The plants of this genus superficially resemble those of *Polygonatum*, SOLOMON'S SEAL. Several species are grown as ornamentals. *Distr.* E North America. *Fam.* Convallariaceae. *Class* M.

U. grandiflora Flowers large, to 5cm long.

U. perfoliata Leaves appearing to pierce stem. Tepals pale yellow, somewhat twisted.

U. sessilifolia STRAWBELL. Young shoots may be eaten like ASPARAGUS.

V

Vaccaria (From Latin *vacca*, cow; it is said to have been used for forage.) A genus of 1 species of annual herb with simple opposite leaves and flowers borne in the axils of the stem branches. The flowers bear 5 sepals and 5 pink petals. The sepals form an inflated tube. *Distr*. Eurasia, Mediterranean, introduced North America. *Fam*. Caryophyllaceae. *Class* D.

V. hispanica COW COCKLE. Occasionally grown as an ornamental.

vacciniifolius, -a, -um: from the genus *Vaccinium*, and *folius*, leaved.

Vaccinium (Latin name of obscure origin, possibly derived from *Hyacinthus* or *bacca*, berry.) A genus of about 450 species of deciduous or evergreen shrubs, small trees, and lianas with simple leaves and 4–5 lobed, white to red flowers. Some species are cultivated as ornamentals or for their edible fruits. *Distr*. N Hemisphere from Arctic circle to tropical mountains; a few species also occur in South Africa. *Fam*. Ericaceae. *Class* D.

V. angustifolium LOW-BUSH BLUEBERRY, LATE SWEET BLUEBERRY, LOW SWEET BLUEBERRY. Fruit much collected from wild plants that are often formed for production of fruit. *Distr*. North America.

V. arctostaphylos (From Greek *arktos*, bear, and *staphyle*, bunch of grapes.) CAUCASIAN WHORTLEBERRY. Deciduous shrub. *Distr*. Caucasus and SE Europe.

V. bracteatum Evergreen shrub to 1m with red fruit. *Distr*. China and Japan.

V. caespitosum DWARF BILBERRY. Small deciduous shrub. *Distr*. N and W North America.

V. corymbosum HIGHBUSH BLUEBERRY, BLUEBERRY, AMERICAN BLUEBERRY, SWAMP BLUEBERRY. Cultivated in North America and W and Central Europe for the edible fruit. *Distr*. E North America.

V. cylindraceum Deciduous or semi-evergreen shrub with blue-black fruit. *Distr*. The Azores.

V. delavayi (After Abbé Jean Marie Delavay (1838–95), French missionary in China who introduced it into cultivation.) Compact evergreen shrub with carmine fruit. *Distr*. SW China and Burma.

V. deliciosum Rapidly spreading, dense shrub. *Distr*. NW North America.

V. donianum See *V. sprengelii*

V. erythrinum Erect compact shrub. *Distr*. Java.

V. erythrocarpum SOUTHERN MOUNTAIN CRANBERRY. Deciduous shrub with sour red fruit. *Distr*. SE USA.

V. glaucoalbum Dense ever-green shrub. *Distr*. Himalaya.

V. macrocarpon CRANBERRY, AMERICAN CRANBERRY, LARGE CRANBERRY. Grown commercially for fruit which is typically eaten with poultry. *Distr*. E North America.

V. membranaceum THIN-LEAVED BILBERRY, BLUE HUCKLEBERRY, MOUNTAIN BLUEBERRY, BILBERRY. Erect deciduous shrub. *Distr*. W North America.

V. moupinense Dense evergreen shrub. *Distr*. W Sichuan, China.

V. myrsinites EVERGREEN BLUEBERRY. More or less procumbent, evergreen shrub. *Distr*. SE USA.

V. myrtillus BILLBERRY, WHORTLEBERRY, BLAEBERRY, WINEBERRY. Evergreen erect shrub. The edible fruit is used in pies and wine-making; the skin is alleged to improve night vision. *Distr*. Eurasia.

V. nummularia Small evergreen shrub. *Distr*. Himalaya.

V. ovatum BOX BLUEBERRY, CALIFORNIA HUCKLEBERRY, EVERGREEN HUCKLEBERRY, SHOT HUCKLEBERRY. Erect bushy shrub. *Distr*. W North America.

V. oxycoccos CRANBERRY, EUROPEAN CRANBERRY, SMALL CRANBERRY. Fruit eaten especially with poultry; somewhat inferior to *V. macrocarpon*. *Distr*. N temperate regions.

V. padifolium MADEIRAN WHORTLEBERRY. Erect evergreen or semi-evergreen shrub. *Distr*. Madeira.

V. parvifolium RED BILBERRY, RED HUCKLEBERRY. Erect deciduous shrub. *Distr.* NE Asia.

V. praestans KAMCHATKA BILBERRY. Erect deciduous shrub. *Distr.* NE Asia.

V. retusum Small evergreen shrub. *Distr.* Himalaya.

V. sprengelii Deciduous or evergreen shrub with dark red-purple fruit. *Distr.* NE India, W and Central China.

V. uliginosum BOG WHORTLEBERRY, BOG BILBERRY, MOORBERRY. Stiffly erect, deciduous shrub. *Distr.* North America and N Europe.

V. virgatum SOUTHERN BLACK BLUEBERRY, RABBIT-EYE BLUEBERRY. Deciduous shrub often forming large colonies. *Distr.* E and SE USA.

V. vitis-idaea COWBERRY, LINGBERRY, FOXBERRY, MOUNTAIN CRANBERRY, CRANBERRY. Evergreen creeping shrub. The fruit is eaten as a cranberry substitute. *Distr.* Europe, Asia, North America.

vagans: wandering, referring to an invasive habit.

Vagaria (From Latin *vagans*, wandering, alluding to the fact that the origin of the first specimen to flower was obscure.) A genus of 2–4 species of bulbiferous perennial herbs with linear white-banded leaves and umbels of white, funnel-shaped flowers. *Distr.* N Africa to Turkey. *Fam.* Amaryllidaceae. *Class* M.

V. parviflora Cultivated as a half-hardy ornamental. *Distr.* Syria and Israel.

vaginalis, -e: with a sheath.

vaginatus, -a, -um: with a sheath.

vagus: uncertain, with no particular direction.

Vahliaceae A family of 1 genus and 5 species of herbs. This family is sometimes included within the family Saxifragaceae. *Distr.* Africa to NW India, China. *Class* D.

valerian *Valeriana* **common** ~ *V. officinalis* **Greek** ~ *Polemonium caeruleum* **red** ~ *Centranthus ruber*.

Valeriana (From Latin *valere*, to be healthy, alluding to the medicinal properties of some species.) VALERIAN. A genus of 150–250 species of perennial herbs and shrubs with opposite, simple to pinnately divided leaves, and heads of small, funnel-shaped or flat flowers. Several species are grown as ornamentals and some have local medicinal uses. *Distr.* N temperate regions, South Africa, Andes. *Fam.* Valerianaceae. *Class* D.

V. officinalis COMMON VALERIAN, GARDEN HELIOTROPE. Leaves pinnately divided or simple. Flowers white or pink. The fragrant rhizome has been widely used medicinally, particularly as a sedative. *Distr.* Eurasia.

Valerianaceae The Spikenard family. 15 genera and 400 species of herbs and shrubs with opposite pinnate leaves and cymes of irregular bisexual flowers. *Distr.* Widespread but chiefly in the N Hemisphere. *Class* D. *See: Centranthus, Patrinia, Valeriana, Valerianella.*

Valerianella (Diminutive of the genus name *Valeriana*.) CORN SALAD. A genus of 50 species of annual and biennial herbs with simple opposite leaves and heads of small tubular flowers. *Distr.* N temperate regions. *Fam.* Valerianaceae. *Class* D.

V. locusta CORN SALAD, LAMB'S LETTUCE, COMMON CORN SALAD. To 40cm high. Flowers pale lilac. Eaten as a salad vegetable. *Distr.* Europe and Mediterranean, N Africa, W Asia.

V. olitoria See *V. locusta*

validus, -a, -um: robust.

Vallea (After Felice Valle (died 1747), Italian botanist.) A genus of 1 species of evergreen shrub with simple, occasionally palmately lobed leaves and cymes of small red flowers that bear 5 petals and numerous stamens. *Distr.* Colombia to Bolivia. *Fam.* Elaeocarpaceae. *Class* D.

V. stipularis Grown as an ornamental.

vallis *Vallisneria*.

Vallisneria (After Antonio Vallisnieri de Vallisnera (1661–1730), professor at the University of Padua.) EEL GRASS, VALLIS. A genus of 8–10 species of submerged, grass-like, perennial herbs with the male and female flowers borne on separate plants. The male flowers are minute and break free to float on the surface where they open, like those of *Elodea canadensis*. *Distr.* Tropical and warm temperate regions. *Fam.* Hydrocharitaceae. *Class* M.

V. americana Leaves to 2m long. *Distr.* E USA.

V. gigantea See *V. americana*

V. spiralis EEL GRASS, TAPE GRASS. A popular oxygenator plant for warm or cold aquaria. *Distr.* S Europe, W Asia.

Vallota See *Cyrtanthus*

Vancouveria (After Captain George Vancouver (1758–98), British explorer.) A genus of about 3 species of rhizo-matous herbs with compound basal leaves and pendulous white-yellow flowers that bear around 12 sepals, 6 petals, and 6 stamens. The smooth black seeds are dispersed by ants. Several species are grown as ornamentals. *Distr.* Coast ranges of the W USA. *Fam.* Berberidaceae. *Class* D.

V. chrysantha Leaves evergreen, leathery. Flowers yellow. *Distr.* S Oregon.

V. hexandra Leaves deciduous. Flowers white. *Distr.* Washington to California.

variabilis, -e: variable.

varicosus, -a, -um: with dilated veins.

variegatus, -a, -um: variegated.

variifolius, -a, -um: with variable, diverse leaves.

varius, -a, -um: diverse.

varnish tree *Koelreuteria paniculata* **Japanese** ~ *Firmiana simplex.*

vegetable hummingbird *Sesbania grandiflora.*

vegetable sponge *Luffa aegyptiaca.*

vegetus, -a, -um: fresh, vigorous.

Veitchia (After James Veitch (1815–69), and his son John Gould Veitch (1839–70), who had nurseries in Exeter and Chelsea and introduced many new plants into cultivation.) CHRISTMAS PALM, MANILA PALM. A genus of 18 species of small to medium-sized palms with feather-shaped leaves to 4m long and branched inflorescences to 1m long. Some of the species are cultivated as ornamentals. *Distr.* Philippines to Fiji. *Fam.* Palmae. *Class* M.

V. merrillii CHRISTMAS PALM, MANILA PALM. Trunk to 5m high, narrowing at base. Leaves ascending. *Distr.* Palawan Island.

Velloziaceae A family of 10 genera and about 300 species of fibrous shrubs with branched stems covered in old leaf bases. *Distr.* E Asia, Africa, Madagascar, South America and China. *Class* M.

Veltheimia (After August Ferdinand von Veltheim (1741–1801), German patron of botany.) A genus of 2 species of bulbiferous herbs with thick strap-shaped leaves and dense racemes of tubular flowers. Cultivated as ornamental greenhouse plants. *Distr.* South Africa. *Fam.* Hyacinthaceae. *Class* M.

V. bracteata Flowers pink-purple. *Distr.* South Africa (Cape Province).

V. capensis Flowers white or pink, marked red. *Distr.* South Africa (Cape Province).

V. viridifolia See *V. capensis*

velutinus, -a, -um: velvety, densely covered with fine short erect hairs.

velvet flower *Amaranthus caudatus.*

velvet leaf *Kalanchoe beharensis*, *Abutilon theophrasti.*

velvet plant *Gynura*, *G. aurantiaca* **purple** ~ *G. aurantiaca* **royal** ~ *G. aurantiaca* **trailing** ~ *Ruellia makoyana.*

× ***Venidio-Arctotis*** See *Arctotis*

Venidium See *Arctotis*

venosus, -a, -um: having many or prominently branched veins.

ventricosus, -a, -um: swollen on one side.

Venus' fly trap *Dionaea muscipula.*

Venus' hair *Adiantum capillus-veneris.*

Venus' slipper *Paphiopedilum.*

venustus, -a, -um: charming.

Veratrum (From Latin *vere*, truly, and *ater*, black, alluding to the colour of the root.) FALSE HELLEBORE. A genus of 20–45 species of rhizomatous herbs of wet places, with pleated leaves and panicles of white or green to purple, star- or bell-shaped flowers. Several species are cultivated as ornamentals and some are used medicinally. *Distr.* N temperate regions. *Fam.* Melanthiaceae. *Class* M.

V. album WHITE HELLEBORE. Lambs born of ewes which have fed on this plant may have a single central eye, suggesting an inspiration for Polyphemus in Homer. *Distr.* Eurasia.

V. nigrum To 1.2m high. Flowers chocolate-purple. *Distr.* Eurasia.

verbascifolius, -a, -um: from the genus *Verbascum*, and *folius*, leaved.

Verbascum (Classical Latin name for these plants.) MULLEIN. A genus of 250–360 species of annual, biennial, and perennial herbs or rarely shrubs with large rosettes of simple leaves and thick-stemmed inflorescences of 5-lobed flowers. A number of species and cultivars are grown as ornamentals. The seeds of some species have been used to poison fish. *Distr.* Eurasia. *Fam.* Scrophulariaceae. *Class* D.

V. blattaria MOTH MULLEIN. Biennial to 2m high. Flowers yellow. *Distr.* Eurasia, naturalized in North America.

V. lychnitis WHITE MAN, WHITE MULLEIN. Biennial to 1.5m high. Inflorescence branched. Flowers white or yellow. *Distr.* W and Central Europe.

V. nigrum DARK MULLEIN. Perennial. Inflorescence simple or branched. Flowers yellow. *Distr.* Europe.

V. thapsus AARON'S ROD, COMMON MULLEIN, FLANNEL PLANT, TORCHES. Biennial to 2m high. Leaves and stem densely white- or grey-hairy. Inflorescence unbranched, flowers yellow. Seeds formerly smoked with tobacco as a treatment for asthma. *Distr.* Eurasia, naturalized in North America.

Verbena (Latin name for the foliage of ceremonial and medicinal plants, possibly applied to *V. officinalis*.) VERVAIN. A genus of 250 species of annual and perennial herbs and subshrubs with simple or pinnately lobed opposite leaves and inflorescences of small, funnel- or salver-shaped, somewhat 2-lipped flowers. A number of species and numerous cultivars are grown as ornamentals. *Distr.* Tropical and temperate America, 2 species in Europe. *Fam.* Verbenaceae. *Class* D.

V. bonariensis PURPLE TOP, SOUTH AMERICAN VERVAIN, TALL VERBENA. Annual or perennial herb to 2m high. Flowers purple to lavender. *Distr.* S America.

V.×hybrida A complex group of hybrids that has given rise to a large number of ornamental cultivars. *Distr.* Garden origin.

V. officinalis COMMON VERVAIN, JUNO'S TEARS, COMMON VERBENA. Perennial herb to 80cm high. Flowers lilac to pale-pink, borne in dense spike-like inflorescences. Formerly used medicinally to treat gallstones and eye diseases. *Distr.* Eurasia, America.

V. patagonica See *V. bonariensis*

verbena: common ~ *Verbena officinalis* **lemon** ~ *Aloysia triphylla* **shrub** ~ *Lantana* **tall** ~ *Verbena bonariensis*.

Verbenaceae The Teak family. 86 genera and about 1750 species of herbs, shrubs, and trees with entire or divided leaves and 4–5 lobed tubular flowers. The family is a source of timber and a number of cultivated ornamentals. *Distr.* Widespread but mostly in tropical regions. *Class* D.
See: Aloysia, Baillonia, Callicarpa, Caryopteris, Citharexylum, Clerodendrum, Duranta, Glandularia, Lantana, Lippia, Petrea, Phyla, Rhaphithamnus, Stachytarpheta, Verbena, Vitex.

veris: of spring.

vernalis, -e: of spring.

vernicosus, -a, -um: varnished.

Vernonia (After William Vernon (died *c.*1711), English botanist who collected in Maryland.) IRONWEED. A genus of 1000 species of annual and perennial herbs, thick-stemmed shrubs and trees with large leaves and flowers in daisy-like heads that lack distinct rays. Several species are grown as ornamentals and some of the trees are a source of timber. *Distr.* Tropical and warm regions to North America. *Fam.* Compositae. *Class* D.

V. lindheimeri WOOLLY IRONWEED. Perennial herb to 80cm high. Stems and leaves densely covered in grey-white hairs. *Distr.* Texas.

vernus, -a, -um: of spring.

Veronica (Probably after St Veronica.) SPEEDWELL, BIRD'S EYE. A genus of 250 species of perennial herbs with simple, opposite and alternate leaves and irregular, 4-lobed, tubular to flat flowers that bear 2 exserted, spreading stamens. Visiting insects grasp the base of the stamens for support, pulling them together and so dusting themselves with pollen. Many species and cultivars are grown as ornamentals. *Distr.* N temperate regions, few species on tropical mountains and in S temperate regions. *Fam.* Scrophulariaceae. *Class* D.

V. beccabunga BROOKLIME, EUROPEAN BROOKLIME. Decumbent perennial. Flowers pale to deep blue, 7mm across, borne in slender racemes. *Distr.* Eurasia, naturalized North America.

V. chamaedrys GERMANDER SPEEDWELL, ANGEL'S EYES, BIRD'S EYE, GOD'S EYE. Rhizomatous perennial. Stems erect. Flowers blue with a white throat, to 10mm across, borne in racemes. *Distr.* Eurasia, naturalized in North America.

V. filiformis Mat-forming annual or perennial. A persistent weed. *Distr.* Turkey, Caucasus.

V. fruticans ROCK SPEEDWELL. Perennial herb with a woody stock. Flowers blue, tinged red in the centre. *Distr.* Europe.

V. incana SILVER SPEEDWELL. Rhizomatous, densely white-hairy perennial. Stems erect, to 60cm high. Flowers to 9mm across, clear blue. *Distr.* E Europe.

V. kellereri See *V. spicata*

V. lyallii See *Parahebe lyallii*

V. officinalis COMMON SPEEDWELL, GYPSYWEED. Procumbent or ascending perennial. Flowers 8mm across, violet-blue. Formerly used as a medicinal tea. *Distr.* N temperate regions.

V. perfoliata DIGGER'S SPEEDWELL. Large perennial herb to 1.5m high. Flowers in elongated racemes, to 15mm across, pale blue with darker lines. *Distr.* Australia (New South Wales, Victoria).

V. prostrata Mat-forming perennial. Flowers deep blue, to 8mm across. *Distr.* Europe.

V. rupestris See *V. prostrata*

V. saxatilis See *V. fruticans*

V. spicata Rhizomatous, glandular-hairy perennial. Stems erect, to 60cm high. Flowers clear blue, borne in spike-like racemes. A protected species in Great Britain. *Distr.* N Europe, Asia.

V. virginica See *Veronicastrum virginicum*

veronica, yellow *Paederota lutea.*

Veronicastrum (From the genus name *Veronica*, and Latin *-aster*, suggesting an inferior resemblance.) A genus of 1–2 species of perennial herbs with simple whorled leaves and spike-like terminal racemes of tubular flowers that have spreading lobes and bear 2 stamens. *Distr.* 1 species NE Asia, 1 species E North America. *Fam.* Scrophulariaceae. *Class* D.

V. virginicum CULVER'S ROOT, BLACK ROOT, BOWMAN'S ROOT. Erect perennial to 1.8m high. Flowers pale blue, borne in long slender spikes. Grown as an ornamental and formerly used medicinally as a purgative. *Distr.* E North America.

verrucosus, -a, -um: warty.

verruculosus, -a, -um: covered with small wart-like lumps.

versicolor: diversely coloured.

verticillatus, -a, -um: whorled.

verus, -a, -um: true.

vervain *Salvia verbenaca, Verbena* **bastard** ~ *Stachytarpheta* **common** ~ *Verbena officinalis* **false** ~ *Stachytarpheta* **South American** ~ *Verbena bonariensis.*

vescus, -a, um: thin.

Vestia (After L. C. de Vest (1776–1840), professor at Graz, Austria.) A genus of 1 species of evergreen shrub with simple, alternate, foetid leaves and cymes of funnel-shaped yellow-green flowers. *Distr.* Chile. *Fam.* Solanaceae. *Class* D.

V. foetida Grown as a tender ornamental.

V. lycioides See *V. foetida*

vestitus, -a, -um: clothed, with a covering of hairs.

vetch *Vicia, V. sativa* **bird** ~ *V. cracca* **crown** ~ *Coronilla varia* **horseshoe** ~ *Hippocrepis comosa* **kidney** ~ *Anthyllis vulneraria* **milk** ~ *Astragalus* **spring** ~ *Vicia sativa* **tuberous** ~ *Lathyrus tuberosus* **tufted** ~ *Vicia cracca.*

vetchling: common ~ *Lathyrus pratensis* **meadow** ~ *L. pratensis* **yellow** ~ *L. pratensis.*

vexans: wounding.

vexillaria: with a standard, one large uppermost petal.

viburnifolius, -a, -um: from the genus *Viburnum*, and *folius*, leaved.

viburnoides: from the genus *Viburnum*, with the ending *-oides*, indicating resemblance.

Viburnum (Classical Latin name for *V. lantana*.) ARROW WOOD, WAYFARING TREE. A genus of about 150 species of deciduous and evergreen shrubs and small trees with simple or occasionally lobed leaves and panicles or umbel-like cymes of small flowers. The outer flowers of the inflorescences are often irregular and more showy than the central ones. The fruit is also often showy. Many species are important garden ornamentals and a few have local medicinal uses. *Distr.* Temperate and warm regions, especially Asia and North America. *Fam.* Caprifoliaceae. *Class* D.

V. acerifolium ARROW WOOD, DOCKMACKIE. Erect deciduous shrub. Flowers white. Fruit red, turning black. The bark has been used medicinally and the wood for arrows. *Distr.* E North America.

V. alnifolium See *V. lantanoides*

V. betulifolium Deciduous shrub. Flowers white. Fruit bright red. *Distr.* W and Central China.

V. × bodnantense (After Bodnant, the garden in Wales where the hybrid originated.) A hybrid of *V. farreri* × *V. grandiflorum*. Deciduous shrub. Flowers white, fragrant, appearing in winter. *Distr.* Garden origin.

V. carlesii (After W. R. Charles, who collected the type specimen.) Deciduous shrubs. Flowers pink at first, then white, very fragrant. *Distr.* Korea, Japan.

V. dentatum ARROW WOOD, SOUTHERN ARROW WOOD. Deciduous shrub. Wood used for arrows.

V. farreri (After Reginald John Farrer (1880–1920), English horticultural writer who collected in China, Burma, and the Alps.) Erect deciduous shrub. Flowers fragrant, appearing in winter. *Distr.* Japan, China.

V. fragrans See *V. farreri*

V. lantana WAYFARING TREE, TWISTWOOD. Erect, deciduous, tree-like shrub. Fruit red, becoming glossy black. *Distr.* Europe, N Africa, W Asia, Caucasus.

V. lantanoides HOBBLE BUSH, MOOSEBERRY, MOOSEWOOD. Straggling shrub. *Distr.* E North America.

V. lentago SHEEPBERRY. Small deciduous tree. Fruit prune-like, cooked by North American Indians. *Distr.* E North America.

V. opulus GUELDER ROSE, CRAMPBARK, EUROPEAN CRANBERRY BUSH. Thicket-forming deciduous shrub. Leaves red in autumn. Fruit bright red. The wood was formerly used for skewers, the fruit is edible and sometimes a substitute for cranberries and the bark has been used medicinally. There are numerous ornamental varieties available. *Distr.* Europe, N Africa, W Asia, Caucasus.

V. plicatum Deciduous shrub with somewhat pleated leaves. *Distr.* Japan, China.

V. prunifolium BLACK HAW. Tall deciduous shrub. Leaves turning red-yellow in winter. Fruit dark-blue. *Distr.* E North America.

V. semperflorens See *V. plicatum*

V. tinus Evergreen shrub. Fruit black, poisonous. *Distr.* Mediterranean region.

V. tomentosum See *V. plicatum*

V. trilobum AMERICAN CRANBERRY BUSH, HIGHBUSH CRANBERRY, CRANBERRY. Deciduous shrub. Fruit bright red. *Distr.* North America.

Vicia (From Latin *vincire*, to bind, alluding to the clasping tendrils.) VETCH, TARE. A genus of 140 species of annual and perennial, climbing or scrambling herbs with clasping tendrils, pinnate leaves, and pea-like flowers that are borne singly or in clusters. Several species are grown as fodder crops and as green manures. *Distr.* N temperate regions, South America, tropical E Africa. *Fam.* Leguminosae. *Class* D.

V. cracca TUFTED VETCH, BIRD VETCH, CANADA PEA. Perennial. Flowers indigo, borne in dense racemes. *Distr.* N temperate regions.

V. faba BROAD BEAN, ENGLISH BEAN, FIELD BEAN. Erect annual. Tendrils absent. Flowers white, marked purple. Seeds eaten as a vegetable. *Distr.* N Africa, SW Asia.

V. sativa VETCH, TARE, SPRING VETCH. Annual to biennial herb. Flowers borne in pairs, purple. *Distr.* Europe, naturalized North America.

viciifolius, -a, -um: from the genus *Vicia*, and *folius*, leaved.

Victoria (After Queen Victoria (1819–1901).) GIANT WATER-LILY, WATER-PLATTER. A genus of 2 species of very large, rhizomatous, aquatic herbs with floating, round, peltate leaves that are strengthened by prominent netted veins below and upturned margins. The large floating flowers bear numerous cream petals and have an intoxicating fragrance. The leaves are often big enough to support the weight of a child. Both species are grown as tender ornamentals. *Distr.* Tropical South America. *Fam.* Nymphaeaceae. *Class* D.

V. amazonica ROYAL WATER-LILY, GIANT WATER-LILY, AMAZON WATER-LILY, WATER MAIZE. Leaves to 2m across, spiny below. Seeds edible if roasted. *Distr.* Amazon Basin, Guyana, French Guiana, Surinam.

V. cruziana SANTA CRUZ WATER-LILY. Leaves to 1.5m across, softly hairy below, not spiny. *Distr.* N Argentina, Paraguay, Bolivia, Brazil.

victoria-regina: Queen Victoria.

Villadia (For Dr Manuel M. Villada, 19th-century Mexican scientist.) A genus of 25–30 species of succulent annual and perennial herbs, typically with prostrate rooting stems, small cylindrical leaves, and clusters of small tubular flowers. Several species are grown as ornamentals. *Distr.* Texas to Peru. *Fam.* Crassulaceae. *Class* D.

V. elongata Perennial. Stems creeping, thin. Leaves to 8mm long. Flowers bell-shaped, pink, 5mm long, borne in a panicle. *Distr.* Mexico.

Villaresia See *Citronella*

Villarsia (After Dominique Villars (1745–1814), professor at Grenoble.) A genus of 16 species of perennial herbs with simple ovate to round leaves and loose erect cymes of white or yellow bell-shaped flowers that bear 5 fused petals. Several species are grown as ornamentals. *Distr.* SE Asia to Australia, 1 species in Africa. *Fam.* Menyanthaceae. *Class* D.

V. bennettii See *Nymphoides peltata*

V. parnassifolia Leaves mostly basal. Flowers yellow. *Distr.* Australia.

villosus, -a, -um: villous, softly hairy.

viminalis, -e: with long slender shoots.

Vinca (After the Latin name *vinca pervinca*, from *vincio*, to bind, alluding to the use of the foliage in wreaths; also the origin of the modern English name through the Middle English *per wynke*.) PERIWINKLE. A genus of 6–7 species of low perennial herbs and subshrubs with simple leaves and large, solitary, funnel-shaped flowers. Several species are grown as ornamentals. *Distr.* Europe to N Africa and Central Asia. *Fam.* Apocynaceae. *Class* D.

V. difformis Evergreen. Flowers blue to white. *Distr.* Morocco, Algeria.

V. herbacea Deciduous herb. Flowers violet-blue, occasionally white. *Distr.* E Europe, W Asia.

V. major GREATER PERIWINKLE. Evergreen, prostrate. Flowers purple-blue. Numerous cultivars available. *Distr.* Mediterranean region.

V. minor LESSER PERIWINKLE. Evergreen, shoots prostrate. Flowers blue, borne on more or less erect shoots. Numerous cultivars available. Used medicinally locally. *Distr.* Europe, W Asia, naturalized British Isles.

Vincetoxicum (From Latin *vincere*, to conquer, and *toxicum*, poison, alluding to their supposed powers as antidotes.) SWALLOW WORT. A genus of 15–80 species of erect and twining herbs and subshrubs with simple leaves and clusters of small inconspicuous flowers. *Distr.* Temperate Eurasia. *Fam.* Asclepiadaceae. *Class* D.

V. nigrum Stems twining. Flowers in simple umbels, green-brown. *Distr.* S Europe.

V. officinale Stems erect. Leaves long, pointed. Flowers green-white. *Distr.* Europe.

vine: Allegheny ~ *Adlumia fungosa* **apricot** ~ *Passiflora incarnata* **Argentine trumpet** ~ *Clytostoma callistegioides* **arrowhead** ~ *Syngonium angustatum, S. podophyllum* **bamboo** ~ *Smilax* **Bengal clock** ~ *Thunbergia grandiflora* **bleeding heart** ~ *Clerodendrum thomsoniae* **blue trumpet** ~ *Thunbergia grandiflora* **bush clock** ~ *T. erecta* **canary bird** ~ *Tropaeolum* **celery** ~ *Phyllocladus* **chalice** ~ *Solandra* **chestnut** ~ *Tetrastigma voinierianum* **chocolate** ~ *Akebia* **clock** ~ *Thunbergia grandiflora* **common grape** ~ *Vitis vinifera* **common matrimony** ~ *Lycium barbarum* **confederate** ~ *Antigonon leptopus* **coral** ~ *A., A. leptopus* **cross**

~ *Bignonia capreolata* **cup and saucer** ~ *Cobaea scandens* **cypress** ~ *Ipomoea quamoclit* **Easter lily** ~ *Beaumontia grandiflora* **firecracker** ~ *Manettia luteo-rubra* **flame** ~ *Pyrostegia venusta* **golden chalice** ~ *Solandra maxima* **guinea gold** ~ *Hibbertia, H. scandens* **heart** ~ *Ceropegia linearis* **kangaroo** ~ *Cissus antarctica* **lemon** ~ *Pereskia aculeata* **lipstick** ~ *Aeschynanthus radicans* **Madeira** ~ *Anredera cordifolia* **maidenhair** ~ *Muehlenbeckia complexa* **matrimony** ~ *Lycium* **mattress** ~ *Muehlenbeckia complexa* **mignonette** ~ *Anredera cordifolia* **necklace** ~ *Muehlenbeckia complexa* **pink trumpet** ~ *Podranea ricasoliana* **Pollyanna** ~ *Soleirolia soleirolii* **potato** ~ *Solanum jasminoides* **purple bell** ~ *Rhodochiton atrosanguineus* **purple passion** ~ *Gynura aurantiaca* **rainbow** ~ *Pellionia pulchra* **sandpaper** ~ *Petrea volubilis* **silk** ~ *Periploca graeca* **silver** ~ *Actinidia polygama, Scindapsus pictus* **silver lace** ~ *Polygonum* **sky** ~ *Thunbergia grandiflora* **snake** ~ *Hibbertia scandens* **tara** ~ *Actinidia arguta* **wax** ~ *Senecio macroglossus* **wire** ~ *Muehlenbeckia complexa* **wonga wonga** ~ *Pandorea pandorana.*

vinegar tree *Rhus glabra.*

vinegar of Sodom *Citrullus colocynthis.*

Viola (Classical Latin name for a scented flower.) PANSY, VIOLET. A genus of about 500 species of annual and perennial herbs and rarely subshrubs with alternate, simple to deeply lobed leaves, and irregular spurred flowers that bear their parts in fives. These plants have been cultivated since antiquity and numerous cultivars have been produced, some of which (pansies) have large flat flowers that bear little resemblance to the wild species. Some species and cultivars are highly valued for their scent and some were formerly important medicinally. *Distr.* Temperate regions. *Fam.* Violaceae. *Class* D.

V. arborescens Subshrub to 30cm high. Flowers white or pale purple. *Distr.* Mediterranean.

V. canina DOG VIOLET, HEATH VIOLET. Perennial herb to 40cm high. Flowers blue or white. *Distr.* Eurasia.

V. cornuta VIOLA, HORNED VIOLET, BEDDING PANSY. Perennial to 30cm high. Flowers large, purple. The pansies used as bedding plants were probably derived from this species. *Distr.* Spain and Pyrenees.

V. curtisii See *V. tricolor*

V. flettii ROCK VIOLET, OLYMPIC VIOLET. Perennial to 15cm high. Flowers small, purple. *Distr.* W North America.

V. glabella STREAM VIOLET. Perennial to 30cm high. Flowers pale yellow with purple veins. *Distr.* NE Asia, W North America.

V. macedonica See *V. tricolor*

V. odorata SWEET VIOLET, GARDEN VIOLET, ENGLISH VIOLET. Perennial. Flowers dark purple to white, very fragrant. This species has been involved in the parentage of many of the cultivars available today; it is also a source of essential oil (100kg of flowers giving 31g of oil). The flowers are sometimes seen crystallized as cake decorations. *Distr.* Eurasia to Africa.

V. palustris MARSH VIOLET, ALPINE MARSH VIOLET. *Distr.* N temperate regions.

V. saxatilis See *V. tricolor*

V. tricolor HEARTSEASE, LOVE IN IDLENESS, JOHNY JUMP UP. Annual to perennial herb. Flowers yellow and purple. The parent of many modern pansies. *Distr.* Europe, naturalized in North America.

viola *Viola cornuta.*

Violaceae The Violet family. 20 genera and about 1000 species of annual and perennial herbs and shrubs with simple, typically alternate leaves. The flowers are usually regular (*Viola* excepted) and have 5 sepals and 5 petals. Some genera are important as ornamentals. *Distr.* Widespread. *Class* D.
See: Melicytus, Viola.

violaceus, -a, -um: violet-coloured.

violescens: becoming violet-coloured.

violet Violaceae, *Viola* **African** ~ *Saintpaulia ionantha*, Gesneriaceae, *Saintpaulia* **Alpine** ~ *Cyclamen* **Alpine marsh** ~ *Viola palustris* **Arabian** ~ *Exacum affine* **bog** ~ *Pinguicula vulgaris* **bush** ~ *Browallia speciosa* **damask** ~ *Hesperis matronalis* **dame's** ~ *H. matronalis* **dog** ~ *Viola canina* **dog's tooth** ~ *Erythronium, E. denscanis*

English ~ *Viola odorata* **false African** ~ *Streptocarpus saxorum* **flame** ~ *Episcia cupreata* **garden** ~ *Viola odorata* **German** ~ *Exacum affine* **heath** ~ *Viola canina* **horned** ~ *V. cornuta* **marsh** ~ *V. palustris* **Olympic** ~ *V. flettii* **Persian** ~ *Exacum affine*, *Cyclamen* **Philippine** ~ *Barleria cristata* **rock** ~ *Viola flettii* **stream** ~ *V. glabella* **sweet** ~ *V. odorata* **Usambara** ~ *Saintpaulia ionantha* **water** ~ *Hottonia palustris* **white dog's tooth** ~ *Erythronium albidum*.

violiflorus, -a -um: with violet flowers.

viper's bugloss *Echium vulgare*.

virens: green.

virescens: becoming green.

virgatus, -a, -um: twiggy, wand-like.

virgaura: a golden rod.

virgin's bower *Clematis virginiana*.

virgineus, -a, -um: virginal, pure white.

virginianus, -a, -um: of Virginia, USA.

virginicus, -a, -um: of Virginia, USA.

viridescens: becoming green.

viridiflorus, -a, -um: with green flowers.

viridifolius, -a, -um: with green leaves.

viridi-glaucescens: becoming glaucous green.

viridis, -e: green.

viridissimus, -a, -um: very green.

viridistriatus, -a, -um: green-striped.

Viscaceae The Mistletoe family. 7 genera and 450 species of semi-parasitic shrubs with simple or scale-like leaves and small regular flowers that are followed by frequently sticky fruits. *Distr*. Widespread. *Class* D.

See: Viscum.

Viscaria vulgaris See *Lychnis viscaria*

viscarius, -a, -um: sticky.

viscidiflorus, -a, -um: with sticky flowers.

viscidifolius, -a, -um: with sticky leaves.

viscidus, -a, -um: sticky.

viscosissimus, -a, -um: very sticky.

Viscum (Classical Latin name for these plants.) MISTLETOE. A genus of 70–100 species of semi-parasitic evergreen shrubs with simple, thick, leathery leaves and clusters of inconspicuous flowers that are followed by 1-seeded berries with a sticky pulp. They grow on a series of deciduous trees, usually high up in the branches. *Distr*. Temperate regions. *Fam*. Viscaceae. *Class* D.

V. album MISTLETOE. Unisexual shrub. Branches pendulous, to 1m long. Berries white. Traditionally brought into the home as a Christmas decoration; originally used as a protection against evil spirits and a fertility symbol (hence kissing under the mistletoe at Christmas). *Distr*. Europe and temperate Asia.

Vitaceae The Grape family. 14 genera and 800 species of climbers with tendrils and a few shrubs and trees. The leaves are simple, palmately or pinnately lobed. The flowers are very small with their parts in fours or fives. The fruit is a berry. The most economically important member of the family is *Vitis vinifera*, the GRAPE, although others are of importance as ornamentals. *Distr*. Tropical to warm-temperate regions. *Class* D.
See: Ampelopsis, Cissus, Parthenocissus, Rhoicissus, Tetrastigma, Vitis.

Vitaliana (After Vitaliano Donati (1717–62), professor of botany at Turin.) A genus of 1 species of tuft-forming perennial herb with simple leaves and solitary yellow flowers that bear their parts in fives. *Distr*. Mountainous regions of Europe. *Fam*. Primulaceae. *Class* D.

V. primuliflora Grown as a rock-garden ornamental.

vitellinus, -a, -um: egg-yolk yellow.

Vitex (Classical Latin name for *V. agnus-castus*.) A genus of 250 species of deciduous shrubs and trees with opposite palmate leaves and panicles or racemes of 2-lipped, typically white flowers. Some species are exploited for their timber and some are grown as ornamentals. *Distr*. Tropical to temperate regions. *Fam*. Verbenaceae. *Class* D.

V. agnus-castus (From the Greek *agnos*, and the Latin *castus*, both meaning chaste.) CHASTE TREE. Small riverside tree. Foliage and flowers fragrant. The young twigs are occasionally used in basketry. Considered a symbol of purity. *Distr.* Mediterranean region to Central Asia.

vitifolius, -a, -um: with leaves like vine leaves.

Vitis (Classical Latin name for the grape vine.) A genus of about 65 species of deciduous woody climbers with simple to palmately lobed leaves and panicles of small male or bisexual flowers that are followed by fleshy berries. Several species are grown for their fruit and as ornamentals. *Distr.* N hemisphere. *Fam.* Vitaceae. *Class* D.

V. inconstans See *Parthenocissus tricuspidata*

V. quinquefolia See *Parthenocissus quinquefolia*

V. vinifera COMMON GRAPE VINE, GRAPE. Vine climbing to 35m high. An extremely important plant with numerous cultivars having been produced for their fruit, the majority of which is used in the production of wine. Some cultivars are also grown for dessert fruits or for fruit juices and the leaves of some are eaten as a vegetable. *Distr.* S and Central Europe.

vitis-idaea: grape of Mount Ida.

Vittaria (From Latin *vitta*, band or ribbon, alluding to the shape of the leaves.) A genus of 50–80 species of epiphytic ferns with linear pendent leaves. Several species are grown as ornamentals. *Distr.* Tropical to warm temperate regions. *Fam.* Vittariaceae. *Class* F.

V. elongata Leaves to 80cm long, tapering to the tip and the base. *Distr.* Malaysia to Australia.

Vittariaceae A family of 8 genera and 122 species of epiphytic ferns with creeping or erect stems that bear simple or rarely divided leaves. The sporangia are arranged in lines or may be in distinct round sori. *Distr.* Tropical to warm-temperate regions. *Class* F.
See: Vittaria.

vittatus, -a, -um: with longitudinal stripes.

vivipara: viviparous, bearing plantlets on the leaves or in the inflorescence.

Vochysiaceae A family of 7 genera and 210 species of trees, shrubs, and climbers with simple leaves and racemes of somewhat irregular flowers. Some species provide timber for construction and boat-building. *Distr.* W Africa, tropical America. *Class* D.

volubilis, -e: twining.

Vriesea (After W. de Vriese (1807–62), Dutch botanist.) A genus of 250–260 species of large epiphytic herbs with tubular rosettes of leathery leaves and short-lived flowers that emerge from between showy bracts on erect condensed spikes. Many species are grown as ornamentals. *Distr.* Central and South America, West Indies. *Fam.* Bromeliaceae. *Class* M.

V. carinata LOBSTER CLAW. Floral bracts bright red with yellow margins. *Distr.* E Brazil.

V. fenestralis Leaves very broadly strap-shaped. Bracts green, flowers yellow-green. *Distr.* Brazil.

V. hieroglyphica KING OF THE BROMELIADS. Leaves strap-shaped, light green, marked with brown-green. Flowers pale yellow. *Distr.* Brazil.

V. psittacina Bracts yellow, red or green, flowers yellow. *Distr.* Brazil.

V. splendens FLAMING SWORD. Leaves with purple cross-banding. Inflorescence flattened, bracts bright red to orange, flowers yellow. *Distr.* Venezuela, Guyana, Trinidad and Tobago.

vulgaris, -e: common.

vulgatus, -a, -um: common.

vulparius, -a, -um: of wolves or foxes.

vulpinus, -a, -um: of wolves or foxes.

W

Wachendorfia (After G. J. Wachendorf (1702–58), Dutch botanist.) A genus of 5 species of tuberous clump-forming herbs with folded leaves and shallow cup shaped flowers. *Distr.* S Africa. *Fam.* Haemodoraceae. *Class* M.

W. thyrsiflora Occasionally grown as an ornamental.

Wahlenbergia (After George Wahlenberg (1780–1851), Swedish professor of botany at Uppsala.) ROCK BELL. A genus of about 150 species of annual and perennial herbs with simple leaves and bell-shaped flowers that are borne singly or in leafy cymes. Several species are grown as ornamentals. *Distr.* Almost cosmopolitan, especially in the S hemisphere. *Fam.* Campanulaceae. *Class* D.

W. albomarginata NEW ZEALAND BLUEBELL. Leaves in basal rosettes from a rhizome. flowers solitary, white to pale blue. *Distr.* New Zealand.

W. congesta Mat-forming perennial. Leaves leathery. Flowers solitary, white to pale-blue. *Distr.* New Zealand.

W. hederaceae IVY-LEAVED BELL-FLOWER. Slender annual or perennial. Leaves round. flowers small, pale blue. *Distr.* W Europe.

W. pumilio See *Edraianthus pumilio*

W. saxicola See *W. albomarginata*

W. tasmanica See *W. saxicola*

wake robin *Arum*, *Trillium*, *T. grandiflorum*.

Waldheimia See *Allardia*

Waldsteinia (After Count Franz Adam Waldstein-Wartenburg (1759–1823), Austrian botanist.) A genus of 6 species of rhizomatous perennial herbs with alternate, 3-lobed or trifoliate leaves and yellow flowers that bear 5 petals and numerous stamens. Several species are grown as ornamentals. *Distr.* N temperate regions. *Fam.* Rosaceae. *Class* D.

W. fragarioides BARREN STRAWBERRY. Mat-forming perennial. Leaves trifoliate. flowers to 2cm across. *Distr.* E USA.

wallflower *Erysimum cheiri* **coastal** *capitatum* **western** *E. asperum*.

wallwort *Sambucus ebulus*.

walnut *Juglans*, *J. regia*, Juglandaceae **African** ~ *Schotia brachypetala* **black** ~ *Juglans nigra* **common** ~ *J. regia* **English** ~ *J. regia* **Japanese** ~ *J. ailantifolia* **Madeira** ~ *J. regia* **Persian** ~ *J. regia* **Texan** ~ *J. microcarpa* **white** ~ *J. cinerea*.

walnut tree, satin *Liquidambar styraciflua*.

wampee *Pontederia cordata*.

wandering jew *Tradescantia*, *T. fluminensis*, *Saxifraga stolonifera*.

wandering sailor *Cymbalaria muralis*.

wandflower *Dierama*, *Sparaxis*, *Galax urceolata*.

wand plant *G. urceolata*.

wangee cane *Phyllostachys nigra*.

wapato *Sagittaria latifolia*.

waratah *Telopea speciosissima* **Tasmanian** ~ *T. truncata*.

wardii after Frank Kingdon-Ward (1885–1958), English botanist and plant collector who introduced many new plants into cultivation.

wasabi *Wasabia japonica*.

Wasabia (From the local Japanese name.) A genus of 2 species of rhizomatous perennial herbs with small white flowers. *Distr.* E Asia. *Fam.* Cruciferae. *Class* D.

W. japonica WASABI, JAPANESE HORSERADISH. Cultivated as the source of the condiment (wasabi) typically eaten with raw fish (sashimi). *Distr.* Japan, Sakhalin.

Washingtonia (After George Washington (1732–99), first President of the United

States.) WASHINGTON PALM. A genus of 2 species of large palms with fan-shaped leaves and arching inflorescences of bisexual flowers. Both species are grown as tender ornamentals and street trees. *Distr.* Arid SW North America. *Fam.* Palmae. *Class* M.

W. filifera DESERT FAN PALM, COTTON PALM, NORTHERN W WASHINGTONIA, CALIFORNIAN WASHINGTONIA. To 15m high. Leaf stalks to 2m long. Fruits and seeds edible, leaf-fibre used for basketry. *Distr.* SW USA.

W. robusta THREAD PALM, SOUTHERN WASHINGTONIA, MEXICAN WASHINGTONIA. To 25m high. Leaf stalks to 1m long. Fruit edible. *Distr.* NW Mexico.

washingtonia: Californian ~ *Washingtonia filifera* **Mexican** ~ *W. robusta* **northern** ~ *W. filifera* **southern** ~ *W. robusta.*

water carpet *Chrysosplenium americanum*

watercress *Rorippa nasturtium-aquaticum.*

water-lily *Nymphaea* **Amazon** ~ *Victoria amazonica* **Australian** ~ *Nymphaea gigantea* **Cape blue** ~ *N. capensis* **fringed** ~ *Nymphoides peltata.* **giant** ~ *Victoria, V. amazonica.* **Indian red** ~ *Nymphaea rubra* **pygmy** ~ *N. tetragona* **royal** ~ *Victoria amazonica* **Santa Cruz** ~ *V. cruziana* **white** ~ *Nymphaea alba* **yellow** ~ *Nuphar lutea.*

water mat *C. americanum.*

water meal *Wolffia.*

waterleaf *Hydrophyllum* **Virginia** ~ *H. virginianum.*

water-platter *Victoria.*

water soldier *Stratiotes aloides.*

waterweed *Elodea, E. canadensis.*

Watsonia (After Sir William Watson (1715–87), British physician and botanist.) A genus of 50–60 species of tender and half-hardy perennial herbs with lance-shaped leaves and spikes of irregular or regular flowers. Several species are cultivated as ornamentals. *Distr.* South Africa, Madagascar. *Fam.* Iridaceae. *Class* M.

W. ardernei See *W. borbonica*

W. beatricis See *W. pillansii*

W. borbonica flowers pink with white lines, fragrant. *Distr.* South Africa (SW Cape).

W. brevifolia See *W. laccata*

W. bulbillifera See *W. meriana.*

W. meriana (After Maria Sybilla Merian (1647–1717), Dutch naturalist and artist.) Clump-forming. flowers pink-red, occasionally yellow. *Distr.* South Africa (Cape Province).

W. pillansii Flowers orange-red, with long curved tubes. *Distr.* South Africa (Cape, Natal, Transkei Provinces).

W. pyramidata See *W. borbonica*

Wattakaka See *Dregea*

W. sinensis See *Dregea sinensis*

wattle *Acacia* **black** ~ *A. mearnsii* **blue** ~ *A. dealbata* **Cootamundra** ~ *A. baileyana* **hedge** ~ *A. paradoxa* **knife-leaf** ~ *A. cultriformis* **Mount Morgan** ~ *A. podalyrifolia* **ovens** ~ *A. pravissima* **Queensland silver** ~ *A. podalyrifolia* **silver** ~ *A. dealbata* **spreading** ~ *A. genistifolia* **swallow** ~ *A. longifolia* **Sydney golden** ~ *A. longifolia.*

waxberry *Gaultheria hispida, Symphoricarpos albus.*

wax flower *Hoya, Stephanotis floribunda* **miniature** ~ *Hoya lanceolata* **Philippine** ~ *Etlingera elatior.*

wax plant *Hoya, H. carnosa.*

wax tree *Rhus succedanea* **white** ~ *Ligustrum lucidum.*

waxwork *Celastrus scandens.*

wayfaring tree *Viburnum, V. lantana.*

weasel's snout *Misopates orontium.*

weather prophet *Dimorphotheca pluvialis.*

wedding flower *Francoa sonchifolia.*

weed, ash *Aegopodium podagraria* **bishop** ~ *A. podagraria* **gout** ~ *A. podagraria.*

Weigela (After Christian Ehrenfried von Weigel (1748–1831), German botanist.) A genus of 10 species of deciduous shrubs with simple opposite leaves and bell-shaped flowers borne singly or in corymbs. Several species and numerous cultivars are grown as ornamentals. *Distr.* E Asia. *Fam.* Caprifoliaceae. *Class* D.

W. florida Stems arching. Flowers pink, solitary. Numerous ornamental cultivars are available. *Distr.* N China, Korea.

W. hortensis flowers clustered, pink. *Distr.* Japan.

W. middendorffiana Flowers yellow, orange within. *Distr.* Japan, N China, Manchuria.

W. praecox Upright shrub. Stems much branched. flowers fragrant, pink, yellow within. *Distr.* NE Asia.

Weinmannia (For Johann Wilhelm Weinmann, 18th-century German apothecary.) A genus of 150–190 species of evergreen shrubs and trees with simple or pinnate leathery leaves and erect racemes or panicles of small unisexual flowers that bear their parts in fours or fives. Some species are grown for their timber or as tender ornamentals. The bark is locally used for tanning. *Distr.* Andes, Madagascar, Mascarenes, Malaysia, Pacific Islands, New Zealand. *Fam.* Cunoniaceae. *Class* D.

W. trichosperma TINEO, MADEN. Large shrub or tree to 20m high. Leaves simple. flowers white. *Distr.* Chile, Argentina.

weld, dyer's *Reseda luteola.*

Weldenia (For L. von Welden (1780–1853), Austrian soldier and Alpine botanist.) A genus of 1 species of tuberous herb with tufts or rosettes of leathery linear leaves and clusters of white tubular flowers. *Distr.* Mountains of Mexico and Guatemala. *Fam.* Commelinaceae. *Class* M.

W. candida Grown as a half-hardy ornamental.

wellingtonia *Sequoiadendron giganteum.*

Welwitschia (After Dr Friedrich Welwitsch (1806–72), who introduced the plant to Britain.) A genus of 1 species of unusual plant with a large, erect, subterranean, woody stem that is approximately carrot-shaped and can grow to 1m in diameter. There are 2 strap-shaped leaves that grow from the top of the stem, close to ground level. These leaves grow continuously and are only limited in length their by being split and worn away at their tips. The flowers are borne in cone-like structures on small branches arising close to the leaf bases. *Distr.* SW African deserts. *Fam.* Welwitschiaceae. *Class* G.

W. mirabilis The only speices.

Welwitschiaceae A family of 1 genus and 1 species. *Class* G.
See: Welwitschia

Westringia (After J. P. Westring (1753–1833), lichen specialist and physician to the king of Sweden.) A genus of 25–27 species of shrubs with simple whorled leaves and typically solitary, 2-lipped, tubular flowers. *Distr.* Australia. *Fam.* Labiatae. *Class* D.

W. angustifolia Leaves 3–4 whorled, narrow. Flowers white, marked yellow or red. *Distr.* Tasmania.

W. fruticosa AUSTRALIAN ROSEMARY. Compact rounded shrub. Leaves 4-whorled. flowers white, spreading. *Distr.* W Australia.

W. rosmariniformis See *W. fruticosa*

wheat *Triticum* **bread** ~ *T. aestivum* **einkorn** ~ *T. monococcum* **emmer** ~ *T. dicoccon* **Inca** ~ *Amaranthus caudatus* **macaroni** ~ *Triticum durum* **semolina** ~ *T. durum.*

wheatgrass *Agropyron* **crested** ~ *A. cristatum* **fairway crested** ~ *A. cristatum.*

whin *Ulex, U. europaeus.*

whitebeam *Sorbus aria* **dwarf** ~ *S. chamaemespilus.*

whitecup *Nierembergia repens.*

white man *Verbascum lychnitis.*

white paint brush *Haemanthus albiflos.*

white sails *Spathiphyllum wallisii.*

white sally *Eucalyptus pauciflora.*

white stemmed filaree *Erodium moschatum.*

white velvet *Tradescantia sillamontana.*

whitewood *Tilia americana, Abies alba* **canary** ~ *Liriodendron tulipifera.*

whitey wood *Acradenia frankliniae.*

whorl flower *Morina longifolia.*

whortleberry *Vaccinium myrtillus* **bog** ~ *V. uliginosum* **Caucasian** ~ *V. arctostaphylos* **Madeiran** ~ *V. padifolium.*

wickup *Epilobium angustifolium.*

Widdringtonia (After Edward Widdrington, late 18th-century British botanist.) AFRICAN CYPRESS. A genus of 3 species of evergreen coniferous shrubs and trees with needle-like juvenile leaves and scale-like adult leaves. *Distr.* Tropical and South Africa. *Fam.* Cupressaceae. *Class* G.

W. nodiflora SAPREE WOOD. Shrub or tree to 45m high. Grown as an ornamental and a source of timber. *Distr.* S Africa.

widdy *Potentilla fruticosa.*

widow: black ~ *Geranium phaeum* **mourning** ~ *G. phaeum.*

widow's tears *Commelina.*

wig tree *Cotinus coggygria.*

Wigandia (After Johannes Wigand (1523–87), German botanist and Bishop of Pomerania.) A genus of 5 species of herbs, shrubs, and trees with very large simple leaves covered in stinging hairs and curled inflorescences of funnel-shaped flowers which bear their parts in fives. Several species are grown as ornamentals, chiefly as annuals for their young foliage. *Distr.* Tropical America. *Fam.* Hydrophyllaceae. *Class* D.

W. caracasana Robust bushy herb. Flowers lilac or white. *Distr.* Central and N South America.

W. urens Shrub. Leaves double-toothed. Flowers indigo to violet. *Distr.* Peru.

willow *Salix*, Salicaceae **Arctic** ~ *Salix arctica* **Babylon weeping** ~ *S. babylonica* **basket** ~ *S. purpurea* **bau** ~ *S. pentandra* **brittle** ~ *S. fragilis* **button** ~ *Cephalanthus occidentalis* **crack** ~ *Salix fragilis* **creeping** ~ *S. repens* **downy** ~ *S. lapponum* **dwarf** ~ *S. herbacea* **eared** ~ *S. aurita* **florist's** ~ *S. caprea* **French** ~ *Epilobium angustifolium* **goat** ~ *Salix caprea* **hoary** ~ *S. candida* **Lapland** ~ *S. lapponum* **laurel** ~ *S. pentandra* **musk** ~ *S. aegyptiaca* **polar** ~ *S. polaris* **prairie** ~ *S. humilis* **purple** ~ *S. purpurea* **pussy** ~ *S. caprea* **sage** ~ *S. candida* **swamp** ~ *S. myrtilloides* **violet** ~ *S. daphnoides* **Virginia** ~ *Itea virginica* **water** ~ *Justicia* **weeping** ~ *Salix* × *sepulcralis* **white** ~ *S. alba* **whortle** ~ *S. myrsinifolia* **woolly** ~ *S. lanata.*

willowherb *Epilobium* **great** ~ *E. angustifolium* **rose bay** ~ *E. angustifolium.*

wineberry *Vaccinium myrtillus*, *Rubus phoenicolasius.*

wine palm, Chilean *Jubaea chilensis.*

wing-nut *Pterocarya* **Caucasian** ~ *P. fraxinifolia.*

Winteraceae A family of 4 genera and about 60 species of shrubs and trees with simple leathery leaves. The flowers bear tepals that are just differentiated into sepals and petals and arranged in spirals along with the numerous stamens. The wood does not contain vessel cells which implies, along with other characteristics, that this is perhaps the most primitive flowering plant family. Some species are grown as ornamentals, others have medicinal uses. *Distr.* Madagascar, Indonesia, and Malaysia to Australasia, Pacific, Central and South America. *Class* D.
See: Drimys, Pseudowintera.

winterberry *Ilex verticillata*, *I. de-cidua* **Japanese** ~ *I. serrata.*

winter cress *Barbarea.*

wintergreen *Gaultheria*, *G. procumbens*, *Trientalis*, *Pyrola* **Alpine** ~ *Gaultheria humifusa* **creeping** ~ *G. procumbens.*

winter's bark *Drimys winteri.*

wintersweet *Chimonanthus praecox.*

wire netting bush *Corokia coton-easter.*

Wisteria (After Caspar Wistar (1761–1818), American professor of anatomy.) A genus of 6–10 species of deciduous woody climbers with alternate pinnate leaves and large pendulous racemes of pea-like flowers. Several species are grown as ornamentals.

Distr. E Asia, North America. *Fam*. Leguminosae. *Class* D.

W. floribunda JAPANESE WISTERIA. Stems twine clockwise. Flowers violet, pink, or blue, fragrant. *Distr*. Japan.

W.* × *formosa A hybrid of *W. floribunda* × *W. sinensis*. Stems twine clockwise. flowers fragrant, violet. *Distr*. Garden origin.

W. sinensis CHINESE WISTERIA. Stems twine anticlockwise. Flowers faintly scented, violet-blue. *Distr*. China.

wisteria: Chinese ~ *Wisteria sinensis* **Japanese** ~ *W. floribunda* **South African** ~ *Bolusanthus speciosus* **wild** *B. speciosus*.

wisteria tree, scarlet *Sesbania grandiflora*.

witch hazel *Hamamelis*, Hamamelidaceae **buttercup** ~ *Corylopsis pauciflora* **Chinese** *Hamamelis mollis* **Japanese** ~ *H. japonica* **Virginian** ~ *H. virginiana*.

witloof *Cichorium intybus*.

woad *Isatis*, *I. tinctoria* **dyer's** ~ *I. tinctoria*.

woadwaxen *Genista*.

wolf's bane *Aconitum*, *A. lycoctonum* **garden** ~ *A. napellus*.

Wolffia (After J. F. Wolff (1778–1806), German physician who wrote about the genus *Lemna*.) WATER MEAL. A genus of about 7 species of minute aquatic herbs consisting of a very small undifferentiated thallus. The rarely produced flowers consist of a single stamen and a ovary of 1 ovule. Thalli reproduce by budding. These are the smallest of all flowering plants, often less than 2mm across. *Distr*. Tropical and warm to temperate regions. *Fam*. Lemnaceae. *Class* M.

W. columbiana COMMON WOLFfiA. Leaves almost spherical, pale green. Flowers occasional, fruit rare. *Distr*. Americas.

wolffia, common *Wolffia columbiana*.

woodbine *Clematis virginiana*, *Lonicera periclymenum* **Italian** ~ *L. caprifolium* **Spanish** ~ *Merremia tuberosa*.

woodland star *Lithophragma*.

woodruff *Asperula*, *Galium odoratum* **dyer's** ~ *Asperula tinctoria* **sweet** ~ *Galium odoratum*.

wood-rush Luzula, field *L. campestris*.

Woodsia (After J. Woods (1776–1864), English author of *The Tourist's Flora*.) A genus of 21 species of medium-sized terrestrial and epiphytic ferns. Some species are cultivated as ornamentals. *Distr*. Temperate and cool-temperate regions. *Fam*. Woodsiaceae. *Class* F.

W. obtusa BLUNT-LOBED WOODSIA, LARGE WOODSIA, COMMON WOODSIA. Terrestrial. Leaves to 60 cm long. *Distr*. North America.

woodsia: blunt-lobed ~ *Woodsia obtusa* **common** ~ *W. obtusa* **large** ~ *W. obtusa*.

Woodsiaceae A family of 20 genera and about 600 species of terrestrial or epiphytic tufted ferns with pinnately divided leaves. The sori are frequently elongate, U- or J-shaped. *Distr*. Temperate to montane tropical regions. *Class* F.

See: Athyrium, Cystopteris, Gymnocarpium, Lunathyrium, Matteuccia, Onoclea, Woodsia.

wood vamp *Decumaria barbara*.

Woodwardia (After Thomas Jenkinson Woodward (1745–1820), English botanist.) CHAIN FERN. A genus of 10–12 species of large deciduous ferns. Several species are grown as ornamentals. *Distr*. Warm regions of the N hemisphere. *Fam*. Blechnaceae. *Class* F.

W. radicans Leaves pinnate, to 2m long. *Distr*. SW Europe.

woolly netbush *Calothamnus villosus*.

worcesterberry *Ribes divaricatum*.

wormwood *Artemisia*, *A. absinthium* **beech** ~ *A. stelleriana* **Roman** ~ *A. pontica* **sea** ~ *Seriphidium maritimum* **sweet** ~ *Artemisia annua*.

Worsleya See *Hippeastrum*

W. rayneri See *Hippeastrum procerum*

woundwort *Stachys* **downy** ~ *S. germanica*.

Wulfenia (After Franz Xavier Freiherr von Wulfen (1728–1805), Austrian botanist.)

A genus of 2–5 species of tufted perennial herbs with simple stalked leaves and spike-like racemes of small, blue or purple, tubular flowers. Several species are grown as rock-garden ornamentals. *Distr.* SE Europe. *Fam.* Scrophulariaceae. *Class* D.

W. amherstiana Leaves in basal rosettes. flowers purple. *Distr.* W Himalaya, Afghanistan.

W. carinthiaca in basal rosettes, dark green. flowers violet-blue. *Distr.* E Alps, Balkan peninsula.

Wurmbea (After F. van Wurmb, Secretary of the Batavian Academy of Sciences about 1800.) A genus of 37 species of perennial herbs with alternate lance-shaped leaves. The regular flowers have 6 fused tepals and 6 stamens and are borne on an erect spike that lacks bracts. Several species are grown as tender ornamentals. *Distr.* South Africa and Australia. *Fam.* Colchicaceae. *Class* M.

W. dioica EARLY NANCY. Flowers white with lilac nectaries. *Distr.* Australia.

X

xanthacanthus, -a, -um: yellow-spined.

xanthinus, -a, -um: yellow.

xanthocalyx: with yellow sepals.

Xanthoceras (From Greek *xanthos*, yellow, and *keras*, horn, alluding to the horn-like appendage borne between the petals.) A genus of 1 species of deciduous shrub or tree with alternate, pinnately divided leaves and racemes or panicles of flowers that bear their parts in fives. *Distr.* N China. *Fam.* Sapindaceae. *Class* D.

X. sorbifolium Grown as a hardy ornamental.

xanthochlorus, -a, -um: yellow-green-coloured.

xanthocodon: a yellow bell.

xantholeucus, -a, -um: yellowish white.

xanthophyllus, -a, -um: with yellow leaves.

Xanthorhiza (From Greek *xanthos*, yellow, and *rhiza*, root.) A genus of 1 species of deciduous shrub with pinnately divided leaves and drooping panicles of small unisexual flowers. *Distr.* E North America. *Fam.* Ranunculaceae. *Class* D.

X. simplicissima YELLOWROOT. Grown as an ornamental and the source of a yellow dye.

Xanthorrhoeaceae The Grass Tree family. 1 genus and 15 species of slow-growing, fire-resistant herbs with very thick stems and thin linear leaves. Some species produce a resin used in varnishes. *Distr.* Australia. *Class* M.

Xanthosoma (From Greek *xanthos*, yellow, and *soma*, body, alluding to the yellow inner tissues of some species.) YAUTIA, TANNIA, MALANGA. A genus of 45–50 species of tuberous or erect herbs with milky sap and very large, simple or palmately lobed leaves. The unisexual flowers are borne on a spadix that is surrounded and exceeded by a green-white spathe. Several species are cultivated as ornamentals and as food-plants. *Distr.* Tropical America. *Fam.* Araceae. *Class* M.

X. sagittifolium TANNIA. Grown for edible tubers which are a staple carbohydrate in parts of Central and South America; the young leaves are cooked and eaten like spinach. Variegated ornamental forms available. *Distr.* Tropical America, West Indies.

xanthostephanus, -a, -um: with a yellow crown.

Xeronema (From Greek *xeros*, dry, and *nema*, thread, alluding to the persistent filaments.) A genus of 2 species of rhizomatous herbs with clumps of narrow, grass-like basal leaves and racemes of red flowers which bear 6 free tepals. *Distr.* New Caledonia, N New Zealand (Poor Knight Island). *Fam.* Phormiaceae. *Class* M.

X. callistemon POOR KNIGHT'S LILY. To 90cm high. flowers densely crowded in 1-sided racemes. Grown as a tender ornamental. *Distr.* Hen and Poor Knight Island, E of Auckland.

Xerophyllum (From Greek *xeros*, dry, and *phyllon*, leaf.) A genus of 2–3 species of rhizomatous herbs with grass-like leaves and cylindrical or pyramidal racemes of star-shaped flowers. *Distr.* North America. *Fam.* Melanthiaceae. *Class* M.

X. asphodeloides TURKEY BEARD, MOUNTAIN ASPHODEL. Flowers 9mm across, fragrant, yellow. Grown as an ornamental. *Distr.* E North America.

X. tenax BEAR GRASS, INDIAN BASKET GRASS, SQUAW GRASS, ELK GRASS. Flowers 1.5cm across, cream-white. Leaves formerly used by Indians to make water-tight baskets, now grown as an ornamental. *Distr.* W North America.

xiphioides: sword-like.

xiphium: the Greek name for the genus *Gladiolus*.

xylocanthus, -a, -um: with woody spines.

xylocarpus, -a, -um: with woody fruit.

xylosteus, -a, -um: with a woody skeleton.

Xyridaceae A family of 5 genera and 260 species of rush-like marsh plants. The flowers are usually yellow and enclosed between bracts. *Distr.* Widespread in tropical and subtropical regions. *Class* M.

Y

yakushimanus, -a, -um: of Yakushima, an island of South Japan.

yam Dioscoreaceae, *Dioscorea* **Chinese** ~ *D. batatas* **cinnamon** ~ *D. batatas* **elephant** ~ *Amorphophallus paeoniifolius* **ornamental** ~ *Dioscorea discolor* **potato** ~ *D. esculenta* **water** ~ *D. alata* **white** ~ *D. alata*.

yampee *D. trifida*.

yangtao *Actinidia deliciosa*.

yargonensis, -e: of the Yargong gorge, Tibet.

yarrow *Achillea, A. millefolium*.

yaupon *Ilex vomitoria*.

yautia *Xanthosoma*.

yegoma oil plant *Perilla frutescens*.

yellow root *Hydrastis canadensis, Xanthorhiza simplicissima*.

yellow wood *Cladrastis*.

yellow wort *Blackstonia perfoliata*.

yerba dulce *Lippia dulcis*.

yesterday today and tomorrow *Brunfelsia pauciflora*.

yew *Taxus, T. baccata*, Taxaceae **California nutmeg** ~ *Torreya californica* **common** ~ *Taxus baccata* **English** ~ *T. baccata* **Japanese nutmeg** ~ *Torreya nucifera* **Japanese plum** ~ *Cephalotaxus harringtonia* **nutmeg** ~ *Torreya* **plum** ~ *Cephalotaxus* **plum-fruited** ~ *Prumnopitys andina* **Prince Albert's** ~ *Saxegothaea conspicua*.

Yorkshire fog *Holcus lanatus*.

youth and old age *Aichryson x domesticum*.

Yucca (From the Caribbean name for CASSAVA (*Manihot esculenta*), originally thought to apply to *Y. gloriosa*.) BEAR GRASS, ADAM'S NEEDLE. A genus of 30–40 species of stout-stemmed shrubs with ridged sword-like leaves and panicles of white flowers that are moth pollinated and most heavily scented at night. Some species are cultivated for their fibres and as ornamentals. *Distr*. Arid regions of the USA, Mexico, Guatemala and the West Indies. *Fam*. Agavaceae. *Class* M.

Y. aloifolia SPANISH BAYONET, DAGGER PLANT. Leaves are a source of fibre used in the manufacture of ropes. *Distr*. S North America, West Indies.

Y. baccata SPANISH BAYONET, BLUE YUCCA, BANANA YUCCA. Leaves flexible near base, twisted. A source of tough fibre and edible fruit and flower buds. *Distr*. SW USA.

Y. brevifolia JOSHUA TREE. Tree to 9m high. Bark scaly. Fibre can be used as newsprint. *Distr*. SW North America.

Y. elata SOAP TREE, SOAP WEED, PALMELLA. Source of fibre and foaming agent. *Distr*. SW USA.

Y. elephantipes SPINELESS YUCCA. Stems densely branched, to 10m high. *Distr*. SW USA, Mexico, Guatemala.

Y. filamentosa ADAM'S NEEDLE, SPOONLEAF YUCCA, NEEDLE PALM. Leaves narrow, like large needles. Commonly cultivated as an ornamental in Europe with several cultivars available. *Distr*. E USA (New Jersey to Florida).

Y. gloriosa SPANISH DAGGER, ROMAN CANDLE, PALM LILY. Perhaps the most widely cultivated ornamental species in Europe with numerous cultivars available. *Distr*. SE USA (N Carolina to Florida).

Y. guatemalensis See *Yucca elephantipes*

Y. rupicola TWISTED LEAF YUCCA. Stem very short. *Distr*. USA (Texas).

Y. smalliana (After John Kunkel Small (1869–1938).) ADAM'S NEEDLE, BEAR GRASS. Leaves very narrow. *Distr*. SE USA.

Y. treculeana SPANISH DAGGER, PALMA PITA. Tree to 5m high. *Distr*. USA (Texas), W Mexico.

Y. whipplei (After Lieutenant Amiel Weeks Whipple (1818–63), who explored North America.) OUR LORD'S CANDLE. A source of strong fibre and edible flowers. *Distr.* USA (California), Mexico (Baja California).

yucca, banana *Yucca baccata* **blue** ~ *Y. baccata* **spineless** ~ *Y. elephantipes* **spoonleaf** ~ *Y. filamentosa* **twisted-leaf** ~ *Y. rupicola.*

yuccifolius, -a, -um: from the genus *Yucca*, and *folius*, leaved.

yuccoides: from the genus *Yucca*, with the ending *-oides*, indicating resemblance.

yulan *Magnolia denudata.*

yunnanensis, -e: of Yunnan province, China.

yusan *Hosta plataginea.*

Yushania See *Sinarundinaria*

Z

Zaluzianskya (After Adam Zaluziansky von Zaluzian (1558–1613), a physician and botanist.) A genus of 35 species of annual and perennial herbs and subshrubs with simple, entire or toothed leaves, and dense spikes of tubular flowers that are fragrant at night. Several species are cultivated as ornamentals. *Distr.* S Africa, E African mountains. *Fam.* Scrophulariaceae. *Class* D.

Z. capensis NIGHT PHLOX. Annual or perennial herb. Flowers few, dark purple outside, white within. *Distr.* S Africa.

Zambac *Jasminum sambac.*

zambesiacus, -a, -um: of the Zambesi river/river basin.

Zamia (From Latin *zamia*, loss.) FLORIDA ARROWROOT. A genus of 30–50 species of tree-like cycads with whorls of large pinnate leaves. The stems are a source of starch which is poisonous until it is cooked. *Distr.* Tropical and warm regions of America. *Fam.* Zamiaceae. *Class* G.

Z. furfuracea CARDBOARD PALM. To 1m high. Grown as an ornamental. *Distr.* Mexico.

Zamiaceae A family of 8 genera and 85 species of palm-like plants with either a subterranean or a tall erect stem and pinnate leaves, each segment having numerous parallel veins. *Distr.* Africa, Australia, America. *Class* G.
See: Dioon, Encephalartos, Lepidozamia, Macrozamia, Zamia.

Zannichellia (After Giovanni Girolamo Zannichelli (1662–1729), Italian botanist and physician.) A genus of 1 species of submerged, perennial, aquatic herb with narrow linear leaves and very small, solitary, unisexual flowers. *Fam.* Zannichelliaceae. *Class* M.

Z. palustris HORNED PONDWEED.

Zannichelliaceae The Horned Pondweed family. 4 genera and 7 species of submerged, rhizomatous, aquatic plants with slender leaves and small, unisexual, water-pollinated flowers. *Distr.* Widespread in brackish and fresh water. *Class* M.
See: Zannichellia.

Zantedeschia (After Francesco Zantedeschi (1773–1846), Italian botanist.) ARUM LILY, CALLA LILY. A genus of 6 species of rhizomatous perennial herbs with simple basal leaves. The small unisexual flowers are borne on a spadix which is surrounded by a funnel-shaped, white to yellow or purple spathe. Several species and cultivars are grown as ornamentals. *Distr.* Tropical and S Africa. *Fam.* Araceae. *Class* M.

Z. aethiopica ARUM LILY, FUNERAL LILY. Spathes white. Frequently used as a cut flower, particularly associated with funerals in Britain. *Distr.* South Africa. Widely naturalized in frost-free regions.

Z. albomaculata BLACK-THROATED ARUM, SPOTTED ARUM. Leaves often white-spotted. Spathe white-yellow, stained purple outside at base. *Distr.* South Africa to tropical E Africa.

Z. elliottiana GOLDEN ARUM, YELLOW ARUM LILY. Leaves spotted white. Spathe yellow. *Distr.* South Africa.

Z. rehmannii PINK ARUM. Spathe more or less tubular, pink to maroon. *Distr.* S Africa.

Zanthoxylum (From Greek *xanthos*, yellow, and *xylon*, wood.) PRICKLY ASH. A genus of 200–250 species of prickly, aromatic, deciduous and evergreen shrubs and trees with aromatic bark, pinnately divided leaves, and cymes or panicles of small flowers. Several species are grown as ornamentals, some are important for their timber and some are used as spices. *Distr.* Widespread. *Fam.* Rutaceae. *Class* D.

Z. piperitum JAPAN PEPPER. Deciduous shrub. Flowers tiny. Fruit used as a condiment in Japan. *Distr.* E Asia.

Zauschneria See *Epilobium*
Z. californica See *Epilobium canum*
Z. microphylla See *Epilobium canum*

Z. villosa See *Epilobium canum*

Zea (From Greek *zea*, cereal.) A genus of 4 species of annual, rarely perennial grasses 2–3m high with broad flat leaves, terminal, tassel-like male inflorescences and axillary, spike-like female inflorescences. *Distr.* C America. *Fam.* Gramineae. *Class* M.

Z. mays MAIZE, CORN, SWEET CORN. The third most important cereal after rice and wheat, grown as animal fodder as well as for human consumption as flour, cornflakes, popcorn, sweet corn and as an oil crop. *Distr.* Cultivated for over 5500 years in the New World, precise wild origin obscure.

zebra plant *Calathea zebrina*, *Cryptanthus zonatus*, *Aphelandra squarrosa*.

zebra wood Connaraceae.

Zebrina See *Tradescantia*

Z. pendula See *Tradescantia zebrina*

zebrinus, -a, -um: striped regularly with white or yellow.

Zelkova (From a local vernacular name in the Caucasus for *Z. carpinifolia*.) A genus of 4–5 species of deciduous shrubs and trees with simple alternate leaves and inconspicuous green flowers that lack petals. Several species are cultivated as ornamentals or a source of timber. *Distr.* Asia, 1 species in Crete. *Fam.* Ulmaceae. *Class* D.

Z. carpinifolia CAUCASIAN ELM. Large shrub or tree to 35m high. *Distr.* Caucasus, N Iran.

Z. serrata KEAKI, KEYAKI, JAPANESE ZELKOVA, SAW-LEAF ZELKOVA. Tree to 35m high. Several ornamental cultivars have been raised from this species and it is often used as a bonsai subject as well as being an important source of timber. *Distr.* E Asia, Japan.

zelkova: Japanese ~ *Zelkova serrata*
saw-leaf ~ *Z. serrata*.

Zenobia (After Zenobia, a Queen of Palmyra around AD 266.) A genus of 1 species of deciduous or semi-evergreen shrub with sweetly scented flowers. *Distr.* SE USA. *Fam.* Ericaceae. *Class* D.

Z. pulverulenta Cultivated ornamentally.

Zephyranthes (From Greek *zephyros*, west wind, and *anthos*, flower; the genus is native to the W hemisphere.) ZEPHYR FLOWER, FAIRY LILY, RAIN LILY. A genus of 60–71 species of bulbiferous, *Colchicum*-like herbs with narrow leaves and funnel-shaped erect flowers borne on a hollow stem. Several species are cultivated as ornamentals. *Distr.* Tropical and warm regions of America. *Fam.* Amaryllidaceae. *Class* M.

Z. atamasco (From the North American Indian name.) ATAMASCO LILY, ZEPHYR LILY. Flowers white. Bulb eaten as an emergency food source. *Distr.* SE USA.

Z. candida FLOWER OF THE WESTERN WIND. Leaves evergreen. Flowers white. *Distr.* Argentina, Uruguay.

Z. citrina Flowers bright lemon-yellow. *Distr.* South America.

Z. grandiflora Most commonly cultivated species, flowers pink. *Distr.* West Indies, Cuba, Central America.

Z. robusta See *Habranthus robustus*

Z. rosea Flowers rose-pink. *Distr.* West Indies, Cuba, Guatemala.

Z. tubispatha BARBADOS SNOWDROP. Flowers white. *Distr.* West Indies, Venezuela, Colombia.

Z. verecunda Flowers white, becoming tinged pink. *Distr.* N Mexico.

zephyranthoides: from the genus *Zephyranthes*, with the ending *-oides*, indicating resemblance.

zephyr flower *Zephyranthes*.

Zigadenus (From Greek *zygos*, yoke, and *aden*, gland, alluding to the paired glands at the base of the sepals.) ALKALI GRASS, DEATH CAMAS. A genus of 15–18 species of rhizomatous or bulbiferous herbs with linear basal leaves and racemes or panicles of star-shaped pale flowers. A few species are cultivated as ornamentals; many are poisonous to livestock. *Distr.* North America, N and E Asia. *Fam.* Melanthiaceae. *Class* M.

Z. elegans WHITE CAMAS, ALKALI GRASS. Flowers green-white, to 25mm across. *Distr.* W North America.

Z. fremontii STAR ZYGADENE, STAR LILY. Flowers to 3cm across, ivory. *Distr.* W North America.

Z. nuttallii DEATH CAMAS, POISON CAMAS, MERRYHEARTS. Flower to 16mm across, pale yellow. *Distr.* W North America.

Zimbabwe climber *Podranea brycei.*

Zingiber (From the classical greek name for these plants, *zingiberi*, which probably came from the Malay name *inchiver*.) A genus of 85–100 species of rhizomatous herbs with reed-like stems, 2-ranked lance-shaped leaves, and spikes of irregular flowers, each of which is subtended by a colourful bract. *Distr.* Tropical Asia, N Australia. *Fam.* Zingiberaceae. *Class* M.

Z. mioga JAPANESE GINGER, MIOGA GINGER. To 1m high. Bracts white or green with red spots. *Distr.* Japan.

Z. officinale GINGER, COMMON GINGER, CANTON GINGER, STEM GINGER. The rhizomes of this species are the ginger of commerce; they are used fresh (green, or root ginger), preserved in syrup, crystallized, or dried and powdered. *Distr.* SE Asia, precise area of origin unknown.

Z. purpureum BENGAL GINGER, CASSUMAR GINGER. To 1.5m high. Bracts purple-brown. *Distr.* India.

Zingiberaceae The Ginger family. 46 genera and 1200 species of perennial aromatic herbs of forests with underground rhizomes and leaves in 2 distinct ranks. The flowers are usually attractive with a large 3-lobed lower lip. The rhizomes are often aromatic. Many species are used as flavourings (e.g. *Zingiber officinale*, ginger) and medicinal plants as well as tender cultivated ornamentals. *Distr.* Tropical regions. *Class* M.
See: Alpinia, Amomum, Cautleya, Curcuma, Elettaria, Etlingera, Globba, Hedychium, Nicolaia, Roscoea, Zingiber.

Zinnia (After Johann Gottfried Zinn (1727–59), professor of botany at Göttingen.) A genus of 20–22 species of annual and perennial herbs and shrubs with simple, opposite or whorled leaves and flowers in daisy-like heads that usually bear distinct rays. Several species are grown as ornamentals. *Distr.* North, Central and South America. *Fam.* Compositae. *Class* D.

Zizania (From the classical Greek name *zizanion*, a weed of wheat fields.) WILD RICE, WATER OATS. A genus of 4 species of annual or perennial aquatic grasses with flat leaves and terminal paniculate inflorescences. Seeds edible. *Distr.* E India to E Asia and N America. *Fam.* Gramineae. *Class* M.

Z. aquatica ANNUAL WILD RICE, WATER RICE, CANADIAN WILD RICE. Annual to over 3m high. Formerly an important food gathered from canoes by N American Indians, now widely cultivated as a grain crop in Canada. *Distr.* North America.

Z. latifolia MANCHURIAN WILD RICE, WATER RICE. Young shoots eaten as green vegetable in China. *Distr.* E Asia.

Zizia A genus of 4 species of perennial herbs with divided leaves and compound umbels of small yellow flowers. *Distr.* India to E Asia, North America. *Fam.* Umbelliferae. *Class* D.

Z. aurea GOLDEN ALEXANDERS. Leaves rounded, sharply toothed. Umbels with unequal rays. Flowers golden yellow. Grown as an ornamental. *Distr.* North America.

zonalis, -e: with different zones or markings.

zonatus, -a, -um: with different zones or markings.

Zosteraceae The Eel Grass family. 3 genera and 17 species of marine herbs living entirely submerged in salt water with grass-like leaves. The dried leaves and stems have been used as packing materials or as strengthening for cement. *Distr.* Coastal areas of N and S temperate regions, rarely tropical. *Class* M.

zosterifolius, -a, -um: from the genus *Zostera*, and *folius*, leaved.

zucchini *Cucurbita pepo.*

zygis: joined, yoked.

zygomeris: with parts joined in pairs.

Zygophyllaceae The Caltrop family. 27 genera and 250 species of shrubs, some herbs and a few trees with fleshy or leathery alternate leaves. The flowers are typically bisexual and 4- or 5-lobed. Some members of the family are a source of timber. *Distr.* Widespread in tropical to warm-temperate regions. *Class* D.
See: Peganum.

GLOSSARY

abortifacient: an agent that causes abortion.
aconitine: a poisonous, alkaloid chemical.
Altai: a mountain range in western Mongolia.
Amboina: an island in Indonesia.
Anatolia: the Asian part of Turkey.
anther: the terminal part of the stamen of a flowering plant. See note below.
Apennines: a mountain range in central Italy.
appressed: lying flat against something.
areole: a small area separated from others by cracks or lines.
aril: a fleshy outgrowth from the seed that is often brightly coloured and aids seed dispersal.
Australasia: Australia, New Zealand, and the islands of the south-western Pacific.
axil: the upper angle where a small stem joins a larger one or where a leaf joins a stem.
axillary: borne in an axil.
Balearic Islands: a group of islands in the western Mediterranean near the east coast of Spain.
bipinnate: a pinnate leaf in which each of the leaflets is pinnately divided. See pinnate.
bract: a specialized leaf that subtends a flower or inflorescence.
bulb: an underground storage organ comprising a short, flattened stem with roots on its lower surface and a number of fleshy, often tightly packed, leaves above. An onion is a bulb.
bulbiferous: possessing a bulb.
bulbils: a small bulb or bulb-like structure.
calcareous: containing calcium carbonate, typically applied to soils which have a high pH and are regarded as alkaline or basic.
calyx: the collective term for the sepals, the outer whorl of flower parts. See note below.
capitulum: a head of small, closely packed, unstalked flowers. Capitula are characteristic of the family Compositae.
carpel: one of the female reproductive parts of a flower. See note below.
catkin: a typically pendulous inflorescence spike, usually made up of small, unisexual flowers.
cladophyll: a branch that has assumed the form and function of a leaf.

compound leaf: a leaf that is made up of a number of smaller leaf-like structures known as leaflets.

coniine: a poisonous alkaloid that paralyses nerves.

corm: an underground storage organ formed from a swollen stem base.

corymb: a raceme in which the lower flowers are on longer stalks than the upper ones so that the whole inflorescence forms a dome-shaped or flat-topped structure.

cultigen: a plant that has arisen in cultivation and does not exists in the wild.

cultivar: a variety of plant which has been produced by horticultural techniques and is not normally found in wild populations.

cupule: a cup-like sheath surrounding or enclosing certain fruits. A good example is the cup at the base of an acorn.

cyme: a type of inflorescence in which each axis ends in a flower; the oldest flowers are in the centre; the younger ones are produced successively from the axils of the bracts.

D: Dicotyledoneae (Dicotyledons). One of the 2 great divisions of the flowering plants. This group is made up of plants that typically bear broad, net-veined leaves, have a 2-leafed embryo, and often bear their floral parts in 5s.

Dalmatia: an area of Adriatic coast north of Albania.

decumbent: lying along the ground but turning up at the tip to become erect.

dehiscent: bursting or splitting open at maturity. Usually used of a fruit that bursts open to release its seeds.

drupe: a fleshy fruit, such as a plum, containing one or few seeds each of which is enclosed in a stony coat.

E: East or eastern.

ephedrine: an alkaloid drug that causes constriction of the blood vessels and widening of the bronchial passages.

epiphyte: a plant that uses another plant, typically a tree, for its physical support but which does not draw nourishment from it. Epiphytes are a conspicuous feature of many kinds of tropical rain forest.

F: Ferns and their allies. A group of vascular plants that do not produce seeds. Adult plants produce spores that are dispersed and then germinate to form a structure known as a prothallus. This prothallus produces the gametes necessary for sexual reproduction which then leads directly to the production of another adult.

fen: a wet, peat-land area that is typically alkaline in nature.

fertilization: the union of the male and female gametes during sexual reproduction.

floret: one of the individual, small flowers of a clustered inflorescence or capitulum.

G: Gymnospermae (Gymnosperms). One of the 2 subdivisions of seed-bearing plants, the other being the Angiospermae or flowering plants. Gymnosperms are typically woody and bear their ovules exposed on cone scales rather than enclosed in a carpel like the flowering plants. See note below.

globose: spherical or nearly so.

herbaceous: not woody.

hesperidium: the technical term for a citrus fruit, e.g. an orange.

indusium: the covering over the sorus of a fern.

inflorescence: a flowering structure that consists of more than a single flower.

infructescence: a fruiting structure that consists of more than a single fruit.

involucre: the bracts below an inflorescence.

liana: a wiry or woody climbing plant.

lithophyte: a plant that grows on rocks.

M: Monocotyledoneae (Monocotyledons). One of the 2 great divisions of the flowering plants. This group is made up of herbs that typically bear narrow, clasping, parallel-veined leaves, have a 1-leafed embryo, and often bear their floral parts in 3s.

Macaronesia: a group of islands in the Atlantic comprising the Azores, Canary Islands, Cape Verde, Madeira, and Selvagens.

Mascarenes: a group of islands in the Indian Ocean comprising Réunion, Mauritius, and Rodriguez.

montane: of or inhabiting mountainous country.

moorland: an area with a peaty, acid soil that is typically high-lying and, in the northern hemisphere, dominated by members of the family Ericaceae.

N: North or northern.

naturalized: applied to a plant that was originally imported from another area but that now behaves like a native and can support itself without further human intervention.

obovate: applied to a leaf or other structure that is ovate but borne the other way round so that the widest point is nearer the apex than the base.

opposite: applied to the arrangement of leaves or flowers when they arise in pairs, one on either side of the stem.

ovate: applied to a leaf or other structure that is egg-shaped in outline, the widest end being the base.

ovule: the female part of a flower or cone that develops into the seed.

palmate leaf: a compound leaf in which the leaflets are all attached to the stalk at more or less the same point so that they spread out like fingers from the palm of the hand. The horse chestnut, *Aesculus hippocastanum*, has palmate leaves.

panicle: an inflorescence which is composed of a number of racemes; this term is often loosely applied to any complex, branching inflorescence.

pantropical: occurring throughout tropical regions of the world.

papilla: a small, rounded projection.

pappus: a tuft of hairs or bristles on a fruit that is derived from the calyx of the flower as in the family Compositae.
Patagonia: a region of southern South America.
pedicel: the stalk of a flower that is part of an inflorescence.
peltate: used for leaves where the stalk is attached to the centre of the blade rather than the margin. The genus *Tropaeolum* (NASTURTIUM) has peltate leaves.
perianth: the outer parts of a flower, used collectively for the sepals and petals. See note below.
petal: one of the inner floral leaves of a flower. See note below.
phyllode: a stem that has assumed the form and function of a leaf.
pinnate leaf: a compound leaf in which the leaflets are borne in 2 ranks along the length of a central axis. An ASH tree, *Fraxinus*, has pinnate leaves.
pinnule: the ultimate division of a fern frond or divided leaf.
pollen: the fine dust-like grains discharged from the male parts of flowers and cones. See note below.
pollinium: a mass of pollen grains that are transported as a single unit. Pollinia are characteristic of the family Orchidaceae.
pome: a fruit-like structure in which the true fruit is surrounded by a swollen, fleshy receptacle. An apple is a pome.
pot-herb: any herb grown in a kitchen garden.
pseudobulbs: bulb-like structures found in the orchid family, Orchidaceae, that are formed from swollen stems.
raceme: an inflorescence in which the central axis continues to grow, producing flowers laterally so that the youngest are apical or at the centre.
rachis: the axis of a compound leaf or inflorescence.
reafforestation: the replanting of former forest land with new trees.
rhizomatous: possessing a rhizome.
rhizome: a horizontal, creeping, underground stem that may bear roots and shoots and which usually persists from season to season.
rootstock: applied to the subterranean parts of a plant, especially when they are woody.
S: South or southern.
Sahel: the belt of arid land bordering the southern edge of the Sahara desert.
salep: a starchy preparation made from the dried tubers of various orchids, used in cooking and formerly medicinally.
saprophyte: a plant that lives on dead or decayed organic matter.
sepal: an outer floral leaf of a flower. See note below.
sessile: attached directly by the base, lacking a stalk.
shrub: a woody plant with short stems that branch close to the ground.
shrublet: a small shrub.
sorus: a cluster of spore-cases (sporangia) on the under-surface of a fern leaf.

spadix: a spike of flowers closely arranged round a swollen axis and typically subtended by a spathe.
spathe: a bract that subtends a spadix.
spike: an inflorescence in which the flowers are attached directly to an unbranched axis.
spikelet: a small spike, typically of grasses but also occurring in reeds and sedges, where it is the fundamental unit of the inflorescence and consists of an axil, 2 bracts, and one or more florets.
sporangium: a receptacle in which spores are formed.
sporocarp: a reproductive structure made up of sporangia surrounded by an indusium.
sporophyll: a leaf that bears sporangia.
stamen: the male organ of the flower. See note below.
stigma: the region of the female parts of the flower that is receptive to pollen. See note below.
stipule: a leaf-like appendage, often occurring in pairs, at the base of a leaf-stalk.
stolon: a short-lived horizontal stem or branch that roots at points along its length forming new plants. Stolons are sometimes called runners. Strawberry plants produce stolons.
stoloniferous: possessing stolons.
strobilus: a cone or cone-like structure.
style: an extension of the carpel or ovary that supports the stigma. See note below.
syncarp: a compound fruit formed from a number of fused carpels.
tepal: a perianth segment where there is no clear distinction between sepals and petals. See note below.
thallus: a type of vegetative plant body in which there is no clear differentiation into stem and leaf. Duckweed (*Lemna*) possesses a thallus.
trifoliate leaf: a compound leaf made up of 3 leaflets.
tripinnate leaf: a compound leaf that is divided pinnately 3 times, i.e. a pinnate leaf that bears pinnately divided leaflets in which the leaflets are pinnately divided.
tuber: a swollen stem or root that functions as an underground storage organ.
tuberous: possessing tubers.
tussock: a mound formed by a clump of vegetation, typically a grass.
umbel: an inflorescence in which all the flower-stalks arise from a single point typically forming an umbrella-shaped structure. Umbels are characteristic of the family Umbelliferae.
understorey: the region below the forest canopy.
vascular plant: a plant that has specialized, internal water-conducting tissues. Most land plants are vascular but not mosses, liverworts, or algae.

vermifuge: a drug that expels intestinal worms.
vesicle: a small bladder, bubble, or hollow structure.
W: West or western.
whorl: the arrangement in which leaves, petals, or other structures all arise at the same point on an axis, encircling it.

A NOTE ON THE STRUCTURE OF FLOWERS

Flower structure and terminology is sometimes difficult to grasp. The paragraphs below give a brief outline of some of the terms used when describing parts of the flower.

All flowers have evolved from a series of whorls of leaves around the end of a branch. As the leaves get closer to the apex of the branch and the centre of the flower, they become more specialized.

In an idealized flower the outer whorl of leaves is termed the calyx and is made up of a number of sepals which are green and leaf-like; inside the calyx is the corolla, made up of a number of petals which are often brightly coloured to attract insects. The individual sepals and petals can be fused to form calyx or corolla tubes which frequently have lobes at the tips representing the original sepals or petals. These two outer whorls together are known as the perianth. In some flowers there is no clear distinction between sepals and petals and so the terms tepal or perianth segment are used.

Within the perianth there are one or more whorls of stamens which are highly specialized for the production of pollen. The stamens represent the male part of the flower and are collectively termed the androecium; they consist of a stalk (the filament) and a head in which the pollen is produced (the anther).

At the tip of the branch and the centre of the flower is the gynoecium, the region that contains the female parts of the flower. In primitive flowers the gynoecium is composed of a whorl of carpels, each carpel somewhat resembling a folded leaf. In advanced flowers the carpels are fused into a more complex structure, the ovary. One part of the ovary is receptive to pollen and this is known as the stigma. The stigma is sometimes borne on a stalk termed the style. After fertilization the ovary swells to form a fruit. All the flower parts are joined at their base to the receptacle which represents the tip of the original branch. In some cases, such as the strawberry, instead of the ovary swelling to form the fruit the receptacle swells and forms what is often termed a 'false' fruit.

FURTHER READING

Brummitt, R. K. (1992) *Vascular Plant Families and Genera*. Royal Botanic Gardens, Kew.
– A list of all the plant families and genera presently recognized by the Royal Botanic Gardens, Kew and the Royal Botanic Garden, Edinburgh.

Huxley, A., *et al.* (1992) *The New Royal Horticultural Society Dictionary of Gardening*. Macmillan, London.
– A comprehensive treatment of garden plants, gardens, and gardening in four volumes.

Mabberley, D. J. (1987) *The Plant-Book*. Cambridge University Press.
– A dictionary of higher plants for the proficient botanist.

Philip, C. (1992) *The Plant Finder*. 6th edition. Moorland Publishing Co.
– A directory of plants available from nurseries in the British Isles, updated yearly.

Stearn, W. T. (1992) *Botanical Latin* 4th edition. David and Charles.
– The standard guide to the use of Botanical Latin.

Walters, S. M. *et al.* (eds.) (1986–) *The European Garden Flora*, Volumes 1 to 3. Cambridge University Press.
– A scientific guide to the garden plants of Europe. This work is being released a volume at a time. Volume 4 will be released in summer 1994. There will be a total of six volumes.